T0280690

Quantum Mechanics

A Modern Development

2nd Edition

Quantum Mechanics

A Modern Development

2nd Edition

Leslie E Ballentine
Simon Fraser University, Canada

NEW JERSEY • LONDON • SINGAPORE • BEIJING • SHANGHAI • HONG KONG • TAIPEI • CHENNAI

Published by

World Scientific Publishing Co. Pte. Ltd.
5 Toh Tuck Link, Singapore 596224
USA office: 27 Warren Street, Suite 401-402, Hackensack, NJ 07601
UK office: 57 Shelton Street, Covent Garden, London WC2H 9HE

Library of Congress Cataloging-in-Publication Data
Ballentine, Leslie E., author.
Quantum mechanics : a modern development / Leslie E. Ballentine, Simon Fraser University, Canada. -- 2nd edition.
pages cm
Includes bibliographical references and index.
ISBN 978-981-4578-57-8 (hardcover : alk. paper) -- ISBN 978-981-4578-58-5 (pbk)
1. Quantum theory. I. Title.
QC174.12.B35 2014
530.12--dc23
2014014232

British Library Cataloguing-in-Publication Data
A catalogue record for this book is available from the British Library.

Printed in Singapore

Contents

Preface

There are many textbooks that deal with the formal apparatus of quantum mechanics (QM) and its application to standard problems, but before the first edition of this book (Prentice-Hall, 1990), none took account of the developments in the foundations of the subject that had taken place in the preceeding few decades. Some specialized treatises on the foundations of QM exist, but they do not integrate those topics with the standard pedagogical material. My objective was to remove that unfortunate dichotomy, which divorced the practical aspects of QM from the interpretation and broader implications of the theory. This emphasis continued in the expanded 1998 World Scientific edition.

The emergence of Quantum Information (QI) as new branch of QM, is the prime motivation for this new edition. QI is now a very active area of research, extending from quantum foundations to new quantum technologies. A whole book would be needed to treat all aspects of QI, and even then it would be incomplete shortly after publication, so rapid is the growth of the research literature. Therefore, I have made no attempt at completeness, and have concentrated on the basic principles. Quantum information is different from "classical" information, presenting both new powers and counter-intuitive limitations. But Shannon's concept of a quantitative measure of information is a common thread, tying together the quantum and classical aspects of information theory. A few key applications are treated, but only at an introductory level. These include quantum cryptography, teleportation of states, and quantum computing. Experts in the field of quantum information will notice the absence of some formal techniques which are commonly used in research. In limiting my coverage, I have been guided by a pedagogical principle: formalism should not be introduced for its own sake, but only when it is needed for some particular problem.

I have paid particular attention to the impact of quantum information theory on the foundations and interpretation of QM. QI provides a new point of view on quantum foundations, and has rejuvenated its study. The spectrum of possible interpretations is much wider than is commonly realized. The interrelations among the categories of *individual* vs *ensemble*, *ontic* vs *epistemic*, and *objective* vs *subjective*, as illustrated in Fig. 21.1, have not previously been studied in detail.

This edition contains minor changes to most chapters, consisting of many new references, improving some discussions with which I was no longer satisfied, and correcting several typographical errors.

This book is intended primarily as a graduate level textbook, but it will also be of interest to physicists and philosophers who study the foundations of quantum mechanics. Parts of it could be used by senior undergraduates. Its evolution can be traced in the prefaces of the earlier editions (from which I now quote, or paraphrase).

The 1990 version introduced several topics that had previously been found in few, if any, textbooks. They included:

- A review of *probability theory* and its relation to the quantum theory.
- Discussion of *state preparation* and *state determination*.
- The Aharonov-Bohm effect.
- Some firmly established results in the theory of *measurement*, which are useful in clarifying the interpretation of quantum mechanics.
- A more complete account of the *classical limit*.
- Introduction of *rigged Hilbert space* as a generalization of the familiar Hilbert space. It allows vectors of infinite norm to be accommodated within the formalism, and eliminates the vagueness that often surrounds the question of whether the operators that represent observables possess a complete set of eigenvectors.
- The *space-time symmetries* of displacement, rotation, and Galilei transformations are exploited to derive the fundamental operators for momentum, angular momentum, and the Hamiltonian.
- A charged particle in a magnetic field (Landau levels).
- Basic concepts of *quantum optics*.
- Modern experiments that illustrate and test the principles of quantum mechanics, such as: the direct measurement of the momentum distribution in the hydrogen atom; the single-crystal neutron interferometer; quantum beats; photon bunching and anti-bunching.
- Bell's theorem and its implications.

The 1998 World Scientific edition added more new material:

- An introduction describing the range of phenomena that quantum theory seeks to explain.
- Feynman's *path integrals.*
- The adiabatic approximation and Berry's phase.
- Expanded treatment of state preparation and determination, including the *no-cloning theorem* and *entangled states.*
- A new treatment of the *energy-time* uncertainty relations.
- The influence of a measurement apparatus on the environment, and vice versa.
- The quantum mechanics of rigid bodies.
- A revised and expanded chapter on the *classical limit.*
- The *phase space* formulation of quantum mechanics.
- Expanded treatment of new interference experiments.
- Optical homodyne tomography as a method of measuring the quantum state of a field mode.
- Bell's theorem without inequalities and probability.

The *Pure State Factor Theorem* of Sec. 8.3 has an interesting history. In his book, *Foundations of Quantum Mechanics* (Addison-Wesley, 1968), the late J. M. Jauch assigns the proof of (essentially) this theorem as Problem 1 (p.182). I do not know whether he possessed a valid proof, since in his next, closely related problem, he asked us to prove something that is *not true!* Reading the relevant part of his chapter, I found the error that misled him into making the false assertion in Problem 2, and the same error could easily have led to an invalid solution of the previous problem. I devised a partial proof for the theorem, but it used perturbation theory, and was not fully general. So, following the professorial stereotype, I assigned it as a problem for my graduate student class (warning them, however, of its origin). No one got it in the first year it was assigned, but in the second year, Bob Goldstein cracked it, and a modified version of his solution appears in this book. I have not seen the *Pure State Factor Theorem* published anywhere else, but perhaps a reader may find an earlier proof of it.

This book is suitable for a two-semester course, with sufficient material to allow the instructor some choice of topics. The introductory chapters are sufficiently novel to deserve some comment.

Chapter 1 contains mathematical topics (vector spaces, operators, and probability), which may be skimmed by mathematically sophisticated readers.

These topics are placed at the beginning, rather than in an appendix, because one needs not only the results but also a coherent overview of their theory, since they form the mathematical language in which quantum theory is expressed. The amount of time spent on this chapter may vary widely, depending on the reader's degree of mathematical preparation. A mathematically advanced reader could proceed directly from the Introduction to Chapter 2, although such a strategy is not recommended.

The *space-time symmetries* of displacement, rotation, and Galilei transformations are used in Chapter 3 to derive the fundamental operators for momentum, angular momentum, and the Hamiltonian. This method replaces the heuristic but inconclusive arguments based upon analogy and wave-particle duality, which so frustrate the serious student. It also introduces *symmetry* concepts and techniques at an early stage, so that they are immediately available for practical applications. No prior knowledge of *group theory* is required. Indeed, a reader who does not know the technical meaning of the word "group", and who interprets the references to "groups" as merely meaning sets of related transformations and operators, will lose none of the essential meaning.

Solutions to some problems are given in Appendix D. The solved problems are those that are particularly novel, and those for which the answer or the method of solution is important for its own sake (rather than merely being an exercise).

At various places, I have segregated in double brackets, $[[\cdots]]$, comments of a historical, comparative, or critical nature. Those remarks would not be needed by a hypothetical reader with no previous exposure to quantum mechanics. They are used to relate my approach, by way of comparison or contrast, to that of earlier writers, and sometimes to show, by means of criticism, the reason for my departure from the older approaches.

Acknowledgments

This book has drawn on a great many published sources, which are acknowledged throughout the text. However, I would like to give special mention to the work of T. F. Jordan, which forms the basis of Chapter 3. Much of the book has been "field-tested" on classes of graduate students at Simon Fraser University. My former student Bob Goldstein discovered a simple proof for *Pure State Factor Theorem* in Sec. 8.3, and his creative imagination was responsible for the paradox that forms the basis of Problem 9.6. The data for Fig. 0.4 was taken by Jeff Rudd of the SFU teaching laboratory staff.

In preparing Sec. 1.5 on probability theory, I benefited from discussions with Prof. C. Villegas. I would like to thank Hans von Baeyer for the key idea in the derivation of the orbital angular momentum eigenvalues in Sec. 7.3, and W. G. Unruh for pointing out interesting features of Example (iii) in Sec. 9.6.

While preparing the chapter on Quantum Information, I have benefited from correspondence or discussions with Joe Emerson, Nathan Wiebe, William Wootters, and Stephen Barnett. I am grateful to Jeremy Hilton and the scientific staff of *D-Wave Systems, Inc.* for explaining to me the workings of their *adiabatic quantum computer*, and in particular to Mark Johnson for supplying useful references.

Leslie E. Ballentine
Professor Emeritus
Simon Fraser University

Introduction

The Phenomena of Quantum Mechanics

Quantum mechanics is a general theory. It is presumed to apply to everything, from subatomic particles to galaxies. But interest is naturally focussed on those phenomena that are most distinctive of quantum mechanics, some of which led to its discovery. Rather than retelling the historical development of quantum theory, which can be found in many books,* I shall illustrate quantum phenomena under three headings: *discreteness*, *diffraction*, and *coherence*. It is interesting to contrast the original experiments, which led to the new discoveries, with the accomplishments of modern technology.

It was the phenomenon of *discreteness* that gave rise to the name "quantum mechanics". Certain dynamical variables were found to take on only a

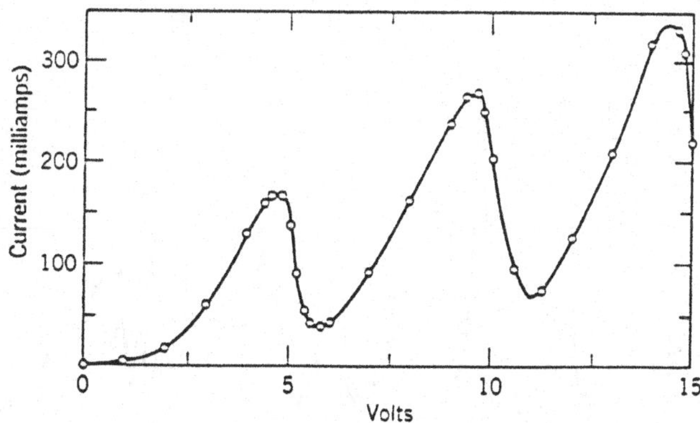

Fig. 0.1 Current through a tube of Hg vapor versus applied voltage, from the data of Franck and Hertz (1914). [Figure reprinted from *Quantum Physics of Atoms, Molecules, Solids, Nuclei and Particles*, R. Eisberg and R. Resnick (Wiley, 1985).]

*See, for example, Eisberg and Resnick (1985) for an elementary treatment, or Jammer (1966) for an advanced study.

discrete, or *quantized*, set of values, contrary to the predictions of classical mechanics. The first direct evidence for discrete atomic energy levels was provided by Franck and Hertz (1914). In their experiment, electrons emitted from a hot cathode were accelerated through a gas of Hg vapor by means of an adjustable potential applied between the anode and the cathode. The current as a function of voltage, shown in Fig. 0.1, does not increase monotonically, but rather displays a series of peaks at multiples of 4.9 volts. Now 4.9 eV is the energy required to excite a Hg atom to its first excited state. When the voltage is sufficient for an electron to achieve a kinetic energy of 4.9 eV, it is able to excite an atom, losing kinetic energy in the process. If the voltage is more than twice 4.9 V, the electron is able to regain 4.9 eV of kinetic energy and cause a second excitation event before reaching the anode. This explains the sequence of peaks.

The peaks in Fig. 0.1 are very broad, and provide no evidence for the sharpness of the discrete atomic energy levels. Indeed, if there were no better evidence, a skeptic would be justified in doubting the discreteness of atomic energy levels. But today it is possible, by a combination of laser excitation and electric field filtering, to produce beams of atoms that are all in the same quantum state. Figure 0.2 shows results of Koch *et al.* (1989), in which

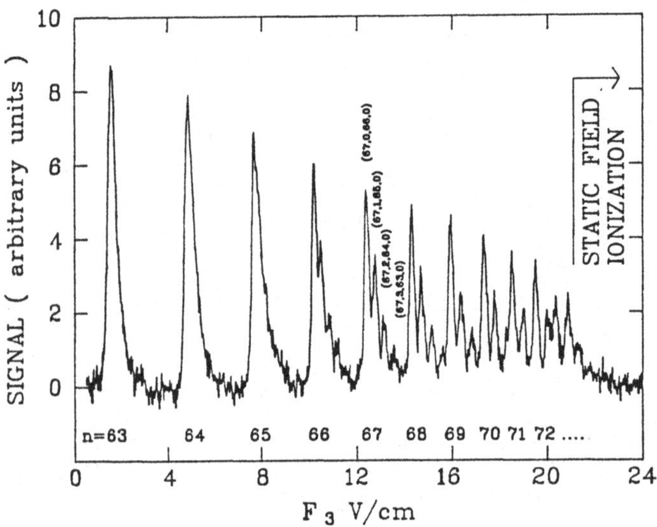

Fig. 0.2 Individual excited states of atomic hydrogen are resolved in this data [reprinted from Koch *et al.*, *Physica Scripta* **T26**, 51 (1989)].

the atomic states of hydrogen with principal quantum numbers from $n = 63$ to $n = 72$ are clearly resolved. Each n value contains many substates that would be degenerate in the absence of an electric field, and for $n = 67$ even the substates are resolved. By adjusting the laser frequency and the various filtering fields, it is possible to resolve different atomic states, and so to produce a beam of hydrogen atoms that are all in the same chosen quantum state. The discreteness of atomic energy levels is now very well established.

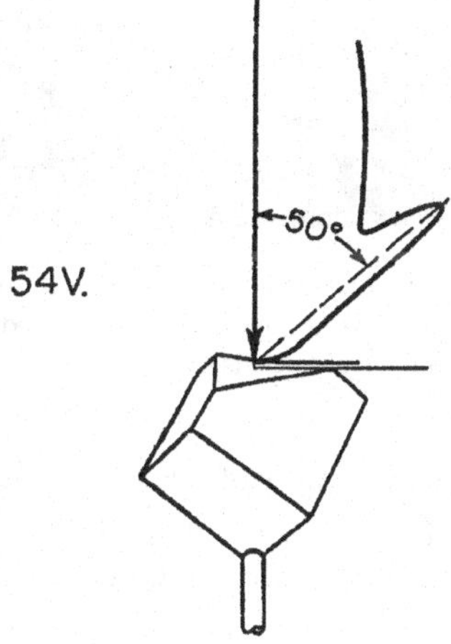

Fig. 0.3 Polar plot of scattering intensity versus angle, showing evidence of electron diffraction, from the data of Davisson and Germer (1927).

The phenomenon of *diffraction* is characteristic of any wave motion, and is especially familiar for light. It occurs because the total wave amplitude is the sum of partial amplitudes that arrive by different paths. If the partial amplitudes arrive in phase, they add constructively to produce a maximum in the total intensity; if they arrive out of phase, they add destructively to produce a minimum in the total intensity. Davisson and Germer (1927), following a theoretical conjecture by L. de Broglie, demonstrated the occurrence of diffraction in the reflection of electrons from the surface of a crystal of nickel. Some of their data is shown in Fig. 0.3, the peak at a scattering angle of 50° being the evidence for electron diffraction. This experiment led to the award of a Noble prize to Davisson in 1937. Today, with improved technology, even an undergraduate can easily produce electron diffraction patterns that are vastly superior to the Nobel prize-winning data of 1927. Figure 0.4 shows an electron

Fig. 0.4 Diffraction of 10 kV electrons through a graphite foil; data from an undergraduate laboratory experiment. Some of the spots are blurred because the foil contains many crystallites, but the hexagonal symmetry is clear.

diffraction pattern from a crystal of graphite, produced in a routine undergraduate laboratory experiment at Simon Fraser University. The hexagonal array of spots corresponds to diffraction scattering from the various crystal planes.

The phenomenon of diffraction scattering is not peculiar to electrons, or even to elementary particles. It occurs also for atoms and molecules, and is a universal phenomenon (see Ch. 5 for further discussion). When first discovered, particle diffraction was a source of great puzzlement. Are "particles" really "waves"? In the early experiments, the diffraction patterns were detected holistically by means of a photographic plate, which could not detect individual particles. As a result, the notion grew that particle and wave properties were mutually incompatible, or *complementary*, in the sense that different measurement apparatuses would be required to observe them. That idea, however, was only an unfortunate generalization from a technological limitation. Today it is possible to detect the arrival of individual electrons, and to see the diffraction pattern emerge as a statistical pattern made up of many small spots (Tonomura *et al.*, 1989). Evidently, quantum particles are indeed particles, but particles whose behavior is very different from what classical physics would have led us to expect.

In classical optics, *coherence* refers to the condition of phase stability that is necessary for interference to be observable. In quantum theory the concept of coherence also refers to phase stability, but it is generalized beyond any analogy with wave motion. In general, a *coherent* superposition of quantum states may have properties than are qualitatively different from a mixture of the properties of the component states. For example, the state of a neutron with its spin polarized in the $+x$ direction is expressible (in a notation that will be developed in detail in later chapters) as a coherent sum of states that are polarized in the $+z$ and $-z$ directions, $|+x\rangle = (|+z\rangle + |-z\rangle)/\sqrt{2}$. Likewise, the state with the spin polarized in the $+z$ direction is expressible in terms of the $+x$ and $-x$ polarizations as $|+z\rangle = (|+x\rangle + |-x\rangle)/\sqrt{2}$.

An experimental realization of these formal relations is illustrated in Fig. 0.5. In part (a) of the figure, a beam of neutrons with spin polarized in the $+x$ direction is incident on a device that transmits $+z$ polarization and reflects $-z$ polarization. This can be achieved by applying a strong magnetic field in the z direction. The potential energy of the magnetic moment in the field, $-\mathbf{B} \cdot \boldsymbol{\mu}$, acts as a potential well for one direction of the neutron spin, but as an impenetrable potential barrier for the other direction. The effectiveness of the device in separating $+z$ and $-z$ polarizations can be confirmed by detectors that measure the z component of the neutron spin.

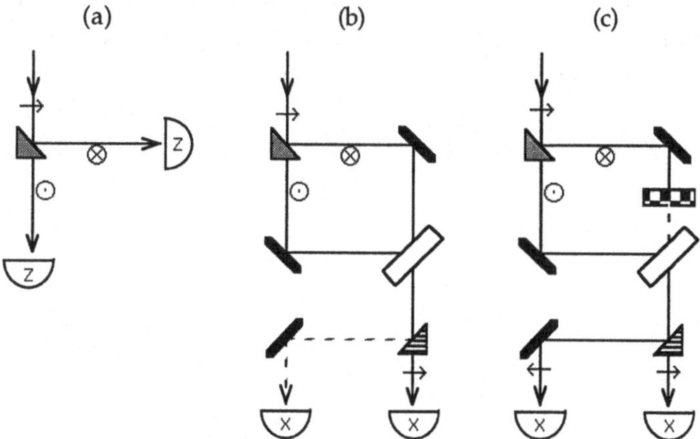

Fig. 0.5 (a) Splitting of a $+x$ spin-polarized beam of neutrons into $+z$ and $-z$ components; (b) coherent recombination of the two components; (c) splitting of the $+z$ polarized beam into $+x$ and $-x$ components.

In part (b) the spin-up and spin-down beams are recombined into a single beam that passes through a device to separate $+x$ and $-x$ spin polarizations. If the recombination is *coherent*, and does not introduce any phase shift between the two beams, then the state $|+x\rangle$ will be reconstructed, and only the $+x$ polarization will be detected at the end of the apparatus. In part (c) the $|-z\rangle$ beam is blocked, so that only the $|+z\rangle$ beam passes through the apparatus. Since $|+z\rangle = (|+x\rangle + |-x\rangle)/\sqrt{2}$, this beam will be split into $|+x\rangle$ and $|-x\rangle$ components.

Although the experiment depicted in Fig. 0.5 is idealized, all of its components are realizable, and closely related experiments have actually been performed.

In this Introduction, we have briefly surveyed some of the diverse phenomena that occur within the quantum domain. Discreteness, being essentially discontinuous, is quite different from classical mechanics. Diffraction scattering of particles bears a strong analogy to classical wave theory, but the element of discreteness is present, in that the observed diffraction patterns are really statistical patterns of the individual particles. The possibility of combining quantum states in coherent superpositions that are qualitatively different from their components is perhaps the most distinctive feature of quantum mechanics, and it introduces a new nonclassical element of continuity. It is the task of quantum theory to provide a framework within which all of these diverse phenomena can be explained.

Chapter 1

Mathematical Prerequisites

Certain mathematical topics are essential for quantum mechanics, not only as computational tools, but because they form the most effective language in terms of which the theory can be formulated. These topics include the theory of linear vector spaces and linear operators, and the theory of probability. The connection between quantum mechanics and linear algebra originated as an apparent by-product of the linear nature of Schrödinger's wave equation. But the theory was soon generalized beyond its simple beginnings, to include abstract "wave functions" in the $3N$-dimensional configuration space of N paricles, and then to include discrete internal degrees of freedom such as spin, which have nothing to do with wave motion. The structure common to all of those diverse cases is that of linear operators on a vector space. A unified theory based on that mathematical structure was first formulated by P. A. M. Dirac, and the formulation used in this book is really a modernized version of Dirac's formalism.

That quantum mechanics does not predict a deterministic course of events, but rather the probabilities of various alternative possible events, was recognized at an early stage, especially by Max Born. Modern applications seem more and more to involve correlation functions and nontrivial statistical distributions (especially in quantum optics), and therefore the relations between quantum theory and probability theory need to be expounded.

The physical development of quantum mechanics begins in Ch. 2, and the mathematically sophisticated reader may turn there at once. But since not only the results, but also the concepts and logical framework of Ch. 1 are freely used in developing the physical theory, the reader is advised to at least skim this first chapter before proceeding to Ch. 2.

1.1 *Linear Vector Space*

A *linear vector space* is a set of elements, called vectors, which is closed under addition and multiplication by scalars. That is to say, if ϕ and ψ are vectors then so is $a\phi + b\psi$, where a and b are arbitrary scalars. If the scalars belong to the field of complex (real) numbers, we speak of a complex (real) linear vector space. Henceforth the scalars will be complex numbers unless otherwise stated.

Among the very many examples of linear vector spaces, there are two classes that are of common interest:

(i) *Discrete vectors*, which may be represented as columns of complex numbers,

$$\begin{pmatrix} a_1 \\ a_2 \\ \vdots \\ \vdots \end{pmatrix}$$

(ii) *Spaces of functions* of some type, for example the space of all differentiable functions.

One can readily verify that these examples satisfy the definition of a linear vector space.

A set of vectors $\{\phi_n\}$ is said to be *linearly independent* if no nontrivial linear combination of them sums to zero; that is to say, if the equation $\sum_n c_n \phi_n = 0$ can hold only when $c_n = 0$ for all n. If this condition does not hold, the set of vectors is said to be *linearly dependent*, in which case it is possible to express a member of the set as a linear combination of the others.

The maximum number of linearly independent vectors in a space is called the *dimension* of the space. A maximal set of linearly independent vectors is called a *basis* for the space. Any vector in the space can be expressed as a linear combination of the basis vectors.

An *inner product* (or scalar product) for a linear vector space associates a scalar (ψ, ϕ) with every ordered pair of vectors. It must satisfy the following properties:

(a) $(\psi, \phi) =$ a complex number,
(b) $(\phi, \psi) = (\psi, \phi)^*$,
(c) $(\phi, c_1\psi_1 + c_2\psi_2) = c_1(\phi, \psi_1) + c_2(\phi, \psi_2)$,
(d) $(\phi, \phi) \geq 0$, with equality holding if and only if $\phi = 0$.

From (b) and (c) it follows that

$$(c_1\psi_1 + c_2\psi_2, \phi) = c_1^*(\psi_1, \phi) + c_2^*(\psi_2, \phi)\,.$$

Therefore we say that the inner product is *linear* in its second argument, and *antilinear* in its first argument.

We have, corresponding to our previous examples of vector spaces, the following inner products:

(i) If ψ is the column vector with elements $a_1, a_2, \ldots$ and ϕ is the column vector with elements $b_1, b_2, \ldots$, then

$$(\psi, \phi) = a_1^* b_1 + a_2^* b_2 + \cdots .$$

(ii) If ψ and ϕ are functions of x, then

$$(\psi, \phi) = \int \psi^*(x)\phi(x)w(x)dx ,$$

where $w(x)$ is some nonnegative weight function.

The inner product generalizes the notions of length and angle to arbitrary spaces. If the inner product of two vectors is zero, the vectors are said to be *orthogonal*.

The *norm* (or length) of a vector is defined as $||\phi|| = (\phi, \phi)^{1/2}$. The inner product and the norm satisfy two important theorems:

Schwarz's inequality,

$$|(\psi, \phi)|^2 \leq (\psi, \psi)(\phi, \phi) . \tag{1.1}$$

The triangle inequality,

$$||(\psi + \phi)|| \leq ||\psi|| + ||\phi|| . \tag{1.2}$$

In both cases equality holds only if one vector is a scalar multiple of the other, i.e. $\psi = c\phi$. For (1.2) to become an equality, the scalar c must be real and positive.

A set of vectors $\{\phi_i\}$ is said to be *orthonormal* if the vectors are pairwise orthogonal and of unit norm; that is to say, their inner products satisfy $(\phi_i, \phi_j) = \delta_{ij}$.

Corresponding to any linear vector space V there exists the *dual space* of *linear functionals* on V. A linear functional F assigns a scalar $F(\phi)$ to each vector ϕ, such that

$$F(a\phi + b\psi) = aF(\phi) + bF(\psi) \tag{1.3}$$

for any vectors ϕ and ψ, and any scalars a and b. The set of linear functionals may itself be regarded as forming a linear space V' if we define the sum of two functionals as

$$(F_1 + F_2)(\phi) = F_1(\phi) + F_2(\phi) . \tag{1.4}$$

Riesz theorem. There is a one-to-one correspondence between linear functionals F in V' and vectors f in V, such that all linear functionals have the form

$$F(\phi) = (f, \phi) , \tag{1.5}$$

f being a fixed vector, and ϕ being an arbitrary vector. Thus the spaces V and V' are essentially isomorphic. For the present we shall only prove this theorem in a manner that ignores the convergence questions that arise when dealing with infinite-dimensional spaces. (These questions are dealt with in Sec. 1.4.)

Proof. It is obvious that any given vector f in V defines a linear functional, using Eq. (1.5) as the definition. So we need only prove that for an arbitrary linear functional F we can construct a unique vector f that satisfies (1.5). Let $\{\phi_n\}$ be a system of orthonormal basis vectors in V, satisfying $(\phi_n, \phi_m) = \delta_{n,m}$. Let $\psi = \sum_n x_n \phi_n$ be an arbitrary vector in V. From (1.3) we have

$$F(\psi) = \sum_n x_n F(\phi_n) .$$

Now construct the following vector:

$$f = \sum_n [F(\phi_n)]^* \phi_n .$$

Its inner product with the arbitrary vector ψ is

$$\begin{aligned}(f, \psi) &= \sum_n F(\phi_n) x_n \\ &= F(\psi) ,\end{aligned}$$

and hence the theorem is proved.

Dirac's bra and ket notation

In Dirac's notation, which is very popular in quantum mechanics, the vectors in V are called *ket* vectors, and are denoted as $|\phi\rangle$. The linear functionals in the dual space V' are called *bra* vectors, and are denoted as $\langle F|$. The numerical value of the functional is denoted as

$$F(\phi) = \langle F|\phi\rangle . \tag{1.6}$$

According to the Riesz theorem, there is a one-to-one correspondence between bras and kets. Therefore we can use the same alphabetic character for the functional (a member of V') and the vector (in V) to which it corresponds, relying on the bra, $\langle F|$, or ket, $|F\rangle$, notation to determine which space is referred to. Equation (1.5) would then be written as

$$\langle F|\phi\rangle = (F, \phi) , \tag{1.7}$$

$|F\rangle$ being the vector previously denoted as f. Note, however, that the Riesz theorem establishes, by construction, an *antilinear* correspondence between bras and kets. If $\langle F| \leftrightarrow |F\rangle$, then

$$c_1^*\langle F_1| + c_2^*\langle F_2| \leftrightarrow c_1|F_1\rangle + c_2|F_2\rangle . \tag{1.8}$$

Because of the relation (1.7), it is possible to regard the "braket" $\langle F|\phi\rangle$ as merely another notation for the inner product. But the reader is advised that there are situations in which it is important to remember that the primary definition of the bra vector is as a linear functional on the space of ket vectors.

[[In his original presentation, Dirac *assumed* a one-to-one correspondence between bras and kets, and it was not entirely clear whether this was a mathematical or a physical assumption. The Riesz theorem shows that there is no need, and indeed no room, for any such assumption. Moreover, we shall eventually need to consideer more general spaces (rigged-Hilbert-space triplets) for which the one-to-one correspondence between bras and kets does not hold.]]

1.2 *Linear Operators*

An *operator* on a vector space maps vectors onto vectors; that is to say, if A is an opetator and ψ is a vector, then $\phi = A\psi$ is another vector. An operator is fully defined by specifying its action on every vector in the space (or in its *domain*, which is the name given to the subspace on which the operator can meaningfully act, should that be smaller than the whole space).

A *linear operator* satisfies

$$A(c_1\psi_1 + c_2\psi_2) = c_1(A\psi_1) + c_2(A\psi_2) . \tag{1.9}$$

It is sufficient to define a linear operator on a set of basis vectors, since every vector can be expressed as a linear combination of the basis vectors. We shall be treating only linear operators, and so shall henceforth refer to them simply as operators.

To assert the *equality* of two operators, $A = B$, means that $A\psi = B\psi$ for *all* vectors (more precisely, for all vectors in the common domain of A and B, this qualification will usually be omitted for brevity). Thus we can define the sum and product of operators,

$$(A + B)\psi = A\psi + B\psi ,$$
$$AB\psi = A(B\psi) ,$$

both equations holding for all ψ. It follows from this definition that operator mulitplication is necessarily *associative*, $A(BC) = (AB)C$. But it need not be *commutative*, AB being unequal to BA in general.

Example (i). In a space of discrete vectors represented as columns, a linear operator is a square matrix. In fact, any operator equation in a space of N dimensions can be transformed into a matrix equation. Consider, for example, the equation

$$M|\psi\rangle = |\phi\rangle . \tag{1.10}$$

Choose some orthonormal basis $\{|u_i\rangle, i = 1 \ldots N\}$ in which to expand the vectors,

$$|\psi\rangle = \sum_j a_j |u_j\rangle , \ |\phi\rangle = \sum_k b_k |u_k\rangle .$$

Operating on (1.10) with $\langle u_i|$ yields

$$\sum_j \langle u_i|M|u_j\rangle a_j = \sum_k \langle u_i|u_k\rangle b_k$$
$$= b_i ,$$

which has the form of a matrix equation,

$$\sum_j M_{ij} a_j = b_i , \tag{1.11}$$

with $M_{ij} = \langle u_i|M|u_j\rangle$ being known as a *matrix element* of the operator M. In this way any problem in an N-dimensional linear vector space, no matter how it arises, can be transformed into a matrix problem.

The same thing can be done formally for an infinite-dimensional vector space if it has a denumerable orthonormal basis, but one must then deal with the problem of convergence of the infinite sums, which we postpone to a later section.

Example (ii). Operators in function spaces frequently take the form of differential or integral operators. An operator equation such as

$$\frac{\partial}{\partial x} x = 1 + x \frac{\partial}{\partial x}$$

may appear strange if one forgets that operators are only defined by their action on vectors. Thus the above example means that

$$\frac{\partial}{\partial x}[x\ \psi(x)] = \psi(x) + x\frac{\partial\psi(x)}{\partial x} \quad \text{for all} \quad \psi(x)\,.$$

So far we have only defined operators as acting to the right on ket vectors. We may define their *action to the left* on bra vectors as

$$(\langle\phi|A)|\psi\rangle = \langle\phi|(A|\psi\rangle) \tag{1.12}$$

for all ϕ and ψ. This appears trivial in Dirac's notation, and indeed this triviality contributes to the practical utility of his notation. However, it is worthwhile to examine the mathematical content of (1.12) in more detail.

A bra vector is in fact a linear functional on the space of ket vectors, and in a more detailed notation the bra $\langle\phi|$ is the functional

$$F_\phi(\cdot) = (\phi, \cdot)\,, \tag{1.13}$$

where ϕ is the vector that corresponds to F_ϕ via the Riesz theorem, and the dot indicates the place for the vector argument. We may define the operation of A on the bra space of functionals as

$$AF_\phi(\psi) = F_\phi(A\psi) \quad \text{for all} \quad \psi\,. \tag{1.14}$$

The right hand side of (1.14) satisfies the definition of a linear functional of the vector ψ (not merely of the vector $A\psi$), and hence it does indeed define a new functional, called AF_ϕ. According to the Riesz theorem there must exist a ket vector χ such that

$$\begin{aligned} AF_\phi(\psi) &= (\chi, \psi) \\ &= F_\chi(\psi)\,. \end{aligned} \tag{1.15}$$

Since χ is uniquely determined by ϕ (given A), there must exist an operator $A^\dagger$ such that $\chi = A^\dagger\phi$. Thus (1.15) can be written as

$$AF_\phi = F_{A^\dagger\phi}\,. \tag{1.16}$$

From (1.14) and (1.15) we have $(\phi, A\psi) = (\chi, \psi)$, and therefore

$$(A^\dagger\phi, \psi) = (\phi, A\psi) \quad \text{for all} \quad \phi \text{ and } \psi\,. \tag{1.17}$$

This is the usual definition of the *adjoint*, $A^\dagger$, of the operator A. All of this nontrivial mathematics is implicit in Dirac's simple equation (1.12)!

The adjoint operator can beformally defined within the Dirac notation by demanding that if $\langle\phi|$ and $|\phi\rangle$ are corresponding bras and kets, then $\langle\phi|A^\dagger \equiv \langle\omega|$ and $A|\phi\rangle \equiv |\omega\rangle$ should also be corresponding bras and kets. From the fact that $\langle\omega|\psi\rangle^* = \langle\psi|\omega\rangle$, it follows that

$$\langle\phi|A^\dagger|\psi\rangle^* = \langle\psi|A|\phi\rangle \quad \text{for all} \quad \phi \text{ and } \psi\,, \tag{1.18}$$

this relation being equivalent to (1.17). Although simpler than the previous introduction of $A^\dagger$ via the Riesz theorem, this formal method fails to prove the existence of the operator $A^\dagger$.

Several useful properties of the adjoint operator that follow directly from (1.17) are

$$\begin{aligned}
(cA)^\dagger &= c^* A^\dagger\,, \quad \text{where } c \text{ is a complex number,}\\
(A+B)^\dagger &= A^\dagger + B^\dagger\,,\\
(AB)^\dagger &= B^\dagger A^\dagger\,.
\end{aligned}$$

In addition to the inner product of a bra and a ket, $\langle\phi|\psi\rangle$, which is a scalar, we may define an *outer product*, $|\psi\rangle\langle\phi|$. This object is an operator because, assuming associative multiplication, we have

$$(|\psi\rangle\langle\phi|)|\lambda\rangle = |\psi\rangle(\langle\phi|\lambda\rangle)\,. \tag{1.19}$$

Since an operator is defined by specifying its action on an arbitrary vector to produce another vector, this equation fully defines $|\psi\rangle\langle\phi|$ as an operator. From (1.18) it follows that

$$(|\psi\rangle\langle\phi|)^\dagger = |\phi\rangle\langle\psi|\,. \tag{1.20}$$

In view of this relation, it is tempting to write $(|\psi\rangle)^\dagger = \langle\psi|$. Although no real harm comes from such a notation, it should not be encouraged because it uses the "adjoint" symbol, †, for something that is not an operator, and so cannot satisfy the fundamental definition (1.16).

A useful characteristic of an operator A is its *trace*, defined as

$$\text{Tr } A = \sum_j \langle u_j|A|u_j\rangle ,$$

where $\{|u_j\rangle\}$ may be any orthonormal basis. It can be shown [see Problem (1.3)] that the value of Tr A is independent of the particular orthonormal basis that is chosen for its evaluation. The trace of a matrix is just the sum of its diagonal elements. For an operator in an infinite-dimensional space, the trace exists only if the infinite sum is convergent.

1.3 *Self-Adjoint Operators*

An operator A that is equal to its adjoint $A^\dagger$ is called *self-adjoint.* This means that it satisfies

$$\langle\phi|A|\psi\rangle = \langle\psi|A|\phi\rangle * \tag{1.21}$$

and that the domain of A (i.e. the set of vectors ϕ on which $A\phi$ is well defined) coincides with the domain of $A^\dagger$. An operator that only satisfies (1.21) is called *Hermitian*, in analogy with a Hermitian matrix, for which $M_{ij} = M_{ji}*$.

[[The distinction between Hermitian and self-adjoint operators is relevant only for operators in infinite-dimensional vector spaces, and we shall make such a distinction only when it is essential to do so. The operators that we call "Hermitian" are often called "symmetric" in the mathematical literature. That terminology is objectionable because it conflicts with the corresponding properties of matrices.]]

The following theorem is useful in identifying Hermitian operators on a vector space with complex scalars.

Theorem 1. If $\langle\psi|A|\psi\rangle = \langle\psi|A|\psi\rangle^*$ for all $|\psi\rangle$, then it follows that $\langle\phi_1|A|\phi_2\rangle = \langle\phi_2|A|\phi_1\rangle^*$ for all $|\phi_1\rangle$ and $|\phi_2\rangle$, and hence that $A = A^\dagger$.

Proof. Let $|\psi\rangle = a|\phi_1\rangle + b|\phi_2\rangle$ for arbitrary a, b, $|\phi_1\rangle$, and $|\phi_2\rangle$.

Then

$$\begin{aligned}\langle\psi|A|\psi\rangle = |a|^2\langle\phi_1|A|\phi_1\rangle + |b|^2\langle\phi_2|A|\phi_2\rangle \\ + a^*b\langle\phi_1|A|\phi_2\rangle + b^*a\langle\phi_2|A|\phi_1\rangle\end{aligned}$$

must be real. The first and second terms are obviously real by hypothesis, so we need only consider the third and fourth. Choosing the arbitrary parameters a and b to be $a = b = 1$ yields the condition

$$\langle\phi_1|A|\phi_2\rangle + \langle\phi_2|A|\phi_1\rangle = \langle\phi_1|A|\phi_2\rangle^* + \langle\phi_2|A|\phi_1\rangle^* .$$

Choosing instead $a = 1$, $b = i$ yields

$$i\langle\phi_1|A|\phi_2\rangle - i\langle\phi_2|A|\phi_1\rangle = -i\langle\phi_1|A|\phi_2\rangle^* + i\langle\phi_2|A|\phi_1\rangle^* .$$

Canceling the factor of i from the last equation and adding the two equations yields the desired result, $\langle\phi_1|A|\phi_2\rangle = \langle\phi_2|A|\phi_1\rangle^*$.

This theorem is noteworthy because the premise is obviously a special case of the conclusion, and it is unusual for the general case to be a consequence of a special case. Notice that the complex values of the scalars were essential in the proof, and no analog of this theorem can exist for real vector spaces.

If an operator acting on a certain vector produces a scalar multiple of that same vector,

$$A|\phi\rangle = a|\phi\rangle , \tag{1.22}$$

we call the vector $|\phi\rangle$ an *eigenvector* and the scalar a an *eigenvalue* of the operator A. The antilinear correspondence (1.8) between bras and kets, and the definition of the adjoint operator $A^\dagger$, imply that the left-handed eigenvalue equation

$$\langle\phi|A^\dagger = a^*\langle\phi| \tag{1.23}$$

holds if the right-handed eigenvalue equation (1.22) holds.

Theorem 2. If A is a Hermitian operator then all of its eigenvalues are real.

Proof. Let $A|\phi\rangle = a|\phi\rangle$. Since A is Hermitian, we must have $\langle\phi|A|\phi\rangle = \langle\phi|A|\phi\rangle^*$. Substitution of the eigenvalue equation yields

$$\langle\phi|a|\phi\rangle = \langle\phi|a|\phi\rangle^* ,$$

$$a\langle\phi|\phi\rangle = a^*\langle\phi|\phi\rangle ,$$

which implies that $a = a^*$, since only nonzero vectors are regarded as nontrivial solutions of the eigenvector equation.

The result of this theorem, combined with (1.23), shows that for a self-adjoint operator, $A = A^\dagger$, the conjugate bra $\langle\phi|$ to the ket eigenvector $|\phi\rangle$ is also an eigenvector with the same eigenvalue a: $\langle\phi|A = a\langle\phi|$.

Theorem 3. Eigenvectors corresponding to distinct eigenvalues of a Hermitian operator must be orthogonal.

Proof. Let $A|\phi_1\rangle = a_1|\phi_1\rangle$ and $A|\phi_2\rangle = a_2|\phi_2\rangle$. Since A is Hermitian, we deduce from (1.21) that

$$\begin{aligned} 0 &= \langle\phi_1|A|\phi_2\rangle - \langle\phi_2|A|\phi_1\rangle^* \\ &= a_1\langle\phi_2|\phi_1\rangle - a_2\langle\phi_1|\phi_2\rangle^* \\ &= (a_1 - a_2)\langle\phi_2|\phi_1\rangle\,. \end{aligned}$$

Therefore $\langle\phi_2|\phi_1\rangle = 0$ if $a_1 \neq a_2$.

If $a_1 = a_2$ ($= a$, say) then any linear combination of the *degenerate* eigenvectors $|\phi_1\rangle$ and $|\phi_2\rangle$ is also an eigenvector with the same eigenvalue a. It is always possible to replace a nonorthogonal but linearly independent set of degenerate eigenvectors by linear combinations of themselves that are orthogonal. Unless the contrary is explicitly stated, we shall assume that such an orthogonalization has been performed, and when we speak of the set of independent eigenvectors of a Hermitian operator we shall mean an orthogonal set.

Provided the vectors have finite norms, we may rescale them to have unit norms. Then we can always choose to work with an *orthonormal* set of eigenvectors,

$$(\phi_i, \phi_j) = \delta_{ij}\,. \tag{1.24}$$

Many textbooks state (confidently or hopefully) that the orthonormal set of eigenvectors of a Hermitian operators is *complete*; that is to say, it forms a basis that spans the vector space. Before examining the mathematical status of that statement, let us see what useful consequences would follow if it were true.

Properties of complete orthonormal sets

If the set of vectors $\{\phi_i\}$ is *complete*, then we can expand an arbitrary vector $|v\rangle$ in terms of it: $|v\rangle = \sum_i v_i|\phi_i\rangle$. From the orthonormality condition (1.24), the expansion coefficients are easily found to be $v_i = \langle\phi_i|v\rangle$. Thus we can write

$$|v\rangle = \sum_i |\phi_i\rangle(\langle\phi_i|v\rangle)$$
$$= \left(\sum_i |\phi_i\rangle\langle\phi_i|\right)|v\rangle \tag{1.25}$$

for an arbitrary vector $|v\rangle$. The parentheses in (1.25) are unnecessary, and are used only to emphasize two ways of interpreting the equation. The first line in (1.25) suggests that $|v\rangle$ is equal to a sum of basis vectors each multiplied by a scalar coefficient. The second line suggests that a certain operator (in parentheses) acts on a vector to produce the same vector. Since the equation holds for all vectors $|v\rangle$, the operator must be the identity operator,

$$\sum_i |\phi_i\rangle\langle\phi_i| = I\,. \tag{1.26}$$

If $A|\phi_i\rangle = a_i|\phi_i\rangle$ and the eigenvectors form a complete orthonormal set — that is to say, (1.24) and (1.26) hold — then the operator can be reconstructed in a useful diagonal form in terms of its eigenvalues and eigenvectors:

$$A = \sum_i a_i|\phi_i\rangle\langle\phi_i|\,. \tag{1.27}$$

This result is easily proven by operating on an arbitrary vector and verifying that the left and right sides of (1.27) yield the same result. One can use the diagonal representation to define *a function of an operator*,

$$f(A) = \sum_i f(a_i)|\phi_i\rangle\langle\phi_i|\,. \tag{1.28}$$

The usefulness of these results is the reason why many authors assume, in the absence of proof, that the Hermitian operators encountered in quantum mechanics will have complete sets of eigenvectors. *But is it true?*

Any operator in a finite N-dimensional vector space can be expressed as an $N \times N$ matrix [see the discussion following Eq. (1.10)]. The condition for a nontrivial solution of the matrix eigenvalue equation

$$M\phi = \lambda\phi\,, \tag{1.29}$$

where M is square matrix and ϕ is a column vector, is

$$\det|M - I| = 0\,. \tag{1.30}$$

The expansion of this determinant yields a polynomial in λ of degree N, which must have N roots. Each root is an eigenvalue to which there must correspond an eigenvector. If all N eigenvalues are distinct, then so must be the eigenvectors, which will necessarily span the N-dimensional space. A more careful argument is necessary in order to handle multiple roots (degenerate eigenvalues), but the proof is not difficult. [See, for example, Jordan (1969), Theorem 13.1.]

This argument does not carry over to infinite-dimensional spaces. Indeed, if one lets N become infinite, then (1.30) becomes an infinite power series in λ, which need not possess any roots, even if it converges. (In fact the determinant of an infinite-dimensional matrix is undefinable except in special cases.) A simple counter-example shows that the theorem is not generally true for an infinite-dimensional space.

Consider the operator $D = -id/dx$, defined on the space of differentiable functions of x for $a \leq x \leq b$. (The limits a and b may be finite or infinite.) Its adjoint, $D^\dagger$, is identified by using (1.21), which now takes the form

$$\begin{aligned} \int_a^b \phi^*(x) D^\dagger \psi(x) dx &= \left\{ \int_a^b \psi^*(x) D\phi(x) dx \right\}^* \\ &= \int_a^b \phi^*(x) D\psi(x) dx + i[\psi(x)\phi^*(x)]|_a^b . \end{aligned} \tag{1.31}$$

The last line is obtained by integrating by parts. If boundary conditions are imposed so that the last term vanishes, then D will apparently be a Hermitian operator.

The eigenvalue equation

$$-i\frac{d}{dx}\phi(x) = \lambda\phi(x) \tag{1.32}$$

is a differential equation whose solution is $\phi(x) = ce^{i\lambda x}, c = \text{constant}$. But in regarding it as an eigenvalue equation for the operator D, we are interested only in eigenfunctions within a certain vector space. Several different vector spaces may be defined, depending upon the boundary conditions that are imposed:

V1. No boundary conditions

All complex λ are eigenvalues. Since D is not Hermitian this case is of no further interest.

V2. $a = -\infty$, $b = +\infty$, $|\phi(x)|$ bounded as $|x| \to \infty$

All real values of λ are eigenvalues. The eigenfunctions $\phi(x)$ are not normalizable, but they do form a complete set in the sense that an arbitrary function can be represented as a Fourier integral, which may be regarded as a continuous linear combination of the eigenfunctions.

V3. $a = -L/2$, $b = +L/2$, periodic boundary conditions $\phi(-L/2) = \phi(L/2)$

The eigenvalues form a discrete set, $\lambda = \lambda_n = 2\pi n/L$, with n being an integer of either sign. The eigenfunctions form a complete orthonormal set (with a suitable choice for c), the completeness being proven in the theory of Fourier series.

V4. $a = -\infty$, $b = +\infty$, $\phi(x) \to 0$ as $x \to \pm\infty$

Although the operator D is Hermitian, it has no eigenfunctions within this space.

These examples suffice to show that *a Hermitian operator in an infinite-dimensional vector space may or may not possess a complete set of eigenvectors*, depending upon the precise nature of the operator and the vector space. Fortunately, the desirable results like (1.26), (1.27) and (1.28) can be reformulated in a way that does not require the existence of well-defined eigenvectors.

The spectral theorem

The outer product $|\phi_i\rangle\langle\phi_i|$ formed from a vector of unit norm is an example of a *projection operator*. In general, a self-adjoint operator p that satisfies $p^2 = p$ is a projection operator. Its action is to project out the component of a vector that lies within a certain subspace (the one-dimensional space of $|\phi_i\rangle$ in the above example), and to annihilate all components orthogonal to that subspace. If the operator A in (1.27) has a degenerate spectrum, we may form the projection operator onto the subspace spanned by the degenerate eigenvectors corresponding to $a_i = a$,

$$P(a) = \sum_i |\phi_i\rangle\langle\phi_i|\delta_{a,a_i} \tag{1.33}$$

and (1.27) can be rewritten as

$$A = \sum_a aP(a)\,. \tag{1.34}$$

The sum on a goes over the eigenvalue spectrum. [But since $P(a) = 0$ if a is not an eigenvalue, it is harmless to extend the sum beyond the spectrum.]

The examples following (1.32) suggest (correctly, it turns out) that the troubles are associated with a continuous spectrum, so it is desirable to rewrite (1.34) in a form that holds for both discrete and continuous spectra. This can most conveniently be done with the help of the *Stieltjes integral*, whose definition is

$$\int_a^b g(x)d\sigma(x) = \lim_{n\to\infty} \sum_{k=1}^{n} g(x_k)[\sigma(x_k) - \sigma(x_{k-1})]\,, \tag{1.35}$$

the limit being taken such that every interval $(x_k - x_{k-1})$ goes to zero as $n \to \infty$. The nondecreasing function $\sigma(x)$ is called the *measure.* If $\sigma(x) = x$, then (1.35) reduces to the more familiar Riemann integral. If $d\sigma/dx$ exists, then we have

$$\int_{\text{(Stieltjes)}} g(x)d\sigma(x) = \int_{\text{(Riemann)}} g(x)\left(\frac{d\sigma}{dx}\right) dx\,.$$

The generalization becomes nontrivial only when we allow $\sigma(x)$ to be discontinuous. Suppose that

$$\sigma(x) = h\theta(x - c)\,, \tag{1.36}$$

where $\theta(x) = 0$ for $x < 0$, $\theta(x) = 1$ for $x > 0$. The only term in (1.35) that will contribute to the integral is the term for which $x_{k-1} < c$ and $x_k > c$. The value of the integral is $hg(c)$.

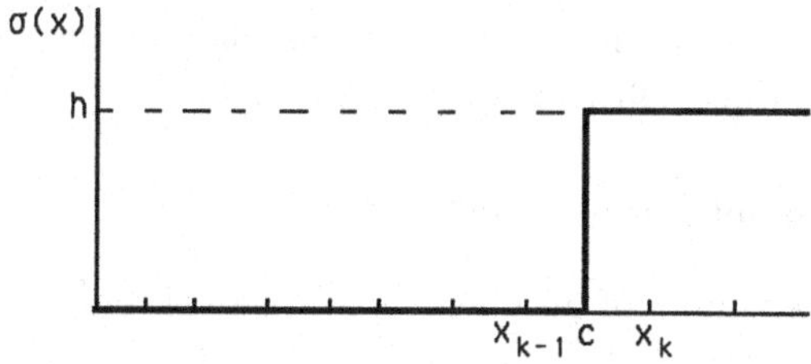

Fig. 1.1 A discontinuous measure function [Eq. (1.36)].

We can now state the *spectral theorem.*

Theorem 4. [For a proof, see Riesz and Sz.-Nagy (1955), Sec. 120.] To each self-adjoint operator A there corresponds a unique family of projection operators, $E(\lambda)$, for real λ, with the properties:

(i) If $\lambda_1 < \lambda_2$ then $E(\lambda_1)E(\lambda_2) = E(\lambda_2)E(\lambda_1) = E(\lambda_1)$
[speaking informally, this means that $E(\lambda)$ projects onto the subspace corresponding to eigenvalues $\leq \lambda$];
(ii) If $\varepsilon > 0$, then $E(\lambda + \varepsilon)|\psi\rangle \to E(\lambda)|\psi\rangle$ as $\varepsilon \to 0$;
(iii) $E(\lambda)|\psi\rangle \to 0$ as $\lambda \to -\infty$;
(iv) $E(\lambda)|\psi\rangle \to |\psi\rangle$ as $\lambda \to +\infty$;
(v) $\int_{-\infty}^{\infty} \lambda dE(\lambda) = A$. (1.37)

In (ii), (iii) and (iv) $|\psi\rangle$ is an arbitrary vector. The integral in (v) with respect to an operator-valued measure $E(\lambda)$ is formally defined by (1.35), just as for a real valued measure.

Equation (1.37) is the generalization of (1.27) to an arbitrary self-adjoint operator that may have discrete or continuous spectra, or a mixture of the two. The corresponding generalization of (1.28) is

$$f(A) = \int_{-\infty}^{\infty} f(\lambda)dE(\lambda)\,. \tag{1.38}$$

Example (discrete case)

When (1.37) is applied to an operator with a purely discrete spectrum, the only contributions to the integral occur at the discontinuities of

$$E(\lambda) = \sum_i |\phi_i\rangle\langle\phi_i|\theta(\lambda - a_i)\,. \tag{1.39}$$

These occur at the eigenvalues, the discontinuity at $\lambda = a$ being just $P(a)$ of Eq. (1.33). Thus (1.37) reduces to (1.34) or (1.27) in this case.

Example (continuous case)

As an example of an operator with a continuous spectrum, consider the operator Q, defined as $Q\psi(x) = x\psi(x)$ for all functions $\psi(x)$. It is trivial to verify that $Q = Q^\dagger$. Now the eigenvalue equation $Q\phi(x) = \lambda\phi(x)$ has the formal solutions $\phi(x) = \delta(x - \lambda)$, where λ is any real number and $\delta(x - \lambda)$ is Dirac's "delta function". But in fact $\delta(x - \lambda)$ is not a well-defined function[a] at all, so strictly speaking there are no eigenfunctions $\phi(x)$.

[a]It can be given meaning as a "distribution", or "generalized function". See Gel'fand and Shilov (1964) for a systematic treatment.

However, the spectral theorem still applies. The projection operators for Q are defined as

$$E(\lambda)\psi(x) = \theta(\lambda - x)\psi(x)\,, \tag{1.40}$$

which is equal to $\psi(x)$ for $x < \lambda$, and is 0 for $x > \lambda$. We can easily verify (1.37) by operating on a general function $\psi(x)$:

$$\begin{aligned}\int_{-\infty}^{\infty} \lambda dE(\lambda)\psi(x) &= \int_{-\infty}^{\infty} \lambda d[\theta(\lambda - x)\psi(x)] \\ &= x\psi(x) = Q\psi(x)\,.\end{aligned}$$

(In evaluating the above integral one must remember that λ is the integration variable and x is constant.)

Following Dirac's pioneering formulation, it has become customary in quantum mechanics to write a formal eigenvalue equation for an operator such as Q that has a continuous spectrum,

$$Q|q\rangle = q|q\rangle\,. \tag{1.41}$$

The orthonormality condition for the continuous case takes the form

$$\langle q'|q''\rangle = \delta(q' - q'')\,. \tag{1.42}$$

Evidently the norm of these formal eigenvectors is infinite, since (1.42) implies that $\langle q|q\rangle = \infty$. Instead of the spectral theorem (1.37) for Q, Dirac would write

$$Q = \int_{-\infty}^{\infty} q|q\rangle\langle q|dq\,, \tag{1.43}$$

which is the continuous analog of (1.27).

Dirac's formulation does not fit into the mathematical theory of *Hilbert space*, which admits only vectors of finite norm. The projection operator (1.40), formally given by

$$E(\lambda) = \int_{-\infty}^{\lambda} |q\rangle\langle q|dq\,, \tag{1.44}$$

is well defined in Hilbert space, but its derivative, $dE(q)/dq = |q\rangle\langle q|$, does not exist within the Hilbert space framework.

Most attempts to express quantum mechanics within a mathematically rigorous framework have restricted or revised the formalism to make it fit within Hilbert space. An attractive alternative is to extend the Hilbert space framework so that vectors of infinite norm can be treated consistently. This will be considered in the next section.

Commuting sets of operators

So far we have discussed only the properties of single operators. The next two theorems deal with two or more operators together.

Theorem 5. If A and B are self-adjoint operators, each of which possesses a complete set of eigenvectors, and if $AB = BA$, then there exists a complete set of vectors which are eigenvectors of both A and B.

Proof. Let $\{|a_n\rangle\}$ and $\{|b_m\rangle\}$ be the complete sets of eigenvectors of A and B, respectively: $A|a_n\rangle = a_n|a_n\rangle$, $B|b_m\rangle = b_m|b_m\rangle$. We may expand any eigenvector of A in terms of the set of eigenvectors of B:

$$|a_n\rangle = \sum_m c_m|b_m\rangle\,,$$

where the coefficients c_m depend on the particular vector $|a_n\rangle$. The eigenvalues b_m need not be distinct, so it is desirable to combine all terms with $b_m = b$ into a single vector,

$$|(a_n)b\rangle = \sum_m c_m|b_m\rangle\delta_{b,b_m}\,.$$

We may then write

$$|a_n\rangle = \sum_b |(a_n)b\rangle\,, \tag{1.45}$$

where the sum is over distinct eigenvalues of B. Now

$$\begin{aligned}(A - a_n)|a_n\rangle &= 0\\ &= \sum_b (A - a_n)|(a_n)b\rangle\,.\end{aligned} \tag{1.46}$$

By operating on a single term of (1.46) with B, and using $BA = AB$,

$$\begin{aligned}B(A - a_n)|(a_n)b\rangle &= (A - a_n)B|(a_n)b\rangle\\ &= b(A - a_n)|(a_n)b\rangle\,,\end{aligned}$$

we deduce that the vector $(A - a_n)|(a_n)b\rangle$ is an eigenvector of B with eigenvalue b. Therefore the terms in the sum (1.46) must be orthogonal, and so are linearly independent. The vanishing of the sum is possible only if each term vanishes separately:

$$(A - a_n)|(a_n)b\rangle = 0\,.$$

Thus $|(a_n)b\rangle$ is an eigenvector of both A and B, corresponding to the eigenvalues a_n and b, respectively. Since the set $\{|a_n\rangle\}$ is complete, the set $\{|(a_n)b\rangle\}$ in terms of which it is expanded must also be complete. Therefore there exists a complete set of common eigenvectors of the commuting operators A and B.

The theorem can easily be extended to any number of mutually commutative operators. For example, if we have three such opeators, A, B and C, we may expand an eigenvector of C in terms of the set of eigenvectors of A and B, and proceed as in the above proof to deduce a complete set of common eigenvectors for A, B and C.

The converse of the theorem, that if A and B possess a complete set of common eigenvectors then $AB = BA$, is trivial to prove using the diagonal representation (1.27).

Let $(A, B, \ldots)$ be a set of mutually commutative operators that possess a complete set of common eigenvectors. Corresponding to a particular eigenvalue for each operator, there may be more than one eigenvector. If, however, there is no more than one eigenvector (apart from the arbitrary phase and normalization) for each set of eigenvalues $(a_n, b_m, \ldots)$, then the operators $(A, B, \ldots)$ are said to be a *complete commuting set* of operators.

Theorem 6. Any operator that commutes with all members of a complete commuting set must be a function of the operators in that set.

Proof. Let $(A, B, \ldots)$ be a complete set of commuting operators, whose common eigenvectors may be uniquely specified (apart from phase and normalization) by the eigenvalues of the operators. Denote a typical eigenvector as $|a_n, b_m, \ldots\rangle$. Let F be an operator that commutes with each member of the set $(A, B, \ldots)$. To say that F is a function of this set of operators is to say, in generalization of (1.28), that F has the same eigenvectors as this set of operators, and that the eigenvalues of F are a function of the eigenvalues of this set of operators. Now since F commutes with $(A, B, \ldots)$, it follows from Theorem 5 that there exists a complete set of common eigenvectors of $(A, B, \ldots, F)$. But since the vectors $|a_n, b_m, \ldots\rangle$ are the unique set of eigenvectors of the complete commuting set $(A, B, \ldots)$, it follows that they must also be the eigenvectors of the augmented set $(A, B, \ldots, F)$. Thus

$$F|a_n, b_m, \ldots\rangle = f_{nm\cdots}|a_n, b_m, \ldots\rangle .$$

Since the eigenvector is uniquely determined (apart from phase and normalization) by the eigenvalues $(a_n, b_m, \ldots)$, it follows that the mapping $(a_n, b_m, \ldots) \to f_{nm\ldots}$ exists, and hence the eigenvalues of F maybe regarded as a function of the eigenvalues of $(A, B, \ldots)$. That is to say, $f_{nm\cdots} = f(a_n, b_m, \ldots)$. This completes the proof that the operator F is a function of the operators in the complete commuting set, $F = f(A, B, \ldots)$.

For many purposes a complete commuting set of operators may be regarded as equivalent to a single operator with a non-degenerate eigenvalue spectrum. Indeed such a single operator is, by itself, a complete commuting set.

1.4 *Hilbert Space and Rigged Hilbert Space*

A linear vector space was defined in Sec. 1.1 as a set of elements that is closed under addition and multiplication by scalars. All finite-dimensional spaces of the same dimension are isomorphic, but some distinctions are necessary among infinite-dimensional spaces. Consider an infinite orthonormal set of basis vectors, $\{\phi_n : n = 1, 2, \ldots\}$. From it we can construct a linear vector space V by forming all possible finite linear combinations of basis vectors. Thus V consists of all vectors of the form $\psi = \sum_n c_n \phi_n$, where the sum may contain any *finite* number of terms.

The space V may be enlarged by adding to it the limit points of convergent infinite sequences of vectors, such as the sums of convergent infinite series. But first we must define what we mean by *convergence* in a space of vectors. The most useful definition is in terms of the norm. We say that the sequence $\{\psi_i\}$ approaches the limit vector χ as $i \to \infty$ if and only if $\lim_{i\to\infty} \|\psi_i - \chi\| = 0$.

The addition of all such limit vectors to the space V yields a larger space, $\mathcal{H}$. For example, the vectors of the form

$$\psi_i = \sum_{n=1}^{i} c_n \phi_n$$

are members of V for all finite values of i. The limit vector as $i \to \infty$ is not a member of V, but it is a member of $\mathcal{H}$ provided $\sum_n |c_n|^2$ is finite. The space $\mathcal{H}$ is called a *Hilbert space* if it contains the limit vectors of all norm-convergent sequences. (In technical jargon, $\mathcal{H}$ is called the completion of V with respect to the norm topology.)

A Hilbert space has the property of preserving the one-to-one correspondence between vectors in $\mathcal{H}$ and members of its dual space $\mathcal{H}'$, composed of continuous linear functionals, which was proved for finite-dimensional spaces in Sec. 1.1. We omit the standard proof (see Jordan, 1969), and proceed instead to an alternative approach that is more useful for our immediate needs, although it has less mathematical generality.

Let us consider our universe of vectors to be the linear space Ξ which consists of all formal linear combinations of the basis vectors $\{\phi_n\}$. A general member of Ξ has the form $\xi = \sum_n c_n\phi_n$, with no constraint imposed on the coefficients c_n. We may think of it as an infinite column vector whose elements c_n are unrestricted in either magnitude or number. Of course the norm and the inner product will be undefined for many vectors in Ξ, and we will focus our attention on certain well-behaved subspaces.

The *Hilbert space* $\mathcal{H}$ is a subspace of Ξ defined by the constraint that $h = \sum_n c_n\phi_n$ is a member of $\mathcal{H}$ if and only if $(h, h) = \sum_n |c_n|^2$ is finite. We now define its *conjugate space*, $\mathcal{H}^\times$, as consisting of all vectors $f = \sum_n b_n\phi_n$ for which the inner product $(f, h) = \sum_n b_n^* c_n$ is convergent for all h in $\mathcal{H}$, and (f, h) is a continuous linear functional on $\mathcal{H}$. It is possible to choose the vector h such that the phase of c_n equals that of b_n, making $b_n^* c_n$ real positive. Thus the convergence of $(f, h) = \sum_n b_n^* c_n$ will be assured if $|b_n|$ goes to zero at least as rapidly as $|c_n|$ in the limit $n \to \infty$, since $\sum_n |c_n|^2$ is convergent. This implies that $\sum_n |b_n|^2$ will also be convergent, and hence the vector f (an arbitrary member of $H^\times$) is also an element of $\mathcal{H}$. Therefore a Hilbert space is identical with its conjugate space,[b] $\mathcal{H} = \mathcal{H}^\times$.

Let us now define a space Ω consisting of all vectors of the form $\omega = \sum_n u_n\phi_n$, with the coefficients subject to the infinite set of conditions:

$$\sum_n |u_n|^2 n^m < \infty \text{ for } m = 0, 1, 2, \ldots .$$

The space Ω, which is clearly a subspace of $\mathcal{H}$, is an example of a *nuclear space*. The conjugate space to Ω, $\Omega^\times$, consists of those vectors $\sigma = \sum_n v_n\phi_n$ such that $(\sigma, \omega) = \sum_n v_n^* u_n$ is convergent for all ω in Ω, and $(\sigma, \cdot)$ is continuous linear functional on Ω. It is clear that $\Omega^\times$ is a much larger space than Ω, since a vector σ will be admissible if its coefficients v_n blow up no faster than a power of n as $n \to \infty$.

Finally, we observe that the space $V^\times$, which is conjugate to V, is the entire space Ξ, since a vector in V has only a finite number of components and so

[b] The conjugate space $\mathcal{H}^\times$ is closely related to the dual space $\mathcal{H}'$. The only important difference is that the one-to-one correspondence between vectors in $\mathcal{H}$ and vectors in $\mathcal{H}'$ is antilinear, (1.8), whereas $\mathcal{H}$ and $\mathcal{H}^\times$ are strictly isomorphic. So one may regard $\mathcal{H}'$ as the complex conjugate of $\mathcal{H}^\times$. Our argument is not quite powerful enough to establish the strict identity of $\mathcal{H}$ with $\mathcal{H}^\times$. Suppose that $c_n \sim n^{-\gamma}$ and $b_n \sim n^{-\beta}$ for large n. The convergence of $\sum_n |c_n|^2$ requires that $\gamma > 1/2$. The convergence of $\sum_n b_n^* c_n$ requires that $\beta + \gamma > 1$. Thus $\beta > 1/2$ is admissible and $\beta < 1/2$ is not admissible. To exclude the marginal case of $\beta = 1/2$ one must invoke the continuity of the linear functional $(f, \cdot)$, as in the standard proof (Jordan, 1969).

no convergence questions arise. Thus the various spaces and their conjugates satisfy the following inclusion relations:

$$V \subset \Omega \subset \mathcal{H} = \mathcal{H}^\times \subset \Omega^\times \subset V^\times = \Xi\,.$$

The important points to remember are:

(a) The smaller or more restricted is a space, the larger will be its conjugate, and

(b) The Hilbert space is unique in being isomorphic to its conjugate.

Of greatest interest for applications is the triplet $\Omega \subset \mathcal{H} \subset \Omega^\times$, which is called a *rigged Hilbert space.* (The term "rigged" should be interpreted as "equipped and ready for action", in analogy with the rigging of a sailing ship.) As was shown in Sec. 1.3, there may or may not exist any solutions to the eigenvalue equation $A|a_n\rangle = a_n|a_n\rangle$ for a self-adjoint operator A on an infinite-dimensional vector space. However, the *generalized spectral theorem* asserts that if A is self-adjoint in $\mathcal{H}$ then a complete set of eigenvectors exists in the extended space $\Omega^\times$. The precise conditions for the proof of this theorem are rather technical, so the interested reader is referred to Gel'fand and Vilenkin (1964) for further details.

We now have two mathematically sound solutions to the problem that a self-adjoint operator need not possess a complete set of eigenvectors in the Hilbert space of vectors with finite norms. The first, based on the spectral theorem (Theorem 4 of Sec. 1.3), is to restate our equations in terms of projection operators which are well defined in Hilbert space, even if they cannot be expressed as sums of outer products of eigenvectors in Hilbert space. The second, based on the generalized spectral theorem, is to enlarge our mathematical framework from Hilbert space to rigged Hilbert space, in which a complete set of eigenvectors (of possibly infinite norm) is guaranteed to exist. The first approach has been most popular among mathematical physicists in the past, but the second is likely to grow in popularity because it permits full use of Dirac's bra and ket formalism.

There are many examples of rigged-Hilbert-space triplets, and although the previous example, based on vectors of infinitely many discrete components, is the simplest to analyze, it is not the only useful example. If Ξ is taken to be the space of functions of one variable, then a *Hilbert space* $\mathcal{H}$ is formed by those functions that are square-integrable. That is, $\mathcal{H}$ consists of those functions $\psi(x)$ for which

$$(\psi, \psi) = \int_{-\infty}^{\infty} |\psi(x)|^2 dx \text{ is finite}\,.$$

A *nuclear space* Ω is made up of functions $\phi(x)$ which satisfy the infinite set of conditions,

$$\int_{-\infty}^{\infty} |\phi(x)|^2 (1+|x|)^m dx < \infty \quad (m = 0, 1, 2, \ldots)\,.$$

The functions $\phi(x)$ which make up Ω must vanish more rapidly than any inverse power of x in the limit $|x| \to \infty$. The *extended space* $\Omega^\times$, which is conjugate to Ω, consists of those functions $\chi(x)$ for which

$$(\chi, \phi) = \int_{-\infty}^{\infty} \chi^*(x)\phi(x) dx \text{ is finite for all } \phi \text{ in } \Omega\,.$$

In addition to the functions of finite norm, which also lie in $\mathcal{H}$, $\Omega^\times$ will contain functions that are unbounded at infinity provided the divergence is no worse than a power of x. Hence $\Omega^\times$ contains e^{ikx}, which is an eigenfunction of the operator $D = id/dx$. It also contains the Dirac delta function, $\delta(x - \lambda)$, which is an eigenfunction of the operator Q, defined by $Q\psi(x) = (x)\psi(x)$. These two examples suffice to show that rigged Hilbert space seems to be a more natural mathematical setting for quantum mechanics than is Hilbert space.

1.5 *Probability Theory*

The mathemetical content of the probability theory concerns the properties of a function $\text{Prob}(A|B)$, which is the *probability* of event A under the conditions specified by event B. In this Section we will use the shortened notation $P(A|B) \equiv \text{Prob}(A|B)$, but in later applications, where the symbol P may have other meanings, we may revert to the longer notation. The meaning or interpretation of the term "probability" will be discussed later, when we shall also interpret what is meant by an "event". But first we shall regard them as mathematical terms defined only by certain axioms.

It is desirable to treat sets of events as well as elementary events. Therefore we introduce certain composite events: $\sim A$ ("not A") denotes the nonoccurrence of A; $A \,\&B$ ("A and B") denotes the occurrence of both A and B; $A \vee B$ ("A or B") denotes the occurrence of at least one of the events A and B. These composite events will also be referred to as events. The three operators ($\sim$, &, $\vee$) are called *negation*, *conjunction*, and *disjunction*. In the evaluation of complex expressions, the negation operator has the highest precedence. Thus $\sim A \,\&B = (\sim A) \,\&B$, and $\sim A \vee B = (\sim A) \vee B$.

The axioms of probability theory can be given in several different but mathematically equivalent forms. The particular form given below is based on the work of R. T. Cox (1961)

Axiom 1: $0 \leq P(A|B) \leq 1$
Axiom 2: $P(A|A) = 1$
Axiom 3a: $P(\sim A|B) = 1 - P(A|B)$
Axiom 4: $P(A\&B|C) = P(A|C)P(B|A\&C)$

Axiom 2 states the convention that the probability of a certainty (the occurrence of A, given the occurrence A) is 1, and Axiom 1 states that no probabilties are greater than the probalitity of a certainty. Axiom 3a expresses the intuitive notion that the probability of nonoccurrence of an event increases as the probabitily of its occurrence decreases. It also implies that $P(\sim A|A) = 0$; that is to say, an impossible event (the nonoccurrence of A given that A occurs) has zero probability. Axiom 4 states that the probability that two events both occur (under some condition C) is equal to the probabitily of occurrence of one of the events multiplied by the probability of the second event given that the first event has already occurred.

The probabilities of negation ($\sim A$) and conjunction ($A\&B$) of events each required an axiom. However, no further axioms are required to treat disjunction because $A \vee B = \sim(\sim A \ \& \ \sim B)$; in other words, "$A$ or B" is equivalent to the negation of "neither A nor B". From Axiom 3a we obtain

$$P(A \vee B|C) = 1 - P(\sim A \ \& \ \sim B|C)\,, \tag{1.47}$$

which can be evaluated from the existing axioms. First we prove a *lemma*, using Axioms 4 and 3a:

$$\begin{aligned} P(X\&Y|C) + P(X\&\sim Y|C) &= P(X|C)P(Y|X\&C) + P(X|C)P(\sim Y|X\&C) \\ &= P(X|C)\{P(Y|X\&C) + P(\sim Y|X\&C)\} \\ &= P(X|C)\,. \end{aligned} \tag{1.48}$$

Using (1.48) with $X = \sim A$ and $Y = \sim B$, we obtain $P(\sim A\&\sim B|C) = P(\sim A|C) - P(\sim A\&B|C)$. Applying Axiom 3a to the first term, and using (1.48) with $X = B$, $Y = A$ in the second term, we obtain $P(\sim A\&\sim B|C) = 1 - P(A|C) - P(B|C) + P(B\&A|C)$, and hence (1.47) becomes

$$P(A \vee B|C) = P(A|C) + P(B|C) - P(A\&B|C)\,. \tag{1.49}$$

If $P(A\,\&B|C) = 0$ we say that the events A and B are *mutually exclusive* on condition C. Then (1.49) reduces to the rule of *addition of probabilities for exclusive events*, which may be used as an alternative to Axiom 3a.

$$\textbf{Axiom 3b:} \qquad P(A \vee B|C) = P(A|C) + P(B|C)\,. \tag{1.49a}$$

The two axiom systems (1, 2, 3a, 4) and (1, 2, 3b, 4), are equivalent. We have just shown that Axioms 3a and 4 imply Axiom 3b. Conversely, since A and $\sim A$ are exclusive events, and $A \vee \sim A$ is a certainty, it is clear that Axiom 3b implies Axiom 3a. Axiom 3a is more elegant, since it applies to all events, however Axiom 3b offers some practical advantages.

Since $A\&B = B\&A$, it follows from Axiom 4 that

$$P(A|C)P(B|A\&C) = P(B|C)P(A|B\&C)\,. \tag{1.50}$$

If $P(A|C) \neq 0$ this leads to *Bayes' theorem*,

$$P(B|A\&C) = P(A|B\&C)P(B|C)/P(A|C)\,. \tag{1.51}$$

This theorem is noteworthy because it relates the probability of B given A to the probability of A given B, and hence it is also known as the principle of *inverse probability*.

Independence. To say that event B is *independent* of event A means that $P(B|A\,\&\,C) = P(B|C)$. That is, the occurrence of A has no influence on the probability of B. Axiom 4 then implies that if A and B are independent (given C) then

$$P(A\,\&B|C) = P(A|C)P(B|C)\,. \tag{1.52}$$

The symmetry of this formula implies that independence is a mutual relationship; if B is independent of A then also A is independent of B. This form of independence is called *statistical* or *stochastic* independence, in order to distinguish it from other notions, such as causal independence.

A set of n events $\{A_k\}(1 < k < n)$ is stochastically independent, given C, if and only if

$$P(A_i\,\&A_j\,\&\cdots\&A_k|C) = P(A_i|C)P(A_j|C)\cdots P(A_k|C) \tag{1.53}$$

holds for all subsets of $\{A_k\}$. It is not sufficient for (1.53) to hold only for the full set of n events; neither is it sufficient only for (1.52) to hold for all pairs.

Interpretations of probability

The abstract probability theory, consisting of axioms, definitions, and theorems, must be supplemented by an *interpretation* of the term "probability". This provides a correspondence rule by means of which the abstract theory can be applied to practical problems. There are many different interpretations of probability because anything that satisfies the axioms may be regarded as a kind of probability.

One of the oldest interpretations is the *limit frequency* interpretation. If the conditioning event C can lead to either A or $\sim A$, and if in n repetitions of such a situation the event A occurs m times, then it is asserted that $P(A|C) = \lim_{n\to\infty}(m/n)$. This provides not only an interpretation of probability, but also a *definition* of probability in terms of a numerical frequency ratio. Hence the axioms of abstract probability theory can be derived as theorems of the frequency theory. In spite of its superficial appeal, the limit frequency interpretation has been widely discarded, primarily because there is no assurance that the above limit really exists for the actual sequences of events to which one wishes to apply probability theory.

The defects of the limit frequency interpretation are avoided without losing its attractive features in the *propensity interpretation*. The probability $P(A|C)$ is interpreted as a measure of the tendency, or propensity, of the physical conditions describe by C to produce the result A. It differs logically from the older limit-frequency theory in that probability is interpreted, but not redefined or derived from anything more fundamental. It remains, mathematically, a fundamental undefined term, with its relationship to frequency emerging, suitably qualified, in a theorem. It also differs from the frequency theory in viewing probability (propensity) as a characteristic of the physical situation C that may potentially give rise to a sequence of events, rather than as a property (frequency) of an actual sequence of events. This fact is emphasized by always writing probability in the conditional form $P(A|C)$, and never merely as $P(A)$. The propensity interpretation of probability is particularly well suited for application to quantum mechanics. It was first applied to statistical physics (including quantum mechanics) by K. R. Popper (1957).

Another application of abstract probabilty theory that is useful in science is the theory of *inductive inference*. The "events", about which we can make probability statements, are replaced by *propositions* that may be either true or false, and the probability $P(\alpha|\gamma)$ is interprtated as the degree of reasonable belief in α given that γ is true. Some of the propositions considered in this theory are trivially related to the events of the propensity theory; proposition α could mean "event A has occurred". But it is also possible to assign probabilities to propositions that do not relate to contingent events, but rather to unknown facts. We can, in this theory, speak of the probability that the electronic charge is between 1.60×10^{-9} and 1.61×10^{-9} coulombs, conditional on some specific experimental data. The theory of inductive inference is useful for testing hypotheses, and for inferring uncertain parameters from statistical data.

The applications of probability theory to physical propensities and to degrees of reasonable belief may be loosely described as *objective* and *subjective* interpretations of probability. (This is an oversimplification, as some theories of inductive inference endeavor to be objective.) A great deal of acrimonious and unproductive debate has been generated over the question of which interpretation is correct or superior. In my view, much of that debate is misguided because the two theories address different classes of problems. Any interpretation of probability that conforms to the axioms is "correct". For example, probability concepts may be employed in number theory. The probability that two integers are relatively prime is $6/\pi^2$. Yet clearly this notion of "probability" refers neither to the propensity for physical variability nor to subjective uncertainty!

Probability and frequency

Suppose that a certain experimental procedure E can yield either of two results, A or $\sim A$, with the probability (propensity) for results A being $P(A|E) = p$. In n independent repetitions of the experiment (denoted as E^n) the result A may occur n_A times $(0 \leq n_A len)$. The probability of obtaining a particular ordered sequence containing A exactly r times and $\sim A$ exactly $n - r$ times is $p^r q^{n-r}$, where $q = 1 - p$. The various different permutations of the sequence are exclusive events, and so we can add their probabilities to obtain

$$P(n_A = r|E^n) = \frac{n!}{r!(n-r)!} p^r q^{n-r} . \tag{1.54}$$

This is known as the *binomial probability distribution.*

The frequency of A in the experiment E^n, $f_n = n_A/n$, is conceptually distinct from the probability p; nevertheless a relationship exists. Consider the average of n_A with respect to the probability distribution (1.54),

$$\langle n_A \rangle = \sum_{r=0}^{n} r P(n_A = r|E^n) .$$

This sum can be easily evaluated by a generating function technique, using the binomial identity,

$$\sum_{n=0}^{\infty} \frac{n!}{r!(n-r)!} p^r q^{n-r} = (p+q)^n .$$

It is apparent that

$$\langle n_A \rangle = p \frac{\partial}{\partial p} (p+q)^n \Big|_{q=1-p} = np .$$

Hence the average frequency of A is

$$\langle f_n \rangle = \frac{\langle n_A \rangle}{n} = p . \tag{1.55}$$

This result provides the first connection between frequency and probability, but it is not sufficient to ensure that the frequency f_n will be close to p.

Consider next a more general experiment than the previous case, with the outcome being the value of a continuous variable X, whose probability density is $P(x < X < x + dx|E) = g(x)dx$. A discrete variable can formally be included by allowing the probability density $g(x)$ to contain delta functions, if necessary.

Lemma. If X is a nonnegative variable [so that $g(x) = 0$ for $x < 0$], then for any $\varepsilon > 0$ we have

$$\begin{aligned} \langle X \rangle = \int_0^\infty g(x) x dx &\geq \int_\varepsilon^\infty g(x) x dx \\ &\geq \varepsilon \int_\varepsilon^\infty g(x) dx = \varepsilon P(X \geq \varepsilon|E) . \end{aligned}$$

Thus $P(X \geq \varepsilon|E) \leq \langle X \rangle / \varepsilon$.

Applying this lemma to the nonnegative variable $|X - c|$, where c is a constant, we obtain

$$P(|X - c| \geq \varepsilon|E) \leq \langle |X - c| \rangle / \varepsilon . \tag{1.56}$$

Furthermore, by considering the nonnegative variable $|X - c|^\alpha$, with $\alpha > 0$, we obtain

$$\begin{aligned} P(|X - c| \geq \varepsilon|E) &= P(|X - c|^\alpha \geq \varepsilon^\alpha|E) \\ &\leq \frac{\langle |X - c|^\alpha \rangle}{\varepsilon^\alpha} . \end{aligned} \tag{1.57}$$

This result is known as *Chebyshev's inequality*. It is most often quoted in the special case for which $\alpha = 2$, $c = \langle X \rangle$ is the mean of the distribution, $\langle |X - c|^2 \rangle = \sigma^2$ is the variance, and $\varepsilon = k\sigma$:

$$P(|X - \langle X \rangle| \geq k\sigma | E) \leq \frac{1}{k^2} \,.$$

The probability of X being k or more standard deviations from the mean is no greater than $1/k^2$, regardless of the form of the probability distribution.

We now return to the experiment E^n (n independent repetitions of a procedure E) to determine a closer relationship between the frequency of occurrence of outcome A and the probability $P(A|E) = p$. We use (1.57) with $\alpha = 2$ and $X = n_A = \sum_{i=1}^{n} J_i$. Here $J_i = 1$ if the outcome of the ith repetition of E is A, and $J_i = 0$ otherwise. We also choose $c = \langle X \rangle$, which is now equal to np, according to (1.55). Thus

$$P(|n_A - np| \geq \varepsilon | E) \leq \frac{\langle (n_A - np)^2 \rangle}{\varepsilon^2} \,.$$

Now we have

$$\langle (n_A - np)^2 \rangle = \left\langle \left\{ \sum_{i=1}^{n} (J_i - p) \right\}^2 \right\rangle = \sum_i \sum_j \langle (J_i - p)(J_j - p) \rangle \,.$$

Since the various repetitions of E are independent, we obtain

$$\langle (J_i - p)(J_j - p) \rangle = \langle J_i - p \rangle \langle J_j - p \rangle = 0 \quad \text{for} \quad i \neq j \,.$$

Hence

$$\langle (n_A - np)^2 \rangle = \left\langle \sum_{i=1}^{n} (J_i - p)^2 \right\rangle \leq n \,.$$

Thus $P(|n_A - np| \geq \varepsilon | E) \leq n/\varepsilon^2$. In terms of the relative frequency of A, $f_n = n_A/n$, this result becomes $P(|f_n - p| \geq \varepsilon/n | E) \leq n/\varepsilon^2$. Putting $\delta = \varepsilon/n$, we see that it becomes

$$P(|f_n - p| \geq \delta | E) \leq \frac{1}{n\delta^2} \,. \tag{1.58}$$

This important result, which is an instance of the *law of large numbers*, asserts that the probability of f_n (the relative frequency of A in n independent repetitions of E) being more than ε away from p converges to zero as $n \to \infty$. It is interesting to note that the proof of this theorem requires the independence condition (1.52) and Axioms 1, 2, and 3b. But it does not require Axiom 4, provided that Axiom 3b is adopted instead of Axiom 3a.

It should be emphasized that the law of large numbers does not assert that f_n ever becomes strictly equal to p, or even that f_n must remain close to p as $n \to \infty$. It merely asserts that deviations of f_n from p become more and more improbable, with the probability of any deviation becoming arbitrarily small for large enough n. From probability theory one derives only statements of probability, not of necessity.

Estimating a probability

In the preceding examples, the propensity p was supposed to be known, and the argument proceeded deductively to obtain other probabilities from it. This is methodologically analogous to quantum theory, many of whose predictions are probabilities. But in order to test those theoretical predictions, we must be able to infer from experimental data some empirical probabilities that may be compared with the theoretical probabilities. For this we need the theory of *inductive inference.*

Suppose that the propensity p for the result A to emerge from the procedure E is unknown. By repeating E independently n times we observe the result A on r occasions. What can we infer about the unknown value of p?

Let us denote $E = (C, p = \theta)$, where C symbolizes all conditions of the experiment except the value of p, and $D = (n_A$ occurrences of A in n repetitions) are the data. Then, using Bayes' theorem (1.51), we obtain

$$P(p = \theta | D\&C) = \frac{P(D|p = \theta, C)P(p = \theta|C)}{P(D|C)} \,.$$

(Strictly speaking, we should consider p to lie within a narrow range δ centered on θ, and should define probability densities in the limit $\delta \to 0$.) Since we are interested only in the relative probabilities for different values of p, we may drop all factors that do not involve θ, obtaining

$$P(p = \theta | D\&C) \propto \theta^r (1 - \theta)^{n-r} P(p = \theta|C) \,. \tag{1.59}$$

As might have been anticipated, this result does not tell us the value of p, but only the probabilities of the various possible values. But there is a further indeterminacy, since we cannot compute the final (or posterior) probability $P(p = \theta|D\&C)$ that is conditioned by the data D until we know the initial (or prior) probability $P(p = \theta|C)$, which represents the degree of reasonable belief that $p = \theta$ *in the absence of the data* D. If we choose the initial probability density to be uniform (independent of θ), then the *most probable* value of p, obtained by maximizing the final probability with respect to θ, is

$$p = \theta_m = \frac{r}{n}\,. \tag{1.60}$$

The justification for the choice of a uniform initial probability is controversial, but it may be noted that if $P(p = \theta|C)$ is any slowly varying function of θ, the location of the maximum of (1.59) will still be close to (1.60) provided n is reasonably large. That is to say, the final reasonable belief about p is dominated by the data, with the initial belief playing a very small practical role. Of course, (1.60) is just equal to the intuitive estimate of the probability p that most persons would make without the help of Bayes' theorem. Even so, the systematic application of Bayes' theorem has advantages:

(a) In addition to yielding the most probable value of p, (1.59) allows us to calculate the probability that p lies within some range. Thus the reliability of the estimate (1.60) can be evaluated.
(b) Depending upon the use that is to be made of the result, the most probable value, θ_m, might not be the most appropriate estimate of p. If, for example, the "cost" of a deviation of the estimate θ from the unknown true value p were proportional to $|\theta - p|$, or to $|\theta - p|^2$, then the best estimates would be, respectively, the median, or the mean, of the final probability density.
(c) Instead of wanting to obtain a purely empirical value of p from the experiment for comparison with a theoretical value, we might want to obtain the best estimate of p, taking into account both an imprecise theoretical calculation of it and a limited set of experimental data. The uncertain theoretical estimate could be expressed in the initial probability density, and the most probable value would be obtained by maximizing the final probability density (1.59).

Further reading for Chapter 1

Full references are given in the Bibliography at the end of the book.

Vectors and operators

Dirac (1958): an exposition of the bra and ket formalism by its originator.
Jauch (1972): a reformulation of Dirac's formalism in the mathematical framework of Hilbert space.

Jordan (1969): a concise account of those aspects of Hilbert space theory that are most relevant to quantum mechanics.
Bohm, A. (1978): the use of rigged Hilbert space in quantum mechanics.

Probability

There are a very large number of books on this subject, of which only a few of special interest are listed here.
Cox (1961): a development of the quantitative laws of probability from more elementary qualitative postulates.
Renyi (1970): a rigorous development of probability theory, based upon its relationship to measure theory.
Fine (1973): a critical analysis of several approaches to probability theory.
Kac (1959): applications of probability to unusual subjects such as number theory.
Jaynes (2003): an exposition of probability as inductive inference.

Problems

1.1 (a) Prove Schwarz's inequality and the triangle inequality from the axioms that define the inner product.
(b) Demonstrate the necessary and sufficient conditions for these inequalities to become equalities.

1.2 Consider the vector space that consists of all possible linear combinations of the following functions: 1, $\sin x$, $\cos x$, $(\sin x)^2$, $(\cos x)^2$, $\sin(2x)$, and $\cos(2x)$. What is the dimension of this space? Exhibit a possible set of basis vectors, and demonstrate that it is complete.

1.3 Prove that the trace of an operator A, $\mathrm{Tr}\,A = \sum_n \langle u_n|A|u_n\rangle$, is independent of the particular orthonormal basis $\{|u_n\rangle\}$ that is chosen for its evaluation.

1.4 Since a linear combination of two matrices of the same shape is another matrix of that shape, it is possible to regard matrices as members of a linear vector space. Show that any 2×2 matrix can be expressed as a linear combination of the following four matrices.

$$I = \begin{bmatrix} 1 & 0 \\ 0 & 1 \end{bmatrix}, \qquad \sigma_x = \begin{bmatrix} 0 & 1 \\ 1 & 0 \end{bmatrix},$$

$$\sigma_y = \begin{bmatrix} 0 & -i \\ i & 0 \end{bmatrix}, \qquad \sigma_z = \begin{bmatrix} 1 & 0 \\ 0 & -1 \end{bmatrix}.$$

1.5 If A and B are matrices of the same shape, show that $(A, B) = \mathrm{Tr}(A^\dagger B)$ has all of the properties of an inner product. Hence show that the four matrices of Problem 1.4 are orthogonal with respect to this inner product.

1.6 Find the eigenvalues and eigenvectors of the matrix

$$M = \begin{bmatrix} 0 & 1 & 0 \\ 1 & 0 & 1 \\ 0 & 1 & 0 \end{bmatrix} .$$

Construct the corresponding projection operators, and verify the spectral theorem for this matrix.

1.7 Show that the symmetrizer S, defined for an arbitrary function $\phi(x)$ as $Sf(x) = \frac{1}{2}[\phi(x) + \phi(-x)]$, and the antisymmetrizer A, defined as $A\phi(x) = \frac{1}{2}[\phi(x) - \phi(-x)]$, are projection operators.

1.8 Using the definition of a function of an operator, $f(A) = \sum_i f(a_i)|a_i\rangle\langle a_i|$, with $A|a_i\rangle = a_i|a_i\rangle$ and $\langle a_i|a_j\rangle = \delta_{ij}$, prove that the power function $f_n(A) \equiv A^n$ satisfies the relation $(A^n)(A^m) = A^{n+m}$.

1.9 (a) Consider a Hilbert space $\mathcal{H}$ that consists of all functions $\psi(x)$ such that

$$\int_{-\infty}^{\infty} |\psi(x)|^2 dx < \infty .$$

Show that there are functions in $\mathcal{H}$ for which $Q\psi(x) \equiv x\psi(x)$ is not in $\mathcal{H}$.

(b) Consider the function space Ω which consists of all $\phi(x)$ that satisfy the infinite set of conditions,

$$\int_{-\infty}^{\infty} |\phi(x)|^2(1 + |x|^n)dx < \infty \quad \text{for} \quad n = 0, 1, 2, \ldots .$$

Show that for any $\phi(x)$ in Ω the function $Q\phi(x) \equiv x\phi(x)$ is also in Ω. (These results are expressed by the statement that the domain of the operator Q includes all of Ω, but not all of $\mathcal{H}$.)

1.10 The extended space $\Omega^\times$ consists of those functions $\chi(x)$ which satisfy the condition

$$(\chi, \phi) = \int_{-\infty}^{\infty} \chi^*(x)\phi(x)dx < \infty \qquad \text{for all } \phi \text{ in } \Omega .$$

The nuclear space Ω and the Hilbert space $\mathcal{H}$ have been defined in the previous problem. Which of the following functions belong to Ω, to $\mathcal{H}$, and/or to $\Omega^\times$? (a) $\sin(x)$; (b) $\sin(x)/x$; (c) $x^2\cos(x)$; (d) $e^{-ax} (a > 0)$; (e) $[\log(1 + |x|)]/(1 + |x|)$; (f) $\exp(-x^2)$; (g) $x^4 e^{-|x|}$.

1.11 What boundary conditions must be imposed on the functions $\{\phi(\mathbf{x})\}$, defined in some finite or infinite volume of space, in order that the Laplacian operator ∇^2 be Hermitian?

1.12 Let $\langle\psi|A|\psi\rangle = \langle\psi|B|\psi\rangle$ for all ψ. Prove that $A = B$, in the sense that $\langle\phi_1|A|\phi_2\rangle = \langle\phi_1|B|\phi_2\rangle$ for all ϕ_1 and ϕ_2.

1.13 The number of stars in our galaxy is about $N = 10^{11}$. Assume that: the probability that a star has planets is $p = 10^{-2}$, the probability that the conditions on a planet are suitable for life is $q = 10^{-2}$, and the probability of life evolving, given suitable conditions, is $r = 10^{-2}$. (These numbers are rather arbitrary.)

(a) What is the probability of life existing in an arbitrary solar system (a star and its planets, if any)?

(b) What is the probability that life exists in at least one solar system? [Note: A naive argument against a purely natural origin of life is sometimes based on the smallness of the probability (a), whereas it is the probability (b) that is relevant.]

1.14 This problem illustrates the law of large numbers.

(a) Assuming the probability of obtaining "heads" in a coin toss is 0.5, compare the probability of obtaining "heads" in 5 out of 10 tosses with the probability of obtaining "heads" in 50 out of 100 tosses.

(b) For a set of 10 tosses and for a set of 100 tosses, calculate the probability that the fraction of "heads" will be between 0.445 and 0.555.

1.15 The probability density for decay of a radioactive nucleus is $P(t) = \alpha e^{-\alpha t}$, where $t \geq 0$ is the (unpredictable) lifetime of the nucleus, and α^{-1} is the mean lifetime for such a decay process. Calculate the probability density for $|t_1 - t_2|$, where t_1 and t_2 are the lifetimes of two such identical independent nuclei.

1.16 Let $X_1, X_2, \ldots, X_n$ be mutually independent random variables, each of which has the probability density

$$\begin{aligned} P_1(x) &= \alpha e^{-\alpha x} \qquad (x \geq 0) \\ &= 0 \qquad\qquad (x < 0) \end{aligned}$$

under some condition C. That is to say, $\text{Prob}(x < X_j < x + dx|C) = P_1(x)dx$ for $1 \leq j \leq n$. Show that the probability density for the sum of these variables, $S = X_1 + X_2 + \cdots + X_n$, is

$$P_n(x) = \alpha(\alpha x)^{n-1} e^{-\alpha x}/(n-1)!\,.$$

Use this result to demonstrate directly (without invoking the law of large numbers) that the mean, S/n, of these variables will probably be close to $\langle X_j \rangle = \alpha^{-1}$ when n is large.

1.17 A source emits particles at an average rate of λ particles per second; however, each emission is stochastically independent of all previous emission events. Calculate the probability that exactly n particles will be emitted within a time interval t.

Chapter 2

The Formulation of Quantum Mechanics

2.1 *Basic Theoretical Concepts*

Every physical theory involves some basic physical concepts, a mathematical formalism, and set of correspondence rules which map the physical objects onto the mathematical representations. The correspondence rules are first used to express a physical problem in mathematical terms. Once the mathematical version of the problem is formulated, it may be solved by purely mathematical techniques that need not have any physical interpretation. The formal solution is then translated back into the physical world by means of the correspondence rules.

Sometimes this mapping between physical and mathematical objects is so obvious that we need not think about it. In classical mechanics the position of a particle (physical concept) is mapped onto a real number or a set of real numbers (mathematical concept). Although the notion of a real number in pure mathematics is not trivial, this correspondence rule can be grasped intuitively by most people, without any risk of confusion. The mathematical formalism of quantum mechanics is much more abstract and less intuitive than that of classical mechanics. The world does not appear to be made up of Hermitian operators and infinite-dimensional state vectors, and we must give careful and explicit attention to the correspondence rules that relate the abstract mathematical formalism to observable reality.

There are two important aspects of quantum theory that require mathematical expression: the mechanical aspect and the statistical aspect.

Mechanical aspect

Certain dynamical variables, which should take on a continuum of values according to classical mechanics, were found to take on only discrete or "quantized" values. Some of the experimental evidence was reviewed in the

Introduction. A good example is provided by atomic spectra. According to classical mechanics and electromagnetism, an electron in an atom should emit radiation at a continuously variable frequency as it loses energy and spirals toward the nucleus. But actually only a discrete set of frequencies is observed. From this fact N. Bohr inferred that a bound electron in an atom can occupy only a discrete set of energy levels, with the frequency of the radiation emitted during a transition between two such allowed energies being proportional to the difference between the energies.

However, energy is not always quantized, since a free electron can take on a continuous range of energies, and when accelerated can emit radiation with a continuous frequency spectrum. Evidently we need some means of calculating the allowed values of dynamical variables, and it should treat the discrete and continuous cases on an unbiased footing. This is accomplished by:

Postulate 1. To each *dynamical variable* (physical concept) there corresponds a *linear operator* (mathematical object), and the possible values of the dynamical variable are the eigenvalues of the operator.

The only justification for this postulate, so far, is that there are operators that possess discrete eigenvalue spectra and continuous spectra, or a combination of discrete and continuous spectra. Thus all possibilities can be accounted for. This postulate will not acquire much content until we obtain rules assigning particular operators to particular dynamical variables.

Statistical aspect

We need some means of calculating the probability, or relative frequency of occurrence, of the various allowed values of the dynamical variables in a specific physical situation. This is also illustrated in atomic spectra, since the observed intensity of a spectral line is proportional to the number of transitions per unit time, which is in turn proportional to the probability of a transition from one energy level to another. However, it is perhaps better illustrated by a scattering experiment.

A particle is subjected to the *preparation* consisting of acceleration and collimation in the apparatus shown schematically at the upper left of Fig. 2.1. It scatters off the target through some angle θ, and is finally detected by one of the detectors at the right of the figure. A single *measurement* consists in the detection of the particle and hence the determination of the angle of scatter, θ. If the same preparation is repeated identically on a similar particle (or even on

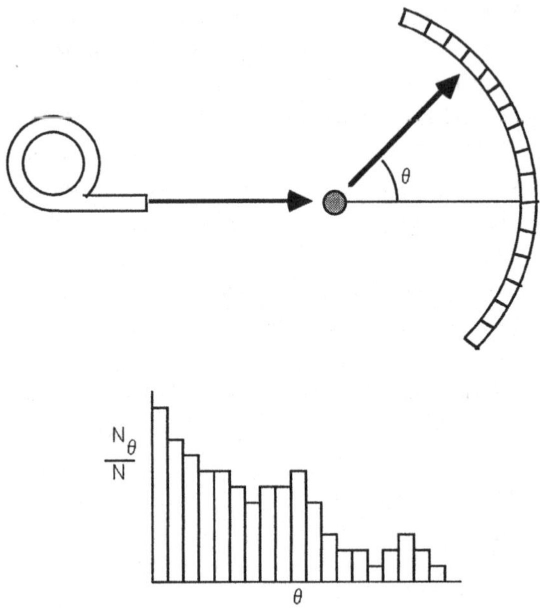

Fig. 2.1 A scattering experiment: apparatus (above); results (below).

the same particle), the angle of scatter that results will, in general, be different. Individual events resulting from identical preparations are not reproducible.[a]

However, in a *statistical experiment*, consisting of a long sequence of identical preparations and measurements, the relative frequencies of the various possible outcomes of the individual measurements usually approach a stable limit. This is illustrated in Fig. 2.1 (bottom), where the relative number of particles counted by each detector is plotted against the angle θ describing the location of the detector. This is the characteristic feature of a statistical experiment: nonreproducibility of individual events but stable limiting frequencies in a long sequence of such events.

Quantum mechanics mirrors this feature of the statistical experiment. It has no means by which to calculate the outcome of an individual event. In the scattering experiment it provides no way to calculate the scattering angle of an individual particle. But it does provide a means to calculate the

[a]Whether this nonreproducibility is due to an indeterminism in nature, or merely to limitations (practical or fundamental) in the preparation procedure, is a question that we cannot, and need not, answer here. The statistical approach is applicable in any case.

probabilities of the various possible outcomes of a scattering event. The fundamental connection between probability and frequency (see Sec. 1.5) allows us to compare the theoretical probabilities with the observed relative frequencies in a statistical experiment.

It is useful to divide the statistical experiment into two phases: *preparation* and *measurement.* In the scattering experiment the preparation consists of passing a particle through the acceleration and collimation apparatus and allowing it to interact with the target. The measurement consists of the detection of the particle and the subsequent inference of the angle of scatter. This subdivision of the experiment is useful because the two phases are essentially independent. For the same preparation one could measure the energy instead of the position of the particle, by means of a different kind of detector. Conversely, the same array of detectors shown in Fig. 2.1 could have been used to measure the positions of particles from some other kind of preparation, involving a different target or even an entirely different preparation apparatus.

Having distinguished preparation from measurement, we need to be more precise about just what is being prepared. At first, one might say that it is the particle (more generally, the object of the subsequent measurement) that is prepared. While this is true in an obvious and trivial sense, it fails to characterize the specific result of the preparation. Two identical objects, each subjected to an identical preparation, may behave differently in the subsequent measurements. Conversely, two objects that yield identical results in measurement could have come from entirely different preparations. In the example of Fig. 2.1, the measurement determines only the direction from which the particle leaves the scatterer. One cannot infer from the result of such a measurement what the direction of incidence onto the target may have been (supposing that the preparation apparatus is not visible). If we want to characterize a preparation by its effect, we must identify that effect with something other than the specific object that has experienced the preparation, because the same preparation could lead to various measurement outcomes, and the same measurement outcome could be a result of various preparations.

A specific preparation determines not the outcome of the subsequent measurement, but the *probabilities* of the various possible outcomes. Since a preparation is independent of the specific measurement that may follow it, the preparation must determine probability distributions for all such possible measurements. This leads us to introduce the concept of a *state*, which is identified with the specification of a probability distribution for each observable. (An *observable* is a dynamical variable that can, in principle, be measured.)

Any repeatable process that yields well-defined probabilities for all observables may be termed a *state preparation procedure.* It may be a deliberate laboratory operation, as in our example, or it may be a natural process not involving human intervention. If two or more procedures generate the same set of probabilities, then these procedures are equivalent and are said to prepare the same *state.*

The empirical content of a probability statement is revealed only in the relative frequencies in a sequence of events that result from the same (or an equivalent) state preparation procedure. Thus, although the primary definition of a *state* is the abstract set of probabilities for the various observables, it is also possible to associate a *state* with an *ensemble* of similarly prepared systems. However, it is important to remember that this ensemble is the conceptual unbounded set of all such systems that may potentially result from the state preparation procedure, and not a concrete set of systems that coexist in space. In the example of the scattering experiment, the system is a single particle, and the ensemble is the conceptual set of replicas of one particle in its surroundings. The ensemble should not be confused with a beam of particles, which is another kind of (many-particle) system. Strictly speaking, the accelerating and collimating apparatus of the scattering experiment can be regarded as a preparation procedure for a one-particle state only if the density of the particle beam is so low that only one particle at a time is in flight between the accelerator and the detectors, and there are no correlations between successive particles.

The mathematical representation of a state must be something that allows us to calculate the probability distributions for all observables. It turns out to be sufficient to postulate only a formula for the average.

Postulate 2. To each state there corresponds a unique *state operator.* The average value of a dynamical variable $\mathbf{R}$, represented by the operator R, in the virtual ensemble of events that may result from a preparation procedure for the state, represented by the operator ρ, is

$$\langle \mathbf{R} \rangle = \frac{\text{Tr}(\rho R)}{\text{Tr}\rho}. \tag{2.1}$$

Here Tr denotes the trace. The state operator is also referred to as the *statistical operator*, and sometimes as the *density matrix*, although the latter term should be restricted to its matrix form in coordinate representation. There are some restrictions on the form that a state operator ρ may have;

these will be developed later. The wording of this postulate is rather verbose because I have deliberately kept separate the physical concepts from the mathematical objects that represent them. When no confusion is likely to occur from a failure to make such explicit distinctions, we may say, "The average of the observable R in the state ρ is $\cdots$ (2.1)."

[[The concept of *state* is one of the most subtle and controversial concepts in quantum mechanics. In classical mechanics the word "state" is used to refer to the coordinates and momenta of an individual system, and so early on it was supposed that the quantum state description would also refer to attributes of an individual system. Since it has always been the goal of physics to give an objective realistic description of the world, it might seem that this goal is most easily achieved by interpreting the quantum state function (state operator, state vector, or wave function) as an element of reality in the same sense that the electromagnetic field is an element of reality. Such ideas are very common in the literature, more often appearing as implicit unanalyzed assumptions than as explicitly formulated arguments. However, such assumptions lead to contradictions (see Ch. 9), and must be abandoned.

The quantum state description may be taken to refer to an ensemble of similarly prepared systems. One of the earliest, and surely the most prominent advocate of the ensemble interpretation, was A. Einstein. His view is concisely expressed as follows [Einstein (1949), quoted here without the supporting argument]:

> "The attempt to conceive the quantum-theoretical description as the complete description of the individual systems leads to unnatural theoretical interpretations, which become immediately unnecessary if one accepts the interpretation that the description refers to ensembles of systems and not to individual systems."

Criticisms of the *ensemble* interpretation have often resulted from a confusion of the ensemble as the virtual unbounded set of similarly prepared systems, with a concrete sequence or assembly of similar systems. These criticisms, misguided though they are, may be alleviated by a slightly more abstract interpretation in which a state is identified with the preparation *procedure* itself. "State" is then an abbreviation for "state preparation procedure". This definition has merit, but it is a bit too operationalistic. It does not, without modification, allow for two procedures to yield the same state. Moreover, it seems to restrict the application of quantum mechanics to laboratory situations, with an experimenter to carry out preparations and measurements. But surely the laws of quantum mechanics must also govern atoms in stars, or on earth before the evolution of life!

By identifying the *state* concept directly with a set of probability distributions, it should be possible to avoid all of the old objections. This approach also makes clear the fact that the interpretation of quantum mechanics is dependent upon choosing a suitable interpretation of probability.]]

2.2 *Conditions on Operators*

Postulates 1 and 2 of the previous section associate an operator with each state and with each dynamical variable, but it is necessary to impose some conditions on these operators in order that they be acceptable.

The first condition imposed on state operators is a conventional normalization,

$$\mathrm{Tr}\ \rho = 1\,, \tag{2.2}$$

which allows us to omit the denominator from (2.1).

The next two conditions are less trivial. Consider a hypothetical observable represented by the projection operator, $P_u = |u\rangle\langle u|$, where $|u\rangle$ is some vector of unit norm. This operator may be regarded as describing some dynamical variable that takes on only the values 0 and 1. Now the average of a variable that takes on only *real* values must certainly be real. Hence $\mathrm{Tr}(\rho P_u) = \langle u|\rho|u\rangle$ must be real. If this requirement is imposed for *all* vectors $|u\rangle$, then by Theorem 1 of Sec. 1.3, we have

$$\rho = \rho^\dagger\,. \tag{2.3}$$

Furthermore, the average of a variable that takes on only *nonnegative* values must itself be nonnegative. Hence

$$\langle u|\rho|u\rangle \geq 0\,. \tag{2.4}$$

If this holds for *all* vectors $|u\rangle$, then ρ is called a *nonnegative operator.*

If we knew that every projection operator onto an arbitrary unit vector corresponds to an observable, then the necessity of (2.3) and (2.4) would be proved. In fact, we have no justification for supposing that all projection operators correspond to observables, so we shall have to be content to introduce Postulate 2a (so labeled because it is a strengthened version of Postulate 2).

Postulate 2a. To each state there corresponds a unique state operator, which must be Hermitian, nonnegative, and of unit trace.

Although this postulate has not been proven to be necessary, it is very strongly motivated, and the possibility of proof remains open if the set of observables turns out to be large enough.

From the fact that the values of dynamical variables are real, and hence any average of them must be real, we can deduce a condition on the operators that correspond to dynamical variables. Consider a special state operator of the form $\rho = |\Psi\rangle\langle\Psi|$, where $|\Psi\rangle$ is a vector of unit norm. Clearly this ρ satisfies the three conditions required of a state operator in Postulate 2a. The average, in this state, of a dynamical variable represented by the operator R is

$$\text{Tr}\,(\rho R) = \text{Tr}\,(|\Psi\rangle\langle\Psi|R) = \langle\Psi|R|\Psi\rangle\,.$$

If this expression is required to be real for all $|\Psi\rangle$, then by Theorem 1 of Sec. 1.3 we have

$$R = R^\dagger\,. \tag{2.5}$$

At this early stage of the theory we cannot justify the assumption that every vector $|\Psi\rangle$ corresponds to a physically realizable state, so we shall introduce a strengthened version of Postulate 1:

Postulate 1a. To each dynamical variable there is a Hermitian operator whose eigenvalues are the possible values of the dynamical variable.

The preceding argument, or some variation of it, is the most common reason given for requiring the operators corresponding to observables to be Hermitian. Unfortunately, the argument has less substance than it might appear to have. The use of real numbers to represent the values of physical quantities is really a convention. Two related physical variables could be represented by a complex number; one physical variable could be described by a set of nested intervals, representing its uncertainty as well as its value. Just because dynamical variables are "real", in the metaphysical sense of "not unreal", it does not follow that they must correspond to "real numbers" in the mathematical sense.

In fact, the property of Hermitian operators that is essential in formulating quantum theory is the existence of a spectral representation, in either the discrete form (1.27) or the continuous form (1.37). The probability calculation in Sec. 2.4 depends essentially on the spectral representation. Whether the eigenvalues are real or complex is incidental and unimportant. Problem 2.1 contains an example of an operator with purely real eigenvalues, but lacking a complete set of eigenvectors, and thus having no spectral representation. If reality of eigenvalues were the only relevant criterion, then that operator would be acceptable. But no consistent statistical interpretation of it is possible because its "average" calculated from (2.1) can be complex, even though all eigenvalues are real.

[[I have been careful to use the term *observable* as a physical concept, meaning a dynamical variable that can, in principle, be measured, and to distinguish it from the mathematical operator to which it corresponds in the formalism. Dirac, to whom we are indebted for so much of the modern formulation of quantum mechanics, unfortunately used the word "observable" to refer indiscriminately to the physical dynamical variable and to the corresponding mathematical operator. This has sometimes led to confusion. There is in the literature a case of an argument about whether or not the electromagnetic vector potential is an observable, one party arguing the affirmative on the grounds that the operator satisfies all of the required conditions, the other party arguing the negative on the grounds that the vector potential cannot be measured.]]

2.3 *General States and Pure States*

As was shown in the preceding section, a mathematically acceptable state operator must satisfy three conditions:

$$\text{Tr}\,\rho = 1\,, \tag{2.6}$$

$$\rho = \rho^\dagger\,, \tag{2.7}$$

$$\langle u|\rho|u\rangle \geq 0 \quad \text{for all} \quad |u\rangle\,. \tag{2.8}$$

Several other useful results can be derived from these. Being a self-adjoint operator, ρ has a spectral representation,

$$\rho = \sum_n \rho_n |\phi_n\rangle\langle\phi_n|\,, \tag{2.9}$$

in terms of its eigenvalues and orthonormal eigenvectors (assumed, for convenience, to be discrete). To each of the three definitive properties of ρ there corresponds a property of the eigenvalues:

$$(2.6) \text{ implies } \sum_n \rho_n = 1\,; \tag{2.10}$$

$$(2.7) \text{ implies } \rho_n = \rho_n^*\,; \tag{2.11}$$

$$(2.8) \text{ implies } \rho_n \geq 0\,. \tag{2.12}$$

Not only does (2.8) imply (2.12), as can be seen by choosing $|u\rangle = |\phi_n\rangle$ in (2.8), but conversely (2.12) implies (2.8). This is proven by using (2.9) to evaluate $\langle u|\rho|u\rangle = \sum_n \rho_n |\langle u|\phi_n\rangle|^2$ for arbitrary $|u\rangle$. The result is clearly nonnegative, provided that all ρ_n are nonnegative. Equation (2.12) provides a more convenient practical test for the nonnegativeness of ρ than does the direct use of (2.8). Combining (2.10) with (2.12), we obtain

$$0 \leq \rho_n \leq 1\,. \tag{2.13}$$

The second inequality holds because no term in a sum of positive terms can exceed the total.

The set of all mathematically acceptable state operators forms a *convex set.* This means that if two or more operators $\{\rho^{(i)}\}$ satisfy the three conditions (2.6)–(2.8), then so does $\rho = \sum_i a_i \rho^{(i)}$, provided that $0 \leq a_i \leq 1$ and $\sum_i a_i = 1$. Such an operator ρ is called a *convex combination* of the set $\{\rho^{(i)}\}$.

Pure states

Within the set of all states there is a special class, called *pure states*, which are distinguished by their simpler properties. A *pure state operator*, by definition, has the form

$$\rho = |\Psi\rangle\langle\Psi|\,, \tag{2.14}$$

where the unit-normed vector $|\Psi\rangle$ is called a *state vector.* The average value of the observable $\mathbf{R}$, in this pure state, is

$$\langle \mathbf{R} \rangle = \mathrm{Tr}\,(|\Psi\rangle\langle\Psi|R) = \langle\Psi|R|\Psi\rangle\,. \tag{2.15}$$

The state vector is not unique, any vector of the form $e^{i\alpha}|\Psi\rangle$ with arbitrary real α being physically equivalent. However, the state operator (2.14) is independent of this arbitrary phase.

A second, equivalent characterization of a pure state is by the condition

$$\rho^2 = \rho\,. \tag{2.16}$$

This condition is necessary because it is satisfied by (2.14). That it is also sufficient may be proven by considering the eigenvalues, which must satisfy

$\rho_n^2 = \rho_n$ if (2.16) holds. The only possible eigenvalues are $\rho_n = 0$ or $\rho_n = 1$. But since, according to (2.10), the sum of the eigenvalues is 1, it must be the case that exactly one of them has the value 1 and all others are 0. Thus the spectral representation (2.9) consists of a single term, and so is of the pure state form (2.14).

A third condition for identifying a pure state, apparently weaker but actually equivalent, is

$$\text{Tr}\,(\rho^2) = 1\,. \tag{2.17}$$

Clearly it is a necessary condition, so we need only prove sufficiency. Because of (2.13) we have $\rho_n^2 \le \rho_n$. Now $\text{Tr}\,(\rho^2) = \sum_n \rho_n^2 \le \sum_n \rho_n = 1$. Thus we have $\text{Tr}\,(\rho^2) \le 1$ for a general state. Equality can hold only if $\rho_n^2 = \rho_n$ for each n. But, by the argument used in proving the second characterization, this can be so only for a pure state.

A fourth way to distinguish a pure state from a general state is by means of the following theorem:

Theorem. A *pure state* cannot be expressed as a nontrivial convex combination of other states, but a *nonpure state* can always be so expressed.

Proof. The latter part of the theorem is trivial, since the spectral representation (2.9) of a nonpure state has the form of a nontrivial convex combination of pure states. To prove the former part we assume the contrary: that a pure state operator ρ may be expressed as a convex combination of distinct state operators,

$$\rho = \sum_i a_i \rho^{(i)}\,, \quad 0 \le a_i \le 1\,, \quad \sum_i a_i = 1\,. \tag{2.18}$$

We shall then use (2.17) to demonstrate a contradiction.

From (2.18) we obtain

$$\text{Tr}\,(\rho^2) = \sum_i \sum_j a_i a_j \text{Tr}\,\{\rho^{(i)} \rho^{(j)}\}\,. \tag{2.19}$$

Now each operator in the sum (2.18) has its own spectral representation, $\rho^{(i)} = \sum_n \rho_n^{(i)} |\phi_n^{(i)}\rangle\langle\phi_n^{(i)}|$. Thus

$$\begin{aligned}
\text{Tr}\,\{\rho^{(i)}\rho^{(j)}\} &= \sum_n \sum_m \rho_n^{(i)} \rho_m^{(j)} \text{Tr}\,\{|\phi_n^{(i)}\rangle\langle\phi_n^{(i)}|\phi_m^{(j)}\rangle\langle\phi_m^{(j)}|\} \\
&= \sum_n \sum_m \rho_n^{(i)} \rho_m^{(j)} |\langle\phi_n^{(i)}|\phi_m^{(j)}\rangle|^2 \\
&\le \sum_n \sum_m \rho_n^{(i)} \rho_m^{(j)} = 1\,.
\end{aligned}$$

Moreover, the inequality becomes an equality if and only if $|\langle\phi_n^{(i)}|\phi_m^{(j)}\rangle| = 1$ for all n and m such that $\rho_n^{(i)}\rho_m^{(j)} \neq 0$. Since the eigenvectors have unit norm, the Schwarz inequality (1.1) implies that $|\phi_n^{(i)}\rangle$ and $|\phi_m^{(j)}\rangle$ differ by at most a phase factor. But each set of eigenvectors is orthogonal, so the foregoing conclusion is impossible unless there is only one n and one m that contributes to the double sum above. The conclusion of this analysis may be stated thus:

Lemma. For any two state operators, $\rho^{(i)}$ and $\rho^{(j)}$, we have

$$0 \leq \text{Tr}\{\rho^{(i)}\rho^{(j)}\} \leq 1\,, \tag{2.20}$$

with the upper limit being reached if and only if $\rho^{(i)} = \rho^{(j)}$ is a pure state operator.

Applying the lemma to (2.19), we obtain

$$\begin{aligned}\text{Tr}\ (\rho^2) &= \sum_i \sum_j a_i a_j \text{Tr}\ \{\rho^{(i)}\rho^{(j)}\}\\ &\leq \sum_i \sum_j a_i a_j = 1\,.\end{aligned}$$

But, by hypothesis, ρ represents a pure state, so according to (2.17) the upper limit of the inequality must be reached. According to the lemma, this is possible only if $\rho^{(i)} = \rho^{(j)}$ for all i and j. This contradicts the assumption that we had a *nontrivial* convex combination of state operators in (2.18); in fact all operators in that sum must be identical. Thus we have proven the theorem that a pure state cannot be expressed as a nontrivial convex combination.

This theorem suggests that the pure states are, in a sense, more fundamental than nonpure states, and that the latter may be regarded as statistical mixtures of pure states. However, this interpretation cannot be taken literally, because the representation of a nonpure state operator as a convex combination of pure state operators is *never unique*. A two-dimensional example suffices to demonstrate this fact. Consider the state operator

$$\rho_a = a|u\rangle\langle u| + (1-a)|v\rangle\langle v|\,, \tag{2.21}$$

where $0 < a < 1$, and where $|u\rangle$ and $|v\rangle$ are orthogonal vectors of unit norm. Define two other vectors,

$$\begin{aligned}|x\rangle &= \sqrt{a}|u\rangle + \sqrt{1-a}|v\rangle\,,\\ |y\rangle &= \sqrt{a}|u\rangle - \sqrt{1-a}|v\rangle\,.\end{aligned}$$

It is easily shown that

$$\rho_a = \tfrac{1}{2}|x\rangle\langle x| + \tfrac{1}{2}|y\rangle\langle y| \,. \tag{2.22}$$

In fact, there are actually an infinite number of ways to represent any nonpure state operator as a convex combination of pure state operators.

The convex set of quantum states is schematically illustrated in Fig. 2.2. The points on the convex boundary represent the pure states, and the interior points represent nonpure states. The nonpure state ρ_a can be mathematically represented as a mixture of pure states u and v, as in (2.21), the relative weights being inversely proportional to the distances of the points u and v from a. It can also be represented as a mixture of the pure states x and y, as in (2.22), or in many other ways.

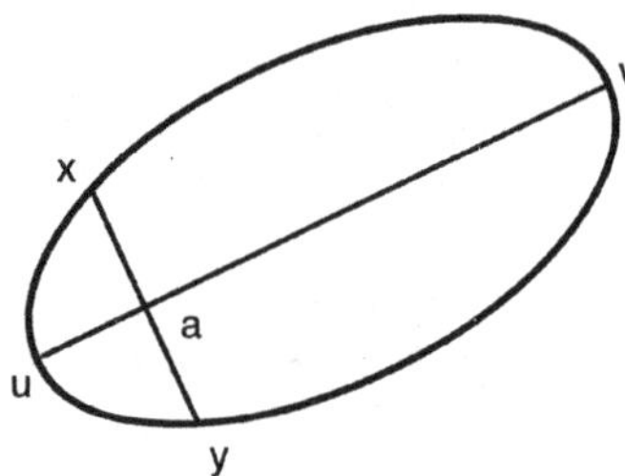

Fig. 2.2 Schematic depiction of pure and nonpure states as a convex set.

Because the pure state content of a "mixture" is not uniquely definable, we shall avoid using the common term "mixed state" for a nonpure state. The physical significance of this nonuniqueness lies in the fact that in quantum mechanics the pure states, as well as the nonpure states, describe statistically nontrivial ensembles. We shall return to this important point in Ch. 9.

Many examples of pure and nonpure states will be studied in the following chapters, but it may be useful to indicate in very broad terms where the two types of state may arise. A nondegenerate energy level of an atom, or indeed of any isolated system, is an example of a pure state. The state of thermal equilibrium is not a pure state, except at the absolute zero of temperature. Polarized monochromatic light produced by a laser can approximate a pure state of the electromagnetic field. Unpolarized monochromatic radiation and black body radiation are examples of nonpure states of the electromagnetic field. Generally speaking, there are fewer fluctuations in a pure state than in a nonpure state. The nature of the information needed to determine the state, and hence to determine whether or not it is pure, will be studied in Ch. 8.

2.4 *Probability Distributions*

According to Postulate 2, the average value in the state represented by ρ, of the observable $\mathbf{R}$ represented by the Hermitian operator R, is equal to

$$\langle \mathbf{R} \rangle = \text{Tr}\,(\rho R)\,. \tag{2.23}$$

We have chosen the state operator ρ to be normalized as in (2.2). This formula for the average is sufficient for us to deduce the entire probability distribution, provided we may assume that the function $F(\mathbf{R})$ is an observable represented by the operator $F(R)$, constructed according to the spectral representation (1.28) or (1.38). This assumption is entirely reasonable because if the physical quantity $\mathbf{R}$ has the value r then a function $F(\mathbf{R})$ must have the value $F(r)$, and precisely this relation is satisfied by the eigenvalues of the operators R and $F(R)$.

Let $g(r)dr$ be the probability that the observable $\mathbf{R}$ lies between r and $r + dr$. Then, by definition,

$$\langle F(\mathbf{R}) \rangle = \int_{-\infty}^{\infty} F(r')g(r')dr'\,. \tag{2.24}$$

But the application of (2.23) to the observable $F(R)$ yields

$$\langle F(\mathbf{R}) \rangle = \text{Tr}\,\{\rho F(R)\}\,. \tag{2.25}$$

By choosing a suitable function $F(\mathbf{R})$, it is possible to use these two equations to extract the probability density $g(r)$. We shall treat separately the cases of discrete and continuous spectra.

Discrete spectrum. Let R be a self-adjoint operator with a purely discrete spectrum. It may be expressed in terms of its eigenvalues r_n and orthonormal eigenvectors $|r_n\rangle$ as

$$R = \sum_n r_n |r_n\rangle\langle r_n|\,.$$

Consider the function $F(\mathbf{R}) = \theta(r - \mathbf{R})$, which is equal to one for $\mathbf{R} < r$ and is zero for $\mathbf{R} > r$. The average of this function, according to (2.24), is

$$\begin{aligned}\langle \theta(r - \mathbf{R}) \rangle &= \int_{-\infty}^{r} g(r')dr' \\ &= \text{Prob}\,(\mathbf{R} < r|\rho)\,.\end{aligned}$$

This is just the probability that the value of observable $\mathbf{R}$ is less than r. But from (2.25) we obtain

$$\begin{aligned}\langle\theta(r-\mathbf{R})\rangle &= \text{Tr}\,\{\rho\theta(r-R)\}\\ &= \text{Tr}\left\{\rho\sum_n \theta(r-r_n)|r_n\rangle\langle r_n|\right\}\\ &= \sum_n \langle r_n|\rho|r_n\rangle\theta(r-r_n)\,.\end{aligned}$$

Hence the probability density is

$$\begin{aligned}g(r) &= \frac{\partial}{\partial r}\text{Prob}\,(\mathbf{R}<r|\rho)\\ &= \sum_n \langle r_n|\rho|r_n\rangle\delta(r-r_n)\,.\end{aligned}$$

The only reason for calculating the probability *density* for a discrete observable is to show that $g(r)=0$ if r is not an eigenvalue. The probability is zero that a dynamical variable will take on a value other than an eigenvalue of the corresponding operator. This is a pleasing demonstration of the consistency of the statistical Postulate 2 with the mechanical Postulate 1.

The probability that the dynamical variable R will have the discrete value r in the virtual ensemble characterized by the state operator ρ is

$$\begin{aligned}\text{Prob}\,(\mathbf{R}=r|\rho) &= \lim_{\varepsilon\to 0}\{\text{Prob}\,(\mathbf{R}<r+\varepsilon|\rho)-\text{Prob}\,(\mathbf{R}<r-\varepsilon|\rho)\}\\ &= \sum_n \langle r_n|\rho|\,r_n\rangle\delta_{r,r_n}\,. \qquad (2.26)\end{aligned}$$

This result can be more concisely expressed in terms of the projection operator $P(r)=\sum_n |r_n\rangle\langle r_n|\delta_{r,r_n}$, which projects onto the subspace spanned by all degenerate eigenvectors with eigenvalue $r_n=r$,

$$\text{Prob}\,(\mathbf{R}=r|\rho) = \text{Tr}\,\{\rho P(r)\}\,. \qquad (2.27)$$

In the special case of a *pure* state, $\rho=|\Psi\rangle\langle\Psi|$, and a non-degenerate eigenvalue r_n, these results reduce to

$$\text{Prob}\,(\mathbf{R}=r_n|\Psi) = |\langle r_n|\Psi\rangle|^2\,. \qquad (2.28)$$

Eigenstates. A particular dynamical variable will have a non-vanishing statistical dispersion in most states. But in the case of a discrete variable it is possible for all of the probability to be concentrated on a single value. If the dynamical variable **R** takes on the unique value r_0 (assumed for simplicity to be a nondegenerate eigenvalue) with probability 1, in some state, then from (2.26) the state operator ρ must satisfy $\langle r_0|\rho|r_0\rangle = 1$. Since any state operator must satisfy $\mathrm{Tr}\rho^2 \leq 1$, we must have

$$\sum_{m,n}\langle r_n|\rho|r_m\rangle\langle r_m|\rho|r_n\rangle = \sum_{m,n}|\langle r_n|\rho|r_m\rangle|^2 \leq 1\,.$$

This limit is exhausted by the single term $\langle r_0|\rho|r_0\rangle = 1$, so all other diagonal and nondiagonal matrix elements of ρ must vanish. Therefore the only state for which **R** takes on the nondegenerate eigenvalue r_0 with probability 1 is the pure state $\rho = |r_0\rangle\langle r_0|$. Such a state, whether described by the state operator ρ or the state vector $|r_0\rangle$, is referred to as an *eigenstate* of the observable **R**.

Continuous spectrum. Let Q be a self-adjoint operator having a purely continuous spectrum:

$$Q = \int q'|q'\rangle\langle q'|dq'\,.$$

Its infinite-length eigenvectors satisfy the orthonormality relation $\langle q'|q''\rangle = \delta(q' - q'')$. Let $g(q)dq$ be the probability that the corresponding observable **Q** lies between q and $q + dq$. As in the previous case, we obtain

$$\begin{aligned}\langle\theta(q - \mathbf{Q})\rangle &= \int_{-\infty}^{q} g(q')dq' \\ &= \mathrm{Prob}\,(\mathbf{Q} < q|\rho)\,,\end{aligned}$$

which is the probability that observable **Q** is less than q. But we also have the relation

$$\begin{aligned}\langle\theta(q - \mathbf{Q})\rangle &= \mathrm{Tr}\,\{\rho\theta(q - Q)\} \\ &= \mathrm{Tr}\left\{\rho\int_{-\infty}^{\infty}\theta(q - q')|q'\rangle\langle q'|dq'\right\} \\ &= \int_{-\infty}^{q}\langle q'|\rho|q'\rangle dq'\,.\end{aligned}$$

Therefore the probability density for the observable **Q** in the virtual ensemble characterized by the state operator ρ is

$$\begin{aligned}g(q) &= \frac{\partial}{\partial q}\mathrm{Prob}\,(\mathbf{Q} < q|\rho) \\ &= \langle q|\rho|q\rangle\,.\end{aligned} \tag{2.29}$$

For the special case of a *pure state*, $\rho = |\Psi\rangle\langle\Psi|$, this becomes

$$g(q) = |\langle q|\Psi\rangle|^2 . \tag{2.30}$$

Although these expressions for probability and probability density have various detailed forms, they always consist of a relation between two factors: one characterizing the state, and one characterizing a portion of the spectrum of the dynamical variable being observed. We shall refer to them as the *state function* and the *filter function*, respectively. In (2.27) these two factors are the state operator ρ and the projection operator $P(r)$. In (2.28) and (2.30) they are the state vector Ψ and an eigenfunction belonging to the observable. The symmetrical appearance of the two factors in these equations should not be allowed to obscure their distinct natures. In particular, the state vector Ψ must be normalized, and so belongs to Hilbert space. But the filter function in (2.30) does not belong to Hilbert space, but rather to the extended space $\Omega^\times$ of the rigged Hilbert space triplet (see Sec. 1.4).

Verification of probability axioms. Several formulas for quantum probabilities have been given in this section. But we are not justified in asserting that a formula expresses a probability unless we can show that it obeys the axioms of probability theory. To do this, it is useful to construct a general probability formula that includes all of the special cases given previously.

Associated with the any dynamical variable **R** and its self-adjoint operator R is a family of projection operators $M_R(\Delta)$ which are related to the eigenvalues and eigenvectors of R as follows:

$$M_R(\Delta) = \sum_{r_n \in \Delta} |r_n\rangle\langle r_n| . \tag{2.31}$$

The sum is over all eigenvectors (possibly degenerate) whose eigenvalues lie in the subset Δ. (In the case of a continuous spectrum the sum should be replaced by an integral.) The probability that the value of **R** will lie within Δ is given by

$$\text{Prob}\,(\mathbf{R} \in \Delta|\rho) = \text{Tr}\,\{\rho M_R(\Delta)\} . \tag{2.32}$$

If the region Δ contains only one eigenvalue, then this formula reduces to (2.26) or (2.27). In the case of a continuous spectrum, (2.32) is equal to the integral of the probability density over the region Δ.

It is easy to verify that (2.32) satisfies the probability axioms 1, 2, and 3 of Sec. 1.5. We note first that since $M_R(\Delta)$ is a projection operator, the trace operation in (2.32) is effectively restricted to the subspace onto which $M_R(\Delta)$ projects. This fact, combined with the normalization (2.6) and nonnegativeness (2.8) of ρ, implies that $0 \leq \mathrm{Tr}\{\rho M_R(\Delta)\} \leq \mathrm{Tr}\rho = 1$. This confirms Axiom 1.

The situation of Axiom 2 is obtained if we choose a state prepared in such a manner that the value of $\mathbf{R}$ is guaranteed to lie within Δ. This will be so for those states which satisfy

$$\rho = M_R(\Delta)\rho M_R(\Delta)\,. \tag{2.33}$$

In the special case where Δ contains only a single eigenvalue, this reduces to the condition that ρ be an eigenstate of R. It is clear that (2.32) becomes identically equal to 1 whenever (2.33) holds.

To verify Axiom 3b (from which Axiom 3a follows) we consider two disjoint sets, Δ_1 and Δ_2, so that $\mathbf{R} \in \Delta_1$ and $\mathbf{R} \in \Delta_2$ are mutually exclusive events. Now $(\mathbf{R} \in \Delta_1) \vee (\mathbf{R} \in \Delta_2)$ is equivalent to $\mathbf{R} \in (\Delta_1 \cup \Delta_2)$, where $\Delta_1 \cup \Delta_2$ denotes the union of the two sets. Since the sets Δ_1 and Δ_2 are disjoint it follows that $M_R(\Delta_1)M_R(\Delta_2) = 0$, and the projection operator corresponding to the union of the sets is just the sum of the separate projection operators, $M_R(\Delta_1 \cup \Delta_2) = M_R(\Delta_1) + M_R(\Delta_2)$. Hence in this case (2.32) becomes

$$\begin{aligned}\mathrm{Prob}\,\{(\mathbf{R} \in \Delta_1) \vee (\mathbf{R} \in \Delta_2)|\rho\} &= \mathrm{Tr}\,\{\rho M_R(\Delta_1 \cup \Delta_2)\} \\ &= \mathrm{Tr}\,\{\rho M_R(\Delta_1)\} + \mathrm{Tr}\,\{\rho M_R(\Delta_2)\}\,,\end{aligned}$$

which satisfies Axiom 3b.

This last calculation may be illuminated by a simple example. Instead of an arbitrary Hermitian operator R, let us consider the operator Q, defined by $Q\Psi(x) = x\Psi(x)$, which will be identified in Ch. 3 as the position operator. Let Δ_1 be the interval $\alpha \leq x \leq \beta$, and let Δ_2 be $\gamma \leq x \leq \delta$. Then the effect of the projection operator $M_Q(\Delta_1)$ is

$$\begin{aligned} M_Q(\Delta_1)\Psi(x) &= \Psi(x) \qquad && \text{for } \alpha \leq x \leq \beta\,, \\ M_Q(\Delta_1)\Psi(x) &= 0 && \text{for } x \leq \alpha \text{ or } \beta \leq x\,. \end{aligned}$$

A similar definition holds for $M_Q(\Delta_2)$, with σ replacing α, and δ replacing β. The projection operator $M_Q(\Delta_1 \cup \Delta_2)$ yields $M_Q(\Delta_1 \cup \Delta_2)\Psi(x) = \Psi(x)$ for $\alpha \leq x \leq \beta$ or $\gamma \leq x \leq \delta$, and $M_Q(\Delta_1 \cup \Delta_2)\Psi(x) = 0$ otherwise. If $\alpha < \beta < \gamma < \delta$ or $\gamma < \delta < \alpha < \beta$, so that Δ_1 and Δ_2 do not overlap, it is clear that $M_Q(\Delta_1)\Psi(x) + M_Q(\Delta_2)\Psi(x) = M_Q(\Delta_1 \cup \Delta_2)\Psi(x)$, and so the above calculation verifying Axiom 3b will be valid. But suppose, on the other hand, that

$\alpha < \gamma < \beta < \delta$, so that the intervals Δ_1 and Δ_2 overlap and the events $x \in \Delta_1$ and $x \in \Delta_2$ are not independent. Then in the region of overlap, $\gamma \le x \le \beta$, we will have $M_Q(\Delta_1)\Psi(x) + M_Q(\Delta_2)\Psi(x) = 2\Psi(x)$, but $M_Q(\Delta_1 \cup \Delta_2)\Psi(x) = \Psi(x)$. Thus the probabilities of nonindependent events will not be additive.

The remaining Axiom 4 will be discussed in Sec. 9.6.

Further reading for Chapter 2

The interpretation of the "state" concept in quantum mechanics, and some related controversies, have been discussed by Ballentine (1970). Ballentine (1986) examines the use of probability in quantum mechanics, and gives examples of erroneous applications of probability theory that have been made in that context. References to many papers on the foundations of quantum mechanics are contained in the "Resource Letter", Ballentine (1987). The infinitely many ways in which a nonpure state operator can be written as a mixture of pure states have been classified by Hughston, Joza and Wootters (1993).

Problems

2.1 (a) Show that the non-Hermitian matrix $M = \left[\begin{smallmatrix} 1 & 1 \\ 0 & 1 \end{smallmatrix}\right]$ has only real eigenvalues, but its eigenvectors do not form a complete set.

(b) Being non-Hermitian, this matrix must violate the conditions of Theorem 1, Sec. 1.3. Find a vector $|v\rangle$ such that $\langle v|M|v\rangle$ is complex. (This example illustrates the need to represent real observables by Hermitian operators, and not merely by operators that have purely real eigenvalues. Since $\langle M\rangle = \langle v|M|v\rangle$ can be complex, it clearly cannot be interpreted as an average of the eigenvalues of M.)

2.2 Show that Tr(AB) = Tr(BA); and, more generally, that the trace of a product of several operators is invariant under cyclic permutation of those operators, $\mathrm{Tr}(ABC \cdots Z) = \mathrm{Tr}(ZABC \cdots)$.

2.3 Prove that $\mathrm{Tr}(|u\rangle\langle v|) = \langle v|u\rangle$.

2.4 The nonnegativeness property, (2.8) or (2.13), of a general state operator ρ implies that $\mathrm{Tr}(\rho^2) \le 1$, as was shown in the course of proving (2.17). Show, conversely, that the condition $\mathrm{Tr}(\rho^2) \le 1$, in conjunction with (2.6) and (2.7), implies that ρ is nonnegative when ρ is 2×2 matrix. Show that these conditions are not sufficient to ensure nonnegativeness of ρ if its dimensions are 3×3 or larger.

2.5 Which of the following are acceptable as state operators? Find state vectors for any of them that represent pure states.

$$\rho_1 = \begin{bmatrix} \frac{1}{4} & \frac{3}{4} \\ \frac{3}{4} & \frac{3}{4} \end{bmatrix}, \quad \rho_2 = \begin{bmatrix} \frac{9}{25} & \frac{12}{25} \\ \frac{12}{25} & \frac{16}{25} \end{bmatrix},$$

$$\rho_3 = \frac{1}{3}|u\rangle\langle u| + \frac{2}{3}|v\rangle\langle v| + \frac{\sqrt{2}}{3}|u\rangle\langle v| + \frac{\sqrt{2}}{3}|v\rangle\langle u|,$$

where $\langle u|u\rangle = \langle v|v\rangle = 1$ and $\langle u|v\rangle = 0$,

$$\rho_4 = \begin{bmatrix} \frac{1}{2} & 0 & \frac{1}{4} \\ 0 & \frac{1}{2} & 0 \\ \frac{1}{4} & 0 & 0 \end{bmatrix}, \quad \rho_5 = \begin{bmatrix} \frac{1}{2} & 0 & \frac{1}{4} \\ 0 & \frac{1}{4} & 0 \\ \frac{1}{4} & 0 & \frac{1}{4} \end{bmatrix}.$$

2.6 Consider a dynamical variable σ that can take only two values, $+1$ or -1. The eigenvectors of the corresponding operator are denoted as $|+\rangle$ and $|-\rangle$. Now consider the following states: the one-parameter family of pure states that are represented by the vectors $|\theta\rangle = \sqrt{\frac{1}{2}}(|+\rangle + e^{i\theta}|-\rangle)$ for arbitrary θ; and the nonpure state $\rho = \frac{1}{2}(|+\rangle\langle +| + |-\rangle\langle -|)$. Show that $\langle\sigma\rangle = 0$ for all of these states. What, if any, are the physical differences between these various states, and how could they be measured?

2.7 It will be shown in Ch. 7 that the matrix operator $\sigma_y = \left[\begin{smallmatrix} 0 & -i \\ i & 0 \end{smallmatrix}\right]$ corresponds to a component of the spin of an electron, in units of $\hbar/2$. For a state represented by the vector $|\Psi\rangle = \left[\begin{smallmatrix} \alpha \\ \beta \end{smallmatrix}\right]$, where α and β are complex numbers, calculate the probability that the spin component is positive.

2.8 Suppose that the operator

$$M = \begin{bmatrix} 0 & 1 & 0 \\ 1 & 0 & 1 \\ 0 & 1 & 0 \end{bmatrix}$$

represents a dynamical variable. Calculate the probability $\text{Prob}(M = 0|\rho)$ for the following state operators:

$$\text{(a)} \;\; \rho = \begin{bmatrix} \frac{1}{2} & 0 & 0 \\ 0 & \frac{1}{4} & 0 \\ 0 & 0 & \frac{1}{4} \end{bmatrix}; \quad \text{(b)} \;\; \rho = \begin{bmatrix} \frac{1}{2} & 0 & \frac{1}{2} \\ 0 & 0 & 0 \\ \frac{1}{2} & 0 & \frac{1}{2} \end{bmatrix}; \quad \text{(c)} \;\; \rho = \begin{bmatrix} \frac{1}{2} & 0 & 0 \\ 0 & 0 & 0 \\ 0 & 0 & \frac{1}{2} \end{bmatrix}.$$

2.9 Let $R = \begin{bmatrix} 6, & -2 \\ -2, & 9 \end{bmatrix}$ represent a dynamical variable, and $|\Psi\rangle = \begin{bmatrix} a \\ b \end{bmatrix}$ be an arbitrary state vector (with $|a|^2 + |b|^2 = 1$). Calculate $\langle R^2 \rangle$ in two ways:

(a) Evaluate $\langle R^2 \rangle = \langle \Psi | R^2 | \Psi \rangle$ directly.

(b) Find the eigenvalues and eigenvectors of R,

$$R|r_n\rangle = r_n|r_n\rangle\,,$$

expand the state vector as a linear combination of the eigenvectors,

$$|\Psi\rangle = c_1|r_1\rangle + c_2|r_2\rangle\,,$$

and evaluate $\langle R^2 \rangle = r_1^2|c_1|^2 + r_2^2|c_2|^2$.

2.10 It was shown by Eqs. (2.21) and (2.22) that any nonpure state operator can be decomposed into a mixture of pure states in at least two ways. Show (by constructing an example depending on a continuous parameter) that this can be done in infinitely many ways.

Chapter 3

Kinematics and Dynamics

The results of Ch. 2 constitute what is sometimes called "the formal structure of quantum mechanics". Although much has been written about its interpretation, derivation from more elementary axioms, and possible generalization, it has by itself very little physical content. It is not possible to solve a single physical problem with that formalism until one obtains *correspondence rules* that identify particular dynamical variables with particular operators. This will be done in the present chapter.

The fundamental physical variables, such as linear and angular momentum, are closely related to space–time symmetry transformations. The study of these transformations serves a dual purpose: a fundamental one by identifying the operators for important dynamical variables, and a practical one by introducing the concepts and techniques of symmetry transformations.

3.1 *Transformations of States and Observables*

The laws of nature are believed to be invariant under certain space–time symmetry operations, including displacements, rotations, and transformations between frames of reference in uniform relative motion. Corresponding to each such space–time transformation there must be a transformation of observables, $A \to A'$, and of states, $|\Psi\rangle \to |\Psi'\rangle$. (We shall consider only pure states, represented by state vectors, since the general case adds no novelty here.) Certain relations must be preserved by these transformations.

(a) If $A|\phi_n\rangle = a_n|\phi_n\rangle$, then after transformation we must have $A'|\phi'_n\rangle = a_n|\phi'_n\rangle$. The eigenvalues of A and A' are the same because A' represents an observable that is essentially similar to A, differing only by transformation to another frame of reference. Since A and A' represent equivalent observables, they must have the same set of possible values.

(b) If a state vector is given by $|\psi\rangle = \sum_n c_n|\phi_n\rangle$, where $\{|\phi_n\rangle\}$ are the eigenvectors of A, then the transformed state vector will be of the form $|\psi'\rangle = \sum_n c'_n|\phi'_n\rangle$, in terms of the eigenvectors of A'. The two state vectors must obey the relations $|c_n|^2 = |c'_n|^2$; that is to say, $|\langle\phi_n|\psi\rangle|^2 = |\langle\phi'_n|\psi'\rangle|^2$. These relations must hold because they express the equality of probabilities for equivalent events in the two frames of reference.

The mathematical character of these transformations is clarified by the following theorem:

Theorem (Wigner). Any mapping of the vector space onto itself that preserves the value of $|\langle\phi|\psi\rangle|$ may be implemented by an operator U:

$$\begin{aligned} |\psi\rangle &\to |\psi'\rangle = U|\psi\rangle\,, \\ |\phi\rangle &\to |\phi'\rangle = U|\phi\rangle\,, \end{aligned} \tag{3.1}$$

with U being either *unitary* (linear) or *antiunitary* (antilinear).

Case (a). If U is *unitary*, then by definition $UU^\dagger = U^\dagger U = I$ is the identity operator. Thus $\langle\phi'|\psi'\rangle = (\langle\phi|U^\dagger)(U|\psi\rangle) = \langle\phi|\psi\rangle$. A unitary transformation preserves the complex value of an inner product, not merely its absolute value.

Case (b). If U is *antilinear*, then by definition $Uc|\psi\rangle = c^*U|\psi\rangle$, where c is a complex number. If U is *antiunitary*, then $\langle\phi'|\psi'\rangle = \langle\phi|\psi\rangle^*$.

An elementary proof of Wigner's theorem has been given by Bargmann (1964).

Only linear operators can describe continuous transformations because every continuous transformation has a square root. Suppose, for example, that $U(\ell)$ describes a displacement through the distance ℓ. This can be done by two displacements of $\ell/2$, and hence $U(\ell) = U(\ell/2)\,U(\ell/2)$. The product of two antilinear operators is linear, since the second complex conjugation nullifies the effect of the first. Thus, regardless of the linear or antilinear character of $U(\ell/2)$, it must be the case that $U(\ell)$ is linear. A continuous operator cannot change discontinuously from linear to antilinear as a function of ℓ, so the operator must be linear for all ℓ. Antilinear operators are needed to describe certain discrete symmetries (see Ch. 13), but we shall have no use for them in this chapter.

The transformation of state vectors, of the form (3.1), is accompanied by a transformation $A \to A'$ of the operators for observables. It must be such that the transformed observables bear the same relationship to the transformed states as did the original observables to the original states. In particular, if $A|\phi_n\rangle = a_n|\phi_n\rangle$, then $A'|\phi'_n\rangle = a_n|\phi'_n\rangle$. Substitution of $|\phi'_n\rangle = U|\phi_n\rangle$, using (3.1), yields $A'U|\phi_n\rangle = a_n U|\phi_n\rangle$, and hence $U^{-1}A'U|\phi_n\rangle = a_n|\phi_n\rangle$. Subtracting this from the original eigenvalue equation yields $(A-U^{-1}A'U)|\phi_n\rangle = 0$. Since this equation holds for each member of the complete set $\{|\phi_n\rangle\}$, it holds for an arbitrary vector, and therefore $(A - U^{-1}A'U) = 0$. Thus the desired transformation of operators that accompanies (3.1) is

$$A \to A' = UAU^{-1}. \tag{3.2}$$

Consider a family of unitary operators, $U(s)$, that depend on a single continuous parameter s. Let $U(0) = I$ be the identity operator, and let $U(s_1 + s_2) = U(s_1)U(s_2)$. It can be shown that it is always possible to choose the parameter in any one-parameter group of operators so that these relations are satisfied. But the proof is not needed here because the operations that we shall treat (displacements, rotations about an axis, Galilei transformations to a moving frame of reference) obviously satisfy them.

If s is allowed to become very small we may express the resultant *infinitesimal* unitary transformation as

$$U(s) = I + \left.\frac{dU}{ds}\right|_{s=0} s + O(s^2).$$

The unitarity condition requires that

$$UU^\dagger = I + s\left.\left[\frac{dU}{ds} + \frac{dU^\dagger}{ds}\right]\right|_{s=0} + O(s^2)$$

should simply be equal to I, independent of the value of s. Hence the coefficient of s must vanish, and we may write

$$\left.\frac{dU}{ds}\right|_{s=0} = iK, \quad \text{with} \quad K = K^\dagger. \tag{3.3}$$

The Hermitian operator K is called the *generator* of the family of unitary operators because it determines $U(s)$, not only for infinitesimal s, but for all s. This can be shown by differentiating

$$U(s_1 + s_2) = U(s_1)U(s_2)$$

with respect to s_2 and using (3.3):

$$\left.\frac{\partial}{\partial s_2}U(s_1 + s_2)\right|_{s_2=0} = U(s_1)\left.\frac{d}{ds_2}U(s_2)\right|_{s_2=0},$$

$$\left.\frac{dU(s)}{ds}\right|_{s=s_1} = U(s_1)\, iK.$$

This first order differential equation with initial condition $U(0) = I$ has the unique solution

$$U(s) = e^{iKs} . \tag{3.4}$$

Thus the operator for any finite transformation is determined by the generator of infinitesimal transformations.

3.2 *The Symmetries of Space–Time*

The symmetries of space–time include rotations, displacements, and transformations between uniformly moving frames of reference. The latter are Lorentz transformations in general, but if we restrict our attention to velocities that are small compared to the speed of light, they may be replaced by Galilei transformations. The set of all such transformations is called the *Galilei group.* The effect of a transformation is

$$\begin{aligned} \mathbf{x} \to \mathbf{x}' &= R\mathbf{x} + \mathbf{a} + \mathbf{v}t , \\ t \to t' &= t + s . \end{aligned} \tag{3.5}$$

Here R is a rotation (conveniently thought of as a 3×3 matrix acting on a three-component vector $\mathbf{x}$), $\mathbf{a}$ is a space displacement, $\mathbf{v}$ is the velocity of a moving coordinate transformation, and s is a time displacement.

Let $\tau_1 = \tau_1(R_1, \mathbf{a}_1, \mathbf{v}_1, s_1)$ denote such a transformation. Let $\tau_3 = \tau_2\tau_1$ be the single transformation that yields the same result as τ_1 followed by τ_2. That is to say, if $\tau_1\{\mathbf{x}, t\} = \{\mathbf{x}', t'\}$ and $\tau_2\{\mathbf{x}', t'\} = \{\mathbf{x}'', t''\}$, then $\tau_3\{\mathbf{x}, t\} = \{\mathbf{x}'', t''\}$. Carrying out these operations, we obtain

$$\begin{aligned} \mathbf{x}'' &= R_2(R_1\mathbf{x} + \mathbf{a}_1 + \mathbf{v}_1 t) + \mathbf{a}_2 + \mathbf{v}_2(t + s_1) , \\ t'' &= t + s_1 + s_2 , \end{aligned}$$

and therefore

$$\begin{aligned} R_3 &= R_2 R_1 , \\ \mathbf{a}_3 &= \mathbf{a}_2 + R_2\mathbf{a}_1 + \mathbf{v}_2 s_1 , \\ \mathbf{v}_3 &= \mathbf{v}_2 + R_2\mathbf{v}_1 , \\ s_3 &= s_2 + s_1 . \end{aligned} \tag{3.6}$$

The laws of physics (in the low velocity, or “nonrelativistic”, limit) are invariant under these transformations, so the quantum-mechanical description of systems that differ only by such transformations must be equivalent. Therefore, corresponding to a space–time transformation τ, there must be a unitary transformation $U(\tau)$ on the state vectors and operators for observables:

$$|\Psi\rangle \to |\Psi'\rangle = U(\tau)|\Psi\rangle\,,$$
$$A \to A' = U(\tau)AU^{-1}(\tau)\,.$$

Since $\tau_2\tau_1$ and τ_3 are the same space–time transformations, we require that $U(\tau_2)U(\tau_1)|\Psi\rangle$ and $U(\tau_3)|\Psi\rangle$ describe the *same state.* This does not mean that they must be the same vector, since two vectors differing only in their complex phases are physically equivalent, but they may differ at most by a phase factor. Thus we have

$$U(\tau_2\tau_1) = e^{i\omega(\tau_2,\tau_1)}U(\tau_2)U(\tau_1)\,. \tag{3.7}$$

One might suppose that the (real) phase $\omega(\tau_2, \tau_1)$ could also depend upon $|\Psi\rangle$. But in that case U would not be a linear operator, and we know from Wigner's theorem (Sec. 3.1) that U must be linear for a continuous transformation.

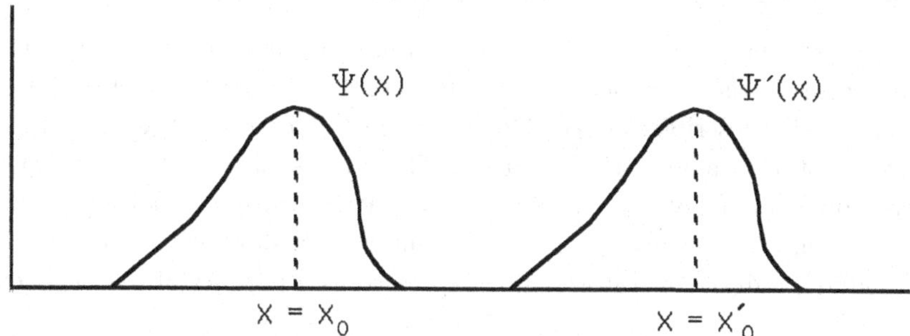

Fig. 3.1 Transformation of a function [Eq. (3.8)].

It is important to be aware that when the abstract vector $|\Psi\rangle$ is represented as a function of space–time coordinates, there is an *inverse relation between transformations on function space and transformations on coordinates.* This is illustrated in Fig. 3.1, where a function $\Psi(x)$ is transformed into a new function, $\Psi'(x) = U(\tau)\Psi(x)$. The original function is located near the point $x = x_0$, and the new function is located near the point $x = x_0'$, where $x_0' = \tau x_0$. The precise relationship between the two functions is $\Psi'(\tau x) = \Psi(x)$; the value of the new function at the transformed point is the same as the value of the original function at the old point. Writing $\tau x = x'$, we have $\Psi'(x') = \Psi(x) = \Psi(\tau^{-1}x')$. But $\Psi'(x') = U(\tau)\Psi(x')$, by definition of $U(\tau)$. Thus we have (dropping the prime from the dummy variable)

$$U(\tau)\,\Psi(x) = \Psi(\tau^{-1}x), \tag{3.8}$$

which exhibits the inverse correspondence between transformations on function space and on coordinates.

The transformation just described is in the *active* point of view, in which the object (in this case a function) is transformed relative to a fixed coordinate system. There is also the *passive* point of view, in which a fixed object is redescribed with respect to a transformed coordinate system. The two points of view are equivalent, and the choice between them is a matter of taste. (The only danger, which must be carefully avoided, is to inadvertently switch from one point of view to the other in the same analysis!) We shall generally adhere to the active point of view in developing the theory. (Exceptions are Sec. 4.3, where the passive point of view is used in a self-contained exercise, and Sec. 7.5, where both active and passive rotations are discussed.)

3.3 *Generators of the Galilei Group*

As was shown in Sec. 3.1, any one parameter group of unitary operators can be expressed as an exponential of a Hermitian *generator*. The set of space–time symmetries described in Sec. 3.2, called the *Galilei group*, has ten parameters: three rotation angles, three space displacements, three velocity components, and one time displacement. The most general transformation of this kind is equivalent to a sequence of ten elementary transformations, and the corresponding unitary operator can be expressed as a product of ten exponentials,

$$U(\tau) = \prod_{\mu=1}^{10} e^{is_\mu K_\mu} . \tag{3.9}$$

Here $s_\mu (\mu = 1, 2, \ldots, 10)$ denotes the ten parameters that define the transformation τ, and $K_\mu = {K_\mu}^\dagger$ are the ten Hermitian generators. The properties of the unitary operators are determined by these generators. Moreover, these generators will turn out to be very closely related to the fundamental dynamical variables, such as momentum and energy.

If we let all the parameters s_μ become infinitesimally small, we obtain a general infinitesimal unitary operator,

$$U = I + i \sum_{\mu=1}^{10} s_\mu K_\mu . \tag{3.10}$$

The multiplication law (3.7) for the U operators expresses itself as a set of commutation relations for the generators.

Consider the following product of two infinitesimal operators and their inverses:

$$e^{i\varepsilon K_\mu}\, e^{i\varepsilon K_\nu}\, e^{-i\varepsilon K_\mu}\, e^{-i\varepsilon K_\nu} = I + \varepsilon^2 (K_\nu K_\mu - K_\mu K_\nu) + O(\varepsilon^3)\,. \tag{3.11}$$

Since any sequence of space–time transformations is equivalent to another transformation in the group, it follows from (3.7) that the operator product in (3.11) must differ at most by a phase factor from some operator of the form (3.9). That is to say, there must be a set of values for the 11 parameters $\{\omega, s_\mu\}$ that will make $e^{i\omega}U(\tau)$ equal to (3.11). It is clear that all 11 parameters must be infinitesimal and of order ε^2 so that (3.11) will be expressible in the form

$$e^{i\omega}U = I + i\sum_{\mu=1}^{10} s_\mu K_\mu + i\omega I\,. \tag{3.12}$$

The equality of (3.11) and (3.12) requires that the commutator of two generators be a linear combination of generators and the identity operator. Hence we can write

$$[K_\mu, K_\nu] = i\sum_\lambda c^\lambda_{\mu\nu}\, K_\lambda + i b_{\mu\nu} I\,. \tag{3.13}$$

The constants $c^\lambda_{\mu\nu}$ are determined from the multiplication rules (3.6) for the space–time transformations $\tau(R, \mathbf{a}, \mathbf{v}, s)$. The multiple of the identity, $b_{\mu\nu}$, arises from the phase factor in (3.7), and would vanish if ω were equal to zero. These general principles will be applied to each specific pair of generators.

It is convenient to introduce a more descriptive notation than (3.9) for the unitary operators that correspond to particular space–time transformations.

Space–Time Transformation	Unitary Operator
Rotation about axis α ($\alpha = 1, 2, 3$)	
$\mathbf{x} \to R_\alpha(\theta_\alpha)\mathbf{x}$	$e^{-i\theta_\alpha J_\alpha}$
Displacement along axis α	
$x_\alpha \to x_\alpha + a_\alpha$	$e^{-ia_\alpha P_\alpha}$
Velocity along axis α	
$x_\alpha \to x_\alpha + v_\alpha t$	$e^{iv_\alpha G_\alpha}$
Time displacement	
$t \to t + s$	e^{isH}

The ten generators $\{-J_\alpha, -P_\alpha, G_\alpha, H\}(\alpha = 1, 2, 3)$ are specific forms of the generic generators $K_\mu(\mu = 1, \ldots, 10)$. The minus signs are introduced only to conform to conventional notations.

Evaluation of commutators

The method for evaluating the commutation relations of the form (3.13) is as follows. We choose a pair of generators and substitute them into (3.11). We then carry out the corresponding sequence of four space–time transformations in order to determine the single transformation that results. The commutator of the chosen pair of generators must therefore differ from the generator corresponding to this resultant transformation by no more than a multiple of the identity.

Several pairs of space–time transformations obviously commute. This is the case for *pure displacements* in space and time, for which Eqs. (3.6) reduce to a form that is independent of the order of transformations τ_1 and τ_2 : $\mathbf{a}_3 = \mathbf{a}_2 = \mathbf{a}_1$, $s_3 = s_2 + s_1$. The commutators of the corresponding generators must vanish, apart from a possible multiple of identity, and hence

$$[P_\alpha, P_\beta] = O + (?)I\,, \tag{3.14}$$

$$[P_\alpha, H] = O + (?)I\,. \tag{3.15}$$

The unknown multiples of identity, $(?)I$, will be dealt with later.

A similar argument applies for space displacements and velocity transformations, for which (3.6) reduce to $\mathbf{a}_3 = \mathbf{a}_2 + \mathbf{a}_1, \mathbf{v}_3 = \mathbf{v}_2 + \mathbf{v}_1$, and hence

$$[P_\alpha, G_\beta] = O + (?)I\,, \tag{3.16}$$

$$[G_\alpha, G_\beta] = O + (?)I\,. \tag{3.17}$$

It is also evident that rotations commute with time displacements, and hence

$$[J_\alpha, H] = O + (?)I\,. \tag{3.18}$$

Furthermore, a rotation commutes with a displacement or a velocity transformation along the rotation axis, and hence

$$[J_\alpha, P_\alpha] = O + (?)I\,, \tag{3.19}$$

$$[J_\alpha, G_\alpha] = O + (?)I\,. \tag{3.20}$$

Consider now a less trivial case,

$$e^{i\varepsilon H}\, e^{i\varepsilon G_1}\, e^{-i\varepsilon H}\, e^{-i\varepsilon G_1} = I + \varepsilon^2 [G_1, H] + \cdots ,$$

which corresponds (from right to left) to a velocity $-\varepsilon$ along the x axis, a time displacement of $-\varepsilon$, and their inverses. The effect of these four successive transformations is

$$\begin{aligned}(x_1, x_2, x_3, t) &\to (x_1 - \varepsilon t, x_2, x_3, t)\\ &\to (x_1 - \varepsilon t, x_2, x_3, t - \varepsilon)\\ &\to (x_1 - \varepsilon t + \varepsilon(t - \varepsilon), x_2, x_3, t - \varepsilon)\\ &\to (x_1 - \varepsilon^2, x_2, x_3, t)\,.\end{aligned}$$

This is just a space displacement by $-\varepsilon^2$ along the x axis, so the product of the four unitary operators must differ by at most a phase factor from

$$e^{i\varepsilon^2 P_1} = I + i\varepsilon^2 P_1 + \cdots .$$

A similar conclusion holds for each of the three axes, so we have

$$[G_\alpha, H] = iP_\alpha + (?)I\,. \tag{3.21}$$

A *rotation* consists of the transformation

$$x_j \to \sum_{k=1}^{3} (R)_{jk} x_k\,.$$

For each of the three axes there is a rotation matrix:

$$R_1(\theta) = \begin{bmatrix} 1 & 0 & 0 \\ 0 & \cos\theta & -\sin\theta \\ 0 & \sin\theta & \cos\theta \end{bmatrix}, \quad R_2(\theta) = \begin{bmatrix} \cos\theta & 0 & \sin\theta \\ 0 & 1 & 0 \\ -\sin\theta & 0 & \cos\theta \end{bmatrix},$$

$$R_3(\theta) = \begin{bmatrix} \cos\theta & -\sin\theta & 0 \\ \sin\theta & \cos\theta & 0 \\ 0 & 0 & 1 \end{bmatrix}.$$

The rotation matrices can be expanded in a power series,

$$R_\alpha(\theta) = I - i\theta M_\alpha + \cdots ,$$

where $M_\alpha = i dR_\alpha/d\theta|_{\theta=0}$:

$$M_1 = \begin{bmatrix} 0 & 0 & 0 \\ 0 & 0 & -i \\ 0 & i & 0 \end{bmatrix}, \quad M_2 = \begin{bmatrix} 0 & 0 & i \\ 0 & 0 & 0 \\ -i & 0 & 0 \end{bmatrix}, \quad M_3 = \begin{bmatrix} 0 & -i & 0 \\ i & 0 & 0 \\ 0 & 0 & 0 \end{bmatrix}.$$

To the second order in the small angle ε, we have

$$\begin{aligned} R_2(-\varepsilon)R_1(-\varepsilon)R_2(\varepsilon)R_1(\varepsilon) &= I + \varepsilon^2(M_1M_2 - M_2M_1) \\ &= I + \varepsilon^2 iM_3 \\ &= R_3(-\varepsilon^2)\,. \end{aligned}$$

The corresponding unitary rotation operators must satisfy a similar relation to within a phase factor,

$$e^{i\varepsilon J_2}\, e^{i\varepsilon J_1}\, e^{-i\varepsilon J_2}\, e^{-i\varepsilon J_1} = e^{i\omega}\, e^{i\varepsilon^2 J_3}\,,$$

and so their generators must satisfy

$$[J_1, J_2] = iJ_3 + (?)I\,.$$

The corresponding relations for other combinations of rotation generators can be deduced by cyclic permutation of the three axes, and by the antisymmetry of the commutator under interchange of its two arguments. This allows us to write

$$[J_\alpha, J_\beta] = i\varepsilon_{\alpha\beta\gamma}J_\gamma + (?)I\,, \tag{3.22}$$

where γ is to be chosen unequal to α or β, with $\varepsilon_{123} = \varepsilon_{231} = \varepsilon_{312} = 1$, $\varepsilon_{213} = \varepsilon_{132} = \varepsilon_{321} = -1$, and $\varepsilon_{\alpha\beta\gamma} = 0$ if $\alpha = \beta$.

Consider next

$$e^{i\varepsilon G_2}\, e^{i\varepsilon J_1}\, e^{-i\varepsilon G_2}\, e^{-i\varepsilon J_1} = I + \varepsilon^2[J_1, G_2] + \cdots\,,$$

which corresponds to a rotation by ε about the x_1 axis, a velocity $-\varepsilon$ along the x_2 axis, and their inverses. The effects of these transformations are

$$\begin{aligned} (x_1, x_2, x_3) &\to (x_1, x_2 \cos\varepsilon - x_3 \sin\varepsilon, x_2 \sin\varepsilon + x_3 \cos\varepsilon) \\ &\to (x_1, x_2 \cos\varepsilon - x_3 \sin\varepsilon - \varepsilon t, x_2 \sin\varepsilon + x_3 \cos\varepsilon) \\ &\to (x_1, x_2 - \varepsilon t \cos\varepsilon, x_3 + \varepsilon t \sin\varepsilon) \\ &\to (x_1, x_2 - \varepsilon t \cos\varepsilon + \varepsilon t, x_3 + \varepsilon t \sin\varepsilon) \\ &\to (x_1, x_2, x_3 + \varepsilon^2 t)\,, \end{aligned}$$

to the second order in ε. This is equivalent to a velocity ε^2 along the x_3 axis, so the operator product must differ by at most a phase factor from

$$e^{i\varepsilon^2 G_3} = I + i\varepsilon^2 G_3 + \cdots$$

Hence we have

$$[J_1, G_2] = iG_3 + (?)I\,.$$

A similar treatment of other components yields

$$[J_\alpha, G_\beta] = i\varepsilon_{\alpha\beta\gamma}G_\gamma + (?)I\,. \tag{3.23}$$

The treatment of a rotation and a space displacement is very similar, and the result is

$$[J_\alpha, P_\beta] = i\varepsilon_{\alpha\beta\gamma}P_\gamma + (?)I\,. \tag{3.24}$$

Multiples of identity

We must now deal with the undetermined *multiples of identity* in (3.14)–(3.24), which result from the unknown phase factor in the operator multiplication law (3.7). The terms of the form $(?)I$ are of three types:

(a) Those that can be determined by consistency conditions;
(b) Those that are arbitrary but may be eliminated by a suitable conventional choice of the phases of certain vectors; and
(c) Those that are irremovable and physically significant.

We shall deal first with the cases of type (a). All commutators are *antisymmetric*,

$$[K_\mu, K_\nu] = -[K_\nu, K_\mu]\,,$$

and satisfy the *Jacobi identity*,

$$[[K_\mu, K_\nu], K_\lambda] + [[K_\nu, K_\lambda], K_\mu] + [[K_\lambda, K_\mu], K_\nu] = 0\,.$$

Antisymmetry implies that every operator commutes with itself, so when $\alpha = \beta$ there can be no multiple of identity in (3.14), in (3.17), or in (3.22).

The Jacobi identity can be written more conveniently as

$$[[K_\mu, K_\nu], K_\lambda] = [[K_\lambda, K_\nu], K_\mu] + [[K_\mu, K_\lambda], K_\nu]\,. \tag{3.25}$$

As an example of its use, let $K_\mu = J_2, K_\nu = P_3$, and $K_\lambda = H$. With the help of (3.24), (3.15), and (3.18), it yields

$$\begin{aligned}
[[J_2, P_3], H] &= [[H, P_3], J_2] + [[J_2, H], P_3]\,,\\
[(iP_1 + ?I), H] &= \quad [?I, J_2] \quad + [?I, P_3]\,,\\
i[P_1, H] &= \qquad 0 \qquad + \quad 0\,.
\end{aligned}$$

A similar result holds for each axis, and hence

$$[P_\alpha, H] = 0\,. \tag{3.26}$$

The way to apply this method in general should be evident from this example. Choose K_μ and K_ν so that their commutator generates one of the operators of interest, and choose K_λ as the other operator. The unknown multiples of identity have no effect inside the commutators. In this way we can obtain the following:

$$[P_\alpha, P_\beta] = 0\,, \tag{3.27}$$

$$[G_\alpha, G_\beta] = 0\,, \tag{3.28}$$

$$[J_\alpha, H] = 0\,. \tag{3.29}$$

We next consider type (b), which consists of those multiples of identity that remain arbitrary but can be transformed to zero by redefining the phases of certain vectors. Such a case is (3.22). Antisymmetry of the commutator implies that the multiple of identity can be expressed thus:

$$[J_\alpha, J_\beta] = i\varepsilon_{\alpha\beta\gamma}J_\gamma + i\varepsilon_{\alpha\beta\gamma}b_\gamma I\,,$$

where $b\gamma(\gamma = 1, 2, 3)$ are real numbers. The multiple of identity can be removed by the substitution $J_\alpha + b\alpha I \to J_\alpha$ for $\alpha = 1, 2, 3$. Then one obtains

$$[J_\alpha, J_\beta] = i\varepsilon_{\alpha\beta\gamma}J_\gamma\,. \tag{3.30}$$

This substitution has the effect of replacing the unitary rotation operator $U(R_\alpha) = e^{-i\theta J_\alpha}$ by $e^{i\theta b_\alpha}e^{-i\theta J_\alpha}$; that is to say, we are replacing $|\Psi'\,\rangle = U|\Psi\,\rangle$ by $e^{i\theta b_\alpha}|\Psi'\,\rangle$. Since the absolute phase of the transformed vector $|\Psi'\,\rangle$ has no physical significance, this redefinition of phase is permitted.

Similar considerations apply to (3.23) and (3.24), although the necessary argument is somewhat longer. Using (3.25) for the generators J_1, J_2 and G_3 yields

$$\begin{aligned}
[[J_1, J_2], G_3] &= [[G_3, J_2], J_1] + [[J_1, G_3], J_2]\,,\\
i[J_3, G_3] &= -i[G_1, J_1] - i[G_2, J_2]\,,\\
[J_3, G_3] &= [J_1, G_1] + [J_2, G_2]\,.
\end{aligned}$$

The latter equation has the form

$$X_3 = X_1 + X_2\,.$$

Since the three axes may be cyclically permuted we also have

$$X_1 = X_2 + X_3 ,$$
$$X_2 = X_3 + X_1 .$$

This set of homogeneous linear equations has only the solution zero, and therefore

$$[J_\alpha, G_\alpha] = 0 .$$

Using (3.25) with J_3, J_1 and G_3 yields

$$[[J_3, J_1], G_3] = [[G_3, J_1], J_3] + [[J_3, G_3], J_1] ,$$
$$i[J_2, G_3] = i[G_2, J_3] + 0 ,$$
$$[J_2, G_3] = -[J_3, G_2] .$$

This result, combined with the previous one, allows us to write

$$[J_\alpha, G_\beta] = -[J_\beta, G_\alpha] .$$

Therefore the multiples of identity in (3.23) must have the form

$$[J_\alpha, G_\beta] = i\varepsilon_{\alpha\beta\gamma} G_\gamma + i\varepsilon_{\alpha\beta\gamma} b_\gamma I .$$

The substitution $G_\alpha + b_\alpha I \to G_\alpha (\alpha = 1, 2, 3)$, which is equivalent to redefining the phase of the transformed vector $|\Psi'\rangle = e^{iv_\alpha G_\alpha}|\Psi\rangle$, then yields

$$[J_\alpha, G_\beta] = i\varepsilon_{\alpha\beta\gamma} G_\gamma . \tag{3.31}$$

A similar calculation yields

$$[J_\alpha, P_\beta] = i\varepsilon_{\alpha\beta\gamma} P_\gamma . \tag{3.32}$$

Having established the above result, we can now evaluate the commutator of G_α and H, by using (3.25) with J_1, G_2 and H:

$$[[J_1, G_2], H] = [[H, G_2], J_1] + [[J_1, H], G_2] ,$$
$$i[G_3, H] = -i[P_2, J_1] + 0$$
$$= -P_3 .$$

Since all three axes are equivalent, we conclude that

$$[G_\alpha, H] = iP_\alpha . \tag{3.33}$$

We now encounter a case of type (c), which involves an irremovable multiple of identity. Using (3.25) with J_1, G_2 and P_1, we obtain

$$[[J_1, G_2], P_1] = [[P_1, G_2], J_1] + [[J_1, P_1], G_2]\,,$$
$$i[G_3, P_1] = 0 + 0\,.$$

Thus $[G_\alpha, P_\beta] = 0$ for $\alpha \neq \beta$. Next we repeat the calculation with P_3 instead of P_1:

$$[[J_1, G_2], P_3] = [[P_3, G_2], J_1] + [[J_1, P_3], G_2]\,,$$
$$i[G_3, P_3] = 0 \; - \; i[P_2, G_2]\,.$$

Thus $[G_\alpha, P_\alpha] = [G_\beta, P_\beta]$. These results may be combined with (3.16) to yield

$$[G_\alpha, P_\beta] = i\delta_{\alpha\beta} M I\,, \tag{3.34}$$

M being a real constant.

The value of M is not determined by any of the equations at our disposal. It cannot be eliminated by adding multiples of I to any of the generators. (That option is, in any case, no longer available to us, since we have already used up such freedom for all generators except H.) It is mathematically irremovable, and will turn out to have a physical significance.

For convenience the final commutation relations are summarized in the following table.

Commutation Relations for the Generators of the Galilei Group of Transformations (3.35)

(a)	$[P_\alpha, P_\beta] = 0$	(f)	$[G_\alpha, P_\beta] = i\delta_{\alpha\beta} M I$
(b)	$[G_\alpha, G_\beta] = 0$	(g)	$[P_\alpha, H] = 0$
(c)	$[J_\alpha, J_\beta] = i\varepsilon_{\alpha\beta\gamma} J_\gamma$	(h)	$[G_\alpha, H] = iP_\alpha$
(d)	$[J_\alpha, P_\beta] = i\varepsilon_{\alpha\beta\gamma} P_\gamma$	(i)	$[J_\alpha, H] = 0$
(e)	$[J_\alpha, G_\beta] = i\varepsilon_{\alpha\beta\gamma} G_\gamma$		

3.4 *Identification of Operators with Dynamical Variables*

In the preceding Section we determined the geometrical significance of the operators $\mathbf{P}, \mathbf{J}, \mathbf{G}$ and H as generators of symmetry transformations in state vector space. However, they have not yet been given any dynamical significance as operators representing observables, although the notation has been suggestively chosen in anticipation of the results that will be established in this section.

The dynamics of a *free particle* are invariant under the full Galilei group of space–time transformations, and this turns out to be sufficient to completely identify the operators for its dynamical variables. The method is based on a paper by T. F. Jordan (1975).

We assume the *position operator* for the particle to be $\mathbf{Q} = (Q_1, Q_2, Q_3)$, where by definition

$$Q_\alpha|\mathbf{x}\rangle = x_\alpha|\mathbf{x}\rangle \quad (\alpha = 1, 2, 3) \tag{3.36}$$

has an unbounded continuous spectrum. Two assumptions are involved here: first, that space is a continuum; and second, that all three components of the position operator are mutually commutative, and so possess a common set of eigenvectors. The first assumption is unavoidable if we are to use continuous transformations. Although it may need revision in a theory that attempts to treat gravitation, and hence space–time, quantum-mechanically, there is no reason to doubt it at the atomic and nuclear levels. The second assumption will be discussed later in the context of composite systems (Sec. 3.5).

We now seek to introduce a *velocity operator* $\mathbf{V}$ such that

$$\frac{d}{dt}\langle\mathbf{Q}\rangle = \langle\mathbf{V}\rangle \tag{3.37}$$

for any state. In particular, for a pure state represented by the vector $|\Psi(t)\rangle$, we want

$$\begin{aligned}\langle\Psi(t)|\mathbf{V}|\Psi(t)\rangle &= \frac{d}{dt}\langle\Psi(t)|\mathbf{Q}|\Psi(t)\rangle \\ &= \left\{\frac{d}{dt}\langle\Psi(t)|\right\}\mathbf{Q}|\Psi(t)\rangle + \langle\Psi(t)|\mathbf{Q}\left\{\frac{d}{dt}|\Psi(t)\rangle\right\}.\end{aligned}$$

Corresponding to the time displacement $t \to t' = t + s$, there is a vector space transformation of the form (3.8),

$$|\Psi(t)\rangle \to e^{isH}|\Psi(t)\rangle = |\Psi(t-s)\rangle\,.$$

Putting $s = t$, we obtain $|\Psi(t)\rangle = e^{-itH}|\Psi(0)\rangle$, and hence

$$\frac{d}{dt}|\Psi(t)\rangle = -iH|\Psi(t)\rangle\,. \tag{3.38}$$

From this result we obtain

$$\begin{aligned}\langle\Psi|\mathbf{V}|\Psi\rangle &= i\langle\Psi|H\mathbf{Q}|\Psi\rangle - i\langle\Psi|\mathbf{Q}H|\Psi\rangle \\ &= i\langle\Psi|[H,\mathbf{Q}]|\Psi\rangle\,.\end{aligned}$$

Therefore

$$\mathbf{V} = i[H, \mathbf{Q}] \tag{3.39}$$

fulfills the role of a velocity operator for a free particle. But it is only expressed in terms of H, whose form is not yet determined.

The *space displacement* $\mathbf{x} \to \mathbf{x}' = \mathbf{x} + \mathbf{a}$ involves a displacement of the localized position eigenvectors,

$$|\mathbf{x}\rangle \to |\mathbf{x}\rangle' = e^{-i\mathbf{a}\cdot\mathbf{P}}|\mathbf{x}\rangle = |\mathbf{x} + \mathbf{a}\rangle\,. \tag{3.40}$$

(As in Fig. 3.1, we are using the active point of view, in which states are displaced with respect to a fixed coordinate system.) The displaced observables bear the same relationship to the displaced vectors as the original observables do to the original vectors, as was discussed in Sec. 3.1. In particular,

$$\mathbf{Q} \to \mathbf{Q}' = e^{-i\mathbf{a}\cdot\mathbf{P}}\mathbf{Q}e^{i\mathbf{a}\cdot\mathbf{P}} \tag{3.41}$$

with

$$Q'_\alpha|\mathbf{x}\rangle' = x_\alpha|\mathbf{x}\rangle' \quad (\alpha = 1, 2, 3)\,. \tag{3.42}$$

But since $|\mathbf{x}\rangle' = |\mathbf{x} + \mathbf{a}\rangle$, a comparison of (3.42) with (3.36) implies that

$$\mathbf{Q}' = \mathbf{Q} - \mathbf{a}I\,. \tag{3.43}$$

In view of (3.42) and (3.43), one may think of the operator $\mathbf{Q}'$ as measuring position with respect to a displaced origin.

Equating terms of first order in $\mathbf{a}$ from (3.43) and (3.41), we obtain $[Q_\alpha, a\cdot\mathbf{P}] = a_\alpha I$, which can hold for arbitrary directions of $\mathbf{a}$ only if

$$[Q_\alpha, P_\beta] = i\delta_{\alpha\beta} I\,. \tag{3.44}$$

A *rotation* through the infinitesimal angle θ about the axis along the unit vector $\hat{\mathbf{n}}$ has the effect $\mathbf{x} \to \mathbf{x}' = \mathbf{x} + \theta\hat{\mathbf{n}} \times \mathbf{x}$. There is a corresponding transformation of the position eigenvectors,

$$|\mathbf{x}\rangle \to |\mathbf{x}\rangle' = e^{-i\theta\hat{\mathbf{n}}\cdot\mathbf{J}}|\mathbf{x}\rangle = |\mathbf{x}'\rangle\,, \tag{3.45}$$

and of the position operator,

$$\begin{aligned} Q_\alpha \to Q'_\alpha &= e^{-i\theta\hat{\mathbf{n}}\cdot\mathbf{J}}Q_\alpha e^{i\theta\hat{\mathbf{n}}\cdot\mathbf{J}} \\ &= Q_\alpha - i\theta[\hat{\mathbf{n}}\cdot\mathbf{J}, Q_\alpha] + O(\theta^2)\,. \end{aligned} \tag{3.46}$$

[The conceptual difference between the notations $|\mathbf{x}\rangle'$and$|\mathbf{x}'\rangle$ is that the former is regarded as an eigenvector of $\mathbf{Q}'$, while the latter is regarded as one of the eigenvectors of $\mathbf{Q}$ in (3.36). The distinction is only for emphasis, since the two vectors are equal.] As in the previous argument for space displacements, we have

$$Q'_\alpha\,|\mathbf{x}\rangle' = x_\alpha\,|\mathbf{x}\rangle' .$$

But also

$$\begin{aligned} Q_\alpha\,|\mathbf{x}'\rangle &= x'_\alpha\,|\mathbf{x}'\rangle = (\mathbf{x} + \theta\hat{\mathbf{n}} \times \mathbf{x})_\alpha\,|\mathbf{x}'\rangle \\ &= (\mathbf{Q}' + \theta\hat{\mathbf{n}} \times \mathbf{Q}')_\alpha\,|\mathbf{x}\rangle' . \end{aligned}$$

Since the vectors $|\mathbf{x}'\rangle = |\mathbf{x}\rangle'$ form a complete set, we have $\mathbf{Q} = \mathbf{Q}' + \theta\hat{\mathbf{n}} \times \mathbf{Q}'$. To the first order in θ this yields

$$\mathbf{Q}' = \mathbf{Q} - \theta\hat{\mathbf{n}} \times \mathbf{Q} . \tag{3.47}$$

Comparing (3.47) with (3.46), we obtain $[\hat{\mathbf{n}}\cdot\mathbf{J}, \mathbf{Q}] = -i\hat{\mathbf{n}} \times \mathbf{Q}$. This result can be written in a more convenient form by taking the scalar product with a unit vector $\hat{\mathbf{u}}$, to obtain

$$[\hat{\mathbf{n}}\cdot\mathbf{J}, \hat{\mathbf{u}}\cdot\mathbf{Q}] = i(\hat{\mathbf{n}} \times \hat{\mathbf{u}})\cdot\mathbf{Q} . \tag{3.48a}$$

Expressed in terms of rectangular components, this becomes

$$[J_\alpha, Q_\beta] = i\varepsilon_{\alpha\beta\gamma}Q_\gamma . \tag{3.48b}$$

This relation of the components of $\mathbf{Q}$ to the generators of rotation is characteristic of the fact that $\mathbf{Q}$ is a *3-vector.*[d]

The operator $\mathbf{G}$ generates a *displacement in velocity space,*

$$e^{i\mathbf{v}\cdot\mathbf{G}}\mathbf{V}e^{-i\mathbf{v}\cdot\mathbf{G}} = \mathbf{V} - \mathbf{v}\,I , \tag{3.49}$$

much as $\mathbf{P}$ generates a displacement in ordinary space [cf. Eq. (3.43)]. The analysis is simplified if we treat only the instantaneous effect of this transformation at $t = 0$. [Since there is nothing special about the instant $t = 0$, there is no real loss of generality in this choice. The general case is treated in Problem

[d] Any object that transforms under rotations in the same way as the coordinate $\mathbf{x}$ or position operator $\mathbf{Q}$ is called a *3-vector*, or simply a *vector* if there is no likelihood of it being confused with a member of the abstract state vector space. Any operator that satisfies (3.48b) in place of $\mathbf{Q}$ is a *3-vector operator.* Thus (3.35c,d,e) imply that $\mathbf{J}, \mathbf{G}$ and $\mathbf{P}$ are *3-vector operators.*

(3.7) at the end of the chapter.] In this case the position will be unaffected by the instantaneous transformation, and hence

$$[G_\alpha, Q_\beta] = 0\,. \tag{3.50}$$

The commutation relations of the *position operator* **Q** (the only operator so far identified with a physical observable) with the *symmetry generators* have now been established.

We shall next obtain more specific forms for the symmetry generators, and their physical interpretations will be deduced from their relation to **Q**. Consider first the generator **G**. In view of (3.44), we see that (3.35f), $[G_\alpha, P_\beta] = i\delta_{\alpha\beta}MI$, will be satisfied by $G_\alpha = MQ_\alpha$, but it is not apparent whether this solution is unique. However, it is apparent that $\mathbf{G} - M\mathbf{Q}$ will commute with **P**, and because of (3.50) it also commutes with **Q**. Further analysis now depends upon whether or not the particle possesses internal degrees of freedom.

Case (i): A free particle with no internal degrees of freedom

In this case the operators $\{\mathbf{Q}, \mathbf{P}\}$ form an *irreducible set*, and according to Schur's lemma any operator that commutes with such a set must be a multiple of the identity. Precise statement and proof of these mathematical assertions are contained in Apps. A and B. Roughly speaking, the argument is as follows. If an operator commutes with Q_α then it must not be a function of P_α, since the commutator of Q_α and P_α never vanishes on any vector. Similarly, if it commutes with P_α it must not be a function of Q_α. If the operator is independent of both **Q** and **P**, and if there are no internal degrees of freedom, then it can only be a multiple of the identity.

Since $\mathbf{G} - M\mathbf{Q}$ commutes with both **Q** and **P**, it must be a multiple of the identity, and hence $G_\alpha = MQ_\alpha + c_\alpha I$. But G_α must satisfy (3.35e); that is to say, it must transform as a component of a 3-vector. Now the term MQ_α transforms as a component of a 3-vector because of (3.48b). But the term $c_\alpha I$ cannot do so because it commutes with J_α, and therefore the multiple c_α must vanish. Thus we must have

$$G_\alpha = MQ_\alpha \tag{3.51}$$

for a particle without internal degrees of freedom.

One can readily verify, by using (3.44), that $\mathbf{J} = \mathbf{Q}\times\mathbf{P}$ satisfies the relations (3.35c,d,e) and (3.48). It then follows from (3.35d) and (3.48b) that $\mathbf{J} - \mathbf{Q}\times\mathbf{P}$ commutes with the irreducible set $\{\mathbf{Q}, \mathbf{P}\}$. Hence Schur's lemma implies that $J_\alpha = (\mathbf{Q}\times\mathbf{P})_\alpha + c_\alpha I$. The constants c_α must vanish in order to satisfy (3.35c). Therefore we must have

$$\mathbf{J} = \mathbf{Q} \times \mathbf{P} \tag{3.52}$$

for a particle without internal degrees of freedom.

The form of the remaining generator H can be determined from (3.35h), which, after we substitute (3.51) for G_α, becomes

$$[Q_\alpha, H] = \frac{iP_\alpha}{M}.$$

It is readily verified that this equation is satisfied by $H = \mathbf{P}\cdot\mathbf{P}/2M$, but this solution may not be unique. However, the above equation implies that $H - \mathbf{P}\cdot\mathbf{P}/2M$ will commute with $\mathbf{Q}$, and (3.35g) implies that it must commute with $\mathbf{P}$, and so by Schur's lemma it is a multiple of the identity. Thus we have

$$H = \frac{\mathbf{P}\cdot\mathbf{P}}{2M} + E_0\,, \tag{3.53}$$

where E_0 is a multiple of the identity.

The velocity operator can now be calculated from (3.39) to be

$$\mathbf{V} = \frac{\mathbf{P}}{M}. \tag{3.54}$$

The appropriate physical interpretations of $\mathbf{P}$, H and $\mathbf{J}$ follow from this result. We now have

$$\mathbf{P} = M\mathbf{V}\,,$$

$$H = \frac{1}{2}M\mathbf{V}\cdot\mathbf{V} + E_0\,,$$

$$\mathbf{J} = \mathbf{Q} \times M\mathbf{V}\,,$$

where $\mathbf{Q}$ and $\mathbf{V}$ are the operators for position and velocity, respectively. If M were the mass of the free particle, these would be the familiar forms of the momentum, the energy, and the angular momentum. But since M is not identified, we can only infer a proportionality:

$$\begin{aligned}\frac{M}{\text{mass}} = \frac{\mathbf{P}}{\text{momentum}} = \frac{H}{\text{energy}} &= \frac{\mathbf{J}}{\text{angular momentum}} \\ &= \text{a fundamental constant} \\ &= \hbar^{-1}, \text{ say}.\end{aligned} \tag{3.55}$$

The parameter $\hbar$ is hereby introduced into the theory as a fundamental constant. Its value can only be determined from experiment. The accepted value, as of 1986, is $\hbar = 1.054573 \times 10^{-34}$ joule-seconds. The first example that we shall consider that permits a measurement of $\hbar$ is the phenomenon of diffraction scattering, or Bragg reflection, of particles by a crystal (see Ch. 5). (The parameter $\hbar$ is sometimes referred to as "Planck's constant", but strictly speaking, Planck's constant is $h = 2\pi\hbar$.) We have now obtained the complete quantum-mechanical description of a free particle without internal degrees of freedom.

Case (ii): A free particle with spin

Internal degrees of freedom, by definition, are independent of the center of mass degrees of freedom, and they must be represented by operators that are independent of both $\mathbf{Q}$ and $\mathbf{P}$. That is to say, they are represented by operators that commute with both $\mathbf{Q}$ and $\mathbf{P}$. The set $\{\mathbf{Q}, \mathbf{P}\}$ is not irreducible in this case because an operator that commutes with that set may be a function of the operators of the internal degrees of freedom.

Spin is, by definition, an internal contribution to the angular momentum, so that instead of (3.52) the rotation generators are of the form

$$\mathbf{J} = \mathbf{Q} \times \mathbf{P} + \mathbf{S} \tag{3.56}$$

with $[\mathbf{Q}, \mathbf{S}] = [\mathbf{P}, \mathbf{S}] = 0$. These operators will be studied in greater detail in Ch. 7.

The operator $\mathbf{J}$ must satisfy (3.35c) in order to be the rotation generator. Since the first term, $\mathbf{Q} \times \mathbf{P}$, satisfies (3.35c), it is necessary that $\mathbf{S}$ must also satisfy it, and hence

$$[S_\alpha, S_\beta] = i\, \varepsilon_{\alpha\beta\gamma}\, S_\gamma\,. \tag{3.57}$$

The relation (3.35f), $[G_\alpha, P_\beta] = i\delta_{\alpha\beta} MI$, is satisfied by $\mathbf{G} = M\mathbf{Q}$, and as in case (i), the three components of $\mathbf{G} - M\mathbf{Q}$ commute with $\mathbf{Q}$ and $\mathbf{P}$. But now there are operators, other than the identity, which commute with $\mathbf{Q}$ and $\mathbf{P}$, namely the operators describing the internal degrees of freedom. Therefore $\mathbf{G} - M\mathbf{Q}$ may be a function of $\mathbf{S}$. The only function of $\mathbf{S}$ that is a 3-vector is a multiple of $\mathbf{S}$ itself. [It follows from (3.57) that $\mathbf{S} \times \mathbf{S} = i\mathbf{S}$, so no new vector operator can be formed by taking higher powers of $\mathbf{S}$.] Therefore $\mathbf{G} = M\mathbf{Q} + c\mathbf{S}$, where c is a real constant. According to (3.35b) the three components of $\mathbf{G}$ commute with each other, and therefore we must have $c = 0$. Hence we obtain $\mathbf{G} = M\mathbf{Q}$ in this case too.

The argument that led to (3.53), $H = \mathbf{P}\cdot\mathbf{P}/2M + E_0$, goes through as in the previous case, except that E_0 may now be a function of $\mathbf{S}$. Because of (3.35i),

$[\mathbf{J}, H] = 0$, we must have $[\mathbf{S}, E_0] = 0$, and so E_0 can only be a multiple of $\mathbf{S}\cdot\mathbf{S}$. This has no effect on the velocity operator, $\mathbf{V} = i[H, \mathbf{Q}]$, (3.39), because $[E_0, \mathbf{Q}] = 0$, so the identification $\mathbf{V} = \mathbf{P}/M$ remains valid. The identification of the momentum and energy operators proceeds as in the previous case, but with E_0 now corresponding to an internal contribution to the energy.

Case (iii): A particle interacting with external fields

For simplicity we shall consider only a spinless particle. The interactions modify the time evolution of the state (and hence the probability distributions of the observables). We shall treat this by retaining the form of the equation of motion for the state vector,

$$\frac{d}{dt}\,|\Psi(t)\rangle = -iH|\Psi(t)\rangle\,, \tag{3.38}$$

but modifying the generator H (now called the *Hamiltonian*) in order to account for the interactions. This means that we must give up the commutation relations (3.35g,h,i), which involve H. The velocity operator is still defined as

$$\mathbf{V} = i[H, \mathbf{Q}]\,, \tag{3.39}$$

since this form was derived from (3.38), but its explicit value may be expected to change when that of H is changed to include interactions.

One may ask why only the time displacement generator H should be changed by the interactions, while the space displacement generators $\mathbf{P}$ are unchanged. If the system under consideration were a self-propelled machine, we could imagine it displacing itself through space under its own power, consuming fuel, expelling exhaust, and dropping worn-out parts along the way. If $\mathbf{P}$ generated that kind of displacement, then the form of the operators $\mathbf{P}$ certainly would be altered by the interactions that were responsible for the displacement. But that is not what we mean by the operation of space displacement. Rather, we mean the purely geometric operation of displacing the system self-congruently to another location. This is the reason why $\mathbf{P}$ and the other generators of symmetry operations are not changed by dynamical interactions. However, H is *redefined* to be the generator of dynamic evolution in time, rather than merely a geometric displacement along the time axis.

The only constraint on H arises from its relation (3.39) to the velocity operator $\mathbf{V}$, whose form we must determine. Now $\mathbf{V}$ transforms as in (3.49) under a transformation to another uniformly moving frame of reference.

Expansion to the first order in the velocity shift parameter $\mathbf{v}$, yields $[i\mathbf{v}\cdot\mathbf{G}, \mathbf{V}] = -\mathbf{v}\,I$, and hence

$$[G_\alpha, V_\beta] = i\,\delta_{\alpha\beta}\,I\,. \tag{3.58}$$

The identification of $G_\alpha = MQ_\alpha$, (3.51), is still valid beacuse its derivation did not make use of any commutators involving H.

Now the earlier result, $V_\alpha = P_\alpha/M$, still represents a possible solution for $\mathbf{V}$, but it is no longer unique. From (3.58) and (3.35f) it follows that $\mathbf{V}-\mathbf{P}/M$ commutes with $\mathbf{G}$. But $\mathbf{G} = M\mathbf{Q}$, and since we have assumed that there are no internal degrees of freedom, the three operators $\mathbf{Q} = (Q_1, Q_2, Q_3)$ form a complete commuting set. Since $\mathbf{V}-\mathbf{P}/M$ commutes with this complete commuting set, it follows from Theorem 6 of Sec. 1.3 that it must be a function of $\mathbf{Q}$. Thus the most general form of the velocity operator is

$$\mathbf{V} = \frac{\mathbf{P}-\mathbf{A}(\mathbf{Q})}{M}\,, \tag{3.59}$$

where $\mathbf{A}(\mathbf{Q})$ is some function of the position operators.

We must now solve (3.39) to obtain H. One possible solution is

$$H_0 = \frac{\{(\mathbf{P}-\mathbf{A})\}^2}{2M}\,,$$

as may be directly verified. From (3.39) it then follows that $[H - H_0, \mathbf{Q}] = 0$. Thus $H - H_0$ commutes with the complete commuting set (Q_1, Q_2, Q_3), and so it must be a function of $\mathbf{Q}$. Therefore the most general form of the time evolution generator, or Hamiltonian, for a spinless particle interacting with external fields is

$$H = \frac{\{(\mathbf{P}-\mathbf{A})\}^2}{2M} + W(\mathbf{Q})\,. \tag{3.60}$$

With this result, we have deduced that the only forms of interaction consistent with invariance under the Galilei group of transformations are a *scalar potential* $W(\mathbf{Q})$ and a *vector potential* $\mathbf{A}(\mathbf{Q})$. Both of these may be time-dependent. As operators they may be functions of $\mathbf{Q}$ but must be independent of $\mathbf{P}$.

It is well known that the electromagnetic field may be derived from a vector potential and a scalar potential, so the electromagnetic interaction has the form demanded by (3.60). But $\mathbf{A}$ and W cannot necessarily be identified with the electromagnetic potentials because $\mathbf{A}(\mathbf{Q})$ and $W(\mathbf{Q})$ are arbitrary functions that need not satisfy Maxwell's equations. For example, the Newtonian gravitational potential can also be included in the scalar W.

Although we have treated only the interaction of a single particle with an external field, this does not restrict the generality of the theory. Interactions between particles can be included by regarding other particles as the sources of fields that act on the particle of interest. Thus the Coulomb interaction between two electrons of charge e is described by the operator $W = e^2/|\mathbf{Q}^{(1)} - \mathbf{Q}^{(2)}|$, where $\mathbf{Q}^{(1)}$ and $\mathbf{Q}^{(2)}$ are the position operators of the two electrons.

Conventional notation adopted

As a final step, we adopt a more conventional notation by *redefining* the symbols M, $\mathbf{P}$, $\mathbf{J}$, and H so that they are equal to the *mass, momentum, angular momentum*, and *energy* of the system, instead of merely being proportional to them as in (3.55). This means that wherever we previously wrote these four symbols, we should henceforth write $M/\hbar$, $\mathbf{P}/\hbar$, $\mathbf{J}/\hbar$, and $H/\hbar$. In particular, unitary operators for space displacement, rotation, and time evolution now become $\exp(-i\mathbf{a}\cdot\mathbf{P}/\hbar)$, and $\exp(-i\theta\hat{\mathbf{n}}\cdot\mathbf{J}/\hbar)$, and $\exp(-itH/\hbar)$. This changed notation will be used in all subsequent sections of this book. When using the equations of Sec. 3.3 in future, one should first perform the substitutions $M \to M/\hbar$, $\mathbf{P} \to \mathbf{P}/\hbar$, $\mathbf{J} \to \mathbf{J}/\hbar$, and $H \to H/\hbar$. Alternatively one may simply think of them as being expressed in units such that $\hbar = 1$.

3.5 *Composite Systems*

Having obtained the operators that represent the dynamical variables of a single particle, we must now generalize those results to composite systems. Consider a system having two components that can be separated. Let the operator $A^{(1)}$ represent an observable of component 1, and let $B^{(2)}$ represent an observable of component 2. If the two components can be separated so that they do not influence each other, then it should be possible to describe one without reference to the other. Moreover, the description of the combined system must be compatible with the separate descriptions. In particular, it must be possible to prepare states for the separate components independently. (This is not to say that all states of the composite system must be of this character, but only that such independent state preparations must be possible.)

It is possible to prepare a state in which the observable corresponding to $A^{(1)}$ has a unique value (with probability 1). As was shown on Sec. 2.4, the appropriate state vector is an eigenvector of the operator $A^{(1)}$. A similar state vector exists for the operator $B^{(2)}$. Since components 1 and 2 can be manipulated independently, there must exist a joint state vector for the combined system that is a common eigenvector of $A^{(1)}$ and $B^{(2)}$, and this must be true for all combinations of eigenvalues of the two operators. That is to say, if $A^{(1)}\,|a_m\rangle^{(1)} = a_m\,|a_m\rangle^{(1)}$ and $B^{(2)}\,|b_n\rangle^{(2)} = b_n\,|b_n\rangle^{(2)}$, then for every m and n there must be a joint eigenvector such that

$$\begin{aligned} A^{(1)}\,|a_m, b_n\rangle &= a_m\,|a_m, b_n\rangle\,, \\ B^{(2)}\,|a_m, b_n\rangle &= b_n\,|a_m, b_n\rangle\,. \end{aligned} \tag{3.61}$$

These equations are satisfied if the joint eigenvectors are of the product form

$$|a_m, b_n\rangle = |a_m\rangle^{(1)}\,|b_n\rangle^{(2)}\,. \tag{3.62}$$

This product is known as the *Kronecker product*, and is often denoted as $|a_m\rangle^{(1)} \otimes |b_n\rangle^{(2)}$. The Kronecker product between a vector in an M-dimensional space and a vector in an N-dimensional space is a vector in an MN-dimensional space. (M and N may be infinite.) If the sets of vectors $\{|a_m\rangle^{(1)}\}$ and $\{|b_n\rangle^{(2)}\}$ span the state vector spaces of the separate components, then the product vectors of the form (3.62) span the state vector space of the composite system.

Let $\{A_i^{(1)}\}$ be the set of operators pertaining to component 1, and let $\{B_j^{(2)}\}$ be the set of operators pertaining to component 2. An operator of the first set acts only on the first factor of (3.62), and an operator of the second set acts only on the second factor:

$$\begin{aligned} A_i^{(1)}\,|a_m, b_n\rangle &= (A_i^{(1)}\,|a_m\rangle^{(1)}) \otimes |b_n\rangle^{(2)}\,, \\ B_j^{(2)}\,|a_m, b_n\rangle &= |a_m\rangle^{(1)} \otimes (B_j^{(2)}\,|b_n\rangle^{(2)})\,. \end{aligned}$$

We can also define a Kronecker product between operators by the relation

$$(A_i \otimes B_j)|a_m, b_n\rangle = (A_i^{(1)}\,|a_m\rangle^{(1)}) \otimes (B_j^{(2)}\,|b_n\rangle^{(2)})\,.$$

In this notation an operator pertaining exclusively to component 1 is denoted as $A_i^{(1)} = A_i \otimes I$, and one pertaining exclusively to component 2 is denoted as $B_j^{(2)} = I \otimes B_j$. It is essential that the notation makes clear which factor of a product vector is acted on by any particular operator. Whether this is done by means of superscripts ($A^{(1)}$, $B^{(2)}$) or by position in a "$\otimes$" product ($A \otimes I, I \otimes B$) is a matter of taste. We shall usually prefer the former notation. Of course, not all important operators have this simple form; in particular, interaction operators act nontrivially on both factors.

The common eigenvectors (3.62) form a complete basis set, and hence the operators $A^{(1)}$ and $B^{(2)}$ must commute. Indeed it must be the case that

$[A_i^{(1)}, B_j^{(2)}] = 0$ for all i and j. In particular, the position, momentum, and spin operators for particle 1 commute with the position, momentum, and spin operators for particle 2.

These properties of operators that pertain to separable components also hold for any operators that pertain to *kinematically independent* (not to be confused with noninteracting) degrees of freedom, even if they are not physically separable. That this is so for the relation between orbital variables (position and momentum) and internal degrees of freedom (spin) emerged naturally in Case (ii) of Sec. 3.4, and indeed it formed a part of the definition of internal degrees of freedom. It is also true for the relation of the three components of position between each other, having been introduced by assumption [see Eq. (3.36)]. In physical terms, this is equivalent to assuming that it is possible to prepare a state that localizes a particle arbitrarily closely to a point. (This assumption may not be acceptable in relativistic quantum theory, but it will not be examined here.)

Corresponding to the state preparations of a two-component system, there must be *joint probability distributions.* If the preparations of components 1 and 2 are independent and do not influence each other, then the joint probability distribution for the observables A and B, corresponding to the operators $A^{(1)}$ and $B^{(2)}$, should obey the condition of statistical *independence* (1.52). If the state is represented by a vector of the factored form, $|\Psi\rangle = |\psi\rangle^{(1)} \otimes |\phi\rangle^{(2)}$, then the joint probability distribution of A and B, obtained from (2.28), is

$$\begin{aligned}\text{Prob}\{(\text{A} = a_m)\&(\text{B} = b_n)|\Psi\} &= |\langle a_m, b_n|\Psi\rangle|^2 \\ &= |\langle a_m|\psi\rangle^{(1)}|^2 \; |\langle b_n|\phi\rangle^{(2)}|^2 \,, \end{aligned} \tag{3.63}$$

which satisfies (1.52). This factorization holds more generally if the state, which need not be pure, is represented by an operator of the factored form $\rho = \rho^{(1)} \otimes \rho^{(2)}$. It should be emphasized that this factored, or *uncorrelated*, state is a very special kind of state, corresponding to independent preparations of the separate components. A full classification of all the possible kinds of states of composite systems is carried out in Sec. 8.3.

3.6 *[[Quantizing a Classical System]]*

[[We may contrast the method of the preceding sections for obtaining the operators that correspond to particular dynamical variables, with an older method based on the *Poisson bracket* formulation of classical mechanics. The Poisson bracket of two functions, $r(q,p)$ and $s(q,p)$, of the generalized coordinates and momenta, q_j and p_j, is defined as

$$\{r, s\} = \sum_j \left[\frac{\partial r}{\partial q_j} \frac{\partial s}{\partial p_j} - \frac{\partial r}{\partial p_j} \frac{\partial s}{\partial q_j} \right] .$$

It possesses many formal algebraic similarities to the commutator of two operators in quantum mechanics. In particular, the classical relation $\{q_\alpha, p_\beta\} = \delta_{\alpha\beta}$ corresponds to the quantum-mechanical relation $[Q_\alpha, P_\beta] = i\hbar\delta_{\alpha\beta}$. This analogy suggests that there should be a rule for assigning to every classical function $r(q,p)$ a quantum-mechanical operator $O(r)$ such that the commutation relation

$$[O(r), O(s)] = i\hbar O(\{r, s\})$$

is obeyed. Such a general substitution rule, $r(q,p) \to O(r)$, is referred to as *quantizing* the classical system.

There are two kinds of objections to this *quantization* program. The first is an epistemological objection. If the quantum-mechanical equations can be obtained from those of classical mechanics by a substitution rule, then the content of quantum theory must be logically contained within classical mechanics, with only a translation key, $r(q,p) \to O(r)$, being required to read it out. This seems implausible. Surely quantum theory is more general in its physical content than is classical theory, with the results of the classical theory being recoverable as a limiting case of quantum theory, but not the other way around.

By way of contrast, our method of obtaining operators for particular dynamical variables is not based on "quantizing" a classical theory. Although analogies with classical expressions for the momentum and kinetic energy of a particle were used to interpret certain operators, the derivation of the operators was based entirely on quantum-mechanical principles, with no use being made of the equations of classical mechanics.

The second objection is of a technical nature. The Poisson bracket equations of classical mechanics are independent of the particular choice of generalized coordinates, so one would like the operator substitution rule to also be independent of the choice of coordinates. Moreover, if $r(q,p)$ is mapped onto the operator $O(r)$, then for consistency one would like a classical function $f(r(q,p))$ to be mapped onto the same function of that operator, $f(O(r))$, as defined by (1.28) or (1.38). That is, one should have $O(f(r)) = f(O(r))$. But there are several theorems proving the impossibility of such general "quantization" rules that satisfy these conditions. For details of these impossibility theorems, see Abraham and Marsden (1978), Arens and Babbitt (1965), and Margenau and Cohen (1967).

Our method, based on the symmetries of space–time, does not yield a general rule for assigning an operator to an arbitrary classical function of coordinates and momenta. But the theory does not appear to suffer from the lack of such a rule, (3.60) being the most general case encountered in practice.]]

3.7 Equations of Motion

Time dependence was introduced into the theory when we defined the velocity operator (3.39), making use of a differential equation of motion, (3.38), for the state vector,

$$\frac{d}{dt}|\Psi(t)\rangle = -\left(\frac{i}{\hbar}\right) H(t)|\Psi(t)\rangle\,. \tag{3.64}$$

If an initial state vector is given at time $t = t_0$, then the solution can be expressed formally by means of a time evolution operator,

$$|\Psi(t)\rangle = U(t,t_0)|\Psi(t_0)\rangle\,. \tag{3.65}$$

It satisfies the same differential equation as does $|\Psi(t)\rangle$,

$$\frac{\partial}{\partial t} U(t,t_0) = -\left(\frac{i}{\hbar}\right) H(t)\,U(t,t_0)\,, \tag{3.66}$$

with the initial condition $U(t_0,t_0)=1$. From (3.66) it follows that

$$\frac{\partial}{\partial t}(U^\dagger U) = \left(\frac{i}{\hbar}\right)(U^\dagger H^\dagger U - U^\dagger H U)\,.$$

If $H(t) = H^\dagger(t)$ then this time derivative vanishes, and we will have $U^\dagger U = 1$ for all time. Thus $U(t,t_0)$ is unitary ($U^\dagger = U^{-1}$) provided that $H(t)$ is Hermitian.

If H is *independent* of t, then

$$U(t,t_0) = e^{-i(t-t_0)H/\hbar}\,.$$

If $H(t)$ is not independent of t, then, in general, no simple closed form can be given for $U(t,t_0)$.

The corresponding equation for a state operator can be obtained directly for the special case of a pure state,

$$\rho(t) = |\Psi(t)\rangle\langle\Psi(t)| = U(t,t_0)|\Psi(t_0)\rangle\langle\Psi(t_0)|U^\dagger(t,t_0)\,.$$

Therefore

$$\rho(t) = U(t,t_0)\, \rho(t_0)\, U^\dagger(t,t_0)\,. \tag{3.67}$$

Differentiating this expression with the help of (3.66) yields the differential equation

$$\frac{d\rho(t)}{dt} = \frac{-i}{\hbar}[H(t), \rho(t)]\,. \tag{3.68}$$

Equations (3.67) and (3.68), which have been derived for pure states, will be assumed to hold also for general states.

[[Some justification can be given for this assumption. If non-pure states are interpreted as mixtures of pure states, then (3.67) and (3.68) must hold because they hold for each pure component of the mixture. But the "mixture" interpretation is ambiguous, as was pointed out in Sec. 2.3, so this argument is suggestive but not fully compelling.]]

Physical significance is attached, not to operators and vectors, but to the probability distributions of observables, and in particular to averages. The time dependence of the average of the observable R, represented by the operator R, is given by

$$\langle \mathrm{R} \rangle_t = \mathrm{Tr}\{\rho(t)R\}\,. \tag{3.69}$$

Substituting (3.67) into (3.69) and using the invariance of the trace of a product with respect to cyclic permutation, we obtain two equivalent expressions:

$$\langle \mathrm{R} \rangle_t = \mathrm{Tr}\{U(t,t_0)\rho_0\, U^\dagger(t,t_0)R\} \tag{3.70a}$$

$$= \mathrm{Tr}\{\rho_0\, U^\dagger(t,t_0)R\, U(t,t_0)\}\,. \tag{3.70b}$$

Here the state operator at time $t = t_0$ is denoted as $\rho(t_0) = \rho_0$. From these two expressions follow two different formalisms for time dependence in quantum theory.

Schrödinger picture. In this approach, which we have been implicitly using all along, the time dependence is carried by the state operator. The first three factors inside the trace in (3.70a) are taken to be the time-dependent state operator, as given by (3.67). The differential equation of motion is (3.68) for the state operator $\rho(t)$, and (3.64) for the state vector $|\Psi(t)\rangle$ in the case of a pure state.

Heisenberg picture. In this approach, we group the last three operators in (3.70b) together and to write

$$\langle \mathrm{R} \rangle_t = \mathrm{Tr}\{\rho_0\, R_H(t)\}\,, \tag{3.71}$$

with $R_H(t)$ defined as

$$R_H(t) = U^\dagger(t, t_0)\, R\, U\,(t, t_0)\,. \tag{3.72}$$

(This is called the *Heisenberg operator* corresponding to the Schrödinger operator R.) In this formalism, the state is independent of time, and the time dependence is carried by the dynamical variables. Differentiating with respect to t and using (3.66), we obtain

$$\frac{dR_H}{dt} = \frac{i}{\hbar}(U^\dagger HRU - U^\dagger RHU) + U^\dagger\,\frac{\partial R}{\partial t}\,U\,.$$

This can be written in the standard form of the Heisenberg equation of motion,

$$\frac{dR_H(t)}{dt} = \frac{i}{\hbar}[H_H(t), R_H(t)] + \left(\frac{\partial R}{\partial t}\right)_H, \tag{3.73}$$

where we have introduced $H_H = U^\dagger HU$ in analogy with (3.72). The last term, $(\partial R/\partial t)_H = U^\dagger(t,t_0)(\partial R/\partial t)U(t,t_0)$, occurs only if the operator R has an intrinsic time dependence. This would be the case if it represented the potential of a variable external field, or if it were the component of a tensor defined with respect to a moving coordinate system.

In the Heisenberg picture the time development is carried by the operators of the dynamical variables, while the state function (ρ or Ψ) describes the initial data provided by state preparation. In the Schrödinger picture the state function must serve both of these roles. The two pictures are equivalent because the physically significant quantity $\langle \mathrm{R}\rangle_t$ depends only on the relative motion of ρ and R. It makes no difference whether ρ moves "forward" (Schrödinger picture) or R moves "backward" (Heisenberg picture). It is the oppositeness of these two possible motions that is responsible for the difference of sign between the commutator terms of (3.68) and (3.73). It should be obvious that those two equations are mutually exclusive and will never be used together. One may use either the Schrödinger picture or the Heisenberg picture, but one must not combine parts of the two.

The rate of change of the average value of an observable has a similar form in the two pictures. From (3.69) of the Schrödinger picture, we obtain

$$\begin{aligned}
\frac{d}{dt}\langle \mathrm{R}\rangle_t &= \mathrm{Tr}\left[\frac{\partial\rho}{\partial t}R + \rho\,\frac{\partial R}{\partial t}\right] \\
&= \mathrm{Tr}\left[\frac{-i}{\hbar}(H\rho R - \rho HR) + \rho\,\frac{\partial R}{\partial t}\right] \\
&= \mathrm{Tr}\left[\frac{-i}{\hbar}(\rho RH - \rho HR) + \rho\,\frac{\partial R}{\partial t}\right].
\end{aligned}$$

Therefore we have

$$\frac{d}{dt}\langle \mathrm{R}\rangle_t = \mathrm{Tr}\left[\frac{i}{\hbar}\rho(t)[H,R] + \rho(t)\,\frac{\partial R}{\partial t}\right] \tag{3.74}$$

in the Schrödinger picture.

On the other hand, from (3.71) of the Heisenberg picture, we obtain

$$\begin{aligned}\frac{d}{dt}\langle \mathrm{R}\rangle_t &= \mathrm{Tr}\left[\rho_0\,\frac{dR_H}{dt}\right] \\ &= \mathrm{Tr}\left[\frac{i}{\hbar}\rho_0[H,R_H(t)] + \rho_0\left(\frac{\partial R}{\partial t}\right)_H\right].\end{aligned} \tag{3.75}$$

For the special case of a *pure state* we can restate these results in terms of the state vector. Let $|\Psi_0\rangle$ be the initial state vector at time $t = t_0$. Then in the Schrödinger picture we have

$$\langle \mathrm{R}\rangle_t = \langle\Psi(t)|R|\Psi(t)\rangle\,, \tag{3.76}$$

with $|\Psi(t)\rangle = U(t,t_0)|\Psi_0\rangle$. Substitution of this expression into (3.76) yields

$$\langle \mathrm{R}\rangle_t = \langle\Psi_0|U^\dagger(t,t_0)RU(t,t_0)|\Psi_0\rangle\,,$$

which can be written in the Heisenberg picture as

$$\langle \mathrm{R}\rangle_t = \langle\Psi_0|R_H(t)|\Psi_0\rangle\,, \tag{3.77}$$

with $R_H(t)$ given by (3.72). The equivalence of the two pictures is obvious.

3.8 *Symmetries and Conservation Laws*

Let $U(s) = e^{isK}$ be a continuous unitary transformation with generator $K = K^\dagger$. [This operator $U(s)$ should not be confused with the time development operator $U(t,t_0)$ of the previous Section.] Several examples of such transformations were discussed in Sec. 3.3. To say that the Hamiltonian operator H is *invariant* under this transformation means that

$$U(s)HU^{-1}(s) = H\,, \tag{3.78}$$

or, equivalently, that $[H,U(s)] = 0$. By letting the parameter s be infinitesimally small, so that $U(s) = 1 + isK + O(s^2)$, the condition for invariance reduces to

$$[H, K] = 0\,. \tag{3.79}$$

The invariance of H under the continuous transformation $U(s)$ for all s clearly implies invariance under the infinitesimal transformation, and hence the commutation relation (3.79). The converse is also true. Since $U(s)$ is a function of K, it follows from (3.79) that H commutes with $U(s)$ for all s. Thus invariance under an infinitesimal transformation implies invariance under the corresponding finite continuous transformations.

In order to draw useful consequences from invariance in cases where $H = H(t)$ is time-dependent, it is necessary for (3.78) and (3.79) to hold for all t. Usually H will be independent of t in the practical cases that we shall encounter.

The Hermitian generators of symmetry transformations often correspond to dynamical variables. The generator of space displacements is the momentum operator, and the generator of rotations is the angular momentum operator. The symmetry generators have no intrinsic time dependence ($\partial K/\partial t = 0$), so (3.74) and the invariance of H under the transformation generated by K imply that the average of the corresponding dynamical variable K is independent of time:

$$\frac{d}{dt}\langle \mathrm{K}\rangle_t = 0\,. \tag{3.80}$$

Since H commutes with K it also commutes with any function $f(K)$, and hence not only is $\langle \mathrm{K}\rangle$ independent of time, but so is $\langle f(\mathrm{K})\rangle$. For the particular function $\theta(x - \mathrm{K})$, which is equal to 1 for positive arguments and is 0 for negative arguments, we have $\langle \theta(x - \mathrm{K})\rangle = \mathrm{Prob}(\mathrm{K} < x|\rho)$. (Similar arguments were used in Sec. 2.4.) Therefore the probability distribution $\mathrm{Prob}(\mathrm{K} < x|\rho)$ is independent of time, regardless of the initial state. Such a quantity K is called a *constant of motion.*

Specific examples of this theorem include the following. Invariance of H under space displacements implies that *momentum* is a constant of motion. Invariance of H under rotations implies that *angular momentum* is a constant of motion. If H is independent of t — or, in other words, if H is invariant under time displacements — then (3.74) implies that H itself represents a conserved quantity, namely *energy* of the system.

The concept of a *constant of motion* should not be confused with the concept of a *stationary state.* Suppose that the Hamiltonian operator H is independent of t, and that the initial state vector is an eigenvector of H, $|\Psi(0)\rangle = |E_n\rangle$ with $H|E_n\rangle = E_n|E_n\rangle$. This describes a state having a unique value of energy E_n. The solution of the equation of motion (3.64) in this case is simply

$$|\Psi(t)\rangle = e^{-iE_n t/\hbar}|E_n\rangle\,. \tag{3.81}$$

From this result it follows that the average of *any* dynamical variable R,

$$\langle \mathrm{R}\rangle = \langle\Psi(t)|R|\Psi(t)\rangle = \langle E_n|R|E_n\rangle\,,$$

is independent of t for such a state. By considering functions of R we can further show that the probability distribution $\mathrm{Prob}(\mathrm{R} < x|\Psi)$ is independent of time. In a *stationary state* the averages and probabilities of *all dynamical variables* are independent of time, whereas a *constant of motion* has its average and probabilities independent of time for *all states.*

Further consequences can be deduced for a constant of motion in a stationary state. If $[K, H] = 0$ then Theorem 5 of Sec. 1.3 implies that the two operators possess a complete set of common eigenvectors. Since the eigenvectors of H describe stationary states, this means that it is possible to prepare stationary states in which both the energy and the dynamical variable described by K have unique values without statistical dispersion. The eigenvalues of such constants of motion are very useful in classifying stationary states.

Further reading for Chapter 3

The principal sources for this chapter are T. F. Jordan (1969) and (1975).

Problems

3.1 Space is invariant under the scale transformation $\mathbf{x} \to \mathbf{x}' = e^c\mathbf{x}$, where c is a parameter. The corresponding unitary operator may be written as e^{-icD}, where D is the dilation generator. Determine the commutator $[D, \mathbf{P}]$ between the generators of dilation and space displacements. (Not all of the laws of physics are invariant under dilation, so this symmetry is less common than displacements or rotations.)

3.2 Use the Jacobi identity to show that there are no multiples of the identity to be added to the commutators $[P_\alpha, P_\beta]$, $[G_\alpha, G_\beta]$, and $[J_\alpha, H]$.

3.3 Prove the following identity, in which A and B are operators, and x is a parameter:

$$e^{xA}B\,e^{-xA} = B + [A,B]x + \frac{[A,[A,B]]x^2}{2} + \frac{[A,[A,[A,B]]]x^3}{6} + \cdots$$

3.4 Prove that $e^{A+B} = e^A e^B e^{-\frac{1}{2}[A,B]}$, provided the operators A and B satisfy $[A,[A,B]] = [B,[A,B]] = 0$.

3.5 Verify the identity $[AB, C] = A[B, C] + [A, C]B$.

3.6 Verify that the operator $\mathbf{Q} \times \mathbf{P}$ satisfies Eqs. (3.35c,d,e) when it is substituted for $\mathbf{J}$.

3.7 The unitary operator $U(\mathbf{v}) = \exp(i\mathbf{v} \cdot \mathbf{G})$ describes the instantaneous ($t = 0$) effect of a transformation to a frame of reference moving at the velocity $\mathbf{v}$ with respect to the original reference frame. Its effects on the velocity and position operators are:

$$U\mathbf{V}U^{-1} = \mathbf{v} - \mathbf{v}I\,, \quad U\mathbf{Q}U^{-1} = \mathbf{Q}\,.$$

Find an operator $\mathbf{G}_t$ such that the unitary operator $U(\mathbf{v}, t) = \exp(i\mathbf{v} \cdot \mathbf{G}_t)$ will yield the full Galilei transformation:

$$U\mathbf{V}U^{-1} = \mathbf{V} - \mathbf{v}I\,, \quad U\mathbf{Q}U^{-1} = \mathbf{Q} - \mathbf{v}tI\,.$$

Verify that $\mathbf{G}_t$ satisfies the same commutation relation with $\mathbf{P}$, $\mathbf{J}$, and H as does $\mathbf{G}$.

3.8 Calculate the position operator in the Heisenberg picture, $\mathbf{Q}_H(t)$, for a free particle.

3.9 Use the equation of motion for a state operator $\rho(t)$ to show that a pure state cannot evolve into a nonpure state, and vice versa.

3.10 If the Hamiltonian is of the form $H = H_0 + H_1$, the so-called *interaction picture* may be obtained by the following transformation of the states and dynamical variables of the Schrödinger picture:

$$\begin{aligned} |\Psi_I(t)\rangle &= \exp\left\{\tfrac{i}{\hbar}(t - t_0)H_0\right\} |\Psi_s(t)\rangle\,, \\ \rho_I(t) &= \exp\left\{\tfrac{i}{\hbar}(t - t_0)H_0\right\} \rho_s(t) \exp\left\{-\tfrac{i}{\hbar}(t - t_0)H_0\right\}\,, \\ R_I(t) &= \exp\left\{\tfrac{i}{\hbar}(t - t_0)H_0\right\} R_s \exp\left\{-\tfrac{i}{\hbar}(t - t_0)H_0\right\}\,. \end{aligned}$$

Find the equation of motion for the state vector $|\Psi_I(t)\rangle$ and the state operator $\rho_I(t)$, and show that their time dependence is due only to the "interaction" term H_1. Show that the time dependence of the average of the observable represented by the operator R, $\langle \mathrm{R} \rangle_t$, is the same in the interaction picture as in the Schrödinger or Heisenberg pictures.

3.11 The Kronecker product of two matrices, $M = A \otimes B$, is defined in terms of their matrix elements as $M_{\alpha\gamma,\beta\delta} = A_{\alpha\beta} B_{\gamma\delta}$, the rows of M being labeled by the pair $\alpha\gamma$ and its columns being labeled by $\beta\delta$. Show that the traces of the matrices satisfy the relation $\text{Tr}M = (\text{Tr}A)(\text{Tr}B)$.

3.12 If the Hamiltonian of a two-component system is of the form $H = H_1 \otimes I + I \otimes H_2$ (i.e. no interaction between the components), show that the time development operator has the form $U(t) = U_1(t) \otimes U_2(t)$, where $U_1(t) = \exp(-itH_1/\hbar)$ and $U_2(t) = \exp(-itH_2/\hbar)$.

Chapter 4

Coordinate Representation and Applications

4.1 *Coordinate Representation*

To form a *representation* of an abstract linear vector space, one chooses a complete orthonormal set of basis vectors $\{|u_i\rangle\}$ and represents an arbitrary vector $|\psi\rangle$ by its expansion coefficients $\{c_i\}$, where $|\psi\rangle = \sum_i c_i|u_i\rangle$. The array of coefficients $c_i = \langle u_i|\psi\rangle$ can be regarded as a column vector (possibly of infinite dimension), provided the basis set is discrete.

Coordinate representation is obtained by choosing as the basis set the eigenvectors $\{|\mathbf{x}\rangle\}$ of the position operator (3.36). Since this is a continuous set, the expansion coefficients define a function of a continuous variable, $\langle \mathbf{x}|\psi\rangle = \psi(\mathbf{x})$. It is a matter of taste whether one says that the set of functions forms a representation of the vector space, or that the vector space consists of the functions $\psi(\mathbf{x})$.

The action of an operator A on the function space is related to its action on the abstract vector space by the rule

$$A\psi(\mathbf{x}) = \langle \mathbf{x}|A|\psi\rangle, \quad \text{where} \quad \psi(\mathbf{x}) = \langle \mathbf{x}|\psi\rangle\,. \tag{4.1}$$

For simplicity of notation we use the same symbol for the corresponding operators on the abstract vector space and on the function space. Application of (4.1) to the *position operator* Q_α, and using (3.36), yields $\langle \mathbf{x}|Q_\alpha|\psi\rangle = x_\alpha\langle \mathbf{x}|\psi\rangle$, so the action of the position operator in function space is merely to multiply by the position coordinate,

$$Q_\alpha\psi(\mathbf{x}) = x_\alpha\psi(\mathbf{x})\,. \tag{4.2}$$

The form of the *momentum operator* is determined from its role as the generator of displacements, $\exp(-i\mathbf{a}{\cdot}\mathbf{P}/\hbar)|\mathbf{x}\rangle = |\mathbf{x}+\mathbf{a}\rangle$, (3.40). (The constant $\hbar$ was introduced at the end of Sec. 3.4.) Thus we have

$$\langle \mathbf{x} + \mathbf{a}|\psi\rangle = \left\langle \mathbf{x} \middle| \exp\left(\frac{i}{\hbar}\mathbf{a}\cdot\mathbf{P}\right) \middle| \psi \right\rangle$$
$$= \left\langle \mathbf{x} \middle| \left(1 + \frac{i}{\hbar}\mathbf{a}\cdot\mathbf{P}\right) \middle| \psi \right\rangle + O(a^2)\,.$$

Therefore, according to (4.1), we have

$$\psi(\mathbf{x} + \mathbf{a}) = \psi(\mathbf{x}) + \left(\frac{i}{\hbar}\mathbf{a}\cdot\mathbf{P}\right)\psi(\mathbf{x}) + O(a^2)\,.$$

Comparing this with the Taylor series in $\mathbf{a}$ for $\psi(\mathbf{x} + \mathbf{a})$ then yields

$$\mathbf{P} = -i\hbar\boldsymbol{\nabla}\,, \quad P_\alpha = -i\hbar\frac{\partial}{\partial x_\alpha} \tag{4.3}$$

as the form of the momentum operator in coordinate representation.

It should be noted that the second expression in (4.3) is valid only in rectangular coordinates. The momentum conjugate to an arbitrary generalized coordinate q is, in general, *not* represented by $-i\hbar\partial/\partial q$. This may be illustrated by expressing the scalar operator $\mathbf{P}\cdot\mathbf{P} = -\hbar^2\nabla^2$ in spherical coordinates and writing it in the form $\mathbf{P}\cdot\mathbf{P} = (P_r)^2 + L^2 r^{-2}$, where the operator P_r involves $\partial/\partial r$ and the operator L^2 involves $\partial/\partial\theta$ and $\partial/\partial\phi$. In this way we obtain $P_r = -i\hbar r^{-1}(\partial/\partial r)r$. The apparently privileged status of the rectangular components of the momentum operator is due to their role as generators of symmetry transformations of space. The commutation relation $[x, P_x] = i\hbar$ is satisfied by any operator of the form $P_x = -i\hbar[g(x)]^{-1}(\partial/\partial x)g(x)$. But the space is invariant under the displacement $x \to x + a$, and so the generator P_x of that transformation should also be form-invariant under $x \to x + a$. Hence we must take $g(x) = 1$. No such argument can be made for an arbitrary generalized coordinate. In particular, for the radial coordinate r the transformation $r \to r + a$ is not a symmetry of space, since it would tear a hole at the origin. Thus there is no reason to expect P_r to have the simple form $-i\hbar\partial/\partial r$.

4.2 *The Wave Equation and Its Interpretation*

The equation of motion for a pure state has the form $H|\Psi(t)\rangle = i\hbar(d/dt)|\Psi(t)\rangle$. The most general form of the Hamiltonian H for a spinless particle, given by (3.60), is the sum of the kinetic and potential energy operators. If there is no vector potential (and hence no magnetic field), the kinetic energy operator for a particle of mass M is $\mathbf{P}\cdot\mathbf{P}/2M = (-\hbar^2/2M)\nabla^2$. For a particle in the scalar potential $W(\mathbf{x})$, the equation of motion in the coordinate representation is *Schrödinger's wave equation*,

$$\left[\frac{-\hbar^2}{2M}\nabla^2 + W(\mathbf{x})\right] \ \Psi(\mathbf{x},t) = i\hbar\frac{\partial}{\partial t} \ \Psi(\mathbf{x},t) \,. \tag{4.4}$$

Because (4.4) has the mathematical form of a wave equation, it is very tempting to interpret the wave function $\Psi(\mathbf{x},t)$ as a physical field or "wave" propagating in real three-dimensional space. Moreover, it may seem plausible to assume that a wave field is associated with a particle, and even that a particle may be identified with a wave packet solution of (4.4). To forestall such misinterpretations we shall immediately generalize (4.4) to many-particle systems.

Coordinate representation for a system of N particles is obtained by choosing as basis vectors the common eigenvectors of the position operators $\mathbf{Q}^{(1)}, \mathbf{Q}^{(2)}, \ldots, \mathbf{Q}^{(N)}$. As was discussed in Sec. 3.5, these eigenvectors have the product form $|\mathbf{x}^{(1)}, \ldots, \mathbf{x}^{(N)}\rangle = |\mathbf{x}^{(1)}\rangle \otimes \cdots |\mathbf{x}^{(N)}\rangle$. The state vector $|\Psi\rangle$ is then represented by a "wave" function of many variables,

$$\langle \mathbf{x}^{(1)}, \ldots, \mathbf{x}^{(N)}|\Psi\rangle = \Psi(\mathbf{x}^{(1)}, \ldots, \mathbf{x}^{(N)}) \,. \tag{4.5}$$

The Hamiltonian is the sum of the single particle kinetic and potential energies plus the interparticle interaction $V(\mathbf{x}^{(1)}, \ldots, \mathbf{x}^{(N)})$. Thus the equation of motion becomes

$$\left[\sum_{n=1}^{N}\frac{-\hbar^2}{2M_n}\nabla_n^2 + \sum_{n=1}^{N} W(\mathbf{x}^{(n)}) + V(\mathbf{x}^{(1)}, \ldots, \mathbf{x}^{(N)})\right]\Psi(\mathbf{x}^{(1)}, \ldots, \mathbf{x}^{(N)}, t)$$
$$= i\hbar\frac{\partial}{\partial t} \ \Psi(\mathbf{x}^{(1)}, \ldots, \mathbf{x}^{(N)}, t) \,. \tag{4.6}$$

The N-particle equation (4.6) does not admit some of the interpretations that may have seemed plausible for (4.4). If a physical wave field were associated with a particle, or if a particle were identified with a wave packet, then corresponding to N interacting particles there should be N interacting waves in ordinary three-dimensional space. But according to (4.6) that is not the case; instead there is one "wave" function in an abstract $3N$-dimensional configuration space. The misinterpretation of Ψ as a physical wave in ordinary space is possible only because the most common applications of quantum mechanics are to one-particle states, for which configuration space and ordinary space are isomorphic.

The correct interpretation of Ψ is as a statistical state function, a function from which probability distributions for all observables may be calculated. In particular, Eq. (2.30) implies that $|\Psi(\mathbf{x}^{(1)}, \ldots, \mathbf{x}^{(N)})|^2$ is the probability density

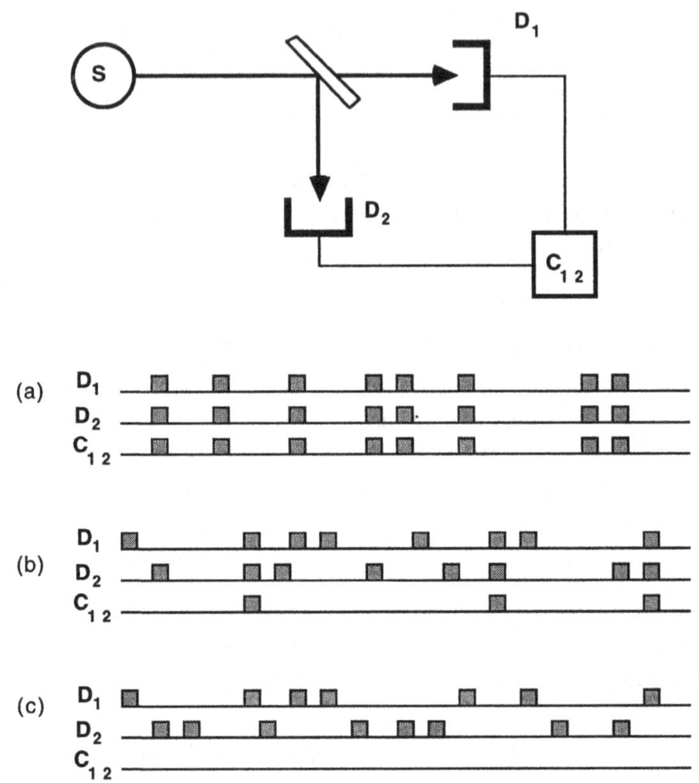

Fig. 4.1 Expected results of a coincidence experiment according to interpretations (a), (b), and (c) discussed in the text.

in configuration space for particle 1 being at the position $\mathbf{x}^{(1)}$, particle 2 being at $\mathbf{x}^{(2)}, \ldots$ and particle N being at $\mathbf{x}^{(N)}$.

The need for a purely statistical interpretation of the wave function can be demonstrated by the experiment shown schematically in Fig. 4.1, in which particles are directed at a semitransparent, semireflecting barrier, with transmitted and reflected particles being counted by detectors D_1 and D_2, respectively. The potential $W(\mathbf{x})$ of the barrier is such that an incident wave packet [a solution of (4.4)] is divided into distinct transmitted and reflected wave packets. [Such a potential is easy to construct in one dimension. Schiff (1968), pp. 106–9, shows a series of computer-generated images of the development of the two wave packets.] For simplicity of analysis we assume that the integrated intensities of the two wave packets are equal.

We shall suppose that the action of the source S can be described by the emission of identical, normalized wave packets at random times, at an average rate of r per second, but no more than one in a small interval Δt. The probability of an emission in any particular interval of duration Δt is $p = r\Delta t$. In every such interval each of the detectors may or may not record a count. If both D_1 and D_2 record counts in a particular interval, then the coincidence counter C_{12} also records a count. We shall now examine the results to be expected in three different interpretations.

(a) Suppose that the wave packet *is* the particle. Then since each packet is divided in half, according to the solution of (4.4), the two detectors will always be simultaneously triggered by the two portions of the divided wave packet. Thus the records of D_1, D_2, and C_{12} will be identical, as shown in (a) of Fig. 4.1.

(b) Suppose the wave function of (4.4) is a physical field in ordinary space, but one that is not directly observable. However, it leads to observable effects through a stochastic influence on the detectors, the probability of a detector recording a count being proportional to the integral of $|\Psi(\mathbf{x},t)|^2$ over the active volume of the detector. Since the emission probability within an interval Δt is $p = r\Delta t$, and since the wave packet divides equally between the transmitted and reflected components, the probability of D_1 recording a count during an interval Δt is $p/2$, as is also the probability for D_2. If the two detectors (and hence also the two wave packets) are sufficiently far apart, the triggering of D_1 and of D_2 should be independent events. Therefore the probability of a coincidence will be

$$c = \frac{p^2}{4}, \tag{4.7}$$

as shown in (b) of Fig. 4.1.

(c) Suppose, finally, that the object emitted by the source S is a single particle, and that $|\Psi(\mathbf{x},t)|^2$ is the probability per unit volume that at time t the particle will be located within some small volume about the point $\mathbf{x}$. Since the particle cannot be in two places at once, the triggering of D_1 and of D_2 at the same time t are mutually exclusive events. Thus the probability of a coincidence will be zero, as shown in (c) of Fig. 4.1.

When the previous (1998) edition of this book was published, this experiment had never been done for any massive particle, although Clauser (1974) had done a similar experiment for photons. Responding to the challenge, Iannuzzi et al. (2000) did it with thermal neutrons. They describe their result as unambiguous evidence in favor of interpretation (c).

4.3 Galilei Transformation of Schrödinger's Equation

Because the requirement of invariance under space–time symmetry transformations was used in Ch. 3 to derive the basic operators and equations of quantum mechanics, there is no doubt that the form of Schrödinger's equation must be invariant under those transformations. Nevertheless it is useful to explicitly demonstrate the invariance of (4.4) under Galilei transformations, and to exhibit the nontrivial transformation undergone by the wave function.

For simplicity we shall treat only one spatial dimension. Let us consider two frames of reference: F with coordinates x and t, and F' with coordinates x' and t'. F' is moving uniformly with velocity v relative to F, so that

$$x = x' + vt', \quad t = t'. \tag{4.8}$$

The potential energy is given by $W(x,t)$ in F, and by $W'(x',t')$ in F', with

$$W'(x',t') = W(x,t). \tag{4.9}$$

In F' the Schrödinger equation (4.4) has the form

$$\frac{-\hbar^2}{2M}\frac{\partial^2\Psi'}{\partial x'^2} + W'(x')\Psi'(x',t') = i\hbar\frac{\partial}{\partial t'}\Psi'(x',t'), \tag{4.10}$$

where $\Psi'(x',t')$ is the wave function in F'. In F the wave function will be $\Psi(x,t)$, and the equation that it satisfies must be determined by substitution of (4.8) and (4.9) into (4.10). If it turns out to be of the same form as (4.10), only expressed in terms of unprimed variables, then we will have demonstrated the invariance of the Schrödinger equation under Galilei transformations.

The probability density at a point in space–time must be the same in the two frames of reference (since the Jacobian of the transformation between coordinate systems is 1),

$$|\Psi(x,t)|^2 = |\Psi'(x',t')|^2, \tag{4.11}$$

and hence we must have

$$\Psi(x,t) = e^{if}\,\Psi'(x',t'), \tag{4.12}$$

where f is a real function of the coordinates. The differential operators transform as

$$\frac{\partial}{\partial x'} = \frac{\partial}{\partial x}, \qquad \frac{\partial}{\partial t'} = \frac{\partial}{\partial t} + v\frac{\partial}{\partial x}, \tag{4.13}$$

so the substitution of (4.9) and (4.12) into (4.10) yields

$$\frac{-\hbar^2}{2M}\frac{\partial^2\Psi}{\partial x^2} + W\Psi + i\hbar\left[\frac{\hbar}{M}\frac{\partial f}{\partial x} - v\right]\frac{\partial\Psi}{\partial x}$$
$$+\left[\frac{i\hbar^2}{2M}\frac{\partial^2 f}{\partial x^2} + \frac{\hbar^2}{2M}\left[\frac{\partial f}{\partial x}\right]^2 - \hbar v\,\frac{\partial f}{\partial x} - \hbar\,\frac{\partial f}{\partial t}\right]\Psi = i\hbar\,\frac{\partial\Psi}{\partial t}. \tag{4.14}$$

It will be possible for this equation to reduce to

$$\frac{-\hbar^2}{2M}\frac{\partial^2\Psi}{\partial x^2} + W(x)\Psi(x,t) = i\hbar\frac{\partial}{\partial t}\Psi(x,t) \tag{4.15}$$

only if there exists a real function $f(x,t)$ such that all of the extra terms vanish from (4.14). Thus three conditions must be satisfied:

$$\frac{\hbar}{M}\frac{\partial f}{\partial x} - v = 0\,, \tag{4.16a}$$

$$\frac{\partial^2 f}{\partial x^2} = 0\,, \tag{4.16b}$$

$$\frac{\hbar}{2M}\left[\frac{\partial f}{\partial x}\right]^2 - v\frac{\partial f}{\partial x} - \frac{\partial f}{\partial t} = 0\,. \tag{4.16c}$$

The first two conditions are both satisfied by $f = Mvx/\hbar + g(t)$, where $g(t)$ is an arbitrary function. The third condition then yields

$$f(x,t) = \frac{Mvx - \frac{1}{2}Mv^2 t}{\hbar}\,, \tag{4.17}$$

apart from an irrelevant constant term. That one function could be found to satisfy all three conditions was possible only because of the Galilean-invariant quality of the Schrödinger equation. For a more general differential equation it would not necessarily have been possible.

Levy–Leblond (1976) has pointed out that these results resolve a minor paradox. If the potential W is identically zero, the solution of (4.10) is

$$\Psi'(x',t') = e^{i(kx'-\omega t')}\,. \tag{4.18}$$

This is an eigenfunction of both the momentum and energy operators, with eigenvalues $P = \hbar k$ and $E = \hbar\omega = \hbar^2k^2/2M$. Now Eq. (4.18) has the form of a wave with wavelength $\lambda = 2\pi/k$, so we obtain Louis de Broglie's relation between momentum and wavelength,

$$P = \frac{2\pi\hbar}{\lambda}\,. \tag{4.19}$$

Upon transforming between the relatively moving frames F' and F, the wavelength of an ordinary wave is unchanged, as can be demonstrated from the relation between the wave amplitudes in F and F',

$$a(x,t) = a'(x',t') = a'(x-vt,t)\,. \tag{4.20}$$

But the momentum undergoes the transformation $P \to P + Mv$. Thus it appears as if the de Broglie relation (4.19) is incompatible with Galilean invariance.

The paradox is resolved by the fact that the Schrödinger wave function is not a classical wave amplitude, and the familiar properties of wave propagation do not necessarily apply to it. In particular, it does not satisfy (4.20), but rather (4.12). When applied to the particular function (4.18), the latter yields

$$\begin{aligned}\Psi(x,t) &= e^{if(x,t)}\;\Psi'(x-vt,t)\\ &= \exp\left[\frac{i}{\hbar}(\hbar k + Mv)x - \frac{i}{\hbar}\frac{(\hbar k + Mv)^2}{2M}t\right],\end{aligned}$$

which is in perfect agreement with the transformation of momentum under a Galilei transformation, $\hbar k \to \hbar k + Mv$. Thus the particle momentum transforms exactly as expected. The wavelength of Ψ is able to transform and to maintain the relation (4.19) precisely because Ψ is not a wave in the ordinary classical sense.

4.4 *Probability Flux*

For a one-particle state the probability of the particle being located within a region Ω is equal to $\int_\Omega |\Psi(\mathbf{x},t)|^2 d^3x$. The rate of change of this probability can be calculated from the rate of change of Ψ as given by (4.4),

$$\begin{aligned}\frac{\partial}{\partial t}\int_\Omega \Psi^*\Psi d^3x &= \int_\Omega\left(\Psi^*\frac{\partial\Psi}{\partial t} + \Psi\frac{\partial\Psi^*}{\partial t}\right)d^3x\\ &= \frac{i\hbar}{2M}\int_\Omega\left(\Psi^*\nabla^2\Psi - \Psi\nabla^2\Psi^*\right)d^3x\\ &= \frac{i\hbar}{2M}\int_\Omega \operatorname{div}(\Psi^*\boldsymbol{\nabla}\Psi - \Psi\boldsymbol{\nabla}\Psi^*)\,d^3x\,.\end{aligned}$$

Since the region Ω is arbitrary, this implies a continuity equation,

$$\frac{\partial}{\partial t}|\Psi(\mathbf{x},t)|^2 + \operatorname{div}\mathbf{J}(\mathbf{x},t) = 0\,, \tag{4.21}$$

where

$$\mathbf{J}(\mathbf{x},t) = \frac{-i\hbar}{2M}(\Psi^*\nabla\Psi - \Psi\nabla\Psi^*)$$
$$= \frac{\hbar}{M}\,\mathrm{Im}(\Psi^*\nabla\Psi) \tag{4.22a}$$

is the *probability flux* vector.

Expressing Ψ in terms of its real amplitude and phase, $\Psi(\mathbf{x},t) = A(\mathbf{x},t)$ $\exp[iS(\mathbf{x},t)/\hbar]$, we obtain another useful form for the flux,

$$\mathbf{J}(\mathbf{x},t) = \frac{A^2\nabla S}{M}\,. \tag{4.22b}$$

Applying the divergence theorem, which relates the volume integral of the divergence of a vector to the surface integral of the vector, we obtain

$$\frac{\partial}{\partial t}\int_\Omega |\Psi(\mathbf{x},t)|^2 d^3x = -\oiint_\sigma \mathbf{J}(\mathbf{x},t)\cdot d\mathbf{s}\,, \tag{4.23}$$

where σ is the bounding surface of Ω. The rate of decrease of probability for the particle to be within Ω is equal to the net outward flux through the surface σ.

The probability flux vector can be expressed conveniently in terms of the velocity operator (3.54), $\mathbf{V} = \mathbf{P}/M = (-i\hbar/M)\nabla$:

$$\mathbf{J}(\mathbf{x},t) = \mathrm{Re}[\Psi^*(\mathbf{x},t)\mathbf{V}\Psi(\mathbf{x},t)]\,. \tag{4.24}$$

If the state function is normalized so that $\int|\Psi|^2 d^3x \equiv \langle\Psi|\Psi\rangle = 1$, then the integral of $\mathbf{J}(\mathbf{x},t)$ over all configuration space is equal to the average velocity in the state

$$\int \mathbf{J}(\mathbf{x},t)\,d^3x = \langle\Psi|\mathbf{V}|\Psi\rangle\,.$$

Since (4.4) omits any vector potential, the expressions (4.22) for the probability flux vector are valid only if there is no magnetic field. However, we shall see in Ch. 11 that (4.24) remains correct in the presence of magnetic fields.

Example (i). Consider the state function $\Psi = Ce^{i\mathbf{k}\cdot\mathbf{x}}$, which is an eigenfunction of the momentum operator (4.3). The probability flux vector, $\mathbf{J} = |C|^2\hbar\mathbf{k}/M$, is equal to the probability density $|C|^2$ multiplied by the velocity. This is analogous to hydrodynamics, in which the fluid flux is equal to the density of the fluid multiplied by its velocity. However this simple interpretation applies only to quantum states that have a unique velocity.

Example (ii). Consider next the superposition state

$$\Psi = C_1 \exp(i\mathbf{k}_1 \cdot \mathbf{x}) + C_2 \exp(i\mathbf{k}_2 \cdot \mathbf{x}) .$$

The corresponding probability flux vector is

$$\mathbf{J} = \frac{\hbar}{M} \big[|C_1|^2 \mathbf{k}_1 + |C_2|^2 \mathbf{k}_2 + (\mathbf{k}_1 + \mathbf{k}_2) \\ \times \{\mathrm{Re}(C_1 C_2^*) \cos[(\mathbf{k}_1 - \mathbf{k}_2)\cdot\mathbf{x}] - \mathrm{Im}(C_1 C_2^*) \sin[(\mathbf{k}_1 - \mathbf{k}_2)\cdot\mathbf{x}]\}\big] .$$

In general the flux is not additive over the terms of a superposition state, although an exception to this rule occurs if $\mathbf{k}_1 = -\mathbf{k}_2$.

4.5 *Conditions on Wave Functions*

The equations of continuity, (4.21) and (4.23), require that the probability flux $\mathbf{J}(\mathbf{x}, t)$ be continuous across any surface, since otherwise the surface would contain sources or sinks. Although this condition applies to all surfaces, implying that $\mathbf{J}(\mathbf{x}, t)$ must be everywhere continuous, its practical applications are mainly to surfaces separating regions in which the potential has different analytic forms.

Let us consider the constraint on Ψ imposed by continuity of $\mathbf{J}$. The wave function $\Psi = C \exp(i\mathbf{k} \cdot \mathbf{x})$ has associated with it a flux vector $\mathbf{J} = |C|^2 \mathbf{k}/M$. Continuity of $\mathbf{J}$ would seem to allow discontinuities in C and $\mathbf{k}$ provided the product $|C|^2 \mathbf{k}$ remained constant. More generally, continuity of (4.22a) would seem to permit compensating discontinuities in Ψ and $\nabla\Psi$. But if we insist on maintaining the relation $\int (\partial\Psi/\partial x)\, dx = \Psi$, then a jump discontinuity in Ψ of magnitude Δ at $x = x_0$ implies a singular contribution of $\Delta\delta(x - x_0)$ to $\partial\Psi/\partial x$. Therefore we need to require continuity of both Ψ and $\nabla\Psi$ in order to keep $\mathbf{J}(\mathbf{x}, t)$ continuous.

An exception occurs if the potential is infinite in some region, which is therefore forbidden to the particle. Consider the simplest example of such a potential in one dimension: $W(x) = 0$ for $|x| < 1$, $W(x) = +\infty$ for $|x| > 1$. In the forbidden region, $|x| > 1$, one must have $\Psi(x) = 0$. The solution to the Schrödinger equation in the allowed region, $|x| < 1$, must match at the boundaries to the necessary result in the forbidden region. Since the differential equation is second order in spatial derivatives, one may impose two boundary conditions, usually chosen to be $\Psi = 0$ at $x = \pm 1$. But it is not possible to also impose the vanishing of $\partial\Psi/\partial x$ at $x = \pm 1$, and it is easy to verify that no solutions exist for which both Ψ and $\partial\Psi/\partial x$ vanish at the boundaries.

This situation can be better understood by considering the infinite potential as the limit of a finite potential: $W(x) = 0$ for $|x| < 1$, $W(x) = V_\infty$ for $|x| > 1$. The bound states in this potential are the stationary state solutions of the Schrödinger equation, satisfying $(-\hbar^2/2M)\partial^2\Psi/\partial x^2 + W(x)\Psi = E\Psi$ with $E > 0$. Because the equation is invariant under the substitution $x \to -x$, it follows that if $\Psi(x)$ is a solution then $\Psi(-x)$ is also a solution for the same E. So are the symmetric and antisymmetric combinations, $\Psi(x) \pm \Psi(-x)$, and hence all linearly independent solutions can be found by considering only even and odd functions.

For simplicity we shall treat only the symmetric case. For $|x| \leq 1$ an even solution has the form $\Psi(x) = A\cos(kx)$, with $E = \hbar^2k^2/2M$. For $|x| \geq 1$ it has the form $\Psi(x) = Be^{-\alpha|x|}$, with $\hbar^2\alpha^2/2M = V_\infty - \hbar^2k^2/2M$. The condition that $\Psi(x)$ be continuous at $x = \pm 1$ implies that $A/B = e^{-\alpha}/\cos(k)$. The continuity of $\partial\Psi/\partial x$ at $x = \pm 1$ implies that $Ak\sin(k) = B\alpha e^{-\alpha}$, which simplifies to $k\tan(k) = \alpha$. The solution of this latter equation determines the allowed values of k, and hence allowed values of the bound state energy E. The lowest energy wave function is shown in Fig. 4.2 for several values of V_∞. (The particular values chosen correspond to k approaching $\pi/2$ from below through equal increments of $\pi/40$.) It is apparent that in the limit $V_\infty \to \infty$ the function $\Psi(x)$ develops a cusp at $x = 1$. Thus $\Psi(x)$ remains continuous, but its derivative becomes discontinuous in the limit $V_\infty \to \infty$. However the vanishing of Ψ at the infinite potential step is sufficient to ensure the continuity and vanishing of the probability flux $\mathbf{J}$. Thus the same physical principle, continuity of the probability flux, governs this limiting case.

Consider next the behavior at a *singular point*, assumed for convenience to be the origin of coordinates. Let S be a small sphere of radius r surrounding the singularity. The probability that the particle is inside S must be finite. Suppose that $\Psi = u/r^\alpha$, where u is a smooth function that does not vanish at $r = 0$. Then we must have $\int |\Psi|^2 d^3r$ convergent at the origin, which implies that $\alpha < 3/2$.

The net outward flow through the surface S is $F = \oint_S \mathbf{J} \cdot d\mathbf{S}$. It must vanish in the limit $r \to 0$, since otherwise the origin would be a point source or sink. Now if $\Psi = u/r^\alpha$, one has $\partial\Psi/\partial r = r^{-\alpha}\partial u/\partial r - \alpha u r^{-\alpha-1}$. The second term does not contribute to the flux (4.22), so we obtain

$$F = r^{2-2\alpha}\left(\frac{-i\hbar}{2M}\right)\oiint\left(u^*\frac{\partial u}{\partial r} - u\frac{\partial u^*}{\partial r}\right)d\Omega\,,$$

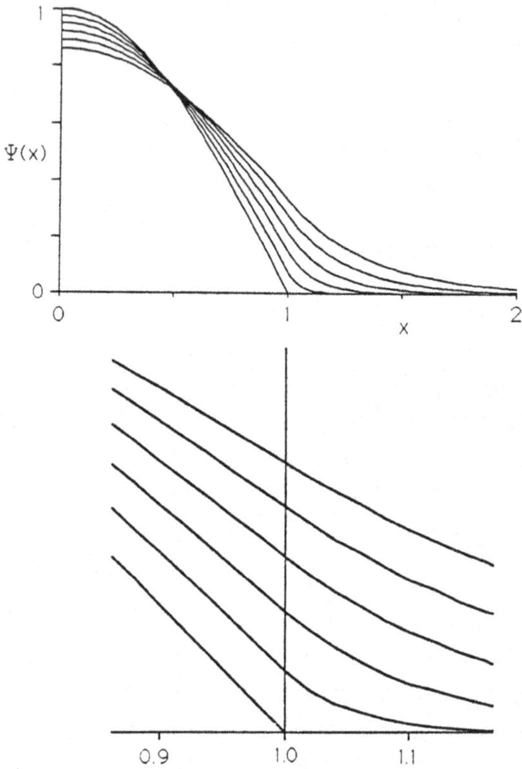

Fig. 4.2 Bound state wave functions in a step potential: $W(x) = 0,\ |x| < 1;\ W(x) = V_\infty$, $|x| > 1$; with an expanded view near the step at $x = 1$. The curves (from top to bottom at $x = 1$) correspond to $V_\infty = 9.48,\ 16.54,\ 32.7,\ 81.7,\ 361.7$, and ∞. (Units: $\hbar = 2M = 1$.)

where the integration is over solid angle. If the integral does not vanish, then we must have $\alpha < 1$ in order for F to vanish in the limit $r \to 0$. This is a stronger condition than that derived from the probability density. But if $u(r)$ is real, or if the above integral vanishes for any other reason, then this argument does not yield a useful condition.

Since $|\Psi|^2$ is a probability density, it must vanish sufficiently rapidly at infinity so that its integral over all configuration space is convergent and equal to 1. The requirement that $\int |\Psi|^2 d^3x \equiv \langle\Psi|\Psi\rangle = 1$ is equivalent to asserting that Ψ lies in Hilbert space (see Sec. 1.4).

The conditions that we have discussed apply to wave functions $\Psi(\mathbf{x})$ which represent *physically realizable states*, but they need not apply to the eigenfunctions of operators that represent observables. Those eigenfunctions, $\chi(\mathbf{x})$, which play the role of *filter functions* in computing probabilities (see Sec. 2.4), are only required to lie in the extended space, $\Omega^\times$, of the rigged-Hilbert-space triplet $(\Omega \subset \mathcal{H} \subset \Omega^\times)$. As was shown in Sec. 1.4, a function $\chi(\mathbf{x})$ belongs to $\Omega^\times$ if $\langle\chi|\phi\rangle \equiv \int \chi^*(x)\,\phi(x)d^3x$ is well defined for all $\phi(\mathbf{x})$ in the nuclear space Ω. Since $|\langle\chi|\Psi\rangle|^2$ is to be interpreted as a probability (2.28) or probability density (2.30), and so should be well defined, it has been suggested that Ψ be restricted to the nuclear space Ω, rather than merely to the Hilbert space $\mathcal{H}$. In many cases this would amount to requiring that Ψ should vanish at infinity more rapidly than any inverse power of the distance, for example like $\exp(-c|x|)$. We shall see that the common examples of bound states do indeed satisfy that condition; however, it is not known whether the condition is satisfied for all physically realizable states.

4.6 *Energy Eigenfunctions for Free Particles*

The calculation of energy eigenfunctions for free particles provides a good illustration of the rigged-Hilbert-space formalism. The energy eigenvalue equation for a free particle, $H|\Psi\rangle = E|\Psi\rangle$, becomes

$$\frac{-\hbar^2}{2M}\nabla^2\Psi(\mathbf{x}) = E\Psi(\mathbf{x}) \tag{4.25}$$

when expressed in the coordinate representation. The solutions of this equation are well known. By separating variables in rectangular coordinates, we obtain a set of solutions of the form

$$\Psi_{\mathbf{k}} = e^{i\mathbf{k}\cdot\mathbf{x}}\,. \tag{4.26}$$

By separating variables in spherical polar coordinates, we obtain another set of solutions (linearly dependent on the first),

$$\Psi_{k\ell m} = j_\ell(kr)Y_\ell^m(\theta,\phi)\,, \tag{4.27}$$

where $j_\ell(kr)$ is a spherical Bessel function and $Y_\ell^m(\theta,\phi)$ is a spherical harmonic, with $\ell = 0, 1, 2, \ldots$, and $m = -\ell, -\ell+1, \ldots, \ell$. The energy eigenvalue is

$$E = \frac{\hbar^2k^2}{2M}\,. \tag{4.28}$$

The problem is that these are mathematically valid solutions of (4.25) for *all complex values* of k, and hence *all complex values* of E. (Solutions for noninteger and complex values of ℓ and m also exist, but they are excluded by the theory of angular momentum in Ch. 6, and so will not be considered.) Moreover, one cannot select the acceptable solutions by imposing the normalization criterion, $\int |\Psi|^2 d^3x = 1$, because the integral is divergent in all cases. Evidently none of the superabundant solutions belong to Hilbert space.

Let us now consider the problem within the broader framework of a rigged-Hilbert-space triplet ($\Omega \subset \mathcal{H} \subset \Omega^\times$). The *nuclear space* Ω is chosen to be the set of functions $\{\phi(\mathbf{x})\}$ which satisfy an infinite set of conditions: that $\int |\phi(\mathbf{x})|^2(1+r^2)^n d^3x$ be convergent for all $n = 0, 1, 2, \ldots$. (Here $r = |\mathbf{x}|$ is the radial distance.) The *Hilbert space* $\mathcal{H}$ consists of the larger set of functions that need only satisfy this condition for the single case of $n = 0$. The *extended space* $\Omega^\times$ consists of those functions $\chi(\mathbf{x})$ such that $\int \chi^*(\mathbf{x})\phi(\mathbf{x})\, d^3x$ is convergent for all ϕ in Ω. Clearly Ω consists of functions that vanish at infinity more rapidly than any inverse power of r, whereas $\Omega^\times$ contains functions that may be unbounded at infinity provided that their divergence is no more rapid than some arbitrary power of r.

The solutions (4.26) and (4.27) are bounded at infinity if the components of $\mathbf{k}$ are all real, and so $\Psi_\mathbf{k}$ and $\Psi_{k\ell m}$ belong to $\Omega^\times$ in this case. But if any component of $\mathbf{k}$ is not real, then $\Psi_\mathbf{k}(\mathbf{x})$ will diverge exponentially at large distances for some directions of $\mathbf{x}$. Similarly, if k is not real, then $j_\ell(kr)$ will diverge exponentially for large r. Such functions do not belong to $\Omega^\times$, and hence they are excluded. Thus we have determined that k must be real, and so the energy (4.28) of a free particle must be nonnegative.

4.7 *Tunneling*

One of the most striking illustrations of the qualitative difference between quantum mechanics and classical mechanics is the phenomenon of tunneling of a particle through a region in which the potential energy function exceeds the total energy of the particle. This would be impossible according to classical mechanics.

We shall consider the simplest example of tunneling through a one-dimensional rectangular potential barrier.

$$\begin{aligned} W(x) &= 0\,, \quad x < 0\,, \\ &= V_0\,, \quad 0 < x < a\,, \\ &= 0\,, \quad a < x\,. \end{aligned} \tag{4.29}$$

If the energy is E, then the wave function $\Psi(x)$ must be a solution of

$$\frac{-\hbar^2}{2M}\frac{d^2\Psi}{dx^2} + W(x)\,\Psi(x) = E\,\Psi(x)\,. \tag{4.30}$$

We shall consider only the case of $0 < E < V_0$, for which crossing through the barrier would be classically forbidden. The solution of (4.30) is of the form

$$\Psi(x) = A_1\,e^{ikx} + B_1\,e^{-ikx}\,,\quad x \le 0\,, \tag{4.31a}$$

$$\Psi(x) = C\,e^{\beta x} + D\,e^{-\beta x}\,,\quad 0 \le x \le a\,, \tag{4.31b}$$

$$\Psi(x) = A_2\,e^{ikx} + B_2\,e^{-ikx}\,,\quad a \le x\,. \tag{4.31c}$$

Here $\hbar^2k^2/2M = E$ and $\hbar^2\beta^2/2M = V_0 - E$. The probability flux (4.22) of this wave function does not vanish at infinity, and so we must imagine that there are distant sources and sinks, not described by (4.30), and that (4.30) really describes the propagation of a particle within some finite region of space between the distant sources and sinks.

The three pairs of unknown constants are restricted by two pairs of equations that impose continuity of Ψ and $d\Psi/dx$ at $x = 0$ and at $x = a$. These can most conveniently be written in matrix form. At $x = 0$ we obtain

$$\begin{bmatrix} 1\,, & 1 \\ ik\,, & -ik \end{bmatrix}\begin{bmatrix} A_1 \\ B_1 \end{bmatrix} = \begin{bmatrix} 1\,, & 1 \\ \beta\,, & -\beta \end{bmatrix}\begin{bmatrix} C \\ D \end{bmatrix}$$

and at $x = a$ we obtain

$$\begin{bmatrix} e^{\beta a}\,, & e^{-\beta a} \\ \beta e^{\beta a}\,, & -\beta e^{-\beta a} \end{bmatrix}\begin{bmatrix} C \\ D \end{bmatrix} = \begin{bmatrix} e^{ika}\,, & e^{-ika} \\ ike^{ika}\,, & -ike^{-ika} \end{bmatrix}\begin{bmatrix} A_2 \\ B_2 \end{bmatrix}.$$

For brevity let us write these two equations as

$$[M_1]\begin{bmatrix} A_1 \\ B_1 \end{bmatrix} = [M_2]\begin{bmatrix} C \\ D \end{bmatrix}, \tag{4.32}$$

$$[M_3]\begin{bmatrix} C \\ D \end{bmatrix} = [M_4]\begin{bmatrix} A_2 \\ B_2 \end{bmatrix}. \tag{4.33}$$

The transmission and reflection characteristics of the potential barrier are given by the transfer matrix $[P]$, defined by

$$\begin{bmatrix} A_1 \\ B_1 \end{bmatrix} = [P]\begin{bmatrix} A_2 \\ B_2 \end{bmatrix},$$

with

$$[P] = [M_1]^{-1}\,[M_2]\,[M_3]^{-1}\,[M_4]\,. \tag{4.34}$$

The elements of the transfer matrix are

$$P_{11} = e^{ika}\left\{\cosh(\beta a) + \frac{1}{2}\,i\,\sinh(\beta a)\left(\frac{\beta}{k} - \frac{k}{\beta}\right)\right\},$$

$$P_{12} = \frac{1}{2}\,i\,e^{-ika}\,\sinh(\beta a)\left(\frac{\beta}{k} + \frac{k}{\beta}\right),$$

$$P_{21} = -\frac{1}{2}\,i\,e^{ika}\,\sinh(\beta a)\left(\frac{\beta}{k} + \frac{k}{\beta}\right),$$

$$P_{22} = e^{-ika}\left\{\cosh(\beta a) - \frac{1}{2}\,i\,\sinh(\beta a)\left(\frac{\beta}{k} - \frac{k}{\beta}\right)\right\}.$$

This transfer matrix method can obviously be generalized to calculate the transmission through any series of potential wells and barriers in one dimension.

There will always be one more pair of constants than the number of equation pairs. The two remaining constants must be determined by the boundary conditions that describe the configuration of the distant sources and sinks. The terms in $\Psi(x)$ proportional to A_1 and B_2 describe, respectively, flux incident from the left and from the right. If there is a source on the left but no source on the right of the potential barrier, then we must have $B_2 = 0$. The transmitted flux on the right ($x > a$) is $|A_2|^2\,\hbar k/M$. The flux on the left ($x < 0$) is $(|A_1|^2 - |B_1|^2)\,\hbar k/M$, with the first term being the incident flux and the second term being the reflected flux. We define the *reflection coefficient* R as

$$R = \left|\frac{B_1}{A_1}\right|^2 = \left|\frac{P_{21}}{P_{11}}\right|^2 \tag{4.35}$$

and the *transmission coefficient* T as

$$T = \left|\frac{A_2}{A_1}\right|^2 = \left|\frac{1}{P_{11}}\right|^2. \tag{4.36}$$

Flux conservation implies that $R+T = 1$, and indeed this can be verified from the specific form of $[P]$ in (4.34).

Some examples of tunneling wave functions are shown in Figs. 4.3a and 4.3b. Contrary to the qualitative sketches that are sometimes seen, the behavior of $\Psi(x)$ inside the barrier is not simply an exponential decay. In Fig. 4.3a the real

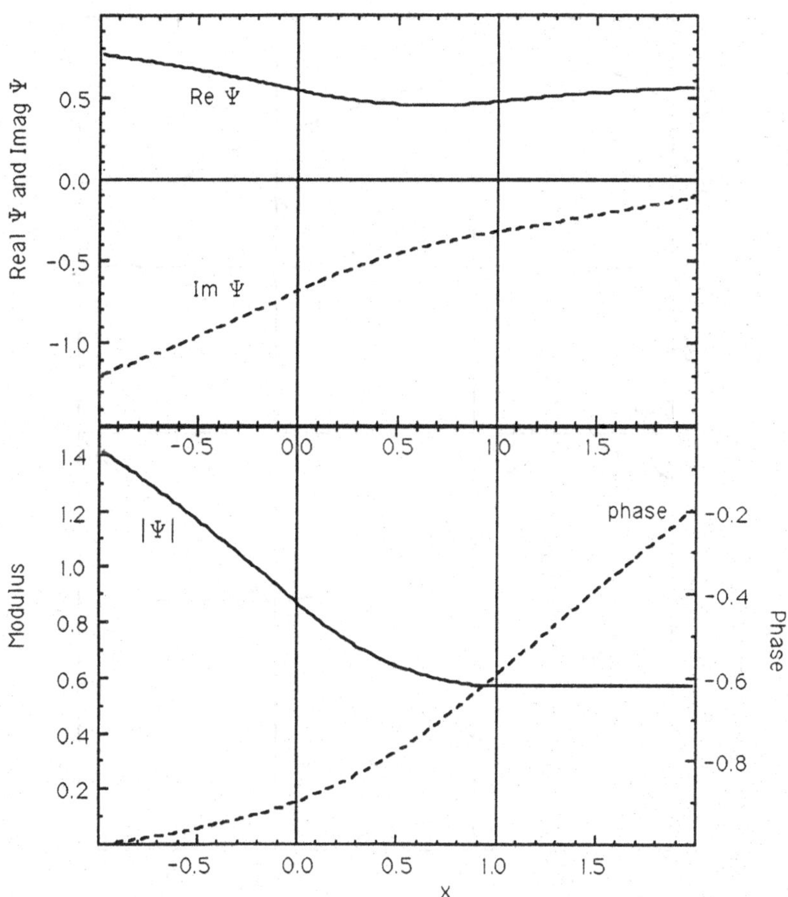

Fig. 4.3a Tunneling wave function for particle energy $E = 0.16$. The potential barrier is $V_0 = 1.0$, $0 < x < 1$. (Units: $\hbar = 2M = 1$.)

part of Ψ first decreases, and then begins to increase. Figure 4.3b shows the case $E = V_0$, for which $\Psi(x)$ varies linearly with x. Nevertheless, $|\Psi|$ always decreases monotonically across the barrier.

The complex nature of $\Psi(x)$ and the progressive increase of its phase are essential for it to carry a nonzero flux. The variation of the amplitude $|\Psi|$ for $x < 0$ is a consequence of interference between the incident and reflected terms. The amplitude for $x > a$ is, of course, constant. Since Ψ and $d\Psi/dx$

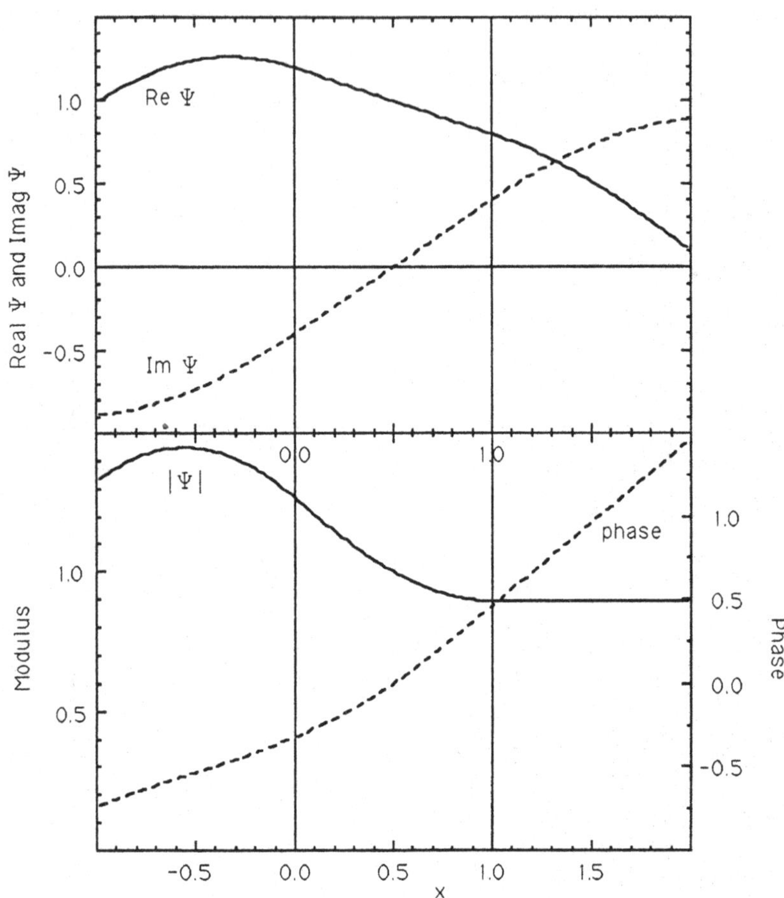

Fig. 4.3b Tunneling wave function for particle energy $E = 1.0$. The potential barrier is $V_0 = 1.0$, $0 < x < 1$. (Units: $\hbar = 2M = 1$.)

are continuous, so are $|\Psi|$ and $d|\Psi|/dx$. Hence it is always the case that $|\Psi|$ has a zero slope at the exit surface of the barrier, which implies that the decay can never be exactly exponential.

The transmission coefficient (4.36) for this potential barrier is

$$T = \left[1 + \frac{V_0^2[\sinh(\beta a)]^2}{4E(V_0 - E)}\right]^{-1} \tag{4.37}$$

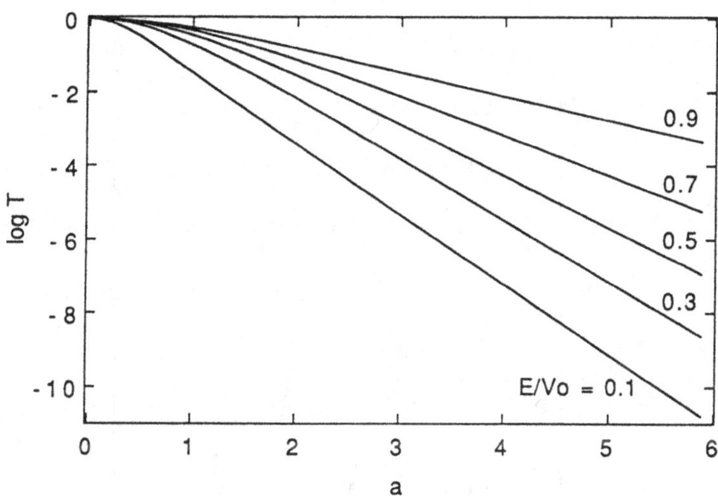

Fig. 4.4 Natural logarithm of the transmission coefficient versus the thickness of the barrier for several energies. (Units: $\hbar = 2M = 1$.)

with $\beta^2 = (V_0 - E)2M/\hbar^2$. For $\beta a \gg 1$ this reduces to

$$T \approx \left[\frac{4E(V_0 - E)}{V_0^2}\right] e^{-2\beta a} . \tag{4.38}$$

This exponential decrease of T with increasing barrier thickness a is evident in Fig. 4.4. The nonexponential variation of T with the barrier thickness for small a would be different for different forms of the barrier potential, but the exponential variation for large values of a can be shown to be independent of the detailed form of the potential.

The exponential dependence of (4.38) on distance has been experimentally confirmed in the phenomenon of *vacuum tunneling* (Binnig *et al.*, 1982). The energy of an electron inside a metal is lower than the energy of a free electron in vacuum. Hence a narrow gap between a sharp metal tip and a metal plate acts as a barrier potential through which electrons may tunnel. The difference between the vacuum potential and the Fermi energy of an electron inside the metal is called the *work function*, and it corresponds to the parameter $V_0 - E = \hbar^2\beta^2/2M$ in our model. Thus the slope of $\log T$ versus a provides a means of measuring the work funciton of the surface. Figure 4.5 illustrates the verification of the exponential distance dependence over four orders of magnitude of T.

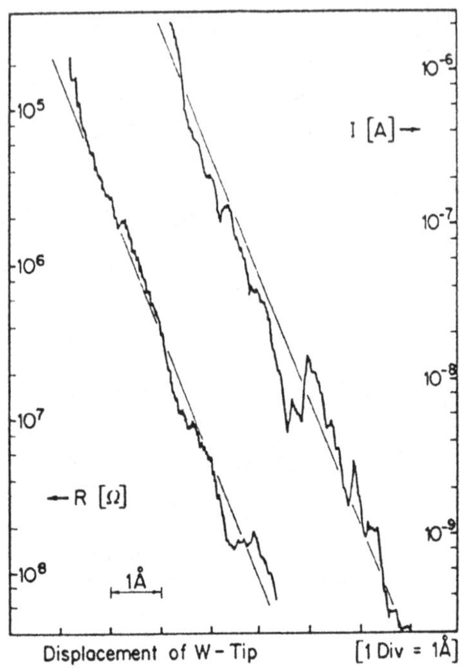

Fig. 4.5 Tunnel resistance and current versus displacement of tip from surface. [From G. Binnig *et al.* (1982), *Physica* **109** & **110B**, 2075–7.]

The very sensitive dependence of the tunneling current on the distance of the metal tip from the surface is utilized in the *scanning tunneling microscope*, which is able to measure surface irregularities as small as 0.1 angstrom (10^{-9} cm) in height.

4.8 *Path Integrals*

The time evolution of a quantum state vector, $|\Psi(t)\rangle = U(t,t_0)|\Psi(t_0)\rangle$, can be regarded as the propagation of an amplitude in configuration space,

$$\Psi(x,t) = \int G(x,t;x',t_0)\ \Psi(x',t_0)dx' \,, \tag{4.39}$$

where

$$G(x,t;x',t_0) = \langle x|U(t,t_0)|x'\rangle \tag{4.40}$$

is often called the *propagator*. As well as giving an explicit solution to the time-dependent Schrödinger equation, the propagator has a direct physical interpretation. If at the initial time t_0 the system were localized about the point x', then the probability of finding it at the point x at a later time t would be proportional to $|G(x,t;x',t_0)|^2$. (It is only proportional, rather than equal, to that probability, because the position eigenvectors are not normalizable state vectors.) Although we use the scalar symbol "x" to label a point in configuration space, and use one-dimensional examples for simplicity, all of the equations can be simply generalized to a configuration space of arbitrary dimension.

R. P. Feynman (1948) showed that the propagator can be expressed as a sum over all possible paths that connect the initial and final states; however, our derivation will not follow his. The first step is to make use of the multiplicative property of the time development operator,

$$U(t_N,t_1) = U(t_N,t_{N-1})U(t_{N-1},t_{N-2})\cdots U(t_3,t_2)U(t_2,t_1)\,, \qquad (4.41)$$

with $t_N > t_{N-1} > \cdots > t_2 > t_1$. It follows that the propagator can be written as

$$G(x,t;x_0,t_0) = \int\cdots\int G(x,t;x_N,t_N)\cdots G(x_2,t_2;x_1,t_1) \times G(x_1,t_1;x_0,t_0)\,dx_1\cdots dx_N\,. \qquad (4.42)$$

The N-fold integration is equivalent to a sum over zigzag paths that connect the initial point (x_0,t_0) to the final point (x,t), as shown in Fig. 4.6. If we now pass to the limit of $N\to\infty$ and $\Delta t\to 0$ (where $\Delta t = t_i - t_{i-1}$), we will have

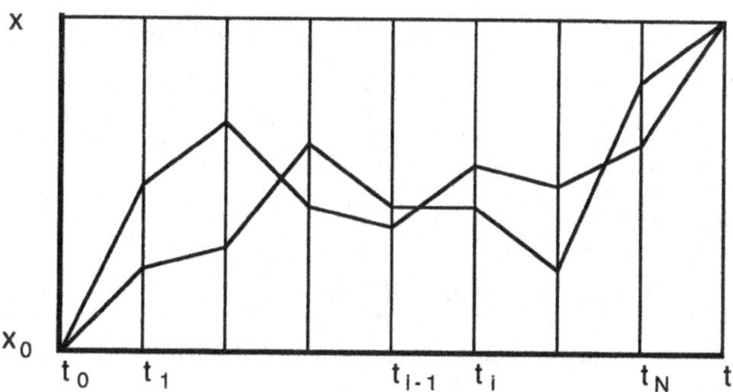

Fig. 4.6 Two paths from (x_0,t_0) to (x,t).

the propagator expressed as a sum (or, rather, as an integral) over all paths that connect the initial point to the final point.

To complete the derivation, we must obtain an expression for the propagator for a short time interval $\Delta t = t_i - t_{i-1}$:

$$\begin{aligned} G(x, t_i; x', t_{i-1}) &= \langle x|U(t_i, t_{i-1})|x'\rangle \\ &= \langle x|e^{-iH\Delta t/\hbar}|x'\rangle \,. \end{aligned} \tag{4.43}$$

(If the Hamiltonian H depends on t it may be evaluated at the midpoint of the interval.) The Hamiltonian is the sum of kinetic and potential energy operators, $H = T + V$, which do not commute. Nevertheless, we can write

$$e^{\varepsilon(T+V)} = e^{\varepsilon T} e^{\varepsilon V} + 0(\varepsilon^2)\,, \tag{4.44}$$

where $\varepsilon = i\Delta t/\hbar$ is a very small number. The error term (which is proportional to the commutator of T and V) will become negligible in the limit $\Delta t \to 0$ because it is of the second order in ε. Thus (4.43) becomes

$$\begin{aligned} \langle x|e^{-iH\Delta t/\hbar}|x'\rangle &\approx \langle x|e^{-iT\Delta t/\hbar} e^{-iV\Delta t/\hbar}|x'\rangle \\ &= \langle x|e^{-iT\Delta t/\hbar}|x'\rangle e^{-iV(x')\Delta t/\hbar}\,, \end{aligned} \tag{4.45}$$

with the error of the approximation[e] vanishing in the limit $\Delta t \to 0$.

The kinetic energy factor of (4.45) can be evaluated by transforming to the momentum representation[f], where the operator is diagonal. Thus we obtain

$$\begin{aligned} \langle x|e^{-iT\Delta t/\hbar}|x'\rangle &= \int \langle x|e^{-iT\Delta t/\hbar}|p\rangle\langle p|x'\rangle\, dp \\ &= \int \langle x|p\rangle e^{-ip^2\Delta t/2m\hbar}\langle p|x'\rangle\, dp \\ &= (2\pi\hbar)^{-1} \int e^{-ip^2\Delta t/2m\hbar} e^{ip(x-x')/\hbar}\, dp\,. \end{aligned} \tag{4.46}$$

Here we have used

$$\langle x|p\rangle = (2\pi\hbar)^{-1/2} e^{ipx/\hbar}\,, \tag{4.47}$$

[e]The error estimates in our limiting processes have been treated very loosely. A rigorous derivation is given in Ch. 1 of Schulman (1981).

[f]See Sec. 5.1 for the derivation of the momentum representation. The author apologizes for this unavoidable forward reference.

which is the one-dimensional version of Eq. (5.4). The Gaussian integral in (4.46) is of a standard form,

$$\int_{-\infty}^{\infty} e^{-ay^2+by}\,dy = (\pi/a)^{1/2}\,e^{b^2/4a}\,, \tag{4.48}$$

so (4.46) simplifies to

$$\langle x|e^{-iT\Delta t/\hbar}|x'\rangle = \{m/(i2\pi\hbar\Delta t)\}^{1/2}\,\exp\left\{\frac{im(x-x')^2}{2\hbar\Delta t}\right\}. \tag{4.49}$$

We now take the limit of (4.42) as N, the number of vertices in the path, becomes infinite:

$$G(x,t;x_0,t_0) = \underset{N\to\infty}{\to}\lim \int\cdots\int \prod_{j=0}^{N} G(x_{j+1},t_{j+1};x_j,t_j)\,dx_1\cdots dx_N\,. \tag{4.50}$$

Here we denote $x = x_{N+1}$, $t = t_{N+1}$. Since Δt becomes infinitesimal in the limit $N\to\infty$, we may substitute (4.43), (4.45), and (4.49) into (4.50). Replacing the product of exponentials by the exponential of a sum, we then obtain

$$\begin{aligned}G(x,t;x_0,t_0) &= \underset{N\to\infty}{\to}\lim \int\cdots\int \left\{\frac{m}{2\pi i\hbar\Delta t}\right\}^{(N+1)/2}\\ &\quad\times\exp\left[\sum_{j=0}^{N}\left\{\frac{im(x_{j+1}-x_j)^2}{2\hbar\Delta t} - V(x_j)\frac{\Delta t}{\hbar}\right\}\right]\,dx_1\cdots dx_N\,.\end{aligned} \tag{4.51}$$

The propagator is now explicitly expressed as an integral over all $(N+1)$-step paths from the initial point to the final point. With a slight transformation, the sum in the argument of the exponential can be interpreted as the Riemann sum of an integral along the path, which remains well defined in the continuum limit $(N\to\infty)$:

$$\begin{aligned}&\frac{i}{\hbar}\sum_{j=0}^{N}\Delta t\left\{\frac{m}{2}\left(\frac{x_{j+1}-x_j}{\Delta t}\right)^2 - V(x_j)\right\}\\ &\qquad\to \frac{i}{\hbar}\int_{t_0}^{t}\left\{\frac{1}{2}m\left(\frac{dx}{d\tau}\right)^2 - V(x)\right\}\,d\tau\,.\end{aligned} \tag{4.52}$$

The integral on the right is over the path $x = x(\tau)$, which is the continuum limit of the $(N+1)$-step zigzag path. Now the integrand is just the Lagrangian function of classical mechanics,

$$\mathcal{L}(x,\dot{x}) = \frac{1}{2}m(\dot{x})^2 - V(x)\,, \tag{4.53}$$

with $\dot{x} = dx/d\tau$. The integral of $\mathcal{L}$ over the path $x = x(\tau)$,

$$S[x(\tau)] = \int_{x(\tau)} \mathcal{L}(x,\dot{x})\,d\tau\,, \tag{4.54}$$

is the *action* associated with the path. Thus in the continuum limit the quantum propagator is expressible as an integral over all possible classical paths that connect the initial and final points, the contribution of each path having the phase factor $\exp(iS[x(\tau)]/\hbar)$.

The result is often expressed in a disarmingly simple form:

$$G(x,t;x_0,t_0) = \int e^{iS[x(\tau)]/\hbar} d[x(\tau)]\,, \tag{4.55}$$

where the integral is a functional integration over all paths $x = x(\tau)$ which connect the initial point (x_0, t_0) to the final point (x, t). Some remarks about this formula are in order.

(a) The class of paths to be included is very large, and includes some very irregular paths. This is evident from the fact that $x_j = x(t_j)$ and $x_{j+1} = x(t_j + \Delta t)$ were treated as independent variables of integration, regardless of how small Δt became.

(b) The measure $d[x(\tau)]$ over the set of all paths is difficult to define in a mathematically rigorous fashion. Furthermore, the convergence of the integral is questionable, since the integrand has absolute value 1 for all paths. In practice, to evaluate (4.55) we must revert to forms like (4.50) and (4.51), which involve discrete approximations to the paths.

Classical limit of the path integral

The classical limit will be discussed in detail in Ch. 14, but it is useful to briefly consider its implications for the path integral formula. Roughly speaking, we may expect classical mechanics to hold to a good approximation when the classical action, S, is much larger than the quantum of action, $\hbar$. Now, in that regime, the phase factor in (4.55) will be very sensitive to small fractional changes in S due to small variations of the path, and so there will be a high degree of cancellation in the integral. An exception occurs if the action is stationary with respect to small variations of a particular path, in which case all paths in a neighborhood of the path of stationary action will contribute to (4.55) with the same phase. Thus the integral will be dominated by a narrow tube of paths.

The condition for $S[x(\tau)]$ to be stationary when the path suffers an infinitesimal variation, $x(\tau) \to x(\tau) + \delta x(\tau)$, is just Hamilton's principle of classical mechanics [see, for example, Goldstein (1980)], which leads to Lagrange's equation for the classical path. Thus, in the limit that the action is large compared to $\hbar$, the path integral is dominated by the contribution of the classical path. This fact is the basis for many useful approximations.

Imaginary time and statistical mechanics

If the Hamiltonian is independent of time then the propagator (4.40) can be written as follows:

$$\begin{aligned} G(x,t;x',0) &= \langle x|e^{-itH/\hbar}|x'\rangle \\ &= \sum_n e^{-itE_n/\hbar}\Psi_n(x)\Psi_n^*(x')\,, \end{aligned} \tag{4.56}$$

where $H\Psi_n(x) = E_n\Psi_n(x)$. Substitution of $t = -i\hbar\beta$ then yields

$$\begin{aligned} G(x,-i\hbar\beta;x',0) &= \sum_n e^{-\beta E_n}\Psi_n(x)\Psi_n^*(x') \\ &= \rho_\beta(x,x')\,. \end{aligned} \tag{4.57}$$

This is the thermal density matrix, $\rho_\beta(x,x')$ (coordinate representation of the state operator), which describes the *canonical ensemble* for a system in equilibrium with a heat reservoir at the temperature T, ($\beta = 1/kT$). It is the starting point for most systematic calculations in quantum statistical mechanics.

In the limit $\beta \to \infty$ (low temperature) the sum is dominated by the ground state term, which allows us to extract the ground state energy and position probability density from the diagonal part of the thermal density matrix,

$$\rho_\beta(x,x) \approx e^{-\beta E_0}|\Psi_0(x)|^2\,. \tag{4.58}$$

Although none of these interesting relations require the path integral formalism, it is possible to evaluate them by path integral summation. Let $t = -iu$ be an imaginary "time" variable. In terms of this imaginary time, the classical Lagrangian becomes $\mathcal{L} = -\frac{1}{2}m(dx/du)^2 - V(x)$, which has the same form as the negative of the classical energy. Equation (4.55) then becomes an integration over imaginary time paths,

$$\rho_\beta(x,x') = \int \exp\left\{-\hbar^{-1}\int_0^{\beta\hbar}\left[\frac{1}{2}m\left(\frac{dx}{du}\right)^2 + V(x)\right]du\right\} d[x(u)] \,. \qquad (4.59)$$

This path integral has a major computational advantage over (4.55) in that there are no subtle cancellations, since all contributions are real and positive. Moreover, the path integral is expected to be convergent because paths of large energy will make only an exponentially small contribution. Gerry and Kiefer (1988) have used this method, along with (4.58), to calculate the ground state energy and position probability density for several simple potentials, and their results compare reasonably well with more accurate solutions obtained from Schrödinger's differential equation.

Discussion of the path integral method

The path integral method has few practical calculational uses in ordinary quantum mechanics. There are some examples that can be solved by that method, but the more traditional solutions, based on operator or differential equation techniques, are usually simpler. Nevertheless, the path integral form of quantum mechanics has some significant merits.

The first is its great generality. Although we have developed the method for a one-dimensional configuration space, it is obvious that for a system with n degrees of freedom we would obtain very similar formulas involving a sum over paths in n-dimensional configuration space. The essence of the formula (4.55) is a *sum over all possible histories* that can connect the initial and final states of the system. Each history carries the phase factor $\exp(iS/\hbar)$, where S is the classical action associated with that particular history. The system need not consist of particles, and need not have a finite number of degrees of freedom. The system could be a field, $\phi(\mathbf{x},t)$, in which case *a history* consists of any continuous sequence of functions, $\{\phi(\mathbf{x})\}$, arranged in a time order. It is clearly impractical to sum over all such histories, each of which has infinitely many degrees of freedom. But it is possible to sum over a representative sample of histories, and with the growth of computer power this is becoming a practical technique.

Perhaps the most important consequence of the path integral formulation is not its potential computational uses, but the point of view that it supports. It is common to all formulations of quantum mechanics that the probability of a process is given by the squared modulus of a complex amplitude. In the path integral form it is clear that if a process can occur through two or more paths, then the amplitudes along each path will generally interfere. Moreover, the phase associated with each partial amplitude is simply related to the action along that path. Since path integrals are often dominated by the classically allowed paths, it is often easy to gain insight into the essential features of an experiment by summing the factor $\exp(iS/\hbar)$ over the classical paths.

Lastly, we raise the question of the physical status of the infinity of Feynman paths (as the possible histories are often called). Does the system really traverse all paths simultaneously? Or does it sample all paths and choose one? Or are these Feynman paths merely a computational device, lacking any physical reality in themselves? In the case of imaginary time path integrals it is clear that they are merely a computational device. This is most likely also true for real time path integrals, although other opinions no doubt exist.

Further reading for Chapter 4

Several detailed calculations of transmission through potential wells and barriers are given by Draper (1979, 1980). The power of the transfer matrix technique is illustrated by Walker and Gathright (1994), who also provide an interactive *Mathematica* notebook.

Two good books about the path integral method are Feynman and Hibbs (1965) and Schulman (1981). The former derives quantum mechanics from path integrals; the latter derives path integrals from quantum theory. Both contain many applications.

Problems

4.1 Show that the commutator of the momentum operator with a function of the position operator is given by $[f(x), P_x] = i\hbar\, \partial f/\partial x$.

4.2 Calculate the energy eigenvalues and eigenfunctions for a particle in one dimension confined by the infinite potential well: $W(x) = 0$ for $0 < x < a$, otherwise $W(x) = \infty$. Calculate the matrices for the position and momentum operators, Q and P, using these eigenfunctions as a basis.

4.3 The simplest model for the potential experienced by an electron at the surface of a metal is a step:

$$\begin{aligned} W(x) &= -V_0 \quad \text{for } x < 0 \text{ (inside the metal)} \\ &= 0 \quad \text{for } x > 0 \text{ (outside the metal)}\,. \end{aligned}$$

For an electron that approaches the surface from the interior, with momentum $\hbar k$ in the positive x direction, calculate the probability that it will escape.

4.4 For a spherical potential of the form $W(r) = C/r^2$, obtain the asymptotic form of the spherically symmetric solutions of the wave equation in the neighborhood of $r = 0$, and hence determine the range of C for which they are physically admissible.

4.5 For a spherical potential of the form $W(r) = C/r^n$, $n > 2$, obtain the asymptotic form of the spherically symmetric solutions of the wave equation in the neighborhood of $r = 0$. For what range of n are they physically admissible? Does the answer depend on the value of C?

4.6 The result (4.22) for the probability flux $\mathbf{J}(\mathbf{x}, t)$ is not uniquely determined by the continuity equation (4.21), since (4.21) is also satisfied by $\mathbf{J}(\mathbf{x}, t) + \mathbf{f}(\mathbf{x}, t)$, where $\text{div}\, \mathbf{f}(\mathbf{x}, t) = 0$ but $\mathbf{f}(\mathbf{x}, t)$ is otherwise arbitrary. Show that if the motion is only in one dimension this formal nonuniqueness has no effect, and so the result (4.22) is practically unique in this case.

4.7 Calculate the transmission and reflection coefficients for an attractive one-dimensional square well potential: $W(x) = -V_0 < 0$ for $0 < x < a$; $W(x) = 0$ otherwise. Give a qualitative explanation for the vanishing of the reflection coefficient at certain energies.

4.8 Use the transfer matrix method to calculate the transmission coefficient for the system of two rectangular barriers shown, for energies in the range $0 < E < V_0$.

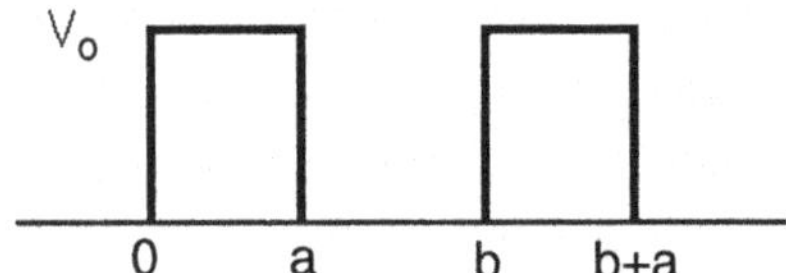

4.9 (a) Determine the condition on the state function $\Psi(x)$ at the one-dimensional deta function potential, $W(x) = c\delta(x)$. [Hint: this can be done directly from the properties of the delta function; alternatively the potential can be considered as the $\varepsilon \to 0$ limit of the finite potential: $W_\varepsilon(x) = c/\varepsilon$ for $|x| < \frac{1}{2}\varepsilon$, $W_\varepsilon(x) = 0$ for $|x| > \frac{1}{2}\varepsilon$.]

(b) Calculate the ground state of a particle in the one-dimensional attractive potential $W(x) = c\delta(x)$ with $c < 0$.

4.10 What is the action associated with the propagation of a free particle along the classical path from (x_1, t_1) to (x_2, t_2)? Use the result to express the Feynman phase factor in Eq. (4.55) in terms of the de Broglie wavelength.

4.11 Show, from its definition (4.40), that the propagator $G(x, t; x', t_0)$ is the Green function of the time-dependent Schrödinger equation,

$$\left\{H_x - i\hbar\left(\frac{\partial}{\partial t}\right)\right\} G(x, t; x', t_0) = -i\hbar\delta(x - x')\,\delta(t - t')\,,$$

where H_x is the Hamiltonian expressed as a differential operator in the x representation. Calculate the propagator for a free particle by this method.

4.12 Use the path integral method to calculate the propagator for a free particle approximately, by including only the classical path. (*Note: It is* not *generally true that this approximation will always yield the exact result.*)

Chapter 5

Momentum Representation and Applications

5.1 *Momentum Representation*

Momentum representation is obtained by choosing as basis vectors the eigenvectors of the three components of the momentum operator,

$$P_\alpha|\mathbf{p}\rangle = p_\alpha|\mathbf{p}\rangle \quad (\alpha = 1, 2, 3)\,. \tag{5.1}$$

Since the eigenvalues form a continuum, the orthonormality condition takes the form

$$\langle\mathbf{p}|\mathbf{p}'\rangle = \delta(\mathbf{p} - \mathbf{p}')\,, \tag{5.2}$$

and the norm of an eigenvector is infinite.

We must now determine the relation between the momentum eigenvectors and the position eigenvectors by evaluating their inner product $\langle\mathbf{x}|\mathbf{p}\rangle$. Using (4.1) and (4.3), and writing the momentum eigenvector as $\mathbf{p} = \hbar\mathbf{k}$, we obtain

$$\begin{aligned} -i\hbar\boldsymbol{\nabla}\langle\mathbf{x}|\hbar\mathbf{k}\rangle &= \langle\mathbf{x}|\mathbf{P}|\hbar\mathbf{k}\rangle \\ &= \hbar\mathbf{k}\langle\mathbf{x}|\hbar\mathbf{k}\rangle\,, \end{aligned}$$

which has the solution

$$\langle\mathbf{x}|\hbar\mathbf{k}\rangle = c(\mathbf{k})e^{i\mathbf{k}\cdot\mathbf{x}}\,. \tag{5.3}$$

The normalization factor $c(\mathbf{k})$ is determined from (5.2):

$$\begin{aligned} \delta(\hbar\mathbf{k} - \hbar\mathbf{k}') = \langle\hbar\mathbf{k}|\hbar\mathbf{k}'\rangle &= \int \langle\hbar\mathbf{k}|\mathbf{x}\rangle\langle\mathbf{x}|\hbar\mathbf{k}'\rangle d^3x \\ &= c^*(\mathbf{k})c(\mathbf{k}') \int \exp\{i(\mathbf{k}' - \mathbf{k})\cdot\mathbf{x}\} d^3x \\ &= c^*(\mathbf{k})c(\mathbf{k}')(2\pi)^3\delta(\mathbf{k} - \mathbf{k}')\,, \end{aligned}$$

whence $c(\mathbf{k}) = (2\pi\hbar)^{-3/2}$. Thus (5.3) becomes

$$\langle \mathbf{x}|\hbar\mathbf{k}\rangle = (2\pi\hbar)^{-3/2} e^{i\mathbf{k}\cdot\mathbf{x}} . \tag{5.4}$$

The coordinate representation of a state vector $|\Psi\rangle$ is a function of $\mathbf{x}$, $\langle \mathbf{x}|\Psi\rangle = \Psi(\mathbf{x})$. In *momentum representation*, the same state vector is represented by

$$\begin{aligned}\langle \hbar\mathbf{k}|\Psi\rangle &= \int \langle \hbar\mathbf{k}|\mathbf{x}\rangle\langle \mathbf{x}|\Psi\rangle \, d^3x \\ &= (2\pi\hbar)^{-3/2} \int e^{-i\mathbf{k}\cdot\mathbf{x}}\Psi(\mathbf{x}) d^3x \\ &= \hbar^{-3/2}\Phi(\mathbf{k}) .\end{aligned} \tag{5.5}$$

Here $\Phi(\mathbf{k}) = (2\pi)^{-3/2} \int e^{-i\mathbf{k}\cdot\mathbf{x}}\Psi(\mathbf{x})d^3x$ is the Fourier transform of $\Psi(\mathbf{x})$.

Since the momentum operator P_α is diagonal in this representation, its effect is simply to multiply $\Phi(\mathbf{k})$ by the eigenvalue $p_\alpha = \hbar k_\alpha$. The effect of position operator Q_α is

$$\begin{aligned}\langle \hbar\mathbf{k}|Q_\alpha|\Psi\rangle &= (2\pi\hbar)^{-3/2} \int e^{-i\mathbf{k}\cdot\mathbf{x}} x_\alpha \Psi(\mathbf{x}) \, d^3x \\ &= \hbar^{-3/2} i \frac{\partial \Phi(\mathbf{k})}{\partial k_\alpha} .\end{aligned}$$

Thus in momentum representation the position operator is equivalent to

$$Q_\alpha = i\frac{\partial}{\partial k_\alpha} = i\hbar \frac{\partial}{\partial p_\alpha} . \tag{5.6}$$

The momentum eigenvectors have infinite norm, and so do not belong to Hilbert space. Although this does not really cause any difficulty, it is sometimes avoided by the device of supposing space to be a large cube of side L, with periodic boundary conditions imposed. If (5.3) is required to be periodic in x_α, then the values of k_α that are permitted by the boundary conditions will be integer multiples of $2\pi/L$. Hence there is one allowed $\mathbf{k}$ value for each $(2\pi/L)^3$ of k space. We shall denote these "box" eigenvectors as $|\hbar\mathbf{k}\rangle_L$. They have unit norm, and satisfy the orthonormality condition

$$_L\langle \hbar\mathbf{k}|\hbar\mathbf{k}'\rangle_L = \delta_{\mathbf{k}',\mathbf{k}} , \tag{5.7}$$

so instead of (5.4) we have

$$\langle \mathbf{x}|\hbar\mathbf{k}\rangle_L = L^{-3/2} e^{i\mathbf{k}\cdot\mathbf{x}} . \tag{5.8}$$

In the limit $L \to \infty$ the results of the "box" method should agree with the results for unbounded space. Now clearly (5.8) does not go into (5.4) in this limit, but that is not necessary. What is required is that the average in state $|\Psi\rangle$ of some observable such as $f(\mathbf{p})$ should have the same value according to both methods of calculation. Now $|\langle\mathbf{p}|\Psi\rangle|^2$ is the probability density in momentum space, according to (2.30), so the first method yields

$$\langle f(\mathbf{p})\rangle = \int f(\mathbf{p})|\langle\mathbf{p}|\Psi\rangle|^2 \, d^3p \, .$$

In the second (box) method, $|_L\langle\hbar\mathbf{k}|\Psi\rangle|^2$ is the probability that the momentum takes on the discrete value $\hbar\mathbf{k}$, where $\mathbf{k}$ is one of the values allowed by the periodic boundary conditions. Hence

$$\langle f(\mathbf{p})\rangle = \sum_{\mathbf{k}} f(\hbar\mathbf{k})|\langle\hbar\mathbf{k}|\Psi\rangle|^2 \, ,$$

where the sum is over the lattice of allowed values. As L becomes large the allowed $\mathbf{k}$ values [one per $(2\pi/L)^3$ of k space] become very dense, and if the summand is a smooth function of $\mathbf{k}$ we may make the replacement

$$\sum_{\mathbf{k}} \to \left[\frac{L}{2\pi}\right]^3 \int d^3k = \left[\frac{L}{2\pi\hbar}\right]^3 \int d^3p \tag{5.9}$$

in the limit of large L. Comparison of (5.8) with (5.4) shows that

$$|_L\langle\hbar\mathbf{k}|\Psi\rangle|^2 = \left(\frac{2\pi\hbar}{L}\right)^3 |\langle\hbar\mathbf{k}|\Psi\rangle|^2 \, ,$$

and so the second method yields the same answer as the first. We shall not make much use of this "box normalization" technique, but it can be helpful if the complications of a continuous eigenvalue spectrum and vectors of infinite norm are not essential to the physics of the problem.

5.2 *Momentum Distribution in an Atom*

According to the theory, the momentum probability distribution in the state represented by $|\Psi\rangle$ is

$$|\langle\mathbf{p}|\Psi\rangle|^2 = (2\pi\hbar)^{-3}\left|\int e^{-i\mathbf{k}\cdot\mathbf{x}}\Psi(\mathbf{x}) \, d^3x\right|^2 \, , \tag{5.10}$$

with $\hbar\mathbf{k} = \mathbf{p}$ being the momentum of the particle. It is desirable to subject this prediction to an experimental test.

The analysis is simplest for the case of the hydrogen atom, which consists of one electron and one proton. The experiment [see Lohmann and Weigold (1981), and Weigold (1982)] involves the ionization of atomic hydrogen by a high energy electron beam, and the measurement of the momentum of the ejected and scattered electrons. Figure 5.1 shows the relative directions of the momentum $\mathbf{p}_0$ of the incident electron, $\mathbf{p}_a$ of the scattered electron, and $\mathbf{p}_b$ of the ejected electron. The equation of momentum conservation is

$$\mathbf{p}_0 + \mathbf{p}_e + \mathbf{p}_N = \mathbf{p}_a + \mathbf{p}_b + \mathbf{p}'_N\,, \tag{5.11}$$

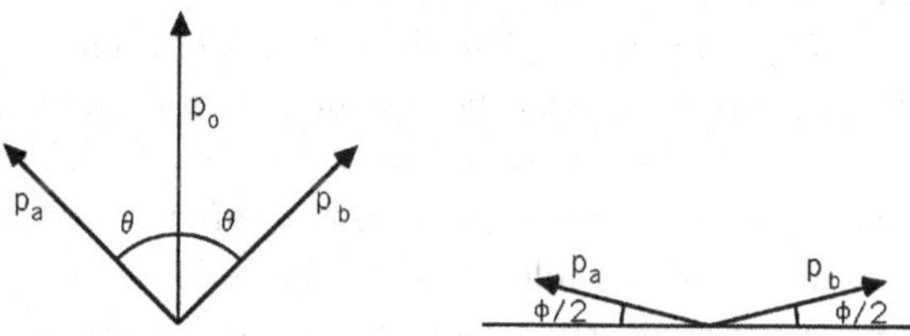

Fig. 5.1 Relative directions of the momenta of the scattered and ejected electrons, $\mathbf{p}_a$ and $\mathbf{p}_b$, and of the incident electron $\mathbf{p}_0$. In the first diagram $\mathbf{p}_0$ lies in the plane of the figure. The second diagram is an end view along $\mathbf{p}_0$.

where $\mathbf{p}_e$ and $\mathbf{p}_N$ are the momentum of the atomic electron and of the nucleus (proton) before the collision, and $\mathbf{p}'_N$ is the final momentum of the nucleus after ionization. The collision of a high energy electron with the atomic electron takes place so quickly (at sufficiently high energies) that the electron is ejected without affecting the nucleus, and so $\mathbf{p}_N = \mathbf{p}'_N$. Thus we can solve for the initial momentum of the atomic electron in terms of measurable quantities,

$$\mathbf{p}_e = \mathbf{p}_a + \mathbf{p}_b - \mathbf{p}_0\,. \tag{5.12}$$

For reasons to be given later, the detectors were arranged so as to select events for which $\mathbf{p}_a$ and $\mathbf{p}_b$ were of equal length and made the same angle θ with respect to the incident momentum $\mathbf{p}_0$. Because $\mathbf{p}_e$ need not be zero, the three vectors $\mathbf{p}_0$, $\mathbf{p}_a$, and $\mathbf{p}_b$ need not be coplanar. The dihedral angle between the plane of $\mathbf{p}_a$ and $\mathbf{p}_0$ and the plane of $\mathbf{p}_0$ and $\mathbf{p}_b$ is $\pi - \phi$. From these geometrical relations, illustrated in Fig. 5.1, we can determine the magnitude of the momentum of the atomic electron to be

$$p_e = \left\{ [2p_a \cos\theta - p_0]^2 + \left[2p_a \sin\theta \sin\left(\frac{\phi}{2}\right)\right]^2 \right\}^{1/2}. \tag{5.13}$$

In the experiment p_e is varied by varying ϕ, the angle θ begin held constant.

The probability of occurrence of such a scattering event is proportional to the electron–electron scattering cross-section, σ_{ee}, for the collision of the incident and atomic electrons, multiplied by the probability that the momentum of the atomic electron will be $\mathbf{p}_e$. Thus the observed detection rate for such events will be proportional to

$$\sigma_{ee}|\langle \mathbf{p}_e|\Psi\rangle|^2\,. \tag{5.14}$$

The scattering cross-section σ_{ee} for electron collision is a function of the energies of the electrons and the scattering angle θ. But all of these are held constant in the experiment, since only ϕ is varied. Thus the detection rate should simply be proportional to the atomic electron momentum distribution, $|\langle \mathbf{p}_e|\Psi\rangle|^2$, and a direct comparison between theory and experiment is possible.

Some further remarks about the experiment are relevant. First, all electrons are identical, so it is not possible to determine which is the scattered electron and which is the ejected electron. But by choosing $|\mathbf{p}_a| = |\mathbf{p}_b|$ and $\theta_a = \theta_b = \theta$ this ambiguity does not complicate the analysis. Second, we have assumed that the electron–atom collision can be regarded as an electron–electron collision, with the proton being a spectator. An electron–proton collision is also possible, but in that case a spectator electron would be left behind with very little energy. By selecting only events with $|\mathbf{p}_a| = |\mathbf{p}_b|$ such unwanted collisions are eliminated.

The theory of the hydrogen atom will be treated in detail in Ch. 10. However, it is easy to verify that $\Psi(r) = ce^{-r/a_0}$ (c is a normalization constant, and $a_0 = \hbar^2/Me^2$) is a stationary state solution of Schrödinger's equation for a particle of mass M in the spherically symmetric potential $W(r) = -e^2/r$. According to (5.10) the momentum probability distribution is proportional to the square of the Fourier transform of $\Psi(r)$, and thus

$$|\langle \mathbf{p}_e|\Psi\rangle|^2 = c'(1 + a_0^2k^2)^{-4}\,, \tag{5.15}$$

where $\mathbf{p}_e = \hbar\mathbf{k}$ is the momentum of the electron in the atom, and c' is another normalization constant.

Figure 5.2 compares the theory with experimental data taken at three different electron energies (all of which are large compared to the hydrogen atom binding energy of 13.6 eV). Since absolute measurements were not obtained, the magnitudes of each of the three sets of data were fitted to the theoretical curve at the low k limit. It is apparent that the experimental confirmation of the theory is very good.

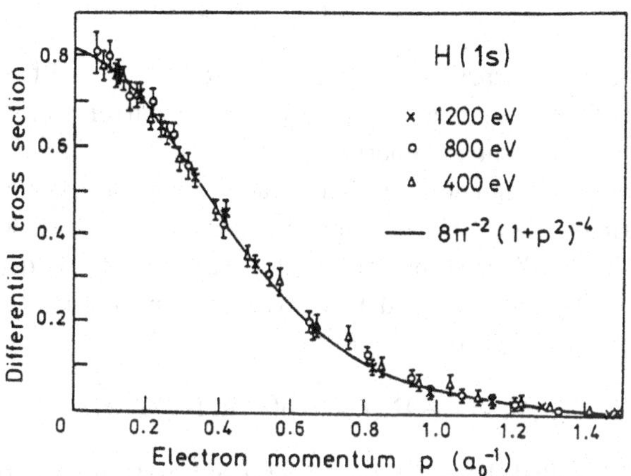

Fig. 5.2 Measured momentum distribution for the hydrogen ground state, for several incident electron energies. [From E. Weigold, *AIP Conf. Proc.* No. 86 (1982), p. 4.]

5.3 *Bloch's Theorem*

This theorem concerns the form of the stationary states for systems that are spatially periodic. It is particularly useful in solid state physics and crystallography. A crystal is unchanged by translation through a vector displacement of the form

$$\mathbf{R}_n = n_1\mathbf{a}_1 + n_2\mathbf{a}_2 + n_3\mathbf{a}_3\,, \tag{5.16}$$

where n_1, n_2, and n_3 are integers, and $\mathbf{a}_1$, $\mathbf{a}_2$, and $\mathbf{a}_3$ form the edges of a unit cell of the crystal. Corresponding to such a translation, there is a unitary operator, $U(\mathbf{R}_n) = \exp(-i\mathbf{p}\cdot\mathbf{R}_n/\hbar)$, which leaves the Hamiltonian of the crystal invariant:

$$U(\mathbf{R}_n)HU^{-1}(\mathbf{R}_n) = H\,. \tag{5.17}$$

These unitary operators for translations commute with each other (as was shown in Sec. 3.3), as well as with H, so according to Theorem 5, Sec. 1.3, there must exist a complete set of common eigenvectors for all of these operators,

$$H|\Psi\rangle = E|\Psi\rangle\,, \tag{5.18a}$$

$$U(\mathbf{R}_n)|\Psi\rangle = c(\mathbf{R}_n)|\Psi\rangle\,. \tag{5.18b}$$

The composition relation of the translation operators, $U(\mathbf{R}_n)U(\mathbf{R}_{n'}) = U(\mathbf{R}_n + \mathbf{R}_{n'})$, implies a similar relation for the eigenvalues, $c(\mathbf{R}_n)c(\mathbf{R}_{n'}) = c(\mathbf{R}_n + \mathbf{R}_{n'})$. This equation is satisfied only by the exponential function,

$$c(\mathbf{R}_n) = \exp(-i\mathbf{k}\cdot\mathbf{R}_n)\,. \tag{5.19}$$

Because $U(\mathbf{R}_n)$ is unitary, we must have $|c(\mathbf{R}_n)| = 1$, and hence the vector $\mathbf{k}$ must be real. These results apply to a system of arbitrary complexity, provided only that it has periodic symmetry.

If the system is a single particle interacting with a periodic potential field, the usual form of *Bloch's theorem* may be obtained by expressing the eigenvector of (5.18) in coordinate representation, $\langle\mathbf{x}|\Psi\rangle = \Psi(\mathbf{x})$. By definition, we have $U(\mathbf{R}_n)\Psi(\mathbf{x}) = \Psi(\mathbf{x} - \mathbf{R}_n)$, and hence the theorem asserts that the common eigenfunctions of (5.18) have the form

$$\Psi(\mathbf{x} - \mathbf{R}_n) = \exp(-i\mathbf{k}\cdot\mathbf{R}_n)\Psi(\mathbf{x})\,. \tag{5.20}$$

The vector $\mathbf{k}$ is called the *Bloch wave vector* of the state. Note that the theorem does not say that all eigenvectors of the periodically symmetric operator H in (5.18a) must be of this form, but rather that they *may be chosen* to also be eigenvectors of (5.18b) and hence have the form (5.20). A linear combination of two eigenfunctions corresponding to the same value of E but different values of $\mathbf{k}$ will satisfy (5.18a), but it will not be of the form (5.20).

Let us now expand a function of the Bloch form (5.20) in a series of plane waves,

$$\Psi(\mathbf{x}) = \sum_{\mathbf{k}'} a(\mathbf{k}')e^{i\mathbf{k}'\cdot\mathbf{x}}\,. \tag{5.21}$$

Substitution of this expansion into (5.20) yields

$$\sum_{\mathbf{k}'} a(\mathbf{k}')e^{-i\mathbf{k}'\cdot\mathbf{R}_n}e^{i\mathbf{k}'\cdot\mathbf{x}} = e^{-i\mathbf{k}\cdot\mathbf{R}_n}\sum_{\mathbf{k}'} a(\mathbf{k}')e^{i\mathbf{k}'\cdot\mathbf{x}}\,,$$

which is consistent if and only if $a(\mathbf{k}')$ vanishes for all values of $\mathbf{k}'$ that do not satisfy the condition $\exp[i(\mathbf{k}' - \mathbf{k})\cdot\mathbf{R}_n] = 1$ for all $\mathbf{R}_n$ of the form (5.16). The vectors that satisfy this condition are of the form

$$\mathbf{k}' - \mathbf{k} = \mathbf{G}_m\,, \tag{5.22}$$

where $\mathbf{G}_m$ is a vector of the *reciprocal lattice*. A detailed theory of lattices and their reciprocals can be found in almost any book on solid state physics [for example, Ashcroft and Mermin (1976)]. For the simplest case, in which the vectors $\{\mathbf{R}_n\}$ of (5.16) form a simple cubic lattice whose unit cell is a cube of side a, the reciprocal lattice vectors $\{\mathbf{G}_m\}$ form a simple cubic lattice whose unit cell is a cube of side $2\pi/a$.

In light of this result, we can rewrite (5.21) in the form

$$\Psi(\mathbf{x}) = \sum_{\mathbf{G}_m} a(\mathbf{k} + \mathbf{G}_m) e^{i(\mathbf{k}+\mathbf{G}_m)\cdot\mathbf{x}} . \tag{5.23}$$

Since an expansion in plane waves is in fact an expansion in momentum eigenfunctions, it follows that the momentum distribution in a state described by (5.23) or (5.20) is discrete, with only momentum values of the form $\hbar(\mathbf{k}+\mathbf{G}_m)$ being present. This result will be important in the next Section.

5.4 *Diffraction Scattering: Theory*

The phenomenon of diffraction-like scattering of particles was very important in the historical development of quantum mechanics, and it remains important as an experimental technique. In this Section we are concerned with the theory of the phenomenon and its implications for the interpretation of quantum mechanics.

Diffraction by a periodic array

Diffraction scattering from a periodic array, such as a grating or a crystal, can be analyzed by two different (though mathematically equivalent) methods, which tend to suggest different interpretations.

(a) Position probability density

The first method is to solve the Schrödinger equation,

$$-\left(\frac{\hbar^2}{2M}\right) \nabla^2\Psi(\mathbf{x}) + W(\mathbf{x})\Psi(\mathbf{x}) = E\Psi(\mathbf{x}) , \tag{5.24}$$

with boundary conditions corresponding to an incident beam from a certain direction, and hence determine the position probability density, $|\Psi(\mathbf{x})|^2$, at the detectors. An exact solution of this equation would be very difficult to obtain, but the most important features of the solution can be found by the method of physical optics. A derivation of optical diffraction theory from a scalar wave equation similar to (5.24) can be found in Ch. 8 of Born and Wolf (1980). We may apply those methods of diffraction theory to (5.24) as a mathematical technique, without necessarily assuming the physical interpretation of the equation and solution to be the same as in optics.

Figure 5.3 depicts an incident beam of particles, each having momentum $p = \hbar k$, which is diffracted by a periodic line of atoms. The source and the detectors are so far away that the rays can be regarded as parallel. The difference in path length from the source to the detector along the two rays shown is $a(\sin\theta_2 - \sin\theta_1)$. If this path difference is an integer multiple, n, of the wavelength of the incident beam, $\lambda = 2\pi\hbar/p = 2\pi/k$, then the amplitudes scattered by the separate atoms will interfere constructively, yielding a large value of $|\Psi|^2$ at the detector. Therefore diffraction maxima in the scattering probability will be observable at angles that satisfy the condition

$$a(\sin\theta_2 - \sin\theta_1) = n\lambda\,. \tag{5.25}$$

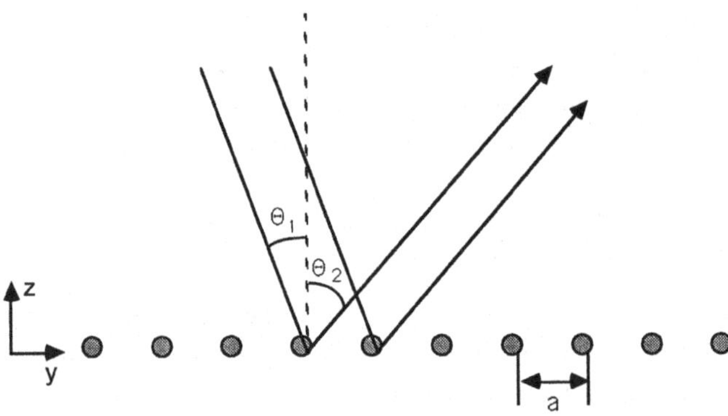

Fig. 5.3 Diffraction by a periodic array of atoms.

The interpretation suggested by this analysis is best described by the phrase *wave–particle duality*. It suggests that there is a wave associated with a particle, although the nature of the association is not entirely clear. Indeed, it suggests that the Schrödinger "wave function" $\Psi(\mathbf{x})$ might be a physical wave in ordinary space. However, as was pointed out in Sec. 4.2, such an interpretation of Ψ does not make sense for other than one-particle states. Moreover, we should not forget that it is only the particles that are observed. To "observe" the diffraction pattern, we actually count the relative numbers of particles that are scattered in various directions.

(b) Momentum probability distribution

The second method is to calculate the momentum probability distribution, since the probability that a particle will have momentum $\mathbf{p}' = \hbar\mathbf{k}'$ is also the probability that it will emerge in the direction of $\mathbf{k}'$. It will not complicate our solution if we now regard the atoms in Fig. 5.3 as constituting a two-dimensional crystal lattice in the xy plane. Moreover, the crystal need not be restricted to a single layer, but rather it may be of arbitrary extent and form in the negative z direction, provided only that it is periodic in the x and y directions.

Because our system is periodic in the x and y directions, the two-dimensional version of Bloch's theorem applies. Hence the solutions of (5.24) may be chosen to have a form that is the two-dimensional analog of (5.23),

$$\Phi_{\mathbf{q}}(\mathbf{x}) = \sum_n e^{i(\mathbf{q}+\mathbf{g}_n)\cdot\mathbf{x}}\, b_n(\mathbf{q}, z)\,. \tag{5.26}$$

Here $\mathbf{g}_n$ is a two-dimensional reciprocal lattice vector, and $\mathbf{q}$ is the two-dimensional analog of the Bloch wave vector $\mathbf{k}$ in (5.23). Both $\mathbf{g}_n$ and $\mathbf{q}$ are confined to the xy plane. Since the periodicity is only in the x and y directions, we can infer the existence, for each fixed value of z, of a solution that is of the Bloch form in its x and y dependences, but nothing can be inferred about the z dependence of Φ.

The general solution of (5.24) is a linear combination of functions of the form (5.26), the particular combination being chosen to fit the boundary conditions. These conditions require that in the region $z > 0$, above the crystal, $\Psi(\mathbf{x})$ should contain an incident wave $e^{i\mathbf{k}\cdot\mathbf{x}}$, with $k_z < 0$. The incident wave is already of the Bloch form provided we identify $\mathbf{q} = \mathbf{k}_{xy}$ (the projection of $\mathbf{k}$ into the xy plane). Therefore it is not necessary to combine functions $\Phi_{\mathbf{q}}(\mathbf{x})$ having different $\mathbf{q}$ values in order to satisfy the boundary condition, since the condition is satisfied by one such function alone. Hence the physical solution $\Psi(\mathbf{x})$ may be taken to have the form (5.26).

Above the crystal, the potential $W(\mathbf{x})$ vanishes, and so the solution must be of the form

$$\Psi(\mathbf{x}) = e^{i\mathbf{k}\cdot\mathbf{x}} + \sum_{\mathbf{k}'} r(\mathbf{k}')e^{i\mathbf{k}'\mathbf{x}}\,, \quad (z > 0)\,, \tag{5.27}$$

where $E = \hbar^2k^2/2M$, and $|\mathbf{k}'| = k$. For the incident wave (first term) we have $k_z < 0$, and for the scattered waves we must have $k'_z > 0$. The probability that a particle is scattered into the direction $\mathbf{k}'$ is proportional to $|r(\mathbf{k}')|^2$.

Now (5.27) must be of the form (5.26). This is possible if we identify $\mathbf{q} = \mathbf{k}_{xy}$, and hence the $n = 0\,(\mathbf{g}_0 = 0)$ term in (5.26) must be identified with the incident wave in (5.27), and the $\mathbf{g}_n \neq 0$ terms in (5.26) must be identified with the scattered waves in (5.27). Therefore in (5.27) we must have

$\mathbf{k}'_{xy} = \mathbf{k}_{xy} + \mathbf{g}_n$, where $\mathbf{g}_n$ is a nonvanishing two-dimensional reciprocal lattice vector. The remaining component, k'_z, is determined by the value of $\mathbf{k}'_{xy}$ and the energy conservation condition $|\mathbf{k}'| = k$. Thus we see that the allowed values of $\mathbf{k}'$ are restricted to a discrete set, and so scattering can occur into only a discrete set of directions.

The reason for the discrete set of scattering directions, according to this analysis, is that the change in the x and y components of the momentum must be a multiple of a two-dimensional reciprocal lattice vector,

$$(\hbar\mathbf{k}' - \hbar\mathbf{k})_{xy} = \hbar\mathbf{g}_n \,. \tag{5.28}$$

Momentum transferred to and from a periodic object (the lattice) *is quantized* in the direction of the periodicity. The z component of momentum is not subject to any such quantization condition because the lattice is not periodic in the z direction. (However, k'_z is fixed by energy conservation.)

For comparison with the result from the first method, we specialize to a one-dimensional array of atoms along the y axis, and we consider only motion in the yz plane. Then the reciprocal lattice vectors lie in the y direction, and their magnitudes are $g_n = 2\pi n/a$, where n is an integer and a is the interatomic separation distance. Thus (5.28) yields

$$\hbar k'_y - \hbar k_y = \frac{2\pi\hbar n}{a} \tag{5.29}$$

for the change of the particle momentum along the direction of the periodicity. In the result (5.25) of the first method, we may substitute $\lambda = 2\pi/k$ and obtain

$$\hbar k(\sin\theta_2 - \sin\theta_1) = \frac{2\pi\hbar n}{a} \,,$$

which is precisely the same as (5.29).

The two methods have thus been shown to yield the same results, but they suggest different interpretations. In particular, the explanation of diffraction scattering by means of quantized momentum transfer to and from a periodic object does not suggest or require any hypothesis that the particle should be literally identified as a wave or wave packet.

> The explanation of diffraction scattering by means of a *hypothesis* of quantized momentum transfer was first proposed by W. Duane in 1923, before quantum mechanics had been formulated by Heisenberg and Schrödinger. That hypothesis is no longer needed, since it has now emerged as a theorem of quantum mechanics. There are three common examples of the relationship between *periodicity* and *quantization*:

(i) Spatial periodicity, of period a, implies that

$$p' - p = \frac{n2\pi\hbar}{a}, \tag{5.30}$$

where p and p' are the initial and final momentum components in the direction of the periodicity, and n is an integer.

(ii) Periodicity in time, of period T, implies that

$$E' - E = \frac{n2\pi\hbar}{T} = n\hbar\omega, \tag{5.31}$$

where $\omega = 2\pi/T$, and E and E' are the initial and final energies. This fact is illustrated by the harmonic oscillator (Ch. 6), and by the effect of a harmonic perturbation (Ch. 12).

(iii) Rotational periodicity about some axis, of period 2π radians, implies that

$$J' - J = \frac{n2\pi\hbar}{2\pi} = n\hbar, \tag{5.32}$$

where J and J' are the initial and final angular momentum components about the axis of rotation. This is demonstrated in Ch. 7.

Some points to note are:

- The size of the quantum is inversely proportional to the period;
- This quantization is not a universal law, but rather it holds only in the presence of an appropriate periodicity [which will always be present in case (iii)];
- Only the *changes* in the dynamical variables are quantized by periodicity, not their absolute magnitudes.

Double slit diffraction

The diffraction of particles by a double slit has become a standard example in quantum mechanics textbooks. In it, we consider the passage of an ensemble of similarly prepared particles through a screen that has two slits. If only one of the slits is open, then the particles that are detected on the downstream side of the screen will have a monotonic spatial distribution whose width is related to the width of the slit. But if both slits are open, the spatial distribution of the detected particles will be modulated by an interference pattern. The positions of the maxima and minima can be calculated by considering the constructive and destructive interference between the partial waves that originate from the two slits.

The double slit diffraction pattern of electrons has been measured by Tonomura *et al.* (1989), using a technique that allows one to see the interference pattern being built up from a sequence of spots as the electrons arrive one at a time. The electron arrival rate is sufficiently low that there is only a negligible chance of more than one electron being present between the source and the detector at any one time. This effectively rules out any hypothetical explanation of the diffraction pattern as being due to electron–electron interaction. Nor can an electron be literally identified with a wave packet, for the positions of the individual electrons are resolved to a precision that is much finer than the width of the interference fringes. The interference pattern is only a statistical distribution of scattered particles.

A remarkable result is that when both slits are open there are places (diffraction minima) where the probability density is nearly zero — particles do not go there — whereas if only one slit were open many particles would go there. This is certainly a remarkable physical phenomenon with interesting theoretical consequences. However, it has unfortunately generated a fallacious argument to the effect that what we are seeing is a violation of "classical" probability theory in the domain of quantum theory. The argument goes as follows:

> Label the two slits as #1 and #2. If only slit #1 is open the probability of detecting a particle at the position X is $P_1(X)$. Similarly, if only slit #2 is open the probability of detection at X is $P_2(X)$. If both slits are open the probability of detections at X is $P_{12}(X)$. Now passage through slit #1 and passage through slit #2 are *exclusive events*, so from the addition rule, Eq. (1.49a), we conclude that $P_{12}(X)$ should be equal to $P_1(X) + P_2(X)$. But these three probabilities are all measurable in the double slit experiment, and no such equality holds. Hence it is concluded that the the addition rule (1.49a) of probability theory does not hold in quantum mechanics.

This would appear to be a very disturbing conclusion, for probability theory is very closely tied to the interpretation of quantum theory, and an incompatibility between them would be very serious. But, in fact, the radical conclusion above is based on an incorrect application of probability theory.

One is well advised to beware of probability statements expressed in the form $P(X)$ instead of $P(X|C)$. The second argument may safely be omitted only if the conditional information C is clear from the context, and is *constant* throughout the problem. But that is not so in the above example. Three distinctly different conditions are used in the argument. Let us denote them as

$$C_1 = (\text{slit } ^{\#}1 \text{ open, } ^{\#}2 \text{ closed, wave function} = \Psi_1)\,,$$
$$C_2 = (\text{slit } ^{\#}2 \text{ open, } ^{\#}1 \text{ closed, wave function} = \Psi_2)\,,$$
$$C_3 = (\text{slits } ^{\#}1 \text{ and } ^{\#}2 \text{ open, wave function} = \Psi_{12})\,.$$

In the experiment we observe that $P(X|C_1) + P(X|C_2) \neq P(X|C_3)$. But probability theory does not suggest that there should be an equality. The inequality of these probabilities (due to interference) may be contrary to classical mechanics, but it is quite compatible with classical probability theory.

This, and other erroneous applications of probability theory in quantum mechanics, are discussed in more detail by Ballentine (1986).

5.5 *Diffraction Scattering: Experiment*

Diffraction scattering from periodic structures (usually crystal lattices) has been observed for many different kinds of particles. Some of the earliest discoveries of this phenomenon are listed in the following table.

Discovery of Diffraction Scattering for Various Particles		
X-ray photons	1912	M. von Laue
Electrons	1927	C. Davisson and L. H. Germer
He atoms	1930	O. Stern
H_2 molecules	1930	O. Stern
Neutrons	1936	D. P. Mitchell and P. N. Powers

Most important, from a theoretical point of view, is the universality of the phenomenon of diffraction. The particle may be charged or uncharged, elementary or composite. The interaction may be electromagnetic or nuclear. (The neutron interacts with the crystal primarily through the strong nuclear interaction.) The effective wavelength λ associated with a particle can be deduced from experiment by means of (5.25), and it is found to be related to the momentum p of the particle by de Broglie's formula, $\lambda = h/p$, where $h = 2\pi\hbar$ is Planck's constant. Hence diffraction experiments provide a means of measuring the universal constant $\hbar$, which was introduced into the theory in Eq. (3.55).

It is conceivable that we might have found different values of the empirical parameter "h" for different particles. Thus we might distinguish h_e, h_n, etc.

for electrons, neutrons, etc. We might also distinguish h_γ for photons by means of the Bohr relation, $h_\gamma \nu = E_2 - E_1$, for the frequency ν of radiation emitted during a transition between two energy levels. Although it is possible to measure these "h" parameters by directly measuring the quantities in their defining equations, more accurate values can be obtained from a combination of indirect measurements [Fischbach, Greene, and Hughes (1991)]. From them, it has been shown that the ratios h_e/h_γ and h_n/h_γ differ from unity by no more than a few parts in 10^{-8}.

The results for He atoms and H_2 molecules are particularly significant, because they demonstrate that the phenomenon of particle diffraction is not peculiar to elementary particles. Diffraction scattering of *composite particles* is also relevant to the interpretation of quantum mechanics. The effective wavelength associated with a particle is $\lambda = h/p$, where p is its total momentum. Thus a particle of mass M_i moving at the speed v_i (small compared to the speed of light) exhibits the wavelength $\lambda_i = h/M_i v_i$ in a diffraction experiment. If there were a real physical wave in space propagating with this wavelength, then one would expect that a composite of several particles would be associated with several waves, and that all of the wavelengths $\{\lambda_i = h/M_i v_i\}$ would appear in the diffraction pattern. But that does not happen. In fact only the single wavelength $\lambda = h/(\sum_i M_i v_i)$, associated with the total momentum of the composite system, is observed. This result would be very puzzling from the point of view of a real wave interpretation. On the other hand, according to interpretation (b) of Sec. 5.4, diffraction scattering is due to quantized momentum transfer to and from the periodic lattice. The size of the quantum is determined entirely by the periodicity of the lattice, and is independent of the nature of the particle being scattered. Thus the observed results for diffraction of composite particles are exactly what one would expect according to this interpretation.

The classic example of diffraction is that of light by a grating, which is a periodic distribution of matter. The inverse of this phenomenon, i.e. the diffraction of matter by light, is known as the *Kapitza–Dirac effect.* Gould, Ruff, and Pritchard (1986) demonstrated that neutral sodium atoms are deflected by a plane standing wave laser field, and have confirmed that the momentum transfer is given by Eq. (5.30). The atom interacts with the field through its electric polarization, the interaction energy being proportional to the square of the electric field. Thus the spatial period in (5.30) is that of the intensity (square of the amplitude), which is half the wavelength of the light from the laser.

Atom interferometry is now a growing field. Many experiments have recently been performed that are the atomic analogs of earlier optical and electron interference experiments. For example, the double slit experiment has been carried out using He atoms with two slits of 1μm width and 8μm separation [Carnal and Mlynek, (1991)]. (One μm is 10^{-6} m.) The de Broglie wavelength in these experiments is typically much smaller than the size of the atom, whereas in electron or neutron diffraction the de Broglie wavelength is much larger than the diameter of the particle. Evidently, the de Broglie wavelength is in no sense a measure of the size of the particle. This is yet another argument against the literal identification of a particle with a wave packet.

In the future, we may expect atomic interferometry to provide new fundamental tests of quantum theory. But, so far, neutrons have been more useful. Single slit and double slit neutron diffraction patterns have been measured, and have accurately confirmed the predictions of diffraction theory [see Zeilinger *et al.* (1988), and Gäler and Zeilinger (1991)].

Very sensitive neutron interference experiments are now possible with the *single crystal interferometer*. It is cut into the shape shown in Fig. 5.4(a) from a crystal of silicon about 10 cm long. The crystal is "perfect" in the sense that it has no dislocations or grain boundaries. (It may contain vacancies but they do not affect the experiment.) The various diffraction beams are shown in Fig. 5.4(b). The incident beam at A is divided into a transmitted beam AC and a diffracted (Bragg-reflected) beam AB. Similar divisions occur at B and C, with transmitted beams leaving the apparatus and playing no further role in the experiment. The diffracted beams from B and C recombine coherently at D, where a further Bragg reflection takes place. The interference of the amplitudes of the two beams is observable by means of the two detectors, D_1 and D_2. The amplitude at D_1 is the sum of the transmitted portion of CD plus the diffracted portion of BD, and similarly the amplitude at D_2 is the sum of the transmitted portion of BD plus the diffracted portion of CD.

To analyze the interferometer, we shall assume that the transmission and reflection coefficients are the same at each of the vertices A, B, C, and D, and that free propagation of plane wave amplitudes takes place between those vertices. As is apparent from the figure, only two distinct propagation directions are involved, and at each diffraction vertex the amplitudes are redistributed between these two modes. Figure 5.4(c) depicts a general diffraction vertex. Because the evolution and propagation are governed by linear unitary operators, it follows that the relation between the amplitudes of the outgoing and incoming waves is of the form

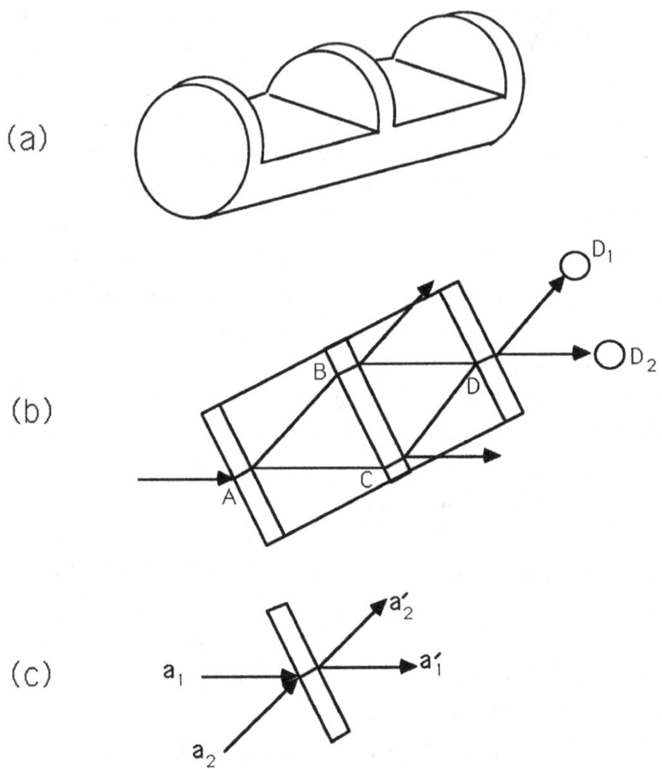

Fig. 5.4 The neutron interferometer.

$$\begin{bmatrix} a'_1 \\ a'_2 \end{bmatrix} = U \begin{bmatrix} a_1 \\ a_2 \end{bmatrix}, \text{ with } U = \begin{bmatrix} t & r \\ s & u \end{bmatrix}. \tag{5.33}$$

Here U is a unitary matrix. The elements t and u are transmission coefficients, and the elements r and s are reflection coefficients.

Several useful relations (not all independent) among the elements of U follow from its unitary nature. For example, $UU^+ = 1$ implies that $|t|^2+|r|^2 = 1$ and $|s|^2 + |u|^2 = 1$. The determinant of a unitary matrix must have modulus 1, and therefore

$$|tu - rs| = 1\,. \tag{5.34}$$

The relation $U^{-1} = U^\dagger$ takes the form

$$\frac{1}{tu - rs}\begin{bmatrix} u & -r \\ -s & t \end{bmatrix} = \begin{bmatrix} t^* & s^* \\ r^* & u^* \end{bmatrix}. \tag{5.35}$$

From (5.34) and (5.35), it follows that $|u| = |t|$ and $|s| = |r|$. Now complex numbers can be regarded as two-dimensional vectors, to which the triangle inequality (1.2) applies. Thus from (5.34) we obtain $|tu| + |rs| \geq 1$. But since $|u| = |t|$ and $|s| = |r|$, it follows that $|tu| + |rs| = 1$. This is compatible with (5.34) only if tu and $-rs$ have the same complex phase, and thus rs/tu must be real and negative.

If the amplitude at A is Ψ_A, then the amplitudes at B and C will be $\Psi_B = \Psi_A r e^{i\phi_{AB}}$ and $\Psi_C = \Psi_A t e^{i\phi_{AC}}$. Here ϕ_{AB} is the phase change occurring during propagation through the empty space between A and B, and ϕ_{AC} is a similar phase change between A and C. The amplitude that emerges toward the detector D_1 is the sum of the amplitudes from the paths $ABDD_1$ and $ACDD_1$:

$$\begin{aligned}\Psi_{D_1} &= \Psi_A(r\, e^{i\phi_{AB}} s\, e^{i\phi_{BD}}\, r + t\, e^{i\phi_{AC}}\, r\, e^{i\phi_{CD}} u) \\ &= \Psi_A\, r(rs\, e^{i\phi_{ABD}} + tu\, e^{i\phi_{ACD}})\,. \end{aligned} \tag{5.36}$$

Similarly the amplitude that emerges toward D_2 is

$$\begin{aligned}\Psi_{D_2} &= \Psi_A(re^{i\phi_{AB}}\, se^{i\phi_{BD}}\, t + te^{i\phi_{AC}} r\, e^{i\phi_{CD}}\, s) \\ &= \Psi_A trs(e^{i\phi_{ABD}} + e^{i\phi_{ACD}})\,. \end{aligned} \tag{5.37}$$

Here we have written $\phi_{ABD} = \phi_{AB} + \phi_{BD}$ and $\phi_{ACD} = \phi_{AC} + \phi_{CD}$.

Any perturbation that has an unequal effect on the phases associated with the two paths, ϕ_{ABD} and ϕ_{ACD}, will influence the intensities of the beams reaching the detectors D_1 and D_2. Since the phase of rs/tu is negative, it follows that if the interference between the two terms in (5.37) is constructive then the interference between the two terms in (5.36) will be destructive, and vice versa. The most convenient way to detect such a perturbation is to monitor the difference between the counting rates of D_1 and D_2.

In one of the most remarkable experiments of this type, Colella, Overhauser, and Werner (1975) detected *quantum interference due to gravity*. The interferometer was rotated about a horizontal axis parallel to the incident beam, causing a difference in the gravitational potential on paths AC and BD, and hence a phase shift of the interference pattern. The phase difference between the two paths is easily calculated from the constancy of the sum of kinetic and gravitational potential energy, $\hbar^2k^2/2M + Mgz = E$, where M is the neutron mass, g is the acceleration of gravity, and z is the elevation relative to the incident beam. The accumulated phase change along any path is given by $\int k ds$, where ds is an element of length along the path. Since the potential energy is small compared to the total energy, we obtain

$$k \approx \frac{\sqrt{2ME}}{\hbar} - \frac{M^2 g z}{\hbar\sqrt{2ME}}\,.$$

The phase difference between the two paths is $\Phi_{\text{ABD}} - \Phi_{\text{ACD}} = \oint k\; ds$, to which the only contribution is from the term above that contains z. Now the integral around a closed path, $\oint z\; ds$, is just the vertical projection of the area bounded by the path. Hence the phase difference is

$$\Phi_{\text{ABD}} - \Phi_{\text{ACD}} = \frac{M^2 g\; A \sin\alpha}{\hbar\sqrt{2ME}} = \frac{\lambda\, M^2 g A \sin\alpha}{2\pi\hbar^2}\,, \tag{5.38}$$

where A is the area of the loop ABDC and α is the angle of its plane with respect to the horizontal. In the second equality, $\lambda = 2\pi\hbar/(2ME)^{1/2}$ is the de Broglie wavelength of the incident neutrons.

The interference pattern shown in Fig. 5.5 was the first demonstration that quantum mechanics applies to the Newtonian gravitational interaction, which has now been shown to function in the Schrödinger equation as does any other potential energy.

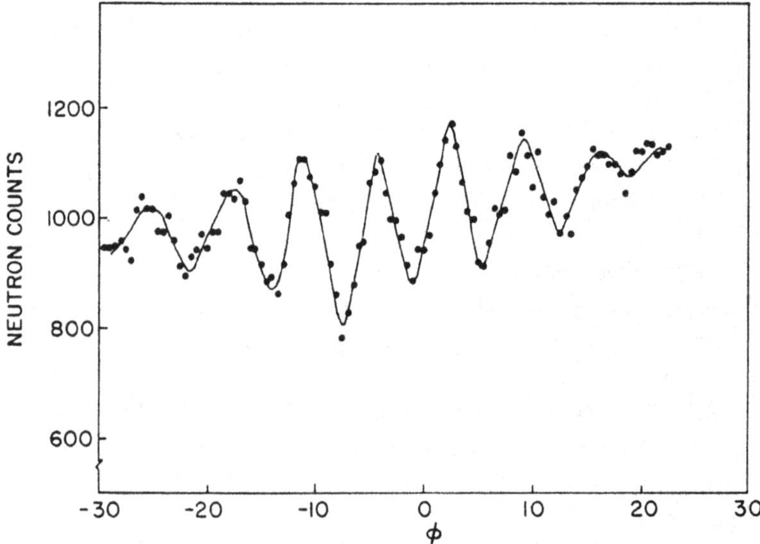

Fig. 5.5 Interference pattern due to the gravitational potential energy of a neutron. Here Φ is the angle (in degrees) of rotation of the interferometer about its horizontal axis. [From Colella, Overhauser, and Werner, *Phys. Rev. Lett.* **34**, 1472 (1975).]

5.6 *Motion in a Uniform Force Field*

Whenever a physical system is invariant under space displacements, we may expect momentum representation to be simpler that coordinate representation. For example, consider the motion of a free particle in one dimension, with some given initial state $|\Psi(0)\rangle$. In coordinate representation, this problem requires the solution of a second order partial differential equation. The momentum representation of the state vector is the one-dimensional version of Eq. (5.5), $\langle\hbar k|\Psi(t)\rangle = \hbar^{-1/2}\Phi(k,t)$, and the Schrödinger equation becomes

$$\frac{\hbar^2 k^2}{2M}\,\Phi(k,t) = i\hbar\,\frac{\partial\Phi(k,t)}{\partial t}\,. \tag{5.39}$$

It has the trivial solution

$$\Phi(k,t) = e^{-it\,\hbar\,k^2/2M}\,\Phi(k,0)\,. \tag{5.40}$$

The coordinate representation of the state function is then obtained by an inverse Fourier transform,

$$\Psi(x,t) = (2\pi)^{-1/2}\int e^{(ikx - it\hbar\,k^2/2M)}\,\Phi(k,0)\,dk\,. \tag{5.41}$$

As an example, we consider a *Gaussian initial state*,

$$\Psi(x,0) = (2\pi a^2)^{-1/4}\,e^{-x^2/4a^2}\,, \tag{5.42a}$$

whose Fourier transform is

$$\Phi(k,0) = \left(\frac{2a^2}{\pi}\right)^{1/4} e^{-a^2k^2}\,. \tag{5.42b}$$

These are normalized so that $\int|\Psi|^2\,dx = \int|\Phi|^2\,dk = 1$. The time-dependent state function, from (5.41), is

$$\Psi(x,t) = \left(\frac{a^2}{2\pi^3}\right)^{-1/4}\int_{-\infty}^{\infty}\exp\left\{ikx - \left(a^2 + \frac{i\hbar t}{2M}\right)k^2\right\}dk\,. \tag{5.43}$$

This integral can be transformed to a standard form by completing the square in the argument of the exponential function. The result is

$$\Psi(x,t) = (2\pi)^{-1/4}\left[a\left(1 + \frac{i\hbar t}{2Ma^2}\right)\right]^{-1/2} e^{-x^2/4\alpha^2}\,, \tag{5.44}$$

where $\alpha^2 = a^2(1 + \frac{i\hbar t}{2Ma^2})$.

Now let us consider a particle in a *homogeneous force field*. Since the components of momentum in the directions perpendicular to the force will remain constant, it is only necessary to treat the motion in the direction of the force, and so the problem becomes essentially one-dimensional. Choose the force to be in the x direction, and so that the potential is $W = -Fx$. The stationary states are described by the eigenvectors of

$$H|\Psi_E\rangle \equiv \left(\frac{P^2}{2M} - Fx\right)|\Psi_E\rangle = E|\Psi_E\rangle\,. \tag{5.45}$$

Even though the force is invariant under the displacement $x \to x + a$, the Hamiltonian is not. However, (5.45) is invariant under the combined transformations

$$x \to x + a\,, \quad E \to E - Fa\,. \tag{5.46}$$

Therefore we need only to calculate one energy eigenfunction, since all energy eigenfunctions can be obtained from one such eigenfunction by displacement.

In momentum representation, (5.45) becomes

$$\frac{\hbar^2 k^2}{2M}\,\Phi(k) - iF\frac{\partial\Phi(k)}{\partial k} = E\,\Phi(k)\,, \tag{5.47}$$

using the form (5.6) for the position operator. This is a first order differential equation, whereas a second order equation would be obtained in coordinate representation. The solution of (5.47) is

$$\Phi(k) = A\,\exp\left\{i\left[k\frac{E}{F} - k^3\frac{\hbar^2}{6MF}\right]\right\}, \tag{5.48}$$

where A is an arbitrary constant. The state function in coordinate representation is obtained by a Fourier transformation,

$$\Psi_E(x) = \int_{-\infty}^{\infty} \exp\left\{i\left[\left(x + \frac{E}{F}\right)k - \frac{\hbar^2}{6MF}k^3\right]\right\} dk\,. \tag{5.49}$$

Since the normalization of this function is arbitrary, we shall drop constant factors in the following analysis. $\Psi_E(x)$ is real because $\Phi(-k) = \Phi^*(k)$. It is apparent that (5.49) is invariant under the transformation (5.46), and hence the eigenfunctions for different energies are related by

$$\Psi_{E+Fa}(x) = \Psi_E(x + a)\,. \tag{5.50}$$

Thus it is sufficient to evaluate (5.49) for $E = 0$.

The function $\Psi_0(x)$ is equivalent to the *Airy function*, apart from scale factors multiplying Ψ and x. It has no closed form expression in terms of simpler functions, but its asymptotic behavior in the limits $x \to \pm\infty$ can be determined by the *method of steepest descent.* Let us write

$$\Psi_0(x) = \int_C e^{i\alpha(k)}\, dk\,, \quad \alpha(k) = kx - k^3 \frac{\hbar^2}{6MF}\,. \tag{5.51}$$

The integrand is an analytic function of k, and so the path of integration C, from $-\infty$ to $+\infty$ along the real axis, may be continuously deformed, provided only that $\text{Im}(k^3) \leq 0$ as $k \to \pm\infty$, so as not to disrupt the convergence of the integral.

As $|x| \to \infty$ the integrand $e^{i\alpha(k)}$ oscillates very rapidly as a function of k, and so its contribution to the integral is nearly zero. An exception to this cancellation occurs at any point where $\partial\alpha/\partial k = 0$. Near such a point of stationary phase, the contributions to the integral from neighboring k values add coherently, rather than canceling, and so the integral will be dominated by such regions. The stationary phase condition $\partial\alpha/\partial k = 0$ has two roots,

$$k_0 = k_0(x) = \pm\left(x\frac{2MF}{\hbar^2}\right)^{1/2}. \tag{5.52}$$

In the neighborhood of one of the points $k = k_0(x)$ we may approximate $\alpha(k)$ by second order Taylor series,

$$\alpha(k) = \frac{2k_0 x}{3} - \left(k_0\frac{\hbar^2}{2MF}\right)(k - k_0)^2\,. \tag{5.53}$$

[Writing $\alpha(k)$ this way, without substituting the explicit value of k_0, yields an expression that is valid for either sign in (5.52).] The contribution to the integral from the neighborhood of this point will be

$$e^{i2k_0x/3} \int_C e^{-a(k-k_0)^2}\, dk\,, \quad \left(a = ik_0\frac{\hbar^2}{2MF}\right). \tag{5.54}$$

The path of integration C should be deformed to pass through $k = k_0$ at an angle such that $a(k - k_0)^2$ is real and positive along C, so that the magnitude of the integrand decreases rapidly from its maximum at $k = k_0$. Then we will have, approximately,

$$\int_C e^{-a(k-k_0)^2}\, dk \approx \int_{-\infty}^{\infty} e^{-az^2\, dz} = \left(\frac{\pi}{a}\right)^{1/2}. \tag{5.55}$$

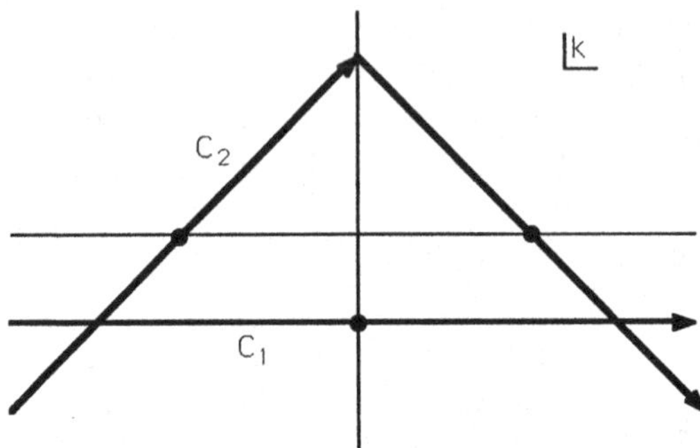

Fig. 5.6 Integration paths in the complex k plane for evaluating Eq. (5.54). C_1 is suitable for $x \to -\infty$, and C_2 is suitable for $x \to +\infty$. The large dots mark the relevant stationary phase points.

For $x < 0$ the stationary phase points (5.52) are located on the imaginary axis. Dropping the integration path in Fig. 5.6 from the real axis to C_1, which passes through $k_0 = -i(|x|2MF/\hbar^2)^{1/2}$, we obtain the following asymptotic behavior:

$$\Psi_0(x) \approx |x|^{-1/4} \exp\left\{-\frac{2}{3}\left(\frac{2MF}{\hbar^2}\right)^{1/2} |x|^{3/2}\right\}, \quad (x \to -\infty)\,. \tag{5.56}$$

For $x > 0$ there are two stationary phase points on the real axis. The path of integration should be distorted to C_2, which intersects the real axis at 45° angles, in order for $a(k - k_0)^2$ to be real positive on the integration path. The contributions to the integral from the two stationary phase points are complex conjugates of each other, and their sum yields the asymptotic limit

$$\Psi_0(x) \approx 2x^{-1/4} \cos\left\{\left[\frac{2}{3}\left(\frac{2MF}{\hbar^2}\right)^{1/2} x^{3/2} - \frac{\pi}{4}\right]\right\}, \quad (x \to +\infty)\,. \tag{5.57}$$

[Although constant factors were dropped between (5.51) and these limiting forms, the two limits (5.56) and (5.57) are mutually consistent in their normalization.]

By solving in momentum representation, we have obtained a unique solution for fixed E. But if we had solved the second order differential equation in coordinate representation, there would be two solutions: the one that we have obtained, and a second solution that diverges exponentially as $x \to -\infty$. This second, physically unacceptable solution is automatically excluded by the momentum representation method. It can be obtained mathematically by diverting the path of integration in (5.54) from the positive real axis to the positive imaginary axis. The stationary phase point at $k_0 = +i(|x|2MF/\hbar^2)^{1/2}$ would then yield a contribution that diverges exponentially as $x \to -\infty$. But this new path cannot be reached by continuous deformation from the real axis through regions in which $\text{Im}(k^3) \leq 0$ at infinite k, and therefore it is excluded as a solution to our problem.

Further reading for Chapter 5

Some of the earliest diffraction experiments were described by Trigg (1971), Ch. 10. These may be contrasted with the capabilities of the modern single crystal neutron interferometer, described by Staudenmann *et al.* (1980), and by Greenberger (1983). The accomplishments and potential of atom interferometry were discussed by Prichard (1992). Since then, interferometry has been extended to large molecules. Brezger, ..., Zeilinger (2002) report results for C_{70} fullerine molecules.

Problems

5.1 Show that the commutator of the position operator with a function of the momentum operator is given by $[Q_x, f(P)] = i\hbar\partial f/\partial P_x$ (cf. Problem 4.1).

5.2 How does the momentum representation of the state vector $|\Psi\rangle$, $\Phi(\mathbf{k}) \equiv \langle\hbar\mathbf{k}|\Psi\rangle$, transform under the Galilei transformation (4.8)?

5.3 A *local* potential is described by an operator W whose matrix in coordinate representation is diagonal, $\langle\mathbf{x}|W|\mathbf{x}'\rangle = \delta(\mathbf{x} - \mathbf{x}')W(\mathbf{x})$. What is the corresponding property of the matrix in momentum representation, $\langle\mathbf{p}|W|\mathbf{p}'\rangle$?

5.4 The Hamiltonian of an electron on a crystal is $H = P^2/2M + W$, where the potential W has the symmetries of the crystal lattice. In particular, it is invariant under displacement by any lattice vector of the form (5.16). Write the eigenvalue equation $H|\Psi\rangle = E|\Psi\rangle$ in momentum representation, and show that it leads naturally to eigenvectors of the Bloch form. Do not invoke Bloch's theorem, since the purpose of the problem is to give an alternative derivation of that theorem.

5.5 For the state function $\Psi(x) = c\exp(iqx - \alpha x^2)$, where c is a normalization constant, and q and α are real parameters, calculate the average momentum in two ways:

(a) Using coordinate representation,

$$\langle P \rangle = \int \Psi^*(x) \left(-i\hbar \frac{\partial}{\partial x} \right) \Psi(x)\, dx\,.$$

(b) Use momentum representation to obtain the momentum probability distribution, and then calculate the average momentum $\langle P \rangle$ from that distribution.

(c) Calculate $\langle P^2 \rangle$ using appropriate generalizations of the methods (a) and (b).

5.6 Use momentum representation to calculate the ground state of a particle in the one-dimensional attractive potential $W(x) = c\,\delta(x)$, $(c<0)$. (Compare this solution with that in coordinate representation, Problem 4.9.)

5.7 Determine the time evolution, $\Psi(x,t)$, of the one-dimensional Gaussian initial state (5.42a) in a constant homogeneous force field.

5.8 (a) For a particle in free space, calculate the time evolution of a Gaussian initial state that has a nonzero average momentum $\hbar q$, $\Psi(x,0) = (2\pi a^2)^{-1/4} e^{-x^2/(2a)^2} e^{iqx}$. Use the method of completing the square, as was done to evaluate the integral in (5.43).

(b) Check your answer by applying a Galilei transformation (Sec. 4.3) to (5.44), which is the solution for $q = 0$.

Chapter 6

The Harmonic Oscillator

A harmonic oscillator is an object that is subject to a quadratic potential energy, which produces a restoring force against any displacement from equilibrium that is proportional to the displacement. The Hamiltonian for such an object whose motion is confined to one dimension is

$$H = \frac{1}{2M}P^2 + \frac{M\omega^2}{2}Q^2\,, \tag{6.1}$$

where P is the momentum, Q is the position, and M is the mass. It is easily shown, by solving the classical equations of motion, that ω is the frequency of oscillation (in radians per unit time). The harmonic oscillator is important because it provides a model for many kinds of vibrating systems, including the electromagnetic field (see Ch. 19). Its solution also illustrates important techniques that are useful in other applications.

6.1 *Algebraic Solution*

The eigenvalue spectrum of the Hamiltonian (6.1) can be obtained algebraically, using *only* the commutation relation

$$[Q, P] = i\hbar \tag{6.2}$$

and the self-adjointness of the operators P and Q,

$$P = P^\dagger\,, \quad Q = Q^\dagger\,. \tag{6.3}$$

We first introduce dimensionless position and momentum operators,

$$q = \left(\frac{M\omega}{\hbar}\right)^{1/2} Q\,, \tag{6.4}$$

$$p = \left(\frac{1}{M\hbar\omega}\right)^{1/2} P\,, \tag{6.5}$$

which satisfy the commutation relation

$$[q,p] = i\,. \tag{6.6}$$

In terms of these new variables the Hamiltonian becomes

$$H = \frac{1}{2}\hbar\omega(p^2 + q^2)\,. \tag{6.7}$$

We next introduce two more operators,

$$a = \frac{q + ip}{\sqrt{2}}\,, \tag{6.8}$$

$$a^\dagger = \frac{q - ip}{\sqrt{2}}\,. \tag{6.9}$$

That these operators are Hermitian conjugates of each other follows from (6.3). From (6.6) it follows that

$$[a, a^\dagger] = 1\,. \tag{6.10}$$

The Hamiltonian (6.7) can be written in several equivalent forms:

$$\begin{aligned} H &= \frac{1}{2}\hbar\omega(aa^\dagger + a^\dagger a) \\ &= \hbar\omega\left(aa^\dagger - \frac{1}{2}\right) \\ &= \hbar\omega\left(a^\dagger a + \frac{1}{2}\right), \end{aligned} \tag{6.11}$$

the last of these being the most useful. The problem of finding the eigenvalue spectrum of H is thus reduced to that of finding the spectrum of

$$N = a^\dagger a\,. \tag{6.12}$$

Using the operator identity, $[AB,C] = A[B,C] + [A,C]B$, along with (6.10), we obtain the commutation relations

$$[N, a] = -a\,, \tag{6.13}$$

$$[N, a^\dagger] = a^\dagger\,. \tag{6.14}$$

The spectrum of N can be easily calculated from these relations.

Let $N|\nu\rangle = \nu|\nu\rangle$, with $\langle\nu|\nu\rangle \neq 0$. Then from (6.13) it follows that

$$Na|\nu\rangle = a(N-1)|\nu\rangle = (\nu-1)a|\nu\rangle\,.$$

Hence $a|\nu\rangle$ is an eigenvector of N with eigenvalue $\nu - 1$, provided only that $a|\nu\rangle \neq 0$. The squared norm of this vector is

$$(\langle\nu|a^\dagger)(a|\nu\rangle) = \langle\nu|N|\nu\rangle = \nu\langle\nu|\nu\rangle\,.$$

Since the norm must be nonnegative, it follows that $\nu \geq 0$, and thus an eigenvalue cannot be negative. By applying the operator a repeatedly, it would appear that one could construct an indefinitely long sequence of eigenvectors having the eigenvalues $\nu-1$, $\nu-2$, $\nu-3, \ldots$. But this would conflict with the fact, just shown above, that an eigenvalue cannot be negative. The contradiction can be avoided only if the sequence terminates with the value $\nu = 0$, since $a|0\rangle = 0$ is the zero vector and further applications of a will produce no more vectors.

From (6.14) it follows that

$$Na^\dagger|\nu\rangle = a^\dagger(N+1)|\nu\rangle = (\nu+1)a^\dagger|\nu\rangle\,.$$

The squared norm of the vector $a^\dagger|\nu\rangle$ is

$$(\langle\nu|a)(a^\dagger|\nu\rangle) = \langle\nu|(N+1)|\nu\rangle = (\nu+1)\langle\nu|\nu\rangle\,,$$

which never vanishes because $\nu \geq 0$. Thus $a^\dagger|\nu\rangle$ is an eigenvector of N with eigenvalue $\nu+1$. By repeatedly applying the operator $a^\dagger$, one can construct an unlimited sequence of eigenvectors, each having an eigenvalue one unit greater than that of its predecessor. The sequence begins with the eigenvalue $\nu = 0$. Therefore the spectrum of N consists of the nonnegative integers, $\nu = n$.

The orthonormal eigenvectors of N will be denoted as $|n\rangle$:

$$N|n\rangle = n|n\rangle\,, \quad n = 0, 1, 2, \ldots\,. \tag{6.15}$$

We have already shown that $a^\dagger|n\rangle$ is proportional to $|n+1\rangle$, and so we may write $a^\dagger|n\rangle = C_n|n+1\rangle$. The proportionality factor C_n can be obtained from the norm of this vector, which was calculated above:

$$|C_n|^2 = (\langle n|a)\,(a^\dagger|n\rangle) = n+1\,.$$

Hence $|C_n| = (n+1)^{1/2}$. The phase of the vector $|n+1\rangle$ is arbitrary because the vector is only defined by (6.15). Thus we are free to choose its phase so that C_n is real and positive, yielding

$$a^\dagger|n\rangle = (n+1)^{1/2}|n+1\rangle\,. \tag{6.16}$$

From this result it follows that

$$|n\rangle = (n!)^{-1/2}(a^\dagger)^n|0\rangle\,. \tag{6.17}$$

From (6.16) and the orthonormality of the eigenvectors, we obtain the matrix elements of $a^\dagger$,

$$\langle n'|a^\dagger|n\rangle = (n+1)^{1/2}\delta_{n',n+1}\,. \tag{6.18}$$

Because a is the adjoint of $a^\dagger$, its matrix elements must be the transpose of (6.18), and may be written as

$$\langle n'|a|n\rangle = n^{1/2}\delta_{n',n-1}\,. \tag{6.19}$$

When written as a matrix, Eq. (6.18) has its nonzero elements one space below the diagonal, and (6.19) has its nonzero elements one space above the diagonal. It follows from (6.19) that

$$\begin{aligned} a|n\rangle &= n^{1/2}|n-1\rangle\,, \quad (n\neq 0)\,, \\ a|0\rangle &= 0\,. \end{aligned} \tag{6.20}$$

Finally we note that the eigenvalues and eigenvectors of the harmonic oscillator Hamiltonian are

$$H|n\rangle = E_n|n\rangle\,,$$

with $E_n = \hbar\omega(n+\frac{1}{2})$, $n = 0, 1, 2, \ldots$. This confirms the assertion in (5.31) that energy transfer to and from a temporally periodic system will be quantized.

6.2 *Solution in Coordinate Representation*

If the eigenvalue equation $H|\psi\rangle = E|\psi\rangle$ for the Hamiltonian (6.1) is written in the coordinate representation, it becomes a differential equation,

$$\frac{-\hbar^2}{2M}\frac{d^2}{dx^2}\psi(x) + \frac{M\omega^2}{2}x^2\psi(x) = E\psi(x)\,. \tag{6.21}$$

The solution of this equation is treated in many standard references [Schiff (1968), Ch. 4, Sec. 13; Merzbacher (1970), Ch. 5], so we shall only treat it briefly.

We first introduce a dimensionless coordinate [as in (6.4)],

$$q = \left(\frac{M\omega}{\hbar}\right)^{1/2} x \tag{6.22}$$

and a dimensionless eigenvalue,

$$\lambda = \frac{2E}{\hbar\omega}. \tag{6.23}$$

If we write $\psi(x) = u(q)$, then (6.21) becomes

$$\frac{d^2u}{dq^2} + (\lambda - q^2)u = 0. \tag{6.24}$$

An estimate of the asymptotic behavior of $u(q)$ for large q can be obtained by neglecting λ compared with q^2 in the second term of (6.24). This yields two solutions, $e^{\frac{1}{2}q^2}$ and $e^{-\frac{1}{2}q^2}$. The first of these is unacceptable, because it diverges so severely as to be outside of both Hilbert space and the extended, or "rigged," Hilbert space (Sec. 1.4). Thus it seems appropriate to seek solutions of the form

$$u(q) = H(q)e^{-\frac{1}{2}q^2}. \tag{6.25}$$

[The traditional notation $H(q)$ for the function introduced here should not be confused with the Hamiltonian, also denoted as H.] Substitution of (6.25) into (6.24) yields

$$H'' - 2qH' + (\lambda - 1)H = 0, \tag{6.26}$$

where the prime denotes differentiation with respect to q.

It is shown in the theory of differential equations that if the equation is written so that the coefficient of the highest order derivative is unity, as in (6.26), then the solution may be singular only at the singular points of the coefficients of the lower order derivatives. Since there are no such singularities in the coefficients of (6.26), it follows that $H(q)$ can have no singularities for finite q, and a power series solution will have infinite radius of convergence. We therefore substitute the power series

$$H(q) = \sum_{n=0}^{\infty} a_n q^n \tag{6.27}$$

into (6.26), and equate the coefficient of each power of q to zero. This yields the recursion formula

$$a_{n+2} = \frac{2n+1-\lambda}{(n+2)(n+1)} a_n, \quad (n \geq 0). \tag{6.28}$$

(The cases of $n = 0$ and $n = 1$ must be treated separately, but it turns out that this formula holds for them too.)

If the series (6.27) does not terminate, then (6.28) yields the asymptotic ratio for the coefficients to be

$$\frac{a_{n+2}}{a_n} \xrightarrow[(n\to\infty)]{} \frac{2}{n}\,. \tag{6.29}$$

This is the same as the asymptotic ratio in the series for $q^k e^{q^2}$ with any positive value of k, and indeed the ratio (6.29) is characteristic of the exponential factor. Such behavior of $H(q)$ would yield an unacceptable divergence of $u(q)$ in (6.25) at large q. The only way that this unacceptable behavior can be avoided is for the series to *terminate*. If

$$\lambda = 2n + 1 \tag{6.30}$$

for some nonnegative integer n, then $H(q)$ will be a polynomial of degree n, and $u(q)$ will tend to zero at large q. Thus we have an eigenvalue condition for λ, and also for E through (6.23),

$$E_n = \hbar\omega\left(n + \frac{1}{2}\right), \qquad n = 0, 1, 2, \ldots\,. \tag{6.31}$$

This is precisely the same result obtained by an entirely different method in the previous section.

For future reference we record the normalized eigenfunctions, which are

$$\psi_n(x) = \left[\frac{\alpha}{\pi^{1/2}2^n n!}\right]^{1/2} H_n(\alpha x)e^{-\frac{1}{2}\alpha^2 x^2}\,. \tag{6.32}$$

Here $\alpha = (M\omega/\hbar)^{1/2}$, and $H_n(z)$ is the Hermite polynomial of degree n. These polynomials and their properties can be obtained from a generating function,

$$\exp(-s^2 + 2sz) = \sum_{n=0}^{\infty} \frac{H_n(z)}{n!}s^n\,. \tag{6.33}$$

Derivations of these results are contained in standard references already cited.

The methods of this section and of the previous section are very different, and yet they lead to precisely the same energy eigenvalue spectrum. It is interesting to inquire why this is so. Part of the equivalence is easy to trace. The differential equation (6.21) is just the eigenvalue equation for the Hamiltonian (6.1) with the operators P and Q expressed in coordinate representation as $-i\hbar d/dx$ and x, respectively. The commutation relation (6.2) is, of course, obeyed in this representation. But it is well known that a differential equation such as (6.21) possesses solutions for *all complex values* of the parameter E.

The eigenvalue restriction came about only by imposing the boundary condition that $\psi(x)$ should tend to zero as $|x| \to \infty$. But there is no direct reference to a boundary condition in Sec. 6.1. However equivalent information is contained in the requirement $P = P^\dagger$. The condition for the operator $-i\hbar d/dx$ to be equal to its Hermitian conjugate is exhibited in (1.31). It is the vanishing of the last term of (1.31), which arose from integration by parts. In unbounded space, it is just the condition $\psi(x) \to 0$ as $|x| \to \infty$ that ensures the vanishing of that term. Thus, in spite of appearances to the contrary, the two methods are closely related.

6.3 *Solution in H Representation*

In the method of Sec. 6.2 the properties of the position operator Q were supposed to be known. By expressing the Hamiltonian H in the representation in which Q is diagonal, we then calculated the eigenvalues of H. The eigenfunctions, $\psi_n(x) = \langle x|n\rangle$, may be thought of as the expansion coefficients of the abstract eigenvectors of H, $|n\rangle$, in terms of the eigenvectors of Q.

The spectrum of H is independently known from the results of Sec. 6.1. So instead of calculating the eigenvalues of H in the representation that diagonalizes Q, was done in Sec. 6.2, we could just as well calculate the eigenvalues of Q in the representation that diagonalizes H. This unusual route will be followed here.

Using (6.4), (6.8), and (6.9), one can express the position operator as

$$Q = \left(\frac{\hbar}{2M\omega}\right)^{1/2} (a + a^\dagger)\,. \tag{6.34}$$

Its matrix elements in the basis formed by the eigenvectors of H, and of $N = a^\dagger a$, can then be obtained from (6.18) and (6.19). Thus

$$\langle n'|Q|n\rangle = \left[\frac{\hbar}{2M\omega}\right]^{1/2} \begin{bmatrix} 0 & \sqrt{1} & 0 & 0 & 0 & \cdots \\ \sqrt{1} & 0 & \sqrt{2} & 0 & 0 & \cdots \\ 0 & \sqrt{2} & 0 & \sqrt{3} & 0 & \cdots \\ 0 & 0 & \sqrt{3} & 0 & \sqrt{4} & \cdots \\ 0 & 0 & 0 & \sqrt{4} & 0 & \cdots \\ \vdots & \vdots & \vdots & \vdots & \vdots & \end{bmatrix}. \tag{6.35}$$

The eigenvalue equation, $Q|x\rangle = x|x\rangle$, now takes the form

$$\sum_n \langle n'|Q|n\rangle\langle n|x\rangle = x\langle n'|x\rangle\,,$$

which upon the use of (6.35) becomes

$$\left(\frac{\hbar}{2M\omega}\right)^{1/2}\left\{\sqrt{n'}\langle n'-1|x\rangle + \sqrt{(n'+1)}\langle n'+1|x\rangle\right\} = x\langle n'|x\rangle\,. \qquad (6.36)$$

This equation may be solved recursively beginning with an arbitrary value for $\langle 0|x\rangle$, from which we can calculate $\langle 1|x\rangle$, and then $\langle 2|x\rangle$, etc. Finally the set $\{\langle n|x\rangle\}$, for all n but fixed x, may be multiplied by a factor so as to achieve some conventional normalization. This recursive solution of (6.36) works for all x, so the eigenvalue spectrum of the operator Q is continuous from $-\infty$ to $+\infty$. (Reality of x is required because of the assumed Hermitian character of Q.)

In this method, we calculate $\langle n|x\rangle$ as a function of n for fixed x, whereas in Sec. 6.2 we calculated $\psi_n(x) = \langle n|x\rangle^*$ as a function of x for fixed n. It is not immediately obvious that the two are in agreement. To demonstrate their agreement, we take the result (6.32) of Sec. 6.2 and substitute it into (6.36). For present purposes (6.32) can be written as

$$\langle n|x\rangle = c(x)(2^n n!)^{-1/2} H_n(\alpha x)\,,$$

where $c(x)$ includes all factors that do not depend on n. Substituting this into (6.36) yields

$$nH_{n-1}(\alpha x) + \frac{1}{2}H_{n+1}(\alpha x) = \alpha x H_n(\alpha x)\,.$$

This is a standard identity satisfied by the Hermite polynomials, and so the results of the two methods are indeed consistent.

Problems

6.1 Calculate the position and momentum operators for the harmonic oscillator in the Heisenberg picture, $Q_H(t)$ and $P_H(t)$.

6.2 Calculate the matrices for the position and momentum operators, Q and P, in the basis formed by the energy eigenvectors of the harmonic oscillator. Square these matrices, and verify that the matrix sum $(1/2M)P^2 + (M\omega^2/2)Q^2$ is diagonal.

6.3 (a) For *finite-dimensional* matrices A and B, show that $\text{Tr}[A,B] = 0$.
(b) *Paradox.* From this result it would seem to follow, by taking the trace of the commutator $[Q,P] = i\hbar$, that one must have $\hbar = 0$. Use the *infinite-dimensional* matrices for Q and P, found in the previous problem, to calculate the matrices QP and PQ, and hence explain in detail why the paradoxical conclusion $\hbar = 0$ is not valid.

6.4 Express the raising operator $a^\dagger$ as a differential operator in coordinate representation. Taking the ground state function $\psi_0(x)$ from (6.32) as given, use $a^\dagger$ to calculate $\psi_1(x)$, $\psi_2(x)$, and $\psi_3(x)$.

6.5 Write the eigenvalue equation $H|\psi\rangle = E|\psi\rangle$ in momentum representation, and calculate the corresponding eigenfunctions, $\langle p|\psi_n\rangle$.

6.6 The Hamiltonian of a three-dimensional isotropic harmonic oscillator is $H = (1/2M)\mathbf{P}\cdot\mathbf{P} + (M\omega^2/2)\mathbf{Q}\cdot\mathbf{Q}$. Solve the eigenvalue equation $H\Psi(x) = E\Psi(x)$ in rectangular coordinates.

6.7 Solve the eigenvalue equation for the three-dimensional isotropic harmonic oscillator in spherical coordinates. Show that the eigenvalues and their degeneracies are the same as was obtained in rectangular coordinates in the previous problem.

6.8 (a) For any complex number z, a vector $|z\rangle$ may be defined by the following expansion in harmonic oscillator energy eigenvectors:

$$|z\rangle = \exp\left(-\frac{1}{2}|z|^2\right)\sum_{n=0}^{\infty}\frac{z^n}{(n!)^{1/2}}|n\rangle\,.$$

Use Eq. (6.20) to show that $|z\rangle$ is a right eigenvector of the lowering operator a.

(b) Show that the raising operator $a^\dagger$ has no right eigenvectors. (*Note*: The vector $|z\rangle$ is called a *coherent state*. It will play an important role in Ch. 19.)

Chapter 7

Angular Momentum

In Ch. 3 we showed that the generators of space–time symmetry operations are related to fundamental dynamical variables. In particular, the generators of rotations are the angular momentum operators (in units of $\hbar$). Rotational symmetry plays a very important role in many physical problems, especially in atomic and nuclear physics. Many useful conclusions can be deduced from the transformation properties of various observables under rotations. Therefore it is useful to develop the theory of angular momentum and rotations in considerable detail.

7.1 *Eigenvalues and Matrix Elements*

In Ch. 3 the commutation relations among the angular momentum operators were found to be

$$[J_x, J_y] = i\hbar J_z \,, \tag{7.1a}$$

$$[J_y, J_z] = i\hbar J_x \,, \tag{7.1b}$$

$$[J_z, J_x] = i\hbar J_y \,. \tag{7.1c}$$

These three operators are self-adjoint:

$$J_x{}^\dagger = J_x \,, \; J_y{}^\dagger = J_y \,, \; J_z{}^\dagger = J_z \,. \tag{7.2}$$

The eigenvalue spectrum of the angular momentum operators can be determined using *only* the above equations.

We first introduce the operator $J^2 = \mathbf{J}\cdot\mathbf{J} = J_x{}^2 + J_y{}^2 + J_z{}^2$. It is easily shown, by means of (7.1), that $[J^2, \mathbf{J}] = 0$. Thus there exists a complete set of common eigenvectors of J^2 and any one component of $\mathbf{J}$. We shall seek the solutions of the pair of eigenvalue equations:

$$J^2|\beta, m\rangle = \hbar^2\beta|\beta, m\rangle \,, \tag{7.3a}$$

$$J_z|\beta, m\rangle = \hbar m|\beta, m\rangle \,. \tag{7.3b}$$

The factors of $\hbar$ have been introduced so that β and m will be dimensionless. Of course the eigenvalue spectrum would be the same if any component of **J** were used instead of J_z, since one component can be transformed into another by a rotation.

From the definition of J^2 we obtain

$$\langle\beta,m|J^2|\beta,m\rangle = \langle\beta,m|{J_x}^2|\beta,m\rangle + \langle\beta,m|{J_y}^2|\beta,m\rangle + \langle\beta,m|{J_z}^2|\beta,m\rangle\,.$$

Now $\langle\beta,m|{J_x}^2|\beta,m\rangle = (\langle\beta,m|{J_x}^\dagger)(J_x|\beta,m\rangle) \geq 0$, since the inner product of a vector with itself cannot be negative. Using a similar condition for ${J_y}^2$, and the eigenvalue conditions (7.3), we obtain

$$\beta \geq m^2\,. \tag{7.4}$$

Thus for a fixed value of β there must be maximum and minimum values for m.

We next introduce two more operators,

$$J_+ = J_x + iJ_y\,, \tag{7.5a}$$

$$J_- = J_x - iJ_y\,. \tag{7.5b}$$

They satisfy the commutation relations

$$[J_z, J_+] = \hbar J_+\,, \tag{7.6a}$$

$$[J_z, J_-] = -\hbar J_-\,, \tag{7.6b}$$

$$[J_+, J_-] = 2\hbar J_z\,, \tag{7.6c}$$

the relations (7.6) and (7.1) being equivalent.

Using (7.6a) we obtain

$$\begin{aligned} J_z J_+|\beta,m\rangle &= (J_+J_z + \hbar J_+)|\beta,m\rangle \\ &= \hbar(m+1)J_+|\beta,m\rangle\,. \end{aligned} \tag{7.7}$$

Therefore, either $J_+|\beta,m\rangle$ is an eigenvector of J_z with the raised eigenvalue $\hbar(m+1)$, or $J_+|\beta,m\rangle = 0$. Now for fixed β there is a maximum value of m, which we shall denote as j. It must be the case that

$$J_+|\beta,j\rangle = 0\,, \tag{7.8}$$

since if it were not zero it would be an eigenvector with eigenvalue $\hbar(j+1)$. But that exceeds the maximum, and so is impossible. The precise relation between β and j can be obtained if we multiply (7.8) by J_- and use

$$\begin{aligned} J_-J_+ &= {J_x}^2 + {J_y}^2 + i(J_xJ_y - J_yJ_x) \\ &= J^2 - {J_z}^2 - \hbar J_z \,. \end{aligned} \tag{7.9}$$

Thus $0 = J_-J_+|\beta, j\rangle = \hbar^2(\beta - j^2 - j)|\beta, j\rangle$, and since the vector $|\beta, j\rangle$ does not vanish we must have

$$\beta = j(j+1)\,. \tag{7.10}$$

By a similar argument using (7.6b), we obtain

$$\begin{aligned} J_zJ_-|\beta, m\rangle &= (J_-J_z - \hbar J_-)|\beta, m\rangle \\ &= \hbar(m-1)J_-|\beta, m\rangle\,. \end{aligned} \tag{7.11}$$

Therefore, either $J_-|\beta, m\rangle$ is an eigenvector of J_z with the lowered eigenvalue $\hbar(m-1)$, or if $m = k$ (the minimum possible value of m) then $J_-|\beta, k\rangle = 0$. In the latter case we have

$$0 = J_+J_-|\beta, k\rangle = \hbar^2(\beta - k^2 + k)|\beta, k\rangle\,.$$

Hence we must have $\beta + k(-k+1) = 0$, which, in the light of (7.10), implies that $k = -j$.

We have thus shown the existence of a set of eigenvectors corresponding to integer spaced m values in the range $-j \le m \le j$. Since the difference between the maximum value j and the minimum value $-j$ must be an integer, it follows that $j =$ integer/2. The allowed values of j and m are

$$\begin{aligned} j &= 0 \quad\,, \quad m = 0\,, \\ j &= 1/2\,, \quad m = 1/2, -1/2\,, \\ j &= 1 \quad\,, \quad m = 1, 0, -1\,, \\ j &= 3/2\,, \quad m = 3/2, 1/2, -1/2, -3/2\,, \\ &\quad \text{etc.} \end{aligned}$$

For each value of j there are $2j+1$ values of m.

Henceforth we shall *adopt the common and more convenient notation* of labeling the eigenvectors by j instead of by $\beta = j(j+1)$. Thus the vector that was previously denoted as $|\beta, m\rangle$ will now be denoted as $|j, m\rangle$.

We have shown above that

$$J_+|j,m\rangle = C\,|j,m+1\rangle\,, \tag{7.12}$$

where C is some numerical factor that may depend upon j and m. The value of $|C|$ can be determined by calculating the norm of (7.12),

$$(\langle j,m|J_-)(J_+|j,m\rangle) = |C|^2\,.$$

With the help of (7.9), this yields

$$|C|^2 = \hbar^2\{j(j+1) - m(m+1)\}\,. \tag{7.13}$$

Notice that this expression vanishes for $m = j$, as it must according to a previous result.

The phase of C is arbitrary because Eqs. (7.3), which define the eigenvectors, do not determine their phases. It is convenient to *choose the phase* of C to be real positive, thereby fixing the relative phases of $|j,m\rangle$ and $|j,m+1\rangle$ in (7.12). Thus we have

$$\begin{aligned} J_+|j,m\rangle &= \hbar\,\{j(j+1) - m(m+1)\}^{1/2}\,|j,m+1\rangle \\ &= \hbar\,\{(j+m+1)(j-m)\}^{1/2}\,|j,m+1\rangle\,. \end{aligned} \tag{7.14}$$

Applying J_- to (7.12) and using (7.9), we obtain

$$C^2|j,m\rangle = CJ_-|j,m+1\rangle\,.$$

Replacing the dummy variable m by $m-1$ then yields

$$\begin{aligned} J_-|j,m\rangle &= \hbar\,\{j(j+1) - m(m-1)\}^{1/2}\,|j,m-1\rangle \\ &= \hbar\,\{(j-m+1)(j+m)\}^{1/2}\,|j,m-1\rangle\,. \end{aligned} \tag{7.15}$$

Explicit *matrix representations* for any component of $\mathbf{J}$ can now be constructed, using as basis vectors the eigenvectors of (7.3). The matrix for J_z is clearly diagonal, $\langle j',m'|J_z|j,m\rangle = \hbar m\,\delta_{j',j}\,\delta_{m',m}$. The matrices for all components of $\mathbf{J}$ must be diagonal in (j',j) since J^2 commutes with all components of $\mathbf{J}$. To prove this we consider

$$\langle j',m'|J^2\mathbf{J}|j,m\rangle = \langle j',m'|\mathbf{J}\,J^2|j,m\rangle\,,$$

$$\hbar^2 j'(j'+1)\langle j',m'|\mathbf{J}|j,m\rangle = \langle j',m'|\mathbf{J}|j,m\rangle\hbar^2 j(j+1)\,.$$

Hence $\langle j', m'|\mathbf{J}|j, m\rangle = 0$ if $j' \neq j$. (A similar argument clearly holds for any other operator that commutes with J^2.) The matrices for $J_x = (J_+ + J_-)/2$ and $J_y = (J_+ - J_-)/2i$ are most easily obtained from those for J_+ and J_-. The matrix $\langle j', m'|J_+|j, m\rangle/\hbar$ is directly obtainable from (7.14), and it has the form

$$
\begin{array}{ll|c|cc|ccc|cccc}
 & & j=0; & \tfrac{1}{2}; & & 1; & & & \tfrac{3}{2} & & & \\
 & & m=0; & \tfrac{1}{2}, & -\tfrac{1}{2}; & 1, & 0, & -1; & \tfrac{3}{2}, & \tfrac{1}{2}, & -\tfrac{1}{2}, & -\tfrac{3}{2} \\
\hline
j'=0,\ m'= & 0 & 0 & & & & & & & & & \\
\hline
j'=\tfrac{1}{2},\ m'= & \tfrac{1}{2} & & 0 & 1 & & & & & & & \\
 & -\tfrac{1}{2} & & 0 & 0 & & & & & & & \\
\hline
j'=1,\ m'= & 1 & & & & 0 & \sqrt{2} & 0 & & & & \\
 & 0 & & & & 0 & 0 & \sqrt{2} & & & & \\
 & -1 & & & & 0 & 0 & 0 & & & & \\
\hline
j'=\tfrac{3}{2},\ m'= & \tfrac{3}{2} & & & & & & & 0 & \sqrt{3} & 0 & 0 \\
 & \tfrac{1}{2} & & & & & & & 0 & 0 & \sqrt{4} & 0 \\
 & -\tfrac{1}{2} & & & & & & & 0 & 0 & 0 & \sqrt{3} \\
 & -\tfrac{3}{2} & & & & & & & 0 & 0 & 0 & 0
\end{array}
\qquad (7.16)
$$

All nonvanishing matrix elements are one position removed from the diagonal. Since $J_- = (J_+)^\dagger$, the matrix for J_- is the transpose of (7.16).

7.2 *Explicit Form of the Angular Momentum Operators*

The angular momentum operators were introduced in Secs. 3.3 and 3.4 as generators of rotations, and their fundamental commutation relations (7.1) derive from their role as rotation generators. The unitary operator corresponding to a rotation through angle θ about an axis parallel to the unit vector $\hat{\mathbf{n}}$ is

$$\mathbf{R}_n(\theta) = e^{i\theta\hat{\mathbf{n}}\cdot\mathbf{J}/\hbar}\,. \qquad (7.17)$$

By examining this rotational transformation in more detail, it is possible to derive a more explicit form for the angular momentum operators.

Case (i): A one-component state function

Let $\Psi(\mathbf{x})$ be a one-component state function in coordinate representation. When it is subjected to a rotation it is transformed into

$$\mathbf{R}\Psi(\mathbf{x}) = \Psi(R^{-1}\mathbf{x})\,, \tag{7.18}$$

where $\mathbf{R}$ is an operator of the form (7.17) and R^{-1} is the inverse of a 3×3 coordinate rotation matrix. The specific form of the matrix R for rotations about each of the three Cartesian axes is given in Sec. 3.3. [Equation (7.18) is a special case of (3.8), written in a slightly different notation.]

For a rotation through angle ε about the z axis, (7.18) becomes

$$\mathbf{R}_z(\varepsilon)\Psi(x,y,z) = \Psi(x\ \cos\varepsilon + y\ \sin\varepsilon,\quad -x\ \sin\varepsilon + y\ \cos\varepsilon, z)\,.$$

If ε is an infinitesimal angle we may expand to the first order, obtaining

$$\mathbf{R}_z(\varepsilon)\Psi(x,y,z) = \Psi(x,y,z) + \varepsilon\left(y\frac{\partial\Psi}{\partial x} - x\frac{\partial\Psi}{\partial y}\right).$$

Comparison of this equation with a first order expansion of (7.17), $\mathbf{R}_z(\varepsilon) = 1 - i\varepsilon J_z/\hbar$, leads to the identification of J_z with $-i\hbar(x\partial/\partial y - y\partial/\partial x)$. This is, of course, just the z component of the orbital angular momentum operator, $\mathbf{L} = \mathbf{Q} \times \mathbf{P}$, with $\mathbf{Q}$ and $\mathbf{P}$ being expressed in the coordinate representation, (4.2) and (4.3).

Case (ii): A multicomponent state function

The rotational transformation of a multicomponent state function can be more complicated than (7.18). In the most general case we may have a transformation of the form

$$\mathbf{R}\begin{bmatrix}\Psi_1(\mathbf{x})\\ \Psi_2(\mathbf{x})\\ \vdots\end{bmatrix} = D\begin{bmatrix}\Psi_1(R^{-1}\mathbf{x})\\ \Psi_2(R^{-1}\mathbf{x})\\ \vdots\end{bmatrix}. \tag{7.19}$$

In addition to the coordinate transformation $R^{-1}\mathbf{x}$, we may have a matrix D that operates on the internal degrees of freedom; that is, it forms linear combinations of components of the state function. Thus the general form of the unitary operator (7.17) will be

$$\mathbf{R}_n(\theta) = e^{i\theta\hat{\mathbf{n}}\cdot\mathbf{L}/\hbar}D_n(\theta)\,. \tag{7.20}$$

The two factors commute because the first acts only on the coordinate $\mathbf{x}$ and the second acts only on the components of the column vector.

The matrix D must be unitary, and so it can be written as

$$D_n(\theta) = e^{i\theta\hat{\mathbf{n}}\cdot\mathbf{S}/\hbar}\,, \tag{7.21}$$

where $\mathbf{S} = (S_x, S_y, S_z)$ are Hermitian matrices. Substituting (7.21) into (7.20) and comparing with (7.17), we see that the angular momentum operator $\mathbf{J}$ has the form

$$\mathbf{J} = \mathbf{L} + \mathbf{S}\,, \tag{7.22}$$

with $\mathbf{L} = \mathbf{Q} \times \mathbf{P}$ and $[L_\alpha, S_\beta] = 0$ $(\alpha, \beta = x, y, z)$. In the particular representation used in this section, we have $\mathbf{L} = -i\hbar\mathbf{x} \times \boldsymbol{\nabla}$, and the components of $\mathbf{S}$ are discrete matrices. The operators $\mathbf{L}$ and $\mathbf{S}$ are called the *orbital* and *spin* parts of the angular momentum. It can be shown by direct calculation (Problem 3.6) that the components of $\mathbf{L}$ satisfy the same commutation relations, (7.1), that are satisfied by $\mathbf{J}$. Therefore it follows that the components of $\mathbf{S}$ must also satisfy (7.1).

The *orbital* part of the angular momentum is the angular momentum due to the motion of the center of mass of the object relative to the origin of coordinates. The *spin* may be identified as the angular momentum that remains when the center of mass is at rest.

7.3 *Orbital Angular Momentum*

The orbital angular momentum operator is

$$\mathbf{L} = \mathbf{Q} \times \mathbf{P}\,, \tag{7.23}$$

where $\mathbf{Q}$ is the position operator and $\mathbf{P}$ is the momentum operator. In the coordinate representation, the action of $\mathbf{Q}$ is to multiply by the coordinate vector, and $\mathbf{P}$ has the form $-i\hbar\boldsymbol{\nabla}$.

It is frequently desirable to express these operators in spherical polar coordinates (r, θ, ϕ), whose relation to the rectangular coordinates (x, y, z) is

$$x = r\ \sin\theta\ \cos\phi\,, \tag{7.24a}$$

$$y = r\ \sin\theta\ \sin\phi\,, \tag{7.24b}$$

$$z = r\ \cos\theta\,, \tag{7.24c}$$

as shown in Fig. 7.1.

The form of the gradient operator in spherical coordinates is

$$\boldsymbol{\nabla} = \hat{\mathbf{e}}_r \frac{\partial}{\partial r} + \hat{\mathbf{e}}_\theta \frac{1}{r}\frac{\partial}{\partial\theta} + \hat{\mathbf{e}}_\phi \frac{1}{r\ \sin\theta}\frac{\partial}{\partial\phi}\,, \tag{7.25}$$

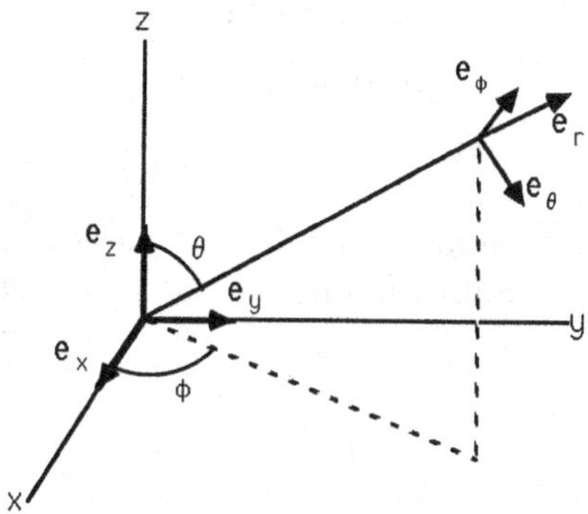

Fig. 7.1 Rectangular and spherical coordinates, showing unit vectors in both systems [see Eq. (7.24)].

where the unit vectors of the spherical coordinate system are given by

$$\hat{\mathbf{e}}_r = \sin\theta\ \cos\phi\ \hat{\mathbf{e}}_x + \sin\theta\ \sin\phi\ \hat{\mathbf{e}}_y + \cos\theta\ \hat{\mathbf{e}}_z\,, \tag{7.26a}$$

$$\hat{\mathbf{e}}_\theta = \cos\theta\ \cos\phi\ \hat{\mathbf{e}}_x + \cos\theta\ \sin\phi\ \hat{\mathbf{e}}_y - \sin\theta\ \hat{\mathbf{e}}_z\,, \tag{7.26b}$$

$$\hat{\mathbf{e}}_\phi = -\sin\phi\ \hat{\mathbf{e}}_x + \cos\phi\ \hat{\mathbf{e}}_y\,, \tag{7.26c}$$

in terms of the unit vectors of the rectangular system. The orbital angular momentum operator then has the form

$$\begin{aligned}\mathbf{L} &= r\hat{\mathbf{e}}_r \times (-i\hbar\boldsymbol{\nabla}) \\ &= (-i\hbar)\left[\hat{\mathbf{e}}_\phi \frac{\partial}{\partial\theta} - \hat{\mathbf{e}}_\theta \frac{1}{\sin\theta}\frac{\partial}{\partial\phi}\right].\end{aligned} \tag{7.27}$$

As in the calculation of Sec. 7.1, we shall seek the eigenvectors of the two commuting operators $L^2 = \mathbf{L}\cdot\mathbf{L}$ and L_z, where

$$L_z = \hat{\mathbf{e}}_z \cdot \mathbf{L} = -i\hbar\frac{\partial}{\partial\phi}. \tag{7.28}$$

In evaluating $\mathbf{L}\cdot\mathbf{L}$ we must remember that the unit vectors $\hat{\mathbf{e}}_\theta$ and $\hat{\mathbf{e}}_\phi$ are not constant, and so the action of the differential operators on them must be included. The result can be written as

$$L^2 = \mathbf{L}\cdot\mathbf{L} = -\hbar^2\left[\frac{1}{\sin\theta}\frac{\partial}{\partial\theta}\left(\sin\theta\frac{\partial}{\partial\theta}\right) + \frac{1}{(\sin\theta)^2}\frac{\partial^2}{\partial\phi^2}\right]. \tag{7.29}$$

We must now solve the two coupled differential equations,

$$L_z Y(\theta,\phi) = \hbar m Y(\theta,\phi)\,, \tag{7.30}$$

$$L^2 Y(\theta,\phi) = \hbar^2 \ell(\ell+1) Y(\theta,\phi)\,. \tag{7.31}$$

(No assumption has been made about the values of ℓ or m, which need not be integers at this point.) Substitution of (7.28) into (7.30) yields $\partial Y/\partial\phi = imY$, and hence

$$\frac{\partial^2 Y}{\partial \phi^2} = -m^2 Y\,. \tag{7.32}$$

This allows us to simplify (7.31) to

$$\sin\theta \frac{\partial}{\partial\theta}\left(\sin\theta\frac{\partial}{\partial\theta}\right) Y + \{(\sin\theta)^2 \ell(\ell+1) - m^2\} Y = 0\,. \tag{7.33}$$

The θ and ϕ dependences, which were coupled in (7.31), are completely separated in (7.32) and (7.33), so it is clear that $Y(\theta,\phi)$ may be taken to be a product of a function of ϕ satisfying (7.32) and a function of θ satisfying (7.33). The solution of (7.32) is obviously $e^{im\phi}$. Equation (7.33) is equivalent to the *associated Legendre equation*, whose solution will be denoted as ${P_\ell}^m(\cos\theta)$. (The standard form of the Legendre equation is obtained by changing from the variable θ to $u = \cos\theta$.) Thus, apart from normalization, we have $Y(\theta,\phi) = e^{im\phi}{P_\ell}^m(\cos\theta)$.

So far nothing has been said about the values of ℓ and m. As is well known, the differential equations (7.32) and (7.33) possess solutions for *all values* of the parameters ℓ and m. Eigenvalue restrictions come about only from the imposition of *boundary conditions*. If we assume that the solution must be *single-valued* under rotation — that is, we assume that $Y(\theta,\phi+2\pi) = Y(\theta,\phi)$ — then it will follow that m must be an integer. If we further assume that it must be *nonsingular* at the singular points of Eq. (7.33), $\theta = 0$ and $\theta = \pi$, then from the standard theory of the Legendre equation it will follow that ℓ must be a nonnegative integer in the range $\ell \geq |m|$. The normalized solutions that result from these assumptions are the well-known *spherical harmonics*,

$${Y_\ell}^m(\theta,\phi) = (-1)^{(m+|m|)/2}\left[\frac{(2\ell+1)(\ell-|m|)!}{4\pi(\ell+|m|)!}\right]^{1/2} e^{im\phi}{P_\ell}^{|m|}(\cos\theta)\,. \tag{7.34}$$

Here ${P_\ell}^m(u)$ is the associated Legendre function. It is derivable from the Legendre polynomial,

$$P_\ell(u) = (2^\ell \ell!)^{-1}\left(\frac{d}{du}\right)^\ell (u^2-1)^\ell\,, \tag{7.35}$$

by the relation

$$P_\ell{}^m(u) = (1-u^2)^{m/2}\left(\frac{d}{du}\right)^m P_\ell(u)\,. \tag{7.36}$$

The arbitrary phase of $Y_\ell{}^m(\theta,\phi)$ has been chosen so that

$$Y_\ell{}^{-m}(\theta,\phi) = (-1)^m\{Y_\ell{}^m(\theta,\phi)\}^*\,, \tag{7.37}$$

which allows it to satisfy (7.14) and (7.15). The spherical harmonics form an orthonormal set:

$$\int_0^{2\pi}\int_0^{\pi}\{Y_\ell{}^m(\theta,\phi)\}^* Y_\ell{}^m(\theta,\phi)\sin\theta\, d\theta\, d\phi = \delta_{\ell,\ell'}\delta_{m,m'}\,. \tag{7.38}$$

The assumptions of *single-valuedness* and *nonsingularity* can be justified in a classical field theory, such as electromagnetism, in which the field is an observable physical quantity. But in quantum theory the state function Ψ does not have such direct physical significance, and the classical boundary conditions cannot be so readily justified. *Why should* Ψ *be single-valued under rotation*? Physical significance is attached, not to Ψ itself, but to quantities such as $\langle\Psi|A|\Psi\rangle$, and these will be unchanged by a 2π rotation even if m is a half-integer and Ψ changes sign. *Why should* Ψ *be nonsingular*? It is clearly desirable for the integral of $|\Psi|^2$ to be integrable so that the total probability can be normalized to one. But $Y = (\sin\theta)^{1/2}e^{i\phi/2}$, which is everywhere finite, satisfies (7.23) and (7.33) for $\ell = m = \frac{1}{2}$. Is it therefore to be admitted as as eigenfunction of orbital angular momentum? It is difficult to give an adequate justification for the conventional boundary conditions in this quantum-mechanical setting.

The orbital angular momentum eigenvalues. The orbital angular momentum operators are Hermitian and satisfy the commutation relation $[L_\alpha, L_\beta] = i\hbar\varepsilon_{\alpha\beta\gamma}L_\gamma$, just as do the components of the total angular momentum operator $\mathbf{J}$. It was shown in Sec. 7.1 that this information by itself implies that an eigenvalue of a component of the angular momentum operator must be $\hbar m$, with m being either an integer or a half-integer. Any further restriction of the orbital angular momentum eigenvalues to only the integer values must come from a special property of the orbital operators that is not possessed by the general angular momentum operators. This special property can only be the relation (7.23) of the orbital angular momentum to the position and momentum operators, $\mathbf{L} = \mathbf{Q}\times\mathbf{P}$, and the commutation relation satisfied by position and momentum.

We shall seek the eigenvalues of the orbital angular momentum operator

$$L_z = Q_x P_y - Q_y P_x \,, \tag{7.39}$$

and shall use only the commutation relations

$$[Q_x, Q_y] = [P_x, P_y] = 0 \,, \quad [Q_\alpha, P_\beta] = i\hbar\delta_{\alpha,\beta} \tag{7.40}$$

and the Hermitian nature of the operators.

For convenience, we shall temporarily adopt a system of units in which Q and P are dimensionless and $\hbar = 1$. We introduce four new operators:

$$q_1 = \frac{Q_x + P_y}{\sqrt{2}} \,, \tag{7.41a}$$

$$q_2 = \frac{Q_x - P_y}{\sqrt{2}} \,, \tag{7.41b}$$

$$p_1 = \frac{P_x - Q_y}{\sqrt{2}} \,, \tag{7.41c}$$

$$p_2 = \frac{P_x + Q_y}{\sqrt{2}} \,. \tag{7.41d}$$

It can be readily verified that

$$[q_1, q_2] = [p_1, p_2] = 0 \,, \quad [q_\alpha, p_\beta] = i\delta_{\alpha,\beta} \,. \tag{7.42}$$

In terms of these new operators, Eq. (7.39) becomes

$$L_z = \frac{1}{2}({p_1}^2 + {q_1}^2) - \frac{1}{2}({p_2}^2 + {q_2}^2) \,. \tag{7.43}$$

Comparing this expression with (6.1), we see that it is the difference of two independent harmonic oscillator Hamiltonians, each having mass $M = 1$ and angular frequency $\omega = 1$. Since the two terms commute, the eigenvalues of L_z are just the differences between the eigenvalues of these two terms. The eigenvalue spectrum of an operator of the form $\frac{1}{2}({p_1}^2 + {q_1}^2)$ was calculated in Sec. 6.1 using only the equivalent of (7.42). From that result we infer that the eigenvalues of (7.43) are equal to

$$\left(n_1 + \frac{1}{2}\right) - \left(n_2 + \frac{1}{2}\right) = n_1 - n_2 \,,$$

where n_1 and n_2 are nonnegative integers. Thus we have shown, directly from the properties of the position and momentum operators, that the orbital angular momentum eigenvalues must be *integer* multiples of $\hbar$, and that half-integer multiples cannot occur. This approach avoids any problematic discussion of boundary conditions on the state function Ψ; it could, however, be regarded as an indirect justification for the conventional boundary conditions that lead to the same result.

7.4 *Spin*

The components of the spin angular momentum$\mathbf{S}$ obey the general angular momentum commutation relations, $[S_x, S_y] = i\hbar S_z$, etc., and so the analysis of Sec. 7.1 applies. The eigenvalue equations for $S^2 = \mathbf{S}\cdot\mathbf{S}$ and S_z,

$$S^2|s,m\rangle = \hbar^2 s(s+1)|s,m\rangle\,, \quad S_z|s,m\rangle = \hbar m|s,m\rangle\,, \tag{7.44}$$

have solutions for $m = s, s-1, \ldots, -s$, with s being any nonnegative integer or half-integer. Because a particular species of particle is characterized by a set of quantum numbers that includes the value of its spin s, it is often sufficient to treat the spin operators S_x, S_y, and S_z as acting on the space of dimension $2s+1$ that is spanned by the eigenvectors of (7.44) for a fixed value of s. We shall treat the most common cases in detail.

Case (i): $s = 1/2$. In this case it is customary to write $\mathbf{S} = \frac{1}{2}\hbar\boldsymbol{\sigma}$, where $\boldsymbol{\sigma} = (\sigma_x, \sigma_y, \sigma_z)$ are called the *Pauli spin operators.* Explicit matrix representations, in the basis formed by the eigenvectors of (7.44), can be deduced for these operators from the 2×2 block of (7.16) and the analogous matrices for other angular momentum components:

$$\sigma_x = \begin{bmatrix} 0 & 1 \\ 1 & 0 \end{bmatrix}, \quad \sigma_y = \begin{bmatrix} 0 & -i \\ i & 0 \end{bmatrix}, \quad \sigma_z = \begin{bmatrix} 1 & 0 \\ 0 & -1 \end{bmatrix}. \tag{7.45}$$

The Pauli spin operators satisfy several important relations:

$${\sigma_x}^2 = {\sigma_y}^2 = {\sigma_z}^2 = \mathbf{1} \tag{7.46}$$

(here $\mathbf{1}$ denotes the identity matrix),

$$\sigma_x\sigma_y = -\sigma_y\sigma_x = i\sigma_z\,, \quad \sigma_y\sigma_z = -\sigma_z\sigma_y = i\sigma_x\,, \quad \sigma_z\sigma_x = -\sigma_x\sigma_z = i\sigma_y\,. \tag{7.47}$$

These relations are operator equalities, which do not depend upon the use of the particular matrix representation (7.45).

The operator corresponding to the component of spin in the direction $\hat{\mathbf{n}} = (\sin\theta\ \cos\phi,\ \sin\theta\ \sin\phi,\ \cos\theta)$ is $\hat{\mathbf{n}}\cdot\mathbf{S} = \frac{1}{2}\hbar\mathbf{n}\cdot\boldsymbol{\sigma}$, with

$$\hat{\mathbf{n}}\cdot\boldsymbol{\sigma} = \begin{bmatrix} \cos\theta, & e^{-i\phi}\sin\theta \\ e^{i\phi}\sin\theta, & -\cos\theta \end{bmatrix}. \tag{7.48}$$

A direct calculation shows that the eigenvalues of this matrix are $+1$ and -1, and that the corresponding (unnormalized) eigenvectors are

$$\begin{bmatrix} e^{-i\phi}\sin\theta \\ 1-\cos\theta \end{bmatrix}, \quad \begin{bmatrix} -e^{-i\phi}\sin\theta \\ 1+\cos\theta \end{bmatrix}.$$

With the help of the trigonometric half-angle formulas, these vectors may be replaced by equivalent normalized eigenvectors:

$$\begin{bmatrix} e^{-i\phi/2}\cos(\theta/2) \\ e^{i\phi/2}\sin(\theta/2) \end{bmatrix}, \quad \begin{bmatrix} -e^{-i\phi/2}\sin(\theta/2) \\ e^{i\phi/2}\cos(\theta/2) \end{bmatrix}. \tag{7.49}$$

Only the relative magnitudes and relative phases of the components of a state vector have physical significance, the norm and the overall phase being irrelevant. Now it is apparent by inspection of the first vector of (7.49) that all possible values of the relative magnitude and relative phase of its two components can be obtained by varying θ and ϕ; and, conversely, the relative magnitude and phase of the components of any two-component vector uniquely determine the values of θ and ϕ. Therefore any pure state vector of an $s = \frac{1}{2}$ system can be associated with a spatial direction $\hat{\mathbf{n}}$ for which it is the $+\frac{1}{2}\hbar$ eigenvector for the component of spin.

We turn now to general states described by a *state operator* ϱ. A 2×2 matrix has four parameters, and so can be expressed as a linear combination of four linearly independent matrices such as $\mathbf{1}$, σ_x, σ_y, and σ_z. Therefore we may write any arbitrary state operator in the form

$$\varrho = \frac{1}{2}(\mathbf{1} + \mathbf{a}\cdot\boldsymbol{\sigma}). \tag{7.50}$$

The factor $\frac{1}{2}$ has been chosen so that Tr $\varrho = 1$. The parameters (a_x, a_y, a_z) must be real to ensure that $\varrho = \varrho^\dagger$. To determine the values and significance of these three parameters, we calculate

$$\begin{aligned}\langle\sigma_x\rangle &= \mathrm{Tr}\,(\varrho\sigma_x)\\ &= \frac{1}{2}\,\mathrm{Tr}\,(\sigma_x + a_x\mathbf{1} - a_y i\sigma_z + a_z i\sigma_y)\\ &= \frac{1}{2} a_x\,\mathrm{Tr}\,\mathbf{1}\\ &= a_x\,.\end{aligned}$$

Here we have used (7.46), (7.47), and the fact that the Pauli operators have zero trace. Similar results clearly hold for the y and z components, and so we have

$$\langle\boldsymbol{\sigma}\rangle = \mathrm{Tr}\,(\varrho\boldsymbol{\sigma}) = \mathbf{a}\,. \tag{7.51}$$

The vector $\mathbf{a}$ is called the *polarization vector* of the state.

Since the eigenvalues of (7.48) are $+1$ and -1, it follows at once that the eigenvalues of ϱ are $\frac{1}{2}(1+|\mathbf{a}|)$ and $\frac{1}{2}(1-|\mathbf{a}|)$. Since an eigenvalue of a state operator cannot be negative, the length of the polarization vector must be restricted to the range $0 \le |\mathbf{a}| \le 1$. The pure states are characterized by the condition $|\mathbf{a}| = 1$, corresponding to maximum polarization. The unpolarized state, having $\mathbf{a} = 0$, is isotropic, and the average of any component of spin is zero in this state. The unpolarized state is the simplest and most common example of a state that cannot be described by a state vector.

Case (ii): $s = 1$. The matrices for the spin operators can be determined from the 3×3 block of (7.16) and related matrices. They are

$$S_x = \hbar\sqrt{\frac{1}{2}}\begin{bmatrix}0 & 1 & 0\\ 1 & 0 & 1\\ 0 & 1 & 0\end{bmatrix},\quad S_y = \hbar\sqrt{\frac{1}{2}}\begin{bmatrix}0 & -i & 0\\ i & 0 & -i\\ 0 & i & 0\end{bmatrix},\quad S_z = \hbar\begin{bmatrix}1 & 0 & 0\\ 0 & 0 & 0\\ 0 & 0 & -1\end{bmatrix}. \tag{7.52}$$

The matrix for the component of spin in the direction $\hat{\mathbf{n}} = (\sin\theta\ \cos\phi,$ $\sin\theta\ \sin\phi,\ \cos\theta)$ is

$$\hat{\mathbf{n}}\cdot\mathbf{S} = \hbar\begin{bmatrix}\cos\theta, & \sin\theta e^{-i\phi}\sqrt{\frac{1}{2}}, & 0\\ \sin\theta e^{i\phi}\sqrt{\frac{1}{2}}, & 0, & \sin\theta e^{-i\phi}\sqrt{\frac{1}{2}}\\ 0, & \sin\theta e^{i\phi}\sqrt{\frac{1}{2}}, & -\cos\theta\end{bmatrix}. \tag{7.53}$$

Its eigenvalues, with the corresponding eigenvectors below them, are

$$\begin{matrix} \hbar & 0 & -\hbar \\ \begin{bmatrix} \frac{1}{2}(1+\cos\theta)e^{-i\phi} \\ \sqrt{\frac{1}{2}}\sin\theta \\ \frac{1}{2}(1-\cos\theta)e^{i\phi} \end{bmatrix} & \begin{bmatrix} -\sqrt{\frac{1}{2}}\sin\theta e^{-i\phi} \\ \cos\theta \\ \sqrt{\frac{1}{2}}\sin\theta e^{i\phi} \end{bmatrix} & \begin{bmatrix} \frac{1}{2}(1-\cos\theta)e^{-i\phi} \\ -\sqrt{\frac{1}{2}}\sin\theta \\ \frac{1}{2}(1+\cos\theta)e^{i\phi} \end{bmatrix} \end{matrix} . \tag{7.54}$$

Unlike the case of $s = \frac{1}{2}$, it is no longer true that every vector must be an eigenvector of the component of spin in some direction. This is so because it requires four real parameters to specify the relative magnitudes and relative phases of the components of a general three-component vector, but the above eigenvectors contain only the two parameters θ and ϕ. Therefore the pure states of an $s = 1$ system need not be associated with a spin eigenvalue in any spatial direction.

A general state is described by a 3×3 state operator ϱ, which depends upon eight parameters after the restriction $\text{Tr}\ \varrho = 1$ is taken into account. The polarization vector $\langle \mathbf{S} \rangle = \text{Tr}(\varrho \mathbf{S})$ provides only three parameters, so it is clear that polarization vector does not uniquely determine the state, unlike the case of $s = \frac{1}{2}$. The additional parameters that are needed to fully determine the state in this case, as well as their physical significance, will be discussed in Ch. 8.

Case (iii): $s = 3/2$. The spin operators for this case will be represented by 4×4 matrices, which we shall not write down explicitly, although they are not difficult to calculate. The sum of the squares of the matrices for the spin components in three orthogonal directions must satisfy the identity

$$S_x{}^2 + S_y{}^2 + S_z{}^2 = \hbar^2 s(s+1)\mathbf{1} . \tag{7.55}$$

This is true, of course, for any value of s. For the case of $s = \frac{1}{2}$, the identity is trivial, since each of the three matrices on the left hand side of (7.55) is a multiple of $\mathbf{1}$. For the case of $s = 1$, the squares of the matrices in (7.52) are not multiples of $\mathbf{1}$, but the 3×3 matrices for $S_x{}^2$, $S_y{}^2$, and $S_z{}^2$ are commutative. Thus there is a complete set of common eigenvectors for the three matrices, and the identity merely expresses a correlation among their eigenvalues: for any such eigenvector, two of the matrices must have the eigenvalue $\hbar^2$ and one must have the eigenvalue zero.

The case of $s = 3/2$ is the simplest one for which the matrices for $S_x{}^2$, $S_y{}^2$, and $S_z{}^2$ are not commutative, and so a set of common eigenvectors does not exist. Therefore the identity (7.55) does not reduce to a relation among eigenvalues. Indeed, it is clear that no such relationship among eigenvalues can hold in this case. This is so because an eigenvalue of each term on the left hand side must be either $9\hbar^2/4$ or $\hbar^2/4$, and no sum of three such numbers can add up to the eigenvalue $15\hbar^2/4$ on the right. This arithmetical fact has implications for the interpretation of quantum mechanics. One might have been inclined to regard quantum-mechanical variables as classical stochastic variables, each of which takes on one of its allowed eigenvalues at any instant of time. (The particular values would, presumably, fluctuate randomly over time in accordance with the quantum-mechanical probability distributions.) But the above example shows that this interpretation cannot be correct, at least not for angular momentum. Questions of interpretation will also arise in later chapters, where they will be treated in more detail.

7.5 *Finite Rotations*

Three parameters are required to describe an arbitrary rotation. For example, they may be the direction of the rotation axis (two parameters) and the angle of rotation, as in (7.17). Another common parameterization is by the *Euler angles*, which are illustrated in Fig. 7.2. From the fixed system of axes $Oxyz$, a new rotated set of axes $Ox'y'z'$ is produced in three steps:

(i) Rotate through angle α about Oz, carrying Oy into Ou;
(ii) Rotate through angle β about Ou, carrying Oz into Oz';
(iii) Rotate through angle γ about Oz', carrying Ou into Oy'.

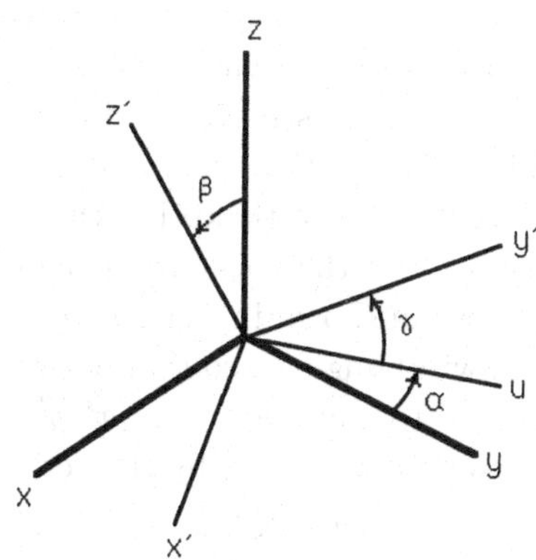

Fig. 7.2 The Euler angles.

At the end of this process Ox will have been carried into Ox'. The corresponding unitary operators for these three rotations are $\mathbf{R}_z(\alpha)$, $\mathbf{R}_u(\beta)$, and $\mathbf{R}_{z'}(\gamma)$, in the notation of (7.17). The net rotation is described by the product of these three operators,

$$\begin{aligned}\mathbf{R}(\alpha,\beta,\gamma) &= \mathbf{R}_{z'}(\gamma)\mathbf{R}_u(\beta)\mathbf{R}_z(\alpha)\\ &= e^{-i\gamma J_{z'}}e^{-i\beta J_u}e^{-i\alpha J_z}\,. \end{aligned} \tag{7.56}$$

(In this section it is convenient to choose units such that $\hbar = 1$.)

This expression for the rotation operator is inconvenient because each of the three rotations is performed about an axis belonging to a different coordinate system. It is more convenient to transform all the operators to a common coordinate system. Applying the formula (3.2) to the rotation (i) above, we obtain $J_u = \mathbf{R}_z(\alpha)J_y\mathbf{R}_z(-\alpha)$, and hence $\mathbf{R}_u(\beta) = \mathbf{R}_z(\alpha)\ \mathbf{R}_y(\beta)\ \mathbf{R}_z(-\alpha)$. Similarly, since $J_{z'}$ is the result of performing rotations (i) and (ii) on J_z, it follows that $\mathbf{R}_{z'}(\gamma) = \mathbf{R}_u(\beta)\ \mathbf{R}_z(\alpha)\ \mathbf{R}_z(\gamma)\ \mathbf{R}_z(-\alpha)\ \mathbf{R}_u(-\beta)$. After substitution of these expressions into (7.56), considerable cancellation becomes possible, and the result is simply

$$\begin{aligned}\mathbf{R}(\alpha,\beta,\gamma) &= \mathbf{R}_z(\alpha)\ \mathbf{R}_y(\beta)\ \mathbf{R}_z(\gamma)\\ &= e^{-i\alpha J_z}\,e^{-i\beta J_y}\,e^{-i\gamma J_z}\,. \end{aligned} \tag{7.57}$$

Now all of the operators are expressed in the original fixed coordinate system Oxyz.

Active and passive rotations

Transformations may be considered from either of two points of view: the *active* point of view, in which the object is rotated with respect to a fixed coordinate system, or the *passive* point of view, in which the object is kept fixed and the coordinate system is rotated. In this book we normally adhere to the active point of view, but both methods are valid, and it is desirable to understand the relation between them. For ease of illustration we shall use two-dimensional examples, but the analysis has more general validity.

Under the *active* rotation shown in Fig. 7.3, the object is rotated through a positive (counterclockwise) angle α, so that a physical point in the object is moved from location (x, y) to a new location (x', y'). The relation between the coordinates of these two points is given by the active rotation matrix,

$$x'_i = \sum_j R^{(a)}_{ij}\,x_j\,. \tag{7.58}$$

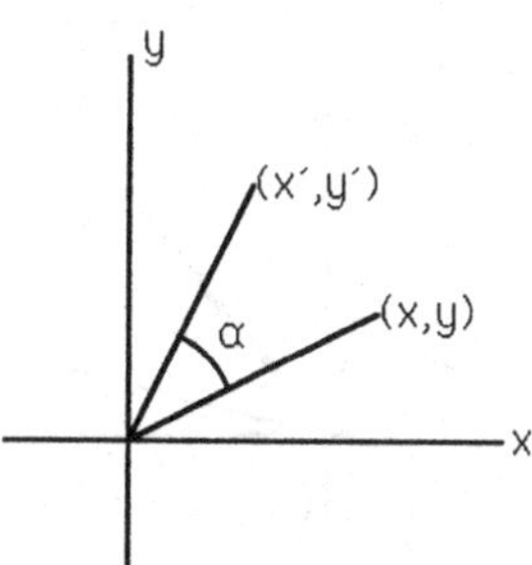

Fig. 7.3 Active rotation [Eq. (7.58)].

The rotation matrix $R_{ij}^{(a)}(\alpha,\beta,\gamma)$ is in general a function of three Euler angles. By inspection of Fig. 7.3, it can be verified that in this case it takes the form

$$R_{ij}^{(a)}(\alpha,0,0) = \begin{bmatrix} \cos\alpha, & -\sin\alpha, & 0 \\ \sin\alpha, & \cos\alpha, & 0 \\ 0, & 0, & 1 \end{bmatrix} . \tag{7.59}$$

Let us take the object of the rotation to be a scalar field, or a state function for a spinless particle, $\Psi(\mathbf{x})$. The rotated function is denoted as $\Psi'(\mathbf{x}) = \mathbf{R}(\alpha,\beta,\gamma)\,\Psi(\mathbf{x})$. By construction, the value of the new function Ψ' at the new point $\mathbf{x}' = (x', y', z')$ is equal to the value of the old function Ψ at the old point $\mathbf{x} = (x, y, z)$, so we have $\Psi'(\mathbf{x}') = \Psi(\mathbf{x}) = \Psi([R^{(a)}]^{-1}\mathbf{x}')$. Thus

$$\mathbf{R}(\alpha,\beta,\gamma)\,\Psi(\mathbf{x}) = \Psi([R^{(a)}(\alpha,\beta,\gamma)]^{-1}\mathbf{x}) , \tag{7.60}$$

this formula being a special case of (3.8).

In a *passive* rotation, the object remains fixed while the coordinate system is rotated. Thus the same physical point P has two sets of coordinates: the old coordinates $\mathbf{x} = (x, y, z)$, and the new coordinates $\mathbf{x}'' = (x'', y'', z'')$. The relation between these two sets of coordinates is given by the passive rotation matrix,

$$x_i'' = \sum_j R_{ij}^{(p)} x_j . \tag{7.61}$$

The matrix $R_{ij}^{(p)}(\alpha,\beta,\gamma)$ takes the following form for the special case shown in Fig. 7.4:

$$R_{ij}^{(p)}(\alpha,0,0) = \begin{bmatrix} \cos\alpha, & \sin\alpha, & 0 \\ -\sin\alpha, & \cos\alpha, & 0 \\ 0, & 0, & 1 \end{bmatrix} . \tag{7.62}$$

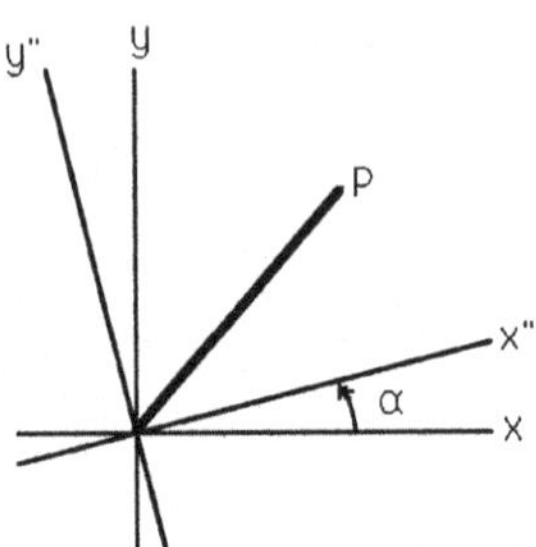

Fig. 7.4 Passive rotation [Eq. (7.61)].

Now the value of the field or state function must be the same at the point P regardless of which coordinate system is used, although its functional form will be different in the two coordinate systems. Thus we must have $\Psi''(\mathbf{x}'') = \Psi(\mathbf{x})$, where Ψ'' and Ψ denote the functions in the new and old coordinates, respectively. From (7.61) we have the relation $\mathbf{x} = [R^{(p)}]^{-1}\mathbf{x}''$, so by a change of dummy variable we obtain

$$\Psi''(\mathbf{x}) = \Psi([R^{(p)}]^{-1}\mathbf{x}) . \tag{7.63}$$

Notice that (7.60) and (7.63) have the same form, involving an inverse matrix acting on the coordinate argument. The difference between the two cases lies only in the differences between the active and passive coordinate rotation matrices (7.59) and (7.62), one being the inverse of the other. This relationship holds also in the general case:

$$R^{(p)}(\alpha,\beta,\gamma) = [R^{(a)}(\alpha,\beta,\gamma)]^{-1} = R^{(a)}(-\gamma,-\beta,-\alpha) .$$

It expresses the intuitive geometrical fact that one may achieve equivalent results by rotating the coordinate system in a positive sense with respect to the object, or by rotating the object in a negative sense with respect to the coordinate system.

[[This discussion of active and passive rotations is based on an article by M. Bouten (1969), in which it is pointed out that the standard reference books by Edmonds (1957) and Rose (1960) contain errors in treating these rotations.]]

Rotation matrices. The matrix representation of the rotation operator (7.57) in the basis formed by the angular momentum eigenvectors,

$$\langle j',m'|\mathbf{R}(\alpha,\beta,\gamma)|j,m\rangle = \delta_{j',j}D^{(j)}_{m',m}(\alpha,\beta,\gamma) , \tag{7.64}$$

gives rise to the *rotation matrices*,

$$D^{(j)}_{m',m}(\alpha,\beta,\gamma) = \langle j,m'|e^{-i\alpha J_z}e^{-i\beta J_y}e^{-i\gamma J_z}|j,m\rangle$$
$$= e^{-i(\alpha m'+\gamma m)}d^{(j)}_{m',m}(\beta)\,, \tag{7.65}$$

where

$$d^{(j)}_{m',m}(\beta) = \langle j,m'|e^{-i\beta J_y}|j,m\rangle\,. \tag{7.66}$$

The matrix (7.64) is diagonal in (j',j) because $\mathbf{R}$ commutes with the operator J^2, and the final simplification in (7.65) takes place because the basis vectors are eigenvectors of J_z.

The matrix element (7.66) is easy to evaluate for the case of $j=\frac{1}{2}$, for which we can replace J_y by $\frac{1}{2}\hbar\sigma_y$. Here σ_y is a Pauli matrix (7.45). From the Taylor series for the exponential function and the identity $(\sigma_y)^2 = \mathbf{1}$ (the 2×2 unit matrix), it follows that

$$e^{-i\beta J_y} = \exp\left(\frac{-i\beta\sigma_y}{2}\right) = \mathbf{1}\,\cos\left(\frac{\beta}{2}\right) - i\sigma_y\,\sin\left(\frac{\beta}{2}\right)\,. \tag{7.67}$$

Hence we obtain

$$d^{(1/2)}(\beta) = \begin{bmatrix}\cos(\beta/2), & -\sin(\beta/2)\\ \sin(\beta/2), & \cos(\beta/2)\end{bmatrix}\,. \tag{7.68}$$

Notice that this matrix is periodic in β with period 4π, but it changes sign when 2π is added to β. This *double-valuedness* under rotation by 2π is a characteristic of the full rotation matrix (7.65) whenever j is a half odd-integer. The matrix is single-valued under rotation by 2π whenever j is an integer.

For the case of $j=1$, we can replace J_y by the 3×3 matrix S_y of (7.52), which satisfies the identity $(S_y)^3 = S_y$. (Recall that we are using $\hbar = 1$ in this section.) By a calculation similar to that leading to (7.67), we obtain

$$\exp(-i\beta S_y) = \mathbf{1} - (S_y)^2(1-\cos\beta) - iS_y\,\sin\beta\,.$$

Hence we have

$$d^{(1)}(\beta) = \begin{bmatrix}\frac{1}{2}(1+\cos\beta), & -(\sqrt{\frac{1}{2}})\sin\beta, & \frac{1}{2}(1-\cos\beta)\\ (\sqrt{\frac{1}{2}})\sin\beta, & \cos\beta, & -(\sqrt{\frac{1}{2}})\sin\beta\\ \frac{1}{2}(1-\cos\beta), & (\sqrt{\frac{1}{2}})\sin\beta, & \frac{1}{2}(1+\cos\beta)\end{bmatrix}\,. \tag{7.69}$$

The matrix element (7.66) was evaluated in the general case by E. P. Wigner. A concise derivation is given in the book by Sakurai (1985). The properties of the rotation matrices can be most systematically derived by means of *group theory* [see Tinkham (1964)]. The only specific result from group theory that we shall use is the *orthogonality theorem*,

$$\int \{D^{(j)}_{\mu,\nu}(R)\}^* D^{(j')}_{\mu',\nu'}(R)\, dR = (2j+1)^{-1}\delta_{j',j}\delta_{\mu',\mu}\delta_{\nu',\nu} \int dR\,. \tag{7.70}$$

Here R denotes the Euler angles of the rotation, (α,β,γ), and $dR = d\alpha \sin\beta d\beta\, d\gamma$. The range of β is from 0 to π. The ranges of α and γ must cover 4π radians, in general, although 2π will suffice if both j and j' are integers.

Rotation of angular momentum eigenvectors

The rotation matrices arise naturally when a rotation operator is applied to the angular momentum eigenvectors:

$$\begin{aligned} \mathbf{R}(\alpha,\beta,\gamma)|j,m\rangle &= \sum_{j',m'} |j',m'\rangle\langle j',m'|\mathbf{R}(\alpha,\beta,\gamma)|j,m\rangle \\ &= \sum_{m'} |j,m'\rangle D^{(j)}_{m',m}(\alpha,\beta,\gamma)\,. \end{aligned} \tag{7.71}$$

The reader can verify that the eigenvectors of an angular momentum component in a general direction, (7.49) and (7.54), are obtainable from this equation.

The $(2j+1)$-dimensional space spanned by the set of vectors $\{|j,m\rangle\}$, for fixed j and all m in the range $(-j \le m \le j)$, is an *invariant irreducible subspace* under rotations. To say that the subspace is *invariant* means that a vector within it remains within it after rotation. This is so because no other values of j are introduced into the linear combination of (7.71). To say that the subspace is *irreducible* means that it contains no smaller invariant subspaces. Proof of irreducibility is left as an exercise for the reader.

Relation to spherical harmonics

The spherical harmonics (7.34), introduced in (7.30) and (7.31) as the eigenfunctions of orbital angular momentum, are by definition the coordinate representation of the angular momentum eigenvectors $|\ell,m\rangle$ for integer values of ℓ and m. Hence they must transform under rotations according to (7.71). From that equation one can derive a useful relation between the spherical harmonics and the rotation matrices. It is slightly more convenient to write the equation for the inverse of the transformation specified in (7.71),

$$\mathbf{R}^{-1}(\alpha,\beta,\gamma)\, Y_\ell{}^m(\theta,\phi) = Y_\ell{}^m(\theta',\phi')$$
$$= \sum_{m'} Y_\ell{}^{m'}(\theta,\phi)\,\{D^{(\ell)}_{m,m'}(\alpha,\beta,\gamma)\}^*\,, \tag{7.72}$$

where the rotation $R(\alpha,\beta,\gamma)$ takes a vector in the direction (θ,ϕ) into the direction (θ',ϕ'). [Note once again the inverse relation between the rotation on function space and on the coordinate arguments of the function, as in (7.60). We use the active point of view.] We have made use of the fact that the rotation matrix D is unitary, and so its inverse is obtained by taking the transpose complex conjugate.

In fact, the spherical harmonics are fully determined by (7.72), except for their conventional normalization. By putting $\beta=\gamma=0$ we obtain

$$\mathbf{R}^{-1}(\alpha,0,0)Y_\ell{}^m(\theta,\phi) = \mathbf{R}(0,0,-\alpha)Y_\ell{}^m(\theta,\phi) = Y_\ell{}^m(\theta,\phi+\alpha)$$
$$= \sum_{m'} Y_\ell{}^{m'}(\theta,\phi)\{D^{(\ell)}_{m,m'}(\alpha,0,0)\}^*$$
$$= e^{i\alpha m}\, Y_\ell{}^m(\theta,\phi)\,.$$

The final step follows from (7.65). Setting $\phi=0$ then yields

$$Y_\ell{}^m(\theta,\alpha) = e^{i\alpha m}\, Y_\ell{}^m(\theta,0)\,, \tag{7.73}$$

which determines the dependence of the spherical harmonic on its second argument. Since the direction $\theta=0$ is the polar axis, continuity of the spherical harmonic requires that $Y_\ell{}^m(0,\alpha)$ be independent of α. Therefore we must have $Y_\ell{}^m(0,0)=0$ for $m\neq 0$, and so we can write

$$Y_\ell{}^m(0,0) = c_\ell\,\delta_{m,0}\,. \tag{7.74}$$

We next put $\theta=0$ in (7.72), so that the direction (θ,ϕ) is the z axis. The equation now reduces to

$$Y_\ell{}^m(\theta',\phi') = \sum_{m'} c_\ell\,\delta_{m',0}\{D^{(\ell)}_{m,m'}(\alpha,\beta,\gamma)\}^*$$
$$= c_\ell\{D^{(\ell)}_{m,0}(\alpha,\beta,\gamma)\}^*\,,$$

where $R(\alpha,\beta,\gamma)$ carries a vector parallel to the z axis into the direction (θ',ϕ'). Clearly this requires $\alpha=\phi'$ and $\beta=\theta'$, with γ remaining arbitrary. But $D^{(\ell)}_{m,0}(\alpha,\beta,\gamma)$ is independent of γ, so we may simply write

$$Y_\ell{}^m(\theta,\phi) = c_\ell\{D^{(\ell)}_{m,0}(\phi,\theta,0)\}^*\,, \tag{7.75}$$

thus obtaining a simple relation between the spherical harmonics and the rotation matrices. Conventional normalization is obtained if we put

$$c_\ell = \left(\frac{2\ell+1}{4\pi}\right)^{1/2} . \tag{7.76}$$

7.6 *Rotation Through* 2π

According to (7.17), the operator for a rotation through 2π about an axis along the unit vector $\hat{\mathbf{n}}$ is $\mathbf{R}_n(2\pi) = e^{-2\pi i\hat{\mathbf{n}}\cdot\mathbf{J}/\hbar}$. Its effect on the standard angular momentum eigenvectors is

$$\mathbf{R}_n(2\pi)|j,m\rangle = (-1)^{2j}|j,m\rangle . \tag{7.77}$$

That is to say, it has no effect if j is an integer, and multiplies by -1 if j is half an odd integer. The truth of (7.77) is self-evident if $\hat{\mathbf{n}}$ points along the z axis because the vectors are eigenvectors of J_z. To show it for an arbitrary direction of $\hat{\mathbf{n}}$, we can express the vector $|j,m\rangle$ as a linear combination of the eigenvectors of $\hat{\mathbf{n}}\cdot\mathbf{J}$, which can be obtained from $|j,m\rangle$ by a rotation as in (7.71). Since different values of j are never mixed by that rotation, it follows that (7.77) also holds for an arbitrary direction of $\hat{\mathbf{n}}$. For this reason we may drop any reference to the axis of rotation, and write simply

$$\mathbf{R}_n(2\pi) = \mathbf{R}(2\pi) . \tag{7.78}$$

We are accustomed to thinking of a rotation through 2π as a trivial operation that leaves everything unchanged. Corresponding to this belief, we shall *assume that all dynamical variables are invariant under* 2π *rotation.* That is, we postulate

$$\mathbf{R}(2\pi)A\,\mathbf{R}^{-1}(2\pi) = A , \quad \text{or} \quad [\mathbf{R}(2\pi), A] = 0 , \tag{7.79}$$

where A may represent *any physical observable* whatsoever. But $\mathbf{R}(2\pi)$ is not a trivial operator (that is, it is not equal to the identity), and so invariance under transformation by $\mathbf{R}(2\pi)$ may have some nontrivial consequences.

It is important to distinguish consequences of invariance of the observable from those that follow from invariance of the state. Since it is difficult to visualize states that are not invariant under 2π rotation, we shall digress to treat the consequences of an arbitrary symmetry. Let U be a unitary operator that leaves a particular dynamical variable F invariant, and hence $[U, F] = 0$. Consider some state that is *not* invariant under the transformation U. If it is a pure state, with state vector $|\Psi\rangle$, then $|\Psi'\rangle = U|\Psi\rangle$ is not equal to $|\Psi\rangle$. The average of the dynamical variable F in the transformed state is

$$\begin{aligned}\langle F\rangle = \langle\Psi'|F|\Psi'\rangle &= \langle\Psi|U^\dagger FU|\Psi\rangle \\ &= \langle\Psi|U^\dagger UF|\Psi\rangle \\ &= \langle\Psi|F|\Psi\rangle\,.\end{aligned}$$

Thus we see that observable statistical properties for F are the same in the two states $|\Psi\rangle$ and $|\Psi'\rangle$, even though the states are not equal. Similarly, for a general state described by the state operator ϱ, for which we assume that $\varrho' = U\varrho U^\dagger$ is not equal to ϱ, the average of F is

$$\begin{aligned}\langle F\rangle = \mathrm{Tr}\,(\varrho' F) = \mathrm{Tr}\,(U\varrho U^\dagger F) &= \mathrm{Tr}\,(\varrho U^\dagger FU) \\ &= \mathrm{Tr}\,(\varrho F)\,.\end{aligned}$$

Thus we see again, in this more general case, that the observable statistical properties for F will be the same in these two symmetry-related but unequal states. Now of course this sort of conclusion holds for the symmetry operation $\mathbf{R}(2\pi)$, but this is not of interest here. We seek rather something that is peculiar to $\mathbf{R}(2\pi)$.

According to (7.77), the operator $\mathbf{R}(2\pi)$ divides the vector space into two subspaces. A typical vector in the first subspace (integer angular momentum), denoted as $|+\rangle$, has the property $\mathbf{R}(2\pi)|+\rangle = |+\rangle$, whereas a typical vector in the second subspace (half odd-integer angular momentum), denoted as $|-\rangle$, has the property $\mathbf{R}(2\pi)|-\rangle = -|-\rangle$. Now, if A represents *any physical observable*, we have

$$\langle +|\mathbf{R}(2\pi)A|-\rangle = \langle +|A\mathbf{R}(2\pi)|-\rangle\,,$$

$$\langle +|A|-\rangle = -\langle +|A|-\rangle\,,$$

and hence it must be the case that $\langle +|A|-\rangle = 0$. Thus no physical observable can have nonvanishing matrix elements between states with integer angular momentum and states with half odd-integer angular momentum. This fact forms the basis of a *superselection rule.*

One statement of this superselection rule is that *there is no observable distinction* among the state vectors of the form

$$|\Psi_\omega\rangle = |+\rangle + e^{i\omega}|-\rangle \tag{7.80}$$

for different values of the phase ω. This is true because for any physical observable A whatsoever we have $\langle\Psi_\omega|A|\Psi_\omega\rangle = \langle+|A|+\rangle + \langle-|A|-\rangle$, since $\langle+|A|-\rangle = \langle-|A|+\rangle = 0$, and this does not depend on the phase ω.

An analogous statement can be made for a general state represented by the state operator $\varrho = \sum_{ij} \varrho_{ij}|i\rangle\langle j|$. The matrix ϱ_{ij} and the similar matrix for a physical observable A can be partitioned into four blocks:

$$[\varrho] = \left[\begin{array}{c|c} \varrho_{++} & \varrho_{+-} \\ \hline \varrho_{-+} & \varrho_{--} \end{array}\right], \qquad [A] = \left[\begin{array}{c|c} A_{++} & 0 \\ \hline 0 & A_{--} \end{array}\right].$$

Now the average of the dynamical variable A in this state is

$$\langle A\rangle = \text{Tr}\,(\varrho A) = \text{Tr}_+\,(\varrho_{++}A_{++}) + \text{Tr}_-\,(\varrho_{--}A_{--})\,,$$

where Tr_+ and Tr_- denote traces over the subspaces. The cross matrix elements ϱ_{+-} and ϱ_{-+} do not contribute to the observable quantity $\langle A\rangle$ because the corresponding matrix elements of the operator A are all zero. This is another, more general way of saying that *interference between vectors of the $|+\rangle$ and $|-\rangle$ types is not observable.*

It is sometimes asserted that states that would be described by vectors of the form (7.80) do not exist. This is equivalent to asserting that the matrix elements ϱ_{+-} and ϱ_{-+} of a state operator must vanish. The correct statement of the superselection rule is that such matrix elements of ϱ, and the phase ω in (7.80), *have no observable consequences.* But since they have no observable consequences, we are free to assume any convenient values, such as $\varrho_{+-} = \varrho_{-+} = 0$. If this assumption is made as an initial condition at $t = 0$, it will remain true for all time because the generator of time evolution H is itself a physical observable and so obeys the invariance condition $[H, \mathbf{R}(2\pi)] = 0$. Therefore the equation of motion decouples into two separate equations in each of the two subspaces, and no cross matrix elements of ϱ between the two subspaces will ever develop.

Superselection versus ordinary symmetry

It is important to understand the difference between a generator of a superselection rule like $\mathbf{R}(2\pi)$, and a symmetry operation that is generated by a universally conserved quantity, such as the displacement operator $e^{-i\mathbf{a}\cdot\mathbf{P}/\hbar}$, which is generated by the total momentum $\mathbf{P}$. Since the Hamiltonian of any closed system must be invariant under both of these transformations, there is an apparent similarity between them. Both give rise to a quantum number that must be conserved in any transition: the eigenvalue ± 1 of the operator $\mathbf{R}(2\pi)$ in the first case, and the total momentum eigenvalue in the second case.

The difference is that there exist observables that do not commute with $\mathbf{P}$, such as the position $\mathbf{Q}$, but there are no observables that fail to commute with $\mathbf{R}(2\pi)$. By measuring the position one can distinguish between states that differ only by a displacement, but there is no way to distinguish between states that differ only by a 2π rotation.

[[The superselection rule generated by $\mathbf{R}(2\pi)$, which separates states of integer angular momentum from states of half odd-integer angular momentum, is the only superselection rule that occurs in the quantum mechanics of stable particles. In quantum field theory, in which particles are regarded as field excitations that can be created and destroyed, it can be shown that the total electric charge operator generates a superselection rule, provided one assumes that all observables are invariant under gauge transformations. (See Sec. 11.2 for the notion of gauge transformation.) In that case, no interference can be observed between states of different total charge because there are no physical observables that do not commute with the charge operator. In a theory of stable particles the charge of each particle, and hence the total charge, is invariable. Therefore the total charge operator is a multiple of the identity. Every operator commutes with it, and so the charge superselection rule becomes trivial.]]

7.7 *Addition of Angular Momenta*

Let us consider a two-component system, each component of which has angular momentum degrees of freedom. For convenience we shall speak of these two components as "particle 1" and "particle 2", and shall denote the corresponding angular momentum operators as $\mathbf{J}^{(1)}$ and $\mathbf{J}^{(2)}$, although they could very well be the orbital and spin degrees of freedom of the same particle.

As was discussed in Sec. 3.5, basis vectors for the composite system can be formed from the basis vectors of the components by taking all binary products of a vector from each set:

$$|j_1, j_2, m_1, m_2\rangle = |j_1, m_1\rangle^{(1)} |j_2, m_2\rangle^{(2)} . \tag{7.81}$$

These vectors are common eigenvectors of the four commutative operators $\mathbf{J}^{(1)} \cdot \mathbf{J}^{(1)}$, $\mathbf{J}^{(2)} \cdot \mathbf{J}^{(2)}$, $\mathbf{J}_z{}^{(1)}$, and $\mathbf{J}_z{}^{(2)}$, with eigenvalues $\hbar^2 j_1(j_1+1)$, $\hbar^2 j_2(j_2+1)$, $\hbar m_1$, and $\hbar m_2$, respectively.

It is often desirable to form eigenvectors of the total angular momentum operators, $\mathbf{J}\cdot\mathbf{J}$ and $\mathbf{J}_z$, where the total angular momentum vector operator is

$$\mathbf{J} = \mathbf{J}^{(1)} + \mathbf{J}^{(2)}\,. \tag{7.82}$$

[A more formal mathematical notation would be $\mathbf{J} = \mathbf{J}^{(1)} \otimes \mathbf{1} + \mathbf{1} \otimes \mathbf{J}^{(2)}$. The essential point of either notation is to indicate that $\mathbf{J}^{(1)}$ operates only on the first factor of (7.81) and that $\mathbf{J}^{(2)}$ operates only on the second factor.] This is useful when the system is invariant under rotation as a whole, but not under rotation of the two components separately, so that the components of $\mathbf{J}$ are constants of motion but the components of $\mathbf{J}^{(1)}$ and $\mathbf{J}^{(2)}$ are not constants of motion.

The eigenvectors of $\mathbf{J}\cdot\mathbf{J}$ and $\mathbf{J}_z$ may be denoted as $|\alpha, J, M\rangle$, where $\hbar^2 J(J+1)$ and $\hbar M$ are the eigenvalues of the two operators and α denotes any other labels that may be needed to specify a unique vector. These eigenvectors can be expressed as linear combinations of the product vectors (7.81). It is easy to verify that the four operators $\mathbf{J}^{(1)}\cdot\mathbf{J}^{(1)}$, $\mathbf{J}^{(2)}\cdot\mathbf{J}^{(2)}$, $\mathbf{J}\cdot\mathbf{J}$, and $\mathbf{J}_z$ are mutually commutative, and hence they possess a complete set of common eigenvectors. Since the set of product vectors of the form (7.81) and the new set of total angular momentum eigenvectors are both eigenvectors of $\mathbf{J}^{(1)}\cdot\mathbf{J}^{(1)}$ and $\mathbf{J}^{(2)}\cdot\mathbf{J}^{(2)}$, the eigenvalues j_1 and j_2 will be constant in both sets. Therefore we may confine our attention to the vector space of dimension $(2j_1+1)(2j_2+1)$ that is spanned by those basis vectors (7.81) having fixed values of j_1 and j_2.

This vector space is invariant under rotation, and so it must be decomposable into one or more *irreducible rotationally invariant subspaces.* (See Sec. 7.5 for the concept of an irreducible invariant subspace.) Now the $2J+1$ vectors $|\alpha, J, M\rangle$, with M in the range $-J \le M \le J$, span an irreducible subspace. Therefore if the vector $|\alpha, J, M\rangle$, for a particular value of M, can be constructed in the space under consideration, then so can the entire set of $2J+1$ such vectors with M in the range $-J \le M \le J$.

For a particular value of J, it might be possible to construct one such set of vectors, two or more linearly independent sets, or none at all. Let $N(J)$ denote the number of independent sets that can be constructed. Let $n(M)$ be the degree of degeneracy, in this space, of the eigenvalue M. The relation between these two quantities is

$$n(M) = \sum_{J \ge |M|} N(J) \tag{7.83}$$

and hence

$$N(J) = n(J) - n(J+1)\,. \tag{7.84}$$

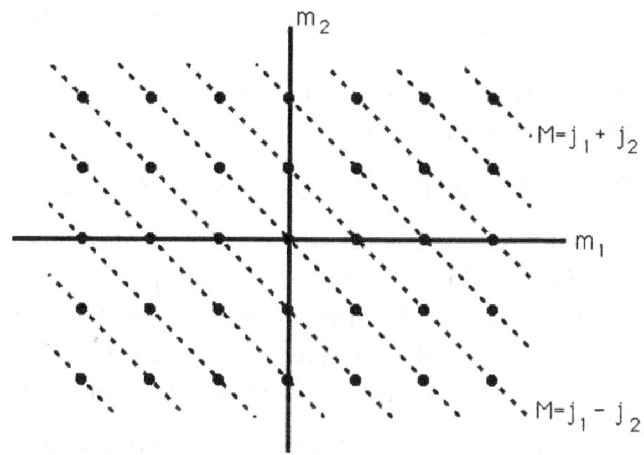

Fig. 7.5 Possible values of $M = m_1 + m_2$, illustrated for $j_1 = 3, j_2 = 2$.

Thus $N(J)$ can be obtained from $n(M)$, which is easier to calculate directly.

The product vectors (7.81) are eigenvectors of the operator $\mathbf{J}_z = \mathbf{J}_z{}^{(1)} + \mathbf{J}_z{}^{(2)}$, with eigenvalue $\hbar M = \hbar(m_1 + m_2)$, and the degree of degeneracy $n(M)$ is equal to the number of pairs (m_1, m_2) such that $M = m_1 + m_2$. This is illustrated in Fig. 7.5, where it is apparent that the number $n(M)$ is equal to the number of dots that lie on the diagonal line $M = m_1 + m_2$. Therefore

$$\begin{aligned} n(M) &= 0 && \text{if} \quad |M| > j_1 + j_2\,, \\ &= j_1 + j_2 + 1 - |M| && \text{if} \quad j_1 + j_2 \geq |M| \geq |j_1 - j_2|\,, \\ &= 2j_{\min} + 1 && \text{if} \quad |j_1 - j_2| \geq |M| \geq 0\,, \end{aligned} \tag{7.85}$$

where $j_{\min}$ is the lesser of j_1 and j_2. It then follows from (7.84) that

$$\begin{aligned} N(J) &= 1 && \text{for} \quad |j_1 - j_2| \leq J \leq j_1 + j_2\,, \\ &= 0 && \text{otherwise}\,. \end{aligned} \tag{7.86}$$

It has turned out that $N(J)$ is never greater that 1, and so the vectors $|\alpha, J, M\rangle$ can be uniquely labeled by the eigenvalues of the four operators $\mathbf{J}^{(1)}\cdot\mathbf{J}^{(1)}$, $\mathbf{J}^{(2)}\cdot\mathbf{J}^{(2)}$, $\mathbf{J}\cdot\mathbf{J}$ and $\mathbf{J}_z$. Henceforth *these total angular momentum eigenvectors will be denoted as* $|j_1, j_2, J, M\rangle$.

Clebsch–Gordan coefficients

The set of all product vectors $\{|j_1, j_2, m_1, m_2\rangle\}$ is a complete basis set, as is the set of all total angular momentum vectors $\{|j_1, j_2, J, M\rangle\}$. Therefore it is possible to express one set in terms of the other:

$$|j_1, j_2, J, M\rangle = \sum_{m_1, m_2} |j_1, j_2, m_1, m_2\rangle\langle j_1, j_2, m_1, m_2|j_1, j_2, J, M\rangle\,. \tag{7.87}$$

The coefficients of this transformation are called the *Clebsch–Gordan coefficients*, and they will be written as

$$(j_1, j_2, m_1, m_2|J, M) = \langle j_1, j_2, m_1, m_2|j_1, j_2, J, M\rangle\,. \tag{7.88}$$

The phases of the CG coefficients are not yet defined because of the indeterminacy of the relative phases of the vectors $|j_1, j_2, J, M\rangle$, which have been defined only as eigenvectors of certain operators. For different values of M but fixed J we adopt the usual phase convention that led to (7.14). This leaves one arbitrary phase for each J value, which we fix by requiring that

$$(j_1, j_2, j_1,\ J - j_1|J, J) \quad \text{be real and positive}\,. \tag{7.89}$$

It can be shown that all of the CG coefficients are now real, although this is not obvious.

The relation (7.87) and its inverse can now be written as

$$|j_1, j_2, J, M\rangle = \sum_{m_1, m_2} |j_1, j_2, m_1, m_2\rangle(j_1, j_2, m_1, m_2|J, M)\,, \tag{7.90}$$

$$|j_1, j_2, m_1, m_2\rangle = \sum_{J, M} |j_1, j_2, J, M\rangle(j_1, j_2, m_1, m_2|J, M)\,. \tag{7.91}$$

Since this is a unitary transformation from one orthonormal set of vectors to another, the coefficients must satisfy the orthogonality relations

$$\sum_{m_1, m_2} (j_1, j_2, m_1, m_2|J, M)(j_1, j_2, m_1, m_2|J', M') = \delta_{J,J'}\delta_{M,M'}\,, \tag{7.92}$$

$$\sum_{J, M} (j_1, j_2, m_1, m_2|J, M)(j_1, j_2, m_1', m_2'|J, M) = \delta_{m_1,m_{1'}}, \delta_{m_2,m_{2'}}\,. \tag{7.93}$$

The CG coefficient $(j_1, j_2, m_1, m_2|J, M)$ vanishes unless the following conditions are satisfied:

$$\text{(a)}\quad m_1 + m_2 = M\,, \tag{7.94}$$

$$\text{(b)}\quad |j_1 - j_2| \leq J \leq j_1 + j_2\,, \tag{7.95}$$

$$\text{(c)}\quad j_1 + j_2 + J = \text{ an integer}\,. \tag{7.96}$$

Conditions (a) and (b) have already been derived above. Condition (b) is actually symmetrical with respect to permutations of (j_1, j_2, J), and it can be re-expressed as requiring that the triad (j_1, j_2, J) should form the sides of a triangle. Condition (c) follows from considering the behavior of (7.87) under rotation by 2π. According to (7.77), the left hand side of (7.87) must be multiplied by $(-1)^{2J}$, whereas the right hand side must be multiplied by $(-1)^{2(j_1+j_2)}$. These two factors will be identical under condition (c), which may now be restated by saying that in the triad (j_1, j_2, J) only an even number of members can be half odd-integers.

Recursion relations

It is possible to calculate the values of the CG coefficients by successive application of the raising or lowering operator to the defining equation,

$$|j_1, j_2, J, M\rangle = \sum_{m_1,m_2} |j_1, m_1\rangle^{(1)} |j_2, m_2\rangle^{(2)} (j_1, j_2, m_1, m_2|J, M) . \qquad (7.97)$$

This is illustrated most simply for the case $j_1 = j_2 = \frac{1}{2}$. When J and M take on their maximum possible values, $J = M = 1$, the sum in (7.97) will contain only one term,

$$|\tfrac{1}{2}, \tfrac{1}{2}, 1, 1\rangle = (\tfrac{1}{2}, \tfrac{1}{2}, \tfrac{1}{2}, \tfrac{1}{2}|1, 1)\ |\tfrac{1}{2}, \tfrac{1}{2}\rangle^{(1)}\ |\tfrac{1}{2}, \tfrac{1}{2}\rangle^{(2)} . \qquad (7.98)$$

The phase condition (7.89) and the fact that the vectors have unit norm imply that $(\frac{1}{2}, \frac{1}{2}, \frac{1}{2}, \frac{1}{2}|\, 1, 1) = 1$. We next apply the lowering operator, $\mathbf{J}_- = \mathbf{J}_-{}^{(1)} + \mathbf{J}_-{}^{(2)}$, to (7.98) and use (7.15) to obtain

$$\sqrt{2}|\tfrac{1}{2}, \tfrac{1}{2}, 1, 0\rangle = |\tfrac{1}{2}, -\tfrac{1}{2}\rangle^{(1)}\ |\tfrac{1}{2}, \tfrac{1}{2}\rangle^{(2)} + |\tfrac{1}{2}, \tfrac{1}{2}\rangle^{(1)} |\tfrac{1}{2}, -\tfrac{1}{2}\rangle^{(2)} .$$

Applying $\mathbf{J}_-$ again to this equation yields

$$|1, -1\rangle = |\tfrac{1}{2}, -\tfrac{1}{2}\rangle^{(1)}\ |\tfrac{1}{2}, -\tfrac{1}{2}\rangle^{(2)} .$$

By comparing these results with (7.97) we obtain the first three columns in the following table.

Values of $(\frac{1}{2}, \frac{1}{2}, m_1, m_2 \mid J, M)$ (7.99)

		(triplet)			(singlet)
	$J, M =$	$1, +1$	$1, 0$	$1, -1$	$0, 0$
$m_1, m_2 =$	$+\frac{1}{2}, +\frac{1}{2}$	1	0	0	0
	$+\frac{1}{2}, -\frac{1}{2}$	0	$\sqrt{\frac{1}{2}}$	0	$\sqrt{\frac{1}{2}}$
	$-\frac{1}{2}, +\frac{1}{2}$	0	$\sqrt{\frac{1}{2}}$	0	$-\sqrt{\frac{1}{2}}$
	$-\frac{1}{2}, -\frac{1}{2}$	0	0	1	0

The fourth column is obtained requiring the *singlet* state $|0,0\rangle$ to be orthogonal to the three *triplet* states, and using the phase condition (7.89).

By applying the lowering operator, $\mathbf{J}_- = \mathbf{J}_-{}^{(1)} + \mathbf{J}_-{}^{(2)}$, to the general case of (7.97), we obtain

$$\begin{aligned}
&[(J-M+1)(J+M)]^{1/2}|j_1,j_2,J,M-1\rangle \\
&\quad = \sum_{m_1,m_2} \Big\{[(j_1-m_1+1)(j_1+m_1)]^{1/2}|j_1,m_1-1\rangle^{(1)}\,|j_2,m_2\rangle^{(2)} \\
&\qquad + [(j_2-m_2+1)(j_2+m_2)]^{1/2}|j_1,m_1\rangle^{(1)}|j_2,m_2-1\rangle^{(2)}\Big\} \\
&\qquad \times (j_1,j_2,m_1,m_2|J,M)\,. \qquad (7.100)
\end{aligned}$$

This may be compared to (7.97) with M replaced by $M-1$:

$$|j_1,j_2,J,M-1\rangle = \sum_{m_1,m_2} |j_1,m_1\rangle^{(1)}\,|j_2,m_2\rangle^{(2)}\,(j_1,j_2,m_1,m_2|J,M-1)\,,$$

which yields

$$\begin{aligned}
&[(J-M+1)(J+M)]^{1/2}\,(j_1,j_2,m_1,m_2|J,M-1) \\
&\quad = [(j_1-m_1)(j_1+m_1+1)]^{1/2}\,(j_1,j_2,m_1+1,m_2|J,M) \\
&\qquad + [(j_2-m_2)(j_2+m_2+1)]^{1/2}\,(j_1,j_2,m_1,m_2+1|J,M)\,. \qquad (7.101)
\end{aligned}$$

A similar calculation using the raising operator $\mathbf{J}_+ = \mathbf{J}_+{}^{(1)} + \mathbf{J}_+{}^{(2)}$ yields

$$\begin{aligned}
&[(J+M+1)(J-M)]^{1/2}\,(j_1,j_2,m_1,m_2|J,M+1) \\
&\quad = [(j_1+m_1)(j_1-m_1+1)^{1/2}\,(j_1,j_2,m_1-1,m_2|J,M) \\
&\qquad + [(j_2+m_2)(j_2-m_2+1)]^{1/2}\,(j_1,j_2,m_1,m_2-1|J,M)\,. \qquad (7.102)
\end{aligned}$$

A useful application of these recursion relations may be made to the case of $j_1 = \ell, j_2 = \frac{1}{2}$, which arises in the study of *spin–orbit coupling*. If we put $m_2 = \frac{1}{2}$, its largest possible value, then the second term on the right hand side of (7.101) will vanish, leaving

$$\begin{aligned}
&[(J-M+1)(J+M)]^{1/2}\,(\ell,\tfrac{1}{2},M-\tfrac{3}{2},\tfrac{1}{2}|J,M-1) \\
&\quad = \left[(\ell-M+\tfrac{3}{2})(\ell+M-\tfrac{1}{2})\right]^{1/2}\,(\ell,\tfrac{1}{2},M-\tfrac{1}{2},\tfrac{1}{2}|J,M)\,.
\end{aligned}$$

Upon the substitution $M \to M+1$ this yields

$$\begin{aligned}
&(\ell, \tfrac{1}{2}, M-\tfrac{1}{2}, \tfrac{1}{2}|J, M) \\
&\quad = \left[\frac{(\ell - M + \frac{1}{2})\,(\ell + M + \frac{1}{2})}{(J-M)(J+M+1)}\right]^{\frac{1}{2}} (\ell, \tfrac{1}{2}, M+\tfrac{1}{2}, \tfrac{1}{2}|J, M+1)\,.
\end{aligned}$$

For the case of $J = \ell + \frac{1}{2}$ we use this equation recursively until the maximum value of M is reached:

$$\begin{aligned}
&(\ell, \tfrac{1}{2}, M-\tfrac{1}{2}, \tfrac{1}{2}|\ell+\tfrac{1}{2}, M) \\
&\quad = \left[\frac{(\ell + M + \frac{1}{2})}{(\ell + M + \frac{3}{2})}\right]^{\frac{1}{2}} (\ell, \tfrac{1}{2}, M+\tfrac{1}{2}, \tfrac{1}{2}|\ell+\tfrac{1}{2}, M+1) \\
&\quad = \left[\frac{(\ell + M + \frac{1}{2})\,(\ell + M + \frac{3}{2})}{(\ell + M + \frac{3}{2})\,(\ell + M + \frac{5}{2})}\right]^{\frac{1}{2}} (\ell, \tfrac{1}{2}, M+\tfrac{3}{2}, \tfrac{1}{2}|\ell+\tfrac{1}{2}, M+2) \\
&\quad\ \ \vdots \\
&\quad = \left[\frac{(\ell + M + \frac{1}{2})}{(2\ell+1)}\right]^{\frac{1}{2}} (\ell, \tfrac{1}{2}, \ell, \tfrac{1}{2}|\ell+\tfrac{1}{2}, \ell+\tfrac{1}{2})\,.
\end{aligned}$$

The final CG coefficient on the right is of the form $(j_1, j_2, j_1, j_2|j_1{+}j_2, j_1{+}j_2) = 1$, so the calculation of $(\ell, \frac{1}{2}, M-\frac{1}{2}, \frac{1}{2}|\ell+\frac{1}{2}, M)$ is complete, and its value is entered in the upper left corner of the following table.

Values of $(\ell, \frac{1}{2}, M - m_s, m_s \mid J, M)$ (7.103)

	$J = \ell + \frac{1}{2}$	$J = \ell - \frac{1}{2}$
$m_s = \frac{1}{2}$	$\left[\frac{(\ell + M + \frac{1}{2})}{(2\ell+1)}\right]^{\frac{1}{2}}$	$-\left[\frac{(\ell - M + \frac{1}{2})}{(2\ell+1)}\right]^{\frac{1}{2}}$
$m_s = -\frac{1}{2}$	$\left[\frac{(\ell - M + \frac{1}{2})}{(2\ell+1)}\right]^{\frac{1}{2}}$	$\left[\frac{(\ell + M + \frac{1}{2})}{(2\ell+1)}\right]^{\frac{1}{2}}$

The lower left entry in this table can be determined similarly from the recursion (7.102) with $m_2 = -\frac{1}{2}$. However its magnitude is more easily determined from the fact that the vector $|J = \ell + \frac{1}{2}, M\rangle$ is normalized, and hence sum of the squares of the entries in the left column of the table must must be 1. The entries in the right column are determined, apart from an overall sign, by requiring the vector $|J = \ell - \frac{1}{2}, M\rangle$ to be normalized and orthogonal to $|J = \ell + \frac{1}{2}, M\rangle$. The phase convention (7.89) determines that it is the $m_s = -\frac{1}{2}$ term that is positive.

The *spin–orbital eigenfunctions* will be denoted as $\mathcal{Y}_\ell{}^{J,M}$. Their explicit form, in the standard matrix representation, is

$$\mathcal{Y}_\ell{}^{\ell+\frac{1}{2},M} = \frac{1}{(2\ell+1)^{1/2}} \begin{bmatrix} (\ell + M + \frac{1}{2})^{1/2}\, Y_\ell{}^{M-\frac{1}{2}}(\theta,\phi) \\ (\ell - M + \frac{1}{2})^{1/2}\, Y_\ell{}^{M+\frac{1}{2}}(\theta,\phi) \end{bmatrix}, \tag{7.104a}$$

$$\mathcal{Y}_\ell{}^{\ell-\frac{1}{2},M} = \frac{1}{(2\ell+1)^{1/2}} \begin{bmatrix} -(\ell - M + \frac{1}{2})^{1/2}\, Y_\ell{}^{M-\frac{1}{2}}(\theta,\phi) \\ (\ell + M + \frac{1}{2})^{1/2}\, Y_\ell{}^{M+\frac{1}{2}}(\theta,\phi) \end{bmatrix}. \tag{7.104b}$$

By construction, they are eigenfunctions of $\mathbf{L}\cdot\mathbf{L}, \mathbf{S}\cdot\mathbf{S}, \mathbf{J}\cdot\mathbf{J}$, and $\mathbf{J}_z$. They are also eigenfunctions of the spin–orbit coupling operator $\mathbf{L}\cdot\mathbf{S}$, since

$$\mathbf{L}\cdot\mathbf{S} = \frac{1}{2}(\mathbf{J}\cdot\mathbf{J} - \mathbf{L}\cdot\mathbf{L} - \mathbf{S}\cdot\mathbf{S}). \tag{7.105}$$

Its eigenvalues are

$$\begin{aligned} \tfrac{1}{2}\hbar^2[j(j+1) - \ell(\ell+1) - 3/4] &= \tfrac{1}{2}\hbar^2\ell && \text{for } j = \ell + \tfrac{1}{2}, \\ &= -\tfrac{1}{2}\hbar^2(\ell+1) && \text{for } j = \ell - \tfrac{1}{2}. \end{aligned} \tag{7.106}$$

3–j symbols

The CG coefficient is related to a more symmetrical coefficient called the 3–j symbol,

$$\begin{pmatrix} j_1 & j_2 & j_3 \\ m_1 & m_2 & m_3 \end{pmatrix} = \frac{(-1)^{j_1-j_2-m_3}}{(2J_3+1)^{1/2}}(j_1, j_2, m_1, m_2|j_3, -m_3). \tag{7.107}$$

Rotenberg *et al.* (1959) have provided extensive numerical tables of these and some more complicated coefficients, as well as listing the principal relations that they satisfy. In accordance with (7.94), (7.95), and (7.96), the 3–j symbol must vanish unless $m_1 + m_2 + m_3 = 0$, (j_1, j_2, j_3) form the sides of a triangle, and $j_1 + j_2 + j_3$ is an integer.

Let us use the abbreviation (1 2 3) to denote the 3–j symbol (7.107). Its principal symmetries are:

(i) Even permutation:
$(1\ 2\ 3) = (2\ 3\ 1) = (3\ 1\ 2)$;

(ii) Odd permutation:
$(3\ 2\ 1) = (2\ 1\ 3) = (1\ 3\ 2) = (1\ 2\ 3)\ \times(-1)^{j_1+j_2+j_3}$;

(iii) Reversal of the signs of m_1, m_2, and m_3 causes the 3–j symbol to be multiplied by $(-1)^{j_1+j_2+j_3}$.

Although symmetry under interchange of 1 and 2 in (7.107) is to be expected from the definition of the CG coefficient in terms of the addition of two angular momenta, the three-fold permutation symmetry would not be expected from that point of view. Other symmetries, whose interpretation is more obscure, have been listed by Rotenberg *et al.*

7.8 *Irreducible Tensor Operators*

One of the reasons why the set of $2j+1$ angular momentum eigenvectors $\{|j,m\rangle : (-j \le m \le j)\}$ are important is that they transform under rotations in the simple manner (7.71), forming the basis for an *invariant irreducible subspace.* A set of $2k+1$ operators $\{T_q{}^{(k)} : (-k \le q \le k)\}$ which transform in an analogous fashion,

$$\mathbf{R}(\alpha,\beta,\gamma)T_q{}^{(k)}\,\mathbf{R}^{-1}(\alpha,\beta,\gamma) = \sum_{q'} T_{q'}{}^{(k)} D^{(k)}_{q',q}(\alpha,\beta,\gamma)\,, \tag{7.108}$$

are said to form an *irreducible tensor of degree k.*

A *scalar* operator S, which is (by definition) invariant under rotations,

$$\mathbf{R}(\alpha,\beta,\gamma)\,S\,\mathbf{R}^{-1}(\alpha,\beta,\gamma) = S\,, \tag{7.109}$$

is an irreducible tensor of degree $k=0$. A *vector* operator $\mathbf{V}$ is an irreducible tensor of degree $k=1$, since the three components of a vector transform into linear combinations of each other under rotation. The *spherical components* of the vector, which satisfy (7.108), are

$$V_{+1} = -(V_x + iV_y)\sqrt{\tfrac{1}{2}}\,,\quad V_0 = V_z\,,\quad V_{-1} = (V_x - iV_y)\sqrt{\tfrac{1}{2}}\,. \tag{7.110}$$

Although an irreducible tensor was defined in terms of its transformation under finite rotations, it may also be characterized by its transformation under infinitesimal rotations. Recalling the definition of the rotation matrix (7.64), we may rewrite (7.108) as

$$\mathbf{R}\,T_q{}^{(k)}\,\mathbf{R}^{-1} = \sum_{q'} T_{q'}{}^{(k)}\,\langle k,q'|\mathbf{R}|k,q\rangle\,.$$

For a rotation through the infinitesimal angle ε about an axis in the direction of the unit vector $\hat{\mathbf{n}}$, the rotation operator is $\mathbf{R} = 1 - i\,\varepsilon\,\hat{\mathbf{n}}\cdot\mathbf{J}/\hbar$, to the first order in ε, and hence the above equation yields

$$[\hat{\mathbf{n}}\cdot\mathbf{J}, T_q{}^{(k)}] = \sum_{q'} T_{q'}{}^{(k)}\ \langle k, q'|\hat{\mathbf{n}}\cdot\mathbf{J}|k, q\rangle\,. \tag{7.111}$$

By applying this result for $\hat{\mathbf{n}}$ in the z, x, and y directions, and using the fundamental properties (7.14) and (7.15) of the operators $J_+ = J_x + iJ_y$ and $J_- = J_x - iJ_y$, we obtain

$$[J_+, T_q^{(k)}] = \hbar\{(k-q)(k+q+1)\}^{1/2}\, T_{q+1}{}^{(k)}\,, \tag{7.112a}$$

$$[J_z, T_q{}^{(k)}] = \hbar q\, T_q{}^{(k)}\,, \tag{7.112b}$$

$$[J_-, T_q{}^{(k)}] = \hbar\{(k+q)(k-q+1)\}^{1/2}\, T_{q-1}{}^{(k)}\,. \tag{7.112c}$$

These commutation relations may be used as an alternative definition of the spherical components of an irreducible tensor, in place of (7.108).

Several useful results follow from the simple rotational properties of the tensor operators. But in order to deduce them, we must first derive some important relations between the rotation matrices and the CG coefficients. Let us apply a rotation to a product vector of the form (7.91):

$$\mathbf{R}|k,q\rangle|j,m\rangle = \sum_J\sum_M \mathbf{R}|k,j,J,M\rangle\ (k,j,q,m|J,M)\,.$$

Using (7.71) then yields

$$\sum_{q'}\sum_{m'} |k,q'\rangle|j,m'\rangle\ D^{(k)}_{q',q}(R)\ D^{(j)}_{m',m}(R)$$

$$= \sum_{M'}\sum_J\sum_M |k,j,J,M'\rangle\ D^{(J)}_{M',M}(R)\ (k,j,q,m|J,M)\,.$$

In the right hand side we now substitute

$$|k,j,J,M'\rangle = \sum_{q'}\sum_{m'} |k,q'\rangle\ |j,m'\rangle\ (k,j,q',m'|J,M')$$

and equate coefficients of $|k,q'\rangle|j,m'\rangle$, obtaining

$$D^{(k)}_{q',q}(R)\ D^{(j)}_{m',m}(R)$$

$$= \sum_{M'}\sum_J\sum_M (k,j,q',m'|J,M')\ D^{(J)}_{M',M}\ (R)(k,j,q,m|J,M)\,. \tag{7.113}$$

(This reduction of a product of rotation matrices into a sum of rotation matrices is sometimes called the *Clebsch–Gordan series.*) With the help of the orthogonality theorem (7.70), we can now evaluate the integral over all rotations of a product of three rotation matrices,

$$\int \{D^{(J)}_{M',M}(R)\}^* \, D^{(k)}_{q',q}(R) \, D^{(j)}_{m',m}(R) \, dR$$
$$= (k,j,q',m'|J,M')(k,j,q,m|J,M)(2J+1)^{-1} \int dR \,. \quad (7.114)$$

Matrix elements of tensor operators

The evaluation of matrix elements of tensor operators can be considerably simplified by means of these results. Let $|\tau, J, M\rangle$ be an eigenvector of (total) angular momentum. The eigenvalue τ represents any other operators that may be combined with $\mathbf{J}\cdot\mathbf{J}$ and J_z to form a complete commutative set. Using (7.71), (7.108), and $\mathbf{R}^\dagger = \mathbf{R}^{-1}$, we obtain

$$\langle \tau', J', M'|T_q{}^{(k)}|\tau, J, M\rangle = \langle \tau', J', M'|\mathbf{R}^\dagger \, \mathbf{R}T_q{}^{(k)}\mathbf{R}^\dagger\mathbf{R}|\tau, J, M\rangle$$
$$= \sum_{\mu'}\sum_{\sigma}\sum_{\mu} \{D^{(J')}_{\mu',M'}(R)\}^* \, D^{(k)}_{\sigma,q}(R) D^{(J)}_{\mu,M}(R)$$
$$\times \langle \tau', J', \mu'|T_\sigma{}^{(k)}|\tau, J, \mu\rangle \,.$$

The left hand side is independent of the rotation R, so we may integrate over all rotations and use (7.114) to obtain

$$\langle \tau', J', M'|T_q{}^{(k)}|\tau, J, M\rangle = \sum_{\mu'}\sum_{\sigma}\sum_{\mu} (2J'+1)^{-1} \, (J,k,\mu,\sigma|J',\mu')$$
$$\times \langle \tau', J', \mu'|T_\sigma{}^{(k)}|\tau, J, \mu\rangle \, (J,k,M,q|J',M') \,.$$

The final CG coefficient may be taken out of the sum as a factor, and the sum over remaining factors depends only upon τ', τ, J', k, and J. Thus the matrix element may be written in the form

$$\langle \tau', J', M'|T_q{}^{(k)}|\tau, J, M\rangle = \langle \tau', J'\|T^{(k)}\|\tau, J\rangle \, (J,k,M,q|J',M') \,, \quad (7.115)$$

which is known as the *Wigner–Eckart theorem.* The quantity $\langle \tau', J'\|T^{(k)}\|\tau, J\rangle$ is called the *reduced matrix element.* It is independent of M', q, and M, with the dependence of the full matrix element on these variables being explicitly given by the CG coefficient.

Example (1). The simplest example of the WE theorem is provided by the *scalar* operator S, of degree $k = 0$. It is evident from the defining equation (7.90) and the phase convention (7.89) that $(J, 0, M, 0|J', M') = \delta_{J',J}\,\delta_{M',M}$, and hence the matrix element

$$\langle \tau', J', M'|S|\tau, J, M\rangle = \langle \tau', J\|S\|\tau, J\rangle\;\delta_{J',J}\;\delta_{M',M} \tag{7.116}$$

is diagonal in angular momentum indices and is independent of $M' = M$.

Example (2). Although the equation leading to (7.115) implicitly provides a closed form expression for the *reduced matrix element*, it is more convenient to deduce it by evaluating the left hand side of (7.115) for one special case. We shall illustrate this procedure for the angular momentum operator $\mathbf{J}$, for which (7.115) becomes

$$\langle \tau', J', M'|{J_q}^{(1)}|\tau, J, M\rangle = \langle \tau', J'\|\mathbf{J}\|\tau, J\rangle\;(J, 1, M, q|J', M')\,. \tag{7.117}$$

But we know that

$$\langle \tau', J', M'|{J_0}^{(1)}|\tau, J, M\rangle \equiv \langle \tau', J', M'|J_z|\tau, J, M\rangle = \hbar M\;\delta_{J',J}\;\delta_{M',M}\;\delta_{\tau',\tau}\,,$$

so it suffices to evaluate (7.117) for the case of $J = M = M' = J'$. The relevant CG coefficient can be found in standard references to be $(J, 1, J, 0|J, J) = \{J/(J+1)\}^{1/2}$, so we obtain $\langle \tau', J\|\mathbf{J}\|\tau, J\rangle = \hbar\{J(J+1)\}^{1/2}\delta_{\tau',\tau}$ for this case. In the general case, the total matrix element vanishes for $J' \neq J$. Since the CG coefficient need not vanish for $J' \neq J$, it follows that the reduced matrix element must do so, and hence

$$\langle \tau', J'\|\mathbf{J}\|\tau, J\rangle = \hbar\{J(J+1)\}^{1/2}\;\delta_{\tau',\tau}\;\delta_{J',J}\,. \tag{7.118}$$

Example (3). The relation (7.75) between spherical harmonics and rotation matrix elements allows us to deduce an integral of a product of three spherical harmonics. From (7.114) we obtain

$$\begin{aligned}&\int\!\!\int\!\!\int \{D^{(L)}_{M,0}(\alpha,\beta,\gamma)\}^*\;D^{(k)}_{q,0}(\alpha,\beta,\gamma)\;D^{(\ell)}_{m,0}(\alpha,\beta,\gamma)\;d\alpha\;\sin\beta\;d\beta\;d\gamma\\ &\quad= (k, \ell, q, m|L, M)\;(k, \ell, 0, 0|L, 0)\;(2L+1)^{-1}\\ &\qquad\times\int\!\!\int\!\!\int d\alpha\;\sin\beta\;d\beta\;d\gamma\,.\end{aligned}$$

Since the right hand side is real, we may formally replace the left hand side by its complex conjugate. Substitution of (7.75) then yields

$$\iint \{Y_L{}^M(\theta,\phi)\}^* \, Y_k{}^q(\theta,\phi) \sin\theta \, d\theta \, d\phi$$

$$= \left[\frac{(2k+1)(2\ell+1)}{4\pi(2L+1)}\right]^{1/2} (\ell,k,0,0|L,0)\,(\ell,k,m,q|L,M)\,, \tag{7.119}$$

which is known as *Gaunt's formula.* This integral can also be regarded as a matrix element of irreducible tensor operator, to which the Wigner–Eckart theorem applies, yielding $\langle L,M|Y_k{}^q|\ell,m\rangle = \langle L\|Y_k\|\ell\rangle(\ell,k,m,q|L,M)$. Thus Gaunt's formula is an instance of the Wigner–Eckart theorem.

Products of tensors

The product of two irreducible tensor operators will usually not be an *irreducible* tensor. However, it is easy to construct irreducible tensors from such a product. Let $X_q{}^{(k)}$ and $Z_m{}^{(\ell)}$ be irreducible tensor operators of rank k and ℓ respectively. Then

$$T_M{}^{(L)} = \sum_q \sum_m (k,\ell,q,m|L,M) X_q{}^{(k)} Z_m{}^{(\ell)} \tag{7.120}$$

is an *irreducible tensor of rank L.* This equation is closely analogous to (7.90). Its proof, using (7.113), is left as an exercise for the reader.

The special case $T_0{}^{(0)}$ is a scalar:

$$T_0{}^{(0)} = \sum_m (j,j,m,-m|0,0)\, X_{-m}{}^{(j)} Z_m{}^{(j)}$$

$$= (2j+1)^{-1/2} \sum_m (-1)^{j-m} X_{-m}{}^{(j)} Z_m{}^{(j)}\,. \tag{7.121}$$

[The CG coefficient here can be obtained from $(j,0,m,0|j,m) = 1$ by using (7.107) and the permutation symmetry of the 3–j symbol.] If $\mathbf{X}$ and $\mathbf{Z}$ are ordinary vectors, then the usual *scalar product* is $\mathbf{X}\cdot\mathbf{Z} = -\sqrt{3}\,T_0{}^{(0)}$.

Example (4). As a final example, we consider the matrix elements of the operator $\mathbf{J}\cdot\mathbf{V}$, where $\mathbf{J}$ is the angular momentum operator and $\mathbf{V}$ is any vector operator. Since $\mathbf{J}\cdot\mathbf{V}$ is a scalar, (7.116) tells us that matrix will be diagonal in angular momentum indices, and hence we need only consider

$$\begin{aligned}\langle \tau', J, M|\mathbf{J}\cdot\mathbf{V}|\tau, J, M\rangle &= \sum_m \langle \tau', J, M|(-1)^m\, J_{-m}{}^{(1)}\, V_m{}^{(1)}|\tau, J, M\rangle \\ &= \sum_{\tau''}\sum_{J''}\sum_{M''}\sum_m (-1)^m \langle \tau', J, M|J_{-m}{}^{(1)}|\tau'', J'', M''\rangle \\ &\quad \times \langle \tau'', J'', M''|V_m{}^{(1)}|\tau, J, M\rangle\,.\end{aligned}$$

The matrix for any component of $\mathbf{J}$ is diagonal in J and τ, so the sums on τ'' and J'' will involve only $\tau'' = \tau'$ and $J'' = J$. Moreover, its value is independent of τ. Using the WE theorem (7.115) for the matrix element of $V_m{}^{(1)}$, we obtain

$$\begin{aligned}\langle \tau', J, M|\mathbf{J}\cdot\mathbf{V}|\tau, J, M\rangle &= \sum_{M''}\sum_m (-1)^m \langle J, M|J_{-m}{}^{(1)}|J, M''\rangle \\ &\quad \times (J, 1, M, m|J, M'')\langle \tau', J\|\mathbf{V}\|\tau, J\rangle \\ &= C_{JM}\langle \tau', J\|\mathbf{V}\|\tau, J\rangle\,. \end{aligned} \tag{7.122}$$

It is clearly apparent that C_{JM} is independent of τ' and τ, and of the particular nature of the vector operator $\mathbf{V}$. Therefore we may evaluate it by substituting $\mathbf{J}$ for $\mathbf{V}$:

$$\langle \tau, J, M|\mathbf{J}\cdot\mathbf{J}|\tau, J, M\rangle = C_{JM}\langle \tau, J\|\mathbf{J}\|\tau, J\rangle\,.$$

Using (7.118), we see that this yields

$$C_{JM} = \hbar\{J(J+1)\}^{1/2}\,. \tag{7.123}$$

This may be substituted into (7.122) to obtain the reduced matrix element of $\mathbf{V}$ (for $J' = J$) in terms of a matrix element of the scalar $\mathbf{J}\cdot\mathbf{V}$, which is often simpler to calculate.

The WE theorem, applied to the operators $\mathbf{V}$ and $\mathbf{J}$, yields

$$\langle \tau', J', M'|V_m{}^{(1)}|\tau, J, M\rangle = \langle \tau', J'\|\mathbf{V}\|\tau, J\rangle\,(J, 1, M, m|J', M')\,,$$

$$\langle \tau, J, M'|J_m{}^{(1)}|\tau, J, M\rangle = \langle \tau, J\|\mathbf{J}\|\tau, J\rangle\,(J, 1, M, m|J, M')\,.$$

In the latter equation, we have taken $\tau' = \tau$ and $J' = J$ to avoid the trivial identity $0 = 0$. Thus we have

$$\langle \tau', J, M'|V_m{}^{(1)}|\tau, J, M\rangle = \frac{\langle \tau', J\|\mathbf{V}\|\tau, J\rangle}{\langle \tau, J\|\mathbf{J}\|\tau, J\rangle}\,\langle \tau, J, M'|J_m{}^{(1)}|\tau, J, M\rangle\,. \tag{7.124}$$

For the case of $J' = J$, the reduced matrix elements can be obtained from (7.118) and (7.122), yielding

$$\langle \tau', J, M'|V_m{}^{(1)}|\tau, J, M\rangle = \frac{\langle \tau', J, M|\mathbf{J}\cdot\mathbf{V}|\tau, J, M\rangle}{\hbar^2 J(J+1)} \langle J, M'|J_m{}^{(1)}|J, M\rangle \,. \quad (7.125)$$

Since the matrix element of the scalar product $\mathbf{J}\cdot\mathbf{V}$ is in fact independent of M, this equation asserts that *for fixed* τ', τ *and* $J = J'$ *the matrix elements of any vector operator are proportional to those of the angular momentum operator.*

Most practical uses of this result are for the case $\tau' = \tau$. If the dynamics of a system are such that the state is approximately confined within a subspace of fixed τ and J, then (7.125) implies that it can be described by a model Hamiltonian involving only angular momentum operators. For example, the *magnetic moment* operator for an atom has the form

$$\boldsymbol{\mu} = \frac{-e}{2m_e c}(g_L\mathbf{L} + g_s\mathbf{S})\,. \quad (7.126)$$

Here the operators $\mathbf{L}$ and $\mathbf{S}$ correspond to the total orbital and spin angular momenta of the atom. The charge and mass of the electron are $-e$ and m_e, and c is the speed of light. The parameters g_L and g_s have approximately the values $g_L = 1$ and $g_s = 2$. Using (7.125), we can write

$$\langle \tau, J, M'|\boldsymbol{\mu}|\tau, J, M\rangle = \frac{-e}{2m_e c}\, g_{\text{eff}}\langle J, M'|\mathbf{J}|J, M\rangle \quad (7.127)$$

where $\mathbf{J} = \mathbf{L} + \mathbf{S}$ is the total angular momentum operator, and

$$\begin{aligned} g_{\text{eff}} &= \frac{\langle \tau, J, M|\boldsymbol{\mu}\cdot\mathbf{J}|\tau, J, M\rangle}{\hbar^2 J(J+1)} \\ &= \frac{\langle \tau, J, M|(g_L\mathbf{L}\cdot\mathbf{J} + g_s\mathbf{S}\cdot\mathbf{J})|\tau, J, M\rangle}{\hbar^2 J(J+1)}\,. \end{aligned}$$

But $\mathbf{L}\cdot\mathbf{J} = \frac{1}{2}(\mathbf{J}\cdot\mathbf{J} + \mathbf{L}\cdot\mathbf{L} - \mathbf{S}\cdot\mathbf{S})$ and $\mathbf{S}\cdot\mathbf{J} = \frac{1}{2}(\mathbf{J}\cdot\mathbf{J} + \mathbf{S}\cdot\mathbf{S} - \mathbf{L}\cdot\mathbf{L})$; hence we have

$$g_{\text{eff}} = \frac{\langle \tau, J, M|\{(g_L + g_s)\mathbf{J}\cdot\mathbf{J} + (g_L - g_s)(\mathbf{L}\cdot\mathbf{L} - \mathbf{S}\cdot\mathbf{S})\}|\tau, J, M\rangle}{2\hbar^2 J(J+1)}\,. \quad (7.128)$$

In the $L - S$ coupling approximation, the atomic state vector is an eigenvector of $\mathbf{L}\cdot\mathbf{L}$ and $\mathbf{S}\cdot\mathbf{S}$, with eigenvalues $\hbar^2 L(L+1)$ and $\hbar^2 S(S+1)$, respectively. With the values $g_L = 1$ and $g_s = 2$, (7.128) then reduces to

$$g_{\text{eff}} = \left[1 + \frac{J(J+1) - L(L+1) + S(S+1)}{2J(J+1)}\right], \tag{7.129}$$

which is known as the *Lande g factor*. These results are useful in atomic spectroscopy and in magnetic resonance.

7.9 Rotational Motion of a Rigid Body

The quantum theory of a many-particle system, such as a polyatomic molecule or a nucleus, is necessarily very complicated. But if the system is tightly bound it will rotate as a rigid body, and the quantum theory of that rotational motion is independent of the other details of the system. The Hamiltonian for such motion is

$$H = \frac{1}{2}\left[\frac{{J_a}^2}{I_a} + \frac{{J_b}^2}{I_b} + \frac{{J_c}^2}{I_c}\right], \tag{7.130}$$

where I_a, I_b, and I_c are the principal moments of inertia, and J_a, J_b, and J_c are the angular momentum operators for the components along the corresponding body-fixed axes.

Although the theory of body-fixed angular momentum operators is not difficult, it presents a couple of surprises. The commutation relation of the angular momentum operators for components along three mutually orthogonal *space-fixed axes* is well known to be

$$[J_x, J_y] = i\hbar J_z\,. \tag{7.131}$$

There is nothing special about the x, y, and z directions, and the orthogonal triplet of axes may have any spatial orientation. Since, at any instant of time, the principle axes a, b, and c form such an orthogonal triplet, one might expect the same commutation relation to hold for J_a, J_b, and J_c. But in fact, the commutation relation for *body-fixed axes* has the opposite sign:

$$[J_a, J_b] = -i\hbar J_c\,. \tag{7.132}$$

The reason for this difference is that a body-fixed component has the form $J_a = \mathbf{J}\cdot\hat{\mathbf{a}}$, where the body-fixed unit vector $\hat{\mathbf{a}}$ represents a dynamical variable of the body, analogous to a position operator. To evaluate the commutator of J_a and J_b, we need the following, easily verified identity:

$$[AB, CD] = A[B,C]D + AC[B,D] + [A,C]DB + C[A,D]B\,. \tag{7.133}$$

Now let **u** and **v** be two position-like vector operators. They commute with each other, and their commutators with the angular momentum operators have the form typical of three-vector operators,

$$[J_\alpha, u_\beta] = i\hbar\varepsilon_{\alpha\beta\gamma}u_\gamma\,, \tag{7.134}$$

where $\varepsilon_{\alpha\beta\gamma}$ is the antisymmetric tensor, introduced in Sec. 3.3. By substitution into (7.133) we obtain

$$\begin{aligned}[u_\alpha J_\alpha, v_\beta J_\beta] &= u_\alpha[J_\alpha, v_\beta]J_\beta + u_\alpha v_\beta[J_\alpha, J_\beta] + [u_\alpha, v_\beta]J_\beta J_\alpha + v_\beta[u_\alpha, J_\beta]J_\alpha \\ &= u_\alpha(i\hbar\varepsilon_{\alpha\beta\gamma}v_\gamma)J_\beta + u_\alpha v_\beta(i\hbar\varepsilon_{\alpha\beta\gamma}J_\gamma) + 0 \\ &\quad + v_\beta(-i\hbar\varepsilon_{\beta\alpha\gamma}u_\gamma)J_\alpha\,.\end{aligned} \tag{7.135}$$

Summing over α, β, and γ, and interchanging dummy indices in some of the terms, we obtain

$$\begin{aligned}[\mathbf{u}\cdot\mathbf{J}, \mathbf{v}\cdot\mathbf{J}] &= \sum_{\alpha\beta\gamma}\{i\hbar\varepsilon_{\alpha\beta\gamma}(-1+1+0-1)u_\alpha v_\beta J_\gamma\} \\ &= \sum_{\alpha\beta\gamma} -i\hbar\varepsilon_{\alpha\beta\gamma}u_\alpha v_\beta J_\gamma\,.\end{aligned} \tag{7.136}$$

If we now choose **u** and **v** to be unit vectors along the principle axes a and b, we obtain Eq. (7.132). If **u** and **v** had been numerically fixed vectors in space, without any operator properties, then only the second term of (7.135) would occur, and we would have obtained the familiar commutation relations (7.131) for space-fixed axes. The change of sign for body-fixed axes comes from the first and fourth terms of (7.135), which are nonzero because the body-fixed vectors are dynamical variables, with nontrivial commutation properties.

The eigenvalue spectrum for body-fixed angular momentum components can be obtained by the methods of Sec. 7.1. It turns out that the eigenvalues are the same as for space-fixed components, but the extra minus sign in (7.132) leads to some sign changes in the matrix elements.

There are several interesting special cases of (7.130), corresponding to different symmetries of the moment-of-inertia tensor. The *spherical top* has all principal moments of inertia equal: $I_a = I_b = I_c \equiv I$. The Hamiltonian takes the form $H = \mathbf{J}\cdot\mathbf{J}/2I$, and its eigenvalues are

$$E_j = \frac{\hbar^2 j(j+1)}{2I}\,. \tag{7.137}$$

Our second surprise concerns the degeneracy of these eigenvalues, which can be determined if we know a complete set of commuting operators for the system. This set consists of:

(i) The magnitude of the angular momentum (the same in both space-fixed and body-fixed coordinates), $\mathbf{J}\cdot\mathbf{J} = {J_x}^2 + {J_y}^2 + {J_z}^2 = {J_a}^2 + {J_b}^2 + {J_c}^2$;
(ii) One space-fixed component, usually chosen to be J_z;
(iii) One body-fixed component, usually chosen to be J_c.

The first and second operators were expected; the third is a surprise, since J_c appears to be an angular momentum component in a direction different from z, which ought not to commute with J_z. But in fact, $J_c = \hat{\mathbf{c}}\cdot\mathbf{J}$ is a scalar operator, which commutes with all the space-fixed rotation generators, including J_z. Thus the degeneracy of the energy eigenvalues is $(2j+1)^2$, rather that only $2j+1$, as might have been expected.

The *symmetric top* has two equal principal moments of inertia, $I_a = I_b \neq I_c$. Since ${J_a}^2 + {J_b}^2 = \mathbf{J}\cdot\mathbf{J} - {J_c}^2$, the Hamiltonian can be written as

$$H = \frac{1}{2}\frac{\mathbf{J}\cdot\mathbf{J}}{I_a} + \frac{1}{2}{J_c}^2\left[\frac{1}{I_c} - \frac{1}{I_a}\right]. \tag{7.138}$$

The energy eigenvalues are

$$E_{jk} = \frac{1}{2}\hbar^2\frac{j(j+1)}{I_a} + \frac{1}{2}\hbar^2\left[\frac{1}{I_c} - \frac{1}{I_a}\right]k^2\,, \tag{7.139}$$

where $\hbar k$ is the eigenvalue of J_c, ranging from $-j$ to j. The degree of degeneracy is $2j+1$, corresponding to the eigenvalues of J_z.

A *linear molecule* can be modeled as the limit $I_c \to 0$. Only the states with $k = 0$ are allowed in this limit, since the energies for $k \neq 0$ become arbitrarily large, and would never be realized. The surviving energy levels have the same values, (7.137), as for the spherical top, but now the degree of degeneracy is only $2j+1$.

The *asymmetric top* has all three principal moments of inertia unequal. No general closed-form expression for the eigenvalues of (7.130) exists. But Landau and Lifshitz (1958, Sec. 101) have given formulas for the energy eigenvalues in the special cases of total angular momentum $j = 1, 2$, and 3.

Further reading for Chapter 7

Quantum Theory of Angular Momentum, edited by L. C. Biedenharn and H. Van Dam (1965), is a collection of interesting reprints and original papers.

The previously unpublished papers by E. P. Wigner and by J. Schwinger are particularly interesting.

Problems

7.1 Find the probability distributions of the orbital angular momentum variables L^2 and L_z for the following orbital state functions:
(a) $\Psi(\mathbf{x}) = f(r)\sin\theta\ \cos\phi$,
(b) $\Psi(\mathbf{x}) = f(r)(\cos\theta)^2$,
(c) $\Psi(\mathbf{x}) = f(r)\sin\theta\ \cos\theta\ \sin\phi$.
Here r, θ, ϕ are the usual spherical coordinates, and $f(r)$ is an arbitrary radial function (not necessarily the same in each case) into which the normalization constant has been absorbed.

7.2 Can there be an *internal linear momentum* analogous to the *internal angular momentum* (known as spin)?
Rationale for the question: The mathematical origin of spin is in the rotational transformation properties of a multicomponent state function, as in (7.19):

$$\mathbf{R}_n(\theta)\begin{bmatrix}\Psi_1(\mathbf{x})\\ \Psi_2(\mathbf{x})\\ \vdots\end{bmatrix} = D_n(\theta)\begin{bmatrix}\Psi_1(R^{-1}\mathbf{x})\\ \Psi_2(R^{-1}\mathbf{x})\\ \vdots\end{bmatrix}.$$

The form of the rotation operator is $\mathbf{R}_n(\theta) = e^{-i\theta\hat{\mathbf{n}}\cdot\mathbf{L}/\hbar}D_n(\theta)$. The first factor produces the coordinate transformation on the argument of each component, and the latter factor is a unitary matrix that replaces the components with linear combinations of each other. Letting θ become infinitesimal, we obtain the rotation generator, which is the angular momentum operator, $\mathbf{J} = \mathbf{L} + \mathbf{S}$, with $\mathbf{L} = -i\hbar\mathbf{x}\times\nabla$ and $\mathbf{S}$ being a matrix derived from $D_n(\theta)$. These two terms of $\mathbf{J}$ are interpreted as the orbital and spin parts of the angular momentum.
Now let us treat space displacement in the same way. The displacement operator is $T(\mathbf{a}) = e^{-i\mathbf{a}\cdot\mathbf{P}/\hbar}$, and its most general effect could be of the form

$$T(\mathbf{a})\begin{bmatrix}\Psi_1(\mathbf{x})\\ \Psi_2(\mathbf{x})\\ \vdots\end{bmatrix} = F(\mathbf{a})\begin{bmatrix}\Psi_1(\mathbf{x}-\mathbf{a})\\ \Psi_2(\mathbf{x}-\mathbf{a})\\ \vdots\end{bmatrix},$$

where $F(\mathbf{a})$ is a matrix. Specializing to an infinitesimal displacement, we find the generator of displacements to be $\mathbf{P} = -i\hbar\nabla + \mathbf{M}$, where the matrix $\mathbf{M}$ is obtained from $F(\mathbf{a}) = e^{-i\mathbf{a}\cdot\mathbf{M}/\hbar}$. The second term, $\mathbf{M}$, would be interpreted as an *internal linear momentum.*

What can you prove about $\mathbf{M}$? How is it related to $\mathbf{S}$? Does $\mathbf{M}$ really exist, or can you prove that necessarily $\mathbf{M} \equiv 0$?

7.3 Let $\mathbf{A}$ and $\mathbf{B}$ be vector operators. This means that they have certain nontrivial commutation relations with the angular momentum operators. Use those relations to prove that $\mathbf{A} \cdot \mathbf{B}$ commutes with J_x, J_y, and J_z.

7.4 Prove that if an operator commutes with any two components of $\mathbf{J}$ it must also commute with the third component.

7.5 The spin matrices for $s = 1$ are given in Eq. (7.52). Show that their squares, ${S_x}^2, {S_y}^2$, and ${S_z}^2$, are commutative. Construct their common eigenvectors. What geometrical significance do those vectors possess?

7.6 Find the $s = 1$ spin matrices in the basis formed by the eigenvectors of Problem 7.5.

7.7 Consider a two-particle system of which one particle has spin s_1 and the other has spin s_2.

(a) If one particle is taken from each of two sources characterized by the state vectors $|s_1, m_1\rangle$ and $|s_2, m_2\rangle$, respectively, what is the probability that the resultant two-particle system will have total spin S?

(b) If the particles are taken from unpolarized sources, what is the probability that the two-particle system will have total spin S?

7.8 Prove the identity $(\sigma \cdot \mathbf{A})(\sigma \cdot \mathbf{B}) = \mathbf{A} \cdot \mathbf{B} + i\sigma \cdot (\mathbf{A} \times \mathbf{B})$, where $(\sigma_x, \sigma_y, \sigma_z)$ are the Pauli spin operators, and $\mathbf{A}$ and $\mathbf{B}$ are vector operators which commute with σ but do not necessarily commute with each other.

7.9 Consider a system of two spin $\frac{1}{2}$ particles. Calculate the eigenvalues and eigenvectors of the operator $\sigma^{(1)} \cdot \sigma^{(2)}$. Use the product vectors $|m_1\rangle \otimes |m_2\rangle$ as basis vectors.

7.10 Two spin $\frac{1}{2}$ particles interact through the spin-dependent potential $V(r) = V_1(r) + V_2(r)\sigma^{(1)} \cdot \sigma^{(2)}$. Show that the equation determining the bound states can be split into two equations, one having the effective potential $V_1(r) + V_2(r)$ and the other having the effective potential $V_1(r) - 3V_2(r)$.

7.11 Prove that the operator defined in (7.120) is indeed an irreducible tensor operator of rank L.

7.12 Use (7.120) to evaluate the spherical tensor components of $\mathbf{L} = \mathbf{Q} \times \mathbf{P}$, regarding the vector operators as tensors of rank 1, and the product as an irreducible tensor product.

7.13 What irreducible tensors can be formed from the nine components of the bivector $A_\alpha B_\beta$ $(\alpha, \beta = x, y, z)$?

7.14 Prove that a nucleus having spin 0 or spin $\frac{1}{2}$ cannot have an electric quadrupole moment. (The "spin" of a nucleus is really the resultant of the spins and relative orbital angular momenta of its constituent nucleons.)

7.15 An electric quadrupole moment couples to the gradient of the electric field, or equivalently to the second derivative of the scalar potential ϕ. If the axes of coordinates are chosen to be the principal axes of the field gradient, the quadrupole Hamiltonian may be taken to be of the form

$$H_q = C\left\{\left(\frac{\partial^2\phi}{\partial x^2}\right) S_x{}^2 + \left(\frac{\partial^2\phi}{\partial y^2}\right) S_y{}^2 + \left(\frac{\partial^2\phi}{\partial z^2}\right) S_z{}^2\right\},$$

where ϕ satisfies the Laplace equation, and the second derivatives are evaluated at the location of the particle.

(a) Show that this Hamiltonian can be written as

$$H_q = A(3S_z{}^2 - \mathbf{S}\cdot\mathbf{S}) + B(S_+{}^2 + S_-{}^2),$$

where A and B are related to C and the derivatives of ϕ.

(b) Find the eigenvalues of H_q for a system with spin $s = 3/2$.

7.16 Consider a system of three particles of spin $\frac{1}{2}$. A basis for the states of this system is provided by the eight product vectors, $|m_1\rangle \otimes |m_2\rangle \otimes |m_3\rangle$, where the m's take on the values $\pm\frac{1}{2}$. Find the linear combinations of these product vectors that are eigenvectors of the total angular momentum operators, $\mathbf{J}\cdot\mathbf{J}$ and J_z, where $\mathbf{J} = \mathbf{S}^{(1)} + \mathbf{S}^{(2)} + \mathbf{S}^{(3)}$.

Chapter 8

State Preparation and Determination

In this chapter we return to the fundamental development of quantum theory. The formal structure of the theory, set forth abstractly in Ch. 2, involves two basic concepts, *dynamical variables* and *states*, and their mathematical representations. In Ch. 3 we determined the particular operators that correspond to particular dynamical variables. It would have been logical to proceed next to the discussion of the mathematical representation of particular physical states. That discussion has been delayed until now because the intervening four chapters have made it possible to treat some specific cases, instead of merely discussing state preparation and state determination in general terms.

Postulate 2 of Sec. 2.1 asserts that "to each state there corresponds a unique state operator". The term "state" was identified with a reproducible preparation procedure that determines a probability distribution for each dynamical variable. Thus we are faced with two problems:

(1) The problem of *state preparation* — what is the procedure for preparing the state that is represented by some chosen state operator (or state vector)?
(2) The problem of *state determination* — for some given situation, how do we determine the corresponding state operator?

8.1 *State Preparation*

If at time $t = t_0$ we have a known pure state represented by the state vector $|\Psi_0\rangle$, then it is possible in principle to construct a time development operator $U(t_1, t_0)$ that will produce any desired pure state, $|\Psi_1\rangle = U(t_1, t_0)|\Psi_0\rangle$, at some later time t_1. But it is not obvious whether this $U(t_1, t_0)$ can be realized in practice, or whether it is only a mathematical construct. In some special cases it clearly can be realized. For example, if we have available the ground state of an atom, then it is possible to prepare an excited state, or a linear combination of excited states, by means of a suitable pulse of electromagnetic radiation from a laser. In the context of quantum optics, Reck *et al.* (1994) have shown how an arbitrary $N \times N$ unitary transformation can be realized from a sequence of beam splitters, phase shifters, and mirrors. Although their constructive proof was done for photon states, analogous methods exist for electrons, neutrons, and atoms. So this demonstrates that, at least for finite-dimensional state spaces, it is indeed possible to produce any desired state from a given initial state.

But these methods rely on having a *known* initial state. The more fundamental problem is how to prepare a particular state *from an arbitrary unknown initial state*. This involves a many-to-one mapping from any arbitrary state to the particular desired state. Since this mapping is not invertible, its realization must involve an irreversible process. Since the fundamental laws of nature are reversible, as far as we know, the effective irreversibility that we need must come about by coupling the system to a suitable apparatus or environment, to which entropy or information can be transferred. Thus, even in a microscopically reversible world, it is possible to achieve an effectively irreversible transformation on the system of interest.

It is possible to prepare the lowest energy state of a system simply by *waiting for the system to decay to its ground state*. The decay of an atomic excited state by spontaneous emission of radiation takes place because of the coupling of the atom to the electromagnetic field. If the survival probability of an excited state decays toward zero (usually exponentially with time, but see Sec. 12.2 for exceptions), then the probability of obtaining the ground state can be made arbitrarily close to 1 by waiting a sufficiently long time. Success of this method is based on the assumption that the energy of the excited state will be radiated away to infinity, never to return. If it is possible for the energy to be reflected back, then the method will not be reliable. It is also necessary to assume that the electromagnetic field is initially in its lowest energy state. If that is not the case, then there will be a nonvanishing probability for the atom to absorb energy from the field and become re-excited. In thermal equilibrium the probability of obtaining an excited atomic state will be proportional to $\exp(-E_x/k_{\mathrm{B}}T)$, where E_x is the lowest excitation energy of the atom, k_{B} is Boltzmann's constant, and T is the effective temperature of the radiation field. This factor is normally quite small, but the presence of cosmic background radiation at a temperature of 3K provides a lower limit unless special shielding and refrigeration techniques are used. This problem will be much more serious if, instead of an atom, we consider a system like a metallic crystal, for which the excitation energy E_x is very small.

If the strategy of waiting can be successfully used to produce a ground state, then a wide variety of states for a spinless particle can be prepared by a generalization of it. Suppose we wish to prepare the state that is represented by the function $\Psi_1(\mathbf{x}) = R(\mathbf{x})e^{iS(\mathbf{x})}$, where $R(\mathbf{x})$ and $S(\mathbf{x})$ are real. The first step is to construct a potential $W_1(\mathbf{x})$ for which the ground state — that is, the lowest energy solution of the equation $(-\hbar^2/2M)\nabla^2\Psi_0 + W_1\Psi_0 = E\Psi_0$ — is $\Psi_0(\mathbf{x}) = R(\mathbf{x})$. We must restrict $R(\mathbf{x})$ to be a nodeless function, otherwise it will not be the ground state. The required potential is

$$W_1(\mathbf{x}) = E + \frac{\hbar^2}{2M}\frac{\nabla^2 R(\mathbf{x})}{R(\mathbf{x})}. \tag{8.1}$$

We then wait until the probability that the system has decayed to its ground state is sufficiently close to 1.

The next step is to apply a pulse potential,

$$W_2(\mathbf{x},t) = -\hbar S(\mathbf{x})\delta_\varepsilon(t), \tag{8.2}$$

where $\delta_\varepsilon(t) = \varepsilon^{-1}$ for the short interval of time, $0 < t < \varepsilon$, and otherwise is equal to zero. The Schrödinger equation (4.4) can be approximated by

$$i\hbar\frac{\partial\Psi}{\partial t} = W_2(\mathbf{x},t)\Psi(\mathbf{x},t)$$

during the short time interval $0 < t < \varepsilon$, because W_2 will overwhelm any other interactions in the limit $\varepsilon \to 0$. Integrating this equation with the initial condition $\Psi(\mathbf{x},0) = R(\mathbf{x})$ yields $\Psi(\mathbf{x},0+\varepsilon) = R(\mathbf{x})e^{iS(\mathbf{x})}$, which is the state that we want to prepare. Of course any realization of this technique will be limited by the kinds of potential fields that can be produced in practice.

Another method of state preparation is *filtering*. A prototype of this technique is provided by the Stern–Gerlach apparatus. It will be analyzed in greater detail in Sec. 9.1, but the principle is very simple. The potential energy of a magnetic moment $\boldsymbol{\mu}$ in a magnetic field $\mathbf{B}$ is equal to $-\mathbf{B}\cdot\boldsymbol{\mu}$. If the magnetic field is spatially inhomogeneous, then the negative gradient of this potential energy corresponds to a force on the particle,

$$\mathbf{F} = \boldsymbol{\nabla}(\mathbf{B}\cdot\boldsymbol{\mu}). \tag{8.3}$$

The magnitude and sign of this force will depend upon the spin state, since the magnetic moment $\boldsymbol{\mu}$ is proportional to the spin $\mathbf{S}$, and hence different spin states will be deflected by this force into sub-beams propagating in different directions. By blocking off, and so eliminating, all but one of the sub-beams (an irreversible process), we can select a particular spin state. The method of preparing particular discrete atomic states [Koch *et al.* (1988)], which was discussed in the Introduction, is similar in principle, although the techniques are much more sophisticated.

No-cloning theorem

A superficially attractive method of state preparation would be to make exact replicas, or "clones", of a prototype of the state (provided that one can be found). This method is common in the domain of classical physics: for example, the duplication of a key, or the copying of a computer file. Surprisingly, the linearity of the quantum equation of motion makes the cloning of quantum states impossible.

In order to clone an arbitrary quantum state $|\phi\rangle$, we would require a device in some suitable state $|\Psi\rangle$ and a unitary time development operator U such that

$$U|\phi\rangle \otimes |\Psi\rangle = |\phi\rangle \otimes |\phi\rangle \otimes |\Psi'\rangle . \tag{8.4}$$

[The dimension of the space of the final device state vector $|\Psi'\rangle$ will be smaller than that of $|\Psi\rangle$, since the overall dimension of the vector space must be conserved. For example, if $|\phi\rangle$ is a one-particle state and $|\Psi\rangle$ is an N-particle state, then $|\Psi'\rangle$ will be an $(N-1)$-particle state.] To prove that such a device is impossible, we assume the contrary. Assume that there are two states, $|\phi_1\rangle$ and $|\phi_2\rangle$, for which (8.4) holds:

$$U|\phi_1\rangle \otimes |\Psi\rangle = |\phi_1\rangle \otimes |\phi_1\rangle \otimes |\Psi'\rangle , \tag{8.5a}$$

$$U|\phi_2\rangle \otimes |\Psi\rangle = |\phi_2\rangle \otimes |\phi_2\rangle \otimes |\Psi''\rangle . \tag{8.5b}$$

(We allow for the possibility that the final device state, Ψ' or Ψ'', may depend on the state being cloned.) Now the linear nature of U implies that for the superposition state $|\phi_s\rangle = (|\phi_1\rangle + |\phi_2\rangle)\sqrt{\frac{1}{2}}$ we must obtain

$$U|\phi_s\rangle \otimes |\mathbf{\Psi}\rangle = |\phi_1\rangle \otimes |\phi_1\rangle \otimes |\Psi'\rangle \sqrt{\tfrac{1}{2}} + |\phi_2\rangle \otimes |\phi_2\rangle \otimes |\mathbf{\Psi}''\rangle \sqrt{\tfrac{1}{2}} .$$

But this contradicts (8.4), according to which we ought to obtain

$$U|\phi_s\rangle \otimes |\Psi\rangle = |\phi_s\rangle \otimes |\phi_s\rangle \otimes |\Psi'''\rangle .$$

Therefore it is impossible to build a device to clone an arbitrary, unknown quantum state.

But classical states are special limiting cases of quantum states, so how is this theorem consistent with the ability to copy an unknown classical state? Clearly, the ability to form linear combinations of quantum states played an essential role in the impossibility proof, which would not apply if we required only that cloning work on a *discrete set* of states $\{|\phi_1\rangle, |\phi_2\rangle, \ldots\}$. In fact, the discrete set of allowed states must also be *orthogonal.* This follows from the fact that the inner product between state vectors is conserved by a unitary transformation. Equating the inner product of the initial and final states in (8.5) yields

$$\langle \phi_1 | \phi_2 \rangle \langle \Psi | \Psi \rangle = \langle \phi_1 | \phi_2 \rangle^2 \langle \Psi' | \Psi'' \rangle . \tag{8.6}$$

Here we have $\langle \Psi | \Psi \rangle = 1$, $|\langle \Psi' | \Psi'' \rangle| \leq 1$, and $|\langle \phi_1 | \phi_2 \rangle| \leq 1$. This will be consistent only if $\langle \phi_1 | \phi_2 \rangle = 0$. Now states that are classically different will certainly be orthogonal, so the no-cloning theorem for quantum states is not in conflict with the well-known possibility of copying classical states.

So far we have discussed only the preparation of pure states. If we can prepare ensembles corresponding to several different pure states, $|\Psi_i\rangle$, then by simply combining them with weights w_i we can prepare the mixed state represented by the state operator $\rho = \Sigma_i w_i |\Psi_i\rangle \langle \Psi_i|$. There is no additional difficulty of principle here. In practice nature usually presents us with states that are not pure, and it is the preparation of pure states that provides the greatest challenge.

8.2 *State Determination*

The problem of state determination may arise in different contexts. We may have an apparatus that has been designed to produce a certain state, and it is necessary to test or calibrate it, in order to determine what state is actually produced. Or we may have a natural process that produces an unknown state. In any case there must be a procedure that is repeatable (whether under the control of an experimenter, or occurring spontaneously in nature), and can yield an ensemble of systems, upon each of which a measurement may be carried out. Because a measurement involves an interaction with the system, it is likely to change the values of some of its attributes, and so will change the state of which it is a representative. Therefore any further measurements on the same system will be of no use in determining the original state. It is necessary to submit a system to the preparation and subsequently to perform a single measurement on it. To obtain more information, it is necessary to repeat the same state preparation before another measurement is performed. Whether the same system is used repeatedly in this *preparation–measurement* sequence, or another identical system is used each time, is immaterial. The objective of this section is to determine what sort of measurements provide sufficient information to determine the state operator ρ associated with the preparation.

Let us first consider the information provided by the measurement of a dynamical variable R, whose operator R has a discrete nondegenerate eigenvalue spectrum, $R|r_n\rangle = r_n|r_n\rangle$. Clearly a single measurement, producing a result such as $\mathrm{R} = r_3$, say, is not very helpful because this result could have come from any state for which there is a nonzero probability of obtaining $\mathrm{R} = r_3$. This result could have come from a state represented by any state vector $|\Psi\rangle$ for which $\langle r_3|\Psi\rangle \neq 0$, or more generally, represented by any state operator ρ for which $\langle r_3|\rho|r_3\rangle \neq 0$. But if we repeat the *preparation–measurement* sequence many times, and determine the relative frequency of the result $\mathrm{R} = r_n$, we will in effect be measuring the probability $\mathrm{Prob}(\mathrm{R} = r_n|\rho)$, where ρ denotes the unknown state operator. [The inference of an unknown probability from frequency data is discussed in Sec. 1.5, in the paragraphs preceding Eq. (1.60).] But, according to (2.26), we have $\mathrm{Prob}(\mathrm{R} = r_n|\rho) = \langle r_n|\rho|r_n\rangle$ for the case of a nondegenerate eigenvalue. Thus the measurement of the probability distribution of the dynamical variable R determines the diagonal matrix elements of the state operator in this representation.

To obtain information about the nondiagonal matrix elements $\langle r_m|\rho|r_n\rangle$, it is necessary to measure some other dynamical variable whose operator does not commute with R. It is not difficult to formally construct a set of operators whose corresponding probability distributions would determine all the matrix elements of ρ. Consider the following Hermitian operators:

$$A_{mn} = \frac{|r_m\rangle\langle r_n| + |r_n\rangle\langle r_m|}{2}\,, \quad B_{mn} = \frac{|r_m\rangle\langle r_n| - |r_n\rangle\langle r_m|}{2i}\,.$$

It follows directly that

$$\mathrm{Tr}(\rho A_{mn}) = \mathrm{Re}\,\langle r_m|\rho|r_n\rangle\,, \quad \mathrm{Tr}(\rho B_{mn}) = -\mathrm{Im}\,\langle r_m|\rho|r_n\rangle\,.$$

Thus if A_{mn} and B_{mn} represent observables, then the measurement of their averages would determine the matrix elements of the state operator ρ. But it is not at all clear that those operators do indeed represent observables, nor is it apparent how one would perform the appropriate measurements, so this formal approach has little practical value. We shall, therefore, examine some particular cases, where the nature of the required measurements is clear.

Spin state $s=\frac{1}{2}$

As was shown in Sec. 7.4, the most general state operator for a system having spin $s = \frac{1}{2}$ is of the form

$$\rho = \frac{1}{2}(\mathbf{1} + \mathbf{a}\cdot\boldsymbol{\sigma})\,. \tag{8.7}$$

Here $\mathbf{1}$ denotes the 2×2 unit matrix, and $\boldsymbol{\sigma} = (\sigma_x, \sigma_y, \sigma_z)$ are the Pauli spin operators (7.45). The state operator depends on three real parameters, $\mathbf{a} = (a_x, a_y, a_z)$, which must be deduced from appropriate measurements. From (7.51), the average of the x component of spin is equal to $\langle S_x \rangle = \text{Tr}(\rho\frac{1}{2}\hbar\sigma_x) = \frac{1}{2}\hbar a_x$, with similar expressions holding for the y and z components. Therefore, in order to fully determine the state operator, it is sufficient to measure the three averages $\langle S_x \rangle$, $\langle S_y \rangle$, and $\langle S_z \rangle$, which can be done by means of the Stern–Gerlach apparatus (introduced in Sec. 8.1, and analyzed in greater detail in Sec. 9.1). More generally, the averages of any three linearly independent (that is, noncoplanar) components of the spin will be sufficient.

Spin state $s=1$

The state operator ρ is a 3×3 matrix that depends on eight independent real parameters. These consist of two diagonal matrix elements (the third being determined from the other two by the condition $\text{Tr}\rho = 1$), and the real and imaginary parts of the three nondiagonal matrix elements above the diagonal (the elements below the diagonal being determined by the condition $\rho_{ji} = \rho_{ij}{}^*$). Three of these parameters can be determined from the averages $\langle S_x \rangle$, $\langle S_y \rangle$, and $\langle S_z \rangle$. Because $\langle \mathbf{S} \rangle$ transforms as a vector, it is clear that no more independent information would be obtained from the average of a spin component in any oblique direction.

The five additional parameters needed to specify the state may be obtained from the averages of the quadrupolar operators, which are quadratic in the spin operators: $(S_\alpha S_\beta + S_\beta S_\alpha)$, where $\alpha, \beta = x, y, z$. Only five of these operators are independent, since $S_x{}^2 + S_y{}^2 + S_z{}^2 = 2\hbar^2$. Using the standard representation (7.52) for the spin operators, in which S_z is diagonal, we can express the most general state operator in terms of eight parameters, $\{a_x, a_y, a_z, q_{xx}, q_{yy}, q_{xy}, q_{yz}$, and $q_{zx}\}$, where

$$
\begin{aligned}
a_\alpha &= \frac{\mathrm{Tr}(\rho S_\alpha)}{\hbar} && (\alpha = x, y, z)\,, \\
q_{\alpha\alpha} &= \frac{\mathrm{Tr}(\rho {S_\alpha}^2)}{\hbar^2} && (\alpha = x, y)\,, \\
q_{\alpha\beta} &= \frac{\mathrm{Tr}\{\rho(S_\alpha S_\beta + S_\beta S_\alpha)\}}{\hbar^2} && (\alpha\beta = xy, yz, zx)\,,
\end{aligned}
\tag{8.8}
$$

$$
\rho = \begin{bmatrix}
1+\frac{1}{2}(a_z - q_{xx} - q_{yy}), & \left(\frac{1}{2}\sqrt{\frac{1}{2}}\right)[a_x + q_{zx} - i(a_y + q_{yz})], & \frac{1}{2}(q_{xx} - q_{yy} - iq_{xy}) \\
\left(\frac{1}{2}\sqrt{\frac{1}{2}}\right)[a_x + q_{zx} + i(a_y + q_{yz})], & -1 + q_{xx} + q_{yy}, & \left(\frac{1}{2}\sqrt{\frac{1}{2}}\right)[a_x - q_{zx} - i(a_y - q_{yz})] \\
\frac{1}{2}(q_{xx} - q_{yy} + iq_{xy}), & \left(\frac{1}{2}\sqrt{\frac{1}{2}}\right)[a_x - q_{zx} + i(a_y - q_{yz})], & 1 - \frac{1}{2}(a_z + q_{xx} + q_{yy})
\end{bmatrix}.
\tag{8.9}
$$

It should be pointed out that the nonnegativeness condition (2.8) imposes some interrelated limits on the permissible ranges of the eight parameters in this expression. For certain values of some parameters, the range of others can even be reduced to a point, so the number of independent parameters may sometimes be less than eight.

Having parameterized the statistical operator, we must next consider how the parameters can be measured. If we use a Stern–Gerlach apparatus with the magnetic field gradient along the x direction to perform measurements on an ensemble of particles that emerge from the state preparation, we will be able to determine the relative frequency of the three possible values of S_x: $\hbar$, 0, and $-\hbar$. We can then calculate the averages $\langle S_x \rangle/\hbar = a_x$ and $\langle {S_x}^2 \rangle/\hbar^2 = q_{xx}$. By means of similar measurements with the field gradient along the y and z directions, we can determine a_y, q_{yy}, and a_z. (The relation $q_{zz} = 2 - q_{xx} - q_{yy}$ can be used as a consistency check on our measurements.)

It is less obvious how we can measure q_{xy}, q_{yz}, and q_{zx}. One method is to make use of the dynamical evolution of our unknown state ρ in a uniform magnetic field $\mathbf{B}$. The Hamiltonian is $H = -\boldsymbol{\mu}\cdot\mathbf{B}$, where the magnetic moment operator $\boldsymbol{\mu}$ is proportional to the spin operator $\mathbf{S}$. If the magnetic field points in the z direction, then the Hamiltonian becomes $H = cS_z$, where the constant c includes the strength of the field and the proportionality factor between the magnetic moment and the spin. Suppose that the initial value of the state operator is ρ, and that the magnetic field is turned on for a time interval t, after which we measure ${S_x}^2$. By doing this many times for each of several values of the interval t, we can estimate $(d/dt)\langle {S_x}^2 \rangle|_{t=0}$ from the data.

From the equation of motion (3.74) we obtain

$$\begin{aligned}
\frac{d}{dt}\left\langle {S_x}^2 \right\rangle\big|_{t=0} &= \frac{i}{\hbar}\,\mathrm{Tr}\left\{\rho\left[H, {S_x}^2\right]\right\} \\
&= \frac{i}{\hbar}\,\mathrm{Tr}\left\{\rho\left[cS_z, {S_x}^2\right]\right\} \\
&= -c\,\mathrm{Tr}\left\{\rho\left(S_xS_y + S_yS_x\right)\right\} \\
&= -c\,\hbar^2\, q_{xy}
\end{aligned} \tag{8.10}$$

when the uniform magnetic field is along the z direction. By measuring ${S_y}^2$ or ${S_z}^2$, with the magnetic field in the x or y direction, respectively, we can similarly obtain q_{yz} or q_{zx}. Thus we have a method, at least in principle, by which an unknown $s = 1$ state can be experimentally determined.

Some generalizations can now be made about the minimum amount of information needed to determine a spin state. Except for the case of $s = \frac{1}{2}$, the average spin $\langle \mathbf{S} \rangle$ is not sufficient, as it provides only three parameters, one for each of the three linearly independent components of the spin vector. If our information is obtained in the form of averages of tensor operators, then we require three dipole (vector) operators for the case of $s = \frac{1}{2}$; three dipole and five quadrupole operators for the case of $s = 1$; three dipole, five quadrupole, and seven octupole operators for $s = 3/2$; etc.

However, our information need not be restricted to the averages of tensor operators, since an ensemble of measurements determines not only the average, but the entire probability distribution of the dynamical variable that is measured. If that variable can take on n different values, then there are $n-1$ independent probabilities, and so the probability distribution will convey more information than does the average (except when $n = 2$, as is the case for $s = \frac{1}{2}$). It is possible to determine all of the matrix elements of the state operator, for any spin s, from the probability distributions of spin components in a sufficiently large number of suitably chosen directions. The results are obtained as the solution of several simultaneous linear equations. For details see the original paper [Newton and Young, (1968)].

Orbital state of a spinless particle

The orbital state of a spinless particle can be described by the coordinate representation of the state operator, $\langle \mathbf{x}|\rho|\mathbf{x}' \rangle$. This function of the two variables, $\mathbf{x}$ and $\mathbf{x}'$, is called the *density matrix*, because its diagonal elements yield the position probability density. It is clear that the determination of the density matrix for an arbitrary state will require the probability distributions for position and one or more variables whose operators do not commute with the position operator.

In 1933 W. Pauli posed the question of whether the position and momentum probability densities are sufficient to determine the state. Several counterexamples are now known, showing that the position and momentum distributions are not sufficient. One such counterexample is a pure state that is an eigenfunction of orbital angular momentum, $\Psi_m(\mathbf{x}) = f(r){Y_\ell}^m(\theta, \phi)$. It is easy to show that Ψ_m and Ψ_{-m} have the same position and momentum distributions. An even simpler, one-dimensional example is provided by any function $\Psi(x)$ that is linearly independent of its complex conjugate $\Psi^*(x)$, and has inversion symmetry, $\Psi(x) = \Psi(-x)$. The two states described by Ψ and Ψ^* have the same position and momentum distributions.

There still remain several interesting but unanswered questions. The simple examples of states with the same position and momentum distributions are all related to discrete symmetries, such as space inversion or complex conjugation. Are there examples that are not related to symmetry? Do such states occur only in discrete pairs, or are there continuous families of states with the same position and momentum distributions? The problem has been reviewed by Weigert (1992), who says that the claims made in the literature are not all compatible. A sufficient set of measurements to determine the orbital state of a particle does not seem to be known. In many papers it is assumed that the unknown state is pure, and hence that there is a wave function. That makes an interesting mathematical problem, but in practice, if you do not know the state you are unlikely to know whether it is pure.

[[Band and Park (1979) have shown that for a *free* particle the infinite set of quantities, $(d/dt)^n \langle (Q_\alpha)^{m+n} \rangle |_{t=0}$, for all positive integers m and n, are sufficient to determine the state at time $t = 0$. Here Q_α is a component of the position operator, and $\alpha = x, y, z$. They claim that their result also holds in the presence of a scalar potential that is an analytic function of position, but that claim cannot be correct. Consider a stationary state, for which all of the time derivative terms with $n \neq 0$ will vanish. Their claim would then be that the state is fully determined by the moments, $\langle (Q_\alpha)^m \rangle$, of the position probability distribution. But we know that the position probability distribution does not fully determine the state.]]

8.3 *States of Composite Systems*

The characterization of the states of composite systems presents some additional problems beyond those that exist for simple systems. Can one define states for the components, or merely states for the system as a whole? Is the state of a composite system determined by the states of its parts?

To answer the first question, we consider a two-component system whose components will be labeled 1 and 2. The state vector space is spanned by a set of product vectors of the form

$$|a_m b_n\rangle = |a_m\rangle \otimes |b_n\rangle\,, \tag{8.11}$$

where $\{|a_m\rangle\}$ is a set of basis vectors for component 1 alone, and similarly $\{|b_n\rangle\}$ is a basis set for component 2 alone. The average of an arbitrary dynamical variable R is given by

$$\langle R\rangle = \mathrm{Tr}(\rho R) = \sum_{m,n,m',n'} \langle a_m b_n|\rho|a_{m'}b_{n'}\rangle\langle a_{m'}b_{n'}|R|a_m b_n\rangle\,. \tag{8.12}$$

If we consider a dynamical variable that belongs exclusively to component 1, then its operator $R^{(1)}$ will act nontrivially on only the first factor of the basis vector (8.11). Therefore in this case (8.12) will reduce to

$$\begin{aligned}\langle R^{(1)}\rangle &= \sum_{m,n,m',n'} \langle a_m b_n|\rho|a_{m'}b_{n'}\rangle\langle a_{m'}|R^{(1)}|a_m\rangle\langle b_{n'}|b_n\rangle\\ &= \sum_{m,m'}\sum_{n}\langle a_m b_n|\rho|a_{m'}b_n\rangle\langle a_{m'}|R^{(1)}|a_m\rangle\\ &= \mathrm{Tr}^{(1)}(\rho^{(1)}R^{(1)})\,,\end{aligned} \tag{8.13}$$

where $\rho^{(1)}$ is an operator in the factor space of component 1, and is defined as

$$\begin{aligned}\rho^{(1)} &= \mathrm{Tr}^{(2)}\rho\,,\\ \langle a_m|\rho^{(1)}|a_{m'}\rangle &= \sum_n \langle a_m b_n|\rho|a_{m'}b_n\rangle\,.\end{aligned} \tag{8.14}$$

$\mathrm{Tr}^{(1)}$ and $\mathrm{Tr}^{(2)}$ signify traces over the factor spaces of components 1 and 2, respectively.

We shall refer to $\rho^{(1)}$ as the *partial state operator* for component 1. (It is sometimes also called a *reduced* state operator.) In order to justify referring to it as a kind of state operator, we must prove that it satisfies the three basic properties, (2.6), (2.7), and (2.8). These follow directly from the definition (8.14) and the fact that the total state operator ρ satisfies those properties.

$$\begin{aligned}\mathrm{Tr}^{(1)}\rho^{(1)} &= \sum_m \langle\, a_m|\rho^{(1)}|a_m\,\rangle = \sum_m\sum_n \langle\, a_m b_n|\rho|a_m b_n\,\rangle \\ &= \mathrm{Tr}\,\rho = 1\,,\end{aligned}$$

and hence $\rho^{(1)}$ satisfies (2.6). The Hermitian property (2.7), $[\rho^{(1)}]^\dagger = \rho^{(1)}$, is evident from the definition (8.14). To prove nonnegativeness (2.8), we assume the contrary and demonstrate that it leads to a contradiction. Suppose that $\langle\, u|\rho^{(1)}|u\,\rangle < 0$ for some vector $|u\,\rangle$. Instead of the basis $\{|a_m\,\rangle\}$ in the factor space of component 1, we use an orthonormal basis $\{|u_m\,\rangle\}$ of which $|u\,\rangle = |u_1\,\rangle$ is the first member. Instead of (8.11), we now use the product vectors $|u_m b_n\,\rangle = |u_m\,\rangle \otimes |b_n\,\rangle$. By hypothesis, we should have

$$0 > \langle\, u_1|\rho^{(1)}|u_1\,\rangle = \sum_n \langle\, u_1 b_n|\rho|u_1 b_n\,\rangle\,,$$

but this is impossible because ρ is nonnegative. Therefore the initial supposition, that $\langle\, u|\rho^{(1)}|u\,\rangle < 0$, must be false, and so $\rho^{(1)}$ must also be nonnegative. We have now shown that $\rho^{(1)}$ satisfies the three basic properties that must be satisfied by all state operators. According to (8.13), $\rho^{(1)}$ suffices for calculating the average of any dynamical variable that belongs exclusively to component 1. Therefore it seems appropriate to call $\rho^{(1)}$ the *partial state operator* for component 1.

Now the partial state operators, $\rho^{(1)} = \mathrm{Tr}^{(2)}\rho$ and $\rho^{(2)} = \mathrm{Tr}^{(1)}\rho$, are sufficient for calculating the averages of any dynamical variables that belong exclusively to component 1 or to component 2. But these two partial state operators are *not sufficient*, in general, for determining the state of the composite system, because they provide no information about *correlations* between components 1 and 2. Let $R^{(1)}$ represent a dynamical variable of component 1, and let $R^{(2)}$ represent a dynamical variable of component 2. If it is the case that

$$\langle\, R^{(1)}R^{(2)}\,\rangle = \langle\, R^{(1)}\,\rangle\,\langle\, R^{(2)}\,\rangle \tag{8.15}$$

for *all* $R^{(1)}$ and $R^{(2)}$, then the state is said to be an *uncorrelated* state. It is easy to show that in this case the state operator must be of the form

$$\rho = \rho^{(1)} \otimes \rho^{(2)}\,. \tag{8.16}$$

This is the only case for which the total state is determined by the partial states of the components.

Classification of states

The partial states and the total state may or may not be pure, several different combinations being possible. In some of those cases the total state may be correlated, and in some it may be uncorrelated. The various possibilities are described in Tables 8.1 and 8.2. For those cases that are marked "yes" (possible) in Table 8.1, we further indicate, in Table 8.2, whether they may exist as correlated states and as uncorrelated states.

Table 8.1

Partial states $\rho^{(1)}$ and $\rho^{(2)}$	Total state ρ	
	pure	nonpure
both pure	yes	no
one pure, one not	no	yes
both not pure	yes	yes

Table 8.2

$\rho^{(1)},\ \rho^{(2)};\ \rho$	Correlated	Uncorrelated
pure, pure; pure	no	yes
pure, nonpure; nonpure	no	yes
nonpure, nonpure; pure	yes	no
nonpure, nonpure; nonpure	yes	yes

To prove that the "yes" cases in both tables are indeed possible, it will suffice to give an example for each "yes" in Table 8.2.

(i) $\rho^{(1)}$ and $\rho^{(2)}$ both pure; ρ pure, uncorrelated:
$\rho^{(1)} = |\phi\rangle\langle\phi|$, $\rho^{(2)} = |\psi\rangle\langle\psi|$, $\rho = |\Psi\rangle\langle\Psi|$, where $|\Psi\rangle = |\phi\rangle \otimes |\psi\rangle$.

(ii) $\rho^{(1)}$ pure, $\rho^{(2)}$ not pure; ρ not pure, uncorrelated:
Let $\rho^{(1)}$ be any pure state, $\rho^{(2)}$ be any nonpure state, and $\rho = \rho^{(1)} \otimes \rho^{(2)}$. Now, according to the pure state criterion (2.17), we have $\mathrm{Tr}^{(1)}[(\rho^{(1)})^2] = 1$ and $\mathrm{Tr}^{(2)}[(\rho^{(2)})^2] < 1$. From the identity

$$\mathrm{Tr}(\sigma \otimes \tau) = \mathrm{Tr}^{(1)}\sigma\, \mathrm{Tr}^{(2)}\tau\,, \tag{8.17}$$

it follows that $\mathrm{Tr}(\rho^2) = \mathrm{Tr}^{(1)}[(\rho^{(1)})^2]\ \mathrm{Tr}^{(2)}[(\rho^{(2)})^2] < 1$, and so ρ is not pure.

(iii) $\rho^{(1)}$ and $\rho^{(2)}$ both not pure; ρ pure, correlated:
Consider two particles each having $s = \frac{1}{2}$. The vector $|m_1, m_2\rangle = |m_1\rangle \otimes |m_2\rangle$ describes a state in which the z component of the spin of particle 1 is equal to $\hbar m_1$ and that of particle 2 is $\hbar m_2$. The singlet state of the two-particle system, having total spin $S = 0$, is described by the state vector $|\Psi\rangle = \left(|+\frac{1}{2}, -\frac{1}{2}\rangle - |-\frac{1}{2}, +\frac{1}{2}\rangle\right)\sqrt{\frac{1}{2}}$, and by the state operator $\rho = |\Psi\rangle\langle\Psi|$. The partial states operators, $\rho^{(1)} = \mathrm{Tr}^{(2)}\rho$ and $\rho^{(2)} = \mathrm{Tr}^{(1)}\rho$, are both of the form $\frac{1}{2}\left(|+\frac{1}{2}\rangle\langle+\frac{1}{2}| + |-\frac{1}{2}\rangle\langle-\frac{1}{2}|\right)$, which represents an unpolarized state. The states $\rho^{(1)}$, $\rho^{(2)}$, and ρ are isotropic. In any chosen direction the spin component of either particle is equally likely to be $+\frac{1}{2}$ or $-\frac{1}{2}$. But the value $+\frac{1}{2}$ for particle 1 is always associated with the value $-\frac{1}{2}$ for particle 2, and vice versa.

(iv) $\rho^{(1)}$ and $\rho^{(2)}$ both not pure; ρ not pure, uncorrelated:
Let $\rho^{(1)}$ and $\rho^{(2)}$ be any nonpure states, and take $\rho = \rho^{(1)} \otimes \rho^{(2)}$. From (8.17) it follows that $\mathrm{Tr}\rho^2 < 1$, and so this is not a pure state.

(v) $\rho^{(1)}$ and $\rho^{(2)}$ both not pure; ρ not pure, correlated:
Take $\rho = \frac{1}{2}\{|+\frac{1}{2}, +\frac{1}{2}\rangle\langle+\frac{1}{2}, +\frac{1}{2}| + |-\frac{1}{2}, -\frac{1}{2}\rangle\langle-\frac{1}{2}, -\frac{1}{2}|\}$. The partial states $\rho^{(1)}$ and $\rho^{(2)}$ will both be unpolarized states, as in example (iii), but the z components of the spins of the two particles are correlated.

Examples (iii) and (v) together illustrate the fact that the total state is not determined by the partial states of the components.

We must now prove the impossibility of the "no" cases in the two tables. That can be done with the help of the following theorem.

Pure state factor theorem. If ρ satisfies the conditions (2.6), (2.7), and (2.8) that are demanded of a state operator, and if $\rho^{(1)} = \mathrm{Tr}^{(2)}\rho$ and $\rho^{(2)} = \mathrm{Tr}^{(1)}\rho$, and if the partial state operator $\rho^{(1)}$ describes a pure state, then it follows that $\rho = \rho^{(1)} \otimes \rho^{(2)}$. In other words, a pure partial state operator must be a factor of the total state operator.

Proof of this theorem is rather subtle, because it is easy to produce what would seem to be a counterexample. Suppose that we have three operators, $\rho^{(1)}$, $\rho^{(2)}$, and ρ, that satisfy the theorem. Now, of the matrix elements $\langle a_m b_n|\rho|a_{m'} b_{n'}\rangle$, only those for $b_{n'} = b_n$ enter into the definition of $\rho^{(1)}$ in (8.14), and only those for $a_{m'} = a_m$ enter into the definition of $\rho^{(2)}$. So it seems that the values of those matrix elements for which $a_{m'} \neq a_m$ and $b_{n'} \neq b_n$ may be changed without affecting $\rho^{(1)}$ or $\rho^{(2)}$, thus invalidating the theorem. In fact, any such change would violate the nonnegativeness condition (2.8), but this is not at all obvious.

Proof. To prove the theorem, we use a representation of ρ that ensures its nonnegativeness. The spectral representation (2.9),

$$\rho = \sum_k \rho_k \, |\phi_k\rangle \langle\phi_k| \, , \tag{8.18}$$

guarantees nonnegativeness provided the eigenvalues ρ_k are nonnegative. The eigenvectors of ρ can be expanded in terms of product basis vectors of the form (8.11),

$$|\phi_k\rangle = \sum_{m,n} c^k_{m,n} \, |a_m b_n\rangle \, , \tag{8.19}$$

and so we have

$$\rho = \sum_k \rho_k \sum_{m,n} \sum_{m',n'} c^k_{m,n} \left(c^k_{m',n'}\right)^* |a_m b_n\rangle \langle a_{m'} b_{n'}| \, . \tag{8.20}$$

Evaluating $\rho^{(1)}$ by means of (8.14) now yields

$$\rho^{(1)} = \sum_{m,m'} \sum_k \rho_k \sum_n c^k_{m,n} \left(c^k_{m',n}\right)^* |a_m\rangle \langle a_{m'}| \, . \tag{8.21}$$

But this is a pure state, and so must be of the form $\rho^{(1)} = |\psi\rangle\langle\psi|$. Since the original basis $\{|a_m\rangle\}$ was arbitrary, we are free to choose it so that $|a_1\rangle = |\psi\rangle$. In that case we will have $\sum_k \rho_k \sum_n c^k_{m,n} (c^k_{m',n})^* = 0$ unless $m = m' = 1$. For $m = m' \neq 1$ this becomes $\sum_k \rho_k \sum_n |c^k_{m,n}|^2 = 0$, which is possible only if $\sum_k \rho_k |c^k_{m,n}|^2 = 0$ for $m \neq 1$. Thus for $m \neq 1$ and any k such that $\rho_k \neq 0$ we must have $c^k_{m,n} = 0$. Therefore (8.20) reduces to

$$\begin{aligned} \rho &= \sum_k \rho_k \sum_n \sum_{n'} c^k_{1,n} \left(c^k_{1,n'}\right)^* |a_1 b_n\rangle \langle a_1 b_{n'}| \\ &= |a_1\rangle\langle a_1| \otimes \sum_k \rho_k \sum_n \sum_{n'} c^k_{1,n} \left(c^k_{1,n'}\right)^* |b_n\rangle\langle b_{n'}| \, , \end{aligned} \tag{8.22}$$

which is the form asserted by the theorem. The first factor in the Kronecker product is $\rho^{(1)}$, and the second factor may be identified with $\rho^{(2)}$.

Using the result of this theorem, we can now prove all of the "no" cases in which at least one of the partial states is pure. Correlated states in the first and second lines of Table 8.2 are impossible because, according to the theorem, the total state operator must be of the product form. The two "no" cases in Table 8.1 are excluded by the relation

$$\mathrm{Tr}[(\rho^{(1)} \otimes \rho^{(2)})^2] = \mathrm{Tr}^{(1)}[(\rho^{(1)})^2]\,\mathrm{Tr}^{(2)}[(\rho^{(2)})^2]\,, \tag{8.23}$$

which is a special case of (8.17). If both $\rho^{(1)}$ and $\rho^{(2)}$ are pure then both factors on the right are equal to 1, and so according to (2.17) the total state operator must also be pure, which proves the first line of Table 8.1. If one of the partial states is not pure then the product of the traces will be less than 1, and so the total state operator cannot be pure, proving the second line of Table 8.1. The third "no" in Table 8.2 is also excluded by this same trace relation. This completes the proofs for the classifications given in Tables 8.1 and 8.2.

In summary, we have shown that partial states for the components of a system can be defined, but the states of the components do not suffice for determining the state of the whole. Indeed, the relation between the states of the components and the state of the whole system is quite complex. Some simplification is provided by the theorem that a pure partial state must be a factor of the total state operator. Since a factorization of the form $\rho = \rho^{(1)} \otimes \rho^{(2)}$ means that there are no correlations between components 1 and 2, this implies that a component described by a pure state can have no correlations with the rest of the system. This may seem paradoxical. Consider a many-particle spin system described by the state vector $|\Psi\rangle = |\uparrow\rangle \otimes |\uparrow\rangle \otimes |\uparrow\rangle \otimes \cdots$. All of the spins are up (in the z direction), and so this seems to be a high degree of correlation. Yet the product form of the state vector is interpreted as an absence of correlation among the particles. The paradox is resolved by noting that the correlation is defined by means of the quantum-mechanical probability distributions. Since $|\Psi\rangle$ is an eigenvector of the z components of the spins, there are no fluctuations in these dynamical variables, and where there is no variability the degree of correlation is undefined. If we consider the components of the spins in any direction other than z, they will be subject to fluctuations and those fluctuation will indeed be uncorrelated in the state $|\Psi\rangle$.

Correlated states of multicomponent systems (also called "entangled states") are responsible for some of the most interesting and peculiar phenomena in quantum mechanics. Figure 8.1 illustrates schematically how such a state could be prepared. A source in the box emits pairs of particles in variable directions, but always with opposite momentum ($\mathbf{k}_b = -\mathbf{k}_a, \mathbf{k}_{b'} = -\mathbf{k}_{a'}$). The two output ports on each side of the box restrict each particle to two possible directions, so the state of the emerging pairs is

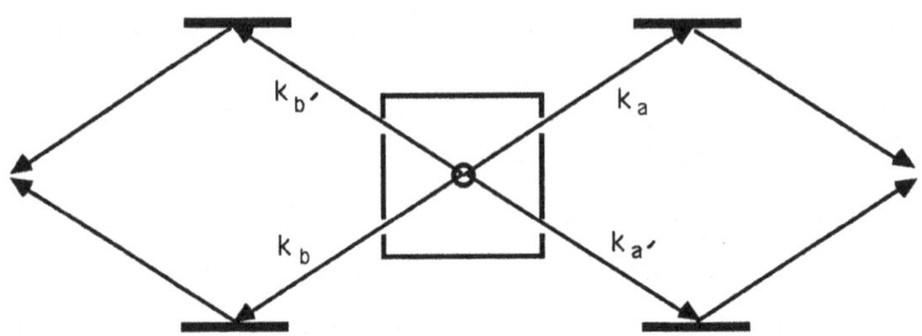

Fig. 8.1 A device to produce a correlated two-particle state.

$$|\Psi_{12}\rangle = (|\mathbf{k}_a\rangle\,|\mathbf{k}_b\rangle + |\mathbf{k}_{a'}\rangle\,|\mathbf{k}_{b'}\rangle)\sqrt{\tfrac{1}{2}}\,. \tag{8.24}$$

The momenta of the two particles are correlated in this state. If particle 1 on the right has momentum $\hbar\mathbf{k}_a$ then particle 2 on the left must have momentum, $\hbar\mathbf{k}_b$, and if 1 has $\hbar\mathbf{k}_{a'}$ then 2 must have $\hbar\mathbf{k}_{b'}$.

By means of appropriately placed mirrors, we can combine beams a and a' on the right, and combine beams b and b' on the left. Looking at only one side of the apparatus, it would appear that the amplitudes from the paths a and a' should produce an interference pattern, as in the double slit experiment (Sec. 5.4), and that a similar interference between paths b and b' should occur on the left. This expectation would be correct if the particles were not correlated, and the state was of the form $|\Psi_{12}\rangle = |\psi_1\rangle\,|\psi_2\rangle$. But, in fact, the correlation between the particles leads to a qualitative difference.

Ignoring normalization, the two-particle configuration space wave function will have the form

$$\Psi_{12}(\mathbf{x}_1,\mathbf{x}_2) \propto e^{i(\mathbf{k}_a\cdot\mathbf{x}_1+\mathbf{k}_b\cdot\mathbf{x}_2)} + e^{i(\mathbf{k}_{a'}\cdot\mathbf{x}_1+\mathbf{k}_{b'}\cdot\mathbf{x}_2)} \tag{8.25}$$

and the position probability density will be

$$|\Psi_{12}(\mathbf{x}_1,\mathbf{x}_2)|^2 \propto \{1+\cos[(\mathbf{k}_a-\mathbf{k}_{a'})\cdot\mathbf{x}_1 + (\mathbf{k}_b-\mathbf{k}_{b'})\cdot\mathbf{x}_2]\}\,. \tag{8.26}$$

(These forms hold only inside the regions where the beams overlap. The wave function falls to zero outside of the beams.) If we ignore particles on the left and place a screen to detect the particles on the right, the detection probability for particle 1 will be given by the integral of (8.26) over $\mathbf{x}_2$, and will be featureless. No interference pattern exists in the single particle probability density. Only in the correlations between particles can interference be observed. For example, if we select only those particles on the right that are detected in coincidence with particles on the left in a small volume near $\mathbf{x}_2 = 0$, then their spatial density will be proportional to $\{1 + \cos[(\mathbf{k}_a - \mathbf{k}_{a'}) \cdot \mathbf{x}_1]\}$.

8.4 *Indeterminacy Relations*

In its most elementary form, the concept of a *state* is identified with the specification of a probability distribution for each observable, as was discussed in Sec. 2.1. However, the probability distributions for different dynamical variables may be interrelated, and cannot necessarily be varied independently. The most important and simplest of these interrelations will now be derived.

Let A and B be two dynamical variables whose corresponding operators are A and B, and let

$$[A, B] = iC\,. \tag{8.27}$$

(The factor of i is inserted so that $C^\dagger = C$.) In an arbitrary state represented by ρ, the mean and the variance of the A distribution are $\langle \mathrm{A} \rangle = \mathrm{Tr}(\rho A)$ and ${\Delta_A}^2 = \langle (\mathrm{A} - \langle \mathrm{A} \rangle)^2 \rangle$, respectively. Similar expressions hold for the B distribution. If we define two operators, $A_0 = A - \langle \mathrm{A} \rangle$ and $B_0 = B - \langle \mathrm{B} \rangle$, then the variances of the two distributions are given by ${\Delta_A}^2 = \mathrm{Tr}(\rho {A_0}^2)$ and ${\Delta_B}^2 = \mathrm{Tr}(\rho {B_0}^2)$.

For any operator T one has the inequality $\mathrm{Tr}(\rho T T^\dagger) \geq 0$. This is most easily proven by choosing the basis in which ρ is diagonal. Now substitute $T = A_0 + i\omega B_0$ and $T^\dagger = A_0 - i\omega B_0$, where ω is any arbitrary real number. The inequality then becomes

$$\mathrm{Tr}\left(\rho T T^\dagger\right) = \mathrm{Tr}\left(\rho {A_0}^2\right) - i\omega\, \mathrm{Tr}(\rho [A_0, B_0]) + \omega^2\, \mathrm{Tr}\left(\rho {B_0}^2\right) \geq 0\,. \tag{8.28}$$

The commutator in this expression has the value $[A_0, B_0] = iC$. The strongest inequality will be obtained if we choose ω so as to minimize the quadratic form. This occurs for the value

$$\omega = -\frac{\mathrm{Tr}(\rho C)}{2\,\mathrm{Tr}\left(\rho {B_0}^2\right)}\,, \tag{8.29}$$

and in this case the inequality may be written as

$$\mathrm{Tr}\left(\rho {A_0}^2\right)\,\mathrm{Tr}\left(\rho {B_0}^2\right) - \frac{\{\mathrm{Tr}(\rho C)\}^2}{4} \geq 0\,,$$

This is equivalent to

$${\Delta_A}^2\, {\Delta_B}^2 \geq \frac{\{\mathrm{Tr}(\rho C)\}^2}{4}\,, \tag{8.30}$$

which may be more compactly written as

$$\Delta_A \Delta_B \geq \frac{1}{2} |\langle C \rangle| \,. \tag{8.31}$$

This result holds for any operators that satisfy (8.27).

Example (i): Angular momentum

In this case the commutator (8.27) becomes $[J_x, J_y] = i\hbar J_z$, and the inequality (8.30) becomes

$$\langle (J_x - \langle J_x \rangle)^2 \rangle \langle (J_y - \langle J_y \rangle)^2 \rangle \geq \left(\tfrac{1}{2}\hbar\right)^2 \langle J_z \rangle^2 \,. \tag{8.32}$$

This result is particularly interesting for the state $\rho = |j, m\rangle\langle j, m|$, which is an *eigenstate* of $\mathbf{J \cdot J}$ and J_z. Because this state is invariant under rotations about the z axis, we have $\langle J_x \rangle = \langle J_y \rangle = 0$ and $\langle {J_x}^2 \rangle = \langle {J_y}^2 \rangle$. Thus (8.32) reduces to $\langle {J_x}^2 \rangle \geq \frac{1}{2}\hbar^2 |m|$. But because this is an eigenstate of $\mathbf{J \cdot J} = {J_x}^2 + {J_y}^2 + {J_z}^2$, we also have the relation $\langle {J_x}^2 \rangle + \langle {J_y}^2 \rangle + (\hbar m)^2 = \hbar^2 j(j+1)$. Hence $\langle {J_x}^2 \rangle = \langle {J_y}^2 \rangle = \frac{1}{2}\hbar^2(j^2 + j - m^2)$. It is apparent that if $|m|$ takes on its largest possible value, j, the inequality becomes an equality.

Example (ii): Position and momentum

For simplicity, only one spatial dimension will be considered. The commutator (8.27) then takes the form $[Q, P] = i\hbar$, and the inequality (8.31) becomes

$$\Delta_Q \Delta_P \geq \tfrac{1}{2}\hbar \,. \tag{8.33}$$

This formula is commonly called "the uncertainty principle", and is associated with the name of Heisenberg.

A state for which this inequality becomes an equality is called a *minimum uncertainty state*. It is apparent from the derivation of (8.31) that this will happen if and only if the nonnegative expression in (8.28) vanishes at its minimum. By appropriately choosing the frame of reference we can have $\langle Q \rangle = 0$ and $\langle P \rangle = 0$, so the condition for a minimum uncertainty state may be taken to be

$$\mathrm{Tr}\left(\rho T T^\dagger\right) = 0 \,, \quad \text{with } T = Q + i\omega P \,, \quad \omega = -\frac{\hbar}{2{\Delta_P}^2} \,. \tag{8.34}$$

As a first step in classifying all minimum uncertainty states, we consider a pure state, $\rho = |\psi\rangle\langle\psi|$, for which the condition (8.34) becomes $\langle \psi | T T^\dagger | \psi \rangle = 0$. This is satisfied if and only if $T^\dagger |\psi\rangle = 0$. In coordinate representation, for which $Q \to x$ and $P \to -i\hbar\partial/\partial x$, the condition becomes $(x+\alpha\partial/\partial x)\psi(x) = 0$, with $\alpha = \hbar^2/2{\Delta_P}^2$. This differential equation has the general solution $\psi(x) = C\exp(-x^2/2\alpha)$, where the constant C could be fixed by normalization. It can easily be shown that ${\Delta_Q}^2 \equiv \langle\psi|x^2|\psi\rangle/\langle\psi|\psi\rangle = \alpha/2 = \hbar^2/4{\Delta_P}^2$, verifying that the minimum uncertainty condition is indeed satisfied. By transforming back to an arbitrary uniformly moving frame of reference in which $\langle Q\rangle = x_0$ and $\langle P\rangle = \hbar k_0$, we obtain the most general minimum uncertainty pure state function,

$$\psi(x) = C\exp\left\{-\frac{(x-x_0)^2}{4{\Delta_Q}^2} + ik_0x\right\}. \tag{8.35}$$

A general state operator can be written in the form $\rho = \sum_n \rho_n|\phi_n\rangle\langle\phi_n|$, where the eigenvalues ρ_n are nonnegative, and the eigenvectors $\{\phi_n\}$ are orthonormal. The minimum uncertainty condition (8.34) then becomes

$$\sum_n \rho_n\langle\phi_n|TT^\dagger|\phi_n\rangle = 0\,. \tag{8.36}$$

This condition can be satisfied only if $T^\dagger|\phi_n\rangle = 0$ for all n such that $\rho_n \neq 0$. Now that is just the condition that determined the minimum uncertainty pure state functions; therefore $\phi_n(x)$ must be of the form (8.35). But it is easily verified that no two functions of the form (8.35) are orthogonal. Since all the eigenvectors $\{\phi_n\}$ must be orthogonal, it is not possible to have more that one distinct term in (8.36). Hence ρ must contain just a single term, $\rho = |\phi_n\rangle\langle\phi_n|$. Thus we have shown that a minimum uncertainty state for position and momentum must be a pure state, a result first proven by Stoler and Newman (1972).

Operational significance

The empirically testable consequence of an indeterminacy relation such as (8.33) is illustrated in Fig. 8.2. One must have a repeatable preparation procedure corresponding to the state ρ which is to be studied. Then on each of a large number of similarly prepared systems, one performs a single measurement (either of Q or of P). The statistical distributions of the results are shown as histograms, and the root-mean-square half-widths of the two distributions, Δ_Q and Δ_P, are indicated in Fig. 8.2. The theory predicts that the product of these two half-widths can never be less that $\hbar/2$, no matter what state is considered.

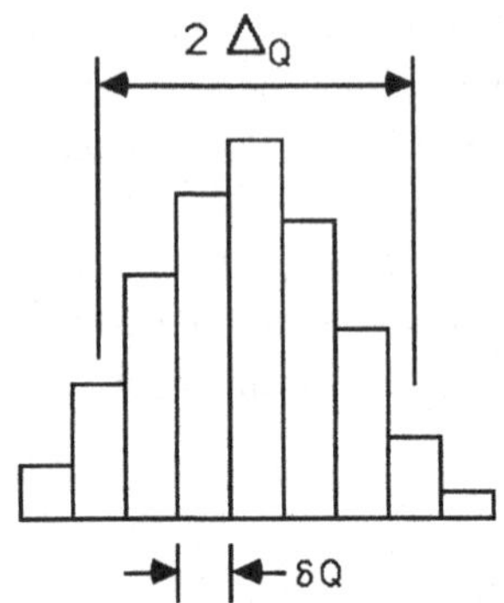

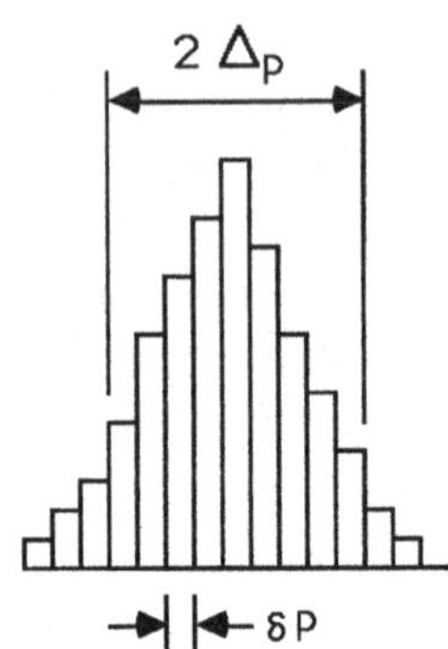

Fig. 8.2 Frequency distributions for the results of independent measurements of Q and P on similarly prepared systems. The standard deviations, which satisfy the relation $\Delta_Q \Delta_P \geq \hbar/2$, must be distinguished from the resolution of the individual measurements, δQ and δP.

[[Because contrary statements abound in the literature, it is necessary to emphasize the following points:

- The quantities Δ_Q and Δ_P are not errors of measurement. The "errors", or preferably the resolution of the Q and P measuring instruments, are denoted as δQ and δP in Fig. 8.2. They are logically unrelated to Δ_Q and Δ_P, and to the inequality (8.33), except for the practical requirement that if δQ is larger than Δ_Q (or if δP is larger than Δ_P) then it will not be possible to determine Δ_Q (Δ_P) in the experiment, and so it will not be possible to test (8.33).
- The experimental test of the inequality (8.33) does not involve *simultaneous measurements* of Q and P, but rather it involves the measurement of one or the other of these dynamical variables on each independently prepared representative of the particular state being studied.

To the reader who is unfamiliar with the history of quantum mechanics, these remarks may seem to belabor the obvious. Unfortunately the statistical quantities Δ_Q and Δ_P in (8.33) have often been misinterpreted as the errors of individual measurements. The origin of the confusion probably lies in the fact that Heisenberg's original paper on the uncertainty principle, published in 1927, was based on early work that predates the systematic formulation and statistical interpretation of quantum theory. Thus the natural derivation and interpretation of (8.33) that is given above was not possible at that time. The statistical interpretation of the indeterminacy relations was first advanced by K. R. Popper in 1934.

It is also sometimes asserted that the quantum lower bound (8.33) on Δ_Q and Δ_P is far below the resolution of practical experiments. That may be so for measurements involving metersticks and stopwatches, but it is not generally true. An interesting example from crystallography has been given by Jauch (1993). The rms atomic momentum fluctuation, Δ_P, is directly obtained from the temperature of the crystal, and hence (8.33) gives a lower bound to Δ_Q, the rms vibration amplitude of an atom. The value of Δ_Q can be measured by means of neutron diffraction, and at low temperatures it is only slightly above its quantum lower bound, $\hbar/2\Delta_P$. Jauch stresses that it is only the rms ensemble fluctuations that are limited by (8.33). The position coordinates of the atomic cell can be determined with a precision that is two orders of magnitude smaller than the quantum limit on Δ_Q.]]

Further reading for Chapter 8

State preparation
Lamb (1969).

State determination
Newton and Young (1968); Band and Park (1970, 1971);
Weigert (1992) — spin states.
Band and Park (1979) — orbital state of a free particle.
Band and Park (1973) — inequalities imposed by nonnegativeness.

Correlated states
Greenberger, Horne, and Zeilinger (1993).

History of the indeterminacy relation
Jammer (1974), Ch. 3.

Problems

8.1 What potentials would be required to prepare each of the following (unnormalized) state functions as ground state? (a) $\Psi(r) = \exp(-\alpha r^2)$; (b) $\Psi(r) = \exp(-ar)$; (c) $\Psi(x) = 1/\cosh(x)$. Cases (a) and (b) are three-dimensional. Case (c) is one-dimensional. What restrictions, if any, must be imposed on the assumed ground state energy E, in order that the potentials be physically reasonable?

8.2 A source of spin $s = 1$ particles is found to yield the following results: $\langle S_x \rangle = \langle S_y \rangle = 0$, $\langle S_z \rangle = a$, with $0 \leq a \leq 1$. This information is not sufficient to uniquely determine the state. Determine all possible state operators that are consistent with the given information. Consider separately the three cases: $a = 0$, $0 < a < 1$, $a = 1$. Identify the pure and nonpure states in the three cases.

8.3 Prove that if for some state of a two-component system one has $\langle R^{(1)} R^{(2)} \rangle = \langle R^{(1)} \rangle \langle R^{(2)} \rangle$ for all Hermitian operators $R^{(1)}$ and $R^{(2)}$, then the state operator must be of the form $\rho = \rho^{(1)} \otimes \rho^{(2)}$. (As usual, the superscript 1 or 2 signifies that the operator acts on the factor space of component 1 or 2, respectively.)

8.4 Consider a system of two particles, each having spin $s = \frac{1}{2}$. The single particle eigenvectors of σ_z are denoted as $|+\rangle$ and $|-\rangle$, and their products serve as basis vectors for the four-dimensional state vector space of this system. The family of state vectors of the form

$$|\Psi_c\rangle = \Big(|+\rangle|-\rangle + c|-\rangle|+\rangle\Big)\sqrt{\tfrac{1}{2}}\,,$$

with $|c| = 1$ but otherwise arbitrary, all share the property

$$\sigma_z{}^{(1)}\sigma_z{}^{(2)}|\Psi_c\rangle = -|\Psi_c\rangle\,.$$

(a) Can two such states, $|\Psi_c\rangle$ and $|\Psi_{c'}\rangle$, be distinguished by any combination of measurements on the two particles separately?

(b) Can they be distinguished by any correlation between the spins of the two particles?

8.5 Show that the three-dimensional single particle state functions $\Psi_m(\mathbf{x}) \equiv f(r)Y_\ell{}^m(\theta,\phi)$ and $\Psi_{-m}(\mathbf{x})$ have the same position and momentum distributions.

8.6 Consider a two-component system that evolves under a time development operator of the form $U(t) = U^{(1)}(t) \otimes I$. (This could describe a system with no interaction between the two components, subject to an external perturbation that acts on component 1 but not on component 2.) Let $\rho(t)$ be an arbitrary correlated state of the two-component system evolving under the action of $U(t)$. Prove that the partial state of component 2, $\rho^{(2)} = \mathrm{Tr}^{(1)}\rho(t)$, is independent of t.

8.7 Consider a system of two particles, each of which has spin $s = \frac{1}{2}$. Suppose that $\langle \boldsymbol{\sigma}^{(1)} \rangle = \langle \boldsymbol{\sigma}^{(2)} \rangle = 0$. (This gives six independent pieces of data, since the spin vectors each have three components.)

(a) Construct a *pure* state consistent with the given data, or prove that none exists.

(b) Construct a *nonpure* state consistent with the given data, or prove that none exists.

8.8 For a system of two particles, each of which has $s = \frac{1}{2}$, suppose that $\langle \sigma_z{}^{(1)} \rangle = \langle \sigma_z{}^{(2)} \rangle = 1$.

(a) Construct a *pure* state consistent with the given data, or prove that none exists.

(b) Construct a *nonpure* state consistent with the given data, or prove that none exists.

8.9 It is possible to prepare a state in which two or more dynamical variables have unique values if the corresponding operators have common eigenvectors. A sufficient condition is that the operators should commute, in which case the set of common eigenvectors forms a complete basis. But a smaller number of common eigenvectors may exist even if the operators do not commute.

(a) For a single spinless particle find all the common eigenvectors of the angular momentum operators, L_x, L_y, and L_z.

(b) Find the common eigenvectors of the angular momentum and linear momentum operators, $\mathbf{L}$ and $\mathbf{P}$.

8.10 Generalize the derivation leading to (8.30) by taking $T = A_0 + i\omega B_0$ with ω *complex*, and then minimizing $\mathrm{Tr}(\rho T T^\dagger)$. This leads to a stronger inequality than (8.30).

Chapter 9

Measurement and the Interpretation of States

A typical experimental run consists of state preparation followed by measurement. The former of these operations was analyzed in the preceding chapter, and the latter will now be treated. An analysis of measurement is required for completeness, but even more important, it turns out to be very useful in elucidating the correct interpretation of the concept of a state in quantum mechanics.

9.1 *An Example of Spin Measurement*

The measurement of a spin component by means of the Stern–Gerlach apparatus is probably the simplest example to analyze. A particle with a magnetic moment passes between the poles of a magnet that produces an inhomogeneous field. The potential energy of the magnetic moment $\boldsymbol{\mu}$ in the magnetic field $\mathbf{B}$ is equal to $-\mathbf{B}\cdot\boldsymbol{\mu}$. The negative gradient of this potential energy corresponds to a force on the particle, equal to $\mathbf{F} = \boldsymbol{\nabla}(\mathbf{B}\cdot\boldsymbol{\mu})$. Since the magnetic moment $\boldsymbol{\mu}$ is proportional to the spin $\mathbf{s}$, the magnitude of this force, and hence the deflection of the particle, will depend on the component of spin in the direction of the magnetic field gradient. Hence the value of that spin component can be inferred from the deflection of the particle by the magnetic field. In practice this method is useful only for neutral atoms or molecules, because the deflection of a charged particle by the Lorentz force will obscure this spin-dependent deflection.

In Fig. 9.1, the velocity of the incident beam is in the y direction, and the magnetic field lies in the transverse xz plane. To simplify the analysis, we shall make some idealizations. We assume that the magnetic field vanishes outside of the gap between the poles. We assume that only the z component of the field is significant, and that within the gap it has a constant gradient in the z direction. Thus, relative to a suitable origin of coordinates (located some distance below the magnet), the components of the magnetic field can

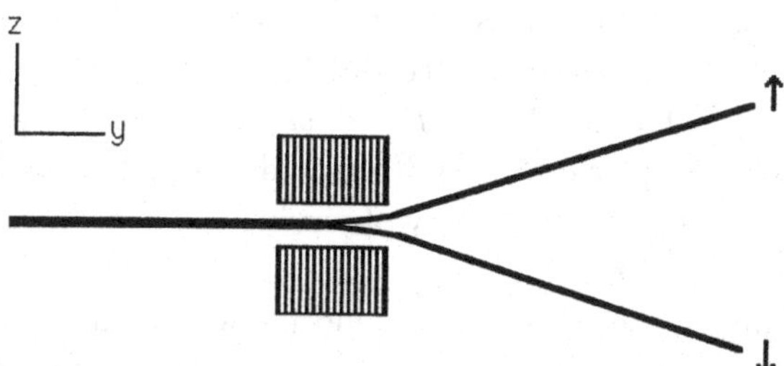

Fig. 9.1 Measurement of spin using the Stern–Gerlach apparatus.

be written as $B_x = B_y = 0$, $B_z = zB'$, where B' is the field gradient. Subject to these idealizations, the magnetic force will be in the z direction, and the y component of the particle velocity will be constant. So it is convenient to adopt a frame of reference moving uniformly in the y direction, with respect to which the incident particle is at rest. In this frame of reference, the particle experiences a time-dependent magnetic field that is nonvanishing only during the interval of time T that the particle spends inside the gap between the magnetic poles. The spin Hamiltonian, $-\boldsymbol{\mu}\cdot\mathbf{B}$, can therefore be written as

$$\begin{aligned} H(t) &= 0 \qquad && (t < 0)\,, \\ &= -cz\sigma_z && (0 < t < T)\,, \\ &= 0 && (T < t)\,. \end{aligned} \tag{9.1}$$

Here σ_z is a Pauli spin operator. The constant c includes the magnetic field gradient B' and magnitude of the magnetic moment.

Since $\boldsymbol{\nabla}\cdot\mathbf{B} = 0$, it is impossible for the magnetic field to have only a z component. A more realistic form for the field would be $B_x = -xB'$, $B_z = B_o + zB'$, which has zero divergence. If B_o is very much larger than B_x, which is true in practice, then any component of the magnetic moment in the xy plane will precess rapidly about the z axis, and the force on the particle due to B_x will average to zero. This has been shown in detail by Alstrom, Hjorth, and Mattuck (1982), using a semiclassical analysis, and by Platt (1992) using a fully quantum-mechanical analysis. Thus the idealized Hamiltonian (9.1) is justifiable.

Suppose that the state vector for $t \leq 0$ is $|\Psi_0\rangle = a|+\rangle + b|-\rangle$, with $|a|^2 + |b|^2 = 1$. Here $|+\rangle$ and $|-\rangle$ denote the spin-up and spin-down eigenvectors of σ_z. Then the equation of motion (3.64) implies that the state vector for $t \geq T$ will be

$$|\Psi_1\rangle = a\, e^{iTcz/\hbar}|+\rangle + b\, e^{-iTcz/\hbar}|-\rangle\,. \tag{9.2}$$

The effect of the interaction is to create a correlation between the spin and the momentum of the particle. According to (9.2), if $\sigma_z = +1$ then $P_z = +Tc$, whereas if $\sigma_z = -1$ then $P_z = -Tc$. Thus the trajectory of the particle will be deflected either up or down, as illustrated in Fig. 9.1, according to whether σ_z is positive or negative, and so by observing the deflection of the particle we can infer the value of σ_z.

In this analysis, the initial state of motion was assumed to be a momentum eigenstate, namely $\mathbf{P} = 0$ in the comoving frame of the incident particle. More realistically, the initial state vector $|\Psi_0\rangle$ should be multiplied by an orbital wave function $\psi(\mathbf{x})$ which has finite beam width and yields an average momentum of zero in the z direction. If the width of the initial probability distribution for P_z is small compared to the momentum $\pm Tc$ which is imparted by the inhomogeneous magnetic field, then it will still be true that a positive value of σ_z will correspond to an upward deflection of the particle and a negative value of σ_z will correspond to a downward deflection. The essential feature of any measurement procedure is the establishment of a correlation between the dynamical variable to be measured (the spin component σ_z in this example), and some macroscopic indicator that can be directly observed (the upward or downward deflection of the beam in this example).

9.2 *A General Theorem of Measurement Theory*

The essential ingredients of a measurement are an *object* (I), an *apparatus* (II), and an interaction that produces a correlation between some dynamical variable of (I) and an appropriate indicator variable of (II). Suppose that we wish to measure the dynamical variable R (assumed to be discrete for convenience) which belongs to the object (I). The corresponding operator R possesses a complete set of eigenvectors,

$$R|r\rangle^{(\mathrm{I})} = r|r\rangle^{(\mathrm{I})} . \tag{9.3}$$

The apparatus (II) has an indicator variable A, and the corresponding operator A has a complete set of eigenvectors,

$$A|\alpha, m\rangle^{(\mathrm{II})} = \alpha|\alpha, m\rangle^{(\mathrm{II})} . \tag{9.4}$$

Here α is the "indicator" eigenvalue, and m labels all the many other quantum numbers needed to specify a unique eigenvector.

The apparatus is prepared in an initial premeasurement state, $|0, m\rangle^{(\mathrm{II})}$, with $\alpha = 0$. We then introduce an interaction between (I) and (II) that produces a unique correspondence between the value r of the dynamical variable R of (I) and the indicator α_r of (II). The properties of the interaction are specified implicitly by defining the effect of the time development operator U, and since only the properties of U enter the analysis, there is no point in discussing the interaction in any more detail. The analysis may be done with various degrees of generality.

Suppose that the initial state of (I) is an eigenstate of the dynamical variable R, $|r\rangle^{(\mathrm{I})}$. Then the initial state of the combined system (I) + (II) will be $|r\rangle^{(\mathrm{I})} \otimes |0, m\rangle^{(\mathrm{II})}$. If we require that the measurement should not change the value of the quantity being measured, then we must have

$$U|r\rangle^{(\mathrm{I})} \otimes |0, m\rangle^{(\mathrm{II})} = |r\rangle^{(\mathrm{I})} \otimes |\alpha_r, m'\rangle^{(\mathrm{II})} . \tag{9.5}$$

Here the value of r is unchanged by the interaction, but the value of m may change. The latter merely represents the many irrelevant variables of the apparatus other than the indicator.

An assumption of the form (9.5) is often made in the formal theory of measurement, but many of the idealizations contained in (9.5) can be relaxed without significantly complicating the argument. There is no reason why the state of the object (I) should remain unaffected by the interaction, and indeed this is seldom true in practice. Nor is it necessary for the state of the apparatus (II) to remain an eigenvector corresponding to a unique value of m'. The most general possibility is of the form

$$\begin{aligned} U|r\rangle^{(\mathrm{I})} \otimes |0, m\rangle^{(\mathrm{II})} &= \sum_{r',m'} u_{r,m}^{r',m'} |r'\rangle^{(\mathrm{I})} \otimes |\alpha_r, m'\rangle^{(\mathrm{II})} \\ &= |\alpha_r\,; (r, m)\rangle , \text{ say}. \end{aligned} \tag{9.6}$$

The labels (r, m) in $|\alpha_r\,; (r, m)\rangle$ do not denote eigenvalues, since the vector is not an eigenvector of the corresponding operators. They are merely labels to indicate the initial state from which this vector was derived.

The only restrictions expressed in (9.6) are that the final state vector be related to the initial state vector by a unitary transformation, and that the particular value of r in the initial state vector should correspond to a unique value of the indicator α_r in the final state vector. The latter restriction is the essential condition for the apparatus to carry out a measurement. The values of α_r that correspond to different values of r should be clearly distinguishable to the observer, and will be referred to as *macroscopically distinct* values.

In the example of Sec. 9.1, the dynamical variable being measured is the spin component σ_z, and the indicator variable is the momentum P_z. This shows that the indicator variable need not be physically separate from the object of measurement; rather, it is sufficient for it to be kinematically independent of the dynamical variable being measured. A more complete analysis of that example would treat the deflected trajectories explicitly, and would use the position coordinate z as the indicator variable.

Consider next a general initial state for the object (I),

$$|\psi\rangle^{(\mathrm{I})} = \sum_r c_r |r\rangle^{(\mathrm{I})} , \tag{9.7}$$

which is not an eigenstate of the dynamical variable R that is being measured. From (9.6) and the linearity of the time development operator U, it follows that the final state of the system will be

$$\begin{aligned} U|\psi\rangle^{(\mathrm{I})} \otimes |0,m\rangle^{(\mathrm{II})} &= \sum_r c_r |\alpha_r\,; (r,m)\rangle \\ &= |\Psi_m{}^f\rangle\,, \ \text{say}\,. \end{aligned} \tag{9.8}$$

This final state is a *coherent superposition* of macroscopically distinct indicator eigenvectors, and this is the theorem referred to in the title of this section.

The probability, in the final state, that the indicator variable A of the apparatus (II) has the value α_r is equal to $|c_r|^2$, just the same as the probability in the initial state (9.7) that the dynamical variable R of the object (I) had the value r. This a consequence of the requirement that there be a faithful mapping from the initial value of r to the final value of α_r.

9.3 *The Interpretation of a State Vector*

The preceding simple theorem, that if the initial state is not an eigenstate of the dynamical variable being measured, then the final state vector for the whole system (object of measurement plus apparatus) must be a coherent superposition of macroscopically distinct indicator eigenvectors, has important implications for the interpretation of the quantum *state* concept. It allows us to discriminate between the two principal classes of interpretations.

A. A pure state $|\Psi\rangle$ provides a complete and exhaustive description of an *individual* system. A dynamical variable represented by the operator Q has a value (q, say) if and only if $Q|\Psi\rangle = q|\Psi\rangle$.

B. A pure state describes the statistical properties of an *ensemble* of similarly prepared systems.

Interpretation A is more common in the older literature on quantum mechanics, although it is often only implicit and not formulated explicitly. Interpretation B has been consistently adopted throughout this book, but it is only now that the reasons for that choice will be examined.

Since the state vector plays a very prominent role in the mathematical formalism of quantum mechanics, it is natural to attempt to give it an equally prominent role in the interpretation. The superficial appeal of interpretation A lies in its attributing a close correspondence between the properties of the world and the properties of the state vector. However, it encounters very serious difficulties when confronted with the measurement theorem. Because the final state (9.8) of the measurement process is not an eigenstate of the indicator variable, one must conclude, according to interpretation A, that the indicator has no definite value. Moreover this is not a microscopic uncertainty, which could be tolerated, but rather a *macroscopic* uncertainty, since the final state vector (9.8) is a coherent superposition of macroscopically distinct indicator eigenvectors. In a typical case, the indicator variable α_r might be the position of a needle on a meter or a mark on a chart recorder, and for two adjacent values of the measured variable, r and r', the separation between α_r and $\alpha_{r'}$ could be several centimeters. It would be apparent to any casual observer that the indicator position α_r is well defined to within at least a millimeter, but the state vector (9.8) would involve a superposition of terms corresponding to values of α_r that differ by several centimeters. Thus the interpretation of (9.8) as a description of an individual system is in conflict with observation. There is no such difficulty with interpretation B, according to which the state vector is an abstract quantity that characterizes the probability distributions of the dynamical variables of an ensemble of similarly prepared systems. Each member system of the ensemble consists of an object and a measuring apparatus.

[[The prototype of the measurement theorem was given by E. Schrödinger in 1935. He considered a chamber containing a cat, a flask of poison gas, a radioactive atom, and an automatic device to release the poison when the atom decays. If the atom were isolated, then after a time equal to one half-life its state vector would be $(|u\rangle + |d\rangle)\sqrt{\frac{1}{2}}$, where the vectors $|u\rangle$ and $|d\rangle$ denote the undecayed and decayed states. But since the atom is coupled to the cat via the apparatus, the state of the system after one half-life is

$$\left(|u\rangle_{\text{atom}}|\text{live}\rangle_{\text{cat}} + |d\rangle_{\text{atom}}|\text{dead}\rangle_{\text{cat}}\right)\sqrt{\tfrac{1}{2}}\,.$$

This is a correlated (or entangled) state like those that were discussed in Sec. 8.3. It is also a superposition of macroscopically distinct states (live cat and dead cat) that is typical of the measurement process. Schrödinger argued that a seemingly plausible interpretation — that an individual quantum system's properties are smeared over the range of values contained in the state vector — cannot be accepted, because it would necessarily imply a macroscopic smearing for classical objects such as the unfortunate cat. Correlated states that involve a superposition of macroscopically distinct terms are often metaphorically called "Schrödinger cat states".]]

The subject could now be concluded, were it not for the persistence of the defenders of the old interpretation (A). In order to save that interpretation, they postulate a further process that is supposed to lead from the state (9.8) to a so-called "reduced state",

$$|\Psi_m{}^f\rangle \to |\alpha_{r_0}\,;(r_0,m)\rangle\,, \tag{9.9}$$

which is an eigenvector of the indicator variable A, with the eigenvalue α_{r_0} being the actual observed value of the indicator position. This postulate of *reduction of the state vector* creates a new problem that is peculiar to interpretation A: namely, how to account for the mechanism of this *reduction* process. Some of the proposed explanations are as follows:

(i) The reduction process (9.9) is caused by an *unpredictable and uncontrollable disturbance* of the object by the measuring apparatus. [Such an argument is offered by Messiah (1964), see p. 140.]

In fact, any interaction between the object (I) and the apparatus (II) that might serve as the cause of such a disturbance is implicitly included in the Hamiltonian from which the time development operator U is constructed. If the interaction satisfies the minimal condition (9.6) for it to be a measurement interaction, then it must lead to the superposition (9.8), and not to the reduced state (9.9). So the disturbance theory is untenable.

(ii) The observer causes the reduction process (9.9) when he reads the result of the measurement from the apparatus.

This is really just a variant of (i) with the observer, rather than the apparatus, causing the disturbance, and it is refuted simply by redefining (II) to include both the apparatus and the observer. But while it circulated, this proposal led to some unfruitful speculations as to whether quantum mechanics can be applied to the consciousness of an observer.

(iii) The reduction (9.9) is caused by the environment, the "environment" being defined as the rest of the universe other than (I) and (II).

This proposal is a bit vague, because it has not been made clear just what part of the environment is supposed to be essential. But it is apparent that if we formally include in (II) all those parts of the environment whose influence might not be negligible, then the same argument that defeated (i) and (ii) will also defeat (iii).

(iv) In proving the measurement theorem, the initial state of the apparatus was assumed to be a definite pure state, $|0, m\rangle^{(\mathrm{II})}$. But in fact m is an abbreviation for an enormous number of microscopic quantum numbers, which are never determined in practice. It is very improbable that they will have the same values on every repetition of the state preparation. Therefore the initial state should not be described as a pure state, but as a mixed state involving a distribution of m values. This, it is hoped, might provide a way around the measurement theorem.

In order to respond to argument (iv), it is necessary to repeat the analysis of Sec. 9.2 using general state operators.

The measurement theorem for general states

Instead of the pure state vector, $|\Psi_m{}^i\rangle = |\psi\rangle^{(\mathrm{I})} \otimes |0, m\rangle^{(\mathrm{II})}$, which was taken to represent the initial state in (9.8), we now assume a more general initial state for the system (I) + (II):

$$\rho^i = \sum_m w_m |\Psi_m{}^i\rangle\langle\Psi_m{}^i| \,. \tag{9.10}$$

Here w_m can be regarded as the probability associated with each of the microscopic states labeled by m, which represents the enormously many quantum numbers of the apparatus other than the indicator α.

The hope of an advocate of interpretation A who defended it by means of argument (iv) is that final state would be a mixture of "indicator" eigenvectors, perhaps of the form

$$\rho^d = \sum_r |c_r|^2 \sum_m v_m |\alpha_r\,; (r, m)\rangle\langle\alpha_r\,; (r, m)| \,, \tag{9.11}$$

but certainly diagonal with respect to α_r. (Any terms that were nondiagonal in α_r would correspond to coherent superpositions of macroscopically distinct "indicator position" eigenvectors, the avoidance of which is essential to the maintenance of interpretation A.) The conjectured achievement of (9.11) as final state of the measurement process is more plausible than (9.9). The latter would have prescribed a unique measurement result, α_{r_0}. But it is universally agreed that quantum mechanics can make only probabilistic predictions, and (9.11) is consistent with a prediction that the result may be α_r with probability $|c_r|^2$.

However, the conjectured form (9.11) is not correct. The actual final state of the measurement process is

$$\rho^f = U\rho^i U^\dagger = \sum_m w_m |\Psi_m{}^f\rangle\langle\Psi_m{}^f| \,, \tag{9.12}$$

where $|\Psi_m{}^f\rangle = U|\Psi_m{}^i\rangle$. From (9.8) it follows that

$$\rho^f = \sum_{r1}\sum_{r2} c_{r1}{}^* c_{r2} \sum_m w_m |\alpha_{r1}\,;(r_1, m)\rangle\langle\alpha_{r2}\,;(r_2, m)| \,. \tag{9.13}$$

The terms with $\alpha_{r1} \neq \alpha_{r2}$ indicate a coherent superposition of macroscopically distinct indicator eigenvectors, just as was the case in (9.8). It is clear that the nondiagonal terms in (9.13) cannot vanish so as to reduce the state to the form of (9.12), and therefore the measurement theorem applies to general states as well as to pure states. In all cases in which the initial state is not an eigenstate of the dynamical variable being measured, the final state must involve coherent superpositions of macroscopically distinct indicator eigenvectors. If this situation is unacceptable according to any interpretation, such as A, then that interpretation is untenable.

Perhaps the best way to conclude this discussion is to quote the words of Einstein (1949):

> "One arrives at very implausible theoretical conceptions, if one attempts to maintain the thesis that the statistical quantum theory is in principle capable of producing a complete description of an individual physical system. On the other hand, those difficulties of theoretical interpretation disappear, if one views the quantum-mechanical description as the description of ensembles of systems."

9.4 *Which Wave Function?*

Once acquired, the habit of considering an individual particle to have its own wave function is hard to break. Even though it has been demonstrated to be strictly incorrect, it is surprising how seldom it leads to a serious error.

This is because the predictions of quantum mechanics that are derived from a wave function consist of probabilities, and the operational significance of a probability is as a relative frequency. Thus one is, in effect, bound to invoke an ensemble of similar systems at the point of comparison with experiment, regardless of how one originally interpreted the wave function. Because so many of the results do not seem to depend in a critical way on the choice of interpretation, some "practical-minded" physicists would like to dismiss the whole subject of interpretation as irrelevant. That attitude, however, is not justified, and a number of practical physicists have been led into unnecessary confusion and dispute because of inattention to the matters of interpretation that we have been discussing.

An interesting case is to be found in the subject of electron interference. Electrons are emitted from a hot cathode, and subsequently accelerated to form a beam, which is then used for various interference experiments. The energy spread of the beam can be accounted for on either of two assumptions (both based on interpretation A of Sec. 9.3):

(a) Each electron is emitted in an energy eigenstate (a plane wave), but the particular energy varies from one electron to the next;
(b) Each electron is emitted as a wave packet which has an energy spread equal to the energy spread of the beam.

One might expect that these two assumptions would lead to quantitatively different predictions about the interference pattern, and so they could be experimentally distinguished.

To simplify the analysis, we shall treat the electron beam as moving from left to right in an essentially one-dimensional geometry. According to assumption (a), each electron has a wave function of the form $\psi_k(x,t) = e^{i(kx-\omega t)}$. The energy of this electron is $\hbar\omega = \hbar^2 k^2/2M$, and the observed energy distribution of the beam allows us to infer the appropriate probability density $W(\omega)$. The state operator corresponding to the thermal emission process will be $\rho = \int |\psi_k\rangle\langle\psi_k| W(\omega) d\omega$. (Remember that k is a function of ω.) In coordinate representation, this becomes

$$\begin{aligned}\rho(x,x') \equiv \langle x|\rho|x'\rangle &= \int \psi_k(x,t)\,\psi_k{}^*(x',t)\,W(\omega)\,d\omega \\ &= \int e^{ik(x-x')}\,W(\omega)\,d\omega\,. \qquad (9.14)\end{aligned}$$

Notice that the time dependence has canceled out, indicating that this is a steady state. All observable quantities, including the interference pattern, can be calculated from the state function $\rho(x,x')$.

According to assumption (b), an individual electron will be emitted in a wave packet state, $\psi_{t_0}(x,t) - \int A(\omega)e^{i[kx-\omega(t-t_0)]}d\omega$, a particular wave packet being distinguished by the time t_0 at which it is emitted. The energy distribution of such a wave packet state is $|A(\omega)|^2 = W(\omega)$. The state function for the emission process is obtained by averaging over the emission time t_0, which is assumed to be uniformly distributed:

$$\langle x|\rho|x'\rangle = \lim_{T\to\infty}\frac{1}{T}\int_{-T/2}^{T/2}\psi_{t_0}(x,t)\psi_{t_0}{}^*(x',t)dt_0$$

$$= \lim_{T\to\infty}\frac{1}{T}\int_{-T/2}^{T/2}\int A(\omega)e^{i[kx-\omega(t-t_0)]}d\omega\int A^*(\omega')e^{-i[k'x'-\omega'(t-t_0)]}d\omega' dt_0\,.$$

Performing the integral over t_0 first and then taking the limit $T\to\infty$ yields zero unless $\omega=\omega'$ (and so also $k=k'$). Therefore the state function reduces to

$$\rho(x,x') = \int e^{ik(x-x')}|A(\omega)|^2 d\omega\,, \tag{9.15}$$

which is the same as the result (9.14) which was obtained from assumption (a). Thus it is apparent that assumptions (a) and (b) do not lead to any observable differences, and the controversy over the form of the supposed wave functions of individual electrons was pointless.

If we now adopt interpretation (B) and regard the state operator ρ as the fundamental description of the state generated by the thermal emission process, which yields an ensemble of systems each of which is a single electron, we can obtain $\rho(x,x')$ directly, without ever speculating about individual wave functions. First we recognize that it is a *steady state* process, so we must have $d\rho/dt = 0$. (The Schrödinger picture is being used here.) Therefore it follows that $[H,\rho]=0$, and so ρ and H possess a complete set of common eigenvectors. These are just the free particle states, $|\psi_\omega\rangle$, having the coordinate representation $\langle x|\psi_\omega\rangle = e^{ikx}$, and satisfying the eigenvalue equation $H|\psi_\omega\rangle = \hbar\omega|\psi_\omega\rangle$. Therefore the state operator must have the form

$$\rho = \int |\psi_\omega\rangle\langle\psi_\omega|W(\omega)d\omega\,, \tag{9.16}$$

where, as before, $W(\omega)$ describes the energy distribution in the beam. In coordinate representation this becomes

$$\rho(x, x') \equiv \langle x|\rho|x'\rangle = \int e^{ik(x-x')} W(\omega) d\omega\,, \tag{9.17}$$

where k is related to ω by the relation $\hbar\omega = \hbar^2 k^2/2M$. The fact that the source was assumed to emit particles from left to right serves as a boundary condition and restricts k to be positive. This is, of course, just the same result as was obtained by the other methods, but it is conceptually superior because it avoids any pointless speculation about the form of any supposed wave function of an individual electron.

[[I am indebted to Professor R. H. Dicke for providing me with some of the historical background for the incident upon which the above example is based. It took place at a conference in 1956. Apparently, many of the participants in the discussion had neglected to perform the calculation leading to the identity of the results (9.14) and (9.15), but had relied on their intuitions about wave functions. Hence they expended considerable effort debating the size and coherence length of the supposed wave packets of the individual electrons. Someone espoused the view that a spread in the energy of the beam leaving the cathode was essential for the occurrence of interference, whereas, in fact, the energy spread tends to wash out the interference pattern. None of the confusion would have occurred were it not for the habit of associating a wave function with an individual electron instead of an ensemble. It goes to show that questions of interpretation in quantum mechanics are not devoid of practical utility.

A similar situation occurred more recently, in which a neutron interference measurement was incorrectly interpreted as providing information about the size of the supposed wave packets of individual neutrons. [See Kaiser *et al.* (1983) and Cosma (1983).]]]

9.5 *Spin Recombination Experiment*

Some evidence that the state vector retains its integrity, and is not subject to any "reduction" process, is provided by the spin recombination experiments that are possible with the single crystal neutron interferometer (see Sec. 5.5). A beam of neutrons with spin polarized in the $+z$ direction is incident from the left (Fig. 9.2). At point A it is split into a transmitted beam AC and a Bragg-reflected beam AB. Similar splittings occur at B and at C, but the

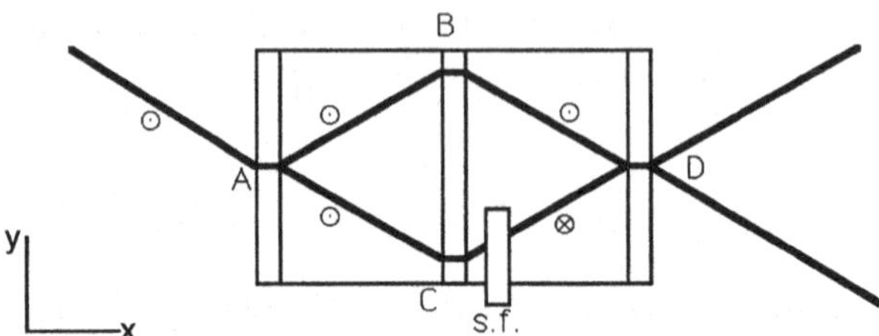

Fig. 9.2 The spin recombination experiment. A spin-flipper (s.f.) overturns the spin of one of the internal beams, which are then recombined.

beams that exit the apparatus at those points play no role in the experiment and are not shown. A spin-flipper is inserted into the beam CD, so that a spin-up and a spin-down beam are recombined at the point D. The spin state of the beams that emerge to the right of the apparatus is then determined.

Let the vectors $|+\rangle$ and $|-\rangle$ denote the spin-up and spin-down eigenvectors of the Pauli spin operator σ_z. It seems reasonable to say that the neutrons at point B have the spin state $|+\rangle$ and the neutrons emerging to the right of the spin-flipper have the spin state $|-\rangle$. What then will be the spin state when the beams recombine at D? Because the beams at B and at C are separated by a distance of a few centimeters, so that their spatial wave functions do not overlap, one might suppose that all coherence has been lost and that no interference will be possible. In that case the spin state should be

$$\rho^{\text{inc}} = \frac{1}{2}(|+\rangle\langle+| + |-\rangle\langle-|)\,, \tag{9.18}$$

which describes an incoherent mixture of spin-up and spin-down. (The two beams are assumed to have equal intensities.) Such a state is also suggested by the "reduction" hypothesis that led to (9.11) in the general theory. If, on the other hand, the coherence is maintained, then the spin state will be of the form

$$\rho^{\text{coh}} = |u\rangle\langle u|\,, \quad \text{with } |u\rangle = \frac{e^{i\alpha}|+\rangle + e^{i\beta}|-\rangle}{\sqrt{2}}\,. \tag{9.19}$$

Both of these state operators predict that $\langle\sigma_z\rangle = 0$; the z component of the spin is equally likely to be positive or negative. But, whereas ρ^{inc} predicts no polarization in any direction, ρ^{coh} predicts the spin to be polarized in some direction within the xy plane. The average x component of spin is given by ρ^{coh} to be $\langle\sigma_x\rangle = \cos(\alpha - \beta)$. Although the phases α and β may not be known in advance, their difference can be systematically varied by placing an additional phase-shifter in one of the beams. The experiment [Summhammer *et al.* (1982)] found a periodic dependence of $\langle\sigma_x\rangle$ on the phase shift, and no such dependence of $\langle\sigma_z\rangle$, confirming that the coherent superposition (9.19) is the correct state.

Let us examine the neutron state function for this experiment in more detail. If both position and spin variables are accounted for, the state function should be written, in place of (9.19), as

$$\psi_+(\mathbf{x})|+\rangle + \psi_-(\mathbf{x})|-\rangle\,. \tag{9.20}$$

The wave functions $\psi_+(\mathbf{x})$ and $\psi_-(\mathbf{x})$ vanish outside of the beams. Along AB, AC, and from B to the left of D, we have $\psi_-(\mathbf{x}) = 0$; from the right of s.f. to the left of D, we have $\psi_+(\mathbf{x}) = 0$. Both components are nonzero to the right of D. The spin state operator, written in the standard basis, is

$$\rho = \begin{bmatrix} |\psi_+|^2\,, & \psi_+{\psi_-}^* \\ \psi_-{\psi_+}^*\,, & |\psi_-|^2 \end{bmatrix}\,. \tag{9.21}$$

At point D the nondiagonal terms are nonzero, indicating the coherent nature of the superposition. The preservation of this coherence over a distance of several centimeters is possible because the neutron spectrometer is cut from a large single crystal of silicon, and the relative separations of the various parts are stable to within the interatomic separation distance.

Suppose, contrary to the conditions of the actual experiment, that the spectrometer were not such a high precision device, and that the relative separations of the points A, B, C and D were subject to random fluctuations that were larger that the de Broglie wavelength of the neutron wave function. This would give rise to random fluctuations in the phases α and β in (9.19), and in the phases of the nondiagonal terms of (9.21). Different neutrons passing through the spectrometer at different times would experience different configurations of the spectrometer, and the observed statistical distributions of spin components would be an average over these fluctuations. If we regard these noise fluctuations as a part of the state preparation process, then ρ should be averaged over the noise. If the phase difference $(\alpha - \beta)$ fluctuates so widely that $(\psi_+{\psi_-}^*)$ is uniformly distributed over a circle in the complex plane, then the average of ρ will be diagonal and will reduce to the incoherent state (9.18). Thus we see that the so-called "reduced" state is physically significant in certain circumstances. But it is only a phenomenological description of an effect on the system (the neutron and spectrometer) due to its environment (the cause of the noise fluctuations), which has for convenience been left outside of the definition of the system. This "reduction" of the state is not a new fundamental process, and, contrary to the impression given in some of the older literature, it has nothing specifically to do with measurement.

Instead of considering the influence of the environment on the spectrometer as an external effect, we may include the environment within the system. Now the neutrons that follow the path ABD will interact differently with the environment than those that follow the path ACD. These interactions will affect the state of the environment, and the final state (9.20) must be generalized to include environment thus:

$$|\Psi\rangle = \psi_+(\mathbf{x})|+\rangle|e_1\rangle + \psi_-(\mathbf{x})|-\rangle|e_2\rangle\,.$$

Here the vector $|e_1\rangle$ is the state of the environment that would result if the neutron followed path ABD, and $|e_2\rangle$ is the environmental state that would result if the neutron followed path ACD. If $|e_1\rangle = |e_2\rangle$ then the formal inclusion of the environment has no effect, and we recover (9.21). (This is a good approximation to the conditions of the actual experiment.) But, in general, the spin state (9.21) must be replaced by

$$\rho = \begin{bmatrix} |\psi_+|^2\,, & \psi_+\psi_-{}^*\langle e_2|e_1\rangle \\ \psi_-\psi_+{}^*\langle e_1|e_2\rangle\,, & |\psi_-|^2 \end{bmatrix}.$$

This ρ is obtained form the total state operator $|\Psi\rangle\langle\Psi|$ by taking the partial trace over the degrees of freedom of the environment. If the difference between the effects of taking paths ABD and ACD on the environment is so great that $|e_1\rangle$ and $|e_2\rangle$ are orthogonal, then the state reduces to the incoherent mixture ρ^{inc} (9.18).

Thus we have two methods of treating the influence, if any, of the environment on the experiment. In the first method, the environment is regarded as a perturbation from outside of the system, which introduces random phases. Coherence will be lost if the phase fluctuations are of magnitude 2π or larger. In the second method, we include the environment within the system. The crucial factor then becomes the action of the apparatus on the environment, rather than the reaction of the environment on the apparatus. Because of the general equality of action and reaction, we may expect these two approaches to be related. Stern, Aharonov, and Imry (1990) have demonstrated their equivalence under rather broad conditions.

9.6 *Joint and Conditional Probabilities*

In the previous discussions, an experimental run was taken to consist of state preparation followed by the measurement of a single quantity. If instead of a single measurement, it involves a sequence of measurements of two or more dynamical variables, then in addition to the probability distributions for the individual quantities, we may also consider correlations between the values of the various quantities. This can be expressed by the joint probability distribution for the results of two or more measurements, or by the probability for one measurement conditional on both the state preparation and the result of another measurement.

These joint and conditional probabilities are related by Axiom 4 of Sec. 1.5,

$$\text{Prob}(A\&B|C) = \text{Prob}(A|C)\ \text{Prob}(B|A\&C)\,. \tag{9.22}$$

It is appropriate here to take the event denoted as C to be the preparation that corresponds to the state operator ρ, and we shall indicate this by writing ρ in place of C. The events A and B shall be the results of two measurements following that state preparation. Let R and S be two dynamical variables with corresponding self-adjoint operators R and S,

$$R|r_n\rangle = r_n|r_n\rangle\,, \quad S|s_m\rangle = s_m|s_m\rangle\,. \tag{9.23}$$

Associated with these operators are the projection operators,

$$M_R(\Delta) = \sum_{r_n \in \Delta} |r_n\rangle\langle r_n|\,, \tag{9.24a}$$

$$M_S(\Delta) = \sum_{s_m \in \Delta} |s_m\rangle\langle s_m|\,, \tag{9.24b}$$

which project onto the subspaces spanned by those eigenvectors whose eigenvalues lie in the interval Δ. Let A denote the event of R taking a value in the range $\Delta_a(\text{R} \in \Delta_a)$, and let B denote the event of S taking a value in the range $\Delta_b(\text{S} \in \Delta_b)$. Suppose the first of these events takes place at time t_a and the second takes place at time t_b. If these times are of interest, then it is convenient to use the Heisenberg picture, and to regard the specification of t_a as implicit in the operators R and $M_R(\Delta)$, and the specification of t_b as implicit in the operators S and $M_s(\Delta)$.

According to the general probability formula (2.32), the first factor on the right hand side of (9.22) is

$$\text{Prob}(A|C) \equiv \text{Prob}(\text{R} \in \Delta_a|\rho) = \text{Tr}\{\rho\, M_R(\Delta_a)\}\,. \tag{9.25}$$

The *joint probability* on the left hand side of (9.22) can be evaluated from the established formalism of quantum mechanics only if we can find a projection operator that corresponds to the compound event $A\&B$. This is possible if the projection operators $M_R(\Delta_a)$ and $M_S(\Delta_b)$ are commutative. In that case the product $M_R(\Delta_a)M_S(\Delta_b)$ is a projection operator that projects onto the subspace spanned by those common eigenvectors of R and S with eigenvalues in the ranges Δ_a and Δ_b, respectively. We then have

$$\begin{aligned}\text{Prob}(A\&B|C) &\equiv \text{Prob}\{(\text{R} \in \Delta_a)\&(\text{S} \in \Delta_b)|\rho\} \\ &= \text{Tr}\{\rho\, M_R(\Delta_a)M_S(\Delta_b)\}\,. \end{aligned} \tag{9.26}$$

This is the joint probability that both events A and B occur on the condition C; that is to say, it is the probability that the result of the measurement of R at time t_a is in the range Δ_a and the result of the measurement of S at time t_b is in the range Δ_b, following the state preparation corresponding to ρ. This calculation will be possible for arbitrary Δ_a and Δ_b if and only if the operators R and S are commutative.

The remaining factor in (9.22),

$$\text{Prob}(B|A\&C) \equiv \text{Prob}\{(S \in \Delta_b)|(\text{R} \in \Delta_a)\&\rho\}\,, \tag{9.27}$$

is the probability for a result of the S measurement, conditional on the state preparation *and* a certain result of the R measurement. This is a kind of probability statement that we have not previously considered, since the general quantum-mechanical probability formula (2.32) involves only conditioning on the state preparation. There are two possibilities open to us. We can regard the preparation of state ρ and the following measurement of R as a composite operation that corresponds to the preparation of a new state ρ'. Alternatively, we can use (9.22) to define $\text{Prob}(B|A\&\rho)$ in terms of the other two factors, both of which are known.

Filtering-type measurements

If we are to regard the initial ρ-state preparation followed by the R measurement as a composite operation that prepares some new state ρ', then we will require a detailed description of the R measurement apparatus and a dynamical analysis of its operation. This can be done for any particular case, but no general treatment seems possible. However, there is one kind of measurement that can be treated quite easily. It is a measurement of the filtering type, in which the ensemble of systems generated by the ρ-state preparation is separated into subensembles according to the value of the dynamical variable R. (The Stern–Gerlach apparatus provides an example of this type of measurement.) If we consider the result of the subsequent S measurement on *only* that subensemble for which R $\in \Delta_a$, and ignore the rest, we shall be determining the conditional probability (9.27). This filtering process, which has the effect of removing all values of R except those for which R $\in \Delta_a$, can be regarded as preparing a new state that is represented by

$$\rho' = \frac{M_R(\Delta_a)\,\rho\,M_R(\Delta_a)}{\text{Tr}\{M_R(\Delta_a)\,\rho\,M_R(\Delta_a)\}} \tag{9.28}$$

and the conditional probability (9.27) can be calculated by means of (2.33):

$$\begin{aligned}\text{Prob}(B|A\&\rho) &\equiv \text{Prob}\{(\text{S} \in \Delta_b)|(\text{R} \in \Delta_a)\&\rho\}\\ &= \text{Prob}\{\text{S} \in \Delta_b)|\rho'\} = \text{Tr}\{\rho' M_S(\Delta_b)\}\,.\end{aligned} \tag{9.29}$$

[[The similarity between *measurement* and *state preparation* in the case of filtering is probably the reason why some of the earlier authors failed to distinguish between these two concepts. Indeed, the statement by Dirac (1958, p. 36) to the effect that the state immediately after an R measurement must be an eigenstate of R, seems perverse unless its application is restricted to filtering-type measurements. But this type of measurement is of a very special kind. A more general measurement, of the sort contemplated in Sec. 9.2, must be expected to have a much more drastic effect on the state, which need not be of the simple form (9.28).]]

We can now calculate a *joint probability for two filtering-type measurements* by substituting (9.25) and (9.29) into (9.22):

$$\begin{aligned}\text{Prob}(A\&B|C) &= \text{Prob}(A|C)\ \text{Prob}(B|A\&C)\\ &= \text{Tr}\{\rho\,M_R(\Delta_a)\}\ \text{Tr}\{\rho'\,M_S(\Delta_b)\}\\ &= \text{Tr}\{\rho\,M_R(\Delta_a)M_S(\Delta_b)M_R(\Delta_a)\}\,.\end{aligned} \tag{9.30}$$

The last line has been simplified by using the cyclic permutation invariance of the trace of an operator product. If the projection operators $M_R(\Delta_a)$ and $M_S(\Delta_b)$ commute, then this expression reduces to (9.26), verifying the consistency of the quantum-mechanical probabilities with Axiom 4 of probability theory.

The joint probability (9.26) was obtained under the condition that the operators R and S be commutative, whereas no such restriction needs to be imposed on the conditional probability (9.29). However, the latter is restricted to filtering-type measurements. We have just seen that these two results are consistent with (9.22) when all of these conditions are satisfied together; nevertheless it seems rather strange that the conditions for evaluating the left and right sides of (9.22) should be different. The answer to this puzzle is that the derivation of (9.26) was implicitly based on the assumption that the measurements of R and S were equivalent to, or at least compatible with, a *joint filtering according to the eigenvalues of both R and S.* That will be possible only if R and S possess a complete set of common eigenvectors, that is, only if R and S commute. In this case, the order of the times t_a and t_b at which the two measurements are performed is irrelevant, as is apparent from symmetry of (9.26) with respect to the two projection operators. (Recall that these operators are in the Heisenberg picture, but their time dependence is not explicitly indicated so as not to complicate the notation.)

If the operators R and S do not commute, then (9.26) does not apply. We can still use (9.22) as a *definition* of the joint probability $\text{Prob}(A\&B|C)$, since the factors on the right hand side are both well defined, in principle. However, it must be remembered that the definition of the event A includes its time of occurrence t_a, and the definition of B includes its time t_b. It is now essential that the time order $t_a < t_b$ be observed, because it is the R measurement that serves as (part of the) state preparation for the S measurement, and not vice versa. This is evident, in the case where the R measurement is of the filtering type, from the lack of symmetry of (9.30) with respect to the two projection operators.

Application to spin measurements

Some of these ideas will now be illustrated for an $s = \frac{1}{2}$ spin system. Consider that a state represented by $|\psi\rangle$ is prepared. It is then subjected to three successive measurements of the filtering type: a measurement of σ_z at time t_1, a measurement of σ_u at time t_2, and a measurement of σ_x at time t_3. The u direction is in the zx plane, making an angle θ with respect to the z axis. These filtering measurements will split the initial beam first into two, then into four, and finally into eight separated subbeams (see Fig. 9.3). Seven Stern–Gerlach machines will be required. For simplicity we assume that the spin vector is a constant of motion between the measurements.

Each of the eight final outcomes of this experiment corresponds to a particular combination of results ($+1$ or -1) for the three $(\sigma_z, \sigma_u, \sigma_x)$ measurements, and the probability of these various outcomes is, in fact, the *joint probability* for the results of the three measurements. The full notation for this joint

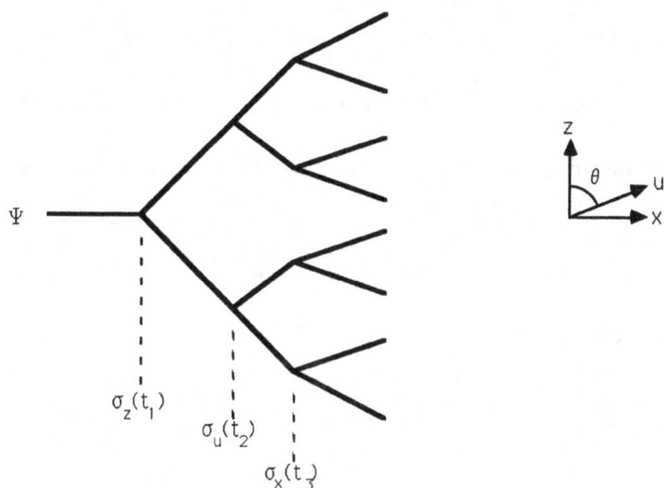

Fig. 9.3 Illustration of three successive spin filtering measurements. The upward sloping lines correspond to a result of +1, and the downward sloping lines correspond to −1.

probability should be $\text{Prob}(\sigma_z = a, \sigma_u = b, \sigma_x = c|\psi\&X)$, with $a = \pm 1$, $b = \pm 1$, and $c = \pm 1$. This probability is conditional on the state preparation (denoted by ψ) and the configuration of the Stern–Gerlach machines (denoted by X). It will be abbreviated as $P(a, b, c|\psi\&X)$, with the order of the three initial arguments corresponding to the unique time ordering of the three measurements. (It should be stressed that this is the joint probability for the results of three actual measurements, and not a joint distribution for hypothetical simultaneous values of three noncommuting observables. Moreover, the various subbeams in this experiment are all separated, and no attempt will be made to recombine them as was done in Sec. 9.5. Therefore questions of relative phase and coherence are not relevant.)

The initial state vector can be written as $|\psi\rangle = \alpha|z+\rangle + \beta|z-\rangle$, in terms of the basis formed by the eigenvectors of σ_z. The amplitudes are divided at each filtering operation, and this division of amplitudes can be calculated from the appropriate projection operators. The absolute squares of these amplitudes yield the probabilities for the outcomes of the various measurements. Following the measurement of σ_z at time t_1 we have, in an obvious notation,

$$\begin{aligned} P_z(a|\psi\&X) = |M_z(a)|\psi\rangle|^2 = \langle\psi|M_z(a)|\psi\rangle &= |\alpha|^2\,, \qquad \text{for } a = +1\,, \\ &= |\beta|^2\,, \qquad \text{for } a = -1\,. \end{aligned} \tag{9.31}$$

The projection operators here are $M_z(+1) = |z+\rangle\langle z+|$, and $M_z(-1) = |z-\rangle\langle z-|$. Similarly, after the measurement of σ_u at time t_2 we have

$$P_{zu}(a, b|\psi\&X) = |M_u(b)M_z(a)|\psi\rangle|^2 = \langle\psi|M_z(a)M_u(b)M_z(a)|\psi\rangle\,. \qquad (9.32)$$

[Notice that this is equivalent to the form (9.30).] Here $M_u(b)$ is a projection operator onto an eigenvector of σ_u,

$$M_u(+1) = |u+\rangle\langle u+|\,, \quad M_u(-1) = |u-\rangle\langle u-|\,,$$

and the eigenvectors are given by (7.49) to be

$$|u+\rangle = \cos\left(\frac{\theta}{2}\right)|z+\rangle + \sin\left(\frac{\theta}{2}\right)|z-\rangle\,,$$

$$|u-\rangle = -\sin\left(\frac{\theta}{2}\right)|z+\rangle + \cos\left(\frac{\theta}{2}\right)|z-\rangle\,.$$

Finally, after completion of the measurement of σ_x at time t_3, we have

$$\begin{aligned} P(a, b, c|\psi\&X) &= |M_x(c)M_u(b)M_z(a)|\psi\rangle|^2 \\ &= \langle\psi|M_z(a)M_u(b)M_x(c)M_u(b)M_z(a)|\psi\rangle\,, \end{aligned} \qquad (9.33)$$

$M_x(c)$ being a projection operator onto an eigenvector of σ_x.

There is an obvious redundancy in explicitly indicating that the probability (9.33) is conditional on the configuration X of the Stern–Gerlach filtering apparatuses, since the mere fact that σ_z, σ_u, and σ_x were measured implies that the appropriate apparatus was in place. The inclusion of X in (9.32) is not entirely redundant, since it indicates the presence of the σ_x filter, as well as the σ_z and σ_u filters that are necessary for the measurement. But to the extent that it is not redundant, it is irrelevant because the results of the σ_z and σ_u measurements cannot be affected by a possible future interaction with the σ_x filter. This follows formally from the fact that $M_x(+1) + M_x(-1) = 1$, and hence

$$\begin{aligned} P_{zu}(a, b|\psi\&X) &= \sum_{c=\pm 1} P(a, b, c)\psi\&X) \\ &= \sum_{c=\pm 1} \langle\psi|M_z(a)M_u(b)M_x(c)M_u(b)M_z(a)|\psi\rangle \\ &= \langle\psi|M_z(a)M_u(b)M_a(a)|\psi\rangle = P_{zu}(a, b|\psi)\,. \end{aligned} \qquad (9.34)$$

The fact that $M_x(c)$ drops out of this expression indicates that the presence or absence of the σ_x filter has no effect on the measurements of σ_z and σ_u. Similarly, we have

$$\begin{aligned} P_z(a|\psi \& X) &= \sum_{b=\pm 1}\sum_{c=\pm 1} \langle\psi|M_z(a)M_u(b)M_x(c)M_u(b)M_z(a)|\psi\rangle \\ &= \langle\psi|M_z(a)|\psi\rangle = P_z(a|\psi)\,, \end{aligned} \tag{9.35}$$

since the presence or absence of the σ_u and σ_x filters has no effect on the measurement of σ_z. However, the explicit notation X, which is redundant or irrelevant here, will be relevant in some later examples.

Several conditional probabilities can be calculated from these joint probability distributions, using the general formula

$$\text{Prob}(B|A\&C) = \frac{\text{Prob}(A\&B|C)}{\text{Prob}(A|C)}\,. \tag{9.36}$$

Example (i): Conditioning on a prior measurement

In (9.36), let us take C to be the preparation of the state ψ, A to be the result $\sigma_z = +1$ for the first measurement, and B to be a result for the measurement of σ_u. Then the probability that the second measurement will yield $\sigma_u = +1$, conditional on *both* the state preparation *and* the result $\sigma_z = +1$ in the first measurement, is

$$\begin{aligned} \text{Prob}\{\sigma_u = +1|(\sigma_z = +1)\&\psi\} &= \frac{P_{zu}(+1,+1|\psi)}{P_z(+1|\psi)} \\ &= \frac{|\alpha\ \cos(\theta/2)|^2}{|\alpha|^2} = \left\{\cos\left(\frac{\theta}{2}\right)\right\}^2. \end{aligned} \tag{9.37}$$

This is clearly equivalent to the probability of obtaining $\sigma_u = +1$ conditional on a new state, $|\psi'\rangle = |z+\rangle$, and indeed this is a special case of (9.28) and (9.29).

Example (ii): Probability distribution for σ_x regardless of σ_z and σ_u

The probability of the result $\sigma_x = +1$ in the final measurement, regardless of the results of the prior measurements, is

$$P_x(+1|\psi\&X) = \sum_a \sum_b P(a,b,+1|\psi\&X)\,. \tag{9.38}$$

From (9.33) we obtain

$$P(+1,+1,+1|\psi\&X) = |\alpha|^2 \left[\cos\left(\frac{\theta}{2}\right)\right]^2 \frac{1}{2}(1+\sin\theta)\,, \tag{9.39a}$$

$$P(-1,+1,+1|\psi\&X) = |\beta|^2 \left[\sin\left(\frac{\theta}{2}\right)\right]^2 \frac{1}{2}(1+\sin\theta)\,, \tag{9.39b}$$

$$P(+1,-1,+1|\psi\&X) = |\alpha|^2 \left[\sin\left(\frac{\theta}{2}\right)\right]^2 \frac{1}{2}(1-\sin\theta)\,, \tag{9.39c}$$

$$P(-1,-1,+1|\psi\&X) = |\beta|^2 \left[\cos\left(\frac{\theta}{2}\right)\right]^2 \frac{1}{2}(1-\sin\theta)\,. \tag{9.39d}$$

These results, which were obtained directly from (9.33), can be understood more intuitively by noting that at each filtering node of Fig. 9.3 an outgoing branch intensity is multiplied by $[\cos(\phi/2)]^2$, where ϕ is the relative angle between the polarization directions of the incoming amplitude and the outgoing branch. Note that $\frac{1}{2}(1+\sin\theta) = [\cos(\theta'/2)]^2$, where $\theta' = \pi/2 - \theta$ is the angle between the x and u directions.

Thus we obtain

$$P_x(+1|\psi\&X) = \frac{1}{2}\{1 + (|\alpha|^2 - |\beta|^2)\sin\theta\cos\theta\}\,. \tag{9.40}$$

This is the probability of obtaining the result $\sigma_x = +1$ with the σ_z and σ_u filters in place, but ignoring the results of the σ_z and σ_u measurements. It is not equal to the probability of obtaining $\sigma_x = +1$ with the σ_z and σ_u filters absent, which is

$$P_x(+1|\psi) = \langle\psi|M_x(+1)|\psi\rangle = \frac{1}{2}|\alpha+\beta|^2\,.$$

The reason why this case differs from (9.35) is clearly that in this case the particle must pass through σ_z and σ_u filters before reaching the σ_x filter, and so the presence of the other filters is relevant.

[[Although these examples are quite simple, they serve as a warning against formal axiomatic theories of measurement that do not explicitly take the dynamical action of the apparatus into account.]]

Example (iii): Conditioning on both earlier and later measurements

We can calculate the probability for a particular result of the intermediate σ_u measurement, conditional on specified results of both the preceding σ_z measurement and the following σ_x measurement. Of course the later measurement can have no causal effect on the outcome of an earlier measurement, but it can give relevant information because of the statistical correlations between the results of successive measurements. In (9.36) we take $C = \psi \& X, A = (\sigma_z(t_1) = +1) \& (\sigma_x(t_3) = +1)$, and $B = (\sigma_u(t_2) = +1)$.

$$\text{Prob}\{(\sigma_u = +1)|(\sigma_z = +1) \& (\sigma_x = +1) \& \psi \& X\} = \frac{p(+1,+1,+1|\psi \& X)}{P_{zx}(+1,+1|\psi \& X)}.$$

The numerator on the right hand side is given by (9.39a), and the denominator is the sum of (9.39a) plus (9.39c). Thus we obtain

$$\text{Prob}\{(\sigma_u = +1)|(\sigma_z = +1) \& (\sigma_x = +1) \& \psi \& X\}$$

$$= \left[\cos\left(\frac{\theta}{2}\right)\right]^2 \frac{1+\sin\theta}{1+\sin\theta\cos\theta}. \tag{9.41}$$

Although the probability distribution for σ_u is well defined for all θ, there is no quantum state ρ' such that (9.41) would be equal to $\text{Prob}(\sigma_u = +1|\rho')$. This is evident from the fact that (9.41) yields probability 1 for both $\theta = 0$ and $\theta = \pi/2$, which is impossible for any quantum state, pure or mixed.

This paradoxical result demonstrates that, although every well-defined quantum-state preparation can yield an ensemble of systems, the converse is not true. Not every abstractly defined ensemble corresponds to a quantum state. The ensemble, in this case, is produced through several steps. First, passage through the σ_z filter. Second, passage through the σ_u filter. Third, passage through the σ_x filter. The probabilities in these examples are all conditional on the *apparatus configuration* X, which includes the angle θ of the σ_u filter, and so might better be written as X_θ. Hence it is a different ensemble for each value of θ. The last step is to condition on the final result $\sigma_x = +1$. This *post-selection* amounts to selecting a subensemble by discarding those cases in which the result was -1. The number of events that get discarded also depends on the angle θ. So the subensemble that we end up counting will depend, in a complicated way, on the angle θ of the σ_u filter. That is why the result (9.41) does not correspond to any possible state for all θ.

In the more usual situation, typified by example (i), all of the specifications correspond to operations performed *before* the measurement of interest. Hence they define an ensemble whose composition does not depend upon what measurement we choose to perform. We then have a well-defined state, which yields a well-defined probability distribution for any subsequent measurement that we may choose to perform (see the discussion in Sec. 2.1). But this is not possible if we define an abstract subensemble by specifying conditional information both *before* and *after* the measurement of interest, as the last example demonstrates. Thus the paradox, which is now resolved, was useful inasmuch as it compelled a more careful attention to the concept of *state preparation.*

Further reading for Chapter 9

The implications of the theory of measurement for the interpretation of QM have been discussed by many authors, including Ballentine (1970), Leggett (1987), Ballentine (1988), Bell (1990), and Peres (1993). Ballentine (1990) shows how the assumption that the state is "*reduced*" by measurment is not necessary to obtain correct predictions, and may even lead to incorrect results. Ballentine (2008) rebutts the notion that environmental decoherence is crucial to the measurement problem. The book by Wheeler and Zurek (1983) contains reprints of older articles on this subject, including Schrödinger's "cat paradox" paper. The resource letter by Ballentine (1987) has further references.

Problems

9.1 Consider the following spin state for a pair of $s = \frac{1}{2}$ particles:

$$|\Psi\rangle = \big(|+\rangle|+\rangle + |-\rangle|-\rangle\big)\sqrt{\tfrac{1}{2}}\,, \qquad \text{where } \sigma_z|\pm\rangle = \pm 1|\pm\rangle\,.$$

(a) Calculate the joint probability distribution for ${\sigma_x}^{(1)}$ and ${\sigma_x}^{(2)}$

(b) Calculate the joint probability distribution for ${\sigma_y}^{(1)}$ and ${\sigma_y}^{(2)}$

(c) Calculate the joint probability distribution for ${\sigma_x}^{(1)}$ and ${\sigma_y}^{(2)}$.

9.2 For the singlet state of a pair of $s = \frac{1}{2}$ particles,

$$|\Psi\rangle = \big(|+\rangle|-\rangle - |-\rangle|+\rangle\big)\sqrt{\tfrac{1}{2}}\,,$$

calculate $\mathrm{Prob}\{({\sigma_x}^{(2)} = +1)|({\sigma_w}^{(1)} = +1)\,\&\,\Psi\}$, which is the probability that x component of spin of particle 2 will be found to be positive on the condition that the w component of spin of particle 1 has been found to be positive. The direction of w is the bisector of the angle between the x and z axes.

9.3 Two physicists who believe that state vectors can be determined for individual particles each take a turn at testing a certain state preparation device for spin $\frac{1}{2}$ particles. Larry performs a series of measurements of σ_z, and concludes that the device is producing a mixture, with 50% of the particles having state vector $|\sigma_z = +1\rangle$ and 50% having state vector $|\sigma_z = -1\rangle$. Moe performs a series of measurements of σ_x, and concludes that it is a mixture of 50% with state vector $|\sigma_x = +1\rangle$ and 50% with state vector $|\sigma_x = -1\rangle$. Is there any measurement that could be done to resolve their argument? Describe it, or alternatively show that it does not exist.

9.4 Let $Q^{(1)}$ be the position of some object that we wish to measure, and let $Q^{(2)}$ and $P^{(2)}$ be the position and momentum of the indicator variable of a measurement apparatus. Show that an interaction of the form $H_{\text{int}} = cQ^{(1)}P^{(2)}\delta(t)$ will induce a correlation between the values of $Q^{(1)}$ and $Q^{(2)}$ such that the value of $Q^{(2)}$ provides a measurement of the value that $Q^{(1)}$ had at $t = 0$.

9.5 The left half of the figure below depicts a double slit diffraction experiment. If the amplitude emerging from the top hole is $\psi_1(x)$ and the amplitude emerging from the bottom hole is $\psi_2(x)$, then the probability density for detecting a particle at a point on the screen where the two amplitudes overlap is $|\psi_1(x)+\psi_2(x)|^2 = |\psi_1(x)|^2+|\psi_2(x)|^2+[\psi_1(x)]^*\psi_2(x)+\psi_1(x)[\psi_2(x)]^*$. The last two terms are responsible for the interference pattern.

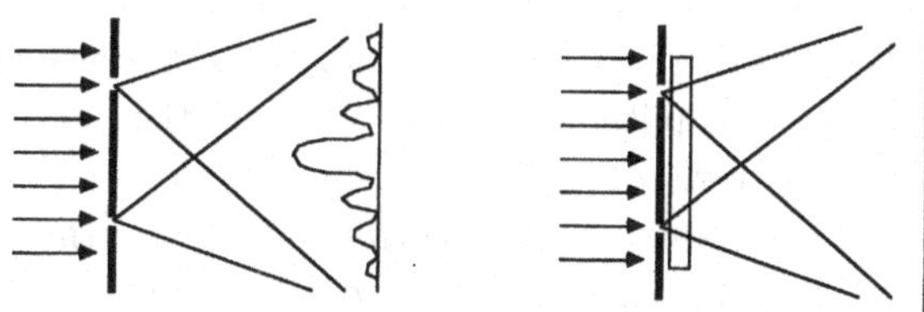

In the right half of the figure, the experiment is modified by the presence, to the right of the holes, of a device whose state is altered by the passage of a particle, but which does not otherwise affect the propagation of the amplitudes. From the state change we may infer (though perhaps not with certainty) which hole a particle has passed through. How will the interference pattern be affected by the presence of this device?

9.6 The figure below depicts a novel proposal for an interference experiment using particles of spin $s = \frac{1}{2}$. The source produces correlated pairs of particles, one of which enters an interferometer on the right, and the other enters a Stern–Gerlach magnet on the left.

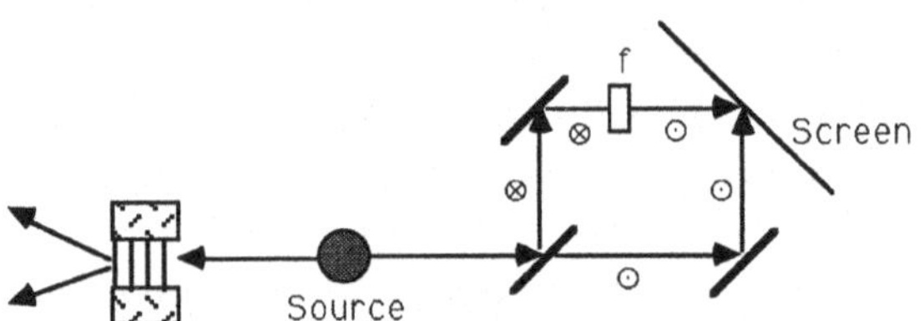

Let us first consider only the right side of the diagram. The first mirror transmits particles whose spin is up (in the z direction) and reflects particles whose spin is down. The other reflectors are of the ordinary spin-independent variety. The upper (spin down) beam passes through a spin-flipper (f), so that both beams have spin up when they reach the screen.

Suppose that the state of the particles emitted to the right of the source were polarized with spin up. Then all particles would take the lower path through the interferometer, and there would be no interference pattern on the screen. Similarly, if the state were polarized with spin down, all particles would take the upper path, and there would be no interference. But if the state were polarized in the x direction, yielding a coherent superposition of spin-up and spin-down components, then there would be amplitudes on both paths of the interferometer, and an interference pattern would be formed on the screen.

Now let the source produce correlated pairs in the singlet spin state, $(|\uparrow\rangle|\downarrow\rangle - |\downarrow\rangle|\uparrow\rangle)/\sqrt{2}$, for which the two particles must have oppositely oriented spins. If we align the magnetic field gradient of the Stern–Gerlach magnet so as to measure the z component of the spin of the particle emitted to the left, we may infer the z component for the particle emitted to the right. Regardless of the result, the above analysis suggests that there should be no interference pattern on the screen. On the other hand, we can rotate the Stern–Gerlach magnet so as to measure the x component of the spin of the particle on the left, and hence infer the value of the x component for the particle on the right. The above analysis now suggests that there should be an interference pattern on the screen. Hence the behavior of the particles going to the right seems to depend on another measurement that may or may not be performed on the particles that go to the left. Resolve this paradox.

Chapter 10

Formation of Bound States

One of the distinctive characteristics of quantum mechanics, in contrast to classical mechanics, is the existence of bound states corresponding to discrete energy levels. Some of the conditions under which this happens will be discussed in this chapter.

10.1 *Spherical Potential Well*

The stationary states of a particle in the potential W are determined by the energy eigenvalue equation, $-(\hbar^2/2M)\nabla^2\Psi + W\Psi = E\Psi$. We shall consider this equation for a spherically symmetric potential $W = W(r)$. The spherical polar coordinates (r, θ, ϕ), shown in Fig. 7.1, are most convenient for this problem. With the well-known spherical form of ∇^2,

$$\nabla^2 = \frac{1}{r^2}\frac{\partial}{\partial r}\left[r^2\frac{\partial}{\partial r}\right] + \frac{1}{r^2\sin\theta}\frac{\partial}{\partial\theta}\left[\sin\theta\frac{\partial}{\partial\theta}\right] + \frac{1}{(r\sin\theta)^2}\frac{\partial^2}{\partial\phi^2},$$

the eigenvalue equation becomes

$$\frac{-\hbar^2}{2M}\frac{1}{r^2}\frac{\partial}{\partial r}r^2\frac{\partial}{\partial r}\Psi + \frac{L^2}{2Mr^2}\Psi + W(r)\Psi = E\,\Psi\,. \tag{10.1}$$

Here the operator L^2, the square of the orbital angular momentum (7.29), arises automatically from the angle derivative terms in ∇^2.

It can be verified by direct substitution that the solution of (10.1) may be chosen to have the factored form

$$\Psi(r, \theta, \phi) = Y_\ell{}^m(\theta, \phi)\frac{u(r)}{r}, \tag{10.2}$$

where the angular factor $Y_\ell{}^m(\theta, \phi)$ is an eigenfunction of L^2 satisfying (7.30) and (7.31). The form $u(r)/r$ for the radial factor is chosen so as to eliminate first order derivatives from the equation. The radial function then satisfies the equation

$$\frac{-\hbar^2}{2M}\frac{d^2u(r)}{dr^2} + \left[\frac{\hbar^2\ell(\ell+1)}{2Mr^2} + W(r)\right]u(r) = E\,u(r)\,. \tag{10.3}$$

This has the same form as the equation for a particle in one dimension, except for two important differences. First, there is a repulsive effective potential proportional to the eigenvalue of L^2, $\hbar^2\ell(\ell+1)$. Second, the radial function must satisfy the boundary condition

$$u(0) = 0\,, \tag{10.4}$$

since $\Psi(r,\theta,\phi)$ would otherwise have an r^{-1} singularity at the origin. It was argued in Sec. 4.5 that such a singularity is unacceptable. The normalization $\langle\Psi|\Psi\rangle = 1$ implies that

$$\int_0^\infty |u(r)|^2\, dr = 1\,. \tag{10.5}$$

Square well potential

The principles may be illustrated by an exact solution for the simplest case, the square well potential. Let the potential be

$$\begin{aligned} W(r) &= -V_0\,, \quad & r < a\,, \\ &= 0\,, & r > a\,. \end{aligned} \tag{10.6}$$

Consider the solution of (10.3) for $\ell = 0$. Inside the potential well, the two linearly independent solutions are $\sin(kr)$ and $\cos(kr)$, with $\hbar^2k^2/2M = E - V_0$. Only $\sin(kr)$ satisfies the boundary condition (10.4), so the solution will be of the form

$$u(r) = N\frac{\sin(kr)}{\sin(ka)}\,, \quad r \le a\,. \tag{10.7}$$

Here N is a normalization factor. The denominator is included for convenience, so that (10.7) and (10.8) will be equal at $r = a$.

Outside of the potential well, the solutions of (10.3) take different forms for positive or negative energies. Bound states may occur in the negative energy region, with the solution of (10.3) being of the form

$$u(r) = N\,e^{-\alpha(r-a)}\,, \quad r \ge a\,, \quad E < 0\,, \tag{10.8}$$

with $\hbar^2\alpha^2/2M = -E$. The other solution, of the form $e^{\alpha r}$, is not allowed because it diverges strongly at infinity.

The wave function and its derivative must be continuous at $r = a$, for reasons that were given in Sec. 4.5. It is more convenient to apply this continuity requirement to the ratio u'/u (with $u' \equiv \partial u/\partial r$), since it is independent of normalization. Equating u'/u evaluated at $r = a$ from (10.8) and from (10.7) yields

$$\alpha = \frac{-k}{\tan(ka)}\,. \tag{10.9}$$

The parameters k and α are also related through the energy, yielding

$$\alpha = \sqrt{\frac{2MV_0}{\hbar^2} - k^2}\,. \tag{10.10}$$

By equating these two expressions for α, we can solve for k, and hence for the energy of the bound state, $E = \hbar^2 k^2/2M - V_0$. This is illustrated in Fig. 10.1.

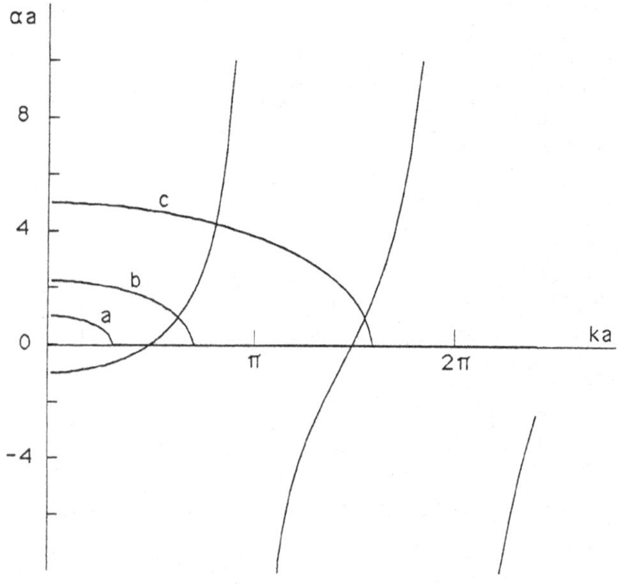

Fig. 10.1 The condition for a bound state in a spherical potential well is the equality of the expressions (10.9) and (10.10), illustrated for three cases: (a) $V_0 = 1$; (b) $V_0 = 5$; (c) $V_0 = 25$. (Units: $\hbar = 2M = a = 1$.)

Equation (10.10) yields a quadrant of an ellipse, which is shown for three different values of V_0. Equation (10.9) yields a curve with infinite discontinuities at $ka = \pi, 2\pi$, etc.

If V_0 is smaller than $(\hbar^2/2Ma^2)(\pi/2)^2$, as in case (a), the two curves do not intersect, and no bound state solution exists. If V_0 lies between

$(\hbar^2/2Ma^2)(\pi/2)^2$ and $(\hbar^2/2Ma^2)(3\pi/2)^2$, as in case (b), there is one intersection, corresponding to a single bound state. If V_0 lies between $(\hbar^2/2Ma^2)(3\pi/2)^2$ and $(\hbar^2/2Ma^2)(5\pi/2)^2$, as in case (c), there are two intersections and two bound states. It is clear that as V_0 increases, the number of bound states increases without limit.

The normalization factor N in (10.7) and (10.8) can now be evaluated by using (10.5). The wave function $u(r)$ is plotted in Fig. 10.2 for several values of the potential well depth V_0. The practical way to do this calculation is to regard k as the independent parameter, and to compute α from (10.9) and then V_0 from (10.10).

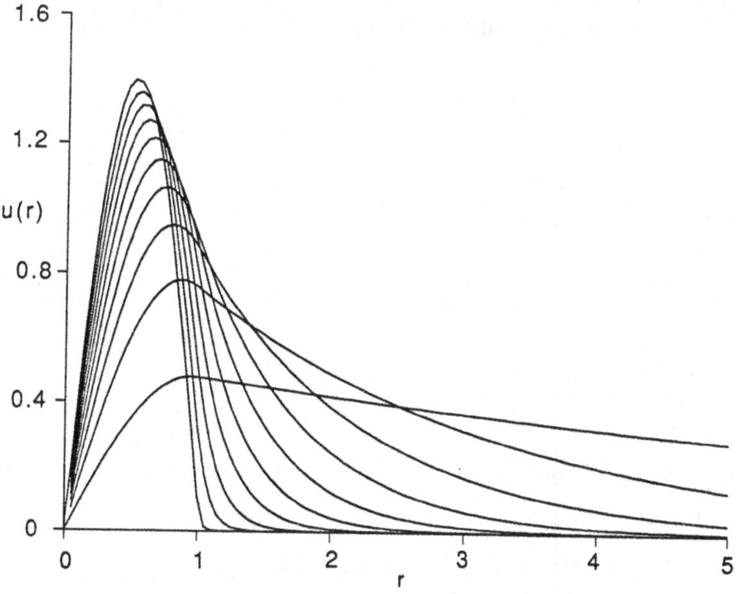

Fig. 10.2 Wave function of the lowest energy bound state, for V_0 ranging from 1524 to 2.737, in units of $\hbar^2/2Ma^2$. The parameter k (rather than V_0) is uniformly spaced from curve to curve. The radius of the potential well is $a = 1$.

For very large V_0, the state is nearly confined within $r < a$. As V_0 becomes smaller, the exponential tail in the region $r > a$ grows longer. As V_0 decreases towards the critical value, $(\hbar^2/2Ma^2)(\pi/2)^2$, the range of the exponential tail diverges, and the state ceases to be bound. No bound states exist for smaller values of V_0.

The size of the bound state may be measured by its root-mean-square radius:

$$\sqrt{\langle r^2 \rangle} = \left\{ \int_0^\infty |u(r)|^2 r^2 \, dr \right\}^{1/2} . \tag{10.11}$$

This can be evaluated analytically, but the expression is messy and uninteresting, and so it will not be reproduced. [To compute the r.m.s. radius for Fig. 10.3, the integral in Eq. (10.11) was evaluated by the computer algebra program REDUCE. It yields the exact formula, expressed in FORTRAN notation, which can then be evaluated numerically.] The radius of the bound state is insensitive to V_0 over a large range because the exponential tail contributes very little to (10.11) when $\alpha a \gg 1$. However, the radius diverges rapidly as V_0 approaches the critical value of $(\hbar^2/2Ma^2)(\pi/2)^2$.

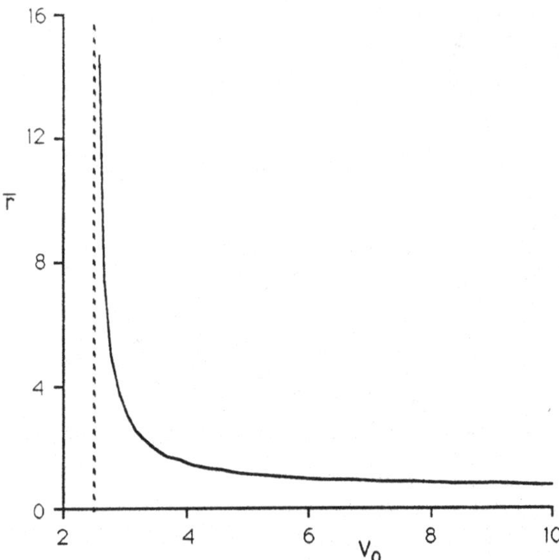

Fig. 10.3 Root-mean-square radius of the lowest energy bound state vs depth of the potential well [Eq. (10.11)]. (Units: $\hbar = 2M = a = 1$.)

These loosely bound large-radius states occur in nuclei, where they are called *halo states* [see Riisager (1994)]. In most nuclei the nucleon density falls off abruptly at some more-or-less well-defined nuclear radius. But in a few cases, such as ^{11}Be, the nucleus consists of a compact core plus one loosely bound neutron. The characteristic of these *nuclear halos* is a component of the density that falls off quite slowly as a function of radial distance, and consequently a nuclear surface that is diffuse and not well defined.

We have just seen how, for $E < 0$, bound states may exist at only a discrete set of energies. It is useful to perform a similar calculation for $E > 0$ in order to show why the energies of unbound states are not similarly restricted. For $r \le a$ the solution of (10.3) and (10.4) is of the form (10.7), for both $E < 0$ and $E > 0$. But for $r \ge a$ the solution of (10.3) has the form

$$u(r) = A'\, e^{-\alpha r} + B'\, e^{\alpha r}\,, \qquad E < 0\,, \tag{10.12a}$$

$$u(r) = A\, \sin(kr) + B\, \cos(kr)\,, \quad E > 0\,, \tag{10.12b}$$

where $\hbar^2 k^2/2M = E$. By matching the value of u'/u at $r = a$ to that from (10.7), we are able to fix the ratios A'/B' and A/B.

It is not possible, in general, to require the eigenfunctions of a self-adjoint operator such as the Hamiltonian to have a finite norm. This fact was discussed in Sec. 1.4, and illustrated for the case of a free particle in Sec. 4.6. In the present case, it is evident that no choice of A and B in (10.12b) will satisfy the normalization condition (10.5). But although the eigenfunctions cannot be restricted to Hilbert space (the space of normalizable functions), they must always lie within the extended space Ω^x, which consists of functions that may diverge at infinity no more strongly than a power of r. This implies, for the case of $E < 0$, that we must have $B' = 0$, thus reducing (10.12a) to (10.8), for which solutions exist for at most a discrete set of eigenvalues E. But in (10.12b) both terms are acceptable, and neither A nor B needs to be eliminated. This extra degree of freedom makes it possible to obtain a solution for any value of $E > 0$.

10.2 *The Hydrogen Atom*

Since the hydrogen atom is treated in almost every quantum mechanics book, our treatment will be brief and will refer to others for derivation of some of the detailed results. Much of our treatment is similar to that of Schiff (1968). For more extensive results, see Bethe and Salpeter (1957).

The hydrogen atom is a two-particle system consisting of an electron and a proton. The Hamiltonian is

$$H = \frac{{P_e}^2}{2M_e} + \frac{{P_p}^2}{2M_p} - \frac{e^2}{|\mathbf{Q}_e - \mathbf{Q}_p|}\,, \tag{10.13}$$

where the subscripts e and p refer to the electron and the proton. The problem is simplified if we take as independent variables the *center of mass* and *relative* coordinates of the particles,

$$\mathbf{Q}_c = \frac{M_e \mathbf{Q}_e + M_p \mathbf{Q}_p}{M_e + M_p}\,, \tag{10.14}$$

$$\mathbf{Q}_r = \mathbf{Q}_e - \mathbf{Q}_p\,. \tag{10.15}$$

The corresponding momentum variables, which satisfy the commutation relations

$$[Q_{c\alpha}, P_{c\beta}] = [Q_{r\alpha}, P_{r\beta}] = i\delta_{\alpha\beta}\mathbf{I}\,,$$
$$[Q_{c\alpha}, P_{r\beta}] = [Q_{r\alpha}, P_{c\beta}] = 0 \qquad (\alpha, \beta = 1, 2, 3)\,,$$

are

$$\mathbf{P}_c = \mathbf{P}_e + \mathbf{P}_p\,, \tag{10.16}$$

$$\mathbf{P}_r = \frac{M_p \mathbf{P}_e - M_e \mathbf{P}_p}{M_e + M_p}\,. \tag{10.17}$$

(This change of variables, which preserves the usual commutation relations, is an example of a *canonical transformation.*) In terms of these center of mass and relative variables, the Hamiltonian becomes

$$H = \frac{{P_c}^2}{2(M_e + M_p)} + \frac{{P_r}^2}{2\mu} - \frac{e^2}{|\mathbf{Q}_r|}\,, \tag{10.18}$$

where μ is called the *reduced mass*, and is defined by the relation

$$\frac{1}{\mu} = \frac{1}{M_e} + \frac{1}{M_p}\,. \tag{10.19}$$

It is apparent that the center of mass behaves as a free particle, and its motion is not coupled to the relative coordinate. Therefore we shall confine our attention to the internal degrees of freedom described by the relative coordinate $\mathbf{Q}_r$.

The Hamiltonian for the internal degrees of freedom is ${P_r}^2/2\mu - e^2/|\mathbf{Q}_r|$, and the energy eigenvalue equation in coordinate representation is

$$\frac{-\hbar^2}{2\mu}\nabla^2\, \Psi(\mathbf{r}) - \frac{e^2}{r}\Psi(\mathbf{r}) = E\, \Psi(\mathbf{r})\,, \tag{10.20}$$

where $\mathbf{r}$ is the position of the electron relative to the proton. This is just the equation for a particle of effective mass μ in a Coulomb potential centered at the origin. In contrast to the spherical potential well studied in Sec. 10.1, the Coulomb potential decays toward zero very slowly at large distances, and we shall see that this is responsible for some qualitatively different features.

Solution in spherical coordinates

When written in spherical coordinates, Eq. (10.20) takes the form of (10.1) with $M = \mu$ and $W(r) = -e^2/r$. We can separate the radial and angular dependences by substituting $\Psi(r,\theta,\phi) = Y_\ell{}^m(\theta,\phi)R(r)$. [For the Coulomb potential it happens to be more convenient not to remove the factor of $1/r$ as was done in (10.2).] The radial equation for angular momentum quantum number ℓ is

$$\frac{-\hbar^2}{2\mu}\frac{1}{r^2}\frac{d}{dr}r^2\frac{d}{dr}R + \frac{\hbar^2\ell(\ell+1)}{2\mu r^2}R - \frac{e^2}{r}R = ER\,. \tag{10.21}$$

We shall be interested in bound state solutions, for which $E = -|E|$.

When solving an equation such as (10.21), it is usually helpful to change to dimensionless variables. Therefore we introduce a dimensionless distance $\rho = \alpha r$, where $\alpha^2 = 8\mu|E|/\hbar^2$, and a dimensionless charge-squared parameter, $\lambda = 2\mu e^2/\alpha\hbar^2 = (\mu e^4/2\hbar^2|E|)^{1/2}$. Equation (10.21) then becomes

$$\frac{1}{\rho^2}\frac{d}{d\rho}\rho^2\frac{d}{d\rho}R + \left[\frac{\lambda}{\rho} - \frac{1}{4} - \frac{\ell(\ell+1)}{\rho^2}\right]R = 0\,. \tag{10.22}$$

The term $1/4$ in the brackets is all that remains of the eigenvalue E, since E was used to define the dimensionless units.

The singular points of this equation at $\rho = 0$ and $\rho = \infty$ require special attention. For very large values of ρ the terms proportional to ρ^{-1} and ρ^{-2} can be neglected compared with $1/4$, and one can easily verify that $R(\rho) = \rho^n e^{\pm\rho/2}$ becomes a solution in the asymptotic limit $\rho \to \infty$. Only the decaying exponential is physically acceptable, and so we shall look for solutions of that form. The possible singularity of the solution at $\rho = 0$ is taken into account by the substitution into (10.22) of

$$R(\rho) = \rho^k\, L(\rho)\, e^{-\rho/2}\,, \tag{10.23}$$

where k may be negative or fractional, and $L(\rho)$ is expressible as a power series, $L(\rho) = \sum_\nu a_\nu \rho^\nu$. (This is equivalent to the standard Frobenius method of substituting ρ^k multiplied by a power series, since the exponential function has a convergent power series.) This yields the equation

$$\rho^2 L''(\rho) + \rho[2(k+1) - \rho]L'(\rho) + [\rho(\lambda - k - 1) + k(k+1) - \ell(\ell+1)]L(\rho) = 0\,,$$

where primes denote differentiation with respect to ρ. When ρ is set equal to zero here, it follows that $k(k+1) - \ell(\ell+1) = 0$, and hence there are two possible roots for $k : k = \ell$ and $k = -(\ell+1)$. It was argued in Sec. 4.5 that the singularities that would correspond to the negative root are unacceptable, so we must have $k = \ell$. The above equation then becomes

$$\rho L''(\rho) + [2(\ell+1) - \rho]L'(\rho) + (\lambda - \ell - 1)L(\rho) = 0\,. \tag{10.24}$$

By substituting the power series for $L(\rho)$ into (10.24) and collecting powers of ρ, we obtain a recurrence relation between successive terms in the series,

$$a_{\nu+1} = \frac{\nu + \ell + 1 - \lambda}{(\nu+1)(\nu+2\ell+2)}\, a_\nu\,. \tag{10.25}$$

If the series does not terminate, the ratio of successive terms will become $a_{\nu+1}/a_\nu \approx 1/\nu$ for large enough values of ν. This is the same asymptotic ratio as in the series for $\rho^n e^\rho$ with any positive value of n. Thus the exponential increase of $L(\rho)$ will overpower the decreasing exponential factor in (10.23), leaving a net exponential increase like $e^{\rho/2}$. Such unacceptable behavior can be avoided only if the power series terminates.

It is apparent from (10.25) that if λ has the integer value

$$\lambda = n = n' + \ell + 1 \tag{10.26}$$

then $L(\rho)$ will be a polynomial of degree n'. Referring back to the definition of λ in terms of E, we see that the energy eigenvalues are

$$E_n = -\frac{\mu e^4}{2\hbar^2 n^2} \quad (n = 1, 2, 3, \ldots)\,. \tag{10.27}$$

The integer n is known as the *principal quantum number* of the state. There are infinitely many bound states within an arbitrarily small energy of $E = 0$. This limit point in the spectrum exists because of the very long range of the Coulomb potential. No such behavior occurs for short range potentials. The degree of degeneracy of an energy eigenvalue with fixed n' and ℓ is $2\ell + 1$, corresponding to the values of $m = -\ell, -\ell+1, \ldots \ell$. Therefore the degeneracy of an eigenvalue E_n is

$$\sum_{\ell=0}^{n-1} (2\ell + 1) = n^2\,.$$

This is the degeneracy of an eigenvalue of Eq. (10.20). The degeneracy of an energy level of a hydrogen atom is greater than this by a factor of 4, which arises from the two-fold orientational degeneracies of the electron and proton spin states. This four-fold degeneracy is modified by the hyperfine interaction between the magnetic moments of the electron and the proton. Those effects of spin will not be considered in this chapter.

The eigenfunctions $L(\rho)$ of (10.24) are related to the *Laguerre polynomials*, which satisfy the equation

$$\rho L_r{}''(\rho) + (1-\rho)L_r{}'(\rho) + rL_r(\rho) = 0\,. \tag{10.28}$$

The rth degree Laguerre polnomial is given by the formula

$$L_r(\rho) = e^{\rho}\frac{d^r}{d\rho^r}(\rho^r\, e^{-\rho})\,. \tag{10.29a}$$

The *associated Laguerre polynomials* are defined as

$$L_r^s(\rho) = \frac{d^s}{d\rho^s}L_r(\rho)\,. \tag{10.29b}$$

[[This is the notation used by Pauling and Wilson (1935) and by Schiff (1968). Messiah (1966) and Merzbacher (1970) used the notation $(-1)^s L_{r-s}^s(\rho)$ for the function defined in (10.29).]]

It satisfies the differential equation

$$\rho L_r^{s''}(\rho) + (s+1-\rho)L_r^{s'}(\rho) + (r-s)L_r^s(\rho) = 0\,. \tag{10.30}$$

Comparing this with (10.24), we see that apart from normalization, the function $L(\rho)$ is equal to $L_{n+\ell}^{2\ell+1}(\rho)$. For more of the mathematical properties of these functions, we refer to the books cited above.

The orthonormal energy eigenfunctions for the hydrogen atom are

$$\Psi_{n\ell m}(r,\theta,\phi) = -\left[\frac{4(n-\ell-1)!}{(na_0)^3 n[(n+\ell)!]^3}\right]^{1/2} \rho^{\ell} L_{n+\ell}^{2\ell+1}(\rho)\, e^{-\rho/2}\, Y_\ell{}^m(\theta,\phi)\,, \tag{10.31}$$

where $\rho = \alpha r = 2r/na_0$, and $a_0 = \hbar^2/\mu e^2$ is a characteristic length for the atom, known as the *Bohr radius*. Detailed formulas and graphs for these functions can be found in Pauling and Wilson (1935). The ground state wave function is

$$\Psi_{100} = (\pi {a_0}^3)^{-1/2}\, e^{-r/a_0}\,, \tag{10.32}$$

a result that can more easily be obtained directly from the eigenvalue equation (10.20) than by specializing the general formula (10.31). It should be emphasized that the infinite set of bound state functions of the form (10.31) is *not* a complete set. To obtain a complete basis set we must include the continuum of unbound state functions for $E \geq 0$.

A measure of the spatial extent of the bound states of hydrogen is given by the averages of various powers of the distance r:

$$\langle r \rangle = \langle \Psi_{n\ell m} | r | \Psi_{n\ell m} \rangle = n^2 a_0 \left\{ 1 + \frac{1}{2} \left[1 - \frac{\ell(\ell+1)}{n^2} \right] \right\}, \tag{10.33}$$

$$\langle r^2 \rangle = \langle \Psi_{n\ell m} | r^2 | \Psi_{n\ell m} \rangle = n^4 {a_0}^2 \left[1 + \frac{3}{2} \left\{ 1 - \frac{\ell(\ell+1) - 1/3}{n^2} \right\} \right], \tag{10.34}$$

$$\langle r^{-1} \rangle = \langle \Psi_{n\ell m} | r^{-1} | \Psi_{n\ell m} \rangle = \frac{1}{n^2 a_0} . \tag{10.35}$$

[These results, as well as formulas for other powers of r, have been given by Pauling and Wilson (1935).] Apparently the characteristic size of a bound state function is of order $n^2 a_0$. This n^2 dependence arises from two sources: the strength of the radial function $L(\rho)$ extends over a region that increases roughly linearly with n; and the scale factor that converts the dimensionless distance ρ into real distance, $r = \alpha^{-1}\rho$, is $\alpha^{-1} = n a_0/2$.

These solutions for the hydrogen atom can be generalized to any one-electron hydrogen-like atom with nuclear charge Ze by substituting $e^2 \to Ze^2$. Thus the energies scale as Z^2, and the lengths (including the Bohr radius a_0) scale as Z^{-1}.

Solution in parabolic coordinates

Equation (10.20) for the energy eigenfunctions and eigenvalues of the hydrogen atom can also be separated in the parabolic coordinates (ξ, η, ϕ), which are related to spherical polar coordinates thus:

$$\begin{aligned} \xi &= r - z = r(1 - \cos\theta) , \\ \eta &= r + z = r(1 + \cos\theta) , \\ \phi &= \phi . \end{aligned} \tag{10.36}$$

The surfaces of constant ξ are a set of confocal paraboloids of revolution about the z axis, opening in the direction of positive z, or $\theta = 0$. The surfaces of constant η are similar confocal paraboloids that open in the direction of negative z, or $\theta = \pi$. All of the paraboloids have their focus at the origin. The surface $\xi = 0$ degenerates into a line, the positive z axis; and the surface $\eta = 0$ degenerates into the negative z axis.

Parabolic coordinates obscure the spherical symmetry of the problem, and so they are less commonly used than are spherical coordinates. But they have the advantage that the equation remains separable in the presence of a uniform electric field along the z axis, which adds to the Hamiltonian the scalar potential $eDz = \frac{1}{2}eD(\eta - \xi)$, with D being the electric field strength and $-e$ being the charge of the electron. We shall solve the equation only for $D = 0$.

The form of (10.20) in parabolic coordinates is

$$\frac{-\hbar^2}{2\mu}\left[\frac{4}{(\xi+\eta)}\left\{\frac{\partial}{\partial\xi}\xi\frac{\partial}{\partial\xi} + \frac{\partial}{\partial\eta}\eta\frac{\partial}{\partial\eta}\right\} + \frac{1}{\xi\eta}\frac{\partial^2}{\partial\phi^2}\right]\Psi - \frac{2e^2}{(\xi+\eta)}\Psi = E\,\Psi\,. \quad (10.37)$$

This may be separated into a set of ordinary differential equations by the substitution $\Psi(\xi,\eta,\phi) = f(\xi)\,g(\eta)\,\Phi(\phi)$. We may anticipate that the third factor will be $\Phi(\phi) = e^{im\phi}$, so that $(\partial^2/\partial\phi^2)\Psi = -m^2\Psi$. Multiplying by $-\mu(\xi+\eta)/2\hbar^2\Psi$ and substituting $E = -|E|$, we obtain

$$\frac{1}{f}\frac{d}{d\xi}\xi\frac{df}{d\xi} + \frac{1}{g}\frac{d}{d\eta}\eta\frac{dg}{d\eta} - \frac{m^2(\xi+\eta)}{4\xi\eta} - \frac{\mu|E|(\xi+\eta)}{2\hbar^2} = \frac{-\mu e^2}{\hbar^2}\,. \quad (10.38)$$

Since $(\xi+\eta)/\xi\eta = \xi^{-1} + \eta^{-1}$, the above equation has the form: (function of ξ) + (function of η) = (constant), and so it separates into a pair of equations,

$$\frac{1}{f}\frac{d}{d\xi}\xi\frac{df}{d\xi} - \left[\frac{m^2}{4\xi} + \frac{\mu|E|\xi}{2\hbar^2}\right] = -K_1\,, \quad (10.39a)$$

$$\frac{1}{g}\frac{d}{d\eta}\eta\frac{dg}{d\eta} - \left[\frac{m^2}{4\eta} + \frac{\mu|E|\eta}{2\hbar^2}\right] = -K_2\,, \quad (10.39b)$$

where $K_1 + K_2 = \mu e^2/\hbar^2$. These two equations are identical in form, so we need only solve one of them.

We shall solve (10.39a) by the same method that was used to solve (10.21). First, introduce a dimensionless length variable $\zeta = \gamma\xi$, where $\gamma^2 = 2\mu|E|/\hbar^2$. This substitution yields

$$\frac{1}{\zeta}\frac{d}{d\zeta}\zeta\frac{df}{d\zeta} + \left[\frac{\lambda_1}{\zeta} - \frac{1}{4} - \frac{m^2}{4\zeta^2}\right] f = 0\,, \quad (10.40)$$

where $\lambda_1 = K_1/\gamma$. It is easily verified that the asymptotic form $f(\zeta) \approx e^{\pm\zeta/2}$ satisfies the equation for very large values of ζ. Therefore, as was done for (10.22), we substitute

$$f(\zeta) = \zeta^k \, L(\zeta) \, e^{-\zeta/2} \,, \tag{10.41}$$

where $L(\zeta)$ is a power series. The two roots for k turn out to be $k = \pm\frac{1}{2}m$. Since the negative root would yield an unacceptable singularity at $\zeta = 0$, we must take $k = \frac{1}{2}|m|$. The resultant equation for $L(\zeta)$ is

$$\zeta L''(\zeta) + (|m| + 1 - \zeta)L'(\zeta) + [\lambda_1 - \tfrac{1}{2}(|m| + 1)]L(\zeta) = 0 \,. \tag{10.42}$$

This is exactly the same form as (10.24) except that now $|m|$ replaces $2\ell + 1$. Therefore we can immediately conclude that the only solutions that do not diverge exponentially at infinity are the associated Laguerre polynomials,

$$L(\zeta) = L^{|m|}_{n_1+|m|}(\zeta) \,,$$

where λ_1 has a value such that

$$n_1 = \lambda_1 - \frac{1}{2}(|m| + 1) \tag{10.43}$$

is a nonnegative integer.

Similarly, an acceptable solution to (10.39b) exists whenever λ_2 has a value such that

$$n_2 = \lambda_2 - \frac{1}{2}(|m| + 1) \tag{10.44}$$

is a nonnegative integer. From the definitions of λ_1 and λ_2, we deduce $\lambda_1 + \lambda_2 = (K_1 + K_2)/\gamma = \mu e^2/\hbar^2\gamma$. The energy eigenvalues are related to γ through its original definition, $E = -|E| = -\hbar^2\gamma^2/2\mu$. Hence we obtain

$$E = -\frac{\mu e^4}{2\hbar^2(\lambda_1 + \lambda_2)^2} \,. \tag{10.45}$$

From (10.43) and (10.44) it follows that

$$\lambda_1 + \lambda_2 = n_1 + n_2 + |m| + 1 \equiv n \tag{10.46}$$

may be any nonnegative integer. Therefore (10.45) agrees with the result (10.27) which was obtained from the solution in spherical coordinates.

The energy eigenfunctions in parabolic coordinates are

$$\Psi_{n_1 n_2 m}(\xi, \eta, \phi) = N \, (\xi\eta)^{|m|/2} \, e^{-\gamma(\xi+\eta)/2} L^{|m|}_{n_1+|m|}(\gamma\xi) \, L^{|m|}_{n_2+|m|}(\gamma\eta) \, e^{im\phi} \,, \tag{10.47}$$

where N is a normalization constant. Here $1/\gamma = a_0(n_1 + n_2 + |m| + 1) = na_0$ is the characteristic length for a state of principal quantum number n.

The unnormalized ground state function Ψ_{000} has the form $e^{-(\xi+\eta)/2a_0} = e^{-r/a_0}$. This agrees with (10.32), which was calculated in spherical coordinates.

In general, however, an eigenfunction from one system of coordinates will be a linear combination of degenerate eigenfunctions from the other system. A parabolic eigenfunction (10.47) with quantum numbers (n_1, n_2, m) is equal to a linear combination of spherical eigenfunctions (10.31) which have the same m and have n given by (10.46), but may have any value of ℓ. Conversely, a spherical eigenfunction with quantum numbers (n, ℓ, m) is equal to a linear combination of parabolic eigenfunctions that have the same m and have n_1+n_2 fixed to give the correct value of n, but may have any value for $n_1 - n_2$.

The sum $n_1 + n_2$ of the parabolic quantum numbers plays a role similar to the radial quantum number n' in (10.26). The significance of the difference $n_1 - n_2$ of the parabolic quantum numbers can be seen by considering the average of the z component of position, $\langle z\rangle = \frac{1}{2}\langle \eta - \xi\rangle = \frac{1}{2}\langle \eta\rangle - \frac{1}{2}\langle \xi\rangle$. Now the average of the dimensionless variable $\zeta = \gamma\xi$ is approximately proportional to n_1, at least for large quantum numbers, because the strength of a Laguerre polynomial extends over a region that increases with n_1. Therefore $\langle \xi\rangle = \langle \zeta\rangle/\gamma$ is proportional to the product nn_1. Similarly, we have $\langle \eta\rangle$ approximately proportional to nn_2. Thus $\langle z\rangle$ is approximately proportional to $n(n_2 - n_1)$. Among all states with fixed values of n and m, the state with the largest value of $n_2 - n_1$ will exhibit the greatest polarization. These state functions are useful in describing a hydrogen atom in an external electric field.

10.3 *Estimates from Indeterminacy Relations*

It is possible to make estimates relating the size and energy of bound states by means of the position–momentum indeterminacy relation, commonly called the uncertainty principle. The indeterminacy relations are precisely defined statistical inequalities, and many arguments that purport to be based on the uncertainty principle are really order-of-magnitude dimensional arguments.

A typical example of a dimensional argument has the following form: $rp \sim \hbar$, where r is a typical dimension of the bound state, p is a typical momentum, and the symbol $\sim$ denotes order-of-magnitude equality. For the hydrogen atom, this argument yields $E = p^2/2\mu - e^2/r \sim \hbar^2/2\mu r^2 - e^2/r$. Minimization of this expression with respect to r yields $E_{\text{min}} = -\mu e^4/2\hbar^2$, at the optimum distance of $r = \hbar^2/\mu e^2$. You should not be impressed by the fact that this crude argument led to the exact ground state energy. All that should be expected is an order-of-magnitude estimate that is neither an upper nor a lower bound. Such dimensional arguments have their uses, but they should not be confused with the indeterminacy relation, which yields a strict inequality.

From the commutation relations satisfied by the relative coordinates and the corresponding momentum of any two-particle system [see (10.15)–(10.17)], it follows from (8.31) that there is an indeterminacy relation of the form

$$\langle (r_\alpha - \langle r_\alpha \rangle)^2 \rangle \, \langle (P_\beta - \langle P_\beta \rangle)^2 \rangle \geq \delta_{\alpha\beta} \, \hbar^2/4 \, .$$

If the state is bound we must have $\langle P_\beta \rangle = 0$, and the origin of coordinates can be chosen so that $\langle r_\alpha \rangle = 0$. Summing over α and β then yields

$$\langle \mathbf{r} \cdot \mathbf{r} \rangle \, \langle \mathbf{P} \cdot \mathbf{P} \rangle \geq \frac{3\hbar^2}{4} \, .$$

If there is no vector potential, this result can be expressed in terms of the relative kinetic energy of the two bound particles, $T_{\text{rel}} = \mathbf{P} \cdot \mathbf{P}/2\mu$:

$$\langle \mathbf{r} \cdot \mathbf{r} \rangle \, \langle T_{\text{rel}} \rangle \geq \frac{3\hbar^2}{8\mu} \, . \tag{10.48}$$

This asserts that the product of the mean square radius of the state and the average kinetic energy is bounded below. The smaller the size of the state, the greater must be its kinetic energy.

A stronger inequality can be obtained for a state whose orbital angular momentum is zero, and which is therefore spherically symmetric. We then have $\langle {r_\alpha}^2 \rangle = \langle \mathbf{r} \cdot \mathbf{r} \rangle / 3$ and $\langle {P_\beta}^2 \rangle = \langle \mathbf{P} \cdot \mathbf{P} \rangle / 3$, from which it follows that

$$\langle \mathbf{r} \cdot \mathbf{r} \rangle \, \langle T_{\text{rel}} \rangle \geq \frac{9\hbar^2}{8\mu} \, . \tag{10.49}$$

This result was first presented by Wolsky (1974).

Examples

As a first example, we apply (10.49) to the ground state of the hydrogen atom (10.32), for which we have

$$\langle \mathbf{r} \cdot \mathbf{r} \rangle = \int r^2 |\, \Psi(\mathbf{r})|^2 \, d^3r = 3 \, {a_0}^2 \, ,$$

$$\langle \mathbf{P} \cdot \mathbf{P} \rangle = \int \hbar^2 \left| \frac{\partial \Psi}{\partial r} \right|^2 \, d^3r = \frac{\hbar^2}{{a_0}^2} \, ,$$

and hence $\langle \mathbf{r} \cdot \mathbf{r} \rangle \, \langle T_{\text{rel}} \rangle = 3\hbar^2/2\mu$. This exceeds the lower bound (10.49) by a factor of 1.333.

As a second example, we consider the deuteron, which is a nuclear particle consisting of a proton and a neutron bound together. From scattering data, it has been deduced that the root-mean-square radius of the deuteron is $(\langle D|\mathbf{r}\cdot\mathbf{r}|D\rangle)^{1/2} = 2.11 \times 10^{-13}$ cm. Here $|D\rangle$ denotes the deuteron state. Then (10.49) implies that $\langle D|T_{\text{rel}}|D\rangle \geq 28.4$ MeV. (One MeV is 10^6 eV. For comparison, the binding energy of the hydrogen atom is 13.6 eV.) Since the binding energy of the deuteron is known to be 2.2 MeV, it follows that the average potential energy must satisfy $\langle D|W|D\rangle \leq -30.6$ MeV. When nuclear forces were not yet understood this was a very valuable piece of information.

10.4 *Some Unusual Bound States*

All of the bound states considered so far have the property that the total energy of the state is less than the value of the potential energy at infinity. The system remains bound because it lacks sufficient energy to dissociate. This same property characterizes classical bound states. However, it is possible in quantum mechanics to have bound states that do not possess this property, and which therefore have no classical analog.

Let us choose the zero of energy so that the potential energy function vanishes at infinity. The usual energy spectrum for such a potential would be a positive energy continuum of unbound states, with the bound states, if any, occurring at discrete negative energies. However, Stillinger and Herrick (1975), following an earlier suggestion by Von Neumann and Wigner, have constructed potentials that have discrete bound states embedded in the positive energy continuum. Bound states are represented by those solutions of the equation $(-\frac{1}{2}\nabla^2 + W)\,\Psi = E\Psi$ for which the normalization integral $\int |\Psi|^2\, d^3x$ is finite. [To follow the notation of Stillinger and Herrick, we adopt units such that $\hbar = 1$ and (particle mass) $= 1$.]

We can formally solve for the potential,

$$W = E + \frac{1}{2}\left(\frac{\nabla^2\Psi}{\Psi}\right). \tag{10.50}$$

For the potential to be nonsingular, the nodes of Ψ must be matched by zeros of $\nabla^2\Psi$. The free particle zero-angular-momentum function $\Psi_0(\mathbf{x}) = \sin(kr)/kr$ satisfies (10.50) with energy eigenvalue $E = \frac{1}{2}k^2$ and with W identically equal to zero, but it is unacceptable because the integral of $|\Psi_0|^2$ is not convergent. This defect can be remedied by taking

$$\Psi(\mathbf{x}) = \Psi_0(\mathbf{x})\, f(r) \tag{10.51}$$

and requiring that $f(r)$ go to zero more rapidly than $r^{-1/2}$ as $r \to \infty$. Substituting (10.51) into (10.50), we obtain

$$W(r) = E - \frac{1}{2}k^2 + k \cot(kr)\frac{f'(r)}{f(r)} + \frac{1}{2}\frac{f''(r)}{f(r)}. \tag{10.52}$$

For $W(r)$ to remain bounded, $f'(r)/f(r)$ must vanish at the poles of $\cot(kr)$; that is, at the zeros of $\sin(kr)$. This can be achieved in many different ways. One possibility is to choose $f(r)$ to be a differentiable function of the variable

$$\begin{aligned} s(r) &= 8k^2 \int_0^r r'\{\sin(kr')\}^2\, dr' \\ &= \frac{1}{2}(2kr)^2 - 2kr \sin(2kr) - \cos(2kr) + 1. \end{aligned} \tag{10.53}$$

The principles guiding this choice (which is far from unique) are: that the integrand must be nonnegative, so that $s(r)$ will be a monotonic function of r; and that the integrand must be proportional to $\sin(kr)$, so that $ds(r)/dr$ will vanish at the zeros of $\sin(kr)$.

We choose

$$f(r) = [A^2 + s(r)]^{-1}, \tag{10.54}$$

where A is an arbitrary real parameter. Ψ decreases like r^{-3} as $r \to \infty$, which ensures its square integrability. The potential (10.52) then becomes

$$W(r) = \frac{64k^4r^2[\sin(kr)]^4}{[A^2 + s(r)]^2} - \frac{4k^2\{[\sin(kr)]^2 + 2kr \sin(2kr)\}}{A^2 + s(r)}. \tag{10.55}$$

At large r we have

$$W(r) \approx -\frac{4k \sin(2kr)}{r}. \tag{10.56}$$

The energy of the bound state is $E = \frac{1}{2}k^2$, independent of A.

Figures 10.4a and 10.4b illustrate the bound state function and the potential. The state function has been arbitrarily normalized so that $\Psi(0) = 1$. The parameter A has been given the value $A = k^4$. In this case the total energy, $E = 4$, is higher than the maximum of the potential $W(r)$, so the classical motion of a particle in such a potential would be unbounded. Clearly this bound state has no classical analog.

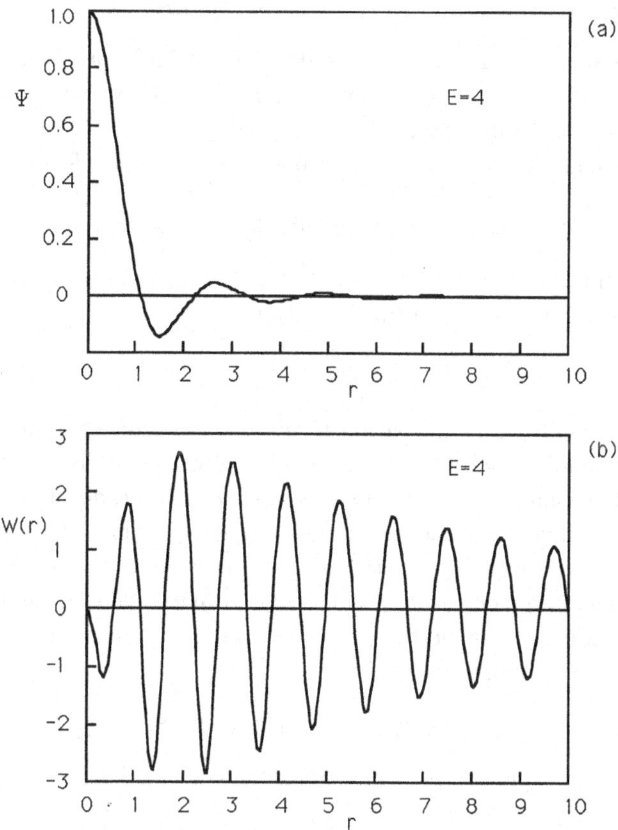

Fig. 10.4 (a) Positive energy bound state function. (b) The potential that supports the bound state in part (a).

Using the analogy of wave propagation to describe the dynamics of the state, it seems that the mechanism which prevents the bound state from dispersing like ordinary positive energy states is the destructive interference of the waves reflected from the oscillations of $W(r)$. Stillinger and Herrick believe that no $f(r)$ that produces a single-particle bound state in the continuum will lead to a potential that decays more rapidly than (10.56). However, they present further calculations and arguments which suggest that nonseparable multiparticle systems, such as two-electron atoms, may possess bound states in the continuum without such a contrived form of potential as (10.56).

10.5 *Stationary State Perturbation Theory*

In this section, and the next, we develop methods for approximate calculation of bound states and their energies. These methods are necessary because many important problems cannot be solved exactly.

Let us consider an energy eigenvalue equation of the form

$$(H_0 + \lambda H_1)|\Psi_n\rangle = E_n|\Psi_n\rangle\,, \tag{10.57}$$

in which the Hamiltonian is of the form $H = H_0 + \lambda H_1$, where the solutions of the "unperturbed" eigenvalue equation

$$H_0|n\rangle = \varepsilon_n|n\rangle \tag{10.58}$$

are known, and the perturbation term λH_1 is small, in some sense that has yet to be made precise. The parameter λ which governs the strength of the perturbation may be a variable such as the magnitude of an external field; it may be a fixed parameter like the strength of the spin–orbit coupling in an atom; or it may be a fictitious parameter introduced for mathematical convenience, in which case we will set $\lambda = 1$ at the end of the analysis.

Our technique will be to expand the unknown quantities E_n and $|\Psi_n\rangle$ in powers of the "small" parameter λ:

$$E_n = E_n{}^{(0)} + \lambda E_n{}^{(1)} + \lambda^2\, E_n{}^{(2)} + \cdots\,, \tag{10.59}$$

$$|\Psi_n\rangle = |\Psi_n{}^{(0)}\rangle + \lambda|\Psi_n{}^{(1)}\rangle + \lambda^2|\Psi_n{}^{(2)}\rangle + \cdots\,. \tag{10.60}$$

Substituting these expansions into (10.57) and collecting powers of λ yields the following sequence of equations:

$$\begin{aligned}
&(0):\ (H_0 - E_n{}^{(0)})|\Psi_n{}^{(0)}\rangle = 0\,,\\
&(1):\ (H_0 - E_n{}^{(0)})|\Psi_n{}^{(1)}\rangle = (E_n{}^{(1)} - H_1)|\Psi_n{}^{(0)}\rangle\,,\\
&(2):\ (H_0 - E_n{}^{(0)})|\Psi_n{}^{(2)}\rangle = (E_n{}^{(1)} - H_1)|\Psi_n{}^{(1)}\rangle + E_n{}^{(2)}|\Psi_n{}^{(0)}\rangle\,,\\
&\qquad\qquad\vdots\\
&(r):\ (H_0 - E_n{}^{(0)})|\Psi_n{}^{(r)}\rangle = (E_n{}^{(1)} - H_1)|\Psi_n{}^{(r-1)}\rangle\\
&\qquad\qquad + E_n{}^{(2)}|\Psi_n{}^{(r-2)}\rangle + \cdots + E_n{}^{(r)}|\Psi_n{}^{(0)}\rangle\,.
\end{aligned} \tag{10.61}$$

The known eigenvectors of H_0 form a complete orthonormal basis, satisfying $\langle n|n'\rangle = \delta_{n',n}$. Therefore we shall express the exact eigenvector of H in terms of these basis vectors,

$$|\Psi_n\rangle = \sum_{n'} |n'\rangle\langle n'|\Psi_n\rangle\,, \tag{10.62}$$

and shall use a similar expansion for each of the terms in the series (10.60).

Nondegenerate case. Suppose, for simplicity, that the eigenvalues of H_0 in (10.58) are nondegenerate; that is to say, $\varepsilon_n \neq \varepsilon_{n'}$ if $n \neq n'$. (The additional complication created by degeneracy will be treated later.) The solution of the zeroth member of the sequence (10.61) is obviously $E_n{}^{(0)} = \varepsilon_n$, $|\Psi_n{}^{(0)}\rangle = |n\rangle$. The zeroth order eigenvector has the usual normalization, $\langle\Psi_n{}^{(0)}|\Psi_n{}^{(0)}\rangle = 1$.

It is more convenient to choose an unusual normalization for the exact eigenvector (10.60):

$$\langle\Psi_n{}^{(0)}|\Psi_n\rangle = \langle n|\Psi_n\rangle = 1\,. \tag{10.63}$$

In view of (10.62), we see that this implies that $\langle\Psi_n|\Psi_n\rangle \geq 1$. It is permissible to choose an arbitrary normalization because the eigenvalue equation (10.57) is homogeneous, and so is not affected by the normalization. It is easier to renormalize $|\Psi_n\rangle$ at the end of the calculation than to impose the standard normalization at each step of the perturbation series. The normalization convention (10.63), when applied to (10.60), yields

$$\begin{aligned}\langle n|\Psi_n\rangle &= \langle n|\Psi_n{}^{(0)}\rangle + \lambda\langle n|\Psi_n{}^{(1)}\rangle + \lambda^2\langle n|\Psi_n{}^{(2)}\rangle + \cdots \\ &= 1 \text{ for all } \lambda\,.\end{aligned}$$

Therefore we obtain

$$\langle n|\Psi_n{}^{(r)}\rangle = 0 \text{ for } r > 0\,. \tag{10.64}$$

This is the reason why the nonstandard normalization (10.63) is so convenient.

To solve the first member of the sequence (10.61), we introduce the expansion

$$|\Psi_n{}^{(1)}\rangle = \sum_{n'\neq n} a_{n'}{}^{(1)}|n'\rangle\,, \tag{10.65}$$

from which the term $n' = n$ may be omitted because of (10.64). Using (10.65) and (10.58) we obtain

$$\begin{gathered}\sum_{n'\neq n} \langle m|(H_0 - E_n{}^{(0)})|n'\rangle\, a_{n'}{}^{(1)} = \langle m|(E_n{}^{(1)} - H_1)|\Psi_n{}^{(0)}\rangle, \\ (\varepsilon_m - \varepsilon_n)a_m{}^{(1)} = E_n{}^{(1)}\,\delta_{m,n} - \langle m|H_1|n\rangle\,.\end{gathered} \tag{10.66}$$

For the case $m = n$, this yields

$$E_n{}^{(1)} = \langle n|H_1|n\rangle \,. \tag{10.67}$$

For the case $m \neq n$ we obtain

$$a_m{}^{(1)} = \frac{\langle m|H_1|n\rangle}{\varepsilon_n - \varepsilon_m} \,,$$

and hence the first order contribution to the eigenvector is

$$\lambda|\Psi_n{}^{(1)}\rangle = \sum_{m\neq n} \frac{|m\rangle\langle m|\lambda H_1|n\rangle}{\varepsilon_n - \varepsilon_m} \,. \tag{10.68}$$

This result suggests that a suitable definition of "smallness" for the perturbation is that the relevant matrix element $\langle m|\lambda H_1|n\rangle$ should be small compared to the corresponding energy denominator $(\varepsilon_n - \varepsilon_m)$.

It is possible to proceed mechanically through the sequence (10.61) and thereby derive formulas for arbitrarily high orders of the perturbation series. That calculation will not be pursued here because the Brillouin–Wigner formulation of perturbation theory provides a more convenient generalization to higher orders. However, a very useful formula can be derived from the general rth member of (10.61). By taking its inner product with the unperturbed bra vector $\langle n|$ we obtain

$$\langle n|(H_0 - E_n{}^{(0)})|\Psi_n{}^{(r)}\rangle = -\langle n|H_1|\Psi_n{}^{(r-1)}\rangle + 0 + \cdots + E_n{}^{(r)}\,\langle n|\Psi_n{}^{(0)}\rangle \,,$$

where all but the first and last terms on the right hand side have vanished because of (10.64). The left hand side of this equation is zero because H_0 can operate to the left to yield the eigenvalue $\varepsilon_n = E_n{}^{(0)}$, and therefore we have

$$E_n{}^{(r)} = \langle n|H_1|\Psi_n{}^{(r-1)}\rangle \quad (r > 0)\,. \tag{10.69}$$

To obtain the energy E_n to order r we need only know $|\Psi_n\rangle$ to order $r - 1$. Taking $r = 2$ and using (10.68), we obtain the second order energy,

$$E_n{}^{(2)} = \sum_{m\neq n} \frac{\langle n|H_1|m\rangle\langle m|H_1|n\rangle}{\varepsilon_n - \varepsilon_m} \,. \tag{10.70}$$

Example (1): Perturbed harmonic oscillator

We wish to calculate the eigenvalues of the Hamiltonian $H = H_0 + H_1$, where the unperturbed Hamiltonian is $H_0 = P^2/2M + \frac{1}{2}M\omega^2 Q^2$, with Q and P being the position and momentum operators for a harmonic oscillator (see Ch. 6), and the perturbation is $H_1 = bQ$, with b a constant. Such a linear perturbation could be due to an external electric field if the oscillator is charged. This problem can be solved exactly, since the linear perturbation merely shifts the position of the minimum of the parabolic potential energy. Thus we may write

$$H = \frac{P^2}{2M} + \frac{1}{2}M\omega^2\left(Q + \frac{b}{M\omega^2}\right)^2 - \frac{b^2}{2M\omega^2}, \tag{10.71}$$

from which it is apparent that the eigenvalues are merely lowered from their unperturbed values by a constant shift of $-b^2/2M\omega^2$. Solving this problem by perturbation theory will only serve to illustrate the technique.

In Sec. 6.1 it was shown that the eigenvalues of $H_0|n\rangle = \varepsilon_n|n\rangle$ are $\varepsilon_n = \hbar\omega(n + \frac{1}{2})$, $(n = 0, 1, 2, \ldots)$. The matrix elements of the perturbation, $b\langle n'|Q|n\rangle$, can most easily be obtained from the relation of Q to the raising and lowering operators, $Q = (\hbar/2M\omega)^{1/2}(a^\dagger + a)$. From (6.16) it follows that

$$\begin{aligned}\langle n|Q|n+1\rangle &= \left(\frac{\hbar}{2M\omega}\right)^{1/2}\sqrt{n+1}\,,\\ \langle n|Q|n-1\rangle &= \left(\frac{\hbar}{2M\omega}\right)^{1/2}\sqrt{n}\,,\end{aligned} \tag{10.72}$$

and all other matrix elements are zero. The perturbed energy eigenvalues are of the form $E_n = \varepsilon_n + E_n{}^{(1)} + E_n{}^{(2)} + \cdots$. From (10.67) we have

$$E_n{}^{(1)} = b\langle n|Q|n\rangle = 0\,.$$

From (10.70) we obtain

$$\begin{aligned}E_n{}^{(2)} &= \frac{b^2|\langle n|Q|n-1\rangle|^2}{\hbar\omega} + \frac{b^2|\langle n|Q|n+1\rangle|^2}{(-\hbar\omega)}\\ &= \frac{-b^2}{2M\omega^2}\,,\end{aligned} \tag{10.73}$$

which is the exact answer.

Example (2): Induced electric dipole moment of an atom

Most atoms have no permanent electric dipole moments. As will be shown in Ch. 13, this follows from space inversion symmetry (Sec. 13.1), or from rotational symmetry combined with time reversal invariance (Sec. 13.3). However, an external electric field will break these symmetries, and will induce a dipole moment that is proportional to the field. The *polarizability* α is defined as the ratio of the induced dipole moment to the electric field, $\langle \mathbf{d} \rangle = \alpha \mathbf{E}$. The potential energy of the polarized atom in the electric field is lowered from that of a free atom by the amount $-\frac{1}{2}\alpha|\mathbf{E}|^2$. (This is the sum of the potential energy of the dipole in the field, $-\langle \mathbf{d} \rangle \cdot \mathbf{E}$, plus the work done by the field on the atom in polarizing it, $\frac{1}{2}\alpha|\mathbf{E}|^2$.) Thus we have two methods to calculate the polarizability α: (a) calculate the energy to the second order in $\mathbf{E}$; or (b) evaluate the perturbed state function to the first order in $\mathbf{E}$ and calculate $\langle \mathbf{d} \rangle$ in the perturbed state. We shall carry out both of these calculations for the ground state of a hydrogen-like atom.

(a) The unperturbed energy levels of the hydrogen atom are determined by the eigenvalue equation (10.20). For a one-electron hydrogen-like atom they will be determined by a similar equation, of the form

$$H_0|n\ell m\rangle = \varepsilon_{n\ell}|n\ell m\rangle\,,$$

where n is the principal quantum number, and ℓ and m are the orbital angular momentum quantum numbers, as in Sec. 10.2. It follows from rotational invariance, and in particular from the fact that H_0 commutes with J_+ and J_-, that the eigenvalue $\varepsilon_{n\ell}$ is independent of m. For the hydrogen atom, the eigenvalue is given by (10.27), which is also independent of ℓ. This is a special property of the Coulomb potential, and it does not hold for any central potential that is not exactly proportional to r^{-1}.

The perturbation due to the electric field is

$$H_1 = -\mathbf{d} \cdot \mathbf{E} = e\mathbf{E} \cdot \mathbf{r}\,, \tag{10.74}$$

where $\mathbf{d} = -e\mathbf{r}$ is the dipole moment operator, $-e$ is the charge of the electron, and $\mathbf{r}$ is the position of the electron relative to the nucleus. It is convenient to choose the direction of $\mathbf{E}$ to be the axis of polar coordinates. Then (10.74) becomes $H_1 = e|\mathbf{E}|r\,\cos\theta$, which is clearly a component of an irreducible tensor of the type ${T_0}^{(1)}$. (See Sec. 7.8

for the definition of an irreducible tensor.) It follows from the Wigner–Eckart theorem that the matrix element $\langle n\ell m|H_1|n'\ell' m'\rangle$ must vanish unless $m = m'$ and three numbers $(\ell, \ell', 1)$ form the sides of a triangle. Moreover, since $\cos\theta$ changes sign under inversion of coordinates $(\mathbf{r} \to -\mathbf{r})$, it is necessary that the two state vectors in the matrix element have opposite parity, one being even and the other odd. Thus we must have $\ell' = \ell \pm 1$ and $m' = m$ in order for the matrix element $\langle n\ell m|H_1|n'\ell' m'\rangle$ to be nonzero.

It follows that the first order (10.67) contribution to the energy vanishes:

$$E_{n\ell m}{}^{(1)} = \langle n\ell m|H_1|n\ell m\rangle = 0\,.$$

The second order (10.68) contribution to the ground state energy is

$$E_{100}{}^{(2)} = \sum_{n'} \frac{|\langle n'10|H_1|100\rangle|^2}{\varepsilon_{10} - \varepsilon_{n'1}}\,. \tag{10.75}$$

Equating this expression to the change in energy of the polarized atom, $-\frac{1}{2}\alpha|\mathbf{E}|^2$, we find the polarizability of the atom in its ground state to be

$$\alpha = 2\sum_{n'} \frac{|\langle n'10|er\ \cos\theta|100\rangle|^2}{\varepsilon_{n'1} - \varepsilon_{10}}\,. \tag{10.76}$$

It must be emphasized that the sum in (10.70) is over *all* of the eigenvectors of H_0 except for the particular state whose perturbed energy is being calculated. Therefore the sum over n' in (10.75) and (10.76) should include an *integral over the continuum* of unbound positive energy states, as well as a sum over the discrete bound states. We shall shortly return to consider this problem, which seriously complicates the evaluation of second order perturbation formulas.

(b) As an alternative to the second order energy calculation, we can evaluate $\langle\mathbf{d}\rangle$ in a first order perturbed state. To the first order, the ground state vector is

$$|\Psi_{100}\rangle = |\Psi_{100}{}^{(0)}\rangle + |\Psi_{100}{}^{(1)}\rangle\,, \tag{10.77}$$

with the first order contribution being [from Eq. (10.68)]

$$|\Psi_{100}{}^{(1)}\rangle = -\sum_{n'} \frac{|n'10\rangle\langle n'10|e|\mathbf{E}|r\ \cos\theta|100\rangle}{\varepsilon_{n'1} - \varepsilon_{10}}\,. \tag{10.78}$$

(The minus sign comes from reversing the sign of the denominator in order to make it positive.) The zeroth order term in (10.77) is even under inversion of coordinates, and the first order term is odd. The dipole moment operator itself is odd, so the average dipole moment in the ground state,

$$\langle \mathbf{d} \rangle \equiv \alpha \mathbf{E} = \frac{\langle \Psi_{100} | \mathbf{d} | \Psi_{100} \rangle}{\langle \Psi_{100} | \Psi_{100} \rangle}\,,$$

evaluated to first order in the electric field, is

$$\langle \mathbf{d} \rangle = \langle {\Psi_{100}}^{(0)} | \mathbf{d} | {\Psi_{100}}^{(1)} \rangle + \langle {\Psi_{100}}^{(1)} | \mathbf{d} | {\Psi_{100}}^{(0)} \rangle\,. \tag{10.79}$$

(The normalization of the perturbed state vector is $\langle \Psi_{100} | \Psi_{100} \rangle = 1 + 0(|\mathbf{E}|^2) \approx 1$ to the first order.) Because of symmetry, we know that the only nonvanishing component of $\langle \mathbf{d} \rangle$ will be directed along the polar axis, and so it is sufficient to evaluate (10.79) for the component $(\mathbf{d})_z = er\,\cos\theta$. Substituting (10.77) into (10.79), we obtain once again the expression (10.76) for the polarizability α. It is no coincidence that these two calculations of α, from the second order energy and from the first order state vector, have led to exactly the same answer. Their compatibility is guaranteed by (10.69), which asserts that the $r-1$ order approximation to the eigenvector contains sufficient information for determining the eigenvalue to the rth order.

Example (3): Second order perturbation energy in closed form

The expression (10.70) for the second order perturbation energy generally involves an infinite summation. In the case of an atom it involves both a sum over the discrete bound states and an integral over the continuum of unbound states. Since these are rather difficult to evaluate, it is sometimes preferable to adopt an alternative method based upon (10.69). Specializing it to a hydrogen-like atom, as in the previous example, we use instead of (10.75),

$${E_{100}}^{(2)} = \langle 100 | H_1 | {\Psi_{100}}^{(1)} \rangle\,. \tag{10.80}$$

This will be useful only if we can somehow obtain the first order correction to the eigenvector.

Of course there is no use trying to obtain it from the perturbation expression (10.68), since we have just seen in the previous example that this would lead to exactly the same computational problem involving the summation and integration over an infinity of states. However, we can make direct use of the first member of the sequence (10.61), which for this case becomes

$$(H_0 - \varepsilon_{10})|\Psi_{100}{}^{(1)}\rangle = (\langle 100|H_1|100\rangle - H_1)|100\rangle\,, \qquad (10.81)$$

where ε_{10} is the ground state eigenvalue of H_0. All quantities on the right hand side of this equation are known, so the problem has been transformed into one of solving an inhomogeneous differential equation. The solution of this equation is not unique, since we can always add to it an arbitrary multiple of the solution of the homogeneous equation $(H_0 - \varepsilon_{10})|100\rangle = 0$. However, uniqueness is restored by the condition (10.64), which requires that $\langle 100|\Psi_{100}{}^{(1)}\rangle = 0$. This method will be effective only if (10.81) is easier to solve than the full eigenvalue equation (10.57). Fortunately there are cases in which this is so.

Let us take H_0 to be the internal Hamiltonian of a hydrogen-like atom with reduced mass μ and a central potential $W(r)$, and H_1 to be the electric dipole interaction (10.74). Then (10.81) becomes

$$\left[\frac{-\hbar^2}{2\mu}\nabla^2 + W(r) - \varepsilon_{10}\right]\Psi^{(1)} = -e|\mathbf{E}|r\ \cos\theta\ \Psi^{(0)}\,. \qquad (10.82)$$

(The state labels $n\ell m = 100$ are omitted to simplify the notation.) The only angular dependence on the right hand side is $\cos\theta$, since the ground state function $\Psi^{(0)}$ is rotationally symmetric. The operator of the left hand side is also rotationally symmetric, and hence it cannot change the angular dependence of $\Psi^{(1)}$, which must therefore be of the form

$$\Psi^{(1)}(r,\theta,\phi) = \cos\theta\ f(r)\,. \qquad (10.83)$$

The subsidiary condition $\langle\Psi^{(0)}|\Psi^{(1)}\rangle = 0$ is automatically satisfied because of the angular dependence. Since (10.83) is an angular momentum eigenfunction with $\ell = 1$, Eq. (10.82) reduces to

$$\frac{-\hbar^2}{2\mu}\frac{1}{r^2}\frac{d}{dr}r^2\frac{d}{dr}f(r) + \left[\frac{\hbar^2}{\mu r^2} + W(r) - \varepsilon_{10}\right]f(r) = -e|\mathbf{E}|\ r\ \Psi^{(0)}(r)\,. \qquad (10.84)$$

Even if this differential equation must be solved approximately, it may be more tractable than the infinite sum in (10.75).

We shall solve (10.84) for the *hydrogen atom*, for which $W(r) = -e^2/r$, $\varepsilon_{10} = -e^2/2a_0$, and $\Psi^{(0)}(r) = (\pi {a_0}^3)^{-1/2} e^{-r/a_0}$, with $a_0 = \hbar^2/\mu e^2$. We anticipate that the solution will be of the form

$$f(r) = p(r)\, e^{-r/a_0}\,, \tag{10.85}$$

where $p(r)$ is a power series. Substituting these expressions into (10.84), we obtain

$$r^2 p''(r) + 2\left(r - \frac{r^2}{a_0}\right) p'(r) - 2p(r) = \frac{2|\mathbf{E}|r^3}{(\pi {a_0}^3)^{1/2} e a_0}\,. \tag{10.86}$$

It is easily verified that the only polynomial solution to this equation is

$$p(r) = -(\pi {a_0}^3)^{-1/2} \left(\frac{|\mathbf{E}|}{e}\right) \left(a_0 r + \frac{1}{2} r^2\right). \tag{10.87}$$

(There is also a solution in the form of an infinite series, but it increases exponentially as $r \to \infty$, and so is unacceptable.) The second order energy can now be evaluated from (10.80):

$$\begin{aligned} {E_{100}}^{(2)} &= \langle \Psi^{(0)} | H_1 | {\Psi_{100}}^{(1)} \rangle \\ &= -|\mathbf{E}|^2 (\pi {a_0}^3)^{-1} \int (\cos\theta)^2 \,(a_0 r^2 + r^3)\, e^{-2r/a_0}\, d^3 r \\ &= -\frac{9}{4} |\mathbf{E}|^2 {a_0}^3 \,. \end{aligned} \tag{10.88}$$

This energy is related to the electric polarizability α by the relation ${E_{100}}^{(2)} = \frac{1}{2}\alpha |\mathbf{E}|^2$, and therefore the polarizability of a hydrogen atom in its ground state is

$$\alpha = \frac{9}{2} {a_0}^3 \,. \tag{10.89}$$

Degenerate case. Formulas such as (10.68) and (10.70) may not be applicable if there are degeneracies among the unperturbed eigenvalues, since that would permit the denominator $\varepsilon_n - \varepsilon_m$ to vanish. The formal perturbation theory must now be re-examined to determine what modifications are needed in the degenerate case.

The formal expansion of the eigenvalues and eigenvectors in powers of the strength of the perturbation is still valid, up to and including (10.62). When we attempt to solve the zeroth member of the sequence (10.61), it is clear that the zeroth order eigenvalue is given by ${E_n}^{(0)} = \varepsilon_n$, as in the nondegenerate case. But we cannot identify the zeroth order eigenvector, except to say that it must be some linear combination of those degenerate eigenvectors of H_0 which all belong to the same eigenvalue ε_n. Because the energy ε_n is not sufficient for identifying a unique eigenvector of H_0, it is necessary to introduce a more detailed notation. Instead of (10.58), we now write

$$H_0|n,r\rangle = \varepsilon_n|n,r\rangle\,, \tag{10.90}$$

where the second label r distinguishes between degenerate eigenvectors. The range of the second label will generally be different for each value of n, corresponding to the degree of degeneracy of that eigenvalue. The zeroth order eigenvectors in the sequence (10.61) must be of the form

$$|\Psi_{n,r}{}^{(0)}\rangle = \sum_{r'} c_{r,r'}|n,r'\rangle\,, \tag{10.91}$$

but the coefficients $c_{r,r'}$ are not yet determined.

The first member of (10.61) will now be written as

$$(H_0 - \varepsilon_n)|\Psi_{n,r}{}^{(1)}\rangle = (E_n{}^{(1)} - H_1)|\Psi_{n,r}{}^{(0)}\rangle\,.$$

Let us consider the inner product of this equation with $\langle n,s|$ for fixed n but for all values of s. Using (10.91) we obtain

$$\langle n,s|(H_0 - \varepsilon_n)|\Psi_{n,r}{}^{(1)}\rangle = \sum_{r'}\langle n,s|(E_n{}^{(1)} - H_1)|n,r'\rangle c_{r,r'}\,,$$

$$(\varepsilon_n - \varepsilon_n)\langle n,s|\Psi_{n,r}{}^{(1)}\rangle = \sum_{r'}\{E_n{}^{(1)}\,\delta_{s,r'} - \langle n,s|H_1|n,r'\rangle\}\,c_{r,r'}\,.$$

Thus we have

$$\sum_{r'}\langle n,s|H_1|n,r'\rangle\,c_{r,r'} = E_n{}^{(1)}\,c_{r,s}\,. \tag{10.92}$$

This has the form of a matrix eigenvector equation that is restricted to the subspace of degenerate unperturbed vectors belonging to the unperturbed energy ε_n. Thus the appropriate choice of zeroth order eigenvectors in (10.91) is those that diagonalize the matrix $\langle n,s|H_1|n,r'\rangle$ in the subspace of fixed n.

If we now use as basis vectors the zeroth order eigenvectors (10.91) with the coefficients determined by (10.92), then the perturbation formulas (10.68) and (10.70) become usable because the diagonalization of (10.92) ensures that $\langle m|H_1|n\rangle = 0$ whenever $\varepsilon_n - \varepsilon_m = 0$. Thus the potentially troublesome terms in the perturbation formulas do not contribute.

Example (4): Linear Stark effect in hydrogen

The shift in the energy levels of an atom in an electric field is known as the *Stark effect.* Normally the effect is quadratic in the field strength, as was shown in Example (2). But the first excited state of the hydrogen atom exhibits an effect that is linear in the field strength. This is due to the degeneracy of the excited state.

If we neglect spin, the stationary states of a free hydrogen atom may be represented by the vectors $|n\ell m\rangle$, where n is the principal quantum number, and ℓ and m are orbital angular momentum quantum numbers. The first excited state is four-fold degenerate, the degenerate states being $|200\rangle, |211\rangle, |210\rangle$, and $|21-1\rangle$. Specialized to this problem, Eq. (10.91) may be written as

$$|\Psi^{(0)}\rangle = c_1|200\rangle + c_2|211\rangle + c_3|210\rangle + c_4|21-1\rangle\,. \tag{10.93}$$

The coefficients are to be determined by diagonalizing the matrix of the perturbation, $H_1 = e\mathbf{E}\cdot\mathbf{r} = e|\mathbf{E}|r\ \cos\theta$, in the four-dimensional subspace spanned by the four degenerate basis vectors.

The matrix element $\langle n\ell m|H_1|n'\ell' m'\rangle$ vanishes unless $m = m'$, and therefore the only nonvanishing elements in the 4×4 matrix in (10.92) are $\langle 210|H_1|200\rangle = \langle 200|H_1|210\rangle^*$. The evaluation of these matrix elements requires only a simple integration over hydrogenic wave functions, yielding the value $\langle 210|H_1|200\rangle = -3e|\mathbf{E}|a_0 = -w$, say. The condition for nontrivial solutions of the eigenvalue equation (10.92) is the vanishing of the determinant

$$\begin{vmatrix} -E^{(1)} & 0 & -w & 0 \\ 0 & -E^{(1)} & 0 & 0 \\ -w & 0 & -E^{(1)} & 0 \\ 0 & 0 & 0 & -E^{(1)} \end{vmatrix} = 0\,. \tag{10.94}$$

It yields four roots: w, $-w$, 0, 0. The corresponding eigenvectors are $\left(\sqrt{\frac{1}{2}}, 0, -\sqrt{\frac{1}{2}}, 0\right)$, $\left(\sqrt{\frac{1}{2}}, 0, \sqrt{\frac{1}{2}}, 0\right)$, (0, 1, 0, 0), and (0, 0, 0, 1). Hence the four-fold degenerate energy level ε_2 of the hydrogen atom is split by the electric field into two perturbed states: $(|200\rangle - 210\rangle)/\sqrt{2}$ with energy $\varepsilon_2 + 3e|\mathbf{E}|a_0$, and $(|200\rangle + |210\rangle)/\sqrt{2}$ with energy $\varepsilon_2 - 3e|\mathbf{E}|a_0$; and two states that remain degenerate at the energy ε_2 : $|211\rangle$ and $|21-1\rangle$.

The two states whose energies depend linearly on the electric field exhibit a *spontaneous electric dipole moment.* The average of the dipole moment in the lowest energy state $|\Psi\rangle = (|200\rangle + |210\rangle)/\sqrt{2}$ has a non-vanishing z component,

$$\begin{aligned}\langle d_z\rangle &= \langle\Psi|(-er\ \cos\theta)|\Psi\rangle \\ &= \frac{1}{2}\langle 200|(-er\ \cos\theta)|210\rangle + \frac{1}{2}\langle 210|(-er\ \cos\theta|)|200\rangle \\ &= 3ea_0\,,\end{aligned}$$

and the corresponding potential energy is $-\langle\mathbf{d}\rangle\cdot\mathbf{E} = -3e|\mathbf{E}|a_0$. The state $(|200\rangle - |210\rangle)/\sqrt{2}$ has a dipole moment of the same magnitude but pointing antiparallel to the electric field, so its energy is raised by $3e|\mathbf{E}|a_0$.

Brillouin–Wigner perturbation theory

The form of perturbation theory described above, which is often called *Rayleigh–Schrödinger* perturbation theory, is based upon an expansion in powers of the perturbation strength parameter. Although it can, in principle, be extended to arbitrarily high orders, the forms of the higher order terms become increasingly complicated as the order increases. The *Brillouin–Wigner* form has the advantage that the generalization to arbitrary order is easy.

Let us put $\lambda = 1$ and rewrite (10.57) as

$$(E_n - H_0)|\Psi_n\rangle = H_1|\Psi_n\rangle\,. \tag{10.95}$$

From the eigenvectors of $H_0|n\rangle = \varepsilon_n|n\rangle$, we construct the projection operators

$$Q_n = \sum_{r\neq n}|r\rangle\langle r| = 1 - |n\rangle\langle n|\,. \tag{10.96}$$

An eigenvector of (10.95), normalized according to (10.63), satisfies

$$|\Psi_n\rangle = |n\rangle + Q_n|\Psi_n\rangle\,. \tag{10.97}$$

It is clear that $Q_nH_0 = H_0Q_n$, since they share the same eigenvectors. Thus we obtain from (10.95)

$$\begin{aligned}Q_n(E_n - H_0)|\Psi_n\rangle &= (E_n - H_0)Q_n|\Psi_n\rangle \\ &= Q_nH_1|\Psi_n\rangle\,,\end{aligned}$$

and hence $Q_n|\Psi_n\rangle = (E_n - H_0)^{-1}Q_nH_1|\Psi_n\rangle$. Substitution of this result into (10.97) yields

$$|\Psi_n\rangle = |n\rangle + R_nH_1|\Psi_n\rangle\,, \tag{10.98}$$

where we have defined

$$R_n = (E_n - H_0)^{-1}\,Q_n = Q_n(E_n - H_0)^{-1}\,. \tag{10.99}$$

Equation (10.98) can be solved by iteration, on the assumption that the perturbation H_1 is small. Neglecting H_1 on the right hand side yields the zeroth order approximation, $|\Psi_n\rangle \approx |n\rangle$. Substitution of this zeroth order approximation on the right then leads to a first order approximation, and so on. Continuing the iteration yields the series

$$|\Psi_n\rangle = |n\rangle + R_nH_1|n\rangle + (R_nH_1)^2|n\rangle + (R_nH_1)^3|n\rangle + \cdots\,. \tag{10.100}$$

We can formally sum this infinite series to obtain

$$|\Psi_n\rangle = (1 - R_nH_1)^{-1}|n\rangle\,, \tag{10.101}$$

however, this exact formal solution seldom has much computational value.

From (10.95) we obtain $\langle n|(E_n - H_0)|\Psi_n\rangle = \langle n|H_1|\Psi_n\rangle$, which yields an expression for the energy eigenvalue,

$$E_n = \varepsilon_n + \langle n|H_1|\Psi_n\rangle\,, \tag{10.102}$$

which becomes a series in powers of H_1 when we substitute (10.100) for $|\Psi_n\rangle$. By introducing the spectral representation of the operator R_n,

$$R_n = \sum_{m\neq n} \frac{|m\rangle\langle m|}{E_n - \varepsilon_m}\,,$$

we obtain a more familiar form of the perturbation expansion,

$$E_n = \varepsilon_n + \langle n|H_1|n\rangle + \sum_{m\neq n} \frac{\langle n|H_1|m\rangle\langle m|H_1|n\rangle}{(E_n - \varepsilon_m)}$$

$$+ \sum_{m\neq n}\sum_{m'\neq n} \frac{\langle n|H_1|m\rangle\langle m|H_1|m'\rangle\langle m'|H_1|n\rangle}{(E_n - \varepsilon_m)\,(E_n - \varepsilon_{m'})} + \cdots\,. \tag{10.103}$$

Notice that the unknown energy E_n appears in the denominators on the right hand side, and hence this is not an explicit expression for E_n. If one wishes to calculate E_n to third order accuracy, then it is sufficient to substitute the zeroth value, $E_n = \varepsilon_n$, into the denominator of the third order term of (10.103); but the first order value, $E_n = \varepsilon_n + \langle n|H_1|n\rangle$, must be used in the denominator of the second order term. A more practical way to compute E_n from (10.103) is to make an estimate of E_n, which is then substituted into all denominators, and the sums are evaluated numerically. The resulting new value of E_n is then substituted into the denominators, and the process is continued iteratively until the result converges to the desired accuracy.

If we formally expand all factors such as $(E_n - \varepsilon_m)^{-1}$ on the right hand side of (10.103) in powers of the strength of the perturbation, we will recover the Rayleigh–Schrödinger perturbation series. In all orders beyond second it will contain many more terms than does (10.103), and so it is much less convenient to handle than is the Brillouin–Wigner perturbation formalism. Some of the higher order terms of Rayleigh–Schrödinger perturbation theory can be found in Ch. 8 of Schiff (1968).

Example (5): Near degeneracy

Consider a simple 2×2 matrix Hamiltonian for which

$$H_0 = \begin{bmatrix} \varepsilon_1 & 0 \\ 0 & \varepsilon_2 \end{bmatrix}, \quad H_1 = \begin{bmatrix} 0 & v \\ v^* & 0 \end{bmatrix}. \tag{10.104}$$

The exact eigenvalues of the equation $(H_0 + H_1)|\Psi\rangle = E|\Psi\rangle$ are given by the vanishing of the determinant

$$\begin{vmatrix} \varepsilon_1 - E & v \\ v^* & \varepsilon_2 - E \end{vmatrix} = 0\,.$$

The expansion of this determinant yields the quadratic equation

$$E^2 - (\varepsilon_1 + \varepsilon_2)E + (\varepsilon_1\varepsilon_2 - |v|^2) = 0\,, \tag{10.105}$$

whose solution is

$$E = \frac{1}{2}(\varepsilon_1 + \varepsilon_2) \pm \frac{1}{2}\sqrt{(\varepsilon_1 - \varepsilon_2)^2 + 4|v|^2}\,. \tag{10.106}$$

In the degenerate limit, $\varepsilon_1 = \varepsilon_2 = \varepsilon$, this becomes

$$E = \varepsilon \pm |v|\,. \tag{10.107}$$

The application of Brillouin–Wigner perturbation theory yields

$$E_1 = \varepsilon_1 + \frac{|v|^2}{E_1 - \varepsilon_2} \tag{10.108}$$

and a similar equation for E_2. Equation (10.108) is equivalent to the exact quadratic equation (10.105), and therefore Brillouin–Wigner perturbation theory yields the exact answer to this problem, in both the degenerate and nondegenerate cases.

For comparison, the application of Rayleigh–Schrödinger perturbation theory to this problem yields

$$E_1 = \varepsilon_1 + \frac{|v|^2}{\varepsilon_1 - \varepsilon_2}\,. \tag{10.109}$$

This is correct to the second order, and will be accurate provided that $|v|/|\varepsilon_1 - \varepsilon_2| \ll 1$. But it is nonsense in the limit $\varepsilon_1 \to \varepsilon_2$.

Rayleigh–Schrödinger perturbation theory provides two distinct formalisms for the degenerate and nondegenerate cases, and so its application to situations of near degeneracy can be problematic. Brillouin–Wigner perturbation theory is superior in such situations. If the degree of degeneracy is greater than 2 the Brillouin–Wigner theory is no longer exact in the degenerate limit; however, it may still form a usable approximation.

10.6 *Variational Method*

The perturbation methods of the previous section rely on there being a closely related problem that is exactly solvable. The variational method is subject to no such restriction, and it is often the method of choice for studying complex systems such as multi-electron atoms and molecules. Although we shall use simple examples to illustrate the method, the overwhelming majority of its practical applications involve numerical computation.

In the variational method, we consider the functional

$$\Lambda(\phi, \psi) = \frac{\langle\phi|H|\psi\rangle}{\langle\phi|\psi\rangle}\,. \tag{10.110}$$

Here H is a linear operator, ϕ and ψ are variable functions. We seek the conditions under which the value of Λ will be stationary with respect to infinitesimal changes in the functions ϕ and ψ. These conditions can be formally expressed as the vanishing of two functional derivatives: $\delta\Lambda/\delta\phi = 0$ and $\delta\Lambda/\delta\psi = 0$. Since functional differentiation may not be a familiar concept to all readers, some explanation is appropriate. Consider the change in Λ when $\langle\phi|$ is replaced by $\langle\phi| + \varepsilon\langle\alpha|$, where ε is a small number and $\langle\alpha|$ is an arbitrary vector. To first order in ε, we obtain

$$\Lambda(\phi + \varepsilon\alpha, \psi) - \Lambda(\phi, \psi) = \varepsilon \left[\frac{\langle\alpha|H|\psi\rangle}{\langle\phi|\psi\rangle} - \frac{\langle\phi|H|\psi\rangle}{\langle\phi|\psi\rangle^2} \langle\alpha|\psi\rangle \right]$$
$$= \varepsilon\langle\alpha| \frac{\{H|\psi\rangle - \Lambda(\phi,\psi)|\psi\rangle\}}{\langle\phi|\psi\rangle} . \qquad (10.111)$$

Formally dividing by $\varepsilon\langle\alpha|$ and letting $\varepsilon \to 0$, we obtain the definition of the functional derivative $\delta\Lambda/\delta\phi$. The condition for (10.111) to vanish to the first order in ε for *arbitrary* $\langle\alpha|$ is equivalent to the eigenvalue equation

$$H|\psi\rangle - \lambda|\psi\rangle = 0 . \qquad (10.112a)$$

Similarly, requiring the functional $\Lambda(\phi, \psi)$ to be stationary under first order variations of ψ leads to the condition

$$\langle\phi|H - \langle\phi|\lambda = 0, \text{ or } H^\dagger|\phi\rangle - \lambda^*|\phi\rangle = 0 . \qquad (10.112b)$$

Thus the conditions for the functional to be stationary are that ϕ and ψ be, respectively, left and right eigenvectors of H, with the eigenvalue λ having the value $\Lambda(\phi, \psi)$.

If $H = H^\dagger$, as is true in most cases of interest, then at the condition of stationarity we will have $\lambda = \lambda^*$ and $|\phi\rangle = |\psi\rangle$. But even in such a case it is useful to regard the variations of the left hand vector ϕ and the right hand vector ψ as being independent, as we shall see in later applications.

If we choose trial functions ϕ and ψ which depend on certain parameters, and vary those parameters to find the stationary points of the functional Λ, we will obtain approximations to the eigenvalues of H. But, in general, those stationary points are neither maxima nor minima, but only inflection points or saddle points in a space of very high dimension (possibly infinite). Such points are not easy to determine numerically, so further developments are needed to make the method useful. Most practical applications are based upon the following theorem.

Variational theorem. If $H = H^\dagger$ and E_0 is the lowest eigenvalue of H, then for any ψ we have the inequality

$$E_0 \le \frac{\langle\psi|H|\psi\rangle}{\langle\psi|\psi\rangle} . \qquad (10.113)$$

Proof. To prove this theorem we use the eigenvector expansion of $|\psi\rangle, |\psi\rangle = \sum_n |\Psi_n\rangle\langle\Psi_n|\psi\rangle$, where $H|\Psi_n\rangle = E_n|\Psi_n\rangle$. Using the orthonormality and completeness of the eigenvectors, we obtain

$$\langle\psi|H|\psi\rangle = \sum_n E_n|\langle\psi|\Psi_n\rangle|^2$$

$$\geq E_0 \sum_n |\langle\psi|\Psi_n\rangle|^2 = E_0\,\langle\psi|\psi\rangle\,,$$

from which the theorem follows at once.

The variational method, applied to the calculation of the lowest eigenvalue, consists of choosing a trial function ψ that depends on one or more parameters, and varying those parameters to obtain the minimum value of the expression on the right hand side of (10.113). Alternatively, one may try several different functions for ψ, based upon whatever information one may have about the problem, or even on intuitive guesses. Regardless of how the trial functions are chosen, the theorem guarantees that the lowest value obtained is the best estimate for E_0.

A common type of variational trial function consists of a linear combination of a *finite subset* of a set of orthonormal basis vectors,

$$|\psi^{\text{var}}\rangle = \sum_{n=1}^{N} a_n|n\rangle\,. \tag{10.114}$$

Stationary values of the functional

$$\Lambda = \frac{\langle\psi^{\text{var}}|H|\psi^{\text{var}}\rangle}{\langle\psi^{\text{var}}|\psi^{\text{var}}\rangle}$$

$$= \frac{\sum_n\sum_m a_n{}^*a_m\langle n|H|m\rangle}{\sum_n a_n{}^*a_n} \tag{10.115}$$

are then sought by varying the parameters $\{a_n\}$. As was explained earlier, the left and right vectors in the functional Λ may be varied independently. This implies, for our current problem, that we may vary $a_n{}^*$ independently of a_n. [It may seem strange to regard $a_n{}^*$ and a_n as independent variables. The strangeness is alleviated if one realizes that the real and imaginary parts of a_n are certainly independent variables. But $a_n{}^*$ and a_n are just two independent linear combinations of $\text{Re}(a_n)$ and $\text{Im}(a_n)$, and so they too are acceptable choices as independent variables.] The set of N conditions, $\partial\Lambda/\partial a_j{}^* = 0$, $(j = 1,\ldots,N)$, yields

$$\sum_m \langle j|H|m\rangle\, a_m = \Lambda\, a_j \quad (j = 1,\ldots,N)\,. \tag{10.116}$$

Because $H = H^\dagger$, the conditions $\partial\Lambda/\partial a_j = 0$ merely lead to the complex conjugate of (10.116), which gives no extra information. Now (10.116) is an $N \times N$ matrix eigenvalue equation. Indeed it is nothing but the original eigenvalue equation, $H|\Psi\rangle = E|\Psi\rangle$, truncated to the N-dimensional subspace in which the trial function (10.114) has been confined. To calculate the eigenvalues of the truncated $N \times N$ matrix as approximations to the true eigenvalues of H is an intuitively natural thing to do. The variational theorem tells us that it is indeed the *best* that can be done with a trial function of the form (10.114).

The variational theorem ensures only that the lowest eigenvalue of the $N \times N$ matrix will be an upper bound to the true E_0. However, for a trial function of the form (10.114), which involves N orthogonal basis functions, it can be shown (Pauling and Wilson, Sec. 26d) that the approximate eigenvalues for successive values of N are interleaved, as shown in Fig. 10.5. Thus, for this particular form of trial function, all approximate eigenvalues must converge from above to their $N \to \infty$ limits.

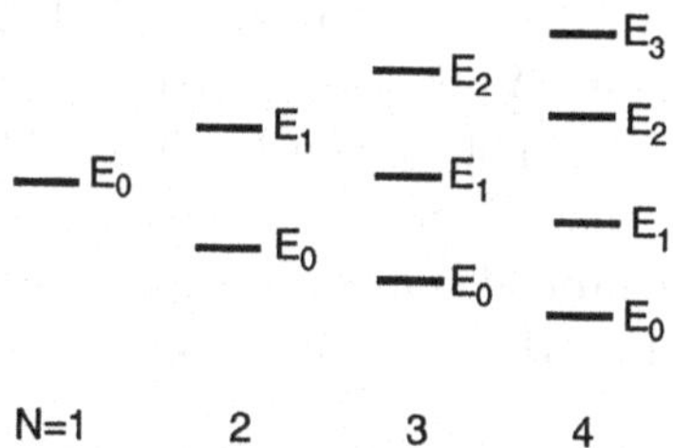

Fig. 10.5 Interleaving of the approximate eigenvalues for trial functions consisting of a linear combination of N basis functions.

The accuracy of a variational approximation to an eigenvalue, ${E_n}^{\text{var}} \equiv \langle \psi^{\text{var}}|H|\psi^{\text{var}}\rangle / \langle \psi^{\text{var}}|\psi^{\text{var}}\rangle$, is clearly determined by the proximity of $|\psi^{\text{var}}\rangle$ to a true eigenvector. Let us write $|\psi^{\text{var}}\rangle = |\Psi_n\rangle + |\varepsilon\rangle$, where $|\varepsilon\rangle$ is a small error vector. Since the value of ${E_n}^{\text{var}}$ is clearly independent of the normalization of the vectors, we shall simplify the algebra by assuming, without loss of generality, that $\langle \psi^{\text{var}}|\psi^{\text{var}}\rangle = \langle \Psi_n|\Psi_n\rangle = 1$. We then have

$$\begin{aligned}
{E_n}^{\text{var}} &= \langle \psi^{\text{var}}|H|\psi^{\text{var}}\rangle \\
&= \langle \Psi_n|H|\Psi_n\rangle + \langle \varepsilon|H|\Psi_n\rangle + \langle \Psi_n|H|\varepsilon\rangle + \langle \varepsilon|H|\varepsilon\rangle \\
&= E_n + E_n\{\langle \varepsilon|\Psi_n\rangle + \langle \Psi_n|\varepsilon\rangle\} + O(\varepsilon^2)\,.
\end{aligned}$$

Although it appears that there are errors of both the first and the second order in ε, that appearance is deceptive. From the normalization condition we have $\langle \psi^{\rm var}|\psi^{\rm var}\rangle = \langle \Psi_n|\Psi_n\rangle + \langle \varepsilon|\Psi_n\rangle + \langle \Psi_n|\varepsilon\rangle + \langle \varepsilon|\varepsilon\rangle$. Since $\langle \psi^{\rm var}|\psi^{\rm var}\rangle = \langle \Psi_n|\Psi_n\rangle = 1$, it follows that $\{\langle \varepsilon|\Psi_n\rangle + \langle \Psi_n|\varepsilon\rangle\} + \langle \varepsilon|\varepsilon\rangle = 0$. Thus the two first order quantities, $\langle \varepsilon|\Psi_n\rangle$ and $\langle \Psi_n|\varepsilon\rangle$, must cancel so that their sum is only of the second order in the magnitude of the error ε. Therefore we have shown that a *first order error in* $|\psi^{\rm var}\rangle$ *leads to only a second order error in* $E_n{}^{\rm var}$. If one's objective is to calculate eigenvalues, this is clearly an advantage. But, on the other hand, one must beware that accurate approximate eigenvalues do not necessarily indicate that the corresponding eigenvectors are similarly accurate.

Example (1): The hydrogen atom ground state

It is useful to test the variational method on an exactly solvable problem. The Hamiltonian for the relative motion of the electron and proton in a hydrogen atom is $H = \mathbf{P}\cdot\mathbf{P}/2\mu - e^2/r$, with μ being the reduced mass. We choose the trial function to be $\psi(r) = e^{-r/a}$, where a is an adjustable parameter. There is no need to normalize the trial function, and it is often more convenient not to do so. The best estimate of the ground state energy is the smallest value of the average energy in the hypothetical state described by ψ,

$$\langle H\rangle = \frac{\langle \psi|H|\psi\rangle}{\langle \psi|\psi\rangle} = \frac{K+P}{N}\,, \tag{10.117}$$

where $K = \langle \psi|\mathbf{P}\cdot\mathbf{P}|\psi\rangle/2\mu$ is the kinetic energy term, $P = -\langle\psi|e^2/r|\psi\rangle$ is the potential energy term, and $N = \langle \psi|\psi\rangle$ is the normalization factor. The value of the normalization factor is

$$N = \int |\psi(r)|^2 d^3r = 4\pi \int_0^\infty e^{-2r/a} r^2 dr = \pi a^3\,.$$

To calculate the kinetic energy, it is often better to evaluate $(\langle \psi|\mathbf{P})\cdot(\mathbf{P}|\psi\rangle)$, rather than $\langle \psi|\nabla^2|\psi\rangle$. Not only is this simpler, requiring only a first derivative, but it is less prone to error. If the trial function happens to have a discontinuity in its derivative (as our trial function does at $r = 0$), then the operator ∇^2 may generate delta function contributions at the discontinuity, which if overlooked will lead to erroneous results. The value of our kinetic energy term is

$$K = \frac{1}{2\mu}(\langle\psi|\mathbf{P})\cdot(\mathbf{P}|\psi\rangle) = \frac{\hbar^2}{2\mu}\int\left|\frac{\partial\psi}{\partial r}\right|^2 d^3r$$
$$= \frac{\hbar^2}{2\mu}\frac{4\pi}{a^2}\int_0^\infty e^{-2r/a}\,r^2\,dr = \frac{\hbar^2\pi a}{2\mu}\,.$$

The value of the potential energy term is

$$P = -\int|\psi|^2\frac{e^2}{r}d^3r$$
$$= -e^2 4\pi\int_0^\infty e^{-2r/a}r\,dr = -\pi e^2 a^2\,.$$

Thus we obtain

$$\langle H\rangle = \frac{\hbar^2}{2\mu a^2} - \frac{e^2}{a}\,. \tag{10.118}$$

The minimum of this expression is determined by the condition $\partial\langle H\rangle/\partial a = 0$, which is satisfied for $a = \hbar^2/\mu e^2$ and corresponds to the energy $\langle H\rangle_{\min} = -\mu e^4/2\hbar^2$. This is the exact value of the ground state energy of the hydrogen atom (10.27). It is unusual for the variational method to yield an exact eigenvalue. In this case it happened because the true ground state function (10.32) happens to be a special case of the trial function, $\psi(r) = e^{-r/a}$, for a particular value of a.

Messiah (1966, Ch. 18) has treated some other trial functions which illustrate the effect on the variational energy of certain errors in the trial functions. Some of his results are summarized in the table below, along with the exact results of the above example. The parameter C in the trial functions is to be chosen so that $\langle\psi|\psi\rangle = 1$. The parameter a is to be varied so as to minimize the energy.

Variational calculations of the hydrogen atom ground state

$\psi(r)$	$C\,e^{-r/a}$	$C\,(r^2+a^2)^{-1}$	$C\,re^{-r/a}$
$\langle H\rangle_{\min}/\lvert E_{100}\rvert$	-1	-0.81	-0.75
$1-\lvert\langle\psi\,\lvert\,\Psi_{100}\rangle\rvert^2$	0	0.21	0.05

The energies in the second row, evaluated for the optimum value of a, are expressed in units of $|E_{100}| = \mu e^4/2\hbar^2$. The last row contains a measure of the overall error in the approximate eigenvector. The first trial function is our example above, which yields the exact ground state. The second trial function decays much too slowly at large r, and is a rather poor overall approximation to the ground state. The third trial function has the correct exponential decay at large r, but is qualitatively incorrect near $r = 0$. However, its overall measure of error in the last row is only 5%. Nevertheless the second trial function, with a 21% overall error, yields a better approximation to the energy than does the apparently more accurate third function. This illustrates the fact that a better approximate energy is no guarantee of a better fit to the state function. In this case the anomaly occurs because the dominant contributions to the potential energy come from small distances, and hence in order to get a good approximate energy it is more important for the state function to be accurate at small distances than at large distances. Although these examples are rather crude, it is more generally true that variational calculations of atomic state functions tend to be least accurate at large distances.

Although the variational theorem (10.113) applies to the lowest eigenvalue, it is possible to generalize it to calculate *low-lying excited states.* In proving that theorem, we formally expressed the trial function as a linear combination of eigenvectors of H, so that $\langle\psi|H|\psi\rangle = \sum_n E_n|\langle\psi|\Psi_n\rangle|^2$. Suppose that we want to calculate the excited state eigenvalue E_m. If we can constrain the trial function $|\psi\rangle$ to satisfy $\langle\psi|\Psi_{n'}\rangle = 0$ for all n' such that $E_{n'} < E_m$, then it will follow that $\langle\psi|H|\psi\rangle \leq E_m \sum_n |\langle\psi|\Psi_n\rangle|^2 = E_m\langle\psi|\psi\rangle$. Hence we can calculate E_m by minimizing $\langle H\rangle \equiv \langle\psi|H|\psi\rangle/\langle\psi|\psi\rangle$ *subject to the constraint that* $|\psi\rangle$ *be orthogonal to all state functions at energies lower than* E_m. This is easy to do if the constraint can be ensured by symmetry. For a central potential one can calculate the lowest energy level for each orbital angular momentum quantum number ℓ, with no more difficulty than is required to calculate the ground state energy. One simply chooses a trial function whose angular dependence is proportional to $Y_\ell{}^m(\theta,\phi)$. If the upper and lower states have the same symmetry, as do the $1s$ and $2s$ atomic states, the orthogonality constraint is not so easy to impose, but the calculation may still be feasible.

As an application of this generalized variational theorem, we prove a theorem on the ordering of energy levels.

Theorem. For any central potential one must have

$$E_\ell{}^{\min} < E_{\ell+1}{}^{\min}, \tag{10.119}$$

where $E_\ell{}^{\min}$ denotes the lowest energy eigenvalue corresponding to orbital angular momentum ℓ.

Proof. Substitute $\Psi(\mathbf{x}) = Y_\ell{}^m(\theta,\phi)\, u_\ell(r)/r$ into the eigenvalue equation $-(\hbar^2/2\mu)\nabla^2\Psi + W(r)\Psi = E\Psi$, as was done in Sec. 10.1, and so obtain another eigenvalue equation,

$$K_\ell\, u_\ell(r) = E\, u_\ell(r)\,, \tag{10.120}$$

where

$$K_\ell = \frac{\hbar^2}{2\mu}\left\{\frac{-d^2}{dr^2} + \frac{\ell(\ell+1)}{r^2}\right\} + W(r)\,. \tag{10.121}$$

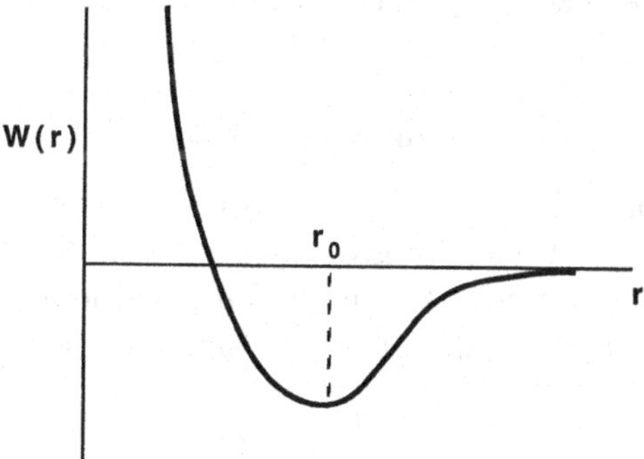

Fig. 10.6 A typical interatomic potential.

Now, at first sight, the theorem (10.119) may seem unsurprising, since the angular momentum term, $\ell(\ell+1)/r^2$, is positive and increases with ℓ. But the situation is really more complicated, since a change in ℓ will change the whole balance between kinetic and potential energy. Consider a central potential of the form shown in Fig. 10.6, which has a strongly repulsive core at short distances and an attractive potential well near $r = r_0$. (The potentials that bind atoms into molecules are of this form.) Near the origin $u_\ell(r)/r$ is proportional to r^ℓ. Thus a particle in an $\ell = 0$ state can penetrate into the region of positive potential energy near the origin, whereas a particle in a state of $\ell > 0$ will tend to be excluded from that region. It seems plausible that the lowest energy would be obtained for a state in which the particle was more-or-less confined in a circular orbit of radius $r = r_0$. This would necessarily correspond to $\ell \neq 0$. The theorem (10.119) proves that this plausible scenario cannot occur.

Continuing with the proof, we apply the variational theorem to (10.121). Using the boundary conditions $u_\ell(0) = u_\ell(\infty) = 0$, it can be shown that $K_\ell = K_\ell{}^\dagger$, and hence the variational theorem applies to (10.121), as well as to the original equation. Let $u_{\ell+1}(r)$ be the true eigenfunction of the operator $K_{\ell+1}$

corresponding to the lowest eigenvalue, ${E_{\ell+1}}^{\min}$. Choosing the normalization $\int_0^\infty |u_{\ell+1}(r)|^2\, dr = 1$, we may write

$$
\begin{aligned}
{E_{\ell+1}}^{\min} = \int_0^\infty u_{\ell+1}(r)\, K_{\ell+1} u_{\ell+1}(r)\, dr &= \int_0^\infty u_{\ell+1}(r)\, K_\ell u_{\ell+1}(r)\, dr \\
&+ \int_0^\infty u_{\ell+1}(r)\, [K_{\ell+1} - K_\ell] u_{\ell+1}(r)\, dr\,.
\end{aligned}
$$

According to the variational theorem, the first term is an upper bound to ${E_\ell}^{\min}$. The second term is equal to $\int_0^\infty |u_{\ell+1}(r)|^2[(\ell+1)(\ell+2)-\ell(\ell+1)]r^{-2}dr$, which is positive. Therefore we conclude that ${E_{\ell+1}}^{\min} > {E_\ell}^{\min}$, which is the theorem (10.119).

Upper and lower bounds on eigenvalues

The variational theorem (10.113) gives a convenient upper bound for the lowest eigenvalue, but does not give any lower bound. It is possible, without a great deal more labor, to obtain both upper and lower bounds to any eigenvalue. To solve, approximately, the eigenvalue equation

$$H|\Psi_k\rangle = \lambda_k|\Psi_k\rangle\,, \tag{10.122}$$

we use a trial function $|\psi\rangle$. It is convenient for this analysis to normalize this function, $\langle\psi|\psi\rangle = 1$, so our approximation to the eigenvalue is

$$\Lambda = \langle\psi|H|\psi\rangle\,. \tag{10.123}$$

To estimate the accuracy of this value, we define an *error vector*,

$$|R\rangle = (H - \Lambda)|\psi\rangle\,, \tag{10.124}$$

which would clearly be zero if the trial vector $|\psi\rangle$ were a true eigenvector. From the spectral representation of H we deduce that

$$
\begin{aligned}
\langle R|R\rangle &= \langle\psi|(H-\Lambda)^2|\psi\rangle \\
&= \sum_j |\langle\psi|\Psi_j\rangle|^2\, (\lambda_j - \Lambda)^2\,.
\end{aligned}
$$

Let λ_k be the closest eigenvalue to Λ. Then we may write

$$\langle R|R\rangle \ge \sum_j |\langle\psi|\Psi_j\rangle|^2\, (\lambda_k - \Lambda)^2 = (\lambda_k - \Lambda)^2\,.$$

Hence we deduce the upper and lower bounds,

$$\Lambda - \Delta \le \lambda_k \le \Lambda + \Delta\,, \tag{10.125}$$

where $\Delta = \sqrt{\langle R|R\rangle}$. The assumption that λ_k is the closest eigenvalue to Λ means that this method can be applied only if we are already sure that our approximate value Λ is closer to the desired eigenvalue λ_k than to any other eigenvalue. If this is not true, then the uncertainty in Λ is so large that there is really no point in calculating upper and lower bounds. It is a feature of all such methods that they cannot be applied blindly, but rather they require certain minimally accurate information in order to be used.

More precise bounds than (10.125) can be deduced, without significantly greater computational effort, by a method due to Kato (1949). It must be assumed, for this method, that we know two numbers, α and β, such that

$$\lambda_j \le \alpha < \beta \le \lambda_{j+1}\,. \tag{10.126}$$

That is to say, we know enough about the eigenvalue spectrum to be sure that there are no eigenvalues between α and β. This is a reasonable requirement, for if the uncertainty in our estimated eigenvalues is greater than the spacing between eigenvalues, then our calculation is too crude to be of any value.

To deduce the bounds we make use of the error vector $|R\rangle$ (10.124), and two auxiliary vectors: $|A\rangle = (H-\alpha)|\psi\rangle$ and $|B\rangle = (H-\beta)|\psi\rangle$. Now we have

$$\begin{aligned}
\langle A|B\rangle &= \langle\psi|(H-\alpha)(H-\beta)|\psi\rangle \\
&= \sum_j \langle\psi|(H-\alpha)|\Psi_j\rangle\langle\Psi_j|(H-\beta)|\psi\rangle \\
&= \sum_j |\langle\psi|\Psi_j\rangle|^2\,(\lambda_j-\alpha)(\lambda_j-\beta)\,.
\end{aligned}$$

Under the hypothesis (10.126), that there is no eigenvalue between α and β, it follows that $\lambda_j - \alpha$ and $\lambda_j - \beta$ have the same sign, and hence

$$\langle A|B\rangle \ge 0\,. \tag{10.127}$$

This inequality can be made more useful by writing

$$\begin{aligned}
\langle A|B\rangle &= \langle\psi|\{(H-\Lambda)-(\alpha-\Lambda)\}\{(H-\Lambda)-(\beta-\Lambda)\}|\psi\rangle \\
&= \langle\psi|(H-\Lambda)^2|\psi\rangle - \langle\psi|(H-\Lambda)|\psi\rangle\,[(\alpha-\Lambda)+(\beta-\Lambda)] \\
&\quad + (\alpha-\Lambda)(\beta-\Lambda) \\
&= \langle R|R\rangle + (\alpha-\Lambda)(\beta-\Lambda) \ge 0\,.
\end{aligned} \tag{10.128}$$

The final inequality comes from (10.127).

Our objective is to calculate the eigenvalue λ_k, so we choose the trial vector $|\psi\rangle$ to approximate $|\Psi_k\rangle$ as best we can, and our estimate will be $\lambda_k \approx \Lambda = \langle\psi|H|\psi\rangle$. To obtain a *lower bound* we set $j = k$ in (10.126) and put $\alpha = \lambda_k$. From (10.128) we then obtain

$$(\lambda_k - \Lambda)\,(\beta - \Lambda) \geq -\langle R|R\rangle\,,$$

and thus if $\beta > \Lambda$ we have

$$\lambda_k \geq \Lambda - \frac{\langle R|R\rangle}{\beta - \Lambda}\,, \quad \text{with } \Lambda < \beta \leq \lambda_{k+1}\,. \tag{10.129}$$

To obtain an *upper bound* we set $j = k - 1$ in (10.126) and put $\beta = \lambda_k$. From (10.128) we obtain

$$(\Lambda - \alpha)(\lambda_k - \Lambda) \leq \langle R|R\rangle\,,$$

and thus if $\alpha < \Lambda$ we have

$$\lambda_k \leq \Lambda + \frac{\langle R|R\rangle}{\Lambda - \alpha}\,, \quad \text{with } \lambda_{k-1} \leq \alpha < \Lambda\,. \tag{10.130}$$

To make these bounds on λ_k as strong as possible, we should choose α as close as possible to the next lower eigenvalue, and β as close as possible to the next higher eigenvalue. The order of magnitude of the error bounds is determined by $\langle R|R\rangle$, and in practice the uncertainties in the choice of α and β are not critical. If λ_k is the lowest eigenvalue then we may let α go to $-\infty$, in which case we recover the upper bound (10.113), which is $\lambda_0 \leq \Lambda$.

Example (2): The screened Coulomb potential

The calculation of the ground state energy of an electron bound in the screened Coulomb potential,

$$W(r) = e^{-\alpha r}\,\frac{e^2}{r}\,,$$

provides a nontrivial test of these methods. We choose the normalized trial function to be

$$\psi(r) = \left(\frac{b^3}{\pi}\right)^{1/2} e^{-br}\,, \tag{10.131}$$

which has the form of the hydrogen atom ground state function. Our answer will be exact in the limit $\alpha = 0$, which is just the hydrogen atom, but the error will grow as α increases. Hence this example will illustrate both the strengths and the limitations of the method.

The average energy for the trial function $\psi(r)$ is

$$\Lambda = \langle\psi|H|\psi\rangle = \frac{\hbar^2 b^2}{2\mu} - \frac{4e^2 b^3}{(\alpha + 2b)^2}, \tag{10.132}$$

the two terms being the kinetic and potential energies, respectively. For each α, the optimum value of b is determined by minimizing Λ, setting $\partial\Lambda/\partial b = 0$. Our best estimate for the lowest energy will then be $E_1 \approx \Lambda$.

For computational purposes, it is convenient to *choose units* such that $\hbar = \mu = e = 1$. Then the unit of length is the Bohr radius, $a_0 = \hbar^2/\mu e^2$, and the lowest energy level of the hydrogen atom is $e^2/2a_0 = 0.5$.

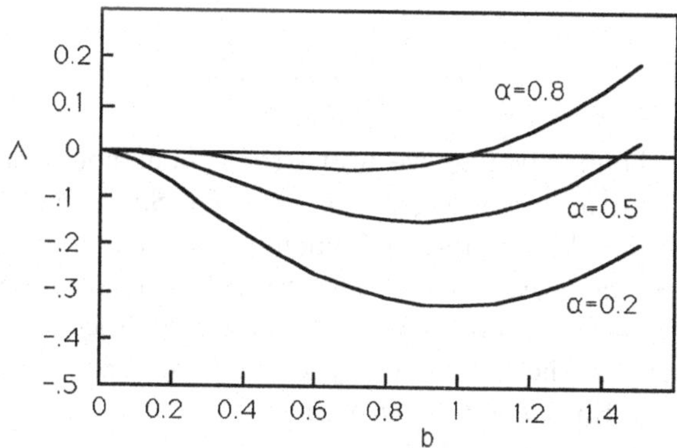

Fig. 10.7 Variational energy for the screened Coulomb potential.

Figure 10.7 shows $\Lambda(b)$ for several values of α. There is a negative minimum of $\Lambda(b)$ when α is in the range $0 \leq \alpha < 1$. As α increases from 0 to 1, the optimum value of b decreases from 1 to 0.5, and the energy increases from -0.5 to 0. Since Λ is an upper bound to the lowest eigenvalue, we can be sure that a bound state exists for all $\alpha < 1$. (More accurate computations yield a critical value of $\alpha_c \approx 1.2$, beyond which the screened Coulomb potential has no bound states.)

To determine lower bounds to the approximate ground state energy, we must evaluate the error vector, $|R\rangle = (H - \Lambda)|\psi\rangle$, (10.124). From the definition of Λ, it follows that

$$\langle R|R\rangle = \langle\psi|H^2|\psi\rangle - \Lambda^2. \tag{10.133}$$

This quantity is a measure of the error in our approximate eigenvector. The function $H\psi(r) = -\frac{1}{2}\nabla^2\psi(r) + W(r)\psi(r)$ can easily be calculated, and from it we obtain

$$\begin{aligned}\langle\psi|H^2|\psi\rangle &= \int |H\psi(r)|^2\, d^3r \\ &= b^3\left\{\frac{5b}{4} + \frac{4b^2}{(\alpha+2b)^2} - \frac{8b}{\alpha+2b} + \frac{2}{\alpha+\beta}\right\}. \qquad (10.134)\end{aligned}$$

We can now evaluate $\langle R|R\rangle$ at the optimum value of b, and compute lower bounds to our approximate ground state energy E_1, for which we already have the upper bound $E_1 \le \Lambda$.

The simplest lower bound is that given by (10.125), which is

$$E_1 \ge E_L = \Lambda - \sqrt{\langle R|R\rangle}\,. \qquad (10.135)$$

To use Kato's bound (10.129), we must estimate a number β that is close to but not higher than the second eigenvalue: $\beta \le E_2$. Since the difference between the screened and the unscreened Coulomb potentials is everywhere positive, it follows that the eigenvalues of the screened Coulomb potential will not be lower than the corresponding eigenvalues of the hydrogen atom. Therefore we shall estimate β as the second energy level of hydrogen, which is $\beta = -1/8$ in our units. Thus Kato's bound becomes

$$E_1 \ge E_K = \Lambda - \frac{\langle R|R\rangle}{-0.125 - \Lambda}, \qquad (10.136)$$

where the denominator must be positive for this expresssion to be valid.

The variational upper bound and these two lower bounds are shown in Fig. 10.8, where the result of a more accurate variational calculation by Lam and Varshni (1971) is also shown. The simple bound E_L is very conservative, seriously overestimating the magnitude of the error. For small α, where our approximation is accurate, Kato's bound E_K provides a good estimate of the error. This can be seen most clearly from the table of numerical values. However, it ceases to be useful for large values of α because the sign of the denominator in (10.136) changes. Even if we had a better estimate for $\beta \le E_2$, Kato's bound would not be very precise because of the large value of $\langle R|R\rangle$. None of our results are accurate for large α, for which a more complicated trial function is needed.

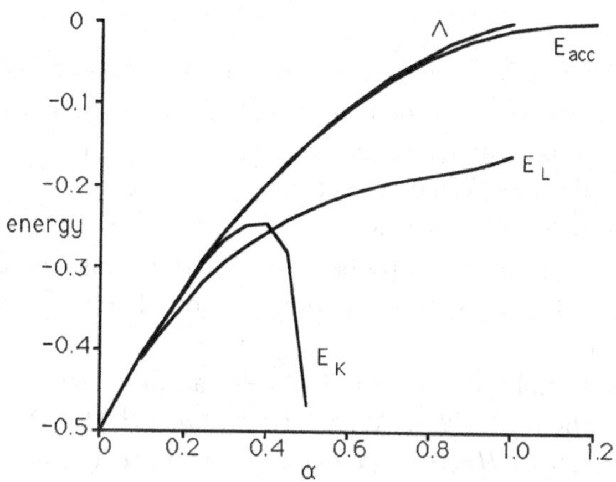

Fig. 10.8 Variational calculation for screened Coulomb potential.

Variational calculations for the screened Coulomb potential

α	b	Λ	$\langle R\,\vert\,R\rangle$	E_L	E_K	$E_{\rm acc}$
0.00	1.0	−0.5	0.0	−0.5	−0.5	−0.5
0.10	0.99333	−0.40705	0.00003	−0.41230	−0.40715	−0.40706
0.20	0.97568	−0.32673	0.00035	−0.34551	−0.32848	−0.32681
0.30	0.94922	−0.25733	0.00141	−0.29481	−0.26795	−0.25763
0.40	0.91502	−0.19758	0.00354	−0.25707	−0.24635	−0.19836
0.50	0.87349	−0.14651	0.00690	−0.22959	−0.46738	−0.14808
0.60	0.82444	−0.10335	0.01139	−0.21009		−0.10608
0.70	0.76698	−0.06750	0.01662	−0.19644		−0.07174
0.80	0.69903	−0.03847	0.02187	−0.18636		−0.04459
0.90	0.61564	−0.01597	0.02588	−0.17684		−0.02418
1.00	0.50000	0.0	0.02604	−0.16137		−0.01016
1.10						−0.00220
1.20						−0.00004

The columns are: screening parameter α; optimum value of b; upper bound to the ground state eigenvalue, Λ; error parameter, $\langle R|R\rangle$; simple lower bound, E_L; Kato's lower bound, E_K; accurate ground state eigenvalue from Lam and Varshni (1971), $E_{\rm acc}$.

Problems

10.1 The attractive square well potential is defined to be $W(r) = -V_0$ for $r < a$, $W(r) = 0$ for $r > a$. In Sec. 10.1 it was shown that in three dimensions there is a minimum value of $V_0 a^2$ below which there are no bound states. What are the corresponding situations in one dimension and in two dimensions?

10.2 For the square well potential in three dimensions, find the minimum value of $V_0 a^2$ needed to produce a bound state of angular momentum $\ell = 1$.

10.3 The Hamiltonian for the hydrogen atom is $H = P^2/2\mu - e^2/r$. Show that the *Runge–Lenz* vector, $\mathbf{K} = (2\mu e^2)^{-1}\,\{\mathbf{L}\times\mathbf{P} - \mathbf{P}\times\mathbf{L}\} + \mathbf{r}/r$, commutes with H. [It is the existence of this extra symmetry and the associated conserved quantity that is responsible for the peculiar degeneracy of the eigenvalues of H, with E_n being independent of ℓ. See Schiff (1968), pp. 236–239 for a full treatment of this topic.]

10.4 For the ground state of the hydrogen atom, calculate the probability that the electron and the proton are farther apart than would be permitted by classical mechanics at the same total energy.

10.5 Calculate explicitly the $n = 2$ (first excited state) functions of the hydrogen atom in parabolic coordinates and in spherical coordinates. Express the parabolic functions as linear combinations of the spherical functions.

10.6 Show that the average momentum vanishes, i.e. $\langle\mathbf{P}\rangle = 0$, in any bound state of the Hamiltonian $H = P^2/2M + W(\mathbf{x})$.

10.7 The following alleged solution to Problem 10.6 is given in a certain texbook.

Since $P_x = (iM/\hbar)\,[H, x]$, it follows that

$$\langle P_x\rangle = \tfrac{iM}{\hbar}(\langle\Psi|Hx|\Psi\rangle - \langle\Psi|x\,H|\Psi\rangle).$$

Using $H|\Psi\rangle = E|\Psi\rangle$ and $\langle\Psi|H = \langle\Psi|E$, we obtain $\langle P_x\rangle = 0$, and hence $\langle\mathbf{P}\rangle = 0$.

This argument, if valid, would establish that $\langle\mathbf{P}\rangle = 0$ for *all* stationary states, bound and unbound.

But the counterexample $W(\mathbf{x}) \equiv 0$, $\Psi(\mathbf{x}) = \exp(i\mathbf{k}\cdot\mathbf{x})$ proves that the argument must be unsound. But just where and why does the argument break down?

10.8 Prove the *virial theorem* for a particle bound in a potential W, $2\langle T\rangle = \langle\mathbf{x}\cdot\nabla W\rangle$, where $T = P^2/2M$ is the kinetic energy. Hence show that if

$W = W(r) \propto r^n$, one has the following relation between the average kinetic and potential energies: $2\langle T\rangle = n\langle W\rangle$.

10.9 Show that in one dimension the bound state energy spectrum must be nondegenerate.

10.10 A harmonic oscillator, which has the unperturbed Hamiltonian $H_0 = P^2/2M + \frac{1}{2}M\omega^2 Q^2$, is given the quadratic perturbation $H_1 = cQ^2$. Evaluate the perturbed energy eigenvalues to the second order in H_1, and compare the result with the exact values.

10.11 Use the variational method to prove that first order perturbation theory always gives an upper bound to the ground state energy of a system, no matter how large the perturbation may be.

10.12 The Hamiltonian for two interacting spins (both $s = \frac{1}{2}$) in a magnetic field B directed along the z axis is

$$H = B(a_1\, {\sigma_z}^{(1)} + a_2 {\sigma_z}^{(2)}) + K\boldsymbol{\sigma}^{(1)}\cdot\boldsymbol{\sigma}^{(2)}\,,$$

where a_1 and a_2 are the negatives of the magnetic moments (assumed to be unequal to avoid degeneracy), and K is the interaction strength.

(a) Use second order perturbation theory to calculate the energy eigenvalues, assuming that B is small.

(b) Use second order perturbation theory to calculate the energy eigenvalues, under the opposite assumption that K is small.

(c) Find the exact energy eigenvalues for this Hamiltonian, and verify the correctness of your answers in parts (a) and (b).

10.13 Use the variational method to obtain an approximate ground state energy for a particle bound in the one-dimensional potential: $W(x) = x$ for $x > 0$, $W(x) = +\infty$ for $x < 0$.

10.14 The three-fold degenerate energy level of the hydrogen atom, with the eigenvectors $|n = 2, \ell = 1, m = \pm 1, 0\rangle$, is subjected to a perturbation of the form $V = b(x^2 - y^2)$. Use degenerate perturbation theory to determine the zero order eigenvectors and the splitting of the energy levels to the first order in b. (You need not evaluate the radial integrals that occur in the matrix elements of V, but you should determine which are zero, which are nonzero, and which nonzero matrix elements are equal.)

10.15 Calculate the quadratic Zeeman effect for the ground state of atomic hydrogen by treating the perturbation of a uniform magnetic field to the second order. By writing the second order energy as $E^{(2)} = -\frac{1}{2}\,\chi B^2$ we see that this yields the *diamagnetic susceptibility* χ.

10.16 Use the variational method to calculate the energy and eigenfunction for the *second* excited state ($n = 2$) of a one-dimensional harmonic oscillator. (Remember that your function must be orthogonal to the eigen functions corresponding to $n = 1$ and $n = 0$.)

10.17 Calculate the shift in atomic energy levels due to the finite size of the nucleus, treating the nucleus as a uniformly charged sphere.

10.18 Use the variational method to calculate the ground state energy of a particle bound in the one-dimensional attractive potential $W(x) = -c\,\delta(x)$ with $c > 0$.

10.19 Apply the variational method to the one-dimensional truncated Coulomb potential, $W(x) = -1/(a + |x|)$. (There are many possible trial functions that could reasonably be used. Your answer should be accurate enough to prove that the lowest energy eigenvalue approaches $-\infty$ in the limit $a \to 0$.)

Chapter 11

Charged Particle in a Magnetic Field

The theory of the motion of a charged particle in a magnetic field presents several difficult and unintuitive features. The derivation of the quantum theory does not require the classical theory; nevertheless it is useful to first review the classical theory in order to show that some of these unintuitive features are not peculiar to the quantum theory, but rather that they are characteristic of motion in a magnetic field.

11.1 *Classical Theory*

The electric and magnetic fields, **E** and **B**, enter the Lagrangian and Hamiltonian forms of mechanics through the vector and scalar potentials, **A** and ϕ:

$$\mathbf{E} = -\nabla\phi - \frac{1}{c}\frac{\partial \mathbf{A}}{\partial t} \tag{11.1a}$$

$$\mathbf{B} = \nabla \times \mathbf{A}\,. \tag{11.1b}$$

(The speed of light c appears only because of a conventional choice of units.) The potentials are not unique. The fields **E** and **B** are unaffected by the replacement

$$\mathbf{A} \to \mathbf{A}' = \mathbf{A} + \nabla\chi\,, \quad \phi \to \phi' = \phi - \frac{1}{c}\frac{\partial\chi}{\partial t}\,, \tag{11.2}$$

where $\chi = \chi(\mathbf{x}, t)$ is an arbitrary scalar function. This change of the potentials, called a *gauge transformation*, has no effect upon any physical result. It thus appears, in classical mechanics, that the potentials are only a mathematical construct having no direct physical significance.

The *Lagrangian* for a particle of mass M and charge q in an arbitrary electromagnetic field is

$$\mathcal{L}(\mathbf{x},\, \mathbf{v},\, t) = \frac{Mv^2}{2} - q\,\phi\,(\mathbf{x}, t) + \frac{q}{c}\,\mathbf{v}\cdot\mathbf{A}\,(\mathbf{x}, t)\,, \tag{11.3}$$

where $\mathbf{x}$ and $\mathbf{v} = d\mathbf{x}/dt$ are the position and velocity of the particle. The significance of (11.3) lies in the fact that Lagrange's equation of motion,

$$\frac{d}{dt}\left[\frac{\partial\mathcal{L}}{\partial v_\alpha}\right] - \frac{\partial\mathcal{L}}{\partial x_\alpha} = 0 \quad (\alpha = 1,2,3)\,, \tag{11.4}$$

leads, after an elementary calculation, to the correct Newtonian equation of motion, $M d\mathbf{v}/dt = q(\mathbf{E} + \mathbf{v}\times\mathbf{B}/c)$.

From the Lagrangian, we can define the *canonical momentum*, $p_\alpha = \partial\mathcal{L}/\partial v_\alpha$. For a particle in a magnetic field it has the form

$$\mathbf{p} = M\mathbf{v} + \frac{q}{c}\mathbf{A}\,. \tag{11.5}$$

Since $\mathbf{p}$, like $\mathbf{A}$, is changed by a gauge transformation, it too lacks a direct physical significance. However, it is of considerable mathematical importance. Lagrange's equation (11.4) can be written as $dp_\alpha/dt = \partial\mathcal{L}/\partial x_\alpha$. Hence it follows that if $\mathcal{L}$ is independent of x_α (or in other words, if $\mathcal{L}$ is invariant under a coordinate displacement of the form $x_\alpha \to x_\alpha + a_\alpha$), then it is the canonical momentum p_α that is conserved, and not the more intuitive quantity Mv_α.

The *Hamiltonian* for a particle in an electromagnetic field is

$$\begin{aligned} H &= \mathbf{v}\cdot\mathbf{p} - \mathcal{L} \\ &= \frac{Mv^2}{2} + q\phi(\mathbf{x},t)\,, \end{aligned} \tag{11.6}$$

with the terms involving $\mathbf{A}$ canceling out of the final expression. Since the magnetic force on a moving particle is perpendicular to the velocity of the particle, the magnetic field does no work and hence does not enter into the expression for the total energy H. How then can the Hamiltonian generate the motion of the particle, which does depend upon the magnetic field, when the magnetic field apparently does not enter into (11.6)? The answer lies in the fact that Hamiltonian is to be regarded as a function of position and momentum, not of position and velocity. Hence it is more proper to rewrite (11.6) using (11.5) as

$$H = \frac{1}{2M}\left(\mathbf{p} - \frac{q}{c}\mathbf{A}\right)^2 + q\phi\,. \tag{11.7}$$

Hamilton's equations, $d\mathbf{p}/dt = -\partial H/\partial\mathbf{x}$ and $d\mathbf{x}/dt = \partial H/\partial\mathbf{p}$, then yield the familiar Newtonian equation of motion.

Two important results from this classical theory, which also hold in the quantum theory, are the relation (11.5) between velocity and canonical momentum, and the fact that the apparently more complicated Hamiltonian (11.7) is really just equal to the sum of kinetic plus potential energy. One should also remember that in the presence of a magnetic field the momentum $\mathbf{p}$ is not an observable quantity, but nevertheless it plays an important mathematical role.

11.2 *Quantum Theory*

It was shown in Sec. 3.4 that the requirement of Galilei invariance restricts the possible external interactions of a particle to a scalar potential and a vector potential. Since the coupling of the particle to the electromagnetic field is proportional to the particle's charge q, the generic form of the Hamiltonian (3.60) should be rewritten as

$$H = \left(\mathbf{P} - \frac{q}{c}\mathbf{A}\right)^2 /2M + q\,\phi \tag{11.8}$$

in this case. (The factor $1/c$ is present only because of a conventional choice of units.) Here $\mathbf{P}$ is the momentum operator of the particle. The vector and scalar potentials, $\mathbf{A} = \mathbf{A}(\mathbf{Q},t)$ and $\phi = \phi(\mathbf{Q},t)$, are operators because they are functions of the position operator $\mathbf{Q}$. Their dependence (if any) on t corresponds to the intrinsic time dependence of the fields. (Here we are using the Schrödinger picture.)

The *velocity operator*, defined in units of $\hbar$ by (3.39), is

$$\begin{aligned} V_\alpha &= \frac{i}{\hbar}[H, Q_\alpha] = \left(\frac{i}{2M\hbar}\right)\left[\left\{P_\alpha - \frac{q}{c}A_\alpha\right\}^2, Q_\alpha\right] \\ &= \frac{1}{M}\left(P_\alpha - \frac{q}{c}A_\alpha\right), \quad (\alpha = x, y, z). \end{aligned} \tag{11.9}$$

As was the case for the classical theory, the momentum $\mathbf{P}$ is mathematically simpler than the velocity $\mathbf{V}$. But the velocity has a more direct physical significance, so it is worth examining its mathematical properties.

The commutator of the position and velocity operators is

$$[Q_\alpha, V_\beta] = i\,\frac{\hbar}{M}\,\delta_{\alpha\beta}\,. \tag{11.10}$$

Apart from the factor of M, this is the same as the commutator for position and momentum. However the commutator of the velocity components with each other presents some novelty:

$$\begin{aligned} [V_x, V_y] &= \frac{[P_x, P_y]}{M^2} + \left(\frac{q}{Mc}\right)^2 [A_x, A_y] \\ &\quad - \frac{q}{M^2c}\{[A_x, P_y] + [P_x, A_y]\}\,. \end{aligned}$$

The first and second terms vanish. The remaining terms can be evaluated most easily by adopting the *coordinate representation* (Ch. 4), in which a vector is represented by a function of the coordinates, $\psi(\mathbf{x})$, and the momentum operator becomes $P_\alpha = -i\hbar\partial/\partial x_\alpha$. Thus we have

$$\begin{aligned}
[V_x, V_y]\psi &= i\,\frac{\hbar q}{M^2c}\left\{A_x\,\frac{\partial}{\partial y} - \frac{\partial}{\partial y}\,A_x + \frac{\partial}{\partial x}\,A_y - A_y\,\frac{\partial}{\partial x}\right\}\psi \\
&= i\,\frac{\hbar q}{M^2c}\left\{-\left(\frac{\partial A_x}{\partial y}\right) + \left(\frac{\partial A_y}{\partial x}\right)\right\}\psi \\
&= i\,\frac{\hbar q}{M^2c}\,(\boldsymbol{\nabla}\times\mathbf{A})_z\psi = i\,\frac{\hbar q}{M^2c}\,B_z\psi\,.
\end{aligned}$$

Since this result holds for an arbitrary function ψ, it may be written as an operator equation, valid in any representation: $[V_x, V_y] = i\,(\hbar q/M^2c)\,B_z$. This may clearly be generalized to

$$[V_\alpha, V_\beta] = i\,\frac{\hbar q}{M^2c}\,\varepsilon_{\alpha\beta\gamma}\,B_\gamma\,, \tag{11.11}$$

where $\varepsilon_{\alpha\beta\gamma}$ is the antisymmetric tensor [introduced in Eq. (3.22)]. The commutator of two components of velocity is proportional to the magnetic field in the remaining direction.

Heisenberg equation of motion

The velocity operator (11.9) is equal to the rate of change of the position operator, calculated from the Heisenberg equation of motion (3.73). Similarly, an acceleration operator can be calculated as the rate of change of the velocity operator. The product of mass times acceleration may naturally be regarded as the force operator,

$$M\frac{dV_\alpha}{dt} = i\,\frac{M}{\hbar}\,[H, V_\alpha] + M\frac{\partial V_\alpha}{\partial t}\,. \tag{11.12}$$

(For simplicity of notation we shall not distinguish between the Schrödinger and Heisenberg operators. This equation should therefore be regarded as referring to the instant of time $t = t_0$ when the two pictures coincide.) To evaluate the commutator in (11.12), it is useful to rewrite the Hamiltonian (11.8) as $H = \frac{1}{2}M\mathbf{V}\cdot\mathbf{V} + q\phi$. Thus we have

$$[H, V_\alpha] = \frac{1}{2} M \sum_\beta [V_\beta^2, V_\alpha] + q\,[\phi, V_\alpha]$$

$$= \frac{1}{2} M \sum_\beta \{V_\beta\,[V_\beta, V_\alpha] + [V_\beta, V_\alpha]\,V_\beta\} + \frac{q}{M}\,[\phi, P_\alpha]$$

$$= \frac{1}{2} i \left(\frac{\hbar q}{Mc}\right) \sum_{\beta,\gamma} (V_\beta \varepsilon_{\beta\alpha\gamma} B_\gamma + \varepsilon_{\beta\alpha\gamma} B_\gamma\, V_\beta) + \frac{q}{M}\, i\hbar\,(\boldsymbol{\nabla}\phi)_\alpha$$

$$= \frac{1}{2} i \left(\frac{\hbar q}{Mc}\right) \sum_{\beta,\gamma} \varepsilon_{\alpha\beta\gamma}(-V_\beta B_\gamma + B_\beta V_\gamma) + i\left(\frac{\hbar q}{M}\right)(\boldsymbol{\nabla}\phi)_\alpha\,.$$

The last term of (11.12) has the value $M\partial V_\alpha/\partial t = -(q/c)\partial A_\alpha/\partial t$. Combining these results and writing (11.12) in vector form, we have

$$M\frac{d\mathbf{V}}{dt} = \frac{1}{2}\left(\frac{q}{c}\right)(\mathbf{V}\times\mathbf{B} - \mathbf{B}\times\mathbf{V}) + q\mathbf{E}\,. \tag{11.13}$$

This is just the operator for the Lorentz force, the only complication being that the magnetic field operator **B** (and also the electric field operator **E**) is a function of **Q**, and so **B** does not commute with **V**.

Coordinate representation

The Hamiltonian (11.8) may be written as

$$H = \frac{\mathbf{P}\cdot\mathbf{P} - (q/c)(\mathbf{P}\cdot\mathbf{A} + \mathbf{A}\cdot\mathbf{P}) + (q/c)^2\mathbf{A}\cdot\mathbf{A}}{2M} + q\phi\,.$$

The difference between $\mathbf{P}\cdot\mathbf{A}$ and $\mathbf{A}\cdot\mathbf{P}$ can be determined by the action of these operators on an arbitrary function $\psi(\mathbf{x})$:

$$\mathbf{P}\cdot\mathbf{A}\,\psi = -i\hbar\,\boldsymbol{\nabla}\cdot\mathbf{A}\psi = -i\hbar\,\mathbf{A}\cdot\boldsymbol{\nabla}\psi - i\hbar\psi\,(\boldsymbol{\nabla}\cdot\mathbf{A})\,.$$

Since ψ is an arbitrary function, this may be written as an operator relation,

$$\mathbf{P}\cdot\mathbf{A} - \mathbf{A}\cdot\mathbf{P} = -\,i\hbar\,\mathrm{div}\mathbf{A}\,, \tag{11.14}$$

which holds in any representation. (It is always possible to choose the vector potential so that $\mathrm{div}\mathbf{A} = 0$, and this is often done.) The general form of the Hamiltonian in coordinate representation is

$$H = -\frac{\hbar^2}{2M}\nabla^2 + \frac{i\hbar q}{Mc}\,\mathbf{A}\cdot\boldsymbol{\nabla} + \frac{i\hbar q}{2Mc}\,(\mathrm{div}\mathbf{A}) + \frac{q^2}{2Mc^2}A^2 + q\phi\,. \tag{11.15}$$

In interpreting this expression, it should be remembered that, in spite of the apparent complexity, the sum of the first four terms is just the kinetic energy, $\frac{1}{2}MV^2$. Sometimes the first term is described as the kinetic energy, and the next three terms are described as paramagnetic and diamagnetic corrections. That is not correct, and indeed the individual terms have no distinct physical significance because they are not separately invariant under gauge transformations. For many purposes, it is preferable not to expand the quadratic term of the Hamiltonian, but rather to write it more compactly as

$$H = \frac{1}{2M}\left(\frac{\hbar}{i}\nabla - \frac{q}{c}\mathbf{A}\right)^2 + q\phi\,. \tag{11.16}$$

Gauge transformations

The electric and magnetic fields are not changed by the transformation (11.2) of the potentials. On the basis of our previous experience, we may anticipate that there will be a corresponding transformation of the state function that will, at most, transform it into a physically equivalent state function. Since the squared modulus, $|\Psi(\mathbf{x},t)|^2$, is significant as a probability density, this implies that only the phase of the complex function $\Psi(\mathbf{x},t)$ can be affected by the transformation. (This is similar to the Galilei transformations, which were studied in Sec. 4.3.)

The Schrödinger equation,

$$\frac{1}{2M}\left(\frac{\hbar}{i}\nabla - \frac{q}{c}\mathbf{A}\right)^2 \Psi + q\,\phi\,\Psi = i\hbar\frac{\partial}{\partial t}\Psi\,, \tag{11.17}$$

is unchanged by the combined substitutions:

$$\mathbf{A} \to \mathbf{A}' = \mathbf{A} + \nabla\chi\,, \tag{11.18a}$$

$$\phi \to \phi' = \phi - \frac{1}{c}\frac{\partial\chi}{\partial t}\,, \tag{11.18b}$$

$$\Psi \to \Psi' = \Psi e^{i(q/\hbar c)\chi}\,, \tag{11.18c}$$

where $\chi = \chi(\mathbf{x},t)$ is an arbitrary scalar function. It is this set of transformations, rather than (11.2), which is called a *gauge transformation* in quantum mechanics. That the transformed equation

$$\frac{1}{2M}\left(\frac{\hbar}{i}\nabla - \frac{q}{c}\mathbf{A}'\right)^2 \Psi' + q\,\phi'\,\Psi' = i\hbar\frac{\partial}{\partial t}\Psi' \tag{11.17$'$}$$

is equivalent to the original (11.17) can be demonstrated in two steps. First, on the right hand side of (11.17′) the time derivative of the phase factor from (11.18c) exactly compensates for the extra term introduced on the left hand side by the scalar potential (11.18b). Second, it is easily verified that

$$\left(\frac{\hbar}{i}\nabla - \frac{q}{c}\mathbf{A}'\right) e^{i(q/\hbar c)\chi}\Psi = e^{i(q/\hbar c)\chi}\left(\frac{\hbar}{i}\nabla - \frac{q}{c}\mathbf{A}\right)\Psi \tag{11.19}$$

since the gradient of the phase factor on the left hand side compensates for the extra term in the vector potential introduced by (11.18a). Hence it follows that (11.17′) differs from (11.17) only by an additional phase factor on both sides of the equation, and so the original and the transformed equations are equivalent.

From (11.19) it follows that the average velocity,

$$\langle\Psi|\mathbf{V}|\Psi\rangle \equiv \left\langle\Psi\left|\left(\frac{\mathbf{P}}{M} - \frac{q\mathbf{A}}{Mc}\right)\right|\Psi\right\rangle,$$

is invariant under gauge transformations, whereas the average momentum $\langle\Psi|\mathbf{P}|\Psi\rangle$ is not. For this reason, the physical significance of a result will usually be more apparent if it is expressed in terms of the velocity, rather than in terms of the momentum.

We can also show that the eigenvalue spectrum of a component of velocity is gauge-invariant, even though the form of the velocity operator, $(\mathbf{P}/M - q\mathbf{A}/Mc)$, depends on the particular choice of vector potential. Suppose that $\psi(\mathbf{x})$ is an eigenvector of V_z,

$$\left(\frac{P_z}{M} - \frac{qA_z}{Mc}\right)\psi(\mathbf{x}) = v_z\psi(\mathbf{x}). \tag{11.20}$$

Consider now another equivalent vector potential, $\mathbf{A}' = \mathbf{A} + \nabla\chi$. From (11.19) and (11.20) we obtain

$$\begin{aligned}\left(\frac{P_z}{M} - \frac{qA'_z}{Mc}\right) e^{i(q/\hbar c)\chi}\psi(\mathbf{x}) &= e^{i(q/\hbar c)\chi}\left(\frac{P_z}{M} - \frac{qA_z}{Mc}\right)\psi(\mathbf{x}) \\ &= v_z e^{i(q/\hbar c)\chi}\psi(\mathbf{x}).\end{aligned} \tag{11.21}$$

Thus the operators $P_z - qA_z/c$ and $P_z - qA'_z/c$ must have the same eigenvalue spectrum.

Probability current density

The probability current density $\mathbf{J}(\mathbf{x},t)$ was introduced in Sec. 4.4 through the continuity equation, $\text{div}\mathbf{J} + (\partial/\partial t)|\Psi|^2 = 0$, which expresses the conservation of probability. In the presence of a nonvanishing vector potential, the expressions (4.22) and (4.24) are no longer equal, and it is the latter that is correct:

$$\mathbf{J}(\mathbf{x},t) = \mathrm{Re}\left\{\Psi^*(\mathbf{x},t)\left[\frac{\mathbf{P}}{M} - \frac{q\mathbf{A}(\mathbf{x},t)}{Mc}\right]\Psi(\mathbf{x},t)\right\}. \tag{11.22}$$

(Proof that this expression satisfies the continuity equation is left for Problem 11.2.) It is apparent from (11.19) that this expression for $\mathbf{J}(\mathbf{x},t)$ is gauge-invariant.

11.3 *Motion in a Uniform Static Magnetic Field*

In this section we treat in detail the quantum theory of a charged particle in a spatially homogeneous static magnetic field. Only the orbital motion will be considered, and any effects of spin or intrinsic magnetic moment will be omitted.

Throughout this section the magnetic field will be of magnitude B in the z direction. There are, of course, many different vector potentials that can generate this magnetic field. Some of the following results will depend only upon the magnetic field, whereas others will depend upon the particular choice of vector potential. Although the vanishing of the electric field requires only that the combination (11.1a) of scalar and vector potentials should vanish, we shall assume that the vector potential is static and that the scalar potential vanishes.

Energy levels

The most direct derivation of the energy levels can be obtained by writing the Hamiltonian (11.8) in terms of the components of the velocity operator (11.9): $H = H_{xy} + H_z$, with $H_{xy} = \frac{1}{2}M(V_x^2 + V_y^2)$ and $H_z = \frac{1}{2}MV_z^2$. Since $B_x = B_y = 0$, it follows from (11.11) that V_z commutes with V_x and V_y. Hence the operators H_{xy} and H_z are commutative, and every eigenvalue of H is just the sum of an eigenvalue of H_{xy} and an eigenvalue of H_z.

By introducing the notations $\gamma = (\hbar|q|B/M^2c)^{1/2}$, $V_x = \gamma Q'$ and $V_y = \gamma P'$, we formally obtain $H_{xy} = \frac{1}{2}(\hbar|q|B/Mc)(P'^2+Q'^2)$ with $Q'P' - P'Q' = i$. (Note that Q' and P' are only formal symbols and do not represent the position and momentum of the particle.) These equations are isomorphic to (6.7) and (6.6) for the harmonic oscillator (Sec. 6.1), and therefore the eigenvalues of H_{xy} must be equal to $(n + \frac{1}{2})\hbar|q|B/Mc$, where n is any nonnegative integer.

The eigenvalue spectrum of H_z is trivially obtained from that of V_z. The spectrum of $V_z \equiv P_z/M - qA_z/Mc$ has been shown to be gauge-invariant. Because the magnetic field is uniform and in the z direction, it is possible to choose the vector potential such that $A_z = 0$. Therefore the spectrum of V_z is continuous from $-\infty$ to ∞, like that of P_z.

Thus the energy eigenvalues for a charged particle in a uniform static magnetic field B are

$$E_n(v_z) = \frac{(n+\frac{1}{2})\hbar|q|B}{Mc} + \frac{1}{2}Mv_z^2\,, \quad (n = 0, 1, 2, \ldots)\,. \tag{11.23}$$

This result is independent of the particular vector potential that may be used to generate the prescribed magnetic field.

This form for the energies is easily interpreted. The motion parallel to the magnetic field is not coupled to the transverse motion, and is unaffected by the field. The classical motion in the plane perpendicular to the field is in a circular orbit with angular frequency $\omega_c = qB/Mc$ (called the *cyclotron frequency*), and it is well known that periodic motions correspond to discrete energy levels whose separation is $|\hbar\omega_c|$.

If we want not only the energies but also the corresponding state functions, it is necessary to choose a particular coordinate system and a particular form for the vector potential.

Solution in rectangular coordinates

Let us choose the vector potential to be $A_x = -yB, A_y = A_z = 0$. One can easily verify that $\nabla \cdot \mathbf{A} = 0$ and that $\nabla \times \mathbf{A} = \mathbf{B}$ is in the z direction. The Hamiltonian (11.8) now becomes

$$H = \frac{(P_x + yqB/c)^2 + P_y^2 + P_z^2}{2M}\,. \tag{11.24}$$

It is apparent that P_x and P_z commute with H (and that P_y does not), so it is possible to construct a complete set of common eigenvectors of H, P_x, and P_z.

In coordinate representation, the eigenvalue equation $H|\Psi\rangle = E|\Psi\rangle$ now takes the form

$$\frac{-\hbar^2}{2M}\nabla^2\Psi - \frac{i\hbar q}{Mc}\,By\,\frac{\partial}{\partial x}\Psi + \frac{q^2B^2}{2Mc^2}\,y^2\,\Psi = E\Psi\,. \tag{11.25}$$

Since Ψ can be chosen to also be an eigenfunction of P_x and P_z, we may substitute

$$\Psi(x, y, z) = \exp\{i(k_x x + k_z z)\}\phi(y)\,, \tag{11.26}$$

thereby reducing (11.25) to an ordinary differential equation,

$$\frac{-\hbar^2}{2M}\frac{d^2\phi(y)}{dy^2} + \frac{\hbar qBk_x}{Mc}\,y\phi(y) + \left[\frac{q^2B^2}{2Mc^2}\,y^2 + \frac{\hbar^2}{2M}\,(k_x^2+k_z^2) - E\right]\phi(y) = 0\,. \tag{11.27}$$

The term linear in y can be removed by shifting the origin to the point $y_0 = -\hbar k_x c/qB$. The equation then takes the form

$$\frac{-\hbar^2}{2M}\frac{d^2\phi(y)}{dy^2} + \left[\frac{M\omega_c^2}{2}\,(y-y_0)^2 - E'\right]\phi(y) = 0\,, \tag{11.28}$$

where $\omega_c = qB/Mc$ is the classical cyclotron frequency, and $E' = E - \hbar^2k_z^2/2M$ is the energy associated with motion in the xy plane. This is just the form of Eq. (6.21) for a simple harmonic oscillator with angular frequency $\omega = |\omega_c|$, whose eigenvalues are $E' = \hbar\omega(n+\frac{1}{2}), n = 0, 1, 2, \ldots$. Thus the energies for the charged particle in the magnetic field must be $E' = \hbar|\omega_c|(n+\frac{1}{2}) + \hbar^2k_z^2/2M$, confirming the result (11.23). The function $\phi(y)$ is a harmonic oscillator eigenfunction, of the form (6.32). Apart from a normalization constant, the eigenfunction (11.26) will be

$$\Psi(x,y,z) = \exp\{i(k_x x + k_z z)\}\, H_n\{\alpha(y-y_0)\}\, \exp\left\{-\frac{1}{2}\alpha^2(y-y_0)^2\right\}, \tag{11.29}$$

with $\alpha = (M|\omega_c|/\hbar)^{1/2} = (|q|B/\hbar c)^{1/2})$, and $y_0 = -\hbar k_x c/qB$. Here H_n is a Hermite polynomial. It is useful to define a characteristic *magnetic length*,

$$a_m = \alpha^{-1} = \left(\frac{\hbar c}{|q|B}\right)^{1/2}. \tag{11.30}$$

In terms of this length, the center of the Hermite polynomial in (11.29) is located at $y_0 = -(q/|q|)k_x a_m^2$.

The interpretation of this state function is far from obvious. The classical motion is a circular orbit in the xy plane. But (11.29) does not reveal such a motion, the x dependence of Ψ being an extended plane wave, while the y dependence is that of a localized harmonic oscillator function. The x and z dependences of Ψ are the same; nevertheless the energy E is independent of k_x while k_z contributes an ordinary kinetic energy term. These puzzles can be resolved by considering the orbit center coordinates.

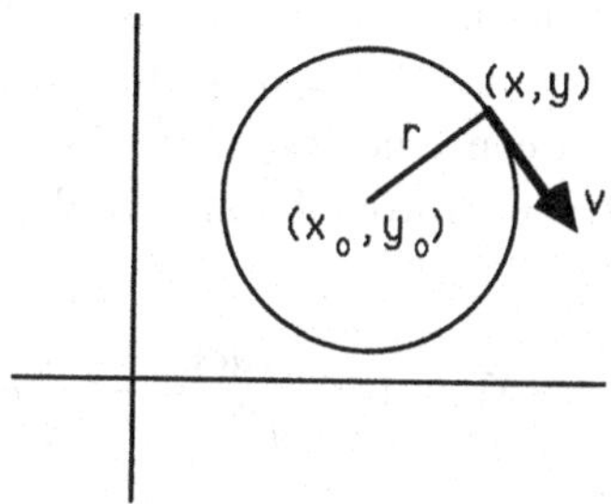

Fig. 11.1 An orbit of a charged particle ($q > 0$) in a magnetic field (directed toward the viewer).

Orbit center coordinates

Consider a classical particle moving with speed v in a circular orbit of radius r, as shown in Fig. 11.1. The magnetic force is equal to the mass times the centripetal acceleration, $qvB/c = Mv^2/r$. The angular frequency, $\omega_c \equiv v/r = qB/Mc$, is independent of the size of the orbit. The equations for the orbital position and velocity of the particle are of the forms

$$\begin{aligned} x - x_0 &= r\cos(\omega_c t + \theta)\,, \qquad y - y_0 = -r\sin(\omega_c t + \theta)\,, \\ v_x &= -\omega_c r\sin(\omega_c t + \theta)\,, \qquad v_y = -\omega_c r\cos(\omega_c t + \theta)\,. \end{aligned} \tag{11.31}$$

(These equations are correct for both positive and negative charges if we take ω_c to have the same sign as the charge q.) Hence the coordinates of the orbit center are $x_0 = x + v_y/\omega_c$ and $y_0 = y - v_x/\omega_c$. We conclude this brief classical analysis with the seemingly trivial remark that *the orbit center coordinates are constants of motion.*

Let us now, by analogy, define quantum-mechanical *orbit center operators*, X_0 and Y_0, in terms of the position operator and the velocity operator (11.9):

$$X_0 = Q_x + \frac{V_y}{\omega_c}\,, \quad Y_0 = Q_y - \frac{V_x}{\omega_c}\,. \tag{11.32}$$

It is easy to verify using (11.10) and (11.11), that if the x and y components of the magnetic field vanish then

$$[H, X_0] = [H, Y_0] = 0\,. \tag{11.33}$$

(This is another case in which it is simpler to express the Hamiltonian in terms of the velocity than in terms of the momentum.) Thus the orbit center coordinates are quantum-mechanical constants of motion, a result that is independent of the particular choice of vector potential.

It is not possible to construct eigenfunctions corresponding to a definite orbit center because the operators X_0 and Y_0 do not commute. A simple calculation yields

$$[X_0, Y_0] = \frac{-i\hbar c}{qB} = -i\frac{q}{|q|}a_m^2 \,. \tag{11.34}$$

In accordance with (8.27) and (8.31), there is an indeterminacy relation connecting the fluctuations of the two orbit center coordinates: $\Delta X_0 \Delta Y_0 \geq \frac{1}{2}a_m^2$. It is possible to construct common eigenfunctions of H and X_0, or of H and Y_0, but not of all three operators.

For the particular vector potential $A_x = -yB$, $A_y = A_z = 0$, the orbit center operators become $X_0 = Q_x + cP_y/qB$, $Y_0 = -cP_x/qB$. Thus it is apparent that the energy eigenfunction (11.29) is also an eigenfunction of Y_0 with eigenvalue $y_0 = -c\hbar k_x/qB$. This result illustrates the nonintuitive nature of the canonical momentum in the presence of a vector potential. The reason why the energy eigenvalue of (11.27) does not depend on k_x is that in this case the momentum component $\hbar k_x$ does not describe motion in the x direction, but rather position in the y direction! Roughly speaking, we may think of the state function (11.29) as describing an ensemble of circular orbits whose centers are distributed uniformly along the line $y = y_0$. (That this picture is only roughly accurate can be seen from the quantum fluctuations in the orbit size, as evidenced by the exponential tails on the position probability density in the y direction.)

Degeneracy of energy levels

Even for fixed n and v_z, the energy eigenvalue (11.23) is highly degenerate. Although the degree of degeneracy must be gauge-invariant, it is easier to calculate it for the particular coordinate system and vector potential corresponding to (11.29). These degenerate energy levels (for fixed n and k_z) are often called *Landau levels*, after Lev Landau, who first obtained the solution (11.29).

With k_z held constant, the problem is effectively reduced to two dimensions. For convenience, we assume that the system is confined to a rectangle of dimension $D_x \times D_y$ and subject to periodic boundary conditions. The allowed values of k_x are $k_x = 2\pi n_x/D_x$, with $n_x = 0, \pm 1, \pm 2, \ldots$. Now the orbit center coordinate, $y_0 = -(q/|q|)k_x a_m^2 = -(q/|q|)a_m^2 2\pi n_x/D_x$, must lie in the range $0 < y_0 < D_y$. In the limit as D_x and D_y become large, we may ignore any problems associated with orbits lying near the boundary, since they will be a negligible fraction of the total. In this limit the number of degenerate states corresponding to fixed n and k_z will be $D_x D_y / 2\pi a_m^2$.

This result suggests a simple geometrical interpretation, namely that each state is associated with an area of magnitude $2\pi a_m^2$ in the plane perpendicular to the magnetic field. The quantity $\Phi_0 = 2\pi\hbar c/q = hc/q$ is a natural unit of magnetic flux. In a homogeneous magnetic field B, the area $2\pi a_m^2$ encloses one unit of flux. Thus the degeneracy factor of a Landau level is simply equal to the number of units of magnetic flux passing through the system.

Orbit radius and angular momentum

It is possible to obtain a more direct description of the circular orbital motion of the particle than that contained implicitly in the state functions of the form (11.29). We may confine our attention to motion in the xy plane, since it is now apparent that the nontrivial aspect of the problem concerns motion perpendicular to the magnetic field. From the position operators, Q_x and Q_y, and the orbit center operators, X_0 and Y_0, we construct an *orbit radius operator*, r_c:

$$r_c^2 = (Q_x - X_0)^2 + (Q_y - Y_0)^2 \,. \tag{11.35}$$

From (11.32) we obtain $r_c^2 = \omega_c^{-2}(V_x^2 + V_y^2)$, and hence the transverse Hamiltonian satisfies the relation

$$H_{xy} \equiv \frac{1}{2}M(V_x^2 + V_y^2) = \frac{1}{2}M\,\omega_c^2 r_c^2 \,, \tag{11.36}$$

a relation that also holds in classical mechanics. From the known eigenvalues of H_{xy}, which are equal to $\hbar|\omega_c|(n+\frac{1}{2})$, we deduce that the eigenvalues of r_c^2 are $a_m^2(2n+1)$, with $n = 0, 1, 2, \ldots$. The degeneracy of the energy levels is due to the fact that the energy does not depend upon the position of the orbit center. Since the operators X_0 and Y_0 commute with H_{xy} but not with each other, it follows that the degenerate eigenvalues of H_{xy} form a one-parameter family (rather than a two-parameter family, as would be the case if the two constants of motion, X_0 and Y_0, were commutative). To emphasize the rotational symmetry of the problem, we introduce the operator

$${R_0}^2 = {X_0}^2 + {Y_0}^2 \,, \tag{11.37}$$

whose interpretation is the square of the distance of the orbit center from the origin. The degenerate eigenfunctions of H_{xy} can be distinguished by the eigenvalues of ${R_0}^2$. [These will not be the particular functions (11.29).]

The set of three operators $\{X_0/a_m, Y_0/a_m, H' \equiv \frac{1}{2}(X_0/a_m)^2 + \frac{1}{2}(Y_0/a_m)^2\}$ are isomorphic in their commutation relations to the position, momentum, and Hamiltonian of a harmonic oscillator (see Sec. 6.1). Hence the eigenvalues of H' are equal to $\ell + \frac{1}{2}$ with $\ell = 0, 1, 2, \ldots$. Thus the eigenvalues of ${R_0}^2$ are equal to $a_m^2(2\ell+1)$, $(\ell = 0, 1, 2, \ldots)$.

Suppose that the system is a cylinder of radius R. Since the orbit center must lie inside the system, we must impose the condition ${R_0}^2 \le R^2$. If we ignore any problems with orbits near the boundary, since they will be a negligible fraction of the total in the limit of large R, then the degeneracy factor of an energy level (the number of allowed values of ℓ) is equal to $\frac{1}{2}(R/a_m)^2 = \pi R^2/2\pi a_m^2$. This agrees with our previous conclusion that each state is associated with an area $2\pi a_m^2$.

The *orbital angular momentum* in the direction of the magnetic field is $L_z = Q_x P_y - Q_y P_x = M(Q_x V_y - Q_y V_x) + (q/c)(Q_x A_y - Q_y A_x)$. It will be a constant of motion if we *choose the vector potential to have cylindrical symmetry.* Hence we take the operator for the vector potential to be $\mathbf{A}(\mathbf{Q}) = \frac{1}{2}\mathbf{B} \times \mathbf{Q}$, which has components $(-\frac{1}{2}BQ_y, \frac{1}{2}BQ_x, 0)$. This yields

$$\begin{aligned} L_z &= M(Q_x V_y - Q_y V_x) + \frac{qB}{2c}(Q_x^2 + Q_y^2) \\ &= \frac{qB}{2c}({R_0}^2 - {r_c}^2)\,. \end{aligned} \tag{11.38}$$

[The second line is obtained by using (11.32) to eliminate the velocity operators.] It is apparent that the angular momentum is indeed a constant of motion, but it is not independent of those already found. Recall that ${r_c}^2$ is proportional to the energy of transverse motion, and that ${R_0}^2$ is an orbit center coordinate that distinguishes degenerate states. Those degenerate states could equally well be distinguished by the orbital angular momentum eigenvalue $\hbar m$.

It is now apparent that the angular momentum can have a very unintuitive significance in the presence of a magnetic field. If we consider it to vary at fixed energy, it has little to do with rotational motion, but is instead related to the radial position of the orbit center. Suppose the radius R of the system becomes infinite. Then for fixed energy (fixed ${r_c}^2$) the allowed values of angular momentum will be bounded in one direction and unbounded in the other, since ${R_0}^2$ is bounded below. (If ${R_0}^2$ is fixed at its minimum value, and the energy and angular momentum are allowed to vary together, then the angular momentum plays a more familiar role.)

It is possible to solve the Schrödinger equation directly in cylindrical coordinates, verifying in detail the results obtained above, and also obtaining explicit eigenfunctions (Problem 11.6). But the interpretation of those eigenfunctions as physical states would be very obscure without knowledge of the relation (11.38).

11.4 *The Aharonov–Bohm Effect*

In classical electrodynamics, the vector and scalar potentials were introduced as convenient mathematical aids for calculating the electric and magnetic fields. Only the fields, not the potentials, were regarded as having physical significance. Since the fields are not affected by the substitution (11.2), it follows that the equations of motion must be invariant under that substitution. In quantum mechanics these changes to the vector and scalar potentials must be accompanied by a change in the phase of the wave function Ψ. The theory is then invariant under the gauge transformation (11.18). Because of its classical origin, it is natural to suppose that the principle of *gauge invariance* merely expresses, in the quantum mechanical context, the notion that only the fields but not the potentials have physical significance. However, Aharonov and Bohm (1959) showed that there are situations in which such an interpretation is difficult to maintain.

They considered an experiment like that shown in Fig. 11.2, which consists of a charged particle source and a double slit diffraction apparatus. A long solenoid is placed perpendicular to the plane of the figure, so that a magnetic field can be created inside the solenoid while the region external to the solenoid remains field-free. The solenoid is located in the unilluminated shadow region so that no particles will reach it, and moreover it may be surrounded by a cylindrical shield that is impenetrable to the charged particles. Nevertheless it can be shown that the interference pattern depends upon the magnetic flux through the cylinder.

Let $\Psi^{(0)}(\mathbf{x},t)$ be the solution of the Schrödinger equation and boundary conditions of this problem for the case in which the vector potential is everywhere zero. Now let us consider the case of interest, in which the magnetic field is nonzero inside the cylinder but zero outside of it. The vector potential $\mathbf{A}$ will not vanish everywhere in the exterior region, even though $\mathbf{B} = \boldsymbol{\nabla} \times \mathbf{A} = 0$ outside of the cylinder. This follows by applying Stokes's theorem to any path surrounding the cylinder: $\oint \mathbf{A} \cdot d\mathbf{x} = \int\int (\boldsymbol{\nabla} \times \mathbf{A}) \cdot d\mathbf{S} = \int\int \mathbf{B} \cdot d\mathbf{S} = \Phi$. If the flux Φ through the cylinder is not zero, then the vector potential must be nonzero on every path that encloses the cylinder. However in any *simply connected* region outside of the cylinder, it is possible to express the vector potential as the gradient of a scalar, from the zero-potential solution by means of a gauge transformation, $\Psi = \Psi^{(0)} e^{i(q/\hbar c)\Lambda}$.

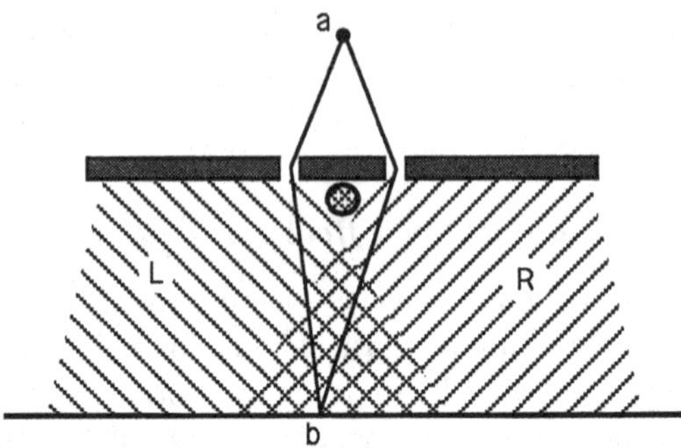

Fig. 11.2 The Aharonov–Bohm experiment. Charged particles from the source **a** pass through the double slit. The interference pattern formed at the bottom screen depends upon the magnetic flux through the impenetrable cylinder.

This technique will now be applied to each of the (overlapping) regions L and R shown in Fig. 11.2. In region L, which contains the slit on the left, the wave function can be written as $\Psi_L = \Psi_L^{(0)} e^{i(q/\hbar c)\Lambda_1}$, where $\Psi_L^{(0)}$ is the zero-potential solution in region L, and $\Lambda_1 = \Lambda_1(\mathbf{x}, t) = \int \mathbf{A} \cdot d\mathbf{x}$, with the integral taken along a path within region L. Since $\boldsymbol{\nabla} \times \mathbf{A} = 0$ inside L, the value of this integral depends only upon the end points of the path, provided, of course, that the path remains within L and does not cross the cylinder of magnetic flux. A similar form can be written for the wave function in the region R, which contains the slit on the right.

At the point $\mathbf{b}$, in the overlap of regions L and R, the wave function is a superposition of contributions from both slits. Hence we have

$$\Psi(\mathbf{b}) = \Psi_L^{(0)} e^{i(q/\hbar c)\Lambda_1} + \Psi_R^{(0)} e^{i(q/\hbar c)\Lambda_2} . \tag{11.39}$$

Here $\Lambda_1 = \int \mathbf{A} \cdot d\mathbf{x}$ with the path of integration running from $\mathbf{a}$ to $\mathbf{b}$ through region L, and $\Lambda_2 = \int \mathbf{A} \cdot d\mathbf{x}$ with the path of integration running from $\mathbf{a}$ to $\mathbf{b}$ through region R. The interference pattern depends upon the relative phase of the two terms in (11.39), $e^{i(q/\hbar c)(\Lambda_1 - \Lambda_2)}$. But $(\Lambda_1 - \Lambda_2)$, the difference between the integrals along paths on either side of the cylinder, is equivalent to an integral around a closed path surrounding the cylinder, $\oint \mathbf{A} \cdot d\mathbf{x} = \Phi$. Therefore the interference pattern is sensitive to the magnetic flux inside of the cylinder, even though the particles never pass through the region in which the magnetic field is nonzero! This prediction, which has been experimentally verified, was very surprising when it was first announced.

Several remarks about this effect are in order. First, the relative phase of the two terms of (11.39) is $(e^{iq\Phi/\hbar c})$. If the magnetic flux Φ were quantized in multiples of $2\pi\hbar c/q$ then this phase factor would be equal to 1, and there would be no observable dependence of the interference pattern upon the flux. This possibility, which would have given quantum-mechanical significance to Faraday's lines of magnetic force, has been experimentally demonstrated to be false. Magnetic flux is not quantized.

[[The phenomenon in superconductivity known as "flux quantization" is that the total flux enclosed by a ring of superconducting material must be a multiple of $\pi\hbar c/e$. The extra factor of $\frac{1}{2}$ occurs because current is carried by correlated pairs of electrons, whose charge is $q = 2e$. This "flux quantization" phenomenon is a peculiar property of the superconducting state, and is not a general property of the electromagnetic field.]]

Second, the existence of the Aharonov–Bohm effect is surprising because the particle never enters the region in which the magnetic field is nonzero. Therefore the classical Lorentz force on the particle is zero, and the classical trajectory would not be deflected by the inaccessible magnetic field in the cylinder. This remains true in quantum mechanics *on the average.* According to (11.13), the ensemble average rate of change of the particle velocity for this state is

$$\left\langle \frac{d\mathbf{V}}{dt} \right\rangle = \frac{1}{2}\left(\frac{q}{Mc}\right)\langle\Psi|(\mathbf{V}\times\mathbf{B} - \mathbf{B}\times\mathbf{V})|\Psi\rangle = 0\,. \tag{11.40}$$

This expression vanishes because $\Psi(\mathbf{x})$ is zero wherever $\mathbf{B}(\mathbf{x})$ is nonzero and vice versa. Although the magnetic flux inside the cylinder affects the motions of the individual particles, it produces zero average deflection. The positions of the fringes within the diffraction pattern shift systematically as the flux Φ is varied, but their intensities change simultaneously, so that the centroid of the diffraction pattern does not move.

Bound state Aharonov–Bohm effect

The above analysis of the AB diffraction experiment was rather schematic. However, it is possible to give a bound state version which can be easily and rigorously analyzed. This example demonstrates more clearly that the AB effect (the influence on charged particles of inaccessible magnetic fields) is an inevitable consequence of the principles of quantum mechanics. Unfortunately, it is not so easy to realize it experimentally.

Consider a particle of charge q confined to the interior of a torus of rectangular cross section. We use cylindrical coordinates (ρ, ϕ, z), whose relations to the rectangular coordinates are $\rho = \sqrt{x^2 + y^2}$ and $\phi = \tan^{-1}(y/x)$. (There should be no confusion between the present use of the symbol ϕ as a coordinate, and its use in previous sections as the electromagnetic scalar potential. No scalar potential will occur in this section.) The z axis is the rotational symmetry axis of the torus. The charged particle is confined within the region defined by the limits

$$a < \rho < b\,, \quad -s < z < s\,. \tag{11.41}$$

The state function $\Psi = \Psi\,(\rho, \phi, z)$ vanishes outside of these limits. A magnetic flux Φ threads the "donut hole" of the torus, but the magnetic field is zero in the region (11.41). The vector potential is necessarily nonzero in the region (11.41), and the cylindrically symmetric potential

$$A_\phi = \frac{\Phi}{2\pi\rho}\,, \quad A_\rho = A_z = 0 \tag{11.42}$$

is consistent with such a magnetic field.

The Hamiltonian (11.15) now takes the form

$$H = -\frac{\hbar^2}{2M}\nabla^2 + \frac{i\hbar q\Phi}{2\pi Mc\rho^2}\frac{\partial}{\partial\phi} + \frac{q^2\Phi^2}{8\pi^2 Mc^2\rho^2}\,, \tag{11.43}$$

with $\nabla^2 = (\partial/\partial\rho)^2 + \rho^{-1}(\partial/\partial\rho) + \rho^{-2}(\partial/\partial\phi)^2 + (\partial/\partial z)^2$ in cylindrical coordinates. The state functions are eigenfunctions of H, satisfying $H\Psi = E\Psi$. It can be verified by direct substitution that the eigenfunctions are of the form

$$\Psi(\rho, \phi, z) = R_n(\rho) e^{im\phi} \sin\left\{\frac{j\pi(z+s)}{2s}\right\}. \tag{11.44}$$

Here j must be a positive integer in order to satisfy the boundary condition $\Psi = 0$ at $z = \pm s$. The value of m must be an integer in order for Ψ to be single-valued under rotation by 2π. (It was shown in Sec. 7.3 that the restriction of m to integer values also follows from the fundamental properties of the orbital angular momentum operators.) The radial function $R_n(\rho)$ satisfies the boundary conditions $R_n(a) = R_n(b) = 0$, and is a solution of the equation

$$\frac{d^2 R_n}{d\rho^2} + \frac{1}{\rho}\frac{dR_n}{d\rho} - \frac{(m-F)^2}{\rho^2}R_n - k_z^2 R_n = \frac{-2M}{\hbar^2}E\,R_n\,, \tag{11.45}$$

where $k_z = j\pi/2s$, and $F = \Phi q/2\pi\hbar c$ is the magnetic flux expressed in natural units. The radial function $R_n(\rho)$ can be given explicitly in terms of Bessel functions, but that is not necessary for present purposes. It is sufficient to note that, according to (11.45), the energy E of the stationary state clearly must depend on the magnetic flux, even though the Schrödinger equation has been solved in the region $a < \rho < b$ and the flux is confined to the inaccessible region $\rho < a$.

The AB effect is a *topological* effect, in that the effect depends on the flux encircled by the paths available to the particle, even though the paths may never approach the region of the flux. Since the magnetic force on the charge q, $q\mathbf{v} \times \mathbf{B}$, vanishes on all possible paths of the particle, one might wonder whether the charge is necessary for the effect. According to the theory, the effect depends on the dimensionless ratio $\Phi q/hc$, which is proportional to the charge, and it has been experimentally confirmed that no AB effect occurs if neutrons are used instead of electrons.

Many analogs of the AB effect have now been observed. One of these, the *Aharonov–Casher* (AC) effect, is interesting because it is the dual of the AB effect. The flux in Fig. 11.2 can be produced by a thin cylinder of magnetized material, which is really a line of magnetic dipoles. So the AB effect can be viewed as the relative phase shift between two charged particle beams that enclose a line of magnetic dipoles. The AC effect is the relative phase shift between two magnetic dipole beams that enclose a line of electric charge. That the AB and AC effects are mathematically exact duals of each other was shown by Hagen (1990). The effect was first demonstrated experimentally by Cimmino *et al.* (1989), using neutrons to form the electrically neutral beam of magnetic dipoles. The effect has been confirmed, with much greater precision, by Sangster *et al.* (1993) using a beam of TlF molecules.

How should we interpret the electromagnetic potentials in light of the Aharonov–Bohm effect? If we adhere to the classical view that only the electric and magnetic fields are physically significant, then we must admit that in quantum mechanics they can have a nonlocal influence. That is to say, they can influence the motions of charged particles even though the particles do not enter any region of space where the fields exist. Alternatively, we may grant that the potentials themselves are physically significant; however, they are subject to the requirement that all observable effects be invariant under gauge transformations. Both points of view are logically tenable. However, the second view seems more natural, since the Hamiltonian and the Schrödinger equation are naturally expressed in terms of the potentials. This view is also more in keeping with the modern non-Abelian gauge theories of fundamental processes.

11.5 *The Zeeman Effect*

The name of this effect is derived from the discovery by P. Zeeman in 1896, that the spectral lines of atoms were often split when the atom was placed in a magnetic field. We shall use the term to refer to the effect of a magnetic field on atomic states and energy levels. To study it mathematically, we must add the spherically symmetric atomic potential $W(r)$ to the Hamiltonian of an electron in a uniform magnetic field.

The Hamiltonian for an electron in the atom is

$$\begin{aligned} H &= \frac{\{\mathbf{P} + (e/c)\mathbf{A}\}^2}{2M} + W(r) \\ &= \frac{P^2}{2M} + \frac{e}{Mc}\mathbf{A}\cdot\mathbf{P} + \frac{e^2A^2}{2Mc^2} + W(r)\,, \end{aligned} \tag{11.46}$$

where the mass of the electron is M, and its charge is $-e = -|e|$. We have used the simplification $\mathbf{P}\cdot\mathbf{A} = \mathbf{A}\cdot\mathbf{P}$, which is valid if $\operatorname{div}\mathbf{A} = 0$ [see Eq. (11.14)]. In order to proceed further, it is necessary to choose a specific form for the vector potential. We shall take it to be $\mathbf{A}(\mathbf{x}) = \frac{1}{2}(\mathbf{B}\times\mathbf{x})$. It then follows that $\mathbf{A}\cdot\mathbf{P} = \frac{1}{2}(\mathbf{B}\times\mathbf{x})\cdot\mathbf{P} = \frac{1}{2}\mathbf{B}\cdot(\mathbf{x}\times\mathbf{P}) = \frac{1}{2}\mathbf{B}\cdot\mathbf{L}$, with $\mathbf{L}$ being the orbital angular momentum operator. The Hamiltonian then becomes

$$H = \frac{P^2}{2M} + \frac{e}{2Mc}\mathbf{B}\cdot\mathbf{L} + \frac{e^2}{8Mc^2}(\mathbf{B}\times\mathbf{x})^2 + W(r)\,. \tag{11.47}$$

For *weak magnetic fields* it is convenient to write this Hamiltonian in the form

$$H = H_a + \frac{e}{2Mc}\mathbf{B}\cdot\mathbf{L} + \frac{e^2}{8Mc^2}(\mathbf{B}\times\mathbf{x})^2\,, \tag{11.48}$$

where $H_a = P^2/2M + W(r)$ is the Hamiltonian of the free atom. Its eigenfunctions are similar in form to those of the hydrogen atom, and they will be denoted as $\Psi_{n\ell m}$, where n is the principal quantum number, and ℓ and m are the orbital angular momentum quantum numbers. The function $\Psi_{n\ell m}$ is a common eigenfunction of the operators H, $\mathbf{L}\cdot\mathbf{L}$, and L_z. If we *neglect the last term* of (11.48), which is of the second order in the magnetic field, and choose the magnetic field to lie along the z axis, then $\Psi_{n\ell m}$ will also be an eigenfunction of H. To the first order in the magnetic field, the atomic energy levels will be displaced by an amount

$$E^{(1)}_{n\ell m} = \frac{e\hbar B}{2Mc}m\,, \tag{11.49}$$

and the eigenfunctions will be unchanged. Thus the degeneracy of the $(2\ell+1)$-fold multiplet of fixed n and ℓ, due to the spherical symmetry of the atom, is broken by the magnetic field.

The term linear in $\mathbf{B}$ in (11.48), which gives rise to (11.49), is equivalent to the potential energy $-\boldsymbol{\mu}_{\mathrm{L}}\cdot\mathbf{B}$ of the orbital magnetic dipole moment $\boldsymbol{\mu}_{\mathrm{L}} = (-e/2Mc)\mathbf{L}$. There is also a magnetic dipole moment associated with the spin,

$\boldsymbol{\mu}_s = (-e/Mc)\mathbf{S}$, and so in practice one must also add the spin term $-\boldsymbol{\mu}_s \cdot \mathbf{B}$ to the Hamiltonian. The net effect of the orbital and spin magnetic moments was treated in an example at the end of Sec. 7.8, and the dynamics of spins are treated in Ch. 12.

Although the calculation leading to (11.49) appears very simple, the approximation that was made is not above suspicion. We have, in effect, neglected the A^2 term in (11.46) compared with the term that is linear in $\mathbf{A}$. But the division between those two terms is not gauge-invariant, and so the effect of neglecting the second order term is ambiguous. Since we first specialized to a particular vector potential, $\mathbf{A} = \frac{1}{2}\mathbf{B} \times \mathbf{x}$, it could be argued that we are really neglecting a term that is second order in the magnetic field strength B. But the second order term in (11.47) becomes arbitrarily large at large distances, no matter how small B may be, and so its neglect is not obviously justified. In this particular problem, we are saved by the fact that the eigenfunctions $\Psi_{n\ell m}$ decay exponentially at large distances, and this overpowers the divergence of $(\mathbf{B} \times \mathbf{x})^2$. But, in general, an expansion in powers of B can be very dangerous, no matter how small B may be. For example, the eigenfunction (11.29) for an unbound particle in a magnetic field has no reasonable limit for $B \to 0$.

For *strong magnetic fields* the term in the Hamiltonian that is proportional to B^2 becomes important. This term is $(e^2/8Mc^2)(\mathbf{B}\times\mathbf{x})^2 = (e^2/8Mc^2)(B\,r\,\sin\theta)^2$, where the angle θ is measured from the axis of cylindrical symmetry, defined by the magnetic field. It has the form of an attractive potential that increases in proportion to the square of the perpendicular distance from the axis of symmetry, $\rho = r\,\sin\theta$. Its effect is to squeeze the atom into a thin elongated shape.

For very strong magnetic fields, the atomic potential $W(r)$ can be treated as a small correction. In this case it is convenient to write the Hamiltonian (11.47) as

$$H = H_\perp + \frac{P_z^2}{2M} + W(r)\,, \tag{11.50}$$

where

$$H_\perp = \frac{(P_x^2 + P_y^2)}{2M} + \frac{e}{2Mc}\mathbf{B}\cdot\mathbf{L} + \frac{e^2}{8Mc^2}(\mathbf{B} \times \mathbf{x})^2 \tag{11.51}$$

is the Hamiltonian for motion in the plane perpendicular to the magnetic field, which is in the z direction. The common eigenfunctions of $H_\perp$ and L_z can be obtained in polar coordinates, and will be denoted as $\psi_{n,m}(\rho,\theta)$. The corresponding energy eigenvalues are given by

$$
\begin{aligned}
H_\perp \psi_{n,m}(\rho,\phi) &= E_\perp \psi_{n,m}(\rho,\phi) \\
&= \hbar\omega_c \left[n + \frac{1}{2}(m + |m|) + \frac{1}{2} \right] \psi_{n,m}(\rho,\phi) . \qquad (11.52)
\end{aligned}
$$

(This result is a part of Problem 11.6.) Here n is the number of radial nodes in the eigenfunction, $\hbar m$ is the orbital angular momentum eigenvalue, and $\omega_c = eB/Mc$ is the cyclotron frequency of the electron.

Now if the atomic potential $W(r)$ is small compared to those terms in (11.51) involving B, the eigenfunctions of the full Hamiltonian (11.50) should closely resemble those of $H_\perp$. Therefore, in the eigenvalue equation

$$
H\Psi(\rho,\phi,z) = E\Psi(\rho,\phi,z) , \qquad (11.53)
$$

we shall seek *approximate* eigenfunctions having the form

$$
\Psi(\rho,\phi,z) \approx \psi_{n,m}(\rho,\theta) f(z) . \qquad (11.54)
$$

When this function is operated on by the Hamiltonian (11.50), we obtain

$$
\begin{aligned}
\left[H_\perp + \frac{P_z^2}{2M} + W(r) \right] \psi_{n,m}(\rho,\theta) f(z) &= E_\perp \psi_{n,m}(\rho,\theta) f(z) \\
&+ \left[-\frac{\hbar^2}{2M} f''(z) + W\left(\sqrt{\rho^2 + z^2}\right) f(z) \right] \psi_{n,m}(\rho,\phi) , \qquad (11.55)
\end{aligned}
$$

where $f''(z)$ is the second derivative of $f(z)$. It is clear that we do not have an exact eigenfunction of H because W depends on both ρ and z. But if we substitute (11.54) into (11.53), multiply by $[\psi_{n,m}(\rho,\phi)]^*$ and integrate over ρ and ϕ, we obtain an equation that determines the function $f(z)$:

$$
\left[\frac{P_z^2}{2M} + V_m(z) \right] f(z) = E_\parallel f(z) , \qquad (11.56)
$$

where

$$
V_m(z) = \iint |\psi_{n,m}(\rho,\phi)|^2 W\left(\sqrt{\rho^2 + z^2}\right) \rho \, d\rho \, d\phi . \qquad (11.57)
$$

Alternatively, we can use (11.54) as a trial function in the variational method (Sec. 10.6). Variation of the unknown function $f(z)$ leads to (11.55) as the condition for minimizing the energy. Thus, in the strong-magnetic-field limit, the problem reduces to that of finding the bound states of an effective one-dimensional potential, (11.57). The corresponding approximate energy eigenvalue for Eq. (11.53) is $E \approx E_\perp + E_\parallel$.

To proceed further we specialize to the *hydrogen atom*, for which $W(r) = -e^2/r$, and calculate the ground state energy. Since the angular momentum quantum number m is exactly conserved, it is no more difficult to calculate the lowest energy level for an arbitrary value of m. The corresponding transverse function, $\psi_{0,m}(\rho,\phi)$, has no radial nodes ($n = 0$), and is of the form (Problem 11.6)

$$\psi_{0,m}(\rho,\phi) = N_\rho{}^{|m|} \exp\left[-\left(\frac{\rho}{2a_m}\right)^2\right] e^{im\phi},$$

where $a_m = (\hbar c/eB)^{1/2}$ is the magnetic length, and N is a normalization factor. The factor $|\psi_{0,m}(\rho,\phi)|^2\rho$ in the integrand of (11.57) is peaked at $\rho = \rho_m \equiv [(2|m|+1)\hbar c/eB]^{1/2}$. If we replace the variable ρ in (11.57) with the dominant value ρ_m, the effective one-dimensional potential will become $V_m(z) \approx -e^2/(\rho_m^2 + z^2)^{1/2}$. It is more convenient, and should be no worse an approximation, to further replace this by

$$V_m(z) \approx \frac{-e^2}{\rho_m + |z|}, \tag{11.58}$$

which agrees with the previous expression at large z and at $z = 0$. This is the so-called *truncated Coulomb potential* in one dimension, whose bound state eigenvalues can be determined exactly [Haines and Roberts (1969)]. Its lowest eigenvalue, in the limit of interest to us, is

$$E_\| = -\frac{\hbar^2}{2Ma_0^2}\left[2\log\left(\frac{a_0}{\rho_m}\right)\right]^2 \quad (\rho_m \ll a_0), \tag{11.59}$$

where $a_0 = \hbar^2/Me^2$ is the Bohr radius. More accurate estimates of the large-magnetic-field limit of the lowest eigenvalue for the true potential (11.57) are similar to (11.59), but with a slightly different numerical factor.

Adding $E_\perp$ and $E_\|$, and substituting the value of ρ_m, we obtain the lowest energy eigenvalue for a fixed of m,

$$\begin{aligned} E_{0,m} &= \frac{1}{2}\hbar\omega_c(m + |m| + 1) \\ &\quad - \frac{\hbar^2}{2Ma_0^2}\left\{\log\left[\frac{eBa_0^2}{\hbar c(2|m|+1)}\right]\right\}^2, \end{aligned} \tag{11.60}$$

this expression being valid for *large magnetic fields*. The ground state is apparently the state with $m = 0$, and its energy is

$$E_{0,0} = \frac{\hbar eB}{2Mc} - \frac{\hbar^2}{2Ma_0^2}\left\{\log\left(\frac{eBa_0^2}{\hbar c}\right)\right\}^2. \tag{11.61}$$

(These expressions omit the contribution from the spin.)

The energies for a hydrogen-like atom, whose potential is $W(r) = -Ze^2/r$, can be obtained from those of the hydrogen atom by the scaling relation $E(Z,B) = Z^2E(1, B/Z^2)$ (see Problem 11.8). Hence the ground state energy of a hydrogen-like atom in a strong magnetic field is

$$E_{0,0}(Z,B) = \frac{\hbar eB}{2Mc} - \frac{Z^2\hbar^2}{2Ma_0^2}\left\{\log\left(\frac{eBa_0^2}{\hbar cZ^2}\right)\right\}^2,$$

the second term being the contribution of the atomic potential. Note that its dependence on the potential strength Z is not analytic, being proportional to $\{Z\log(Z)\}^2$. This is a consequence of the singular character of the Coulomb potential.

Although both the low and high field limits can be treated analytically, there is no simple theory for the spectrum of the hydrogen atom at intermediate field strengths. As the field is varied from zero to near infinity, the energy levels must continuously rearrange themselves from the familiar hydrogenic multiplets (Sec. 10.2) into the Landau level structure (Sec. 11.3). Between these two, relatively simple limits, the spectrum displays great complexity, for which no analytic formula is known. For a review of the theory, which is still a subject of active research, see Friedrich and Wintgen (1989).

Further reading for Chapter 11

Peshkin (1981) gives a good discussion of the Aharonov–Bohm effect, including its close relation to the conservation of the total angular momentum for the particle and electromagnetic field. The first experimental confirmation of the AB effect was by Chambers (1960). Even more striking experimental conformation has been obtained by Tonomura *et al.* (1983), using the technique of electron holography. Silverman (1995) discusses many ingenious interference experiments, several of which involve the AB effect.

Problems

11.1 (a) Evaluate Lagrange's classical equation of motion for a charged particle in an arbitrary electromagnetic field, and show that it leads to Newton's equation of motion.

(b) Do the same for Hamilton's equations.

11.2 Show that the formula (11.22) for the probability current $\mathbf{J}(\mathbf{x}, t)$ satisfies the continuity equation, $\text{div}\mathbf{J} + \partial|\Psi|^2/\partial t = 0$.

11.3 Generalize Eq. (4.22b) so that it becomes correct in the presence of a vector potential and a magnetic field.

11.4 Determine the energy spectrum and wave functions corresponding to a charged particle in uniform crossed electric and magnetic fields, with **B** in the z direction and **E** in the x direction. (*Hint:* This problem may be easy or difficult, depending upon the vector potential that is chosen.)

11.5 Evaluate the average velocity for the states of Problem 11.4.

11.6 Use cylindrical coordinates to solve the Schrödinger equation for a charged particle in the magnetic field generated by the vector potential $\mathbf{A} = \frac{1}{2}\mathbf{B} \times \mathbf{x}$. Note particularly the allowed values of angular momentum corresponding to a particular energy eigenvalue.

11.7 Consider the bound state Aharonov–Bohm effect (Sec. 11.4) for a particle confined to a thin cylindrical shell ($b - a \ll a < s$). Identify the quantum numbers of the ground state and the first excited state, and determine their energies as a function of the magnetic flux threading the center of the shell.

11.8 An energy eigenvalue for a hydrogen-like atom with nuclear charge Ze in a magnetic field B may be denoted as $E(Z,B)$. Show that $E(Z,B) = Z^2E(1, B/Z^2)$, and that hence it is sufficient to consider only $Z = 1$.

11.9 Find the probability current density for the eigenstates of Problem 11.6.

11.10 The Hamiltonian for a charged particle in a homogeneous magnetic field is

$$H = \frac{1}{2M}\left(\frac{\hbar}{i}\nabla - \frac{q}{c}\mathbf{A}(\mathbf{x})\right)^2,$$

with $\nabla \times \mathbf{A}(\mathbf{x}) = \mathbf{B}$ being a constant. Although the physical situation (a homogeneous magnetic field) is translationally invariant, it is apparent that the operator H is *not* translationally invariant. Show, however, that the displaced Hamiltonian H', obtained by the transformation $\mathbf{x} \to \mathbf{x} + \mathbf{a}$, is related to H by a gauge transformation.

11.11 Consider a hydrogen atom in a very strong magnetic field, such that the magnetic-field-dependent terms are much stronger than the atomic potential. Formally treat the atomic potential, $-e^2/r$, as a perturbation, and show in detail why perturbation theory fails.

Chapter 12

Time-Dependent Phenomena

Because of their obvious importance, stationary states (energy eigenstates) have played a prominent role in most of the cases treated so far. But there are many phenomena in which the time dependence is the most interesting feature.

In the simplest case of a pure state and a time-independent Hamiltonian, it is possible to formally express the time dependence in terms of energy levels and the corresponding stationary states. The equation of motion for this case is $(d/dt)|\Psi(t)\rangle = -(i/\hbar)H|\Psi(t)\rangle$. If the eigenvalues and eigenvectors of H are known, $H|E_n\rangle = E_n|E_n\rangle$, and if we can expand the initial state vector as a series of these eigenvectors, $|\Psi(0)\rangle = \sum_n a_n|E_n\rangle, a_n = \langle E_n|\Psi(0)\rangle$, then the time-dependent state vector is simply given by $|\Psi(t)\rangle = \sum_n a_n e^{-iE_nt|E_n\rangle}$. This method has important uses, but it is not adequate in all cases, and it is necessary to devise methods that treat time-dependent states in their own right, without attempting to reduce them to stationary states.

12.1 *Spin Dynamics*

Many particles (such as electrons, neutrons, atoms, and nuclei) possess an intrinsic angular momentum, or spin. The properties of the spin operator $\mathbf{S}$ and the spin states were discussed in Sec. 7.4. A particle that has a nonzero spin also has a magnetic moment proportional to the spin, $\boldsymbol{\mu} = \gamma\mathbf{S}$. The magnetic moment interacts with any applied magnetic field $\mathbf{B}$, yielding a Hamiltonian of the form $H = -\boldsymbol{\mu}\cdot\mathbf{B} = -\gamma\mathbf{B}\cdot\mathbf{S}$. The quantity γ may be of either sign, depending on the particle. For an electron it is approximately equal to $\gamma_e = -e/M_ec$, where $-e$ is the electronic charge, M_e is the electronic mass, and c is the speed of light. For a proton it is $\gamma_p = 2.79e/M_pc$, and for a neutron it is $\gamma_n = -1.91e/M_pc$, where M_p is the proton mass.

Spin precession

The simplest case, a particle of spin $\frac{1}{2}$ in a constant magnetic field, can be solved by the method described in the introduction to this chapter. Choose the magnetic field to point in the z direction with magnitude B_0. The spin operator can be written in terms of the Pauli operators as $\mathbf{S} = \frac{1}{2}\hbar\boldsymbol{\sigma}$. The Hamiltonian then becomes $H = -\frac{1}{2}\gamma\hbar B_0\sigma_z$. The eigenvalues of σ_z are ± 1, and the corresponding eigenvectors may be denoted as $|+\rangle$ and $|-\rangle$. The vector $|+\rangle$ corresponds to the energy $E_1 = -\frac{1}{2}\gamma\hbar B_0$, and $|-\rangle$ corresponds to the energy $E_2 = \frac{1}{2}\gamma\hbar B_0$. A time-dependent state vector has the form

$$|\Psi(t)\rangle = a_1 e^{i\omega_0 t/2}|+\rangle + a_2 e^{-i\omega_0 t/2}|-\rangle\,, \tag{12.1}$$

where $\omega_0 = \gamma B_0 = (E_2 - E_1)/\hbar$, and the constants a_1 and a_2 are determined by the initial conditions. In the 2×2 matrix notation of (7.45), this state vector would become

$$|\Psi(t)\rangle \to \begin{bmatrix} a_1 & e^{i\omega_0 t/2} \\ a_2 & e^{-i\omega_0 t/2} \end{bmatrix}. \tag{12.2}$$

Suppose that $a_1 = a_2 = \sqrt{\frac{1}{2}}$. Then a simple calculation shows the average magnetic moment in this state to be $\langle\mu_x\rangle = \frac{1}{2}\hbar\gamma\cos(\omega_0 t)$, $\langle\mu_y\rangle = -\frac{1}{2}\hbar\gamma\sin(\omega_0 t)$, $\langle\mu_z\rangle = 0$. The magnetic moment is precessing at the rate of ω_0 radians per second about the axis of the static magnetic field.

A much more general treatment is possible in the *Heisenberg picture*, in which the states are independent of time and the time dependence is carried by the operators that represent dynamical variables. In Sec. 3.7 the time-dependent Heisenberg operators were distinguished by a subscript, H. That notation would be cumbersome here because of other subscripts, so we shall use a "prime" notation, $S_x{}'$ being equivalent to $(S_x)_H$ in the original notation of (3.72). The equation of motion (3.73) for the x component of spin in an arbitrary magnetic field now becomes

$$\frac{d}{dt}S_x{}' = \frac{i}{\hbar}[H, S_x{}'] = \frac{i}{\hbar}[-\gamma\mathbf{B}\cdot\mathbf{S}, S_x{}'] = \gamma(-S_z{}'B_y + S_y{}'B_z)\,.$$

This result clearly generalizes to

$$\frac{d}{dt}\mathbf{S}' = \mathbf{S}' \times \gamma\mathbf{B}\,. \tag{12.3}$$

This equation is valid for arbitrary time-dependent magnetic fields.

If we specialize to a constant field of magnitude B_0 in the z direction, as in the simple example above, it is easily verified that the solution is

$$\begin{aligned} S_x{}'(t) &= S_x{}'(0)\cos(\omega_0 t) + S_y{}'(0)\sin(\omega_0 t)\,, \\ S_y{}'(t) &= S_y{}'(0)\cos(\omega_0 t) - S_x{}'(0)\sin(\omega_0 t)\,, \\ S_z{}'(t) &= S_z{}'(0)\,, \end{aligned} \tag{12.4}$$

where $\omega_0 = \gamma B_0$. It is apparent that the magnetic moment precesses at the rate of ω_0 radians per second about the direction of the magnetic field, regardless of the magnitude of the spin, and regardless of the initial state (which need not be a pure state). This rotating magnetic moment can, in principle, be detected with an induction coil, although it might not be practical to do so for a single particle.

Spin resonance

As long as only a static magnetic field is present, there can be no gain or loss of energy by the particle, and hence no transitions between the energy levels. Real transitions become possible if a time-dependent field is applied. Let us consider the effect of a magnetic field of the form $\mathbf{B}_0 + \mathbf{B}_1(t)$, where $\mathbf{B}_0$ is a static field in the z direction, and $\mathbf{B}_1(t)$ is a rotating field in the xy plane:

$$\mathbf{B}_1(t) = \hat{\mathbf{i}}\, B_1 \cos(\omega t) + \hat{\mathbf{j}}\, B_1 \sin(\omega t)\,. \tag{12.5}$$

Here $\hat{\mathbf{i}}$ and $\hat{\mathbf{j}}$ are unit vectors along the x and y axes, respectively. [This rotating field is the simplest to analyze. Another common case, an oscillating field along, say, the x direction, can be represented as a superposition of $\mathbf{B}_1(t)$ and a counterrotating field, $\hat{\mathbf{i}}\, B_1 \cos(\omega t) - \hat{\mathbf{j}}\, B_1 \sin(\omega t)$.]

With this combination of magnetic fields, the Hamiltonian is $H = -\gamma \mathbf{S} \cdot \{\mathbf{B}_0 + \mathbf{B}_1(t)\} = -\gamma B_0 S_z - \gamma B_1 S_u$, where S_u is the component of spin in the direction of $\mathbf{B}_1$. The u direction is obtained from the x axis by a rotation through the angle ωt. The corresponding rotation operator, $e^{-i\omega t J_z/\hbar}$, may be replaced by $e^{-i\omega t S_z/\hbar}$ since only spin operators occur in this problem. Thus the Hamiltonian can be written as

$$H = -\gamma B_0 S_z - \gamma B_1\, e^{-i\omega t S_z/\hbar} S_x e^{i\omega t S_z/\hbar}\,. \tag{12.6}$$

We shall now solve the dynamical problem using the *Schrödinger picture*, in which the dynamics is carried by the state function. (The solution in the Heisenberg picture is the subject of Problem 12.3.) The equation of motion is

$$i\hbar \frac{\partial}{\partial t} |\Psi\rangle = H |\Psi\rangle\,. \tag{12.7}$$

Because of the rotating magnetic field, both H and Ψ are now time-dependent. But since the axis of rotation is the direction of $\mathbf{B}_0$, it is possible to eliminate the time dependence of H by a compensating rotation of the system so as to bring $\mathbf{B}_1$ to rest. Applying that rotation to (12.7) yields

$$
\begin{aligned}
i\hbar e^{i\omega t S_z/\hbar} \frac{\partial}{\partial t}|\Psi\rangle &= e^{i\omega t S_z/\hbar} H e^{-i\omega t S_z/\hbar} e^{i\omega t S_z/\hbar} |\Psi\rangle \\
&= -\{\gamma B_0 S_z + \gamma B_1 S_x\}|\Phi\rangle \,, \qquad (12.8)
\end{aligned}
$$

where $|\Phi\rangle = e^{i\omega t S_z/\hbar}|\Psi\rangle$, or, equivalently,

$$
|\Psi\rangle = e^{-i\omega t S_z/\hbar}|\Phi\rangle \,. \qquad (12.9)
$$

Evaluating the time derivative of $|\Psi\rangle$ in (12.8) then yields

$$
\begin{aligned}
i\hbar\frac{\partial}{\partial t}|\Phi\rangle &= -\{(\gamma B_0 + \omega)S_z + \gamma B_1 S_x\}|\Phi\rangle \\
&= H_{\text{eff}}|\Phi\rangle \,, \text{ say}. \qquad (12.10)
\end{aligned}
$$

Thus we have shown that, with respect to a frame of reference rotating along with $\mathbf{B}_1$ at the rate ω, the motion of the magnetic moment is the same as it would be in an *effective static magnetic field* whose x component is B_1 and whose z component is $B_0 + \omega/\gamma$. The solution to the equation of motion in the rotating frame (12.10) is formally given by $|\Phi(t)\rangle = \exp(-itH_{\text{eff}}/\hbar)|\Phi(0)\rangle$, from which the solution in the static frame may be obtained from (12.9).

Example: spin $\frac{1}{2}$

The formal solution above will now be applied to a specific problem. A particle of spin $\frac{1}{2}$ in the static magnetic field $\mathbf{B}_0$ is initially prepared so that $S_z = +\frac{1}{2}\hbar$. (This corresponds to the lowest energy state if $\gamma > 0$.) The rotating field (12.5) is applied during the time interval $0 \le t \le T$, after which it is removed. What is the probability that the particle will absorb energy from (or, if $\gamma < 0$, emit energy to) the rotating field and flip its spin?

According to the specifications of the problem, the initial state at $t = 0$ is $|\Psi(0)\rangle = |\Phi(0)\rangle = |+\rangle$. The final state at $t = T$, in the rotating frame of reference is $|\Phi(T)\rangle = \exp(-iTH_{\text{eff}}/\hbar)|+\rangle$. In the original static frame of reference, it is $|\Psi(T)\rangle = \exp(-i\omega T S_z/\hbar)\exp(-iTH_{\text{eff}}/\hbar)|+\rangle$ from (12.9). If we substitute $\mathbf{S} = \frac{1}{2}\hbar\boldsymbol{\sigma}$, and denote $\gamma B_0 = \omega_0$, $\gamma B_1 = \omega_1$, then the effective Hamiltonian in the rotating frame can be written as

$$
H_{\text{eff}} = -\tfrac{1}{2}\hbar\{(\omega_0 + \omega)\sigma_z + \omega_1\sigma_x\} \,. \qquad (12.11)
$$

A reminder of the meanings of the frequency parameters may be in order: ω is the angular frequency of the rotating field in the xy plane;

ω_0 is the frequency at which the magnetic moment would precess if only the static field B_0 were present; ω_1 is the frequency at which the magnetic moment would rotate if only a static field of magnitude B_1 were present. (We shall see that at resonance it actually does oscillate at the frequency ω_1.)

Since σ_z and σ_x are components of the vector $\boldsymbol{\sigma}$, it is apparent that H_{eff} is proportional to some component, σ_a, at an intermediate direction in the zx plane. Thus we have

$$H_{\text{eff}} = -\tfrac{1}{2}\hbar\alpha\sigma_a \,, \tag{12.12}$$

with

$$\sigma_a = \frac{(\omega_0 + \omega)\sigma_z + \omega_1\sigma_x}{\alpha} \,, \tag{12.13}$$

$$\alpha = \{(\omega_0 + \omega)^2 + {\omega_1}^2\}^{1/2} \,. \tag{12.14}$$

Note that ${\sigma_a}^2 = 1$, as a consequence of (7.47). Therefore we have the identity $\exp(ix\sigma_a) = \cos(x) + i\sigma_a \sin(x)$, which was first used in (7.67). This makes it easy to express the time development operator in 2×2 matrix form:

$$\begin{aligned}
&\exp\left(-\frac{itH_{\text{eff}}}{\hbar}\right) \\
&= 1\cos\left(\tfrac{1}{2}\alpha t\right) + i\,\sigma_a \sin\left(\tfrac{1}{2}\alpha t\right) \\
&= \begin{bmatrix} \cos\left(\tfrac{1}{2}\alpha t\right) + \frac{i(\omega_0+\omega)}{\alpha}\sin\left(\tfrac{1}{2}\alpha t\right), & \frac{i\omega_1}{\alpha}\sin\left(\tfrac{1}{2}\alpha t\right) \\ \frac{i\omega_1}{\alpha}\sin\left(\tfrac{1}{2}\alpha t\right), & \cos\left(\tfrac{1}{2}\alpha t\right) - \frac{i(\omega_0+\omega)}{\alpha}\sin\left(\tfrac{1}{2}\alpha t\right) \end{bmatrix} .
\end{aligned} \tag{12.15}$$

The initial state vector $|+\rangle$ corresponds to the two component vector, $\binom{1}{0}$. Hence for the duration of the field $\mathbf{B}_1(t), 0 \le t \le T$, the time-dependent state vector in the rotating frame is

$$|\Phi(t)\rangle = \exp\left(-\frac{itH_{\text{eff}}}{\hbar}\right)|+\rangle = a_1(t)|+\rangle + a_2(t)|-\rangle \,,$$

where the coefficients $a_1(t)$ and $a_2(t)$ are the elements of the first column of (12.15). The state vector in the static frame is given by

$$\begin{aligned}
|\Psi(t)\rangle &= \exp\left(-\frac{i\omega t\sigma_z}{2}\right)|\Phi(t)|\rangle \\
&= e^{-i\omega t/2}a_1(t)|+\rangle + e^{i\omega t/2}a_2(t)|-\rangle \,,
\end{aligned} \tag{12.16}$$

valid for $0 \le t \le T$. For $t \ge T$, after the rotating field has been removed and only the static field remains, the amplitudes of $|+\rangle$ and $|-\rangle$ remain constant with only their phase changing, as in (12.1). Thus, for $t \ge T$, we have

$$|\Psi(t)\rangle = e^{i\omega_0(t-T)/2}e^{-i\omega T/2}a_1(T)|+\rangle + e^{-i\omega_0(t-T)/2}e^{i\omega T/2}a_2(T)|-\rangle\,. \tag{12.17}$$

The probability that the spin S_z will have the value $-\frac{1}{2}\hbar$ at any time $t \ge T$ is

$$|\langle -|\Psi(t\ge T)\rangle|^2 = |a_2(T)|^2 = \left|\omega_1 \frac{\sin\left(\frac{1}{2}\alpha T\right)}{\alpha}\right|^2\,. \tag{12.18}$$

This is the probability that a particle prepared at $t = 0$ in the spin-up state and subjected to the rotating field during the interval $0 \le t \le T$ will, at the end of the experiment, be found to have its spin flipped down.

If we consider the spin flip probability (12.18) as a function of the duration T of the rotating field, then its maximum possible value will be

$$\frac{\omega_1^2}{\alpha^2} = \frac{\omega_1^2}{(\omega_0+\omega)^2 + {\omega_1}^2} \le 1\,.$$

This expression achieves its maximum value when $\omega_0 + \omega = 0$ or, equivalently, $\omega = -\gamma B_0$, which is known as the *resonance* condition. Now γ may be of either sign, and so may ω. A positive value of ω means that the field $\mathbf{B}_1$ in (12.5) rotates in the positive (counterclockwise) direction in the xy plane. Unfortunately the situation is complicated by the following fact: from (12.3) it follows that a positive magnetic moment in a positively directed magnetic field will precess in the negative sense, and hence a positive value of ω_0 implies precession in the negative (clockwise) direction.

At *resonance* the spin flip probability (12.18) becomes

$$|\langle -|\Psi(t\ge T)\rangle|^2 = \left|\sin\left(\tfrac{1}{2}\omega_1 T\right)\right|^2\,. \tag{12.19}$$

The origin of this result can be most easily seen by reference to the effective Hamiltonian (12.11) in the rotating frame. When $\omega_0 + \omega = 0$, the effect of the rotation exactly cancels the effect of the static field B_0 in the z direction, and the magnetic moment precesses around field

B_1 which points along the x axis of the rotating frame. By choosing a suitable value of the product $\omega_1 T = \gamma B_1 T$, one can rotate the magnetic moment through any desired angle with respect to the z axis. This technique is very useful in nuclear magnetic resonance experiments.

The only property of a spin $\frac{1}{2}$ system that was essential to this analysis is that the state space be two-dimensional, and hence analogous results will hold for any two-state system. One such useful analog is the so-called *two-state atom*, in which the excitation mechanism is such that only the ground state and one excited state are significantly populated. If we denote those state vectors as $|E_1\rangle$ and $|E_2\rangle$, then we can define spin-like operators σ_1, σ_2, and σ_3 such that $\sigma_3|E_1\rangle = |E_1\rangle$, $\sigma_3|E_2\rangle = -|E_2\rangle$, $\sigma_1|E_1\rangle = |E_2\rangle$, $\sigma_1|E_2\rangle = |E_1\rangle$, $\sigma_2|E_1\rangle = i|E_2\rangle$, and $\sigma_2|E_2\rangle = -i|E_1\rangle$. These are just the same relations that are satisfied by the spin angular momentum operators σ_x, σ_y, and σ_z on the eigenvectors of σ_z. Hence the formalism of spin resonance can also be applied to a two-state atom or to any two-state system.

12.2 *Exponential and Nonexponential Decay*

There are many examples of spontaneous decay of unstable systems: radioactive disintegration of nuclei and the decay of atomic excited states are the most familiar cases. These decay processes are commonly found to be describable by an exponential formula. The survival probability of an undecayed state is an exponentially decreasing function of time, or, in the case of a large number of noninteracting unstable systems, the number of surviving systems decreases exponentially with time.

The exponential decay law

The exponential decay law can be derived from a simple plausible argument. Denote by the symbol $u(t)$ the event of the system being in the undecayed state at time t. Then $P(t_2, t_1) = \text{Prob}\{u(t_2)|u(t_1)\}$, $(t_2 > t_1)$, is the probability that the system remains undecayed at time t_2 conditional on its having been undecayed at t_1. Implicit in this notation is the *assumption* that the probability in question depends only on the information specified at time t_1, and that the history earlier than t_1 is irrelevant. Since the laws of nature do not depend on the particular epoch, the probability should not depend separately on t_2 and t_1, but only on their difference, and hence $P(t_2, t_1) = P(t_2 - t_1)$. Now, for any three times, $t_1 < t_2 < t_3$, it must be the case that

$$\text{Prob}\{u(t_3)\,\&\,u(t_2)|u(t_1)\} = \text{Prob}\{u(t_3)|u(t_1)\}\,,$$

since if it was undecayed at time t_3 it must also have been undecayed at the earlier time t_2. But a fundamental rule of probability theory (Sec. 1.5, Axiom 4) implies that

$$\begin{aligned}&\text{Prob}\{u(t_3)\ \&\ u(t_2)|u(t_1)\}\\&\quad= \text{Prob}\{u(t_3)|u(t_2)\ \&\ u(t_1)\} \times \text{Prob}\{u(t_2)|u(t_1)\}\,.\end{aligned}$$

Thus, by combining these two equations, we obtain

$$P(t_3 - t_1) = P(t_3 - t_2)\,P(t_2 - t_1)\,.$$

The only continuous solution to this functional equation is the exponential, $P(t) = e^{-\lambda t}$, with $\lambda > 0$ so that P does not exceed 1.

From this result, we can calculate the probability that the system will decay within an arbitrary time interval from t_1 to t_2. This is the probability that it is decayed at t_2 conditional on its having been undecayed at t_1,

$$\begin{aligned}\text{Prob}\{\sim u(t_2)|u(t_1)\} &= 1 - \text{Prob}\{u(t_2)|u(t_1)\}\\&= 1 - \exp\{-\lambda(t_2 - t_1)\}\,,\end{aligned}$$

which is approximately equal to $\lambda(t_2 - t_1)$ if $(t_2 - t_1)$ is very small. Thus the decay probability per unit time, for a small time interval, is just equal to the constant λ. This is another way in which the exponential decay law may be characterized.

We have seen that the exponential decay law follows necessarily from the one assumption above, namely that the probability of survival from t_1 to t_2, $\text{Prob}\{u(t_2)|u(t_1)\}$, depends only on the condition (undecayed rather than decayed) of the system at t_1, and does not depend on the previous history. Although this assumption is very plausible, the existence of nonexponential decays must be interpreted as exceptions to the validity of that assumption.

The decay probability in quantum mechanics

Suppose that a system is prepared at time $t = 0$ in a state Ψ_u. This might be an atomic state that has been excited by a laser pulse, or it might be the naturally occurring state of a radioactive nucleus. A system that has been subjected to this preparation, be it artificial or natural, will exhibit some distinctive characteristic, which we will designate by "u" (for "undecayed"). (For the radioactive nucleus the property "u" would be the existence of the nucleus as a single particle, rather than in several fragments.) At time t, the state vector (in the Schrödinger picture) will evolve to be $e^{-iHt/\hbar}|\Psi_u\rangle$, and the probability that the system retains the property "u" at time t will be

$$P_{\mathrm{u}}(t) = |A(t)|^2\,, \quad \text{with } A(t) = \langle \Psi_{\mathrm{u}}|e^{-iHt/\hbar}|\Psi_{\mathrm{u}}\rangle\,. \tag{12.20}$$

We are interested in spontaneous decays, so it is appropriate to assume that the Hamiltonian H is independent of time, since any time dependence in H would be due to some external force.

It is easy to show that the quantum-mechanical decay law (12.20) is not exactly exponential. For small t we may use a Taylor expansion of the amplitude, $A(t) = 1 - i\langle H\rangle t/\hbar - \langle H^2\rangle t^2/2\hbar^2 + \cdots$, and hence the survival probability is

$$\begin{aligned} P_{\mathrm{u}}(t) &= 1 - \frac{\langle H\rangle t^2}{\hbar^2} + \frac{\langle H\rangle^2 t^2}{\hbar^2} \\ &= 1 - \frac{\langle (H - \langle H\rangle)^2\rangle t^2}{\hbar^2}\,, \end{aligned} \tag{12.21}$$

where the averages are taken in the state Ψ_{u}. Therefore, if $\langle H\rangle$ and $\langle H\rangle^2$ are finite, it follows that $P_{\mathrm{u}}(t)$ must be parabolic in form, rather than exponential, for short times.

Further insight into the nonexponential component of the decay law can be obtained by writing the time-independent state vector in the form

$$e^{-iHt/\hbar}|\Psi_{\mathrm{u}}\rangle = A(t)|\Psi_{\mathrm{u}}\rangle + |\Phi(t)\rangle\,, \tag{12.22}$$

where $|\Phi(t)\rangle$ is orthogonal to the undecayed state,

$$\langle \Psi_{\mathrm{u}}|\Phi(t)\rangle = 0\,, \tag{12.23}$$

and so should be interpreted as describing decay products. Now, by applying the operator $e^{-iHt'/\hbar}$ to (12.22) and taking the inner product with $\langle \Psi_{\mathrm{u}}|$, we obtain

$$A(t + t') = A(t)A(t') + \langle \Psi_{\mathrm{u}}|e^{-iHt'/\hbar}|\Phi(t)\rangle\,. \tag{12.24}$$

If the last term were to vanish, then the amplitude $A(t)$ would necessarily be an exponential function. The deviations from exponential decay are therefore due to the fact that upon further evolution from time t to $t + t'$ the previous state of the decay products, described by the vector $|\Phi(t)\rangle$, does not remain orthogonal to the original state $|\Psi_{\mathrm{u}}\rangle$. In other words, the undecayed state is being at least partially regenerated.

The survival probability as a function of time is fully determined by the *energy spectral content* of the initial state. To see this, we expand the initial state vector in terms of the energy eigenvectors,

$$|\Psi_{\mathrm{u}}\rangle = \sum_n |E_n\rangle\langle E_n|\Psi_{\mathrm{u}}\rangle\,, \tag{12.25}$$

where $H|E_n\rangle = E_n|E_n\rangle$. The amplitude in (12.20) may then be written as

$$A(t) = \sum_n |\langle E_n|\Psi_u\rangle|^2 e^{-iE_nt/\hbar}$$
$$= \int \eta(E)e^{-iE_nt/\hbar}dE\,, \tag{12.26}$$

where

$$\eta(E) = \sum_n |\langle E_n|\Psi_u\rangle|^2\delta(E - E_n) \tag{12.27}$$

is the *spectral function* for the state Ψ_u. If the spectrum of H is continuous, then the sum in (12.27) should be an integral, and $\eta(E)$ may be a continuous function. Since $A(t)$ and $n(E)$ are related by a Fourier transform, and the only restriction on $\eta(E)$ is that it cannot be negative, it appears that there can be no universal decay law for unstable states. Every different choice for the spectral content of the initial state leads to a different time dependence of the decay. Attention must therefore be shifted to the nature of the unstable states that are likely to occur in practice.

Suppose that the spectral function of the state were to have the form of a Lorentzian distribution,

$$\eta(E) = \frac{\frac{1}{2}\lambda\hbar/\pi}{(E - E_0)^2 + (\frac{1}{2}\lambda\hbar)^2}\,. \tag{12.28}$$

Then (12.26) could be evaluated as a contour integral, yielding

$$A(t) = \exp\left(-\frac{1}{2}\lambda t - \frac{iE_0t}{\hbar}\right), \tag{12.29}$$

and an exponential survival probability, $P(t) = e^{-\lambda t}$. Since the Fourier transform relation between $\eta(E)$ and $A(t)$ is a one-to-one correspondence, it follows that only a state with a Lorentzian spectral function can have an exactly exponential decay law. The short time result (12.21) is not applicable to the Lorentzian distribution because it has no finite moments. That is to say, the integrals $\langle H^n\rangle \equiv \int E^n\eta(E)dE$ $(n > 0)$ are not convergent at their infinite upper and lower limits. Therefore the Taylor series expansion (12.21) does not exist in this case.

Any real physical system has a lower bound to its energy spectrum, and so we must have $\eta(E) = 0$ for $E < E_{\text{min}}$. This will clearly cause some deviation from exponential decay. Indeed, it can be shown (Khalfin, 1958) that the existence of a lower bound to the spectrum implies that at sufficiently long times the decay must be slower than any exponential.

This analysis has shown that the familiar exponential "law" of decay must, in fact, be an approximation whose validity depends on special properties of the unstable states that occur commonly. One such example, the *virtual bound state* of resonant scattering, will be discussed in Sec. 16.6. Winter (1961) has analyzed an exactly solvable model of an unstable state that decays by tunneling through a potential barrier. The analysis is too lengthy to reproduce here, but if the results may be taken as typical, then the course of a decay is as follows. For a short time the decay will be nonexponential, initially parabolic as is (12.21). This phase lasts a relatively short time, and in many cases, such as natural radioactivity, it will have escaped observation. The second phase is one of approximately exponential decay. It lasts several times as long as the characteristic exponential "lifetime", λ^{-1}, and during this phase the undecayed population decreases by many orders of magnitude. This is followed by a final phase of slower-than-exponential decay. The radioactive intensity is by now so small that, although observable in principle, it may escape observation because it is so weak.

The "watched pot" paradox

This paradox is amusing, but also instructive since it has implications for the interpretation of quantum mechanics. The paradox arises within the interpretation (A) in Sec. 9.3, according to which a state vector is attributed to each individual system. If any system is observed to be undecayed, it will be assigned the state vector $|\Psi_u\rangle$, within that interpretation. Although that interpretation has superficial plausibility, and was once widely accepted, it has been rejected in this book. Some of the reasons were given in Ch. 9, and this paradox provides further evidence against it.

Suppose that an unstable system, initially in the state $|\Psi_u\rangle$, is observed n times in a total interval of duration t; that is to say, it is observed at the times $t/n, 2t/n, \ldots, t$. Since t/n is very small, the probability that the system remains undecayed at the time of the first observation is given by (12.21) to be $P_u(t/n) = 1 - (\sigma t/n\hbar)^2$, where $\sigma^2 = \langle (H - \langle H\rangle)^2\rangle$. Now, according to the interpretation (A), whose consequences are being explored, the observation of no decay at time t/n implies that the state is still $|\Psi_u\rangle$. Thus the probability of survival between the first and the second observation will also be equal to $P_u(t/n)$, and so on for each successive observation. The probability of survival in state $|\Psi_u\rangle$ at the end of this sequence of n independent observations is the product of the probabilities for surviving each of the short intervals, and thus

$$P_u(t) = [P_u(t/n)]^n = \left[1 - \left(\frac{\sigma t}{n\hbar}\right)^2\right]^n . \tag{12.30}$$

We now pass to the limit of continuous observation by letting n become infinite. The limit of the logarithm of (12.30) is

$$\begin{aligned} \log P_u(t) &= n \log \left[1 - \left(\frac{\sigma t}{\hbar n}\right)^2\right] \\ &= n\left[-\left(\frac{\sigma t}{\hbar n}\right)^2 - O(n^{-4})\right] \to 0 \quad \text{as} \quad n \to \infty . \end{aligned}$$

Thus we obtain $P_u(t) = 1$ in the limit of continuous observation. Like the old saying "A watched pot never boils," we have been led to the conclusion that a continuously observed system never changes its state!

This conclusion is, of course, false. The fallacy clearly results from the assertion that if an observation indicates no decay, then the state vector must be $|\Psi_u\rangle$. Each successive observation in the sequence would then "reduce" the state back to its initial value $|\Psi_u\rangle$, and in the limit of continuous observation there could be no change at all. The notion of "reduction of the state vector" during measurement was criticized and rejected in Sec. 9.3. A more detailed critical analysis, with several examples, has been given by Ballentine (1990). Here we see that it is disproven by the simple empirical fact that continuous observation does not prevent motion. It is sometimes claimed that the rival interpretations of quantum mechanics differ only in philosophy, and cannot be experimentally distinguished. That claim is not always true, as this example proves.

12.3 *Energy–Time Indeterminacy Relations*

The rms half-widths of a function $f(t)$ and its Fourier transform $g(\omega) = \int e^{i\omega t} f(t)dt$ are related by the classical inequality $\Delta_f \Delta_g \geq \frac{1}{2}$. Since the position and momentum representations are connected by a Fourier transform, this classical inequality can be used to derive the position–momentum indeterminacy relation (8.33), $\Delta_Q \Delta_P \geq \frac{1}{2}\hbar$. The putative identification of ω with $E/\hbar$ would then lead, by analogy, to an energy–time indeterminacy relation. But the analogy between (P, Q) and (E, t) breaks down under closer scrutiny. The meaning, and even the existence, of an energy–time indeterminacy relation has long been a subject of confusion. [Peres (1993, Sec. 12.8) gives a lucid analysis of the controversy.]

The derivation of an energy–time relation by analogy with the properties of Fourier transforms is unsound because the relation between frequency and energy is not $\omega = E/\hbar$, but rather $\omega = (E_1 - E_2)/\hbar$. A frequency is not associated with an energy level, but with the difference between two energy levels. The significance of this distinction is apparent from the fact that a frequency is directly measurable, whereas the energy can be altered by the addition of an arbitrary constant without producing observable effects.

The position–momentum indeterminacy relation, $\Delta_Q\Delta_P \geq \frac{1}{2}\hbar$, asserts that the product of the rms half-widths of the position and the momentum probability distributions cannot be less than the constant $\frac{1}{2}\hbar$. But there is no probability distribution for time, which is not a dynamical variable. (Indeed, the term *dynamical variable* refers to something that can vary as a function of time, thereby excluding *time* itself.) So any analogy between (P, Q) and (E, t) can only be superficial.

In the formalism of quantum theory, time enters as a parameter, and is not represented by an operator. One might want to restore the symmetry between space and time by introducing a time operator T, which would be required to satisfy the commutation relation $[T, H] = i\hbar$. However, it was shown by W. Pauli in 1933 that the operator T does not exist if the eigenvalue spectrum of H is bounded below. Suppose that a self-adjoint operator T satisfying the desired commutation relation were to exist. Then the unitary operator $e^{i\alpha T}$ would generate a displacement in energy, just as the operator $e^{i\beta Q}$ produces a displacement in momentum and $e^{i\gamma P}$ produces a displacement in position. Thus if $H|E\rangle = E|E\rangle$, then we should have $He^{i\alpha T}|E\rangle = (E + \alpha\hbar)\, e^{i\alpha T}|E\rangle$ for arbitrary real α. But this is inconsistent with the existence of a lower bound to the spectrum of H, and so the initial supposition that the operator T exists must be false.

In practice, the quantitative determination of the passage of time is not obtained by measuring a special *time* variable, but rather by observing the variation of some other dynamical variable, which can serve as a clock. This suggests another approach to the problem. Let us apply the general form of the indeterminacy relation (8.31), which applies to an arbitrary pair of dynamical variables, to the Hamiltonian H and any other dynamical variable whose operator R does not commute with H. From (8.31) we have $\Delta_R\Delta_E \geq \frac{1}{2}|\langle [R, H]\,\rangle|$, where Δ_R and Δ_E are the rms half-widths of the probability distributions for R and for energy, respectively, in the state under consideration. With the help of (3.74), this inequality can be written as

$$\Delta_R \Delta_E \geq \tfrac{1}{2}\hbar \frac{d\langle R \rangle}{dt} \,. \tag{12.31}$$

This is often called the *Mandelstam–Tamm inequality.*

We can define a characteristic time for the variation of R,

$$\tau_R = \Delta_R \left\{ \frac{d\langle R \rangle}{dt} \right\}^{-1} , \tag{12.32}$$

from which we obtain an energy–time indeterminacy relation of the form

$$\tau_R \Delta_E \geq \tfrac{1}{2}\hbar \,. \tag{12.33}$$

In this relation, Δ_E is a standard measure of the statistical spread of energy in the state, but τ_R is not an indeterminacy or statistical spread in a time variable. It is, rather, a characteristic time for variability of phenomena in this state. Note that τ_R depends on both the particular dynamical variable R and the state, and that it may vary with time.

Several other useful results can be derived from (12.31). [See Uffink (1993) for a survey.] Let the initial state at $t = 0$ be pure, and substitute its projection operator, $|\psi(0)\rangle\langle\psi(0)|$, for R in (12.31). Then we obtain $(\Delta_R)^2 \equiv \langle R^2 \rangle - \langle R \rangle^2 = \langle R \rangle (1 - \langle R \rangle)$, with $\langle R \rangle = |\langle \psi(t)|\psi(0)\rangle|^2$. This is the *survival probability* of the initial state, introduced in (12.20), which we shall here denote as

$$P(t) = |\langle \psi(t)|\psi(0)\rangle|^2 \,. \tag{12.34}$$

From (12.31), we now obtain $\{P(1-P)\}^{1/2}\Delta_E \geq \frac{1}{2}\hbar|dP/dt|$. Solving for Δ_E and integrating with respect to t then yields

$$\Delta_E t \geq \frac{1}{2}\hbar \int_0^t \{P(1-P)\}^{-1/2} \frac{dP}{dt} dt = \hbar \cos^{-1}(\sqrt{P}) \,. \tag{12.35}$$

The shortest time at which the survival probability drops to $\frac{1}{2}$ is called the *half-life*, $\tau_{1/2}$. From (12.35) we obtain the inequality

$$\Delta_E \tau_{1/2} \geq \frac{\pi\hbar}{4} \,. \tag{12.36}$$

The shortest time required for $|\psi(t)\rangle$ to become orthogonal to the initial state, denoted as τ_0, is the minimum time for destruction of the initial state to be complete. It is restricted by

$$\Delta_E \tau_0 \geq \frac{\pi\hbar}{2} \,. \tag{12.37}$$

These inequalities are useful if the second moment of the energy distribution, $(\Delta_E)^2$, is finite. But there are cases, such as the Lorentzian distribution (12.28), that have no finite moments, and so Δ_E is infinite. Therefore another approach that does not rely on moments is needed. Consider an initial state vector $|\psi(0)\rangle$ with an arbitrary energy distribution $|\langle E|\psi(0)\rangle|^2$, which will be independent of time. Define $W(\alpha)$ to be the size of the shortest interval W such that

$$\int_W |\langle E|\psi(0)\rangle|^2 dE = \alpha\,.$$

A reasonable measure of the width of the energy distribution is $W(\alpha)$ for some value of α, such as $\alpha = 0.9$. Let τ_β be the minimal time for the survival probability (12.34) to fall to the value β. Let P_W be the projection operator onto the subspace spanned by those energy eigenvectors in the energy range W, and let $P_{W\perp}$ be the projector onto the complementary subspace. We can then write the state vector as

$$\begin{aligned}|\psi(t)\rangle &= P_W|\psi(t)\rangle + P_{W\perp}|\psi(t)\rangle \\ &= \sqrt{\alpha}\,|\psi_W(t)\rangle + \sqrt{(1-\alpha)}\,|\psi_{W\perp}(t)\rangle\,. \end{aligned} \tag{12.38}$$

The vectors $|\psi_W(t)\rangle$ and $|\psi_{W\perp}(t)\rangle$ are orthogonal, and are chosen to have unit norms. Since $P_W P_{W\perp} = 0$, the inner product of (12.38) with $\langle\psi(0)|$ is $\langle\psi(0)|\psi(t)\rangle = \alpha\langle\psi_W(0)|\psi_W(t)\rangle + (1-\alpha)\langle\psi_{W\perp}(0)|\psi_{W\perp}(t)\rangle$, from which we obtain the inequality

$$|\langle\psi(0)|\psi(t)\rangle| + (1-\alpha)|\langle\psi_{W\perp}(0)|\psi_{W\perp}(t)\rangle| \geq \alpha|\langle\psi_W(0)|\psi_W(t)\rangle|\,.$$

We evaluate this expression for $t = \tau_\beta$, so the first term has the value $\sqrt{\beta}$. The absolute value in the second term is bounded by 1, so we obtain

$$\frac{1-\alpha+\sqrt{\beta}}{\alpha} \geq |\langle\psi_W(0)|\psi_W(\tau_\beta)\rangle|\,. \tag{12.39}$$

Now the inequality (12.35) can be applied to the survival probability of $|\psi_W(0)\rangle$, instead of $|\psi(0)\rangle$, yielding

$$\cos^{-1}\left(|\langle\psi_W(0)|\psi_W(\tau_\beta)\rangle|\right) \leq \frac{\Delta_W \tau_\beta}{\hbar}\,.$$

Here Δ_W is the rms half-width of the energy distribution of the state ψ_W. But, by construction, its absolute width is $W(\alpha)$. Therefore $\Delta_W \leq \frac{1}{2}W(\alpha)$. Taking the inverse cosine of (12.39), we obtain

$$W(\alpha)\tau_\beta \geq 2\hbar\cos^{-1}\left(\frac{1-\alpha+\sqrt{\beta}}{\alpha}\right), \tag{12.40}$$

with the restriction $\sqrt{\beta} \le 2\alpha - 1$, since the argument of the inverse cosine cannot exceed 1. This result can be applied to all states, regardless of whether their energy distributions have finite moments. If, for illustration, we take $\beta = \frac{1}{2}$ and $\alpha = 0.9$, then it yields the inequality $W(\alpha)\tau_{1/2} \ge 0.917\hbar$.

Perhaps it is now possible to resolve the long-standing controversy over energy–time indeterminacy relations with the following conclusion. There is no energy–time relation that is closely analogous to the well-known position–momentum indeterminacy relation (8.33). However, there are several useful inequalities relating some measure of the width of the energy distribution to some aspect of the time dependence. But none of these inequalities has such a priority as to be called *the* energy–time indeterminacy relation.

12.4 *Quantum Beats*

If a coherent superposition of two or more discrete energy states is excited, the resulting nonstationary state will exhibit a characteristic time dependence at the frequencies corresponding to the differences between those energy levels. The resulting modulations of observable phenomena at those frequencies are known as *quantum beats.*

The time dependence of such nonstationary states can be observed in neutron interferometry. Strictly speaking this is not an example of quantum beats, but it is a simpler case that exhibits similar phenomena. The experimental setup is similar to that of Fig. 9.2. A neutron beam with spin polarized in the z direction is split by Bragg reflection into two spatially separated beams. The spin of one beam is flipped, and the two beams are then recombined. But whereas in Sec. 9.5 the spin flip was accomplished by precession in a static magnetic field, it is now accomplished by spin resonance. The entire apparatus is immersed in a static magnetic field of magnitude B_0 in the z direction, and a small radio frequency (r.f.) coil supplies a perturbation to one of the beams at the resonant frequency $\omega_0 = \gamma B_0$. After spin flip the energies of the two beams will differ by $\Delta E = \hbar\gamma B_0$. The spin state of the recombined beam will now be of the form

$$|\Psi(t)\rangle = \frac{e^{i\omega_0 t/2}|+\rangle + e^{-i\omega_0 t/2}|-\rangle}{\sqrt{2}}, \tag{12.41}$$

where the vectors $|+\rangle$ and $|-\rangle$ are eigenvectors of the z components of spin. This is a nonstationary state with the spin polarization rotating in the xy plane: $\langle\sigma_x\rangle = \cos(\omega_0 t), \langle\sigma_y\rangle = -\sin(\omega_0 t), \langle\sigma_z\rangle = 0$. If the x component of spin is analyzed, the probability of obtaining a positive value will have an oscillatory dependence on the time elapsed since the particle emerged from the r.f. coil. This prediction has been experimentally confirmed by Badurek *et al.* (1983).

A similar time-dependent effect can be observed in atomic spectroscopy. Consider an atom that has a ground state $|a\rangle$, and two closely spaced excited states $|b\rangle$ and $|c\rangle$ (Fig. 12.1).

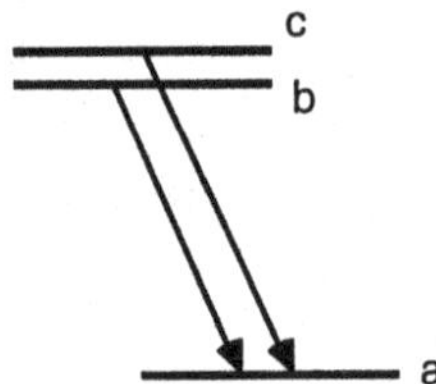

Fig. 12.1 Quantum beats. The atom is excited into a coherent superposition of the two upper states. Its spontaneous emission intensity will be modulated at the beat frequency $(E_c - E_b)/\hbar$.

A short laser pulse can excite the atom into a coherent superposition of the two upper states. This will be a nonstationary state, of the form

$$|\Psi(t)\rangle = \alpha e^{-iE_b t}|b\rangle + \beta e^{-iE_c t}|c\rangle \,. \tag{12.42}$$

The atom will decay from this excited state by spontaneous emission of radiation.

If the spontaneous emission radiation could be treated classically, the radiation field would be described as a sum of two components whose frequencies are $\omega_{ba} = (E_b - E_a)\hbar$ and $\omega_{ca} = (E_c - E_a)/\hbar$. In view of the identity

$$\sin(\omega_{ca}t) + \sin(\omega_{ba}t) = 2\sin\left[\frac{1}{2}(\omega_{ca} + \omega_{ba})t\right]\cos\left[\frac{1}{2}(\omega_{ca} - \omega_{ba})t\right],$$

we should expect a radiation field at the mean frequency, $\frac{1}{2}(\omega_{ca}+\omega_{ba})$, with its amplitude modulated at the frequency $\frac{1}{2}(\omega_{ca}-\omega_{ba}) = \frac{1}{2}(E_c - E_b)\hbar$. The radiation intensity is the square of the field, and so the intensity will be modulated at twice that frequency, $\omega_{cb} = (E_c - E_b)\hbar$.

A complete theory of this quantum beat effect requires that the radiation field be treated quantum-mechanically. This will be done in Ch. 19. However, a qualitative description can be obtained if we recognize that the measured intensity of a classical radiation field is proportional to the probability of the detector absorbing a photon from the field. Thus the probability of detecting a photon will not be a monotonic function of the time elapsed since the atom was excited; rather, it will be modulated at the quantum beat frequency $\omega_{cb} = (E_c - E_b)\hbar$, as illustrated in Fig. 12.2. The smaller the spacing of the energy levels, E_c and E_b, the longer will be the period of the modulation T_m,

$$T_m = \frac{2\pi}{\omega_{cb}} = \frac{2\pi\hbar}{E_c - E_b}. \tag{12.43}$$

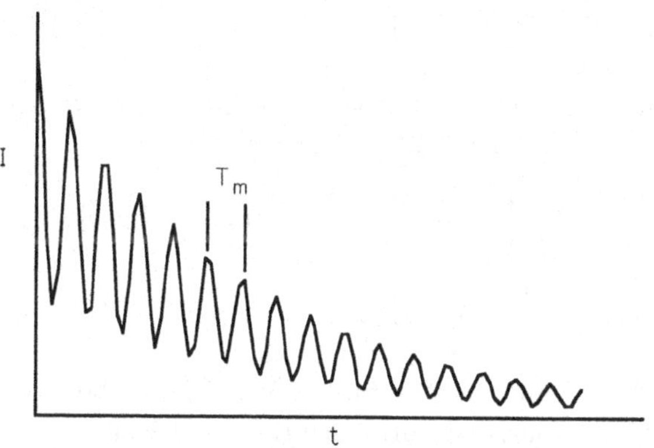

Fig. 12.2 Intensity versus time for a quantum beat signal.

This fact has made possible the technique of *quantum beat spectroscopy*, which can resolve two very closely spaced energy levels, so close that it would be impossible to resolve the separate radiation frequencies, ω_{ca} and ω_{ba}.

12.5 *Time-Dependent Perturbation Theory*

It is possible to solve time-dependent problems by a form of perturbation theory. Consider a time-dependent Schrödinger equation,

$$i\hbar\frac{d|\Psi(t)\rangle}{dt} = [H_0 + \lambda H_1(t)]|\Psi(t)\rangle\,, \tag{12.44}$$

in which the Hamiltonian is of the form $H = H_0 + \lambda H_1(t)$, with the time dependence confined to the perturbation term $\lambda H_1(t)$. We may anticipate that the perturbation must be small, but it is not yet obvious what the appropriate condition of smallness might be.

The eigenvalues and eigenvectors of H_0 are assumed to be known:

$$H_0|n\rangle = \varepsilon_n|n\rangle\,. \tag{12.45}$$

Since the set of eigenvectors $\{|n\rangle\}$ is complete, it can be used as a basis for expansion of $|\Psi(t)\rangle$:

$$|\Psi(t)\rangle = \sum_n a_n(t) e^{-i\varepsilon_n t/\hbar}|n\rangle\,. \tag{12.46}$$

If $\lambda = 0$ the general solution of (12.44) is of the form (12.46) with the coefficients a_n being constant in time. Therefore, if λ is nonzero but small, we expect the time dependence of $a_n(t)$ to be weak, or, in other words, $da_n(t)/dt$ should be small. This is the intuitive idea that motivates time-dependent perturbation theory.

Substituting (12.46) into (12.44), performing the differentiation, and using the eigenvalue equation (12.45), we obtain

$$\sum_n \left\{ i\hbar \frac{da_n(t)}{dt} + \varepsilon_n\, a_n(t) \right\} e^{-i\varepsilon_n t/\hbar}|n\rangle$$
$$= \sum_n \{\varepsilon_n\, a_n(t) + \lambda H_1 a_n(t)\} e^{-i\varepsilon_n t/\hbar}|n\rangle\,,$$

where the second and third terms cancel. The orthonormality of the basis vectors leads to a matrix equation for the coefficients,

$$i\hbar\, \frac{da_m(t)}{dt} = \lambda \sum_n \langle m|H_1(t)|n\rangle e^{i\omega_{mn}t}\, a_n(t)\,, \tag{12.47}$$

where $\omega_{mn} = (\varepsilon_m - \varepsilon_n)/\hbar$. This equation, which is exact, shows that the time dependence of $a_n(t)$ is entirely due to the perturbation λH_1, confirming our intuitive notions. The phase factors in (12.46) have absorbed all of the time dependence due to H_0.

The perturbation approximation is introduced by expanding the coefficients in powers of λ,

$$a_n(t) = a_n{}^{(0)} + \lambda^1\, a_n{}^{(1)} + \lambda^2 a_n{}^{(2)} + \cdots\,, \tag{12.48}$$

substituting this expansion into (12.47), and collecting powers of λ. In zeroth order we merely recover the known result $da_n{}^{(0)}/dt = 0$. In the first order we obtain

$$i\hbar\, \frac{da_m{}^{(1)}(t)}{dt} = \sum_n \langle m|H_1(t)|n\rangle e^{i\omega_{mn}t}\, a_n{}^{(0)}\,, \tag{12.49}$$

and in order $r+1$ we obtain

$$i\hbar\, \frac{da_m{}^{(r+1)}(t)}{dt} = \sum_n \langle m|H_1(t)|n\rangle e^{i\omega_{mn}t}\, a_n{}^{(r)}(t)\,. \tag{12.50}$$

The zeroth order coefficients ${a_n}^{(0)}$ are obtained from the initial state, $|\Psi(0)\rangle = \sum_n {a_n}^{(0)}|n\rangle$, which must be given in any particular problem. These are fed into (12.49), which can then be integrated to obtain the first order coefficients ${a_n}^{(1)}(t)$. The first order coefficients can then be used to calculate the second order coefficients, and so on.

A *typical problem* is of the following form. For times $t \leq 0$ the Hamiltonian is H_0, and the system is in a state of energy ε_i, described by the stationary state vector $|\Psi(t)\rangle = e^{-i\varepsilon_i t/\hbar}|i\rangle$. The perturbation $\lambda H_1(t)$ is applied during the time interval $0 \leq t \leq T$, during which the coefficients $a_n(t)$ in (12.46) will be variable. For times $t \geq T$ the perturbation vanishes, and the coefficients will retain the constant values $a_n(t) = a_n(T)$. The probability that, as a result of this perturbation, the energy of the system will become ε_f, is equal to $|a_f(T)|^2$. (We assume for simplicity that the eigenvalue ε_f is nondegenerate.) The required amplitude is obtained to the first order from (12.49):

$$a_f(T) \approx \lambda {a_f}^{(1)}(T) = (i\hbar)^{-1} \int_0^T \langle f|\lambda H_1(t)|i\rangle e^{i\omega_{fi}t}\, dt \quad (f \neq i)\,. \tag{12.51}$$

Notice that it involves only the Fourier component of the perturbation at the frequency corresponding to the difference between the final and initial energies, $\omega_{fi} = (\varepsilon_f - \varepsilon_i)/\hbar$.

The amplitude $a_i(T)$ will be diminished from its initial value of $a_i(0) = 1$. Although it can also be calculated from (12.49), its magnitude is more easily obtained from the normalization of the state vector (12.46),

$$1 = \langle\Psi(t)|\Psi(t)\rangle = |a_i(t)|^2 + \sum_{n\neq i} |a_n(t)|^2\,.$$

Now $|a_n(t)| = O(\lambda)$ for $n \neq i$, so we have

$$|a_i(t)| = [1 - O(\lambda^2)]^{1/2} = 1 - O(\lambda^2)\,. \tag{12.52}$$

To the first order the perturbation affects $a_i(t)$ only in its phase.

[[When problems of this sort are discussed formally, it is common to speak of the perturbation as causing *transitions* between the eigenstates H_0. If this means only that the system has absorbed from the perturbing field (or emitted to it) the energy difference $\hbar\omega_{fi} = \varepsilon_f - \varepsilon_i$, and so has changed its energy, there is no harm in such language. But if the statement is interpreted to mean that the state has changed from its initial value of $|\Psi(0)\rangle = |i\rangle$ to a final value of $|\Psi(T)\rangle = |f\rangle$, then it is incorrect. The perturbation leads to a final state $|\Psi(t)\rangle$, for $t \geq T$, that is of the form (12.46) with

$a_n(t)$ replaced by $a_n(T)$. It is not a stationary state, but rather it is a coherent superposition of eigenstates of H_0. The interference between the terms in (12.46) is detectable, though of course it has no effect on the probability $|a_f(T)|^2$ for the final energy to be $E = \varepsilon_f$. The spin-flip neutron interference experiments of Badurek *et al.* (1983), which were discussed in Sec. 12.4, provide a very clear demonstration that *the effect of a time-dependent perturbation is to produce a nonstationary state*, rather than to cause a jump from one stationary state to another. The ambiguity of the informal language lies in its confusion between the two statements, "the energy is ε_f" and "the state is $|f\rangle$". If the state vector $|\Psi\rangle$ is of the form (12.46) it is correct to say that the probability of the energy being ε_f is $|a_f|^2$. In the formal notation this becomes $\text{Prob}(E = \varepsilon_f|\Psi) = |a_f|^2$, which is a correct formula of quantum theory. But it is nonsense to speak of the probability of the state being $|f\rangle$ when in fact the state is $|\Psi\rangle$.]]

Harmonic perturbation

Further analysis is possible only if we choose a specific form for the time dependence of the perturbation. We shall now specialize to a sinusoidal time dependence, since it is often encountered in practice. We shall put $\lambda = 1$, since λ was only a bookkeeping device for the derivation of (12.49) and (12.50). The perturbation is taken to be

$$H_1(t) = H'e^{-i\omega t} + H'^{\dagger}e^{i\omega t} \quad (0 \le t \le T)\,, \tag{12.53}$$

and to vanish outside the interval $0 \le t \le T$. Both positive and negative frequency terms must be included in order that the operator $H_1(t)$ be Hermitian. At any time $t \ge T$, the first order amplitude of the component $|f\rangle$ $(f \ne i)$ in (12.46) will be

$$\begin{aligned} a_f{}^{(1)}(T) &= (i\hbar)^{-1}\,\langle f|H'|i\rangle \int_0^T e^{i(\omega_{fi}-\omega)t}\,dt \\ &\quad + (i\hbar)^{-1}\,\langle f|H'^{\dagger}|i\rangle \int_0^T e^{i(\omega_{fi}+\omega)t}\,dt \\ &= \frac{\langle f|H'|i\rangle}{\hbar}\,\frac{1-e^{i(\omega_{fi}-\omega)T}}{\omega_{fi}-\omega} + \frac{\langle f|H'^{\dagger}|i\rangle}{\hbar}\,\frac{1-e^{i(\omega_{fi}+\omega)T}}{\omega_{fi}+\omega}\,. \end{aligned} \tag{12.54}$$

The square of this amplitude, $|a_f{}^{(1)}(T)|^2$, is the probability that the final energy of the system will be ε_f, on the condition that the initial energy was ε_i (assuming nondegenerate energy eigenvalues).

Example: Spin resonance

The problem of spin resonance, which was solved exactly in Sec. 12.1, will now be used to illustrate the conditions for the accuracy of time-dependent perturbation theory. The system is a particle of spin $s = \frac{1}{2}$ and a static magnetic field B_0 in the z direction. The unperturbed Hamiltonian is $H_0 = -\frac{1}{2}\hbar\gamma B_0\sigma_z$. The perturbation is due to a magnetic field, of magnitude B_1, rotating in the xy plane with angular velocity ω. The perturbation term in the Hamiltonian has the form

$$H_1(t) = -\tfrac{1}{2}\hbar\gamma B_1[\sigma_x \cos(\omega t) + \sigma_y \sin(\omega t)] \,.$$

In the standard basis formed by the eigenvectors of σ_z, and using the notation

$$|+\rangle = \begin{bmatrix} 1 \\ 0 \end{bmatrix}, \qquad |-\rangle = \begin{bmatrix} 0 \\ 1 \end{bmatrix},$$

the perturbation becomes

$$H_1(t) = -\tfrac{1}{2}\hbar\gamma B_1 \begin{bmatrix} 0 & e^{-i\omega t} \\ e^{i\omega t} & 0 \end{bmatrix} . \tag{12.55}$$

The initial state at $t = 0$ was chosen to be $|i\rangle = |+\rangle$, which corresponds to spin up and energy $\varepsilon_i = -\frac{1}{2}\hbar\gamma B_0$. At the end of the interval $0 \le t \le T$ during which the perturbation acts, the probability that the spin will be down and the energy will be $\varepsilon_f = \frac{1}{2}\hbar\gamma B_0$ is given by (12.18) to be

$$|a_f(T)|^2 = \left\{\sin\left(\frac{1}{2}\alpha T\right)\right\}^2 \left(\frac{\omega_1}{\alpha}\right)^2 \tag{12.56}$$

where $\alpha^2 = (\omega_0 + \omega)^2 + {\omega_1}^2$, $\omega_0 = \gamma B_0$, and $\omega_1 = \gamma B_1$. The lowest order perturbation approximation for this probability can be obtained from (12.54). Comparing (12.53) with (12.55), we see that $\langle f|H'|i\rangle = \langle -|H'|+\rangle = 0$, $\langle f|H'^{\dagger}|i\rangle = \langle -|H'^{\dagger}|+\rangle = -\frac{1}{2}\hbar\gamma B_1 = -\frac{1}{2}\hbar\omega_1$, and $\omega_{fi} \equiv (\varepsilon_f - \varepsilon_i)/\hbar = \omega_0$. Thus the square of (12.54) reduces to

$$|a_f{}^{(1)}(T)|^2 = \left\{\sin\left(\frac{1}{2}(\omega_0 + \omega)T\right)\right\}^2 \frac{{\omega_1}^2}{(\omega_0 + \omega)^2} \,. \tag{12.57}$$

The *conditions for validity* of the perturbation approximation can be determined by comparing the exact and approximate answers. If the exact probability (12.56) is expanded to the lowest order in ω_1, which is proportional to the strength of the perturbation, then we obtain (12.57), the approximation being accurate if $|\omega_1/(\omega_0+\omega)| \ll 1$. This condition can be satisfied by making the perturbing field B_1 sufficiently weak, provided that $\omega_0 + \omega \neq 0$.

At *resonance* we have $\omega_0 + \omega = 0$, and the above condition cannot be satisfied. The exact answer (12.56) then becomes

$$|a_f(T)|^2 = \left\{\sin\left(\frac{1}{2}\alpha T\right)\right\}^2 , \tag{12.58}$$

and the result of perturbation theory (12.57) becomes

$$|a_f^{(1)}(T)|^2 = \frac{(\omega_1 T)^2}{4} . \tag{12.59}$$

It is apparent that perturbation theory will be accurate at resonance only if $|\omega_1 T| \ll 1$. No matter how weak the perturbing field may be, perturbation theory will fail if the perturbation acts for a sufficiently long time.

There is another condition under which perturbation theory is accurate. If (12.56) and (12.57) are expanded to the lowest order in T, both expressions reduce to $(\omega_1 T)^2/4$. So perturbation theory is correct for very short times, no matter how strong the perturbation may be. The reason for this surprising result is that the effect of the perturbation depends, roughly, on the product of its strength and duration. This effect can be made small if the perturbation is allowed to act for only a very short time.

Harmonic perturbation of long duration

Let us now consider the behavior of the amplitude (12.54) in the limit $|\omega T| \gg 1$. Provided the denominators do not vanish, this amplitude remains bounded as T increases. But if $\omega_{fi} - \omega \to 0$ the first term of (12.54) will grow in proportion to T, and if $\omega_{fi} + \omega \to 0$ the second term will grow in proportion to T. These are both conditions for *resonance.* If $\omega > 0$ then the first of them, $\hbar\omega = \varepsilon_f - \varepsilon_i$, is the condition for *resonant absorption* of energy by the system; and the second, $\hbar\omega = \varepsilon_i - \varepsilon_f$, is the condition for *resonant emission* of energy. Near a resonance it is permissible to retain only the dominant term. By analogy with the example of spin resonance, we infer that the validity of perturbation theory at resonance is assured only if $|\langle f|H'|i\rangle T|$ and $|\langle f|H'^{\dagger}|i\rangle T|$ are small. We shall assume that the matrix elements are small enough to ensure these conditions.

Let us consider the case $\varepsilon_f - \varepsilon_i > 0$, and retain only the *resonant absorption term* of (12.54). Then the absorption probability is given by

$$|a_f^{(1)}(T)|^2 = \hbar^{-2}|\langle f|H'|i\rangle|^2 \frac{|1 - e^{i(\omega_{fi}-\omega)T}|^2}{(\omega_{fi}-\omega)^2}$$
$$= \frac{1}{\hbar^2}|\langle f|H'|i\rangle|^2 \left[\frac{\sin[\frac{1}{2}(\omega-\omega_{fi})T]}{\frac{1}{2}(\omega-\omega_{fi})}\right]^2 . \qquad (12.60)$$

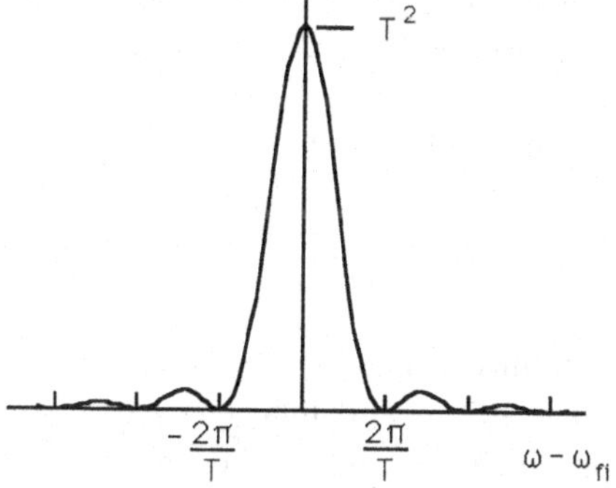

Fig. 12.3 The function $\{\sin[\frac{1}{2}(\omega-\omega_{fi})T]/\frac{1}{2}(\omega-\omega_{fi})\}^2$.

The last factor of this expression is plotted in Fig. 12.3. The height of the peak is T^2, its width is proportional to $1/T$, and the area under the curve is $2\pi T$. Most of the area is under the central peak, and by neglecting the side lobes, we may say that the absorption probability is significant only if $|\varepsilon_f - \varepsilon_i - \hbar\omega| < 2\pi\hbar/T$.

[[Landau and Lifshitz (1958), Ch. 44, use the condition $|\varepsilon_f - \varepsilon_i - \hbar\omega| < 2\pi\hbar/T$ to argue that energy conservation holds only to an accuracy of $\Delta E \approx 2\pi\hbar/T$. (More precisely, they claim that conservation of energy can be verified by two measurements separated by a time T only to this accuracy, but in their terms of reference the former statement means the same as the latter.) Their opinion is questionable. There are strong reasons for believing that energy conservation is exact. In this case it requires that an energy quantum of magnitude $\hbar\omega' = \varepsilon_f - \varepsilon_i$ be absorbed by the system from the perturbing field, with $\omega' \neq \omega$ but $|\omega' - \omega| < 2\pi/T$. It was pointed out in connection with (12.51) that only the Fourier component of the time-dependent perturbation that has the frequency ω' will be effective in inducing this transition. Although our perturbation has the *nominal* frequency ω, its duration is restricted to the finite time interval $0 < t < T$. Its Fourier transform is peaked at the frequency ω, but it is nonzero at other frequencies. Indeed, the function shown in Fig. 12.3 arose from the Fourier transform of the perturbation. Thus the reason why our perturbation can induce transitions for which $\varepsilon_f - \varepsilon_i \equiv \hbar\omega' \neq \hbar\omega$ is simply that it has components at the frequency ω', which is required for energy conservation.]]

Formally passing to the limit $T \to \infty$, we can define a "transition rate" or, more correctly, a *transition probability per unit time.* For $T \to \infty$ we obtain $\{\sin[\frac{1}{2}(\omega - \omega_{fi})T]/\frac{1}{2}(\omega - \omega_{fi})\}^2 \to 2\pi T\delta(\omega - \omega_{fi})$, and thus

$$\lim_{T\to\delta} T^{-1}|a_f^{(1)}(T)|^2 = \hbar^{-2}|\langle f|H'|i\rangle|^2 2\pi\delta(\omega - \omega_{fi})$$

$$= \frac{2\pi}{\hbar}|\langle f|H'|i\rangle|^2\delta(\hbar\omega - \varepsilon_f + \varepsilon_i)\,. \qquad (12.61)$$

This expression is infinite whenever it is not zero, indicating that it cannot be applied if both the initial and final energies belong to the discrete point spectrum. But suppose that we want to calculate the transition rate *from the discrete initial energy level* ε_i *to an energy* ε_f *in the continuum.* The eigenvalue ε_f will now be highly degenerate, and we must integrate over all possible degenerate final states.

Let $n(E)$ be the density of states per unit energy in the continuum. That is to say, $n(E)dE$ is the number of states in the energy range of E to $E + dE$. Then the total transition rate from the discrete energy level ε_i by absorption of an energy quantum $\hbar\omega$ will be

$$R = \int \frac{2\pi}{\hbar}|\langle f|H'|i\rangle|^2\delta(\hbar\omega - \varepsilon_f + \varepsilon_i)n(\varepsilon_f)d\varepsilon_f$$

$$= \frac{2\pi}{\hbar}|\langle f|H'|i\rangle|^2 n(\varepsilon_i + \hbar\omega)\,. \qquad (12.62)$$

This result is known as *Fermi's rule* for transition rates. It has proven to be very useful, in spite of its humble origin as a lowest order perturbation approximation.

12.6 *Atomic Radiation*

One of the earliest applications of time-dependent perturbation theory was to study the absorption and emission of radiation by matter. In this section we shall develop the theory of a single charged particle interacting with a classical electromagnetic field. Correlations and cooperative effects among the electrons will not be considered here. The electromagnetic field will be treated as a quantum-mechanical system in Ch. 19, but not in this section.

The Hamiltonian describing an electron in an atom interacting with a radiation field is

$$H = \frac{\{\mathbf{P} - (q/c)\mathbf{A}\}^2}{2M} + q\phi + W\,, \tag{12.63}$$

where $q = -e$ is the charge of the electron, W is the potential energy that binds it to the atom, and $\mathbf{A}$ and ϕ are the vector and scalar potentials that generate the electric and magnetic fields of the radiation:

$$\mathbf{E} = -\boldsymbol{\nabla}\phi - \frac{1}{c}\frac{\partial \mathbf{A}}{\partial t}\,, \quad \mathbf{B} = \boldsymbol{\nabla} \times \mathbf{A}\,.$$

To use perturbation theory, the Hamiltonian (12.63) is written as

$$H = H_0 + H_1\,,$$

where

$$H_0 = \frac{P^2}{2M} + W \tag{12.64}$$

is the Hamiltonian of the free atom, and

$$H_1 = \frac{q}{2Mc}(\mathbf{P}\cdot\mathbf{A} + \mathbf{A}\cdot\mathbf{P}) + \frac{q^2}{2Mc^2}(\mathbf{A}\cdot\mathbf{A}) + q\phi \tag{12.65}$$

describes the interaction of the atom with the radiation field.

The gauge problem

The electromagnetic potentials are not unique. As was discussed in Sec. 11.2, the electromagnetic fields and the time-dependent Schrödinger equation (12.44) are invariant under a *gauge transformation* of the form (11.18):

$$\mathbf{A} \to \mathbf{A}' = \mathbf{A} + \boldsymbol{\nabla}\chi\,, \tag{12.66a}$$

$$\phi \to \phi' = \phi - \frac{1}{c}\frac{\partial \chi}{\partial t}\,, \tag{12.66b}$$

$$\Psi \to \Psi' = \Psi e^{i(q/\hbar c)\chi}\,, \tag{12.66c}$$

where $\chi = \chi(\mathbf{x}, t)$ is an arbitrary scalar function. This leads to ambiguities in applying the methods of Sec. 12.5. The first step was to expand the state vector in terms of the eigenvectors of H_0,

$$|\Psi(t)\rangle = \sum_n c_n(t)|n\rangle\,, \tag{12.67}$$

and to interpret $|c_n|^2$ as a probability. [The coefficients $c_n(t)$ are equal to $a_n(t)e^{-i\varepsilon_n t/\hbar}$ in the notation of (12.46).] Suppose that we use different potentials, related to the old potentials by the gauge transformation (12.66). The transformed state vector (12.66c) can also be expanded,indexRadiation

$$|\Psi'(t)\rangle = \sum_n c'_n(t)|n\rangle\,, \tag{12.68}$$

and the relation between the new coefficients and the old is

$$c'_n(t) = \sum_m \left\langle n \left| \exp\left(\frac{iq\chi}{\hbar c}\right) \right| m \right\rangle c_m(t)\,. \tag{12.69}$$

Since $\chi(\mathbf{x},t)$ is an arbitrary function, it is clear that $|c'_n|^2$ and $|c_n|^2$ need not be equal. Then $|c_n|^2$ cannot be physically meaningful, in general, because it is not gauge-invariant. The solution to this gauge problem is discussed in detail by Kobe and Smirl (1978).

We shall restrict our attention to the effects of perturbing fields that act *only during the finite time interval* $0 < t < T$. For $t > T$, when the field vanishes, it is natural to take $\mathbf{A} = \phi = 0$, although any potentials of the form $\mathbf{A} = \boldsymbol{\nabla}\chi$, $\phi = -(1/c)\partial\chi/\partial t$ would be consistent with vanishing electromagnetic fields. Provided we choose $\chi(\mathbf{x},t) = 0$ for $t > T$, we shall have $|\Psi'(t)\rangle = |\Psi(t)\rangle$ for $t > T$, and the interpretation of $|c_n|^2$ as the probability that the final energy is ε_n will be unambiguous. So, for our restricted class of problems, we shall slightly restrict the kind of gauge transformations permitted.

Although, in our problem, an exact calculation of $c_n(t)$ would yield gauge-invariant probabilities for $t > T$, this need not be true in any finite order of perturbation theory, such as (12.50), because the form of the perturbation (12.65) is not gauge-invariant. So there still remains a practical problem of choosing an appropriate gauge.

The electric dipole approximation

Because the wavelength of atomic radiation is very much longer than the diameter of an atom, we may neglect the variation of the fields throughout the volume of an atom. Although the magnetic and electric components of a radiation field are of equal magnitude (in Gaussian units), the magnetic force on an electron with speed v is smaller than the electric force by a factor of v/c. Thus the magnetic effects are usually negligible compared with the electric effects. The so-called *electric dipole* approximation can be derived under the conditions that (a) the variation of the electric field over the size of the atom is negligible, and (b) the magnetic field can be neglected.

If the magnetic field is negligible, then the fields $\mathbf{E} = \mathbf{E}(\mathbf{x}, t)$ and $\mathbf{B} = 0$ can be generated by the potentials

$$\mathbf{A} = 0\,, \quad \phi = -\int_0^{\mathbf{x}} \mathbf{E}(\mathbf{x}', t) \cdot d\mathbf{x}'\,. \tag{12.70}$$

The integral is independent of the path because $\nabla \times \mathbf{E} = -(1/c)\partial\mathbf{B}/\partial t = 0$. It is easy to verify that any other potentials that generate the same electric and magnetic fields can be gauge-transformed into the form (12.70). If the spatial variation of the electric field can be neglected, these potentials can be simplified to

$$\mathbf{A} = 0\,, \quad \phi = -\mathbf{x} \cdot \mathbf{E}(0, t)\,. \tag{12.71}$$

The atomic nucleus is here assumed to be located at the origin. These potentials are valid whenever the conditions for the electric dipole approximation hold, and are almost always the most convenient choice. The electric field need not be weak for (12.71) to be valid, but if it is weak then the potential may be treated as a perturbation, with the perturbation Hamiltonian (12.65) being

$$H_1 = -q\mathbf{x} \cdot \mathbf{E}(0, t)\,. \tag{12.72}$$

[[Another common approach is to treat (12.65) as the perturbation, and to expand in powers of the potentials. Since $\nabla \cdot \mathbf{E} = 0$ for a radiation field, it is possible to choose $\phi = 0$ and $\nabla \cdot \mathbf{A} = 0$ (by a gauge transformation, if necessary). The perturbation expansion is then in powers of $\mathbf{A}$. This is always a hazardous thing to do, because $\mathbf{A}$ is not gauge-invariant. The first order term of (12.65), $H_1{}' = -(q/2Mc)(\mathbf{P} \cdot \mathbf{A} + \mathbf{A} \cdot \mathbf{P}) = -(q/Mc)\mathbf{A} \cdot \mathbf{P}$, yields the so-called "$\mathbf{A} \cdot \mathbf{P}$" form of the interaction.

Let us compare this approach with the recommended method based on (12.72). When the electric dipole approximation is valid, the fields $\mathbf{E} = \mathbf{E}(\mathbf{x}, t) \approx \mathbf{E}(0, t)$ and $\mathbf{B} = 0$ can be generated by the alternative potentials $\mathbf{A}(\mathbf{x}, t) = -c \int \mathbf{E}(\mathbf{x}, t) dt$ and $\phi(\mathbf{x}, t) = 0$. If the time dependence of the electric field is taken to be $e^{-i\omega t}$, then we may write $\mathbf{A} = c\mathbf{E}/i\omega$. Using the relation $\mathbf{P}/M = (i/\hbar)[H_0, \mathbf{x}]$, we obtain $H_1{}' = -(q/\omega\hbar)[H_0, \mathbf{x}] \cdot \mathbf{E}$. The matrix elements of this operator in the basis formed by the eigenvectors of H_0 are

$$\begin{aligned} \langle m|H_1'|n\rangle &= -\frac{q}{\omega\hbar}(\varepsilon_m - \varepsilon_n)\langle m|\mathbf{x} \cdot \mathbf{E}|n\rangle \\ &= \frac{\omega_{mn}}{\omega}\langle m|H_1|n\rangle\,, \end{aligned}$$

where $\omega_{mn} = (\varepsilon_m - \varepsilon_n)/\hbar$. Since the matrix element of the $\mathbf{A}\cdot\mathbf{P}$ interaction, $H_1{}'$, differs from the matrix element of (12.72) by the factor (ω_{mn}/ω), it follows that transition probabilities calculated to the lowest order in H_1' will be incorrect, except at resonance $(\omega = \omega_{mn})$. One reason why the $\mathbf{A}\cdot\mathbf{P}$ interaction gives incorrect results is that we have assumed the perturbation to be zero for $t < 0$ and $t \geq T$. No difficulty is caused by H_1 [Eq. (12.72)] jumping discontinuously to zero. But if the $\mathbf{A}\cdot\mathbf{P}$ interaction jumps discontinuously to zero, then the relation $\mathbf{E} = -c^{-1}\partial\mathbf{A}/\partial t$ generates a spurious delta function impulse electric field.

That calculations based upon (12.72) agree with the experimental shape of the resonance curve, whereas those based on the $\mathbf{A}\cdot\mathbf{P}$ interaction do not, was noted by W. E. Lamb in 1952. The relation between the two forms of interaction has been studied in greater detail by Milonni *et al.* (1989).]]

Induced emission and absorption

In order to use the analysis of a harmonic perturbation in Sec. 12.5, we assume that the time dependence of the perturbation (12.72) is of the form

$$\begin{aligned} H_1(t) &= -q\mathbf{x}\cdot\mathbf{E}_0\left(e^{-i\omega t} + e^{i\omega t}\right) \quad (0 < t < T) \\ &= 0 \quad (t < 0 \ \text{or} \ t > T)\,. \end{aligned} \tag{12.73}$$

Here $\mathbf{E}_0$ is a constant vector, giving the strength and polarization of the radiation field. This form is appropriate for describing monochromatic laser radiation. In the notation of (12.53), we have $H' = H'^\dagger = -q\mathbf{x}\cdot\mathbf{E}_0$. If the initial state at $t = 0$ is an eigenstate of the atomic Hamiltonian H_0 [Eq. (12.64)] with energy ε_i, then the probability that at any time $t \geq T$ the atom will have the final energy ε_f is equal to $|a_f^{(1)}(T)|^2$, where the amplitude $a_f^{(1)}(T)$ is given by (12.54). If $\varepsilon_f > \varepsilon_i$ this is the probability of absorbing radiation; if $\varepsilon_f < \varepsilon_i$ it is the probability of emitting radiation.

The theory of transitions between two discrete atomic states is very similar to the theory of spin resonance. In the so-called *rotating wave* approximation, we retain only one of the terms of (12.73). Then the two-level atom problem becomes identical to the spin resonance problem with a rotating magnetic field, and this analogy leads to the term "rotating wave" approximation. The dependence of the transition probability on the duration T of the perturbation is quite complicated, as was seen in Sec. 12.1.

If, instead of monochromatic laser radiation, we have *incoherent radiation* with a continuous frequency spectrum, a different analysis is appropriate. We consider the case of near-resonant absorption ($\omega \approx \omega_{fi}$) and retain only the first term of (12.54), and hence

$$\left|a_f^{(1)}(T)\right|^2 = \frac{e^2}{\hbar^2}|\langle f|\mathbf{x}\cdot\mathbf{E}_0|i\rangle|^2 \left[\frac{\sin\left[\frac{1}{2}(\omega-\omega_{fi})T\right]}{\frac{1}{2}(\omega-\omega_{fi})}\right]^2, \tag{12.74}$$

where we have substituted $q = -e$. This expression applies to radiation of a single angular frequency ω. But we actually have a continuous spectrum of radiation whose energy density in the angular frequency range $\Delta\omega$ is $u(\omega)\Delta\omega$. Strictly speaking, we should integrate the amplitude $a_f^{(1)}(T)$ over the frequency spectrum, and then square the integrated amplitude. But if the radiation is *incoherent*, the cross terms between different frequencies will average to zero, and the correct result will be obtained by integrating the probability (12.74) over the frequency spectrum. If the radiation is *unpolarized*, we may average over the directions of $\mathbf{E}_0$, and so replace $|\langle f|\mathbf{x}\cdot\mathbf{E}_0|i\rangle|^2$ with $(1/3)|\mathbf{E}_0|^2|\langle f|\mathbf{x}|i\rangle|^2 = (1/3)|\mathbf{E}_0|^2\langle f|\mathbf{x}|i\rangle\cdot\langle f|\mathbf{x}|i\rangle^*$. The instantaneous energy density in a radiation field is $|\mathbf{E}|^2/4\pi$, including equal contributions from the electric and magnetic fields. The electric field in (12.73) is $\mathbf{E} = 2\mathbf{E}_0\cos(\omega t)$, so the average of $|\mathbf{E}|^2$ over a cycle of the oscillation is $2|\mathbf{E}_0|^2$. Therefore it is appropriate to replace $|\mathbf{E}_0|^2$ by $2\pi u(\omega)d\omega$, where $u(\omega)$ is the time average energy density per unit ω. In the limit of very large T, we replace $\{\sin[\frac{1}{2}(\omega-\omega_{fi})T]/\frac{1}{2}(\omega-\omega_{fi})\}^2$ by $2\pi T\delta(\omega-\omega_{fi})$, as was done in deriving (12.61). In this way we obtain the *transition rate for absorption* of radiation at the angular frequency ω_{fi}:

$$\begin{aligned} R_a &= T^{-1}\int |a_f{}^{(1)}(T)|^2 \\ &= \frac{4\pi^2}{3}\left(\frac{e}{\hbar}\right)^2 u(\omega_{fi})|\langle f|\mathbf{x}|i\rangle|^2 . \end{aligned} \tag{12.75}$$

An almost identical calculation yields the same transition rate for *stimulated emission* of radiation.

Spontaneous emission

It is well known that an atom in an excited state will spontaneously emit radiation and return to its ground state. That phenomenon is not predicted by this version of the theory, in which only matter is treated as a quantum-mechanical system, but the radiation is treated as an external classical field. If no radiation field is present, then $H_1 \equiv 0$ and all eigenstates of H_0 are stationary.

When the electromagnetic field is also treated as a dynamical system, it has a Hamiltonian H_{em}. If there were no coupling between the atom and the electromagnetic field, the total Hamiltonian would be $H = H_{\text{at}} + H_{\text{em}}$, where H_{at} is the atomic Hamiltonian (previously denoted as H_0). The two terms H_{at} and H_{em} act on entirely different degrees of freedom, and so the operators commute. The stationary state vectors would be of the form $|\,\text{atom}\,\rangle \otimes |\,\text{em}\,\rangle$ which are common eigenvectors of H_{at} and H_{em}. Thus the atomic excited states would be stationary and would not decay. But of course there is an interaction between the atomic and electromagnetic degrees of freedom. The total Hamiltonian is of the form $H = H_{\text{at}} + H_{\text{em}} + H_{\text{int}}$, where the interaction term H_{int} does not commute with other two terms. Now an eigenvector of H_{at} is not generally an eigenvector of H, since H_{at} and H do not commute. Therefore the excited states of the atom are not stationary, and will decay spontaneously.

A calculation based upon these ideas requires a quantum theory of the electromagnetic field, some aspects of which will be developed in Ch. 19. Nevertheless, Einstein was able to deduce some of the most important features of spontaneous emission in 1917, when most of the quantum theory was unknown. The argument below, derived from Einstein's ideas, is based on the principle that the radiation mechanism must preserve the statistical equilibrium among the excited states of the atoms.

Let the number of atoms in state n be $N(n)$, and consider transitions between states i and f involving the emission and absorption of radiation at the frequency $\omega_{fi} = (\varepsilon_f - \varepsilon_i)/\hbar > 0$. We have calculated the probability per unit time for an atom to absorb radiation [Eq. (12.75)]. It has the form $B_{if}u(\omega_{fi})$, where $u(\omega_{fi})$ is the energy density of the radiation at the angular frequency ω_{fi}. Therefore the rate of excitation of atoms by absorption of radiation will be $B_{if}u(\omega_{fi})N(i)$. Einstein assumed that the rate of de-excitation of atoms by emitting radiation is of the form $B_{fi}u(\omega_{fi})N(f) + A_{fi}N(f)$. The first term corresponds to induced emission, which we have shown how to calculate. The second term, which is independent of the presence of any radiation, describes *spontaneous emission*. At equilibrium the rates of excitation and de-excitation must balance, so we must have

$$B_{fi}u(\omega_{fi})N(f) + A_{fi}N(f) = B_{if}u(\omega_{fi})N(i)\,. \tag{12.76}$$

Invoking the principle of *detailed balance*, Einstein assumed that the probabilities of induced emission and absorption should be equal: $B_{fi} = B_{if}$. This relation was confirmed by our quantum–mechanical calculation above. Therefore we may solve for energy density of the radiation field at equilibrium:

$$u(\omega_{fi}) = \frac{A_{fi}N(f)}{B_{fi}[N(i) - N(f)]}\,.$$

But we know that in thermodynamic equilibrium the relative occupation of the various atomic states is given by the Boltzmann distribution, so we must have $N(i)/N(f) = \exp[(\varepsilon_f - \varepsilon_i)/kT] = \exp(\hbar\omega_{fi}/kT)$, where T is the temperature. Therefore the energy density of the radiation field can be written as

$$u(\omega_{fi}) = \frac{A_{fi}/B_{fi}}{\exp(\hbar\omega_{fi}/kT) - 1}\,. \tag{12.77}$$

Except for the numerator, this has the form of the Planck distribution for black body radiation. Since A_{fi} and B_{fi} are elementary quantum-mechanical probabilities, they do not depend on temperature. Therefore it is sufficient to equate the low frequency, high temperature limit of $u(\omega_{fi})$ to the classical Rayleigh–Jeans formula, $u(\omega_{fi}) = (\omega_{fi}^2/\pi^2c^3)kT$, which says that in this limit the energy density is equal to kT per normal mode of the field. We thus obtain

$$A_{fi} = \frac{\hbar\omega_{fi}^3}{\pi^2c^3}B_{fi}\,, \tag{12.78}$$

which relates the spontaneous emission probability to the induced emission probability that has already been calculated. This relation, derived before most of quantum mechanics had been formulated, remains valid in modern quantum electrodynamics.

12.7 *Adiabatic Approximation*

The perturbation theory of Sec. 12.5 was based on the assumed small magnitude of the time-dependent part of the Hamiltonian. The adiabatic approximation is based, instead, on the assumption that the time dependence of H is slow.

Suppose that the Hamiltonian $H(R(t))$ depends on time through some parameter or parameters $R(t)$. The state vector evolves through the Schrödinger equation,

$$i\hbar\frac{d|\Psi(t)\rangle}{dt} = H(R(t))|\Psi(t)\rangle\,. \tag{12.79}$$

Now the time-dependent Hamiltonian has instantaneous eigenvectors $|n(R)\rangle$, which satisfy

$$H(R)|n(R)\rangle = E_n(R)|n(R)\rangle\,. \tag{12.80}$$

It is intuitively plausible that if $R(t)$ varies sufficiently slowly and the system is prepared in the initial state $|n(R(0)\rangle$, then the time-dependent state vector should be $|n(R(t)\rangle$, apart from a phase factor.

To give this intuition a firmer footing, we use the instantaneous eigenvectors of (12.80) as a basis for representing a general solution of (12.79),

$$|\Psi(t)\rangle = \sum_n a_n(t) e^{i\alpha_n(t)} |n(R(t))\rangle \,. \tag{12.81}$$

Here the so-called *dynamical phase*,

$$\alpha_n(t) = -\frac{1}{\hbar} \int_0^t E_n(R(t'))dt' \,, \tag{12.82}$$

has been introduced, generalizing the phase that would be present for a time-independent Hamiltonian. Substituting (12.81) into (12.79), we obtain

$$\sum_n \dot{a}_n e^{i\alpha_n} |n\rangle + \sum_n a_n e^{i\alpha_n} |\dot{n}\rangle = 0 \,. \tag{12.83}$$

[Here, for simplicity, we do not indicate the implicit time dependences of the various quantities, and we denote the time derivatives of $a_n(t)$ and $|n(R(t))\rangle$ by $\dot{a}_n$ and $|\dot{n}\rangle$, respectively.] Taking the inner product of (12.83) with another instantaneous eigenvector, $\langle m| = \langle m(R(t))|$, yields

$$\dot{a}_m = -\sum_n a_n e^{i(\alpha_n - \alpha_m)} \langle m|\dot{n}\rangle \,. \tag{12.84}$$

Now the time derivative of the eigenvalue equation (12.80) yields

$$\dot{H}|n\rangle + H|\dot{n}\rangle = \dot{E}_n|n\rangle + E_n|\dot{n}\rangle \,, \tag{12.85}$$

where $\dot{H} = dH/dt$, etc. The inner product with $\langle m|$ then yields

$$\langle m|\dot{n}\rangle(E_n - E_m) = \langle m|\dot{H}|n\rangle \quad (m \neq n) \,, \tag{12.86}$$

which may be substituted into (12.84) to obtain

$$\dot{a}_m = \sum_n a_n e^{i(\alpha_n - \alpha_m)} \langle m|\dot{H}|n\rangle (E_m - E_n)^{-1} \quad (m \neq n) \,. \tag{12.87}$$

Let us now choose the *initial state* to be one of the instantaneous eigenvectors, $|\Psi(0)\rangle = |n(\mathrm{R}(0))\rangle$, so that $a_n(0) = 1$ and $a_m(0) = 0$ for $m \neq n$. Then for $m \neq n$ we will have, approximately,

$$\dot{a}_m \approx e^{i(\alpha_n - \alpha_m)} \, \langle m|\dot{H}|n\rangle \, (E_m - E_n)^{-1} \,, \tag{12.88}$$

which can be integrated (bearing in mind the implicit time dependences in all quantities) to obtain $a_m(t)$. To estimate the magnitude of the excitation probability $|a_m(t)|^2$, we assume that the time dependences of $\langle m|\dot{H}|n\rangle$ and $E_m - E_n$ are slow. Then the most important time dependence will be in the exponential, which can be approximated by $e^{i(\alpha_n - \alpha_m)} \approx e^{i(E_m - E_n)t/\hbar}$. Neglecting the other slow time dependences then yields

$$a_m(t) \approx -i\hbar\langle m|\dot{H}|n\rangle \, (E_m - E_n)^{-2}\{e^{i(E_m - E_n)t/\hbar} - 1\}, \tag{12.89}$$

which will be small provided the rate of variation of $H(\mathrm{R}(t))$ is slow compared to the transition frequency $\omega_{mn} = (E_m - E_n)/\hbar$. In fact, this simple estimate is often much too large. If the time dependence of $H(R(t))$ is sufficiently smooth, and characterized by a time scale τ, then $a_m(t)$ may be only of order $e^{-\omega_{mn}\tau}$. An example is given in Problem 12.11.

The Berry phase

In the adiabatic limit, where excitation to other instantaneous eigenvectors is negligible, the choice of initial state $|\Psi(0)\rangle = |n(\mathrm{R}(0))\rangle$ will imply that $|a_n(t)| = 1$, $a_m(t) = 0$ for $m \neq n$. Then Eq. (12.84) will reduce to $\dot{a}_n = -a_n\langle n|\dot{n}\rangle$. If we write $a_n = e^{i\gamma_n(t)}$, we obtain

$$\dot{\gamma}_n(t) = i\langle n(R(t))|\dot{n}(R(t))\rangle \,, \tag{12.90}$$

and the adiabatic evolution of the state vector becomes

$$|\Psi_n(t)\rangle = e^{i[\alpha_n(t)+\gamma_n(t)]}|n(R(t))\rangle \,. \tag{12.91}$$

Now the vector $|n(R)\rangle$ is defined only by the eigenvalue equation (12.80), so its phase is arbitrary and can be modified to have any continuous dependence on the parameter $R(t)$. Hence the phase $\gamma(t)$ is not uniquely defined, and many older books assert that it can be transformed to zero. However, M. V. Berry (1984) showed that not to be so.

Equation (12.90) can be written as

$$\dot{\gamma}_n(t) = i\langle n(R(t))|\boldsymbol{\nabla}_R \; n(R(t))\rangle \cdot \dot{R}(t) \,, \tag{12.92}$$

where the gradient is taken in the space of the parameter R, and $\dot{R}(t)$ is the time derivative of R. Now suppose that $R(t)$ is carried around some closed curve C in parameter space, such that $R(0) = R(T)$. The net change in the phase $\gamma_n(t)$ will be

$$\gamma_n(T) - \gamma_n(0) = \oint_C \dot{\gamma}_n(t)dt$$

$$= i \oint_C \langle n(R)|\boldsymbol{\nabla}_R\, n(R)\rangle \cdot dR\,. \qquad (12.93)$$

This net phase change depends only on the closed path C in parameter space that is traversed by $R(t)$, but not on the rate at which it is traversed. It is therefore called a *geometrical phase*, or often a *Berry phase*, after its discoverer.

The vector in the integrand of (12.93) depends on the arbitrary phase of the vector $|n(R)\rangle$, but the integral around C is independent of those phases. To show this, we use Stoke's theorem to transform the path integral into an integral over the surface bounded by C,

$$\oint_C \langle n(R)|\boldsymbol{\nabla}_R \dot{n}(R)\rangle \cdot d(R) = \int\int_c [\boldsymbol{\nabla} \times \langle n(R)|\boldsymbol{\nabla}_R\, n(R)\rangle] \cdot d(\mathrm{S})\,. \qquad (12.94)$$

(For convenience we take the parameter space to be three-dimensional, but the results can be generalized to any number of dimensions.) Now, if we introduce an arbitrary change of the phases of the basis vectors, $|n\rangle \to e^{i\chi(R)}|n\rangle$, then $\langle n|\boldsymbol{\nabla} n\rangle \to \langle n|\boldsymbol{\nabla} n\rangle + i\boldsymbol{\nabla}\chi$. But $\boldsymbol{\nabla} \times \boldsymbol{\nabla}\chi = 0$, so the net phase change (12.93) is an invariant quantity that depends only on the geometry of the path C.

The Aharonov–Bohm effect (Sec. 11.4) can be viewed as an instance of the geometrical phase. Consider a tube of magnetic flux near a charged system that is confined within a box. Although the magnetic flux does not penetrate into the box, the vector potential $\mathbf{A}(\mathbf{r})$ will be non-zero inside the box. Let $\mathbf{r}$ be the position operator of a charged particle inside the box, and $\mathbf{R}$ be the position of the box, as shown in Fig. 12.4. In the absence of a vector potential,

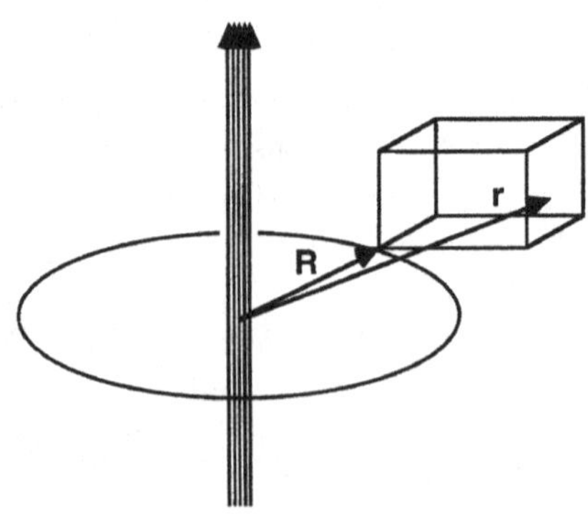

Fig. 12.4 Aharonov–Bohm effect in a box transported around a flux tube.

the Hamiltonian of the charged particle would be a function of its position and momentum: $H = H(\mathbf{p}, \mathbf{r} - \mathbf{R})$. In the presence of the vector potential, it would have the form $H = H(\mathbf{p} - q\mathbf{A}(\mathbf{r})/c, \mathbf{r} - \mathbf{R})$.

The box is then transported adiabatically along a closed path encircling the flux tube, with $\mathbf{R}$ playing the role of the parameter that is carried around a closed path. The geometrical phase change of the wave function turns out to be equal to the Aharonov–Bohm phase, $q\Phi/\hbar c$, where Φ is the magnetic flux in the tube (Problem 12.12).

Further reading for Chapter 12

The following are extensive reviews of certain topics in this chapter.

Theory of the decay of unstable quantum systems: Fonda, Ghirardi and Rimini (1978).

Applications and generalizations of the geometrical phase: Shapere and Wilczek (1989).

Generalized energy–time indeterminacy relations: Pfeifer (1995).

A deeper insight into the energy–time indeterminacy relations, and the controversies surrounding them, was provided by Aharonov, Massar and Popescu (2002). They showed that if the Hamiltonian of the system is known, then it is possible, in principle, to measure the energy to an arbitrary precision in an arbitrarily short time. But if the Hamiltonian is not known, and the measurement of the energy must be done by observing the motion of the system, then the minimum time Δt required to determine the energy to an accuracy ΔE is limited by $\Delta T\,\Delta E \simeq \hbar$.

Problems

12.1 Suppose the complex coefficients in Eq. (12.1) have the values $a_1 = ae^{i\alpha}$, $a_2 = ae^{i\beta}$. Evaluate the three components of the average magnetic moment $\langle \boldsymbol{\mu} \rangle$ as a function of time. What are the polar angles of the instantaneous direction in which this vector points?

12.2 The most general state operator for a spin $\frac{1}{2}$ system has the form given in Eq. (7.50), $\rho = \frac{1}{2}(1 + \mathbf{a} \cdot \boldsymbol{\sigma})$, where $\mathbf{a}$ is a vector whose length is not greater than 1. If the system has a magnetic moment $\boldsymbol{\mu} = \frac{1}{2}\gamma\hbar\boldsymbol{\sigma}$ and is in a constant magnetic field $\mathbf{B}$, calculate the time-dependent state operator $\rho(t)$ in the Schrödinger picture. Describe the result geometrically in terms of the variation of the vector $\mathbf{a}$.

12.3 To treat a magnetic moment acted on by a static magnetic field $\mathbf{B}_0$ in the z direction and a field $\mathbf{B}_1$ rotating in the xy plane at the rate of ω radians per second, it is useful to treat the problem in the rotating coordinate system defined by orthogonal unit vectors $\mathbf{u}, \mathbf{v}$, and $\mathbf{k}$. Here $\mathbf{u} = \mathbf{i}\cos(\omega t) + \mathbf{j}\sin(\omega t), \mathbf{v} = -\mathbf{i}\sin(\omega t) + \mathbf{j}\cos(\omega t)$, with $\mathbf{i}, \mathbf{j}$, and $\mathbf{k}$ being the unit vectors of the static coordinate system. Obtain the Heisenberg equations of motion for the spin components $S_u \equiv \mathbf{S}\cdot\mathbf{u}$ and $S_v \equiv \mathbf{S}\cdot\mathbf{v}$. Show that they are equivalent to the equations of motion for S_x and S_y in an effective static magnetic field.

12.4 Show that the following set of nine operators forms a complete orthonormal set for an $s = 1$ spin system. This means that $\text{Tr}({R_i}^\dagger R_j) = \delta_{ij}$, and that any operator on the three-dimensional state vector space can be written as a linear combination of the operators $R_j (j = 0, \ldots, 8)$. The operators are $R_0 = I/\sqrt{3}$, $R_1 = S_x/\hbar\sqrt{2}, R_2 = S_y/\hbar\sqrt{2}, R_3 = S_z/\hbar\sqrt{2}$, $R_4 = [3(S_z/\hbar)^2 - 2]/\sqrt{6}, R_5 = (S_xS_z + S_zS_x)/\hbar^2\sqrt{2}, R_6 = (S_yS_z + S_zS_y)/\hbar^2\sqrt{2}, R_7 = (S_x^2 - S_y^2)/\hbar^2\sqrt{2}, R_8 = (S_xS_y + S_yS_x)/\hbar^2\sqrt{2}$. Of these, R_0 is a scalar, the next three are components of a vector, and the last five are components of a tensor of rank 2.

12.5 The state operator for an $s = 1$ system can be written in terms of the nine operators defined in Problem 12.4: $\rho(t) = \sum_j c_j(t)R_j$. Determine the time dependence of the coefficients $c_j(t)$ for the magnetic dipole Hamiltonian, $H = -\gamma B_0 S_z$.

12.6 Repeat the previous problem for the axially symmetric quadrupole Hamiltonian, $H = A(3{S_z}^2 - 2)$. [Notice how vector and tensor terms of $\rho(t)$ become mixed as time progresses.]

12.7 The spin Hamiltonian for a system of two $s = \frac{1}{2}$ particles is $H = {\sigma_x}^{(1)}{\sigma_x}^{(2)} + {\sigma_y}^{(1)}{\sigma_y}^{(2)}$. Find the time dependence of the state vector $|\Psi(t)\rangle$ if its initial value is $|\Psi(0)\rangle = |+\rangle^{(1)}|-\rangle^{(2)}$, and hence evaluate the time dependence of the spin correlation function $\langle{\sigma_z}^{(1)}{\sigma_z}^{(2)}\rangle$.

12.8 Show that for a charged particle with spin in a spatially uniform, time-varying magnetic field, the time-dependent state function can be separated into the product of a position-dependent factor and a spin function. (It is assumed, of course, that the initial conditions are compatible with this separation.)

12.9 Evaluate $W(\alpha)$ and τ_β of Sec. 12.3 for the Lorentzian distribution (12.28). Compare the inequality (12.40) with the exact values, by evaluating the various quantities for several representative values of α and β.

12.10 Consider a one-dimensional harmonic oscillator of angular frequency ω_0 that is perturbed by the time-dependent potential $W(t) = bx\cos(\omega t)$, where x is the displacement of the oscillator from equilibrium. Evaluate $\langle x \rangle$ by time-dependent perturbation theory. Discuss the validity of the result for $\omega \approx \omega_0$ and for ω far from ω_0.

12.11 A hydrogen atom is placed in a time-dependent homogeneous electric field, of magnitude $|\mathbf{E}(t)| = A\tau/(t^2+\tau^2)$. (Note that the total impulse of the force is independent of τ.) If at $t = -\infty$ the atom is in its ground state, calculate the probability that at $t = +\infty$ it has been excited to the first excited state.

12.12 Calculate the geometrical phase of the wave function of a charged particle in a box when the box is adiabatically transported around a magnetic flux tube that does not enter the box. (See Fig. 12.4.)

Chapter 13

Discrete Symmetries

When symmetry transformations were first considered in Sec. 3.1, it was pointed out that a theorem due to Wigner proves that such a transformation may, in principle, be implemented by a *unitary* (linear) operator or by an *antiunitary* (antilinear) operator. An operator U is said to be unitary or antiunitary if the mapping $|\Psi\rangle \to U|\Psi\rangle = |\Psi'\rangle$ is one-to-one and $|\langle\Psi|\Psi\rangle| = |\langle\Psi'|\Psi'\rangle|$. A *linear* operator L, by definition, satisfies the relation

$$L(c_1|\Psi_1\rangle + c_2|\Psi_2\rangle) = c_1 L|\Psi_1\rangle + c_2 L|\Psi_2\rangle\,, \tag{13.1}$$

whereas an *antilinear* operator A satisfies

$$A(c_1|\Psi_1\rangle + c_2|\Psi_2\rangle) = {c_1}^* A|\Psi_1\rangle + {c_2}^* A|\Psi_2\rangle\,. \tag{13.2}$$

In previous chapters we considered only continuous symmetry transformations, which must be represented by linear operators. However, in this chapter we will need both possibilities.

13.1 *Space Inversion*

The space inversion transformation is $\mathbf{x} \to -\mathbf{x}$. The corresponding operator on state vector space is usually called the *parity* operator. It will be denoted by Π (since the symbol P is already in use for momentum, and also for probability). By definition, the parity operator reverses the signs of the position operator and the momentum operator,

$$\Pi\mathbf{Q}\Pi^{-1} = -\mathbf{Q}\,, \tag{13.3}$$

$$\Pi\mathbf{P}\Pi^{-1} = -\mathbf{P}\,. \tag{13.4}$$

It follows that the orbital angular momentum, $\mathbf{L} = \mathbf{Q}\times\mathbf{P}$, is unchanged by the parity transformation. This property is extended, by definition, to any angular momentum operator,

$$\Pi\mathbf{J}\Pi^{-1} = \mathbf{J}\,. \tag{13.5}$$

We must next determine whether the operator Π should be linear or antilinear. Under the operation of Π the commutator of the position and momentum operators, $Q_\alpha P_\alpha - P_\alpha Q_\alpha = i\hbar$, becomes

$$\Pi Q_\alpha \Pi^{-1} \Pi P_\alpha \Pi^{-1} - \Pi P_\alpha \Pi^{-1} \Pi Q_\alpha \Pi^{-1} = \Pi i\hbar \Pi^{-1} .$$

By the use of (13.3) and (13.4), this becomes $Q_\alpha P_\alpha - P_\alpha Q_\alpha = \Pi i \Pi^{-1}\hbar$, which is compatible with the original commutation relation provided that $\Pi i \Pi^{-1} = i$. This will be true if Π is linear, but not if Π is antilinear. Therefore the parity operator is a unitary operator, and cannot be an antiunitary operator. Hence $\Pi^{-1} = \Pi^\dagger$.

Since two consecutive space inversions produce no change at all, it follows that the states described by $|\Psi\rangle$ and by $\Pi^2|\Psi\rangle$ must be the same. Thus the operator Π^2 can differ from the identity operator by at most a phase factor. This phase factor is left arbitrary by the defining equations (13.3)–(13.5), since any phase factor in Π would be canceled by that in Π^{-1}. It is most convenient to choose that phase factor to be unity, and hence we have

$$\Pi = \Pi^{-1} = \Pi^\dagger . \tag{13.6}$$

The effect of the parity operator on vectors and wave functions will now be determined. Consider its effect on an eigenvector of position, $\Pi Q_\alpha|\mathbf{x}\rangle = \Pi x_\alpha|\mathbf{x}\rangle = x_\alpha \Pi|\mathbf{x}\rangle$. Now from (13.3) we have $\Pi Q_\alpha|\mathbf{x}\rangle = \Pi Q_\alpha \Pi^{-1}\Pi|\mathbf{x}\rangle = -Q_\alpha \Pi|\mathbf{x}\rangle$, and thus $Q_\alpha(\Pi|\mathbf{x}\rangle) = -x_\alpha(\Pi|\mathbf{x}\rangle)$. But we know that $Q_\alpha|-\mathbf{x}\rangle = -x_\alpha|-\mathbf{x}\rangle$, and that these eigenvectors are unique. Therefore the vectors $\Pi|\mathbf{x}\rangle$ and $|-\mathbf{x}\rangle$ can differ at most by a phase factor, which may conveniently be chosen to be unity. Hence we have

$$\Pi|\mathbf{x}\rangle = |-\mathbf{x}\rangle . \tag{13.7}$$

The effect of Π on a wave function, $\Psi(\mathbf{x}) \equiv \langle \mathbf{x}|\Psi\rangle$, is now easily determined. From (4.1), (13.6), and (13.7), we obtain

$$\Pi\Psi(\mathbf{x}) \equiv \langle \mathbf{x}|\Pi|\Psi\rangle = \langle -\mathbf{x}|\Psi\rangle = \Psi(-\mathbf{x}) . \tag{13.8}$$

[If instead of $\Pi^2 = 1$, we had chosen some other phase, say $\Pi^2 = e^{i\theta}$, then we would have obtained $\Pi\Psi(\mathbf{x}) = e^{i\theta/2}\Psi(-\mathbf{x})$. This would only be a complicating nuisance, without any physical significance.]

From the fact that $\Pi^2 = 1$, it follows that Π has eigenvalues ± 1. Any even function, $\Psi_e(\mathbf{x}) = \Psi_e(-\mathbf{x})$, is an eigenfunction on Π with eigenvalue $+1$, and any odd function, $\Psi_0(\mathbf{x}) = -\Psi_0(-\mathbf{x})$, is an eigenfunction of Π with eigenvalue

-1. A function corresponding to parity $+1$ is also said to be of *even* parity, and a function corresponding to parity -1 is said to be of *odd* parity.

Example (i). Orbital angular momentum

Under space inversion, $\mathbf{x} \to -\mathbf{x}$, the spherical harmonic (7.34) undergoes the transformation

$$Y_\ell{}^m(\theta, \Phi) \to Y_\ell{}^m(\pi - \theta, \phi + \pi) = (-1)^\ell Y_\ell{}^m(\theta, \phi)\,. \tag{13.9}$$

Hence the single particle orbital angular momentum eigenvector $|\ell, m\rangle$ is also an eigenvector of parity,

$$\Pi|\ell, m\rangle = (-1)^\ell |\ell, m\rangle\,. \tag{13.10}$$

This vector is said to have parity equal to $(-1)^\ell$.

The same result does not extend to the eigenfunctions of total angular momentum for a multiparticle system. For example, according to (7.90) a total orbital angular momentum eigenvector for a two-electron atom is of the form

$$|\ell_1, \ell_2, L, M\rangle = \sum_{m_1, m_2} (\ell_1, \ell_2, m_1, m_2|L, M)|\ell_1, m_1\rangle \otimes |\ell_2, m_2\rangle\,.$$

It is apparent that

$$\Pi|\ell_1, \ell_2, L, M\rangle = (-1)^{\ell_1+\ell_2}|\ell_1, \ell_2, L, M\rangle\,,$$

and that $(-1)^{\ell_1+\ell_2} \neq (-1)^L$. Thus we see that, in general, the parity of an angular momentum state is not determined by its total angular momentum.

Example (ii). Permanent electric dipole moments

The electric dipole moment operator for a multiparticle system has the form $\mathbf{d} = \sum_j q_j \mathbf{Q}_j$, where q_j and $\mathbf{Q}_j$ are the charge and position operator of the jth particle. Thus it follows from (13.3) that the operator $\mathbf{d}$ has odd parity: $\Pi\mathbf{d}\Pi^{-1} = -\mathbf{d}$. If, in the absence of any external electric field, a stationary state $|\Psi\rangle$ has a nonzero average dipole moment, $\langle\mathbf{d}\rangle = \langle\Psi|\mathbf{d}|\Psi\rangle$, we say that the state has a *permanent* or *spontaneous* dipole moment.

Consider now the implications of space inversion symmetry on the average dipole moment:

$$\langle\Psi|\mathbf{d}|\Psi\rangle = \langle\Psi|\Pi^{-1}\Pi\mathbf{d}\Pi^{-1}\Pi|\Psi\rangle = -\langle\Psi'|\mathbf{d}|\Psi'\rangle\,, \tag{13.11}$$

where $|\Psi'\rangle = \Pi|\Psi\rangle$. We are considering a stationary state, and so $H|\Psi\rangle = E|\Psi\rangle$. Now assume that the Hamiltonian is invariant under space inversion, $\Pi H \Pi^{-1} = H$. Then we can make the following derivation:

$$H|\Psi\rangle = E|\Psi\rangle\,,$$
$$\Pi H \Pi^{-1}\Pi|\Psi\rangle = E\Pi|\Psi\rangle\,,$$
$$H|\Psi'\rangle = E|\Psi'\rangle\,.$$

Thus both $|\Psi\rangle$ and $|\Psi'\rangle \equiv \Pi|\Psi\rangle$ describe stationary states with the same energy, E. If this energy level is nondegenerate then these two eigenvectors cannot be independent, and hence we must have $\Pi|\Psi\rangle = c|\Psi\rangle$. The constant c must be equal to one of the parity eigenvalues, $c = \pm 1$. Equation (13.11) for the average dipole moment then yields

$$\langle\Psi|\mathbf{d}|\Psi\rangle = -\langle\Psi'|\mathbf{d}|\Psi'\rangle = -c^2\langle\Psi|\mathbf{d}|\Psi\rangle = -\langle\Psi|\mathbf{d}|\Psi\rangle\,,$$

and hence $\langle\Psi|\mathbf{d}|\Psi\rangle = 0$. Therefore we have proven that *if the Hamiltonian is invariant under space inversion, and if the state is nondegenerate, then there can be no spontaneous electric dipole moment in that state.*

The second condition of nondegeneracy must not be forgotten, because the theorem fails if it does not hold. This is illustrated by Example (4) of Sec. 10.5. The atomic states of hydrogen, denoted $|n,\ell,m\rangle$, have parity $(-1)^\ell$, and so, by the above argument, they should have no spontaneous dipole moment. However, the first excited state ($n = 2$) is fourfold degenerate, and one can easily verify that the eigenvector $(|2,0,0\rangle + |2,1,0\rangle)/\sqrt{2}$ has a nonvanishing average dipole moment. Thus hydrogen in its first excited state can exhibit a spontaneous electric dipole moment.

The necessary condition for a state to exhibit a spontaneous electric dipole moment is that it be a linear combination of even parity and odd parity components. This can be seen most easily for a single particle state function, $\Psi(\mathbf{x})$. If $\Psi(\mathbf{x})$ has definite parity, whether even or odd, the probability density $|\Psi(\mathbf{x})|^2$ is inversion-symmetric, and so the average dipole moment, $q\int|\Psi(\mathbf{x})|^2\mathbf{x}d^3x$, is zero. But if $\Psi(\mathbf{x})$ is a linear combination of even and odd terms, $\Psi(\mathbf{x}) = a\Psi_e(\mathbf{x}) + b\Psi_0(\mathbf{x})$, then $|\Psi(\mathbf{x})|^2$ will not have inversion symmetry, and the average dipole moment will not be zero.

13.2 *Parity Nonconservation*

If the parity operator Π commutes with the Hamiltonian H, then parity eigenvalue ± 1 is a conserved quantity. In that case an even parity state can never acquire an odd parity component, and an odd parity state can never acquire an even parity component. This will be true regardless of the complexity of H, provided only that $\Pi H = H\Pi$.

For a long time it was believed that the fundamental laws of nature were invariant under space inversion, and hence that parity conservation was a fundamental law. This is equivalent to saying that if a process is possible, its mirror image is also possible. In rather loose language, one could say that nature does not distinguish between left-handedness and right-handedness. However, in 1956 an experiment was performed which showed that nature does not obey this symmetry.

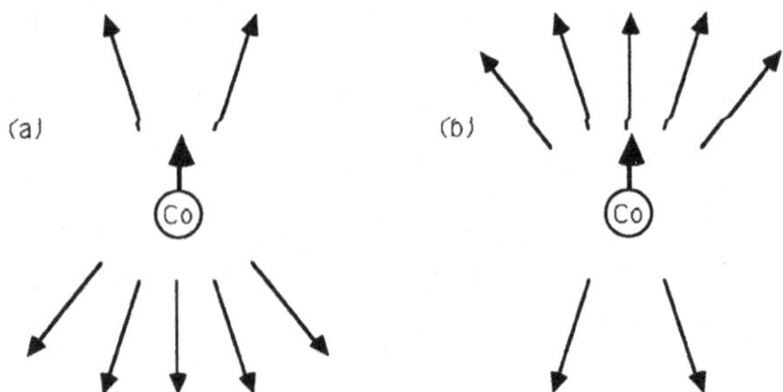

Fig. 13.1 (a) In the actual experiment electrons are emitted preferentially into the hemisphere opposite the nuclear spin. (b) Under space inversion the electron momentum is reversed but the nuclear spin is unchanged.

The radioactive nucleus ^{60}Co undergoes β decay. This is essentially a process whereby a neutron within the nucleus decays into a proton plus an electron plus a neutrino. Only the emitted electron can be readily detected. The nuclei have nonzero spin and magnetic moment, and hence their spins can be aligned at low temperatures by means of a magnetic field. It was found that the electrons were emitted preferentially in the hemisphere opposite to the direction of the nuclear spin, as shown in Fig. 13.1(a). The operation of space inversion reverses the electron momentum but does not change the direction of the nuclear spin, as shown in Fig. 13.1(b). These two processes, (a) and (b), are images of each other with respect to space inversion, yet one happens in nature but the other does not. Thus it appears that nature is not indifferent to left-handedness and right-handedness.

The argument can be formulated more mathematically. Let **S** be the nuclear spin operator, and **P** be the electron momentum operator. Part (a) of Fig. 13.1 illustrates a state for which $\langle\Psi|\mathbf{S}\cdot\mathbf{P}|\Psi\rangle < 0$, whereas part (b) illustrates a state for which $\langle\Psi'|\mathbf{S}\cdot\mathbf{P}|\Psi'\rangle > 0$. The relation between the two states is $|\Psi'\rangle = \Pi|\Psi\rangle$. Now, if it were true that $\Pi H = H\Pi$ then it would follow that $|\Psi\rangle$ and $|\Psi'\rangle$ must be either degenerate states or the same state. However, they cannot be the same state, for this would require that $\langle\Psi|\mathbf{S}\cdot\mathbf{P}|\Psi\rangle = 0$, which is contrary to observation. Therefore it is possible to maintain space inversion symmetry, $\Pi H = H\Pi$, only if the spin-polarized state of the radioactive ^{60}Co nucleus is degenerate. This hypothesis is not supported by detailed theories of nuclear structure.

If we are to entertain the hypothesis that two degenerate states, $|\Psi\rangle$ and $|\Psi'\rangle$, both exist, we need to account for the observation that one of them occurs in nature while its inversion symmetry image does not. The observed parity *asymmetry* (the state $|\Psi\rangle$ is common but the state $\Pi|\Psi\rangle$ is not) does not obviously imply parity *nonconservation.* Most humans have their heart on the left side of their body. Why is it that asymmetries such as this were never advanced as evidence for parity nonconservation, but a similar asymmetry in the nucleus ^{60}Co was taken as overthrowing the supposed law of parity conservation? On the face of it, these asymmetries seem compatible with either of two explanations: (1) parity nonconservation ($\Pi H \neq H\Pi$); or (2) parity conservation ($\Pi H = H\Pi$) with a nonsymmetric initial state, involving components of both parities. Let us examine the second possible explanation, using a highly simplified model as an analog of the more complicated systems of interest.

The potential shown in Fig. 13.2 has a symmetric ground state Ψ_s, with energy E_s. The first excited state is the antisymmetric Ψ_a, at a slightly higher energy E_a. If the barrier separating the two potential minima were infinitely high, these two states would be degenerate. The energy difference, $E_a - E_s$, is nonzero only because of the possibility of tunneling through the barrier.

From these two stationary states we can construct two nonstationary states:

$$\begin{aligned}\Phi &= \frac{\Psi_s + \Psi_a}{\sqrt{2}}\,,\\ \Phi' &= \frac{\Psi_s - \Psi_a}{\sqrt{2}} = \Pi\Phi\,.\end{aligned} \tag{13.12}$$

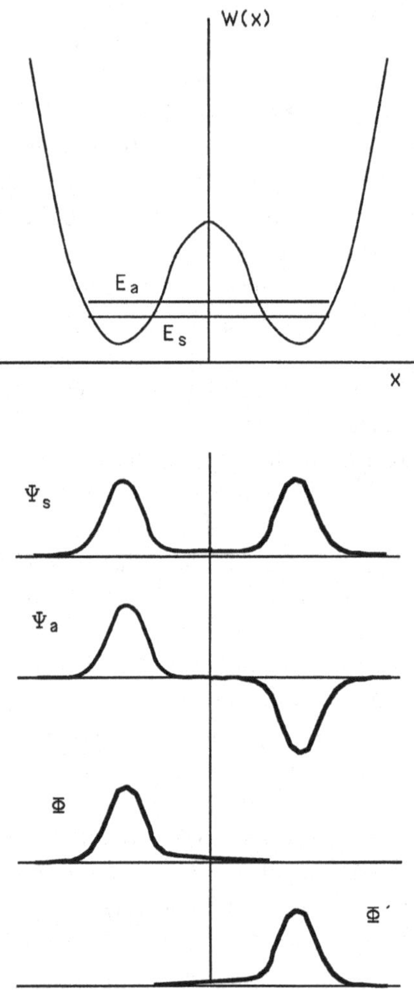

Fig. 13.2 The double minimum potential supports a symmetric ground state Ψ_s, and an antisymmetric excited state Ψ_a. From these one can construct the nonstationary states: $\Phi = (\Psi_s + \Psi_a)/\sqrt{2}$ and $\Phi' = \Pi\Phi$.

These states are not stationary, but would become stationary in the limit of an infinitely high barrier separating the two potential minima. Suppose that at time $t = 0$ the state function is $|\Psi(0)\rangle = |\Phi\rangle$. Then the time-dependent state function will be

$$|\Psi(t)\rangle = (e^{-iE_s t/\hbar}|\Psi_s\rangle + e^{-iE_a t/\hbar}|\Psi_a\rangle)/\sqrt{2}\,. \tag{13.13}$$

This nonstationary state can be described as oscillating back and forth between $|\Phi\rangle$ and $|\Phi'\rangle$ at the frequency $\omega = (E_a - E_s)/\hbar$.

Now, for one's heart, which is initially on the left side of the body, the barrier against tunneling to the right side is very large, and the energy difference $E_a - E_s$ is extremely small. (It can be shown to be an exponentially decreasing function of the barrier height.) Hence the tunneling time from left to right, $\pi\hbar/(E_a - E_s)$, is enormously large, even when compared to the age of the universe. Therefore the observed parity asymmetry in the location of the heart in the body can be explained by an unsymmetric initial state, and does not require nonconservation of parity for its explanation.

We can formally carry this line of argument over to the case of β decay of a nucleus or a neutron. But in such a case the supposed tunneling barrier would be very much smaller, and the tunneling time between the analogs of the "left" and "right" states should be quite short. We would therefore expect to find the "left-handed" state of Fig. 13.1(a) and the "right-handed" state of Fig. 13.1(b) to be equally common. Since this is contrary to observation, we are led to prefer explanation (1), according to which the weak interaction responsible for β decay does not conserve parity. We see from this analysis that the logical path form the observed parity *asymmetry* to the inferred *nonconservation* of parity in β decay is considerably more complex than the popular presentations would indicate.

It should be emphasized that the violation of inversion symmetry, and the related nonconservation of parity, occur only for the weak interactions that are responsible for phenomena such as β decay. There is still a large domain of physics in which inversion symmetry holds to a very good approximation.

13.3 *Time Reversal*

One might suppose that time reversal would be closely analogous to space inversion, with the operation $t \to -t$ replacing $\mathbf{x} \to -\mathbf{x}$. In fact, this simple analogy proves to be misleading at almost every step.

In the first place, the term "time reversal" is misleading, and the operation that is the subject of this section would be more accurately described as *motion reversal*. We shall continue to use the traditional but less accurate expression "time reversal", because it is so firmly entrenched. The effect of the time reversal operator T is to reverse the linear and angular momentum while leaving the position unchanged. Thus we require, by definition,

$$TQT^{-1} = \mathbf{Q}\,, \tag{13.14}$$

$$TPT^{-1} = -\mathbf{P}\,, \tag{13.15}$$

$$TJT^{-1} = -\mathbf{J}\,. \tag{13.16}$$

Consider now the effect that T has on the commutator of the position and momentum operators, $Q_\alpha P_\alpha - P_\alpha Q_\alpha = i\hbar$:

$$TQ_\alpha T^{-1}TP_\alpha T^{-1} - TP_\alpha T^{-1}TQ_\alpha T^{-1} = Ti\hbar T^{-1}\,.$$

According to (13.14) and (13.15), this becomes $Q_\alpha(-P_\alpha) + P_\alpha Q_\alpha = TiT^{-1}\hbar$, which is compatible with the original commutation relation provided that $TiT^{-1} = -i$. Therefore it is necessary for T to be an *antilinear* operator. This same conclusion will be reached if we consider the commutation relations between the components of $\mathbf{J}$, or between components of $\mathbf{P}$ and $\mathbf{J}$.

Properties of antilinear operators

An antilinear operator is one that satisfies (13.2). It is similar to a linear operator except that it takes the complex conjugate of any complex number on which it acts. Hence we have

$$Ac = c^* A\,, \tag{13.17}$$

where A is any antilinear operator and c is any complex number. The product of two antilinear operators A_2 and A_1, defined by the relation

$$(A_2 A_1)|u\rangle = A_2(A_1|u\rangle) \quad \text{for all } |u\rangle\,,$$

is a linear operator, since the second operation of complex conjugation undoes the result of the first.

An operator A is *antiunitary* if it is antilinear, its inverse A^{-1} exists, and it satisfies $\|\,|u\rangle\,\| = \|\,A|u\rangle\,\|$ for all $|u\rangle$. It follows from this definition (see Problem 13.4) that if $|u'\rangle = A|u\rangle$ and $|v'\rangle = A|v\rangle$, then $\langle u'|v'\rangle = \langle v|u\rangle \equiv \langle u|v\rangle^*$. The time reversal operator T is antiunitary.

The action of an antilinear operator to the right on a ket vector is defined by (13.2), but no action to the left on a bra vector has yet been defined. In fact this cannot be done in the simple way that was used in Sec. 1.2 to allow linear operators to act to the left. Recall that a bra vector $\langle\xi|$ is defined as a *linear functional* on the space of ket vectors; that is to say, it must satisfy the relation $\langle\xi|(a|u\rangle + b|v\rangle) = a\langle\xi|u\rangle + b\langle\xi|v\rangle$. For any *linear* operator L we can define another linear functional, $\langle\eta| \equiv \langle\xi|L$ (with L operating to the left), by requiring it to satisfy the relation $\langle\eta|\phi\rangle = \langle\xi|(L|\phi\rangle)$ or, equivalently, $(\langle\xi|L)|\phi\rangle = \langle\xi|(L|\phi\rangle) = \langle\xi|L|\phi\rangle$. But this is possible only because this expression is indeed a *linear functional* of $|\phi\rangle$, satisfying $\langle\xi|L(a|u\rangle + b|v\rangle) = a\langle\xi|L|u\rangle + b\langle\xi|L|v\rangle$. Thus $\langle\xi|L$ really does satisfy the definition of a bra vector.

If we attempt to carry out the same construction using an *antilinear* operator A in place of the linear operator L, we formally obtain $\langle\xi|A(a|u\rangle+b|v\rangle) = a^*\langle\xi|A|u\rangle+b^*\langle\xi|A|v\rangle$. Thus if we were to define $(\langle\xi|A)|\phi\rangle = \langle\xi|(A|\phi\rangle)$ we would not obtain a *linear* functional of $|\phi\rangle$, and therefore the object $\langle\xi|A$ so defined would not be a bra vector. We shall deal with this complication by adopting the convention that *antilinear operators act only to the right, and never to the left.* Because of this convention, we shall not make use of the adjoint, $A^\dagger$, of an antilinear operator.

[[This convention is not the only way of dealing with the problem. Messiah (1966) allows antilinear operators to act either to the left or to the right, but as a consequence he must caution his readers that $(\langle\xi|A)|\phi\rangle \neq \langle\xi|(A|\phi\rangle)$, and hence the common expression $\langle\xi|A|\phi\rangle$ becomes undefined. Both his approach and ours impose a certain inconvenience on the reader, which is ultimately not the fault of either author, but rather a reflection of the fact that antilinear operators do not fit into the bra–ket notation as neatly as do linear operators.]]

The *complex conjugation* operator is the simplest example of an antilinear operator. Unlike a linear operator, it is not independent of the phases of the basis vectors in terms of which it is defined. Consider an orthonormal set of basis vectors, $\{|n\rangle\}$, and an arbitrary vector, $|\psi\rangle = \sum_n a_n|n\rangle$. The complex conjugation operator in this n-basis, $K_{(n)}$, is defined by the equation

$$K_{(n)}|\psi\rangle = \sum_n a_n{}^*\,|n\rangle\,. \tag{13.18}$$

Consider next some other orthonormal set of basis vectors, $\{|\nu\rangle\}$, in terms of which the same vector is given by $|\psi\rangle = \sum_\nu \alpha_\nu|\nu\rangle$. In this ν-basis the complex conjugation operator $K_{(\nu)}$ is defined by

$$K_{(\nu)}|\psi\rangle = \sum_\nu \alpha_\nu{}^*\,|\nu\rangle\,. \tag{13.19}$$

To determine whether these two complex conjugation operators are equivalent, we shall express the ν-basis vectors in terms of the n-basis:

$$|\nu\rangle = \sum_n |n\rangle\langle n|\nu\rangle\,. \tag{13.20}$$

Thus we obtain $|\psi\rangle = \sum_\nu \alpha_\nu \sum_n |n\rangle\langle n|\nu\rangle$, and so the relation between the two sets of coefficients is $a_n = \sum_\nu \alpha_\nu \langle n|\nu\rangle$. Substitution of (13.20) into (13.19) yields $K_{(\nu)}|\psi\rangle = \sum_\nu \alpha_\nu{}^* \sum_n |n\rangle\langle n|\nu\rangle = \sum_n \sum_\nu \alpha_\nu{}^* \langle n|\nu\rangle |n\rangle$. This is not equal to (13.18) unless the inner product $\langle n|\nu\rangle$ is real for all n and ν. Thus we have shown that the complex conjugate operators defined with respect to two different sets of basis vectors are, in general, not equivalent. This is true, in particular, if the two basis sets are identical except for the complex phases of the vectors.

Time reversal of the Schrödinger equation

Contrary to what is suggested by the name, the application of the time reversal operator T to the Schrödinger equation,

$$H|\Psi(t)\rangle = i\hbar \frac{\partial}{\partial t}|\Psi(t)\rangle\,, \tag{13.21}$$

does not change t into $-t$. Indeed, since t is merely a parameter it cannot be directly affected by an operator, and so the connection between the action of T and the parameter t can only be indirect. The time reversal transformation of (13.21) yields

$$\begin{aligned} THT^{-1}T|\Psi(t)\rangle &= Ti\hbar \frac{\partial}{\partial t}|\Psi(t)\rangle \\ &= -i\hbar \frac{\partial}{\partial t} T|\Psi(t)\rangle\,. \end{aligned} \tag{13.22}$$

Suppose that $THT^{-1} = H$, or, in words, that H is invariant under time reversal. If we rewrite (13.22) with the dummy variable t replaced by $-t$, then it is apparent that $T|\Psi(-t)\rangle$ is also a solution of (13.21). Whereas the invariance of H under a linear transformation gives rise to a conserved quantity (the parity eigenvalue in the case of space inversion), there is no such conserved quantity associated with invariance under the antilinear time reversal transformation. Instead, the solutions of the Schrödinger equation occur in pairs, $|\Psi(t)\rangle$ and $T|\Psi(-t)\rangle$.

So far we have not obtained the explicit form of the time reversal operator, except that it is antilinear and so must involve complex conjugation. Indeed the explicit form of T depends upon the basis, and so we shall consider separately the most common cases.

In *coordinate representation* the Schrödinger equation takes the form

$$\left[\frac{-\hbar^2}{2M}\nabla^2 + W(\mathbf{x})\right]\Psi(\mathbf{x},t) = i\hbar\frac{\partial}{\partial t}\Psi(\mathbf{x},t)\,.$$

Its complex conjugate is

$$\left[\frac{-\hbar^2}{2M}\nabla^2 + W^*(\mathbf{x})\right]\Psi^*(\mathbf{x},t) = -i\hbar\frac{\partial}{\partial t}\Psi^*(\mathbf{x},t)\,.$$

The condition for the Hamiltonian to be invariant under complex conjugation is that the potential be real: $W^* = W$. In that case it is apparent that if $\Psi(\mathbf{x},t)$ is a solution then so is $\Psi^*(\mathbf{x},-t)$. This suggests that we may identify the time reversal operator with the complex conjugation operator in this representation,

$$T = K_0\,, \tag{13.23}$$

where, by definition, $K_0\Psi(\mathbf{x},t) = \Psi^*(\mathbf{x},t)$. In this case T is its own inverse.

In coordinate representation the effect of the position operator is merely to multiply by $\mathbf{x}$, and therefore (13.14) is satisfied. The momentum operator has the form $-i\hbar\boldsymbol{\nabla}$. Its sign is reversed by complex conjugation and so (13.15) is satisfied. It is also apparent that (13.16) holds for the orbital angular momentum operator, $\mathbf{L} = \mathbf{x}\times(-i\hbar\boldsymbol{\nabla})$. Therefore (13.23) is valid in coordinate representation for spinless particles.

The formal expression for an arbitrary vector in coordinate representation is $|\Psi\rangle = \int \Psi(\mathbf{x})|\mathbf{x}\rangle d^3x$, where the basis vector $|\mathbf{x}\rangle$ is an eigenvector of the position operator. Since T is equal to the complex conjugation operator, its effect is simply $T|\Psi\rangle = \int \Psi^*(\mathbf{x})|\mathbf{x}\rangle d^3x$, with $T|\mathbf{x}\rangle = |\mathbf{x}\rangle$. [cf. (13.18).]

In *momentum representation* an arbitrary vector can be written as $|\Psi\rangle = \int \Psi(\mathbf{p})|\mathbf{p}\rangle d^3p$, where the basis vector $|\mathbf{p}\rangle$ is a momentum eigenvector (5.1). The effect of the time reversal operator is $T|\Psi\rangle = \int \Psi^*(\mathbf{p})T|\mathbf{p}\rangle d^3p$, and so T will be completely defined as soon as we determine its effect on a momentum eigenvector. To do this we transform back to coordinate representation, where the form of T is already known.

$$\begin{aligned}T|\mathbf{p}\rangle &= T\int |\mathbf{x}\rangle\langle\mathbf{x}|\mathbf{p}\rangle d^3\mathbf{x} = \int T|\mathbf{x}\rangle e^{i\mathbf{p}\cdot\mathbf{x}/\hbar}(2\pi\hbar)^{-3/2}d^3\mathbf{x}\\ &= \int |\mathbf{x}\rangle e^{-i\mathbf{p}\cdot\mathbf{x}/\hbar}(2\pi\hbar)^{-3/2}d^3\mathbf{x} = \int T|\mathbf{x}\rangle\langle\mathbf{x}|-\mathbf{p}\rangle d^3\mathbf{x}\\ &= |-\mathbf{p}\rangle\,.\end{aligned}$$

Therefore the time reversal operator in momentum representation is not merely complex conjugation; rather, its effect is given by

$$T|\Psi\rangle = \int \Psi^*(\mathbf{p})|-\mathbf{p}\rangle d^3p\,. \tag{13.24}$$

Time reversal and spin

The time reversal operator must reverse the angular momentum, as is asserted by (13.16). This condition has been shown to hold for the orbital angular momentum, and for consistency it must be imposed on the spin angular momentum:

$$T\mathbf{S}T^{-1} = -\mathbf{S}\,. \tag{13.25}$$

Since the form of the time reversal operator is representation-dependent, we choose coordinate representation for orbital variables, and the standard representation of the spin operators in which S_z is diagonal. In this representation the matrices for S_x and S_z are real, and the matrix for S_y is imaginary. This was shown explicitly for $s = \frac{1}{2}$ in (7.45), and for $s = 1$ in (7.52). That it is true in general may be shown from the argument leading up to (7.16), which demonstrates that the matrices for $S_+ \equiv S_x + iS_y$ and $S_- \equiv S_x - iS_y$ are real, and hence S_x must be real and S_y must be imaginary. The time reversal operator T cannot be equal to the complex conjugation operator K_0 in this representation, since the effect of the latter is

$$K_0 S_x K_0 = S_x\,, \quad K_0 S_y K_0 = -S_y\,, \quad K_0 S_z K_0 = S_z\,. \tag{13.26}$$

(Note that ${K_0}^{-1} = K_0$.)

Let us write the time reversal operator as $T = YK_0$, where Y is a linear operator because it is the product of two antilinear operators, TK_0. To satisfy (13.25), Y must have the following properties:

$$YS_xY^{-1} = -S_x\,, \quad YS_yY^{-1} = S_y\,, \quad YS_zY^{-1} = -S_z\,. \tag{13.27}$$

The correct transformation of the orbital variables is produced by the complex conjugation operator K_0 by itself, and so in order that Y should not spoil this situation, we must have

$$Y\mathbf{Q}Y^{-1} = \mathbf{Q}\,, \quad Y\mathbf{P}Y^{-1} = \mathbf{P}\,. \tag{13.28}$$

Thus Y must operate only on the spin degrees of freedom. It is apparent that (13.27) and (13.28) are satisfied by the operator $Y = e^{-i\pi S_y/\hbar}$, whose effect is to rotate spin (and only spin) through the angle π about the y axis. Therefore the explicit form of the time reversal in this representation is

$$T = e^{-i\pi S_y/\hbar} K_0\,. \tag{13.29}$$

Time reversal squared

Two successive applications of the time reversal transformation, i.e. two reversals of motion, leave the physical situation unchanged. Therefore the vectors $|\Psi\rangle$ and $T^2|\Psi\rangle$ must describe the same state, and hence we must have

$$T^2|\Psi\rangle = c|\Psi\rangle \tag{13.30}$$

for some c that satisfies $|c| = 1$. To determine the possible values of c, we evaluate $T^3|\Psi\rangle$ using the associative property of operator multiplication and (13.30), obtaining

$$T^2(T|\Psi\rangle) = T(T^2|\Psi\rangle) = T(c|\Psi\rangle) = c^*T|\Psi\rangle\,. \tag{13.31}$$

Now an equation of the form (13.30) must hold for every state vector, so we must have $T^2(|\Psi\rangle + T|\Psi\rangle) = c'(|\Psi\rangle + T|\Psi\rangle)$ for some c'. But from (13.30) and (13.31) we obtain $T^2(|\Psi\rangle + T|\Psi\rangle) = c|\Psi\rangle + c^*T|\Psi\rangle$, which is consistent with the previous requirement only if $c' = c = c^*$. Thus we must have $c = \pm 1$, and hence (13.30) can be rewritten as

$$T^2|\Psi\rangle = \pm|\Psi\rangle\,. \tag{13.32}$$

Although two successive time reversals is a seemingly trivial transformation, the corresponding operator T^2 is not the identity operator. In Sec. 7.6 we encountered a similar operator, $R(2\pi)$, which corresponds to rotation through a full circle. In fact the relation between these two operators is much stronger than analogy. From (13.29) and (13.26) we obtain

$$\begin{aligned} T^2 &= e^{-i\pi S_y/\hbar}\, K_0\, e^{-i\pi S_y/\hbar}\, K_0 \\ &= e^{-i\pi S_y/\hbar}\, e^{+i\pi(-S_y)/\hbar} \\ &= e^{-i2\pi S_y/\hbar}\,. \end{aligned}$$

This may equivalently be written as

$$T^2 = e^{-i2\pi J_y/\hbar}\,,$$

since J_y is the sum of two commutative operators, $J_y = L_y + S_y$, and $e^{-i2\pi L_y/\hbar} = 1$. Thus the operator T^2 is equal to the rotation operator for a full revolution about the y axis, and so we have an identity

$$T^2 = R(2\pi)\,. \tag{13.33}$$

These two identical operators have eigenvalue $+1$ for any state of integer total angular momentum, and have eigenvalue -1 for any state of half odd-integer total angular momentum.

Example (i). Kramer's theorem

It has been shown that invariance of the Hamiltonian under the unitary operator of space inversion gives rise to a conserved quantity, the *parity* of the state. Invariance under the antiunitary time reversal operator does not produce a conserved quantity, but it sometimes increases the degree of degeneracy of the energy eigenstates. Let us consider the energy eigenvalue equation, $H|\Psi\rangle = E|\Psi\rangle$, for a time-reversal-invariant Hamiltonian, $TH = HT$. Then $HT|\Psi\rangle = TH|\Psi\rangle = ET|\Psi\rangle$, and so both $|\Psi\rangle$ and $T|\Psi\rangle$ are eigenvectors with energy eigenvalue E. There are two possibilities: (a) $|\Psi\rangle$ and $T|\Psi\rangle$ are linearly dependent, and so describe the same state, or (b) $|\Psi\rangle$ and $T|\Psi\rangle$ are linearly independent, and so describe two degenerate states. It will now be shown that case (a) is not possible in certain circumstances.

Suppose that (a) is true, in which case we must have $T|\Psi\rangle = a|\Psi\rangle$ with $|a| = 1$. A second application of T yields $T^2|\Psi\rangle = Ta|\Psi\rangle = a^*T|\Psi\rangle = a^*a|\Psi\rangle = |\Psi\rangle$. Thus case (a) is possible only for those states that satisfy $T^2|\Psi\rangle = |\Psi\rangle$. But for those states that satisfy $T^2|\Psi\rangle = -|\Psi\rangle$ it is necessarily true that $|\Psi\rangle$ and $T|\Psi\rangle$ are linearly independent, degenerate states. This result is known as *Kramer's theorem*: any system for which $T^2|\Psi\rangle = -|\Psi\rangle$, such as an odd number of $s = \frac{1}{2}$ particles, has only degenerate energy levels.

In many cases the degeneracy implied by Kramer's theorem is merely the degeneracy between states of spin up and spin down, or something equally obvious. The theorem is nontrivial for a system with spin–orbit coupling in an unsymmetrical electric field, so that neither spin nor angular momentum is conserved. Kramer's theorem implies that no such field can split the degenerate pairs of energy levels. However, the degeneracy can be broken by an external magnetic field, which couples to the magnetic moment and contributes a term in the Hamiltonian like $\gamma\mathbf{S}\cdot\mathbf{B}$, which is not invariant under time reversal.

Example (ii). Permanent electric dipole moments

It was shown in Sec. 13.1 that there can be no permanent electric dipole moment in a nondegenerate state if the Hamiltonian is invariant under space inversion. However, it is known that the weak interactions, which are responsible for β decay, are not invariant under space inversion, so this raises the possibility that elementary particles might have electric dipole moments. Such moments, if they exist, are very small and have not been detected. It can be shown that electric dipole moments are excluded by invariance under both rotations and time reversal.

If the Hamiltonian is rotationally invariant it must commute with the angular momentum operators, which are the generators of rotations. Thus there is a complete set of common eigenvectors of H, $\mathbf{J}\cdot\mathbf{J}$, and J_z, which we shall denote as $|E,j,m\rangle$. We assume that the only degeneracy of these energy eigenvectors is associated with the $2j+1$ values of m. The electric dipole moment operator $\mathbf{d}$ is an irreducible tensor operator of rank 1, so we may invoke (7.125) to write the average dipole moment in one of these states as

$$\langle E,J,m|\mathbf{d}|E,j,m\rangle = C_{E,j}\langle E,j,m|\mathbf{J}|E,j,m\rangle\,, \tag{13.34}$$

where $C_{E,j}$ is a scalar that does not depend on m.

It is shown in Problem 13.4 that if $|u'\rangle = T|u\rangle$ and $|v'\rangle = T|v\rangle$ then $\langle u'|v'\rangle = \langle u|v\rangle^*$. Let us take $|u\rangle = |\Psi\rangle$ and $|v\rangle = d_\alpha|\Psi\rangle$, where d_α is a component of the electric dipole operator. Then $|u'\rangle = |\Psi'\rangle \equiv T|\Psi\rangle$ and $|v'\rangle = Td_\alpha|\Psi\rangle = d_\alpha T|\Psi\rangle \equiv d_\alpha|\Psi'\rangle$. Thus $\langle\Psi'|d_\alpha|\Psi'\rangle = \langle\Psi|d_\alpha|\Psi\rangle^*$. But d_α is a Hermitian operator, so we may write

$$\langle\Psi'|\mathbf{d}|\Psi'\rangle = \langle\Psi|\mathbf{d}|\Psi\rangle\,, \qquad \text{where } |\Psi'\rangle = T|\Psi\rangle\,. \tag{13.35}$$

A similar calculation may be performed using J_α (a component of angular momentum) in place of d_α, except that because of (13.16) we now have $TJ_\alpha = -J_\alpha T$, and so we obtain

$$\langle\Psi'|\mathbf{J}|\Psi'\rangle = -\langle\Psi|\mathbf{J}|\Psi\rangle\,. \tag{13.36}$$

From the relation $J_z|E,j,m\rangle = \hbar m|E,j,m\rangle$, we obtain

$$TJ_zT^{-1}T|E,j,m\rangle = \hbar mT|E,j,m\rangle\,,$$

and hence $J_z(T|E,j,m\rangle) = -\hbar m(T|E,j,m\rangle)$. Under the previous assumption restricting the degeneracy, the vector $T|E,j,m\rangle$ can differ from $|E,j,-m\rangle$ by at most a phase factor. Therefore, by taking $|\Psi\rangle = |E,j,m\rangle$, we see from (13.35) that $\langle E,j,-m|\mathbf{d}|E,j,-m\rangle = \langle E,j,m|\mathbf{d}|E,j,m\rangle$, and from (13.36) we obtain $\langle E,j,-m|\mathbf{J}|E,j,-m\rangle = -\,\langle E,j,m|\mathbf{J}|E,j,m\rangle$. But these two results imply that under the substitution $m \to -m$, the right hand side of (13.34) changes sign while the left hand side does not. This is possible only if both sides vanish. Hence *the spontaneous dipole moment of the state must vanish under the combined assumptions of rotational invariance, time reversal invariance, and the degeneracy of the state being only that due to* m.

This would suffice to prove that elementary particles cannot have electric dipole moments, but for the fact that there is indirect evidence for a superweak interaction that violates time reversal invariance. Thus experiments to detect very small electric dipole moments are of considerable interest. The best upper bound on the magnitude of any electric dipole moment of the electron is $|d_e| < 1.05 \times 10^{-27}\, e\,\text{cm}$ (Hudson *et al.*, 2011).

Further reading for Chapter 13

The discrete symmetries of parity and time reversal (and also charge conjugation, not treated in this book) find many of their applications in nuclear and particle physics. Many such applications are discussed in Ch. 3 of the book by Perkins (1982).

Problems

13.1 Show that mirror reflection is equivalent to the combined effect of space inversion and a certain rotation.

13.2 Show in detail that if $\Pi H = H\Pi$ and if the initial state $|\Psi(0)\rangle$ has definite parity (either even or odd), then the state vector $|\Psi(t)\rangle$ remains a pure parity eigenvector at all future times.

13.3 An unstable particle whose spin is $\mathbf{S}$ decays, emitting an electron and possibly other particles. Consider the angular distribution of the electrons emitted from a spin-polarized sample of such particles. It may depend upon $\mathbf{S}$, $\boldsymbol{\sigma}$, and $\mathbf{p}$, where $\boldsymbol{\sigma}$ and $\mathbf{p}$ are the electron spin and momentum. (a) Write down the most general distribution function that is consistent with space inversion symmetry. (b) Write down the most general distribution function that is consistent with time reversal symmetry. (c) Write down the most general distribution consistent with both symmetries.

13.4 An operator A is *antiunitary* if it is antilinear, its inverse A^{-1} exists, and it satisfies $\|\,|u\rangle\,\| = \|\,A|u\rangle\,\|$ for all $|u\rangle$. Prove from this definition that if $|u'\rangle = A|u\rangle$ and $|v'\rangle = A|v\rangle$ then $\langle u'|v'\rangle = \langle u|v\rangle^*$.

13.5 Kramer's theorem states that if the Hamiltonian of a system is invariant under time reversal, and if $T^2|\Psi\rangle = -|\Psi\rangle$ (as is the case for an odd number of electrons), then the energy levels must be at least doubly degenerate. In fact the degree of degeneracy must be even. Show explicitly that threefold degeneracy is not possible.

13.6 In Example (ii) of Sec. 13.3 it was proved, under certain assumptions, that a state of the form $|\Psi\rangle = |E, j, m\rangle$ cannot possess a permanent electric dipole moment. Since the $2j+1$ states having different m values are degenerate, one can also have stationary states of the form $|\Psi\rangle = \sum_m c_m|E, j, m\rangle$. Prove, under the same assumptions, that a state of this more general form cannot have a permanent electric dipole moment.

13.7 Suppose that the Hamiltonian is invariant under time reversal: $[H, T] = 0$. Show that, nevertheless, an eigenvalue of T is *not* a conserved quantity.

13.8 Use the explicit form of the time reversal operator T for a particle of spin $\frac{1}{2}$ to evaluate $T\begin{bmatrix}\alpha\\ \beta\end{bmatrix}$, where the vector is expressed in the standard representation in which σ_z is diagonal.

13.9 The probability of tunneling through a potential barrier from left to right is clearly equal to the probability of tunneling from right to left if the barrier potential possesses mirror reflection symmetry. (In one dimension this is the same as space inversion.) But if the barrier potential is asymmetric, having no mirror reflection symmetry, it is not apparent that these two probabilities should be equal. Use time reversal invariance to prove that the left-to-right and right-to-left tunneling probabilities must be equal, even if the barrier potential is asymmetric.

Chapter 14

The Classical Limit

Classical mechanics has been verified in a very wide domain of experience, so if quantum mechanics is correct it must agree with classical mechanics in the appropriate limit. Ideally we would like to exhibit quantum mechanics as a broader theory, encompassing classical mechanics as a special limiting case. Loosely stated, the limit must be one in which $\hbar$ is negligibly small compared with the relevant dynamical parameters. However, the matter is quite subtle. One cannot merely define $\hbar \to 0$ to be the classical limit. That limit is not well defined mathematically unless one specifies what quantities are to be held constant during the limiting process. Moreover, there are conceptual problems that are at least as important as the mathematical problem.

It is useful to first examine the manner in which special relativity reduces to classical Newtonian mechanics in the limit where the speed of light c becomes infinite. Consider a typical formula of relativistic mechanics, such as the kinetic energy of a particle of mass M moving at the speed v: KE $= Mc^2[(1-v^2/c^2)^{-1/2}-1]$. In the limit $c \to \infty$, this formula reduces to the Newtonian expression KE $= \frac{1}{2}Mv^2$. More generally, all the results of classical Newtonian mechanics are recovered in the limit where $v/c \ll 1$ or, equivalently, in the limit where kinetic and potential energies are small compared to the rest energy Mc^2. In this limit, the trajectories predicted by relativistic mechanics merge with those predicted by Newtonian mechanics, and it is quite correct to say that relativistic mechanics includes Newtonian mechanics as a special limiting case.

Bohr and Heisenberg stressed the analogy between the limits $c \to \infty$ and $\hbar \to 0$, both of which supposedly lead back to the familiar ground of Newtonian mechanics, in an attempt to convert Einstein to their view of quantum mechanics. Einstein was unmoved by such arguments, and indeed the proposed analogy seriously oversimplifies the problem. Newtonian mechanics and relativistic mechanics are formulated in terms of the same concepts: the continuous trajectories of individual particles through space–time. Those trajectories differ quantitatively between the two theories, but the differences vanish in the limit $c \to \infty$. But quantum mechanics is formulated in terms of probabilities, and does not refer directly to trajectories of individual particles. A conceptual difference is much more difficult to bridge than a merely quantitative difference.

14.1 *Ehrenfest's Theorem and Beyond*

The term *classical limit of quantum mechanics* will be used, broadly, to refer to the predictions of quantum mechanics for systems whose dynamical magnitudes are large compared to $\hbar$. Often these will be macroscopic systems whose dimensions and masses are of the order of centimeters and grams. The concepts of *classical* and *macroscopic* systems are distinct, as the existence of macroscopic quantum phenomena (such as superconductivity) demonstrates, but the behavior of most macroscopic systems can be described by classical mechanics. Throughout this book, we have stressed that quantum theory does not predict the individual observed phenomenon, but only the probabilities of the possible phenomena. This fact is particularly relevant in studying the classical limit, where we will see that, in a generic case, *the classical limit of a quantum state is an ensemble of classical trajectories, not a single classical trajectory.*

If quantum mechanics were to yield an individual trajectory in its classical limit, it would be necessary for the probability distributions to become arbitrarily narrow as $\hbar \to 0$. The indeterminacy relation, $\Delta x\, \Delta p \geq \hbar/2$, allows the possibility that the widths of position and momentum distributions might both vanish as $\hbar \to 0$. But whether or not this actually happens depends on the particular state. Some special states behave in that way, but there are many physically realistic states that do not. A good example is provided by a measurement process (see Ch. 9), in which a correlation is established between the eigenvalue r of the measured dynamical variable R and the indicator variable α of the measuring apparatus. The indicator is a macroscopic object, such as the position of a pointer on an instrument. If the initial state is not an eigenstate of the measured variable R, but is rather a state in which two (or more) eigenvalues, r_1 and r_2, have comparable probability, then in the final state there will be two (or more) indicator positions, α_1 and α_2, that have comparable probability. The values α_1 and α_2 are macroscopically distinct, being perhaps centimeters apart, and hence the probability distribution for the indicator variable will be spread over a macroscopic range.

Even though the indicator may be an ordinary classical object, like a pointer on an instrument, its quantum-mechanical description will be a broad probability distribution, quite unlike any classical trajectory. Therefore we should not expect to recover an individual classical trajectory when we take the classical limit of quantum mechanics. Rather, we should expect the probability distributions of quantum mechanics to become equivalent to the probability distributions of an ensemble of classical trajectories.

Ehrenfest's theorem

This theorem is the first step in relating quantum probabilities to classical mechanics. It is sufficient for our purposes to consider only the simplest example, a single particle in one dimension, whose Hamiltonian is $H = P^2/2M + W(Q)$. Using the Heisenberg picture (Sec. 3.7), in which the operators for dynamical variables are time-dependent and the states are time-independent, the equations of motion for the position and momentum operators are

$$\frac{dQ}{dt} = \frac{i}{\hbar}[H, Q] = \frac{P}{M}, \tag{14.1}$$

$$\frac{dP}{dt} = \frac{i}{\hbar}[H, P] = F(Q), \tag{14.2}$$

where $F(Q) = -\partial W(Q)/\partial Q$ is the force operator. [The result of Problem 4.1 has been used in deriving (14.2).] Taking averages in some state, we obtain

$$\frac{d\langle Q\rangle}{dt} = \frac{\langle P\rangle}{M}, \tag{14.3}$$

$$\frac{d\langle P\rangle}{dt} = \langle F(Q)\rangle. \tag{14.4}$$

Now, *if* we can approximate the average of the function of position with the function of the average position,

$$\langle F(Q)\rangle \approx F(\langle Q\rangle), \tag{14.5}$$

then (14.4) may be replaced by

$$\frac{d\langle P\rangle}{dt} = F(\langle Q\rangle). \tag{14.6}$$

Equations (14.3) and (14.6) together say that the quantum-mechanical averages, $\langle Q\rangle$ and $\langle P\rangle$, obey the classical equations of motion. The approximation (14.5) is exact only if the force $F(Q)$ is a linear function of Q, as is the case for a harmonic oscillator or a free particle. But if the width of the position probability distribution is small compared to the typical length scale over which the force varies, then the centroid of the quantum-mechanical probability distribution will follow a classical trajectory. This is Ehrenfest's theorem.

It is sometimes asserted that the conditions for classical behavior of a quantum system are just those required for Ehrenfest's theorem. But, in fact, Ehrenfest's theorem is *neither necessary nor sufficient* to define the classical regime (Ballentine, Yang, and Zibin, 1994). Lack of sufficiency — that a system may obey Ehrenfest's theorem but not behave classically — is proved by the example of the harmonic oscillator. It satisfies (14.6) exactly for all states. Yet a quantum oscillator has discrete energy levels, which make its thermodynamic properties quite different from those of the classical oscillator. Lack of necessity — that a system may behave classically even when Ehrenfest's theorem does not apply — will be demonstrated below.

Corrections to Ehrenfest's theorem

Let us introduce operators for the deviations from the mean values of position and momentum,

$$\delta Q = Q - \langle Q \rangle \,, \tag{14.7}$$

$$\delta P = P - \langle P \rangle \,, \tag{14.8}$$

and expand (14.1) and (14.2) in powers of these deviation operators. Taking the average in some chosen state then recovers (14.3), and yields, in place of (14.4),

$$\frac{dp_0}{dt} = F(q_0) + \frac{1}{2}\langle(\delta Q)^2\rangle \, \frac{\partial^2}{\partial {q_0}^2} \, F(q_0) + \cdots \,, \tag{14.9}$$

where the average position and momentum are $q_0 = \langle Q \rangle$ and $p_0 = \langle P \rangle$. If $\langle(\delta Q)^2\rangle$ and higher order terms are negligible, we recover Ehrenfest's theorem, with q_0 and p_0 obeying the classical equations.

The terms in (14.9) beyond $F(q_0)$ are corrections to Ehrenfest's theorem. But they are not essentially quantum-mechanical in origin, as is evidenced by the fact that they do not depend explicitly on $\hbar$. Indeed, $\langle(\delta Q)^2\rangle$ is just a measure of the width of the position probability distribution, which need not vanish in the classical limit. The proper interpretation of these correction terms can be found by comparison with a suitable classical ensemble.

Let $\rho_c(q, p, t)$ be the probability distribution in phase space for a classical ensemble. It satisfies the Liouville equation, which describes the flow of probability in phase space,

$$\frac{\partial}{\partial t}\,\rho_c(q,p,t) = -\dot{q}\,\frac{\partial}{\partial q}\,\rho_c(q,p,t) \,-\, \dot{p}\,\frac{\partial}{\partial p}\,\rho_c(q,p,t)$$

$$= -\frac{p}{M}\,\frac{\partial}{\partial q}\,\rho_c(q,p,t) - F(q)\frac{\partial}{\partial p}\,\rho_c(q,p,t)\,. \qquad (14.10)$$

From it, we can calculate the classical averages,

$$q_c = \iint q\,\rho_c(q,p,t)\,dq\,dp\,, \qquad (14.11)$$

$$p_c = \iint p\,\rho_c(q,p,t)\,dq\,dp\,. \qquad (14.12)$$

Differentiating these expressions with respect to t, using (14.10), and integrating by parts as needed, we obtain

$$\frac{dq_c}{dt} = \frac{p_c}{m}\,, \qquad (14.13)$$

$$\frac{dp_c}{dt} = \iint F(q)\,\rho_c(q,p,t)\,dq\,dp\,, \qquad (14.14)$$

which are the classical analogs of (14.3) and (14.4). Expanding (14.14) in powers of $\delta q = q - q_c$ then yields

$$\frac{d}{dt}p_c = F(q_c) + \frac{1}{2}\,\langle(\delta q)^2\rangle_c\,\frac{\partial^2}{\partial {q_c}^2}F(q_c) + \cdots\,, \qquad (14.15)$$

where $\langle(\delta q)^2\rangle_c = \iint(\delta q)^2\rho_c(q,p,t)\,dq\,dp$ is a measure of the width of the classical probability distribution.

The significance of the terms involving δq is now clear. The centroid of a classical ensemble need not follow a classical trajectory if the width of the probability distribution is not negligible. The quantal equation (14.9) has exactly the same form as the classical (14.15), and its appropriate interpretation is simply that the centroid of the quantal probability distribution does not follow a classical trajectory unless it is very narrow.

Example: Particle between reflecting walls

Consider a particle confined to move between two impenetrable walls, at $x = 0$ and $x = L$. A general time-dependent state function can be expanded in terms of the energy eigenfunctions,

$$\psi(x,t) = \sum_{n=1}^{\infty} c_n\,\sin(k_n x)\,\exp\left(-\frac{iE_n t}{\hbar}\right)\,, \qquad (14.16)$$

where $k_n = n\pi/L$ and $E_n = (\hbar^2 \pi^2/2mL^2)n^2$. Because all the frequencies in (14.16) are integer multiples of the lowest frequency, it follows that $\psi(x,t)$ is periodic, but its period,

$$T_{\rm qm} = \frac{4mL^2}{\pi\hbar}\,, \tag{14.17}$$

bears no relation to the classical period of a particle with speed v, $T_{\rm cl} = 2L/v$. The failure of (14.16) to oscillate with the classical period would be a problem if, in the classical limit, the wave function were supposed to describe the orbit of a single particle. But there is no difficulty if it is compared to an ensemble of classical orbits, since the motion of the ensemble need not be periodic. The quantum recurrence period $T_{\rm qm}$ diverges to infinity as $\hbar \to 0$, and so becomes irrelevant in the classical limit.

Consider an initial wave function of the form

$$\psi(x,0) = A(x)\, e^{ikx}\,, \tag{14.18}$$

where $A(x)$ is a real amplitude function. The mean velocity of this state is $v = \hbar k/m$. The motion of this quantum state will be compared to that of a classical ensemble whose initial position and momentum distributions are equal to those of the quantum state (14.18), the initial phase space distribution being the product of the position and momentum distributions. We choose a Gaussian amplitude,

$$A(x) = C\,\exp\left[-\left(\frac{x-x_0}{2a}\right)^2\right]. \tag{14.19}$$

This initial state has rms half-width $\Delta x = a$, and its mean position is taken to be $x_0 = L/2$. Results for $a = 0.1$, $v = 20$ (units: $\hbar = m = L = 1$) are shown in Fig. 14.1. The average position of the quantum state, $\langle x\rangle = \langle\psi(x,t)|x|\psi(x,t)\rangle$, exhibits a complex pattern of decaying and recurring oscillations that repeat with period $T_{\rm qm}$. The average position of the classical ensemble closely follows the first quantum oscillations, but it decays to a constant value, $\langle x\rangle = L/2$, where it remains. The decay of the classical oscillation is due to the distribution of velocities in the ensemble, which causes it to spread and eventually cover the range $(0,\, L)$ uniformly. The initial spreading of the quantum wave function is essentially equivalent to the spreading of the classical ensemble. The later

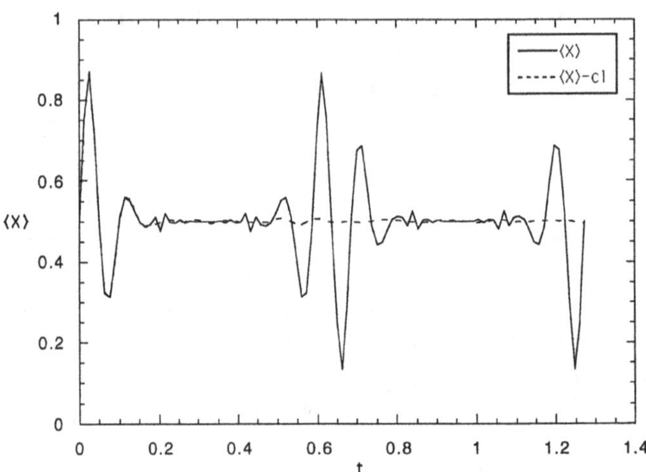

Fig. 14.1 Average position for a particle confined to the unit interval, according to quantum theory (solid line) and classical ensemble theory (dotted line).

periodic recurrences of the quantum state are due to the interference of reflected waves and to the discreteness of the quantum spectrum, which are essentially nonclassical.

The time interval during which the classical and quantum theories agree well is shown in more detail in Fig. 14.2. Ehrenfest's theorem, which predicts $\langle x \rangle$ to follow a classical trajectory, is very inaccurate, even before the first reflection. But the failure of Ehrenfest's theorem does not indicate nonclassical behavior; the quantum state and the classical ensemble are in close agreement, even though Ehrenfest's theorem is not applicable. The lower half of Fig. 14.2 shows that $\Delta x = (\langle x^2 \rangle - \langle x \rangle^2)^{1/2}$ is also correctly given by the classical theory for $t \leq 0.14$. The nonmonotonic behavior of Δx is caused by the folding of the ensemble upon itself when it is reflected from a wall. Indeed, for $t = 0.025$ the value of Δx is smaller than it was for the original minimum uncertainty wave function. For large t, the rms half-width of the classical ensemble approaches the limit $\Delta x \to L(2\sqrt{3})^{-1} \approx 0.2887L$, which is the value for a uniform distribution.

14.2 *The Hamilton–Jacobi Equation and the Quantum Potential*

The Schrödinger equation, $-(\hbar^2/2M)\,\nabla^2\Psi + W\Psi = i\hbar\,\partial\Psi/\partial t$, takes on an interesting form when Ψ is expressed in terms of its real amplitude and phase,

$$\Psi(\mathbf{x},t) = A(\mathbf{x},t)\, e^{iS(\mathbf{x},t)/\hbar}\,. \tag{14.20}$$

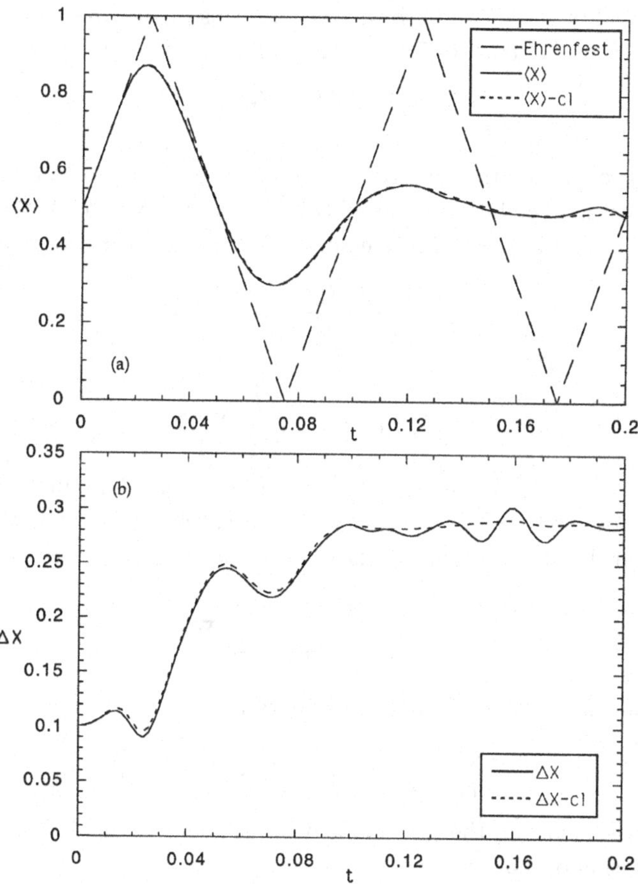

Fig. 14.2 (a) Average position: quantum (solid line), classical (dotted line), Ehrenfest's theorem (sawtooth curve). (b) Rms half-width of position probability distribution.

Making this substitution in the Schrödinger equation and separating the real and imaginary parts, we obtain two equations,

$$-\frac{\hbar^2}{2M}\nabla^2 A + \frac{1}{2M}\,A(\boldsymbol{\nabla} S)^2 + WA = -A\frac{\partial S}{\partial t}\,, \tag{14.21a}$$

$$\frac{-1}{2M}\{A\,\nabla^2 S + 2(\boldsymbol{\nabla} A)\cdot(\boldsymbol{\nabla} S)\} = \frac{\partial A}{\partial t}\,. \tag{14.21b}$$

The second of these can be rewritten in terms of the *probability density*, $P \equiv |\Psi|^2 = A^2$, as $\partial P/\partial t + \{P\nabla^2 S + (\boldsymbol{\nabla} P)\cdot(\boldsymbol{\nabla} S)\}/M = 0$, or, equivalently,

$$\frac{\partial P}{\partial t} + \frac{\boldsymbol{\nabla}\cdot(P\boldsymbol{\nabla}S)}{M} = 0\,. \tag{14.22}$$

This is the continuity equation (4.21) for conservation of probability, since it was shown in Sec. 4.4 that the *probability flux* is given by $\mathbf{J} = P\boldsymbol{\nabla}S/M$.

Equation (14.21a) can conveniently be written in the form

$$\frac{\partial S}{\partial t} + \frac{(\boldsymbol{\nabla}S)^2}{2M} + W + W_Q = 0\,, \tag{14.23}$$

where

$$W_Q = -\frac{\hbar^2}{2M}\,\frac{\nabla^2 A}{A} \tag{14.24}$$

is called the *quantum potential*, because it enters the equation in the same way as does the ordinary potential W. Equation (14.23) has the form of the Hamilton–Jacobi equation of classical mechanics. If we introduce a velocity field

$$\mathbf{v}(\mathbf{x},t) = \frac{\mathbf{J}}{P} = \frac{\boldsymbol{\nabla}S}{M} \tag{14.25}$$

and take the gradient of (14.23), we obtain

$$M\frac{\partial \mathbf{v}}{\partial t} + M(\mathbf{v}\cdot\boldsymbol{\nabla})\,\mathbf{v} + \boldsymbol{\nabla}(W + W_Q) = 0\,.$$

A particle following the flow defined by the velocity field (14.25) would obey the equation of motion

$$M\frac{d\mathbf{v}}{dt} = -\boldsymbol{\nabla}\,(W + W_Q)\,. \tag{14.26}$$

Therefore, if $W_Q \to 0$ in the limit as $\hbar \to 0$, the particle trajectories will obey Newton's law of motion.

There are two major logical steps involved in demonstrating, on the basis of this result, that quantum mechanics has the correct classical limit. One is to show that the quantum potential vanishes in the limit $\hbar \to 0$, which is not trivial in spite of its formal proportionality to $\hbar^2$. This problem will be examined later. The other is a deeper conceptual question regarding the meaning of the state function Ψ and its relation to Hamilton's principal function.

We have shown that *the phase of* Ψ and *Hamilton's principal function* in classical mechanics (both denoted by the symbol S) obey the same mathematical equation in the limit of vanishing W_Q. Now the physical significance of Hamilton's principal function is as a *generator of trajectories* through its gradient in (14.25). The classical version of (14.23) is

$$\frac{(\nabla S)^2}{2M} + W = -\frac{\partial S}{\partial t}\,. \tag{14.27}$$

If W does not depend explicitly on t, then this equation has a solution for which S is linear in t, so that $\partial S/\partial t = -Et$, and E may be any constant not less than the minimum value of W. Then, in view of (14.25), the classical equation for S becomes $\frac{1}{2}Mv^2 + W = E$. Thus it is apparent that in classical mechanics the function S determines the set of all possible trajectories for a particle with energy E. To make contact between classical mechanics and quantum mechanics through this route, it seems necessary to *interpret the phase of the state function Ψ as a generator of trajectories*. But no such interpretation has been given to the phase function $S(\mathbf{x}, t)$ in the usual interpretations of quantum mechanics, where Ψ is interpreted only as a probability amplitude. The relation (4.22b) between $S(\mathbf{x}, t)$ and the probability flux is compatible with the interpretation of $S(\mathbf{x}, t)$ as a generator of trajectories, and this suggests a possible generalization of quantum mechanics, which will be discussed in the next section.

We now return to the behavior of the quantum potential W_Q in the limit $\hbar \to 0$. This evidently may depend on the nature of the particular state function. There does not seem to have been a systematic study of this problem, so we shall consider only a couple of simple examples that illustrate the essential features.

The first example is a free particle in one dimension whose initial state is a Gaussian wave packet of half-width a, $\Psi(x, 0) \propto \exp(x^2/4a^2)$. The time-dependent wave function $\Psi(x, t)$ was calculated in Sec. 5.6, and is given by (5.44). Because the quantum potential (14.24) is independent of the normalization of Ψ, we may drop all factors from (5.44) that do not depend on x. In dropping such factors, the real amplitude of $\Psi(x, t)$ takes the form $A = \exp(-x^2/4\beta^2)$, where $\beta^2 = a^2[1 + (\hbar t/2Ma^2)^2]$. Therefore the quantum potential is

$$W_Q = \frac{\hbar^2}{4M\beta^2}\left(1 - \frac{x^2}{\beta^2}\right)\,. \tag{14.28}$$

Taking the limit $\hbar \to 0$ with a fixed, we find that the quantum potential does indeed vanish. Therefore (14.26) does indeed reduce to the classical equation in that limit.

Roughly speaking, the criterion for smallness of the quantum potential is that $\hbar^2/M\beta^2$ be small compared to other energy terms. Here $\beta(t) =$

$\langle (x - \langle x \rangle)^2 \rangle^{1/2}$ is the half-width of the state. Thus the broader the position probability density, the more accurate will be the approximation provided by the classical limit. This is a very different perspective on the classical limit from that suggested by Ehrenfest's theorem, which attempted to obtain classical behavior by concentrating the probability in the neighborhood of a single classical trajectory, and so would require a small value of β. The approach via the Hamilton–Jacobi equation is more powerful because it recognizes that a quantum state generally corresponds to an ensmble of classical trajectories rather than to a single trajectory.

The important features of $\Psi(x,t)$ in this example, which will also apply to a much broader class of states, are: (a) a very rapid oscillation in the phase of $\Psi(x,t)$ on a scale that vanishes with $\hbar$; and (b) an amplitude $A(x,t)$ varying smoothly on a scale that is not sensitive to $\hbar$. If a state satisfies these conditions, then it will obey the correct classical limit.

However, there are many quantum states that do not satisfy these conditions. Consider the state function $\Psi(x) = \sin(px/\hbar)$, which can describe a particle of energy $E = p^2/2M$ confined between reflecting walls at $x = 0$ and $x = L$, with $L = n\hbar\pi/p$ for some large integer n. The quantum potential for this state is $W_Q = p^2/2M$, which does not vanish when we take the limit $\hbar \to 0$ with p fixed. Moreover Eq. (14.25) would yield a velocity $v = M^{-1}\,\partial S/\partial x = 0$, contrary to the classical velocity $v = p/M$. The failure of this method to yield the expected classical limit in this case is clearly due to the formation of a standing wave, which is a manifestation of the quantum-mechanical phenomenon of interference between the leftward- and rightward-reflected waves that make up Ψ.

14.3 *Quantal Trajectories*

In the previous section, the classical limit of a pure quantum state was obtained as an ensemble of classical trajectories. The Hamilton–Jacobi equations (14.23) and (14.25) are formally capable of generating trajectories from the total potential $W + W_Q$, and there is no apparent reason to restrict their application to the cases where W_Q vanishes. This suggests that quantum mechanics can be extended beyond its purely statistical role, to describe microscopic trajectories of individual particles. The continuity equation (14.22) guarantees that if a probability density is assigned on this ensemble of quantal trajectories such that at some initial time it agrees with the quantum probability postulate, $P = |\Psi|^2$, then the motion along the trajectories will preserve this agreement for all time. Thus this model of deterministic trajectories of individual particles is consistent with the statistical predictions of quantum mechanics. Only if the quantum potential vanished would the quantal trajectories be the same as classical trajectories.

This extension of quantum mechanics was proposed by David Bohm (1952). In such a distinctively quantum-mechanical problem as two-slit diffraction (Philippidis, Dewdney, and Hiley, 1979) it yields an intuitively reasonable set of trajectories, the bunching of the trajectories into diffraction peaks being due to the force produced by the quantum potential.

It is less satisfactory for bound states. Time reversal invariance (Sec. 13.3) of the Hamiltonian implies that stationary bound state functions can always be chosen to have form $\Psi(\mathbf{x},t) = \psi(\mathbf{x})\, e^{iEt/\hbar}$, with $\psi(\mathbf{x})$ real. Therefore $\boldsymbol{\nabla} S = 0$, and so Eq. (14.25) implies that the particle is at rest, in neutral equilibrium through an exact cancellation between $\boldsymbol{\nabla} W$ and $\boldsymbol{\nabla} W_Q$. Thus we see that although Bohm's theory yields the same position probability distribution as does quantum mechanics, the momentum distribution is very different. (For Bohm's response to this difficulty through an analysis of the measurement process, see the references at the end of this chapter.)

The source of this trouble may lie in an ambiguity in the interpretation of the velocity (14.25), which is defined as the ratio of the probability flux vector to the probability density. To see the problem in its simplest form, we consider a surface across which the probability flux $\mathbf{J}$ is zero. This could occur because no particles cross the surface, or alternatively it could occur because, on average, equal numbers cross from left to right as from right to left. In the general case where $\mathbf{J}$ is not zero, the two alternatives are to interpret (14.25) as the velocity of an *individual* particle trajectory, or to interpret it as the *average* velocity of a web of intersecting trajectories. The analogy with the classical Hamilton–Jacobi equation encouraged us to choose the first alternative, and that was done in Bohm's theory. But the unnatural picture which emerges, of particles in bound states begin motionless, suggests that perhaps the wrong alternative was chosen. If so, it follows that the approach to classical limit via the Hamilton–Jacobi theory of Sec. 14.2, though helpful, cannot be regarded as definitive.

Another unintuitive feature of the quantum potential is that it need not vanish at infinite distances. Consider the ground state function of a hydrogen-like atom, which has the form $\Psi(r) = A(r) = e^{-\alpha r}$. For this state the quantum potential (14.24) is $W_Q(r) = (\hbar^2\alpha^2/2M)\,[(2/\alpha r) - 1]$, which does not go to zero as the interparticle separation r becomes infinite. This *nonseparability* was for a long time regarded as fatal defect of Bohm's theory. However, it has been discovered through the study of Bell's theorem that nonseparability is not peculiar to Bohm's specific model, but rather it seems to be inherent in quantum mechanics. (This matter will be discussed in Ch. 20.)

The most important consequence of Bohm's theory is its demonstration that, contrary to previous belief, it is logically possible to give a more detailed account of microscopic phenomena than that given by the statistical quantum theory. The significance and utility of the resulting quantal trajectories, however, remain controversial.

14.4 *The Large Quantum Number Limit*

The attempts in the preceding sections to establish full dynamical equivalence between classical mechanics and quantum mechanics in the limit $\hbar \to 0$ have met with partial success. The approach in this section is to examine specific quantum-mechanical results in the limit where classical mechanics is expected to be valid. This is the limit in which dynamical variables such as angular momentum, energy, etc. are large compared to the relevant quantum unit, and thus it is the limit of large quantum numbers.

We first consider the example of a particle in one dimension confined between reflecting walls at $x = 0$ and $x = L$, for which the method based on the Hamilton–Jacobi equation (Sec. 14.2) failed most drastically. The normalized stationary state function is $\Psi(x) = (2/L)^{1/2} \sin(k_n x)$, where $k_n = n\pi/L$, and the quantum number n is a positive integer. In this quantum state the energy is $E = \hbar^2 {k_n}^2/2M$, and the two values of momentum $p = \pm\hbar k_n$ are equally probable. These values are the same as in a stationary classical statistical ensemble. But whereas the classical position probability density would be uniform on the interval $0 < x < L$, the probability density in the quantum state is $|\Psi(x)|^2 = 2[\sin(k_n x)]^2/L$. The rapid oscillations are a manifestation of quantum-mechanical interference. It is clear that the quantal probability density does not converge *pointwise* to the classical value in any limit. But if we calculate the probability that the particle is in some small *interval*,

$$\text{Prob}\,(a < x < a + \Delta x|\Psi) = \int_a^{a+\Delta x} |\Psi(x)|^2\, dx\,,$$

it will converge to the classical value, $\Delta x/L$, in the limit $n \to \infty$.

The conclusion suggested by this example is that, strictly speaking, quantum mechanics does not converge to classical mechanics, but that in the classical limit the distinctive quantum phenomena like interference fringes become so finely spaced as to be practically undetectable. Any real measurement involves some kind of coarse-grained average which will eventually obscure the quantum effects, and it is this average that obeys classical mechanics. The fact that this example failed to yield the correct classical limit by the Hamilton–Jacobi method should, therefore, not be regarded as evidence for a failure of classical mechanics or of quantum mechanics; nor does it constitute an unbridgeable gap between the two theories. Rather, it indicates that in the previous sections we did not adequately characterize the subtle nature of the classical limit.

We shall apply the lesson of this simple model to a wider class of problems. The property of the state function that we shall exploit is the existence, in the large quantum number limit, of two length scales: a very rapid fine scale oscillation modulated by a slowly varying envelope. The local wavelength of the fine scale oscillation decreases as the quantum number increases, whereas the envelope varies on the scale of the potential. Unfortunately, the mathematical technique that is most convenient for such a problem is applicable only to ordinary differential equations, so we shall treat only one-dimensional problems.

The Schrödinger equation for stationary states in one dimension has the form

$$\frac{d^2\psi}{dx^2} + k^2(x)\,\psi(x) = 0\,, \tag{14.29}$$

with

$$k^2(x) = \frac{[E - W(x)]2M}{\hbar^2}\,. \tag{14.30}$$

It is most convenient to first obtain the two complex linearly independent solutions of this equation without regard to boundary conditions, and then to form the actual state function as a linear combination of them. Hence we substitute $\psi(x) = e^{i\Phi(x)}$, so that (14.29) becomes

$$-\left(\frac{d\Phi}{dx}\right)^2 + i\left(\frac{d^2\Phi}{dx^2}\right) + k^2(x) = 0\,. \tag{14.31}$$

If the potential W were constant, the solution would be $\Phi(x) = \pm kx$ with k constant, and so we would have $d^2\Phi/dx^2 = 0$. If $W(x)$ is not constant but changes very little over the distance of the local wavelength, $\lambda = 2\pi/k(x)$, we may expect $d^2\Phi/dx^2$ to be small compared with the other terms in (14.31). Since λ decreases as E increases, this approximation should be valid in the large quantum number limit. The approximation scheme based upon this idea is called the *WKB method* (after Wentzel, Kramers, and Brillouin).

As a first approximation we drop $d^2\Phi/dx^2$ from (14.31), obtaining

$$\frac{d\Phi}{dx} \approx \pm k(x)\,. \tag{14.32}$$

To obtain the second approximation, we substitute $d^2\Phi/dx^2 = \pm dk(x)/dx$ into (14.31), obtaining

$$\begin{aligned}
\left(\frac{d\Phi}{dx}\right)^2 &= k^2(x) \pm i\left(\frac{dk}{dx}\right),\\
\frac{d\Phi}{dx} &= \pm k(x)\left\{1 \pm \frac{i(dk/dx)}{2k^2(x)}\right\}\\
&= \pm k(x) + \frac{i(dk/dx)}{2k(x)},\\
\Phi(x) &= \pm \int k(x)\,dx + i\,\frac{1}{2}\log[k(x)]\,.
\end{aligned}$$

Thus the approximate complex solutions of (14.29) are

$$\psi(x) = e^{i\Phi(x)} = [k(x)]^{-1/2}\exp\left\{\pm i\int k(x)\,dx\right\}, \tag{14.33}$$

and the real bound state functions will have the form

$$\Psi(x) = \frac{c}{[k(x)]^{1/2}}\cos\left\{\int k(x)\,dx + \phi\right\}. \tag{14.34}$$

The constant ϕ will be determined by the boundary conditions, and c will be determined by normalization. The rapid fine scale oscillations and the smooth envelope are exhibited explicitly in this form. The average of $|\Psi(x)|^2$ over a short distance Δx yields the coarse-grained probability density $\frac{1}{2}|c|^2/k(x)$. The higher the energy, the smaller will be the wavelength of the fine scale oscillation, and hence smaller may Δx be chosen.

The *classical* position probability density is proportional to the time that the particle spends in an interval Δx, and so is proportional to $dt/dx = 1/v$, with the constant of proportionality being determined by normalizing the total probability to 1. Since $\frac{1}{2}Mv^2 = E - W(x)$, it is apparent from (14.30) that the classical velocity is equal to $v = \hbar k(x)/M$, and so the coarse-grained quantal probability density agrees with the classical probability density. The coarse-graining length Δx may become arbitrarily small in the limit of high enough energy. But even though classical and quantal position probability densities become indistinguishable in the high energy (large quantum number) limit, quantum mechanics need not become identical with classical mechanics in this limit. In the example of a particle between reflecting walls, the allowed energies remain discrete and the separation between energy levels does not go to zero.

The approach of the position probability to its classical limit has been studied explicitly in simple systems. Pauling and Wilson (1935, p. 76) illustrate it for the harmonic oscillator. Rowe (1987) has examined the hydrogen atom, illustrating the large n limit for the cases of minimum angular momentum ($\ell = 0$) and maximum angular momentum ($\ell = n-1$). The $\ell = 0$ states correspond to narrow ellipses that have degenerated into straight lines through the center of the orbit, and the radial position probability density is broad. The states of maximum ℓ correspond to circular orbits, and the radial position probability density sharpens about the classical orbit radius in the limit $n \to \infty$. This can be deduced from the formulas (10.33) and (10.34) for $\langle r \rangle$ and $\langle r^2 \rangle$. The mean radius $\langle r \rangle$ in the atomic state $|n\ell m\rangle$ is of order $n^2 a_0$, where a_0 is the Bohr radius and n is the principal quantum number. The mean square fluctuation in the radial variable is

$$\langle r^2 \rangle - \langle r \rangle^2 = \frac{{a_0}^2(n^4 + 2n^2 - \ell^4 - 2\ell^3 - \ell^2)}{4},$$

with both terms on the left being of order $n^4 a_0$. But for $\ell = n-1$ the mean square fluctuation reduces to $\langle r^2 \rangle - \langle r \rangle^2 = \frac{1}{2}{a_0}^2\, n^3\, (1 + 1/2n)$, and thus in the limit $n \to \infty$ the relative fluctuation, $(\langle r^2 \rangle - \langle r \rangle^2)/\langle r \rangle^2$, vanishes like n^{-1}. The angular dependence of the probability density is given by $|Y_\ell{}^m(\theta,\phi)|^2 \propto |P_\ell{}^m(\cos\theta)|^2$. It is apparent from Eq. (7.36) that when m has its maximum value, $m = \ell$, the angular density reduces to $|Y_\ell{}^\ell(\theta,\phi)|^2 \propto (\sin\theta)^\ell$, which becomes arbitrarily sharp about $\theta = \pi/2$ in the limit $\ell \to \infty$. Thus we see that in the limit $n \to \infty$ the position probability distribution of the atomic state with $m = \ell = n-1$ approximates an equatorial circular orbit. Since the width of the probability distribution is a vanishing fraction of the mean radius $\langle r \rangle$, the classical limit of this quantum state appears to be a single orbit. This example is not typical because of its high degree of symmetry. It is more common for the limit of a quantum state to be an ensemble of classical orbits, and a coarse-grained smoothing of the probability density is usually required.

Further reading for Chapter 14

The classical limit for a particle confined between reflecting walls was used by Einstein to demonstrate the need for an *ensemble* interpretation of quantum state functions. This led to an inconclusive correspondence with Max Born, who seems to have missed Einstein's point. The debate was eventually concluded through the mediation of W. Pauli, who endorsed most of Einstein's specific arguments, and yet dissented from his conclusion. See items 103–116 of *The Born–Einstein Letters* (Born, 1971). This debate is discussed in a broader context by Ballentine (1972).

D. Bohm's quantum potential theory was published in *Phys. Rev.* **85**, 166–193 (1952), and is reprinted in Wheeler and Zurek (1983).

The use of the WKB method as a calculational tool is treated in more detail in the textbooks by Merzbacher (1970) and Messiah (1966).

Detailed numerical studies [Ballentine (2004), Wiebe and Ballentine (2005)] show that the emergence of classical behavior from quantum mechanics is very complex, with probability distributions converging to the classical limit more slowly than do coarse averages of observables. In some cases, the amplitudes of fine-scale interference patterns do not tend to zero in the classical limit, but their wavelength becomes arbitrarily small. A small amount of coarse graining is sufficient to obliterate them. That smoothing may result from the limited resolution of the measurement apparatus, or it may result from environmental perturbations (called *decoherence*). The role of *decoherence* in the emergence of classicality has been exagerated by some of its advocates, leading even to the dubious claim that decoherence is responsible for the appearence of the classical world! The two sides of this controversy are debated by Schlosshauer (2008) and Ballentine (2008).

Problems

14.1 (a) The initial state function (not normalized) for a free particle in one dimension is $\Psi(x,0) = \exp(-x^2/2a)$. Calculate $\Delta x = \langle (x - \langle x \rangle)^2 \rangle^{1/2}$ as a function of time.

(b) Construct a classical probability distribution in phase space, $P(x,p)$, which has the same position and momentum distributions at $t = 0$ as does quantum state in part (a). From it calculate the classical variation of Δx as a function of t, and compare with the quantum-mechanical result.

14.2 According to quantum mechanics, the frequency of radiation emitted by a system is given by $\omega = (E_n - E_{n-1})/\hbar$, where E_n is an energy eigenvalue. According to classical mechanics, ω should be equal to the frequency of some periodic motion, such as an orbit. Show that the quantum-mechanical value of ω for the hydrogen atom approaches the classical value in the large quantum number limit, and calculate the order of magnitude of the difference.

14.3 Do the same calculations as in the previous problem for: (a) a particle confined between reflecting walls in one dimension; (b) a spherically symmetric rotator whose Hamiltonian is $H = J^2/2I$, where $\mathbf{J}$ is the angular momentum and I is the moment of inertia.

14.4 Apply the WKB method (Sec. 14.4) to the linear potential of Sec. 5.6, and show that it yields the correct asymptotic forms at $x \to \infty$ and $x \to -\infty$.

14.5 The position probability density for the hydrogen atom state with quantum numbers $m = \ell = n - 1$, in the limit $n \to \infty$, is concentrated in a toriodal tube in the equatorial plane. (This was shown in Sec. 14.4.) The thickness of the tube, $\Delta r = (\langle r^2 \rangle - \langle r \rangle^2)^{1/2}$, diverges in the limit $n \to \infty$. (It was shown, however, that the fractional thickness, $\Delta r / \langle r \rangle$, vanishes in that limit.) Modify the theory of the hydrogen atom so as to describe two objects bound together by gravity, and estimate the principal quantum number n for the earth–moon system. Supposing it to be described by an $m = \ell = n - 1$ quantum state, calculate the magnitude of the quantum fluctuation Δr in the radius of the moon's orbit.

Chapter 15

Quantum Mechanics in Phase Space

15.1 *Why Phase Space Distributions?*

In the previous chapter, we studied probability distributions in configuration space, and showed how they can approach the classical limit. Similar calculations could be done for the momentum probability distribution, and for the probability distributions of other dynamical variables. But even if the probability distribution of each dynamical variable were shown to have an appropriate classical limit, this would not constitute a complete classical description. In classical mechanics we also have correlations between dynamical variables, such as position and momentum, and these are described by a joint probability distribution in phase space, $\rho_c(q,p)$. If a full classical description is to emerge from quantum mechanics, we must be able to describe quantum systems in phase space.

It would be desirable if, for each state ρ, there were a quantum phase space distribution $\rho_Q(q,p)$ with the following properties: its marginal distributions should yield the usual position and momentum probability distributions,

$$\int \rho_Q(q,p)dp = \langle q|\rho|q\rangle\,, \tag{15.1}$$

$$\int \rho_Q(q,p)dq = \langle p|\rho|p\rangle\,, \tag{15.2}$$

and it should be nonnegative,

$$\rho_Q(q,p) \geq 0\,, \tag{15.3}$$

so as to permit a probability interpretation.

It is sometimes said that such a quantum phase space distribution cannot exist because of the indeterminacy principle (Sec. 8.4), but that is not true. In order to satisfy the Heisenberg inequality (8.33), it is sufficient that $\rho_Q(q,p)$ should have an effective area of support in phase space of order $2\pi\hbar$ (the numerical factor depends on the shape of the area), so that the product of the rms half-widths of (15.1) and (15.2) is not less than $\frac{1}{2}\hbar$. In fact, for any ρ, there are infinitely many functions $\rho_Q(q,p)$ which satisfy the three equations above (Cohen, 1986). The problem is that no principle has been found to single out any one of them for particular physical significance.

To obtain a unique form for $\rho_Q(q,p)$, one may try imposing additional conditions. For a pure state, $\rho = |\psi\rangle\langle\psi|$, the familiar probability formulas are bilinear in $|\psi\rangle$ and $\langle\psi|$, having the form $\langle\psi|P|\psi\rangle$, where P is a projection operator. For example, the position probability density is $\langle\psi|q\rangle\langle q|\psi\rangle$. Hence one might require the phase space distribution to be expressible in the form $\rho_Q(q,p) = \langle\psi|\, M(q,p)\, |\psi\rangle$, where $M(q,p)$ is some self-adjoint operator. Wigner (1971) has proven that any such $\rho_Q(q,p)$ could not satisfy (15.1), (15.2), and (15.3). The bilinearity condition is mathematically attractive, but it lacks physical motivation. However, the theorem has been generalized (Srinivas, 1982), with the bilinearity condition being replaced by the *mixture property*. This is motivated by the fact that the representation of a nonpure state operator as a mixture of pure states is not unique (Sec. 2.3). If ρ is not a pure state, it can be written in the form $\rho = \sum_i w_i|\psi_i\rangle\langle\psi_i|$ in infinitely many ways. The mixture property is the requirement that the phase space distribution should depend only on the state operator ρ, and not on the particular way that it is represented as a mixture of some set of pure states $\{\psi_i\}$.

In view of these negative results, two approaches have been pursued. The Wigner function of Sec. 15.2 satisfies the mixture property, but not (15.3). It cannot be interpreted as a probability, but it is still useful for calculations. The Husimi distribution of Sec. 15.3 has a probability interpretation, but does not satisfy (15.1) and (15.2).

15.2 *The Wigner Representation*

The state operator ρ can be given several matrix representations, the position representation $\langle q|\rho|q'\rangle$ and the momentum representation $\langle p|\rho|p'\rangle$ being the most common. The Wigner representation is, in a sense, intermediate between these two. For a single particle in one dimension, it is defined as

$$\rho_w(q,p) = (2\pi\hbar)^{-1}\int_{-\infty}^{\infty} \langle q - \tfrac{1}{2}y|\rho|q + \tfrac{1}{2}y\rangle\, e^{ipy/\hbar}dy\,. \tag{15.4}$$

[The generalization to N particles in three dimensions is straightforward, and is given in the original paper (Wigner, 1932). It requires that all variables be interpreted as $3N$-dimensional vectors, and that the factor $(2\pi\hbar)^{-1}$ become $(2\pi\hbar)^{-3N}$.] The Wigner representation can also be obtained from the momentum representation,

$$\rho_w(q,p) = (2\pi\hbar)^{-1} \int_{-\infty}^{\infty} \langle p - \tfrac{1}{2}k|\rho|p + \tfrac{1}{2}k\rangle \, e^{-iqk/\hbar} dk \,, \tag{15.5}$$

showing that it is, indeed, intermediate between the position and momentum representations. It follows directly from these two relations that the Wigner function satisfies (15.1) and (15.2):

$$\int_{-\infty}^{\infty} \rho_w(q,p)dp = \langle q|\rho|q\rangle \,, \tag{15.6}$$

$$\int_{-\infty}^{\infty} \rho_w(q,p)dq = \langle p|\rho|p\rangle \,. \tag{15.7}$$

The three basic properties of the state operator, (2.6), (2.7), and (2.8), can all be expressed in the Wigner representation. From the definition (15.4), it follows that the trace of ρ is given by

$$\iint \rho_w(q,p)dqdp = \int \langle q|\rho|q\rangle dq = \mathrm{Tr}(\rho) \,. \tag{15.8}$$

The first property, Tr $\rho = 1$, becomes $\iint \rho_w(q,p)dqdp = 1$. The second property, $\rho = \rho^\dagger$, corresponds to the fact that $\rho_w(q,p)$ is real. The third, $\langle u|\rho|u\rangle \geq 0$, however, does *not* imply nonnegativity for $\rho_w(q,p)$. We shall return to it after certain necessary results have been obtained.

The Wigner representation for any operator R, *other than* ρ, is defined as

$$R_w(q,p) = \int_{-\infty}^{\infty} \langle q - \tfrac{1}{2}y|R|q + \tfrac{1}{2}y\rangle \, e^{ipy/\hbar} dy \,. \tag{15.9}$$

The omission of the factor $(2\pi\hbar)^{-1}$, as compared with (15.4), is done to simplify the normalization in the case of a function of q only, such as a potential energy $V(q)$, for which $\langle x|V|x'\rangle = V(x)\delta(x-x')$. Equation (15.9) then yields

$$V_w(q,p) = \int V\left(q - \tfrac{1}{2}y\right) \delta(y) e^{ipy/\hbar} dy = V(q) \,.$$

Similarly, the Wigner representation for a function of p only, $K(p)$, is simply $K_w(q,p) = K(p)$.

The average of the dynamical variable R in the state ρ is $\langle R\rangle = \mathrm{Tr}(\rho R)$. Thus we need to express the trace of a product of two operators in terms of their Wigner representation. To do this, we first write the trace in the position representation:

$$\text{Tr}(\rho R) = \iint \langle q|\rho|q'\rangle\langle q'|R|q\rangle\, dq\, dq' \,.$$

We next express the position representations of R and ρ in terms of the Wigner representation, using the Fourier inverse of (15.4),

$$\int e^{-ipy/\hbar}\rho_w(q,p)dp = \langle q - \tfrac{1}{2}y|\rho|q + \tfrac{1}{2}y\rangle \,,$$

and a similar Fourier inverse of (15.9). The resulting expression for $\text{Tr}(\rho R)$ is initially cumbersome, but it simplifies to

$$\langle R\rangle = \text{Tr}(\rho R) = \iint \rho_w(q,p)R_w(q,p)\, dq\, dp \,. \tag{15.10}$$

The similarity of this formula to a classical phase space average is responsible for much of the intuitive appeal and practical utility of the Wigner representation. It should be stressed, however, that the Wigner function $\rho_w(q,p)$ is *not a probability distribution* because it typically takes on both positive and negative values.

We now return to the nonnegativeness property (2.8) of the state operator ρ, and its consequences in the Wigner representation. It is convenient to replace this property with a generalization (2.20), which states that for any pair of state operators that satisfy (2.6), (2.7), and (2.8), the trace of their product obeys the inequality $0 \le \text{Tr}(\rho\rho') \le 1$. If we put $\rho' = |u\rangle\langle u|$, this yields the nonnegativeness condition (2.8), $0 \le \langle u|\rho|u\rangle \le 1$. Substituting $R_w(q,p) = 2\pi\hbar\rho'_w(q,p)$ into (15.10), we obtain

$$\text{Tr}(\rho\rho') = 2\pi\hbar \iint \rho_w(q,p)\, \rho'_w(q,p)\, dq\, dp \,, \tag{15.11}$$

and hence (2.20) implies that

$$0 \le \iint \rho_w(q,p)\rho'_w(q,p)\, dq\, dp \le (2\pi\hbar)^{-1} \,. \tag{15.12}$$

The special case $\rho = \rho'$, for which

$$\iint \{\rho_w(q,p)\}^2\, dq\, dp \le (2\pi\hbar)^{-1} \,, \tag{15.13}$$

is particularly interesting, since it implies that the Wigner function cannot be too sharply peaked. Suppose, for example, that $\rho_w(q,p)$ approximately vanishes outside of some region in phase space, of area A, and has the value A^{-1} inside that region. Then the integral in (15.13) would be equal to A^{-1}. Therefore the area of support cannot be too small: $A \geq 2\pi\hbar$, a result that is related to the indeterminacy principle.

These results have several interesting consequences if we specialize to *pure states*, $\rho = |\psi\rangle\langle\psi|$ and $\rho' = |\phi\rangle\langle\phi|$. Then (15.11) becomes

$$|\langle\phi|\psi\rangle|^2 = 2\pi\hbar \iint \rho_w(q,p)\, \rho'_w(q,p)\, dq\, dp\,. \tag{15.14}$$

Both sides must vanish if the two state vectors are orthogonal, which proves that the Wigner functions take on both positive and negative values, and so cannot be probabilities. The derivation of (2.20) also implies that upper limit of (15.12) is achieved if and only if $\rho = \rho'$ is a pure state. Therefore we have

$$\iint \{\rho_w(q,p)\}^2\, dq\, dp = (2\pi\hbar)^{-1} \tag{15.15}$$

for a pure state, $\rho = |\psi\rangle\langle\psi|$. This corresponds to the property (2.17), $\mathrm{Tr}(\rho^2) = 1$ for a pure state.

Here are some simple examples of Wigner functions. For a pure state, Eq. (15.4) can be written as

$$\rho_w(q,p) = (2\pi\hbar)^{-1} \int_{-\infty}^{\infty} \Psi\left(q - \tfrac{1}{2}y\right) \Psi^*\left(q + \tfrac{1}{2}y\right) e^{ipy/\hbar} dy\,, \tag{15.16}$$

where $\Psi(q) = \langle q|\Psi\rangle$ is the wave function of the state.

Example (i): Gaussian wave packet

Consider first a Gaussian wave packet of the form

$$\Psi(q) = (2\pi a^2)^{-1/4} e^{-q^2/4a^2}\,. \tag{15.17}$$

From (15.16), its Wigner function is

$$\begin{aligned} \rho_w(q,p) &= \frac{e^{-q^2/2a^2}}{2\pi\hbar(2\pi a^2)^{1/2}} \int_{-\infty}^{\infty} \exp\left(\frac{ipy}{\hbar} - \frac{y^2}{8a^2}\right) dy \\ &= \frac{1}{\pi\hbar} e^{-q^2/[2(\Delta q)^2]} e^{-p^2/[2(\Delta p)^2]}\,. \end{aligned} \tag{15.18}$$

The values of the rms half-widths of the position and momentum distributions, $\Delta q = a$ and $\Delta p = \hbar/2a$, have been introduced in the last line to better show the symmetry between q and p.

The most general Gaussian wave function is obtained by displacing the centroid of the state to an arbitrary point in phase space, $\langle q\rangle = q_0$ and $\langle p\rangle = p_0$:

$$\Psi(q) = (2\pi a^2)^{-1/4} e^{-(q-q_0)^2/4a^2} e^{ip_0 q/\hbar} . \tag{15.19}$$

The Wigner function becomes

$$\rho_w(q,p) = \frac{1}{\pi\hbar} e^{-(q-q_0)^2/2(\Delta q)^2} e^{-(p-p_0)^2/2(\Delta p)^2} . \tag{15.20}$$

This is just the product of the position and momentum distributions, and is everywhere positive. Unfortunately, such a simple result is not typical. It has been proven (Hudson, 1974) that Gaussian wave functions are the *only* pure states with nonnegative Wigner functions.

Example (ii): Separated Gaussian wave packets

Consider next a superposition of two Gaussian packets centered at $q = \pm c$:

$$\Psi(q) = \frac{N}{2^{1/2}(2\pi a^2)^{1/4}} \left\{ e^{-(q-c)^2/4a^2} + e^{-(q+c)^2/4a^2} \right\} . \tag{15.21}$$

The normalization factor N occurs because the two Gaussians are not orthogonal: $N = [1 + e^{-c^2/2a^2}]^{-1/2}$. When the Wigner function is evaluated from (15.16), there will be four terms: the Wigner functions of the two separate Gaussian packets, and two interference terms. The result is

$$\rho_w(q,p) = \frac{N^2}{2\pi\hbar} e^{-p^2/2(\Delta p)^2} \left\{ e^{-(q-c)^2/2(\Delta q)^2} + e^{-(q+c)^2/2(\Delta q)^2} + 2e^{-q^2/2(\Delta q)^2} \cos\frac{2cp}{\hbar} \right\} . \tag{15.22}$$

Here again we use $\Delta q = a$ and $\Delta p = \hbar/2a$. In addition to the expected peaks at $q = \pm c$, there is another peak at $q = 0$. It is multiplied by an oscillatory factor that represents interference between the two Gaussian packets. Clearly this Wigner function takes both positive and negative values, and so cannot be interpreted as a probability distribution. Moreover, it retains this character in the macroscopic limit, in which the separation c between the packets becomes macroscopically large. As $c \to \infty$ the amplitude of the interference term does not diminish, so the *Wigner function does not approach a classical phase space probability distribution even in the macroscopic limit.*

This does not prevent it from yielding the expected two-peak position distribution, since the interference term averages to zero upon integration over momentum.

Time dependence of the Wigner function

The time evolution of the Wigner function can be deduced from that of the state vector, or, more generally, of the state operator (3.68), $d\rho/dt = (i/\hbar)(\rho H - H\rho)$. Since the Hamiltonian, $H = P^2/2M + V$, is the sum of kinetic and potential energies, it is convenient to write

$$\frac{d\rho}{dt} = \frac{\partial_K \rho}{\partial t} + \frac{\partial_V \rho}{\partial t}, \tag{15.23}$$

where

$$\frac{\partial_K \rho}{\partial t} = \frac{i}{2M\hbar}(\rho P^2 - P^2\rho), \tag{15.24}$$

$$\frac{\partial_V \rho}{\partial t} = \frac{i}{\hbar}(\rho V - V\rho). \tag{15.25}$$

It is most convenient to evaluate (15.24) in the momentum representation, where it becomes

$$\begin{aligned}\frac{\partial_K}{\partial t}\langle p|\rho|p'\rangle &= \frac{i}{2M\hbar}\langle p|\rho|p'\rangle(p'^2 - p^2)\\ &= \frac{i}{2M\hbar}\langle p|\rho|p'\rangle(p' + p)(p' - p). \end{aligned}\tag{15.26}$$

Using (15.5) to transform to the Wigner representation, we obtain

$$\frac{\partial_K}{\partial t}\rho_w(q,p,t) = \frac{i}{\hbar M}\int_{-\infty}^{\infty}\langle p - \tfrac{1}{2}k|\rho|p + \tfrac{1}{2}k\rangle\, pk\, e^{-iqk/\hbar}dk.$$

We may replace the factor k inside the integral with the operation $(-\hbar/i)\partial/\partial q$ outside the integral, obtaining

$$\frac{\partial_K}{\partial t}\rho_w(q,p,t) = -\frac{p}{M}\frac{\partial}{\partial q}\rho_w(q,p,t). \tag{15.27}$$

Equation (15.25) is most easily evaluated in position representation:

$$\frac{\partial_V}{\partial t}\langle x|\rho|x'\rangle = \frac{i}{\hbar}\langle x|\rho|x'\rangle[V(x) - V(x')].$$

Using (15.4) to transform to the Wigner representation, we obtain

$$\frac{\partial_V}{\partial t}\rho_w(q,p,t) = \frac{i}{\hbar(2\pi\hbar)}\int_{-\infty}^{\infty}\langle q-\tfrac{1}{2}y|\rho|q+\tfrac{1}{2}y\rangle \times \left[V\left(q+\tfrac{1}{2}y\right)-V\left(q-\tfrac{1}{2}y\right)\right]e^{ipy/\hbar}dy\,.$$

If $V(x)$ is analytic, it can be expressed as a Taylor series,

$$V\left(q+\tfrac{1}{2}y\right)-V\left(q-\tfrac{1}{2}y\right) = \sum_{n=\text{odd}}\frac{2}{n!}\left(\tfrac{1}{2}y\right)^n\frac{d^nV(q)}{dq^n}\,. \tag{15.28}$$

When this series expansion is substituted into the integral above, we may replace the factor $(\frac{1}{2}y)^n$ inside the integral with the operation $[(\hbar/2i)(\partial/\partial p)]^n$ outside the integral. This yields

$$\frac{\partial_V}{\partial t}\rho_w(q,p,t) = \sum_{n=\text{odd}}\frac{1}{n!}\left(-\tfrac{1}{2}i\hbar\right)^{n-1}\frac{d^nV(q)}{dq^n}\frac{\partial^n}{\partial p^n}\rho_w(q,p,t)\,. \tag{15.29}$$

The sum of (15.27) and (15.29) yields the equation for time evolution of the Wigner function.

There are several points worth noting about this result. First, the factor $i=\sqrt{-1}$ in (15.29) appears to an even power, so all terms are real. Second, the sum is a formal power series in $\hbar$, which suggests that this equation should have a simple classical limit. Combining (15.27) and (15.29), we obtain

$$\frac{\partial}{\partial t}\rho_w(q,p,t) = -\frac{p}{M}\frac{\partial}{\partial q}\rho_w(q,p,t)+\frac{dV}{dq}\frac{\partial}{\partial p}\rho_w(q,p,t)+O(\hbar^2)\,. \tag{15.30}$$

If the correction $O(\hbar^2)$ can be neglected, this is just the classical Liouville equation (14.10). But the form of this equation is misleading. The correction terms, formally of order $\hbar^n$, also involve an nth order derivative of $\rho_w(q,p,t)$ with respect to p. This can generate factors of $1/\hbar$, and so cancel the explicit $\hbar$ factors. Equation (15.22) is an example of a Wigner function that behaves in this way. In such cases the corrections terms in (15.30) do not vanish in the limit $\hbar\to 0$. This is very similar to the possible nonvanishing of the quantum potential (Sec. 14.2) in the limit $\hbar\to 0$, in spite of its formal proportionality to $\hbar$.

The harmonic oscillator is an interesting special case. Since the third and higher derivatives of $V(q)$ vanish, the terms in (15.29) for $n>1$, which explicitly contain $\hbar$, are all zero. Hence its Wigner function satisfies the classical Liouville equation exactly, even if the state is not nearly classical.

This is analogous to the situation noted in Sec. 14.1, that the harmonic oscillator satisfies Ehrenfest's theorem exactly, even for states that are not nearly classical. The harmonic oscillator is a very special case, and its approach to the classical limit is not typical.

In summary, the Wigner representation has the virtue of providing information about the state of the system in phase space. This contrasts with the more conventional representations, which may provide information about position only, or about momentum only, but not both together. It can be a useful calculational tool. In the original paper, Wigner (1932) used it to calculate the quantum corrections to the equation of state of a gas of interacting atoms. But one must remember that the Wigner function is *not* a probability distribution, being both positive and negative, and in general it does not become equal to the classical phase space distribution function in the classical limit. In spite of some attractive formal properties of the Wigner representation, it does not seem to provide a good approach to the classical limit.

15.3 *The Husimi Distribution*

The Husimi distribution is defined in a manner that guarantees it to be nonnegative, and gives it a probability interpretation. To motivate its definition, we first recall how the configuration space distribution is constructed. This is done by introducing the position eigenvectors $\{|q\rangle\}$, which satisfy the orthonormality relation $\langle q|q'\rangle = \delta(q-q')$, and the completeness relation $\int |q\rangle\langle q|dq = 1$. The position probability density for the state ρ is then given by $\langle q|\rho|q\rangle$, which becomes $|\langle q|\Psi\rangle|^2$ in the special case of a pure state ($\rho = |\Psi\rangle\langle\Psi|$).

The obstacle to constructing a phase distribution is, apparently, the nonexistence of eigenvectors of both position and momentum together. But although exact eigenvectors do not exist, we can use the next best thing — a set of minimum uncertainty states localized in phase space. We shall denote these vectors as $|q,p\rangle$. In position representation, they have the form (15.19), apart from a slight change of notation,

$$\langle x|q,p\rangle = (2\pi s^2)^{-1/4} e^{-(x-q)^2/4s^2} e^{ipx/\hbar} . \tag{15.31}$$

This function is centered at the point (q,p) in phase space, with Gaussian distributions in both position and momentum, and with rms half-widths $\delta q = s$ and $\delta p = \hbar/2s$. The parameter s is arbitrary, and each choice of s yields a different basis function set $\{|q,p\rangle\}$. In the following discussion, we shall regard the parameter s as having been fixed.

The functions (15.31) are clearly not orthogonal. They form an overcomplete set, satisfying the completeness relation

$$\int |q,p\rangle\langle q,p|dq\,dp = 2\pi\hbar\,. \tag{15.32}$$

[This identity is equivalent to (19.68), whose proof is given in detail.] For the state operator ρ, *the Husimi distribution is defined as*

$$\rho_H(q,p) = (2\pi\hbar)^{-1}\langle q,p|\rho|q,p\rangle\,. \tag{15.33}$$

For the special case of a pure state, $\rho = |\Psi\rangle\langle\Psi|$, it becomes

$$\rho_H(q,p) = (2\pi\hbar)^{-1}|\langle q,p|\Psi\rangle|^2\,. \tag{15.34}$$

The normalizing factor in the definition is necessary because of the factor on the right hand side of (15.32). It ensures that $\int \rho_H(q,p)\,dq\,dp = 1$.

The Husimi distribution, $\rho_H(q,p)$, can be interpreted as the probability density for the system to occupy a fuzzy region in phase space, of half-widths $\delta q = s$ and $\delta p = \hbar/2s$, centered at (q,p). In the limit $s \to 0$ the minimum uncertainty function (15.31) becomes vanishingly narrow in position, and so approximates a position eigenfunction. Alternatively, in the limit $s \to \infty$ it approximates a momentum eigenfunction. Thus the Husimi representation, like the Wigner representation, is intermediate between the position and momentum representations.

The Husimi distribution is also equal to a Gaussian smoothing of the Wigner function. To see this, we write $\rho_H(q,p) = (2\pi\hbar)^{-1}\langle q,p|\rho|q,p\rangle = (2\pi\hbar)^{-1}\,\mathrm{Tr}(|q,p\rangle\langle q,p|\rho)$, and then use (15.11) to express the trace as an integral of two Wigner functions:

$$\rho_H(q,p) = \iint \rho_{qpw}(q',p')\,\rho_w(q',p')\,dq'\,dp'\,. \tag{15.35}$$

Here $\rho_w(q',p')$ is the Wigner function for the state ρ, and $\rho_{qpw}(q',p')$ is the Wigner function for the minimum uncertainty state $|q,p\rangle$. From (15.20), the latter is

$$\rho_{qpw}(q',p') = (\pi\hbar)^{-1}e^{-(q'-q)^2/2s^2}\,e^{-(p'-p)^2(2s^2/\hbar^2)}\,.$$

Thus the Husimi distribution $\rho_H(q,p)$ is derivable from the Wigner function $\rho_w(q,p)$ by a Gaussian smoothing in both position and momentum. This property of the Wigner function may explain why it has been found to provide a qualitatively useful description of phase space structures, even though it has no probability interpretation. Any strongly pronounced feature of the Husimi distribution will also show up in the Wigner function, although the latter may also contain unphysical structures (as in Example ii of the previous section).

The Husimi distribution does not obey (15.1) and (15.2); the momentum integral of $\rho_H(q,p)$ does not yield the quantal position probability distribution, and the position integral does not yield the momentum distribution. We shall demonstrate this for a pure state, $\rho = |\Psi\rangle\langle\Psi|$, the extension to a general state operator being straightforward. From (15.34) we find the Husimi position distribution to be

$$\begin{aligned} P_H(q) &= \int \rho_H(q,p)\,dp = (2\pi\hbar)^{-1}\int |\langle q,p|\Psi\rangle|^2 dp \\ &= (2\pi\hbar)^{-1}\int dp \iint \langle q,p|x\rangle\langle x|\Psi\rangle\langle\Psi|x'\rangle\langle x'|q,p\rangle\,dx\,dx' \,. \end{aligned} \tag{15.36}$$

Now use (15.31) to substitute for $\langle q,p|x\rangle$ and $\langle x'|q,p\rangle$. The dependence of the integrand on p is exponential, and upon integration yields a factor $\delta(x-x')$. Thus we obtain

$$\begin{aligned} P_H(q) &= \int |\langle x|q,p\rangle|^2 |\langle x|\Psi\rangle|^2 dx \\ &= \int (2\pi s^2)^{-1/2} e^{-(x-q)^2/2s^2} |\langle x|\Psi\rangle|^2 dx \,. \end{aligned} \tag{15.37}$$

This is a Gaussian-broadened version of the quantal position probability distribution $|\langle x|\Psi\rangle|^2$, which approaches $|\langle q|\Psi\rangle|^2$ in the limit $s \to 0$.

Similarly, one can show that the Husimi momentum distribution, $P_H(p) = \int \rho_H(q,p)\,dq$, is a Gaussian-broadened version of the quantal momentum distribution $|\langle p|\Psi\rangle|^2$, and it approaches $|\langle p|\Psi\rangle|^2$ in the limit $s \to \infty$.

Indeterminacy relation for the Husimi distribution

In general, averages calculated from the Husimi distribution will differ from standard quantum state averages because of the above-noted broadening of the probabilities. Nevertheless, the Husimi averages are of some interest.

For a normalized state vector $|\Psi\rangle$, we define the average position and momentum to be $\langle q\rangle = \langle\Psi|Q|\Psi\rangle$ and $\langle p\rangle = \langle|\Psi|P|\Psi\rangle$. We may also define averages for the Husimi distribution, $\langle q\rangle_H = \int qP_H(q)\,dq$ and $\langle p\rangle_H = \int pP_H(p)\,dp$. In fact, the Husimi averages of q and p are equal to the quantum averages. To show this, notice that $P_H(q)$ has the form of a convolution,

$$P_H(q) = \int f(x)g(q-x)\,dx \,, \tag{15.38}$$

with $f(x) = |\langle x|\Psi\rangle|^2$ and $g(q-x) = |\langle x|q,p\rangle|^2 = (2\pi s^2)^{-1/2} e^{-(x-q)^2/2s^2}$. Thus we have

$$\begin{aligned}\langle q\rangle_H &= \iint q f(x)\, g(q-x)\, dx\, dq \\ &= \int f(x) \left[\int q g(q-x)\, dq\right] dx \\ &= \int f(x)\, x\, dx = \langle q\rangle\,. \end{aligned} \tag{15.39}$$

The third equality follows because $g(q-x)$ is symmetric about $q = x$. A similar argument shows that

$$\langle p\rangle_H = \langle p\rangle\,. \tag{15.40}$$

The variances of the quantum position and momentum distributions are $(\Delta q)^2 = \langle\Psi|(Q - \langle q\rangle)^2|\Psi\rangle$ and $(\Delta p)^2 = \langle\Psi|(P - \langle p\rangle)^2|\Psi\rangle$. The variances for the Husimi distribution are

$$(\Delta q)^2_H = \int (q - \langle q\rangle)^2 P_H(q)\, dq\,, \tag{15.41}$$

$$(\Delta p)^2_H = \int (p - \langle p\rangle)^2 P_H(p)\, dp\,. \tag{15.42}$$

We may expect these to be larger than the quantum state variances because of the Gaussian broadening of the probabilities. For simplicity, and without loss of generality, we displace the state so that $\langle q\rangle = 0$. Then $(\Delta q)^2 = \langle\Psi|Q^2|\Psi\rangle$, and

$$\begin{aligned}(\Delta q)^2_H &= \int q^2 P_H(q)\, dq = \iint q^2 f(x) g(q-x)\, dx\, dq \\ &= \int f(x) \left[\int q^2 g(q-x)\, dq\right] dx \\ &= \int f(x)[x^2 + (\delta q)^2]\, dx \\ &= (\Delta q)^2 + (\delta q)^2\,. \end{aligned} \tag{15.43}$$

Here δq is the rms half-width of the basis state $|q,p\rangle$, whose position probability density is $g(q-x)$, and Δq is the rms half-width of the quantum state $|\Psi\rangle$. By a similar argument, we can show that

$$(\Delta p)^2_H = (\Delta p)^2 + (\delta p)^2\,. \tag{15.44}$$

The indeterminacy product for the Husimi distribution is

$$(\Delta q)^2_H(\Delta p)^2_H = [(\Delta q)^2 + (\delta q)^2][(\Delta p)^2 + (\delta p)^2]$$
$$= (\Delta q)^2(\Delta p)^2 + (\delta q)^2(\delta p)^2 + (\Delta q)^2(\delta p)^2 + (\delta q)^2(\Delta p)^2 \,.$$

Since $|q,p\rangle$ is a Gaussian state, we have $\delta q = s, \delta p = \hbar/2s$. This yields

$$(\Delta q)^2_H(\Delta p)^2_H = (\Delta q)^2(\Delta p)^2 + \frac{\hbar^2}{4} + \frac{(\Delta q)^2\hbar^2}{4s^2} + s^2(\Delta p)^2 \,.$$

The first term on the right is bounded below by $\hbar^2/4$, according to the standard indeterminacy relation (8.33). Minimizing the last two terms with respect to s then yields $(\Delta q)^2_H(\Delta p)^2_H \geq \hbar^2$. Thus the indeterminacy product for the Husimi distribution,

$$(\Delta q)_H(\Delta p)_H \geq \hbar \,, \tag{15.45}$$

has twice as large a lower bound as that for a quantum state,

$$\Delta q \Delta p \geq \frac{\hbar}{2} \,. \tag{15.46}$$

A physical interpretation of this result will be suggested later.

We now consider the Husimi distributions for the same examples that were treated for the Wigner function in the previous section.

Example (i): Gaussian wave packet

The Gaussian wave packet centered at the origin is (15.17):

$$\Psi(x) = (2\pi a^2)^{-1/4} e^{-x^2/4a^2} \,,$$

from which the Husimi distribution is calculated using (15.34):

$$\rho_H(q,p) = (2\pi\hbar)^{-1} \left| \int \langle q,p|x\rangle\langle x|\Psi\rangle dx \right|^2 \,. \tag{15.47}$$

The result is

$$\rho_H(q,p) = \frac{a\, s}{\pi\hbar(a^2+s^2)} e^{-q^2/2(a^2+s^2)} e^{-2p^2/(a^{-2}+s^{-2})\hbar^2} \,. \tag{15.48}$$

This is similar in form to the Wigner function (15.18), but with Δq and Δp replaced by $(\Delta q)_H$ and $(\Delta p)_H$, as would be expected from the Husimi distribution being equivalent to a broadened Wigner function (15.35).

Example (ii): Separated Gaussian wave packets

Consider next a superposition of two Gaussian packets (15.21) centered at $x = \pm c$,

$$\Psi(x) = \frac{N}{2^{1/2}(2\pi a^2)^{1/4}}\{e^{-(x-c)^2/4a^2} + e^{-(x+c)^2/4a^2}\}.$$

Using (15.47) to calculate the Husimi distribution, we obtain

$$\begin{aligned}\rho_H(q,p) = \frac{N^2 as}{2\pi\hbar(a^2+s^2)} \exp\left[\frac{-2p^2}{\hbar^2(a^{-2}+s^{-2})}\right] &\left\{\exp\left[\frac{-(q-c)^2}{2(a^2+s^2)}\right]\right. \\ + \exp\left[\frac{-(q+c)^2}{2(a^2+s^2)}\right] &+ 2\exp\left[\frac{-(q^2+c^2)}{2(a^2+s^2)}\right] \\ \times \cos\left[\frac{2cps^2}{\hbar(a^2+s^2)}\right]&\left.\right\}. \qquad (15.49)\end{aligned}$$

This consists of the Husimi distributions for the two Gaussian packets at $q = \pm c$ plus an interference term centered at $q = 0$. But, in contrast with the Wigner function (15.22), the amplitude of the interference term vanishes rapidly in the limit $c \to \infty$. Thus the macroscopic limit of the Husimi distribution is a proper classical phase space distribution.

It is possible to derive an equation of motion for the Husimi distribution $\rho_H(q,p,t)$ [O'Connell and Wigner, 1981; O'Connell, Wang, and Williams, 1984]. It will not be given here, since the derivation and the form of the equation are rather complicated. In practice it is usually more efficient to solve the Schrödinger equation for the state vector or state operator, and then calculate the Husimi distribution from (15.33) or (15.34).

In summary, the Husimi distribution is a true phase space probability density, representing the probability that the system occupies a certain area of magnitude $2\pi\hbar$ in phase space. The boundaries of this area are fuzzy, being defined by a Gaussian function in both position and momentum. The shape of the fuzzy region is elliptical, with its semimajor axes being $\delta q = s$ and $\delta p = \hbar/2s$. In the limit $s \to 0$ the quantal position probability density is resolved without broadening, but no information is given about momentum. In the opposite limit of $s \to \infty$ the momentum probability density is resolved faithfully, but no information is given about position. By varying the parameter s, we can get a variety of complementary images of the phase space structures.

A nearly classical description will be obtained if it is possible to choose s such that δq is small compared to the significant structures in position space, and δp is small compared to the significant structures in momentum space. Whether this is possible depends on both the system Hamiltonian and the state.

The notion that the parameter s governs the degree of position resolution vs momentum resolution suggests an interpretation of the indeterminacy principle (15.45) for the Husimi distribution. It suggests that the vector $|q,p\rangle_s$, for some value of s, is a highly idealized state vector for a measuring apparatus that performs simultaneous but imprecise measurements of position and momentum. The extra factor of 2 in (15.45), compared with the standard indeterminacy relation (15.46), is then due to the fact that both the system and the measuring apparatus are subject to quantum indeterminacies, each contributing a minimum of $\hbar/2$. This idea can be made precise. Stenholm (1992) has given a detailed analysis of the simultaneous coupling of a system to idealized position- and momentum-measuring devices. If the initial states of these devices are chosen optimally, the joint distribution of the measurement outcomes is just the Husimi distribution for the state of the system.

Further reading for Chapter 15

K. Husimi (1940) first introduced the phase space distribution that bears his name, although it was not widely recognized for several years. A review of the Wigner representation is given by Hillery, O'Connell, Scully, and Wigner (1984). Lee (1995) reviews the relations among the Wigner, Husimi, and other phase space functions that have been defined.

Problems

15.1 Carry out the derivation of Eq. (15.10), which states that $\mathrm{Tr}(\rho R) = \iint \rho_w(q,p) R_w(q,p)\, dq\, dp$.

15.2 Calculate the Wigner function for the first excited state of a harmonic oscillator. Notice that it takes on both positive and negative values.

15.3 Show that the Husimi momentum distribution, $P_H(p) = \int \rho_H(q,p)\, dq$, is a Gaussian broadening of the quantal momentum distribution $|\langle p|\Psi\rangle|^2$.

15.4 Calculate the Wigner and Husimi functions for the state $\Psi(x) = A\sin(kx)$. (Normalization may be ignored, since this state function is not normalizable over $-\infty < x < \infty$.) Compare the interference terms in the two phase-space functions.

Chapter 16

Scattering

The phenomenon of scattering was first mentioned in Sec. 2.1 of this book as an illustration of the fact that quantum mechanics does not predict the outcome of an individual measurement, but rather the statistical distribution or probabilities of all possible outcomes. Scattering is even more important than that illustration would indicate, much of our information about the interaction between particles being derived from scattering experiments. Entire books have been written on the subject of scattering theory, and this chapter will cover only the basic topics.

16.1 *Cross Section*

The angular distribution of scattered particles in a particular process is described in terms of a *differential cross section*. Suppose that a flux of J_i particles per unit area per unit time is incident on the target. The number of particles per unit time scattered into a narrow cone of solid angle $d\Omega$, centered about the direction whose polar angles with respect to the incident flux are θ and ϕ, will be proportional to the incident flux J_i and to the angular opening $d\Omega$ of the cone. Hence it may be written as $J_i\,\sigma(\theta,\phi)\,d\Omega$. The proportionality factor $\sigma(\theta,\phi)$ is known as the differential cross section.

Suppose that a particle detector is located in the direction (θ,ϕ), at a sufficiently large distance r from the target so as to be outside of the incident beam.

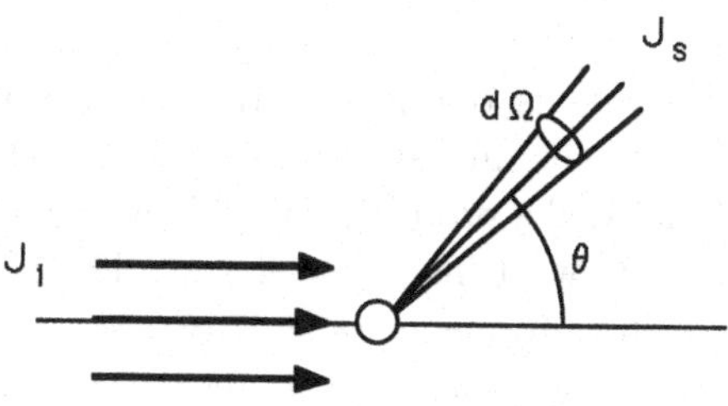

Fig. 16.1 Defining the differential cross section [Eq. (16.1)].

If it subtends the solid angle $d\Omega$ it will receive $J_i\,\sigma(\theta,\phi)\,d\Omega$ scattered particles per unit time. Dividing this number by the area of the detector, we obtain the flux of scattered particles at the detector, $J_s = J_i\,\sigma(\theta,\phi)\,d\Omega/r^2\,d\Omega$. Thus the differential cross section can be written as

$$\sigma(\theta,\phi) = \frac{r^2\,J_s}{J_i}\,, \tag{16.1}$$

from which it is apparent that it has the dimensions of an area. Its value is independent of the distance r from the target to the detector because J_s is inversely proportional to r^2. This expression is convenient because the fluxes J_s and J_i are measurable quantities, and can also be calculated theoretically.

By integrating over all scattering directions we obtain the *total cross section*,

$$\begin{aligned}\sigma &= \int \sigma(\theta,\phi)\,d\Omega \\ &= \int_0^{2\pi}\int_0^{\pi} \sigma(\theta,\phi)\,\sin\theta\,d\theta\,d\phi\,.\end{aligned} \tag{16.2}$$

Laboratory and center-of-mass frames

In defining $\sigma(\theta,\phi)$ above, we have reasoned as if the target were fixed at rest. This is never exactly true, because the total momentum of the projectile and the target particles is conserved. For theoretical analysis it is most convenient to use a frame of reference in which the center of mass (CM) of the two particles is at rest. The description of the scattering event is then symmetric between the projectile and the target. The distance r and the direction (θ,ϕ) in the above expressions refer to the relative separation of the projectile from the target. However, (θ,ϕ) is also the direction of the scattered projectile from the fixed CM, and the recoil of the target particle is in the opposite direction $(\pi-\theta,\phi+\pi)$. Scattering cross sections are almost always calculated in this CM frame.

Experimental results are obtained in the laboratory frame of reference, in which the target particle is initially at rest (see Fig. 16.2). In the laboratory frame we have, initially, the projectile particle with mass M_1 and velocity $\mathbf{v}_1$, and the target particle with mass M_2 at rest. The velocity of the CM with respect to the laboratory frame is $\mathbf{V}_0 = M_1\,\mathbf{v}/(M_1+M_2)$.

To transform from laboratory coordinates to the frame of reference in which the CM is at rest, we must subtract $\mathbf{V}_0$ from all velocities. Thus in the CM

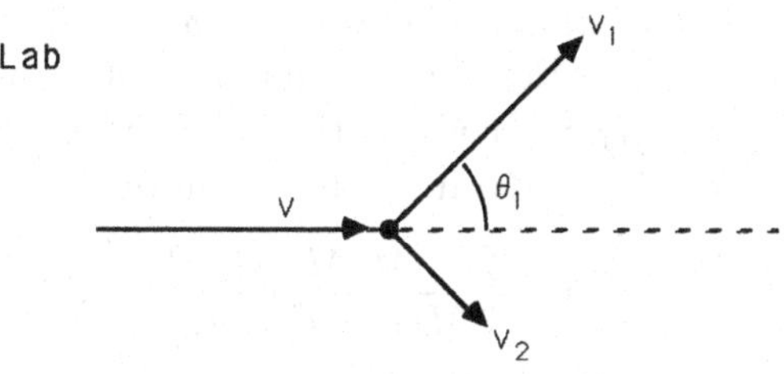

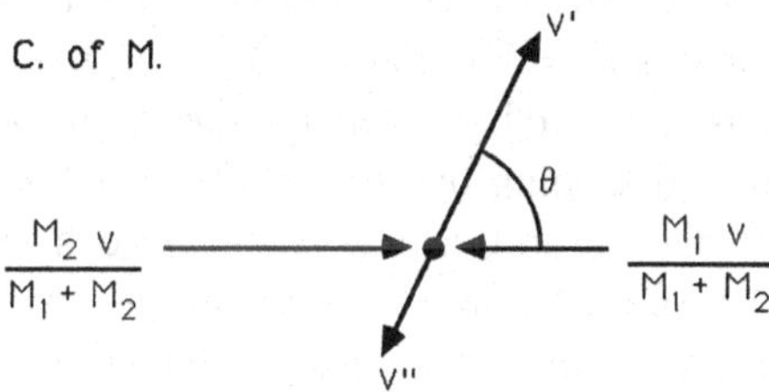

Fig. 16.2 Scattering event in the laboratory frame (top), and in the center-of-mass frame (bottom).

frame the initial velocity of the projectile is $\mathbf{v} - \mathbf{V}_0 = M_2\,\mathbf{v}/(M_1 + M_2)$, and the initial target velocity is $-M_1\,\mathbf{v}/(M_1 + M_2)$. It is apparent from Fig. 16.2 that the final velocity and direction of the projectile in the two frames of reference are related by

$$\begin{aligned} v_1 \cos\theta_1 &= v' \cos\theta + V_0\,, \\ v_1 \sin\theta_1 &= v' \sin\theta\,. \end{aligned} \tag{16.3}$$

Taking the ratio of these two equations, we obtain

$$\tan\theta_1 = \frac{\sin\theta}{\cos\theta + \beta}\,, \quad \text{with } \beta = \frac{\mathrm{V}_0}{\mathrm{v}'}\,. \tag{16.4}$$

In an *elastic collision* the speeds of the particles relative to the CM are unchanged, and so $v' = M_2\,v/(M_1 + M_2)$. Thus $\beta = M_1/M_2$ in this case.

In a general *inelastic collision* the internal energy of the particles may change, and so the total kinetic energy need not be conserved. In a *rearrangement collision* between composite particles there is a transfer of mass between the particles. (Examples are nuclear reactions, chemical reactions between molecules, and charge exchange between atoms and ions.) Suppose that the masses of the incoming particles are M_1 and M_2, and the masses of the outgoing particles are M_3 and M_4. Since we are treating only nonrelativistic kinematics, we have $M_1 + M_2 = M_3 + M_4$. It can be shown that

$$\beta = \left\{ \frac{M_1\, M_3\, E}{M_2\, M_4\, (E+Q)} \right\}^{1/2} . \tag{16.5}$$

Here M_1 is the mass of the projectile, M_2 is the initial mass of the target, M_3 is the mass of the detected particle (whose direction is θ in the CM frame), and M_4 is the mass of the (usually undetected) recoil particle. The initial kinetic energy in the CM frame is $E = M_1\, M_2\, v^2/2(M_1 + M_2)$, and Q is the amount of internal energy that is converted into kinetic energy in the reaction. The limit of an elastic collision is obtaining by putting $Q = 0, M_1 = M_3$, and $M_2 = M_4$.

The relation between the differential cross sections in the laboratory and CM frames can be determined from the fact that the number of particles scattered into a particular cone must be the same in the two coordinate systems. The incident flux (particles per unit area per unit time) is the same in the two frames, so we have

$$\sigma_1(\theta_1, \phi_1)\, \sin\theta_1\, d\theta_1\, d\phi_1 = \sigma(\theta,\phi)\, \sin\theta\, d\theta\, d\phi\,. \tag{16.6}$$

The relation between θ_1 and θ is given by (16.4), and it is clear that $\phi_1 = \phi$. After some algebra it follows that the cross section in the laboratory frame is given by

$$\sigma_1(\theta_1, \phi_1) = \sigma(\theta,\phi)\, \frac{(1+\beta^2+2\beta\, \cos\theta)^{3/2}}{|1+\beta\, \cos\theta|} . \tag{16.7}$$

The total cross section must be the same in the two frames of reference because the total number of scattered particles is independent of the coordinate system.

The quantum state function in scattering

No specific reference was made to quantum mechanics in the previous discussions, which were concerned with the formal definitions of cross sections in terms of numbers of scattered particles, and with the transformation of velocity vectors between frames of reference. Those results are independent of the differences between classical and quantum mechanics. We must now relate those definitions to the quantum state function.

The Hamiltonian for a system of two interacting particles is

$$H = -\frac{\hbar^2}{2M_1}\, {\nabla_1}^2 - \frac{\hbar^2}{2M_2}\, {\nabla_2}^2 + V(\mathbf{x}_1 - \mathbf{x}_2)\,, \tag{16.8}$$

where the first term involves derivatives with respect to $\mathbf{x}_1$ and the second term involves derivatives with respect to $\mathbf{x}_2$. (For simplicity, the internal degrees of freedom of the two particles are not indicated). The first step is to change variables from the coordinates of the individual particles to the CM and relative coordinates,

$$\mathbf{X} = \frac{M_1 \mathbf{x}_1 + M_2 \mathbf{x}_2}{M_1 + M_2}, \tag{16.9a}$$

$$\mathbf{r} = \mathbf{x}_1 - \mathbf{x}_2 . \tag{16.9b}$$

This transformation was performed in Sec. 10.2 by a canonical transformation of the position and momentum operators, but it can be done simply by introducing the new variables (16.9) into the differential operators of (16.8). The result is

$$H = -\frac{\hbar^2}{2(M_1 + M_2)} \nabla_X{}^2 - \frac{\hbar^2}{2\mu} \nabla^2 + V(r) . \tag{16.10}$$

The first term is the kinetic energy of the CM, and is of no present interest. The second term is the kinetic energy of relative motion, with the derivatives taken with respect to the relative coordinate $\mathbf{r}$. It involves the *reduced mass*, $\mu = M_1 M_2/(M_1 + M_2)$. The eigenfunctions of H can be chosen to have the separated form $\Psi(\mathbf{X}, \mathbf{r}) = \Phi(\mathbf{X})\, \psi(\mathbf{r})$. The second factor satisfies

$$-\frac{\hbar^2}{2\mu} \nabla^2 \psi(\mathbf{r}) + V(r)\, \psi(\mathbf{r}) = E\, \psi(\mathbf{r}) , \tag{16.11}$$

where E is the energy associated with the *relative motion* of the two particles in the CM frame.

The appropriate *boundary condition* for (16.11) is determined from the experimental conditions, shown in Fig. 16.1, that are used to define the differential cross section (16.1). There must be an incident flux J_i directed from the source to the target, and a scattered flux J_s radiating outward in all directions. The particle source is not included in Eq. (16.11), so the value of the incident flux must be imposed as a boundary condition. Therefore we require that the solution of (16.11) be of the form

$$\psi(\mathbf{r}) = \psi_i(\mathbf{r}) + \psi_s(\mathbf{r}) , \tag{16.12}$$

where the "incident wave" $\psi_i(\mathbf{r})$ represents the flux of the incident beam, and $\psi_s(\mathbf{r})$ is an outgoing "scattered wave".

The quantum state does not describe the position of the incident particle, but rather it gives the probability density, $|\psi(\mathbf{r})|^2$, for it to be a distance $\mathbf{r}$ from the target. Similarly the state does not describe the actual flux of particles, but rather the *probability flux*, which is the net probability per unit time that a particle crosses a unit area. Applying (4.22) to our problem, we can write the probability flux in (16.11) as

$$\mathbf{J} = \frac{\hbar}{\mu}\,\mathrm{Im}(\psi^* \boldsymbol{\nabla}\psi)\,. \tag{16.13}$$

The incident beam can be described by $\psi_i = A\,e^{i\mathbf{k}\cdot\mathbf{r}}$, for which the flux, $\mathbf{J}_i = |A|^2\,\hbar\mathbf{k}/\mu$, is uniform. If the scattering potential $V(r)$ is of finite range, then for large values of r, Eq. (16.11) will reduce to the free particle equation, $-(\hbar^2/2\mu)\,\nabla^2\,\psi(\mathbf{r}) = E\,\psi(\mathbf{r})$, and we may expect $\psi(\mathbf{r})$ to become asymptotically equal to some solution of the free particle equation.[g] An outgoing spherical wave at large r has the asymptotic form $\psi_s \sim A\,f(\theta,\phi)\,e^{ikr}/r$, where the angular function $f(\theta,\phi)$ is not yet specified. The radial component of the flux for this function is $(\mathbf{J}_s)_r = (\hbar/\mu)\,\mathrm{Im}(\psi_s^*\partial\psi_s/\partial r) = |A|^2\,(\hbar k/\mu)\,|f(\theta,\psi)|^2/r^2$. Therefore we seek a solution of (16.11) that satisfies the asymptotic boundary condition

$$\psi(\mathbf{r}) \sim A\left\{e^{i\mathbf{k}\cdot\mathbf{r}} + \frac{f(\theta,\phi)e^{ikr}}{r}\right\} \tag{16.14}$$

in the limit of large r. Substituting the fluxes of these two terms into (16.1) yields the differential cross section

$$\sigma(\theta,\phi) = |f(\theta,\phi)|^2\,. \tag{16.15}$$

Thus the solution to a scattering problem is reduced to determining the asymptotic behavior of the scattering state function. The amplitude A is irrelevant, and is usually set equal to 1 for convenience. Since we have neglected any internal degrees of freedom of the particles, we have implicitly restricted our solution to the case of *elastic scattering*. The result (16.15) will be modified when we treat inelastic scattering.

The alert reader will have noticed that we did not calculate the flux by substituting (16.14) into (16.13). Instead, we calculated separate fluxes for the two terms of (16.14), and thereby apparently omitted certain cross terms. This is not an error; rather, it is a recognition of the fact that the incident beam must be of finite width, as is indicated in Fig. 16.1. Thus, strictly speaking, we have $\psi_i = A\,e^{i\mathbf{k}\cdot\mathbf{r}}$ within the incident beam, and $\psi_i = 0$ elsewhere. At the detector we will have $\psi(\mathbf{r}) \equiv \psi_i(\mathbf{r}) + \psi_s(\mathbf{r}) = \psi_s(\mathbf{r})$, provided the detector

[g]This is not true for the Coulomb potential, which goes to zero very slowly, being proportional to $1/r$. See the references at the end of this chapter for the special treatment that it needs.

is located outside of the incident beam. This is always done in practice, for obvious reasons, but if it were not the case then the cross terms involving both $\psi_i(\mathbf{r})$ and $\psi_s(\mathbf{r})$ in $\mathbf{J}$ would have to be included.

Most of scattering theory is concerned with the solution of (16.11), subject to a boundary condition like (16.12) or (16.14). When concentrating on the technical details of that solution, one may be inclined to think of $\psi(\mathbf{r})$ as describing the motion of the scattered particle, and regard the target particle as a fixed force center supporting the potential $V(r)$. Strictly speaking, that interpretation is not correct. Equation (16.11) describes the *relative* motion of the two particles, and the description is completely symmetrical with respect to the two particles (except for the change of sign of $\mathbf{r}$ when they are interchanged). Thus, even though calculations can be done as if the target were fixed at the origin, we are in fact working in the CM frame of reference, and the state function $\psi(\mathbf{r})$ describes the quantum-mechanical properties of the target particle as well as of the incident particle.

16.2 *Scattering by a Spherical Potential*

In this section we consider in detail the scattering of particles which interact by a potential $V(r)$ that depends only on the magnitude of the distance between the particles. Since no internal degrees of freedom of the particles can be excited by such an interaction, they may be ignored. Only *elastic scattering* is possible in this model.

Equation (16.11), which governs the state of relative motion of the two particles, will, for convenience, be rewritten as

$$\nabla^2\psi(\mathbf{r}) + [k^2 - U(r)]\,\psi(\mathbf{r}) = 0\,, \tag{16.16}$$

where $k = (2\mu E/\hbar^2)^{1/2}$ and $U(r) = (2\mu/\hbar^2)\,V(r)$. The relative velocity of the particles is $\hbar k/\mu$. The solution of (16.16) can be written as a series of partial waves,

$$\psi(\mathbf{r}) = \sum_{\ell m} a_{\ell m}\, Y_\ell{}^m(\theta,\phi)\,\frac{u_\ell(r)}{r}\,, \tag{16.17}$$

where the radial functions satisfy the equation

$$\frac{d^2\,u_\ell(r)}{dr^2} + \left[k^2 - U(r) - \frac{\ell(\ell+1)}{r^2}\right] u_\ell(r) = 0\,. \tag{16.18}$$

[A substitution like (16.17) was previously used in Sec. 10.1.]

We must now determine the asymptotic behavior of the radial function $u_\ell(r)$. At sufficiently large r, the U and ℓ terms will be negligible compared with k^2, suggesting that the asymptotic form of the radial function should be $e^{\pm ikr}$. To verify this intuitive but nonrigorous estimate, we put

$$u_\ell(r) = e^{h(r)}\, e^{\pm ikr}\,, \tag{16.19}$$

and we expect the first exponential to be slowly varying at large r, compared to the second exponential. Substitution of (16.19) into (16.18) yields a differential equation for $h(r)$,

$$h''(r) + [h'(r)]^2 \pm 2ikh'(r) = U(r) + \frac{\ell(\ell+1)}{r^2}\,, \tag{16.20}$$

where the primes indicate differentiation. This is really a first order differential equation for $h'(r)$, since $h(r)$ does not appear in it. If $U(r)$ falls off at large r at least as rapidly as r^{-2}, then the third term on the left will be dominant, with $h'(r)$ going to zero like r^{-2}. Then at large r we will have $h(r) = b + c/r$, which becomes a constant in the limit $r \to \infty$. In this case our intuitive estimate is correct, and $u_\ell(r)$ does indeed go as $e^{\pm ikr}$ for large r, which is compatible with the asymptotic form (16.14).

We can also see from this argument why the Coulomb potential, for which $U(r) \propto r^{-1}$, requires special treatment. In that case, we see from (16.20) that $h'(r)$ falls off like r^{-1} at large r, and hence $h(r)$ goes as $\log(r)$ and does not have a finite limit as $r \to \infty$. Therefore the first exponential in (16.19) does not become constant at large r, and consequently the asymptotic form (16.14) cannot be obtained. References to the special treatment needed for the Coulomb potential are given at the end of this chapter. Henceforth our discussion will be confined to *short range* potentials, which means potentials that fall off at large r at least as rapidly as r^{-2}.

Phase shifts

It is convenient to consider a more restricted class of potentials which vanish [or at least become negligible compared to the $\ell(\ell+1)/r^2$ term] for $r > a$. Later we will indicate how the principal results can be generalized to any short range potential.

Let us return to the problem of solving (16.16) with the boundary condition (16.14). If the scattering potential were identically zero, the unique (apart from normalization) solution would be

$$\psi_i = e^{i\mathbf{k}\cdot\mathbf{r}} = \sum_\ell (2\ell+1)\, i^\ell\, j_\ell(kr)\, P_\ell(\cos\theta)\,, \tag{16.21}$$

where j_ℓ is a spherical Bessel function, P_ℓ is a Legendre polynomial, and θ is the angle between $\mathbf{k}$ and $\mathbf{r}$. Let us write the solution of (16.16) with the scattering potential in the form

$$\psi = \sum_\ell (2\ell + 1)\, i^\ell\, A_\ell\, R_\ell(r)\, P_\ell(\cos\theta)\,, \tag{16.22}$$

where the radial function $R_\ell(r)$ satisfies the partial wave equation,

$$\frac{1}{r^2}\frac{d}{dr}r^2\frac{d}{dr}R_\ell(r) + \left[k^2 - U(r) - \frac{\ell(\ell+1)}{r^2}\right]R_\ell(r) = 0\,. \tag{16.23}$$

If the potential $U(r)$ were not present, this would be the equation satisfied by the spherical Bessel functions, $j_\ell(kr)$ and $n_\ell(kr)$. [For the many formulas and identities satisfied by these functions, see Morse and Feshbach (1953).] Hence, for $r > a$, where $U(r) = 0$, the solution of (16.23) must be a linear combination of these two Bessel functions, which we write as

$$R_\ell(r) = \cos(\delta_\ell)\, j_\ell(kr) - \sin(\delta_\ell)\, n_\ell(kr)\,,\ \ (r \geq a)\,. \tag{16.24}$$

Since the differential equation is real, the solution $R_\ell(r)$ may be chosen real, and so δ_ℓ is real.

The asymptotic forms of the Bessel functions in the limit $kr \to \infty$ are

$$j_\ell(kr) \to \frac{\sin(kr - \frac{1}{2}\pi\ell)}{kr}\,, \tag{16.25a}$$

$$n_\ell(kr) \to \frac{-\cos(kr - \frac{1}{2}\pi\ell)}{kr}\,, \tag{16.25b}$$

and therefore the corresponding limit of $R_\ell(r)$ is

$$R_\ell(r) \to \frac{\sin(kr - \frac{1}{2}\pi\ell + \delta_\ell)}{kr}\,. \tag{16.26}$$

If the scattering potential were exactly zero, the form (16.24) would be valid all the way in to $r = 0$. But the function $n_\ell(kr)$ has an r^{-1} singularity at $r = 0$, which is not allowed in a state function, so we must have $\delta_\ell = 0$ if $U(r) = 0$ for all r. Comparing the asymptotic limit of the zero scattering solution, $j_\ell(kr)$, with (16.26), we see that the only effect of the short range scattering potential that appears at large r is the *phase shift* of the radial function by δ_ℓ. Since the differential cross section is entirely determined by the asymptotic form of the state function, it follows that the cross section must be expressible in terms of these phase shifts.

If we substitute the series (16.22) and (16.21) into (16.14) with $A = 1$, and replace the Bessel functions with their asymptotic limits, we obtain

$$\sum_{\ell} (2\ell+1)\, i^{\ell}\, P_{\ell}(\cos\theta)\, A_{\ell}\, \frac{\sin(kr - \frac{1}{2}\pi\ell + \delta_{\ell})}{kr}$$

$$= \sum_{\ell} (2\ell+1)\, i^{\ell}\, P_{\ell}(\cos\theta)\, \frac{\sin(kr - \frac{1}{2}\pi\ell)}{kr} + f(\theta,\phi)\, \frac{e^{ikr}}{r}\,. \qquad (16.27)$$

We next express the sine functions in terms of complex exponentials, using $\sin(x) = (e^{ix} + e^{-ix})/2i$. Equating the coefficients of e^{-ikr} in the above equation yields

$$\sum_{\ell} (2\ell+1)\, i^{\ell}\, P_{\ell}(\cos\theta)\, A_{\ell}\, \exp\left(i\tfrac{1}{2}\pi\ell - i\delta_{\ell}\right)$$

$$= \sum_{\ell} (2\ell+1) i^{\ell}\, P_{\ell}(\cos\theta)\, \exp\left(i\tfrac{1}{2}\pi\ell\right)\,.$$

This equality must hold term by term, since the Legendre polynomials are linearly independent, and so we have

$$A_{\ell} = \exp(i\delta_{\ell})\,. \qquad (16.28)$$

Equating the coefficients of e^{ikr} in (16.27) and using (16.28) then yields

$$\begin{aligned} f(\theta,\phi) &= f(\theta) \\ &= (2ik)^{-1} \sum_{\ell} (2\ell+1) i^{\ell} \exp\left(-\tfrac{1}{2}i\pi\ell\right) [\exp(2i\delta_{\ell}) - 1]\, P_{\ell}(\cos\theta) \\ &= (2ik)^{-1} \sum_{\ell} (2\ell+1) [\exp(2i\delta_{\ell}) - 1]\, P_{\ell}(\cos\theta) \\ &= k^{-1} \sum_{\ell} (2\ell+1) \sin(\delta_{\ell}) \exp(i\delta_{\ell})\, P_{\ell}(\cos\theta)\,. \end{aligned} \qquad (16.29)$$

Notice that this expression is unchanged by the substitution $\delta_{\ell} \to \delta_{\ell} + \pi$, and hence all scattering effects are periodic in δ_{ℓ} with period π (rather than 2π, as might have been expected). The differential cross section is now given by (16.15) to be

$$\sigma(\theta,\phi) = \sigma(\theta) = |f(\theta)|^2\,. \qquad (16.30)$$

This is independent of ϕ because the potential is spherically symmetric, and we have measured the angle θ from the direction $\mathbf{k}$ of the incident beam.

The total elastic cross section is obtained by integrating $\sigma(\theta)$ over all directions, as in (16.2). Because the Legendre polynomials are orthogonal, there are no terms involving different values of ℓ, and we have

$$\sigma = \frac{4\pi}{k^2} \sum_{\ell} (2\ell + 1)[\sin(\delta_\ell)]^2 \,. \tag{16.31}$$

Calculation of phase shifts

The phase shift δ_ℓ for a scattering potential $U(r)$ that may be nonzero for $r < a$ but vanishes for $r > a$, is obtained by solving (16.23) for the radial function $R_\ell(r)$ in the region $r \leq a$ and matching it to the form (16.24) at $r = a$. There are two linearly independent solutions to (16.23), but only one of them remains finite at $r = 0$, and so the solution for $R_\ell(r)$ in the interval $0 \leq r \leq a$ is unique except for normalization. It can be obtained numerically, if necessary. Although the boundary condition at $r = a$ is that both R_ℓ and dR_ℓ/dr must be continuous (see Sec. 4.5), it is sufficient for our purposes to impose continuity on the so-called *logarithmic derivative*,

$$\gamma_\ell = \frac{d \log(R_\ell)}{dr} = \frac{1}{R_\ell} \frac{dR_\ell}{dr} \,, \tag{16.32}$$

which is independent of the arbitrary normalization. This yields the condition

$$\gamma_\ell = \frac{k[\cos(\delta_\ell)\, j'_\ell(ka) - \sin(\delta_\ell)\, n'_\ell(ka)]}{\cos(\delta_\ell)\, j_\ell(ka) - \sin(\delta_\ell)\, n_\ell(ka)} \,, \tag{16.33}$$

where a prime indicates differentiation of a function with respect to its argument, and γ_ℓ now denotes the logarithmic derivative evaluated from the interior at $r = a$. The phase shift is then given by

$$\tan(\delta_\ell) = \frac{kj'_\ell(ka) - \gamma_\ell\, j_\ell(ka)}{kn'_\ell(ka) - \gamma_\ell\, n_\ell(ka)} \,. \tag{16.34}$$

If the scattering potential is not identically zero for $r > a$, but is still of *short range*, it is still possible to define phase shifts as the limit of (16.34) as $a \to \infty$, remembering of course that γ_ℓ depends on a. This limit will exist provided the potential falls off more rapidly than r^{-1}.

It can be shown that, for sufficiently large values of ℓ, the phase shift δ_ℓ decreases as the reciprocal of a factorial of ℓ. [See Schiff (1968), Sec. 19.] This is a very rapid decrease, being faster than exponential, and it ensures that the series (16.29) is convergent. However, this "sufficiently large" value of ℓ may be very large, and the phase shift series is practical only if it converges in a small number of terms. To estimate the condition under which this will occur, let us suppose the potential $U(r)$ to be identically zero. Then we would have $R_\ell(r) = j_\ell(kr)$, which is proportional to $(kr)^\ell$ in the regime $kr \ll \ell$, and is very small in that regime for large ℓ. We now reintroduce the potential $U(r)$, of range a, into (16.23). If $ka \ll \ell$ then $U(r)$ will be multiplied by the small quantity $R_\ell(r)$, and so will have little effect on the solution. By this rather loose argument, we can see that the phase shift δ_ℓ will be small provided that $ka \ll \ell$. Therefore the phase shift series will be most useful when ka is small.

Example: hard sphere scattering

Consider scattering by the potential $V(r) = +\infty$ for $r < a, V(r) = 0$ for $r > a$. Then the boundary condition becomes $R_\ell(a) = 0$, from which (16.24) yields

$$\tan(\delta_\ell) = \frac{j_\ell(ka)}{n_\ell(ka)}, \tag{16.35}$$

a result that also follows from (16.34) by taking the limit $\gamma_\ell \to \infty$.

In the *low energy limit*, for which $ka \ll 1$, we may use the approximate values $j_\ell(ka) \approx (ka)^\ell/1{\cdot}3{\cdot}5\cdots(2\ell{+}1)$ and $n_\ell(ka) \approx -1{\cdot}3{\cdot}5\cdots(2\ell{-}1)/(ka)^{\ell+1}$, whence (16.35) becomes $\tan(\delta_\ell) \approx -(ka)^{2\ell+1}/[1\cdot 3\cdot 5\cdots(2\ell-1)]^2(2\ell+1)$. This proportionality of $\tan(\delta_\ell)$, and hence also of δ_ℓ to $(ka)^{2\ell+1}$ in the low energy limit, is actually a general feature of short range potentials. Therefore in this limit the phase shift series converges in only a few terms. When $k \to 0$ only the $\ell = 0$ term contributes to (16.31), so the zero energy limit of the total elastic scattering cross section is $\sigma = 4\pi a^2$, four times the geometric cross section. The de Broglie wavelength, $\lambda = 2\pi/k$, is very large in this limit, so the difference of σ from the classical value should not be surprising.

The *high energy limit* of the cross section is more difficult to evaluate because it involves a very large number of phase phase shifts (see Sakuri, 1982). The result is $\sigma = 2\pi a^2$, twice the geometric area. This is a surprise, since we expect the classical limit to be recovered when the de Broglie wavelength is very small. A simple explanation that applies to all rigid scatterers was given by Keller (1972). The scattered wave is equal to the total wave function minus the incident wave:

$$\psi(\mathbf{r}) - \psi_i(\mathbf{r}) = \psi_s(\mathbf{r})\,. \tag{16.36}$$

The flux lines associated with the three terms are depicted pictorially in Fig. 16.3. These lines become straight trajectories in the limit $\lambda \to 0$. Since the flux associated with ψ_i is not affected by changing the sign of ψ_i, the subtraction of this term gives ψ_s an ongoing flux in the region of the geometric shadow of the scatterer. It is apparent that the total flux associated with ψ_s consists of a reflected flux and a "shadow" flux, each equal in magnitude to the incident flux, and so the conventional definition of the scattering cross section, (16.1) and (16.2), results in σ being equal to twice the geometric cross section of the scatterer. One would not normally count an undeviated flux as being scattered, so the definition of σ seems unreasonable in this case. However, if $\lambda = 2\pi/k$ is small but nonzero the "shadow" flux lines will be slightly deflected as a result of diffraction, and so the "shadow scattering" really must be counted as a part of the scattering cross section.

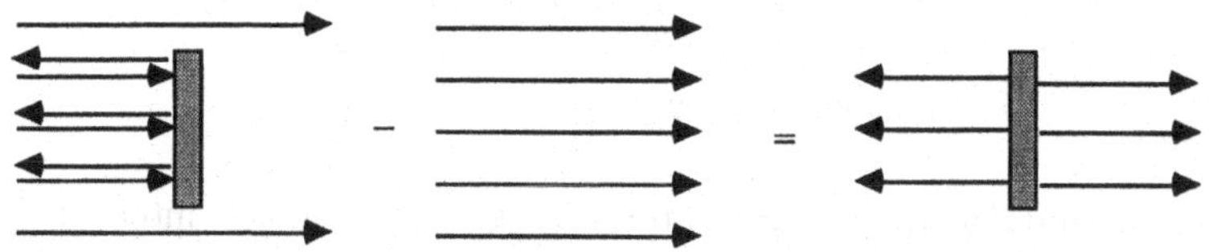

Fig. 16.3 Flux lines of the total, incident, and scattered waves for a rigid scatterer.

16.3 *General Scattering Theory*

In the previous section we treated only scattering by a central potential, which cannot change the internal states of the particles, and so produces only elastic scattering. We shall now consider a more general interaction between the two particles. It may depend on variables other that the separation distance, such as spin and orientation, and may change the internal states of the particles.

Collision events may be classified into several types:

(a) *Elastic scattering*, symbolized as $A + X \to A + X$, in which the internal states of the particles are unchanged.

(b) *Inelastic scattering*, $A + X \to A' + X'$, which involves a change of the internal states.
(c) *Rearrangement collisions*, $A + X \to B + Y$, in which matter is exchanged and the outgoing particles may be of different species from the incoming particles. Nuclear and chemical reactions are examples.
(d) *Particle production*, $A + X \to B + Y + Z + \cdots$, in which there are three or more particles in the final state.

Each mode of breakup of the system into a set of separate particles is termed a *channel.* Instead of having only one differential cross section, we must now define differential cross sections for each channel.

We shall treat only the elastic and inelastic channels (a) and (b). The theory of (c) and (d) presents considerably greater technical difficulties, and would be too lengthy for his chapter. It may be found in the specialized references at the end of the chapter.

As in the previous section, the CM variable is of no interest, and we may consider only the Hamiltonian for the relative motion of the particles, as well as their internal degrees of freedom. It has the form

$$\begin{aligned} H &= \frac{\hbar^2}{2\mu} \nabla^2 + h_1 + h_2 + V \\ &= H_0 + V \,. \end{aligned} \tag{16.37}$$

Here h_1 and h_2 are the Hamiltonian operators for the internal degrees of freedom of the two particles (labeled 1 and 2), and μ is the reduced mass ($\mu^{-1} = {M_1}^{-1} + {M_2}^{-1}$). The differential operator ∇^2 acts on the relative coordinate $\mathbf{r}$. For reasons discussed in the previous section, we consider only interactions V that decrease at large separations more rapidly than r^{-1}, however the dependence of V on internal variables such as spin is unrestricted.

It would make for a very cumbersome notation if all of the labels for particles and states were always explicitly indicated, so we shall keep the notation as compact as possible. We shall write

$$(h_1 + h_2) w_a = e_a \, w_a \,, \tag{16.38}$$

where w_a is a state vector for the internal degrees of freedom of *both* particles, and e_a is the sum of their internal energies. Frequently, but not always, w_a will be factorable into a product of state vectors for each particle; however, this will not be indicated in the notation. If the initial and final internal states are w_a and w_b, the condition of energy conservation will be

$$E = \frac{\hbar^2 {k_a}^2}{2\,\mu} + e_a = \frac{\hbar^2 {k_b}^2}{2\mu} + e_b\,. \tag{16.39}$$

The collision will be called *elastic* if the internal states w_a and w_b are the same. If the internal states are different it will be called *inelastic*, even if there is no change in the internal and kinetic energies. The kinetic energy terms in (16.39) are, of course, the kinetic energies of relative motion in the CM frame.

Scattering cross sections

We seek steady state solutions of the Schrödinger equation,

$$H\,\Psi_{\mathbf{a}}^{(+)} = E\,\Psi_{\mathbf{a}}^{(+)}\,. \tag{16.40}$$

The boldfaced label $\mathbf{a}$ is a composite of internal and motional state labels, $\mathbf{a} = (\mathbf{k}_a, a)$, with $\hbar\mathbf{k}/\mu$ being the velocity of the incident particle relative to the target, and with a being the internal state label. The solution of (16.40) must satisfy an asymptotic boundary condition analogous to (16.14), but now, instead of a single outgoing scattered wave, we may have an outgoing wave with components for each of the possible internal states that may be produced in the collision. Therefore we shall write the state function as

$$\Psi_{\mathbf{a}}^{(+)} = \sum_{b'} {\psi_{\mathbf{a},b'}}^{(+)}(\mathbf{r})\,w_{b'}\,, \tag{16.41}$$

where the wave functions have the following asymptotic behavior as the separation r becomes very large:

$$ {\psi_{\mathbf{a},a}}^{(+)}(\mathbf{r}) \sim A\left\{e^{i\mathbf{k}_a\cdot\mathbf{r}} + {f_{\mathbf{a}a}}^{(+)}(\Omega_r)\,\frac{e^{ik_a r}}{r}\right\}, \tag{16.42a}$$

$$ {\psi_{\mathbf{a},b}}^{(+)}(\mathbf{r}) \sim A\,{f_{\mathbf{a}b}}^{(+)}(\Omega_r)\,\frac{e^{ik_b r}}{r}\,, \quad (b \neq a)\,. \tag{16.42b}$$

Here Ω_r denotes the angles of the vector $\mathbf{r}$. The first of these expressions, corresponding to the elastic scattering channel, contains the incident beam and an outgoing scattered wave. The second expression, describing inelastic scattering, contains only an outgoing wave since there is no incident beam corresponding to the internal state w_b with $b \neq a$.

As was discussed in Sec. 16.1, the magnitude of the flux of the incident beam is $J_i = |A|^2\,\hbar k_a/\mu$. The scattered flux corresponding to the internal state w_b is $J_b = |A|^2(\hbar k_b/\mu)|f_{\mathbf{a}b}{}^{(+)}(\Omega_r)|^2/r^2$. Therefore, according to the definition (16.1), we have the differential cross section

$$\sigma_{a\to b}(\Omega) = \frac{k_b}{k_a}\,|f_{\mathbf{a}b}{}^{(+)}(\Omega)|^2 \tag{16.43}$$

for the collision process involving the change of internal state $a \to b$. Here Ω denotes the angular direction of the detector from the target, relative to the direction $\mathbf{k}_a$ of the incident beam. If $a = b$ then (16.43) reduces to the elastic scattering cross section (16.15).

Scattering amplitude theorem

Although the scattering cross sections depend only on the asymptotic limit of the state function at large distance, through the angular functions $f_{\mathbf{a}b}{}^{(+)}(\Omega)$ known as *scattering amplitudes*, the values of those amplitudes are determined by the scattering interaction at short distances. We will now derive a theorem relating the asymptotic limit of the scattering state function to an integral involving the interaction.

As a technical step in the derivation, we must introduce new scattering-like functions, $\Psi_{\mathbf{b}}{}^{(-)}$, which are eigenfunctions of H,

$$H\,\Psi_{\mathbf{b}}{}^{(-)} = E\,\Psi_{\mathbf{b}}{}^{(-)}\,, \tag{16.44}$$

but which satisfy different asymptotic boundary condition from (16.41):

$$\Psi_{\mathbf{b}}{}^{(-)} = \sum_{a'} \psi_{\mathbf{b},a'}{}^{(-)}(\mathbf{r})\,\mathbf{w}_{a'}\,, \tag{16.45}$$

with

$$\psi_{\mathbf{b},b}{}^{(-)}(\mathbf{r}) \sim A\left\{e^{i\mathbf{k}_b\cdot\mathbf{r}} + f_{\mathbf{b}b}{}^{(-)}(\Omega_r)\,\frac{e^{-ik_b r}}{r}\right\}, \tag{16.46a}$$

$$\psi_{\mathbf{b},a}{}^{(-)}(\mathbf{r}) \sim A\,f_{\mathbf{b}a}{}^{(-)}(\Omega_r)\,\frac{e^{-ik_a r}}{r}\,, \quad (a \neq b)\,. \tag{16.46b}$$

These new functions consist of an incident beam plus *incoming* spherical waves. Although they do not correspond to any state that is likely to be produced in an experiment, they play an essential mathematical role in the theory.

The two sets of functions, $\{\Psi_{\mathbf{a}}{}^{(+)}\}$ and $\{\Psi_{\mathbf{b}}{}^{(-)}\}$, each span the subspace of positive energy eigenfunctions of H. If the interaction V is strong enough to produce bound states of the two particles, then these bound states must be added to each of the sets to make them complete. The existence of two complete sets of eigenfunctions of H can be understood from the fact that an eigenvector problem is fully determined by an operator plus boundary conditions. Thus (16.40) plus b.c. (16.42) is one such well-defined problem, and (16.44) with b.c. (16.46) is another, and each problem has its own complete set of eigenfunctions. One of the sets, $\{\Psi_{\mathbf{a}}{}^{(+)}\}$ or $\{\Psi_{\mathbf{b}}{}^{(-)}\}$, can be expressed as linear combinations of the other.

None of the scattering functions has a finite norm, and thus none belongs to Hilbert space (see Sec. 1.4). The internal state functions are properly normalized, $\langle w_a | w_a \rangle = 1$, but the wave functions are not square-integrable because of their behavior as $r \to \infty$. Thus $\langle \Psi_{\mathbf{a}}{}^{(+)} | \Psi_{\mathbf{a}}{}^{(+)} \rangle = \infty$ for all scattering states. This unnormalizability is an essential part of their nature, and not merely a technical detail.

[[Some writers try to avoid this essential unnormalizability of the scattering functions by supposing the universe to be a large cube, subject to periodic boundary conditions, with the length of its sides being allowed to approach infinity at the end of the calculation. All eigenfunctions of H are then normalizable. But if periodic boundary conditions are imposed on a finite space, then the *incoming and outgoing wave solutions of* (16.40) *and* (16.44) *do not exist*, and all eigenfunctions of H must be standing waves. Although the users of the "box normalization" technique seldom derive wrong answers from it, their method is fundamentally inconsistent at its outset. Therefore we shall not use it in scattering theory.]]

The operator ∇^2 is Hermitian only within a space of functions that satisfies certain boundary conditions (see Problem 1.11) that are violated by the scattering state functions. (This situation is not anomalous; indeed, the generalized spectral theorem, discussed in Sec. 1.4, applies to operators like ∇^2 that are self-adjoint in Hilbert space but whose eigenvectors belong to a larger space. This is the normal situation for operators that have a continuous eigenvalue spectrum.) This fact will prove to be crucial in deriving the scattering amplitude theorem.

To derive our theorem we shall compare two Hamiltonians, both of the form (16.37), but with different scattering potentials: $H = H_0 + V$ and

$H' = H_0 + V'$. We consider an outgoing wave eigenfunction of H and an incoming wave eigenfunction of H', for the same energy:

$$H\,\Psi_{\mathbf{a}}{}^{(+)} = E\,\Psi_{\mathbf{a}}{}^{(+)}\,, \quad H'\,\Psi'_{\mathbf{b}}{}^{(-)} = E\,\Psi'_{\mathbf{b}}{}^{(-)}\,. \tag{16.47}$$

The wave vectors of the incident beams in these functions are, respectively, $\mathbf{k}_a$ and $\mathbf{k}_b$.

Because the scattering functions have infinite norms, it is useful to define a *partial inner product*, $\langle\,\Psi'|\Psi\,\rangle_R$, which comprises an ordinary inner product for the internal degrees of freedom such as spin, and an integration of the relative coordinate $\mathbf{r}$ over the finite sphere $|\mathbf{r}| \le R$. The ordinary inner product would be given by $\langle\,\Psi'|\Psi\,\rangle = \lim_{R\to\infty}\langle\,\Psi'|\Psi\,\rangle_R$ *provided the limit exists*, which cannot be assumed in advance. Equations (16.47) can be rewritten as

$$\frac{\hbar^2}{2\mu}\,\nabla^2\,\Psi_{\mathbf{a}}{}^{(+)} = (h_1 + h_2 + V - E)\Psi_{\mathbf{a}}{}^{(+)}\,,$$

$$\frac{\hbar^2}{2\mu}\,\nabla^2\,\Psi'_{\mathbf{b}}{}^{(-)} = (h_1 + h_2 + V' - E)\Psi'_{\mathbf{b}}{}^{(-)}\,.$$

We now form the partial inner products between $\Psi'_{\mathbf{b}}{}^{(-)}$ and the first of these equations, between the second equation and $\Psi_{\mathbf{a}}{}^{(+)}$, and substract the results, obtaining

$$\frac{\hbar^2}{2\mu}\left\{\langle\,\Psi'_{\mathbf{b}}{}^{(-)}|\nabla^2\,\Psi_{\mathbf{a}}{}^{(+)}\,\rangle_R - \langle\,\nabla^2\,\Psi'_{\mathbf{b}}{}^{(-)}|\,\Psi_{\mathbf{a}}{}^{(+)}\,\rangle_R\right\}$$
$$= \langle\,\Psi'_{\mathbf{b}}{}^{(-)}\,|(V - V')|\,\Psi_{\mathbf{a}}{}^{(+)}\,\rangle_R\,. \tag{16.48}$$

The left hand side of this equation would vanish if the operator ∇^2 were Hermitian in the space of scattering functions, but that is not the case. To evaluate it, we use (16.41) and (16.45):

$$\frac{\hbar^2}{2\mu}\left\{\langle\,\Psi'_{\mathbf{b}}{}^{(-)}|\nabla^2\,\Psi_{\mathbf{a}}{}^{(+)}\,\rangle_R - \langle\,\nabla^2\,\Psi'_{\mathbf{b}}{}^{(-)}|\Psi_{\mathbf{a}}{}^{(+)}\,\rangle_R\right\}$$
$$= \frac{\hbar^2}{2\mu}\sum_{a'}\sum_{b'}\int_{r\le R}\Big\{[\psi'_{\mathbf{b},a'}{}^{(-)}]^*\,\nabla^2\,\psi_{\mathbf{a},b'}{}^{(+)}$$
$$- [\nabla^2\,\psi'_{\mathbf{b},a'}{}^{(-)}]^*\,\psi_{\mathbf{a},b'}{}^{(+)}\Big\}\,d^3\mathbf{r}\,\langle\,w_{a'}|w_{b'}\,\rangle\,. \tag{16.49}$$

Because the internal states vectors are orthonormal, we can reduce the double sum to a single sum over the dummy variable c. The volume integrals can be transformed into surface integrals by using the divergence theorem and the identity $\boldsymbol{\nabla}\cdot(f\boldsymbol{\nabla}g) = (\boldsymbol{\nabla}f)\cdot\boldsymbol{\nabla}g + f\nabla^2 g$. This yields

$$\frac{\hbar^2}{2\mu}\sum_c \oiint_{r=R} \left\{[\psi'_{\mathbf{b},c}{}^{(-)}]^*\,\boldsymbol{\nabla}\,\psi_{\mathbf{a},c}{}^{(+)} - [\boldsymbol{\nabla}\,\psi'_{\mathbf{b},c}{}^{(-)}]^*\,\psi_{\mathbf{a},c}{}^{(+)}\right\}\cdot d\mathbf{S}\,. \tag{16.50}$$

We now let the radius R of the sphere be sufficiently large so that the asymptotic forms, (16.42) and (16.46), can be substituted for the wave functions. Three types of terms will arise: those involving two spherical waves, those involving two plane waves, and those involving a plane wave and a spherical wave. Apart from a constant factor, the terms of the first type yield

$$\sum_c \oiint [f'_{\mathbf{b}c}{}^{(-)}(\Omega)]^*\,f_{\mathbf{a}c}{}^{(+)}(\Omega)\,d\Omega\;\,2ik_c\;\,e^{i2k_cR}\,.$$

The integration is over all directions on the spherical surface. As $R\to\infty$, the exponential will oscillate infinitely rapidly as a function of k_c, and may be regarded as averaging to zero. This can be justified by observing that any physical state will contain a distribution of energies over some range, however narrow it may be, and so it will always be necessary to integrate over some small range of k_c, which will contain infinitely many oscillations in the limit $R\to\infty$. A similar argument can be used to eliminate the term involving two plane waves. Alternatively, one can transform the surface integral back into a volume integral, going back from (16.50) to (16.49) for this term, and use the orthogonality of plane waves when $R\to\infty$. Finally, we must evaluate the terms involving a plane wave and a spherical wave, which turn out to be nonzero. The plane wave term of $\Psi'_{\mathbf{b}}{}^{(-)}$ gives rise to the integrals

$$\oiint_{r=R} [e^{i\mathbf{k}_b\cdot\mathbf{r}}]^*\,f_{\mathbf{a}b}{}^{(+)}(\Omega_r)\left[\frac{\partial}{\partial r}\frac{e^{ik_br}}{r}\right] r^2\,d\Omega_r$$
$$-\oiint_{r=R}\left[\frac{\partial}{\partial r}\,e^{i\mathbf{k}_b\cdot\mathbf{r}}\right]^*\,f_{\mathbf{a}b}{}^{(+)}(\Omega_r)\,\frac{e^{ik_br}}{r}\,r^2\,d\Omega_r\,. \tag{16.51}$$

To evaluate them, we use the spherical harmonic expansion of a plane wave, which is equivalent to (16.21):

$$e^{i\mathbf{k}\cdot\mathbf{r}} = 4\pi\sum_\ell\sum_m [Y_\ell{}^m(\Omega_r)]^*\,Y_\ell{}^m(\Omega_k)\,i^\ell\,j_\ell(kr)\,. \tag{16.52}$$

Here Ω_k and Ω_r denote the angles of $\mathbf{k}$ and $\mathbf{r}$, respectively. For $r\to\infty$ we can use the asymptotic form (16.25a),

$$j_\ell(kr) \sim \frac{\sin(kr - \frac{1}{2}\pi l)}{kr} = \frac{1}{2ikr}\left(i^{-l}e^{ikr} - i^\ell e^{-ikr}\right), \tag{16.53}$$

to obtain an asymptotic expansion for a plane wave in the limit $r \to \infty$:

$$e^{i\mathbf{k}\cdot\mathbf{r}} \sim \frac{2\pi}{ikr} \sum_{\ell} \sum_{m} [Y_\ell{}^m(\Omega_r)]^* \, Y_\ell{}^m(\Omega_k)[e^{ikr} - (-1)^\ell e^{-ikr}] \,. \tag{16.54}$$

It consists of outgoing and incoming spherical waves. When it is substituted into (16.51), the incoming (e^{-ikr}) terms of (16.54) will produce a factor e^{i2k_bR}, which oscillates infinitely rapidly as $R \to \infty$, and so may be regarded as averaging to zero. Therefore we may substitute only the e^{ikr} terms of (16.54) into (16.51), which then yields

$$-4\pi \sum_{\ell} \sum_{m} [f_{\mathbf{ab}}{}^{(+)}]_{\ell m} \, Y_\ell{}^m(\Omega_{kb}) + O(R^{-1}) \,, \tag{16.55}$$

where

$$[f_{\mathbf{ab}}{}^{(+)}]_{\ell m} = \oint\!\!\oint [Y_\ell{}^m(\Omega_r)]^* \, f_{\mathbf{ab}}{}^{(+)}(\Omega_r) \, d\Omega_r \,.$$

But this is just a coefficient in the spherical harmonic expansion of the scattering amplitude,

$$f_{\mathbf{ab}}{}^{(+)}(\Omega_{kb}) = \sum_{\ell} \sum_{m} [f_{\mathbf{ab}}{}^{(+)}]_{\ell m} \, Y_\ell{}^m(\Omega_{kb}) \,,$$

so (16.55) is equal to $-4\pi \, f_{\mathbf{ab}}{}^{(+)}(\Omega_{kb})$ in the limit $R \to \infty$. Here Ω_{kb} denotes the angles of the direction $\mathbf{k}_b$. The plane wave term of $\Psi_{\mathbf{a}}{}^{(+)}$ yields a contribution to (16.50) that is similar in form to (16.51), and can be similarly evaluated. Its value, in the limit $R \to \infty$, is $4\pi[f'_{\mathbf{ba}}{}^{(-)}(-\Omega_{ka})]^*$, where $-\Omega_{ka}$ denotes the angles of the direction of $-\mathbf{k}_a$. Combining these results, restoring the constant factors that have been omitted for brevity, and taking the limit $R \to \infty$ in (16.48), we finally obtain

$$\begin{aligned} &\frac{2\pi\hbar^2}{\mu} \, |A|^2 \left\{ [f'_{\mathbf{ba}}{}^{(-)}(-\Omega_{ka})]^* - f_{\mathbf{ab}}{}^{(+)}(\Omega_{kb}) \right\} \\ &\qquad = \langle \, \Psi'_{\mathbf{b}}{}^{(-)} | (V - V') | \Psi_{\mathbf{a}}{}^{(+)} \, \rangle \,, \end{aligned} \tag{16.56}$$

which is the *scattering amplitude theorem* that we have been seeking. The right hand side is finite because V and V' go to zero more rapidly than r^{-1} as $r \to \infty$. The normalization factor A, defined in (16.42) and (16.46), is arbitrary, but does not affect (16.56) because it is implicitly contained as a factor in the scattering state functions on the right. The most common (but not universal) choice is $A = 1$.

The theorem (16.56) has several useful applications. As its first application, we put $V = V'$, so that also $f' = f$. It then follows that

$$[f_{\mathbf{b}a}{}^{(-)}(-\Omega_{ka})]^* = f_{\mathbf{a}b}{}^{(+)}(\Omega_{kb})\,. \tag{16.57}$$

We shall use this result to eliminate the amplitude $f^{(-)}$, which corresponds to a physically unrealistic incoming spherical wave state, in favor of the more intuitively meaningful scattering amplitude $f^{(+)}$. With this understanding, we may *simplify the notation* by writing $f_{\mathbf{a}b} = f_{\mathbf{a}b}{}^{(+)}$.

For the second application, we put $V' = 0$, so that $H' = H_0$ and $H = H_0 + V$. The eigenfunctions in the absence of any scattering potential are given by

$$H_0\,\Phi_{\mathbf{b}} = E\,\Phi_{\mathbf{b}}\,, \quad \Phi_{\mathbf{b}} = A\,e^{i\mathbf{k}_b\cdot\mathbf{r}}\,w_b\,. \tag{16.58}$$

Since there is no scattered wave, there is no distinction between $\Phi^{(+)}$ and $\Phi^{(-)}$. Then (16.56), with $A = 1$, yields

$$f_{\mathbf{a}b}{}^{(+)}(\Omega_{kb}) = -\frac{\mu}{2\pi\hbar^2}\,\langle\,\Phi_{\mathbf{b}}|V|\Psi_{\mathbf{a}}{}^{(+)}\,\rangle\,. \tag{16.59}$$

Although Ω_{kb} (the angles of $\mathbf{k}_b$) appears explicitly as a variable, this amplitude also depends implicity on the fixed direction of $\mathbf{k}_a$ (the direction of the incident beam), as is indicated by the subscript $\mathbf{a} = (\mathbf{k}_a, a)$. If the interaction is spherically symmetric, then the amplitude will depend only on the relative direction of $\mathbf{k}_a$ and $\mathbf{k}_b$. This important formula expresses the scattering amplitude in terms of an integral over the scattering potential and the state function. It is a remarkable result, since it relates the asymptotic behavior of the state function at large distance (16.42) to the scattering interaction, which is a short range quantity.

16.4 *Born Approximation and DWBA*

Two useful approximations can be derived from the scattering amplitude theorem of the previous section. The first of them, called the *Born approximation*, is derived by observing that if the scattering potential is weak, then the difference between the operators H_0 and $H = H_0 + V$ is small, and so the difference between their eigenfunctions, $\Phi_{\mathbf{a}}$ and $\Psi_{\mathbf{a}}{}^{(+)}$, should be small too. Hence, from (16.59), we obtain an approximation for the scattering amplitude by replacing $\Psi_{\mathbf{a}}{}^{(+)}$ with $\Phi_{\mathbf{a}}$:

$$f_{\mathbf{a}b}(\Omega_{kb}) \approx -\frac{\mu}{2\pi\hbar^2}\,\langle\,\Phi_{\mathbf{b}}|V|\Phi_{\mathbf{a}}\,\rangle\,, \tag{16.60}$$

where $\Phi_{\mathbf{a}} = e^{i\mathbf{k}_a\cdot\mathbf{r}}\,w_a$ and $\Phi_{\mathbf{b}} = e^{i\mathbf{k}_b\cdot\mathbf{r}}\,w_b$ are eigenvectors of H_0.

The second approximation, called the *distorted wave Born approximation* (DWBA), is useful when the scattering potential can be written as $V = V_1 + V_2$, and V_2 is small. We shall apply the theorem (16.56) to the Hamiltonians $H = H_0 + V_1$ and $H' = H + V_2$. The (16.56) becomes

$$\frac{2\pi\hbar^2}{\mu}\left\{[f'_{\mathbf{ba}}{}^{(-)}(-\Omega_{ka})]^* - f_{\mathbf{ab}}{}^{(+)}(\Omega_{kb})\right\} = -\langle\, \Psi'_{\mathbf{b}}{}^{(-)}|V_2|\Psi_{\mathbf{a}}{}^{(+)}\,\rangle\,.$$

But (16.57) can be used to transform the first of the scattering amplitudes, yielding

$$f'_{\mathbf{ab}}{}^{(+)}(\Omega_{kb}) - f_{\mathbf{ab}}{}^{(+)}(\Omega_{kb}) = -\frac{\mu}{2\pi\hbar^2}\langle\, \Psi'_{\mathbf{b}}{}^{(-)}|V_2|\Psi_{\mathbf{a}}{}^{(+)}\,\rangle\,. \tag{16.61}$$

This is an exact expression for the change in the scattering amplitude due to the extra scattering potential V_2. If V_2 is small, we may replace the eigenfunction of H', $\Psi'_{\mathbf{b}}{}^{(-)}$, by the corresponding eigenfunction of H, and so obtain the approximation

$$f'_{\mathbf{ab}}{}^{(+)}(\Omega_{kb}) \approx f_{\mathbf{ab}}{}^{(+)}(\Omega_{kb}) - \frac{\mu}{2\pi\hbar^2}\langle\, \Psi_{\mathbf{b}}{}^{(-)}|V_2|\Psi_{\mathbf{a}}{}^{(+)}\,\rangle\,, \tag{16.62}$$

where the right hand side involves only quantities corresponding to the Hamiltonian $H = H_0 + V_1$. This is the DWBA. It is useful if the scattering problem for H is simpler than that for H'.

Example (1): Spin–spin interaction

As an example of a problem involving both elastic and inelastic scattering, we consider two particles of spin $s = \frac{1}{2}$ whose interaction is of the form $V = V_0(r) + V_s(r)\,\boldsymbol{\sigma}^{(1)}\cdot\boldsymbol{\sigma}^{(2)}$. Both the orbital and spin interactions are of short range in the interparticle separation r. To use the Born approximation (16.60) we evaluate the matrix element

$$\langle\, \Phi_{\mathbf{b}}|V|\Phi_{\mathbf{a}}\,\rangle = v_0(\mathbf{k}_b - \mathbf{k}_a)\,\langle\, w_b|w_a\,\rangle + v_s(\mathbf{k}_b - \mathbf{k}_a)\,\langle\, w_b|\boldsymbol{\sigma}^{(1)}\cdot\boldsymbol{\sigma}^{(2)}|w_a\,\rangle \tag{16.63}$$

where

$$v_0(\mathbf{q}) = \int \exp\left(-i\mathbf{q}\cdot\mathbf{r}\right) V_0(r)\, d^3\mathbf{r}\,,$$

and a similar definition relates $v_s(\mathbf{q})$ to $V_s(r)$. The internal state vector $|w_a\,\rangle$ is a two-particle spin state vector: $|\uparrow\uparrow\rangle, |\uparrow\downarrow\rangle, |\downarrow\uparrow\rangle$, or $|\downarrow\downarrow\rangle$, where the arrows refer to the z components of spin of the two particles. (A more formal mathematical notation would be $|\uparrow\downarrow\rangle = |\uparrow\rangle \otimes |\downarrow\rangle$, etc.)

The eigenvectors of the operator $\boldsymbol{\sigma}^{(1)}\cdot\boldsymbol{\sigma}^{(2)}$ are the *triplet* $\{|\uparrow\uparrow\rangle, (|\uparrow\downarrow\rangle + |\downarrow\uparrow\rangle)/\sqrt{2}, |\downarrow\downarrow\rangle\}$ and the *singlet* $(|\uparrow\downarrow\rangle - |\downarrow\uparrow\rangle)/\sqrt{2}$, and its eigenvalues are 1 and -3, respectively. (The triplet and singlet vectors are eigenvectors of total spin, with $s = 1$ and $s = 0$, respectively. These were discussed in Sec. 7.7, using a slightly different notation.) Therefore if in the initial state the spins of the two particles are parallel, there can be no change of spin state, since it is an eigenstate of the Hamiltonian. Hence if $|w_a\rangle = |\uparrow\uparrow\rangle$ the scattering will be purely elastic, and the amplitude in the Born approximation is

$$f_{\uparrow\uparrow,\uparrow\uparrow}(\theta) = -\frac{\mu}{2\pi\hbar^2}\left[v_0(\mathbf{k}_b - \mathbf{k}_a) + v_s(\mathbf{k}_b - \mathbf{k}_a)\right], \tag{16.64}$$

where θ denotes the angle between $\mathbf{k}_b$ and $\mathbf{k}_a$.

For antiparallel spin states we obtain, after a simple calculation,

$$\langle\uparrow\downarrow|\boldsymbol{\sigma}^{(1)}\cdot\boldsymbol{\sigma}^{(2)}|\uparrow\downarrow\rangle = -1, \quad \langle\uparrow\downarrow|\boldsymbol{\sigma}^{(1)}\cdot\boldsymbol{\sigma}^{(2)}|\downarrow\uparrow\rangle = 2.$$

Therefore the elastic scattering amplitude, in the Born approximation, is

$$f_{\uparrow\downarrow,\uparrow\downarrow}(\theta) = -\frac{\mu}{2\pi\hbar^2}\left[v_0(\mathbf{k}_b - \mathbf{k}_a) - v_s(\mathbf{k}_b - \mathbf{k}_a)\right], \tag{16.65}$$

and the inelastic, or spin flip, amplitude is

$$f_{\uparrow\downarrow,\downarrow\uparrow}(\theta) = -2\frac{\mu}{2\pi\hbar^2}\, v_s(\mathbf{k}_b - \mathbf{k}_a). \tag{16.66}$$

No energy change takes place in the flipping of the spins, so the kinetic energy does not change and we have $k_b = k_a$. The scattering cross sections are equal to the absolute squares of these amplitudes.

Example (2): Spin–orbit interaction

In this example we consider the scattering of an electron by a spinless target, but we include the small spin–orbit interaction. The scattering interaction is taken to be $V = V_o(r) + V_{so}$. The physical origin of the spin–orbit interaction is discussed by Fisher (1971). If the other interactions are spherically symmetric, it has the form

$$V_{so} = \frac{\hbar}{4{m_e}^2c^2}\,\frac{1}{r}\,\frac{dV_0(r)}{dr}\,\mathbf{L}\cdot\boldsymbol{\sigma}, \tag{16.67}$$

where m_e is the mass of the electron and c is the speed of light.

We shall assume that the phase shifts of the central potential $V_0(r)$ are known, and shall treat V_{so} as a perturbation, using the DWBA. The scattering amplitude due to $V_0(r)$ alone is of the form (16.29):

$$f_{\mathbf{ab}}(\theta) = \frac{1}{k}\,\delta_{a,b} \sum_{\ell} (2\ell+1)\,\sin(\delta_\ell)\,e^{i\delta_\ell}\,P_\ell(\cos\theta)\,, \tag{16.68}$$

where a and b refer to the spin states, and θ is the angle between $\mathbf{k}_b$ and $\mathbf{k}_a$. The Kronecker delta $\delta_{a,b}$ indicates that there is no change of spin. The outgoing and incoming wave eigenfunctions of $H = H_0 + V_0(r)$ are

$$\Psi_{\mathbf{a}}^{(\pm)} = 4\pi \sum_{\ell,m} i^\ell\, e^{\pm i\delta_\ell}\, R_\ell(r)\,[Y_\ell{}^m(\Omega_{k_a})]^*\, Y_\ell{}^m(\Omega_r)|a\,\rangle\,, \tag{16.69}$$

where Ω_{k_a} and Ω_r denote the angles of $\mathbf{k}_a$ and $\mathbf{r}$, and $|a\,\rangle$ is an electron spin state. The radial function $R_\ell(r)$ is real and has the asymptotic form

$$R_\ell(r) \sim \cos(\delta_\ell)\,j_\ell(kr) - \sin(\delta_\ell)\,n_\ell(kr)$$

in the limit $r \to \infty$. It is left as an exercise to verify that (16.69) does indeed have the asymptotic limit (16.42) for $\Psi^{(+)}$ and (16.46) for $\Psi^{(-)}$.

The additional scattering amplitude due to the spin–orbit interaction is given by the DWBA to be

$$g_{\mathbf{ab}}{}^{(+)}(\Omega_{kb}) = -\frac{\mu}{2\pi\hbar^2}\,\langle\,\Psi_{\mathbf{b}}{}^{(-)}|V_{so}|\Psi_{\mathbf{a}}{}^{(+)}\,\rangle\,. \tag{16.70}$$

This amplitude will now be evaluated by substituting (16.67) and (16.69):

$$g_{\mathbf{ab}}{}^{(+)}(\Omega_{kb}) = -\frac{\mu}{2\pi\hbar^2}\,(4\pi)^2 \sum_{\ell',m'}\sum_{\ell,m} i^{(\ell-\ell')}\,\exp[i(\delta_\ell+\delta_{\ell'})]\,\Lambda_\ell$$
$$\times\, Y_{\ell'}{}^{m'}(\Omega_{k_b})\,[Y_\ell{}^m(\Omega_{k_a})]^* \int [Y_{\ell'}{}^{m'}(\Omega_r)]^*\,\mathbf{L}\cdot(\boldsymbol{\sigma})_{ba}\,Y_\ell{}^m(\Omega_r)\,d\Omega_r\,. \tag{16.71}$$

The radial functions are contained in Λ_ℓ, which we define below. The notation $(\boldsymbol{\sigma})_{ba}$ denotes the standard 2×2 matrix representation of the Pauli spin operators. The orbital angular momentum operator, $\mathbf{L} = -i\hbar\mathbf{r}\times\boldsymbol{\nabla}$, is the generator of rotations, so it does not produce any new ℓ values when it operates on the spherical harmonic to its right.

Hence the angular integral over Ω_r will vanish unless $\ell' = \ell$. We have anticipated this result in defining the radial integral only for $\ell' = \ell$:

$$\Lambda_\ell = \frac{\hbar}{4{m_e}^2 c^2} \int_0^\infty \frac{1}{r} \frac{dV_0(r)}{dr} [R_\ell(r)]^2 r^2 \, dr \,. \tag{16.72}$$

This integral is convergent at the upper limit provided $V_0(r)$ is of short range. At the lower limit, where usually $V_0(r) \propto r^{-1}$, it is convergent for $\ell \geq 1$. There is no term with $\ell = 0$ because the operator $\mathbf{L}$ yields zero in that case.

Using the identity

$$4\pi \sum_m [{Y_\ell}^m(\Omega_k)]^* \, {Y_\ell}^m(\Omega_r) = (2\ell + 1) \, P_\ell \left(\frac{\mathbf{k} \cdot \mathbf{r}}{kr} \right) \tag{16.73}$$

we can simplify (16.71) to

$$\begin{aligned} {g_{\mathbf{ab}}}^{(+)}(\Omega_{kb}) &= -\frac{\mu}{2\pi\hbar^2} \sum_\ell e^{i2\delta_\ell} \, \Lambda_\ell \, (2\ell+1)^2 \\ &\quad \times \int P_\ell \left(\frac{\mathbf{k}_b \cdot \mathbf{r}}{k_b r} \right) \mathbf{L} \cdot (\boldsymbol{\sigma})_{ba} \, P_\ell \left(\frac{\mathbf{k}_a \cdot \mathbf{r}}{k_a r} \right) d\Omega_r \,. \end{aligned} \tag{16.74}$$

Since $P_\ell(\mathbf{k}_a \cdot \mathbf{r}/k_a r)$ depends only on the relative direction of $\mathbf{k}_a$ and $\mathbf{r}$, a rotation of $\mathbf{r}$ is equivalent to the opposite rotation of $\mathbf{k}_a$. Now $\mathbf{L} = i\hbar\mathbf{r} \times \boldsymbol{\nabla}$ is the generator of rotations in $\mathbf{r}$-space, and so $\mathbf{L}^{(k)} = -i\hbar\mathbf{k} \times \partial/\partial\mathbf{k}$ generates similar rotations in $\mathbf{k}$-space, therefore we have the relation $\mathbf{L} \, P_\ell(\mathbf{k} \cdot \mathbf{r}/kr) = -\mathbf{L}^{(k)} \, P_\ell(\mathbf{k} \cdot \mathbf{r}/kr)$. Using this relation and the identity (16.73), we can simplify the integral in (16.74):

$$\begin{aligned} &\int P_\ell \left(\frac{\mathbf{k}_b \cdot \mathbf{r}}{k_b r} \right) \mathbf{L} \cdot (\boldsymbol{\sigma})_{ba} \, P_\ell \left(\frac{\mathbf{k}_a \cdot \mathbf{r}}{k_a r} \right) d\Omega_r \\ &\quad = -(\boldsymbol{\sigma})_{ba} \cdot \mathbf{L}^{(k_a)} \int P_\ell \left(\frac{\mathbf{k}_b \cdot \mathbf{r}}{k_b r} \right) P_\ell \left(\frac{\mathbf{k}_a \cdot \mathbf{r}}{k_a r} \right) d\Omega_r \\ &\quad = -(\boldsymbol{\sigma})_{ba} \cdot \mathbf{L}^{(ka)} \left(\frac{4\pi}{2\ell+1} \right)^2 \\ &\qquad \times \int \sum_{m'} {Y_\ell}^{m'}(\Omega_{kb}) \, [{Y_\ell}^{m'}(\Omega_r)]^* \sum_m [{Y_\ell}^m(\Omega_{ka})]^* \, {Y_\ell}^m(\Omega_r) \, d\Omega_r \end{aligned}$$

$$= -(\boldsymbol{\sigma})_{ba} \cdot \mathbf{L}^{(ka)} \left(\frac{4\pi}{2\ell + l}\right)^2 \sum_m Y_\ell^{m'}(\Omega_{kb}) \, [Y_\ell^m(\Omega_{ka})]^*$$

$$= -4\pi(2\ell + 1)^{-1} \, \mathbf{L}^{(k_a)} \, P_\ell \left(\frac{\mathbf{k}_a \cdot \mathbf{k}_b}{k_a k_b}\right) . \tag{16.75}$$

Let us introduce the scattering angle θ, defined by

$$\cos\theta = \frac{\mathbf{k}_a \cdot \mathbf{k}_b}{k_a k_b} . \tag{16.76}$$

Then we have

$$\begin{aligned} \mathbf{L}^{(k_a)} \, P_\ell(\cos\theta) &= -i\hbar \mathbf{k}_a \times \frac{\partial}{\partial \mathbf{k}_a} \, P_\ell(\cos\theta) \\ &= -i\hbar \mathbf{k}_a \times \frac{\partial(\cos\theta)}{\partial \mathbf{k}_a} \, \frac{dP_\ell(\cos\theta)}{d(\cos\theta)} \\ &= -i\hbar \left(\frac{\mathbf{k}_a \times \mathbf{k}_b}{k_a k_b}\right) \frac{dP_\ell(\cos\theta)}{d(\cos\theta)} . \end{aligned} \tag{16.77}$$

Substituting this sequence of results back to (16.74), we finally obtain the additional scattering amplitude due to the spin–orbit interaction:

$$\begin{aligned} g_{\mathbf{ab}}{}^{(+)}(\Omega_{kb}) &= -i\hbar(\boldsymbol{\sigma})_{ba} \cdot (\mathbf{k}_a \times \mathbf{k}_b)(k_a k_b)^{-1} \frac{2\mu}{\hbar^2} \\ &\quad \times \sum_\ell \exp(i2\delta_\ell) \, \Lambda_\ell \, (2\ell + 1) \frac{dP_\ell(\cos\theta)}{d(\cos\theta)} . \end{aligned} \tag{16.78a}$$

The total scattering amplitude is the sum of (16.68) and (16.78), and the scattering cross section is

$$\sigma_{a \to b}(\mathbf{k}_a, \mathbf{k}_b) = |f_{ab} + g_{ab}|^2 , \tag{16.78b}$$

since $k_a = k_b$. The most interesting part of the spin–orbit amplitude (16.78a) is the factor $(\boldsymbol{\sigma})_{ba} \cdot (\mathbf{k}_a \times \mathbf{k}_b)$. We shall choose the spin states to be the eigenvectors of σ_z, denoted as $|\uparrow\rangle$ and $|\downarrow\rangle$. If the initial and final spin states are the same, then only the z component of $\boldsymbol{\sigma}$ contributes, and the factor becomes $(\boldsymbol{\sigma})_{\uparrow\uparrow} \cdot (\mathbf{k}_a \times \mathbf{k}_b) = (\mathbf{k}_a \times \mathbf{k}_b)_z$, or $(\boldsymbol{\sigma})_{\downarrow\downarrow} \cdot (\mathbf{k}_a \times \mathbf{k}_b) = -(\mathbf{k}_a \times \mathbf{k}_b)_z$. If the spin states are different, then σ_x and σ_y contribute. The spin flip cross section is equal to $\sigma_{\uparrow \to \downarrow} = |g_{\uparrow\downarrow}|^2$. Of greater interest is the non-spin flip, or elastic scattering cross section, $\sigma_{\uparrow \to \uparrow}(\mathbf{k}_a, \mathbf{k}_b) = |f_{\uparrow\uparrow} + g_{\uparrow\uparrow}|^2$. It is apparent that the potential scattering amplitude $f_{\uparrow\uparrow}$ is symmetric under interchange of $\mathbf{k}_a$ and $\mathbf{k}_b$, whereas the spin–orbit amplitude $g_{\uparrow\uparrow}$ is antisymmetric. Therefore the probabilities of the scattering events $\mathbf{k}_a \to \mathbf{k}_b$ and $\mathbf{k}_a \to \mathbf{k}_b$ will not be equal. This inequality is known as *skew scattering*, and it causes the principle of detailed balance to fail. It is caused by the interference between the two amplitudes $f_{\uparrow\uparrow}$ and $g_{\uparrow\uparrow}$.

It is worthwhile to examine the symmetries of the operator $i\boldsymbol{\sigma}\cdot(\mathbf{k}_a \times \mathbf{k}_b)$, which is responsible for the antisymmetry of the spin–orbit amplitude, and hence for the existence of skew scattering. It is a scalar product of three vectors, and hence is rotationally invariant. It is invariant under space inversion, with both $\mathbf{k}_a$ and $\mathbf{k}_b$ changing sign. It is invariant under time reversal, under which all four factors change sign. It is not easy to construct a function that obeys all of these symmetries and yet is not symmetric under interchange of $\mathbf{k}_a$ and $\mathbf{k}_b$, and so there are not many examples of skew scattering. Skew scattering can be important in the *Hall effect*, which is a phenomenon that accompanies electrical conduction in crossed electric and magnetic fields (Ashcroft and Mermin, 1976; Ballentine and Huberman, 1977).

Suppose that we had calculated the scattering cross section by means of the Born approximation instead of the DWBA. Then, apart from a constant factor, the cross section would have been given by $|\langle \Phi_{\mathbf{b}}|(V_0+V_{so})|\Phi_{\mathbf{a}}\rangle|^2$. Since both terms of the scattering interaction are Hermitian, we have $\langle \Phi_{\mathbf{a}}|(V_0 + V_{so})|\Phi_{\mathbf{b}}\rangle = \langle \Phi_{\mathbf{b}}|(V_0+V_{so})|\Phi_{\mathbf{a}}\rangle^*$, and so the cross section would be symmetric under interchange of $\mathbf{k}_a$ and $\mathbf{k}_b$. Therefore the phenomenon of skew scattering, and the consequent failure of detailed balance, cannot be detected by the Born approximation.

16.5 *Scattering Operators*

The theory of the preceding sections has relied heavily on the coordinate representation of the Schrödinger equation. That is a natural thing to do, because the asymptotic conditions at large separation play an essential role. However, it is possible to formulate an elegant and general theory in terms of operators, avoiding for the most part any need to invoke detailed representations. This section presents an outline of the operator formalism.

As in the preceding sections, we consider a Hamiltonian of the form

$$H = H_0 + V\,, \tag{16.79}$$

where V is the scattering interaction, and H_0 is translationally invariant so that momentum would be conserved were it not for V. We introduce two resolvent operators,

$$G(z) = (z - H)^{-1}, \quad G_0(z) = (z - H_0)^{-1}. \tag{16.80}$$

It is necessary to take the energy parameter z to be complex, in general, because the inverse operator does not exist when z is equal to an eigenvalue of H or H_0, respectively. It will be shown that the operators (16.80) have different limits when z is allowed to approach the positive real axis from above and from below in the complex plane.

We now define an operator $T(z)$ by the relation

$$G(z) = G_0(z) + G_0(z)\, T(z)\, G_0(z). \tag{16.81}$$

$T(z)$ is called, rather unimaginatively, the *t matrix*, which is short for "transition matrix". Its properties and its relation to scattering will now be demonstrated. From the definition, $G_0\, T\, G_0 = G - G_0$, we deduce that

$$\begin{aligned} T &= G_0^{-1} G\, G_0^{-1} - G_0^{-1} \\ &= (z - H_0)(G\, G_0^{-1} - 1) \\ &= (z - H_0)(G\, G_0^{-1} - G\, G^{-1}) \\ &= (z - H_0) G\, V. \end{aligned}$$

Because the first line of this calculation reads the same from right to left as from left to right, a mirror image sequence of steps will lead to

$$T = V\, G(z - H_0).$$

This left–right symmetry exists even though the factors do not commute. From these results we obtain

$$G_0 T = G\, V, \quad T\, G_0 = V\, G. \tag{16.82}$$

Hence (16.81) may be written as

$$G(z) = G_0(z) + G(z)\, V\, G_0(z) = G_0(z) + G_0(z)\, V\, G(z). \tag{16.83}$$

Now, from (16.82), or from the intermediate results above it, we obtain

$$\begin{aligned} T - V &= G_0^{-1} G V - V \\ &= (G_0^{-1} G - 1) V \\ &= (G_0^{-1} - G^{-1}) G V \\ &= V G V , \end{aligned}$$

and hence

$$T = V + V G V . \tag{16.84}$$

Equation (16.83) can easily be solved iteratively to obtain a formal perturbation series,

$$G = G_0 + G_0 V G_0 + G_0 V G_0 V G_0 + \cdots , \tag{16.85}$$

which can be substituted into (16.84) to obtain

$$T = V + V G_0 V + V G_0 V G_0 V + \cdots . \tag{16.86}$$

These series can be used as the basis for systematic approximations.

From the first equation (16.83) we have $G(z) = [1 + G(z) V] G_0(z)$, and hence

$$G(z) G_0(z)^{-1} = 1 + G(z) V . \tag{16.87}$$

Rewriting the second equation (16.83) as $G_0(z) = G(z) - G_0(z) V G(z) = [1 - G_0(z) V] G(z)$, we obtain

$$G_0(z) G(z)^{-1} = 1 - G_0(z) V . \tag{16.88}$$

Multiplying (16.87) and (16.88) yields

$$[1 + G(z) V][1 - G_0(z) V] = [1 - G_0(z) V][1 + G(z) V] = 1 . \tag{16.89}$$

One must remember that these relations hold for z not on the real axis, and that any use of them for $z = E$ (real) must be done as a limiting process, either from above or from below the real axis.

To relate the t matrix to scattering, we introduce the scattering eigenvectors, rewriting (16.40) as

$$(E - H_0)|\Psi_{\mathbf{a}}{}^{(+)}\rangle = V|\Psi_{\mathbf{a}}{}^{(+)}\rangle . \tag{16.90}$$

From this we obtain the *Lippmann–Schwinger* equation,

$$|\Psi_{\mathbf{a}}{}^{(+)}\rangle = |\Phi_{\mathbf{a}}\rangle + G_0(E^+) V|\Psi_{\mathbf{a}}{}^{(+)}\rangle . \tag{16.91}$$

Here $|\Phi_{\mathbf{a}}\rangle$ satisfies the homogeneous equation $(E - H_0)|\Phi_{\mathbf{a}}\rangle = 0$. The notation E^+ in $G_0(E^+)$ signifies that we are to take the limit of $G_0(E + i\varepsilon)$ as $\varepsilon \to 0$ through positive values. The compatibility of (16.91) with (16.90) can be verified by operating with $E - H_0$, to prove that any $|\Psi_{\mathbf{a}}{}^{(+)}\rangle$ that is a solution of (16.91) will also satisfy (16.90). However, (16.91) contains more information than does (16.90). In coordinate representation Eq. (16.90) would be a differential equation, for which boundary conditions must be specified to make the solution unique. But (16.91) is an inhomogeneous equation, and the relevant asymptotic boundary conditions are already built into it. If we put $V = 0$ in (16.91) the solution becomes $|\Phi_{\mathbf{a}}\rangle$, which is an eigenvector of the free particle Hamiltonian H_0, and represents the incident beam. Thus the solution of (16.91) will be of the following form: incident beam + scattered wave. By evaluation $G_0(E^+)$ in the limit from the positive imaginary side of the real axis, we ensure that the scattered wave is *outgoing* rather than incoming. (This will be demonstrated below.)

The Lippmann–Schwinger equation (16.91) can be rewritten as

$$[1 - G_0(E^+)\,V]\,|\Psi_{\mathbf{a}}{}^{(+)}\rangle = |\Phi_{\mathbf{a}}\rangle\,,$$

which can be formally solved by means of (16.89):

$$|\Psi_{\mathbf{a}}{}^{(+)}\rangle = [1 + G(E^+)\,V]|\Phi_{\mathbf{a}}\rangle\,. \tag{16.92}$$

Therefore we have $V|\Psi_{\mathbf{a}}{}^{(+)}\rangle = [V + V\,G(E^+)\,V]|\Phi_{\mathbf{a}}\rangle = T(E^+)|\Phi_{\mathbf{a}}\rangle$, where the last step used (16.84). Finally we obtain the connection between the t matrix and the scattering amplitude through (16.59):

$$\begin{aligned}\langle\Phi_{\mathbf{b}}|T(E^+)|\Phi_{\mathbf{a}}\rangle &= \langle\Phi_{\mathbf{b}}|V|\Psi_{\mathbf{a}}{}^{(+)}\rangle \\ &= -\frac{2\pi\hbar^2}{\mu}\,f_{\mathbf{a}b}{}^{(+)}(\Omega_{kb})\,.\end{aligned} \tag{16.93}$$

Outgoing waves and the limit $E + i\varepsilon$

The operator $G_0(z)$ has two limits on the positive real axis, $G_0(E^+)$ and $G_0(E^-)$, obtained from the two limits $z \to E \pm i\varepsilon$ as the nonnegative quantity ε vanishes. We stated that the choice of $G_0(E^+)$ in the Lippmann–Schwinger equation corresponds to outgoing scattered waves, and this will now be demonstrated. For simplicity we shall ignore the internal degrees of freedom of the particles for this calculation. The Lippmann–Schwinger equation (16.91) can then be rewritten in coordinate representation as

$$\Psi_{\mathbf{a}}{}^{(+)}(\mathbf{r}) = \Phi_{\mathbf{a}}(\mathbf{r}) + \int G_0(\mathbf{r},\mathbf{r}';E^+)\, V(\mathbf{r}')\, \Psi_{\mathbf{a}}{}^{(+)}(\mathbf{r}')\, d^3\mathbf{r}\,, \tag{16.94}$$

where $G_0(\mathbf{r},\mathbf{r}';E^+) = \langle\, \mathbf{r}|(E^+ - H_0)^{-1}|\mathbf{r}'\,\rangle$ is called a *Green's function.* We place the superscript (+) on $\Psi_{\mathbf{a}}{}^{(+)}$ in anticipation of the result that we shall obtain, even though we have not yet determined its asymptotic form.

The free particle Hamiltonian is $H_0 = P^2/2\mu$, and its eigenvectors are the momentum eigenvectors, $\mathbf{P}|\hbar\mathbf{k}\,\rangle = \hbar\mathbf{k}|\hbar\mathbf{k}\,\rangle$. The Green's function can be constructed from the spectral representation of the resolvent. Let ε be an arbitrary small positive quantity that will be allowed to vanish at the end of the calculation. Then we have

$$\begin{aligned}
G_0(\mathbf{r},\mathbf{r}';E+i\varepsilon) &= \int \frac{\langle\, \mathbf{r}|\hbar\mathbf{k}\,\rangle\ \langle\, \hbar\mathbf{k}|\mathbf{r}'\,\rangle}{E+i\varepsilon-\hbar^2k^2/2\mu}\, d^3\mathbf{k} \\
&= \frac{2\mu}{\hbar^2(2\pi)^3}\int \frac{\exp[i\mathbf{k}\cdot(\mathbf{r}-\mathbf{r}')]}{\mathcal{K}^2+i\varepsilon-k^2}\, d^3\mathbf{k}\,,\quad \left(\frac{\hbar^2\mathcal{K}^2}{2\mu}=E\right), \\
&= \frac{2\mu}{\hbar^2(2\pi)^3}\int_0^\infty 2\pi \int_0^\pi \frac{\exp(ikR\cos\theta)}{\mathcal{K}^2+i\varepsilon-k^2}\sin\theta\, d\theta\, k^2\, dk\,,\ (R=|\mathbf{r}-\mathbf{r}'|)\,, \\
&= \frac{2\mu}{\hbar^2 4\pi^2 R}\int_0^\infty \frac{e^{ikR}-e^{-ikR}}{i(\mathcal{K}^2+i\varepsilon-k^2)}\, k\, dk\,.
\end{aligned}$$

The last integrand is an even analytic function of k, so we may change the lower limit to $-\infty$, multiply by $\frac{1}{2}$, and use the residue theorem to evaluate the integral. The first term, involving e^{ikR}, vanishes as the imaginary part of k approaches $+\infty$, so we may close the contour of integration from $-\infty$ to ∞ with an infinite semicircle in the upper half of the complex k plane. In the limit $\varepsilon \to 0$ this contour encloses a simple pole at $k = \mathcal{K}$. For the second term we must close the contour with an infinite semicircle in the lower half of the k plane, and it will enclose a pole at $k = -\mathcal{K}$. The final result for the Green's function is

$$G_0(\mathbf{r},\mathbf{r}';E^+) = -\frac{2\mu}{\hbar^2}\,\frac{\exp(i\mathcal{K}|\mathbf{r}-\mathbf{r}'|)}{4\pi|\mathbf{r}-\mathbf{r}'|}\,. \tag{16.95}$$

Had we chosen ε to be negative we would have obtained

$$G_0(\mathbf{r},\mathbf{r}';E^+) = -\frac{2\mu}{\hbar^2}\,\frac{\exp(-i\mathcal{K}|\mathbf{r}-\mathbf{r}'|)}{4\pi|\mathbf{r}-\mathbf{r}'|}\,.$$

We now substitute (16.95) into the Lippmann–Schwinger equation (16.94) and choose the incident wave to be $\Phi_{\mathbf{a}}(\mathbf{r}) = e^{i\mathbf{k}_a\cdot\mathbf{r}}$, where $\mathbf{a}$ now denotes $\mathbf{k}_a$, since internal states are not considered. Then the energy will be $E = \hbar^2 k_a{}^2/2\mu$, and so $\mathcal{K} = k_a$.

$$\Psi_{\mathbf{a}}{}^{(+)}(\mathbf{r}) = e^{i\mathbf{k}_a \cdot \mathbf{r}} - \frac{\mu}{2\pi\hbar^2} \int \frac{\exp(ik_a|\mathbf{r}-\mathbf{r}'|)}{|\mathbf{r}-\mathbf{r}'|} V(\mathbf{r}')\, \Psi_{\mathbf{a}}{}^{(+)}(\mathbf{r}')\, d^3r' \,. \quad (16.96)$$

Now the integration variable r' is effectively confined within the range of the iteration V, so in the limit of large r we can use the approximation $k_a|\mathbf{r}-\mathbf{r}'| \approx kr - \mathbf{k}_b\!\cdot \mathbf{r}'$, where we define $\mathbf{k}_b = k_a\mathbf{r}/r$ to have the magnitude of k_a but the direction of $\mathbf{r}$. Therefore the asymptotic limit of (16.96) for large r is

$$\Psi_{\mathbf{a}}{}^{(+)}(\mathbf{r}) \sim e^{i\mathbf{k}_a \cdot \mathbf{r}} - \frac{\mu}{2\pi\hbar^2} \int e^{-i\mathbf{k}_b \cdot \mathbf{r}'}\, V(\mathbf{r}')\, \Psi_{\mathbf{a}}{}^{(+)}(\mathbf{r}')\, d^3r' \, \frac{e^{ik_a r}}{r} \,, \quad (16.97)$$

which consists of the incident wave plus an outgoing scattered wave. Thus we have shown that the $E + i\varepsilon$ prescription does indeed yield outgoing scattered waves, as was claimed earlier. The coefficient of $e^{ik_a r}/r$ is, by definition, the scattering amplitude. It is of the form $-(\mu/2\pi\hbar^2)\langle\, \Phi_{\mathbf{b}}|V|\Psi_{\mathbf{a}}{}^{(+)}\,\rangle$, so we have also rederived (16.59) for this case of purely elastic scattering.

Properties of the scattering states

We have just shown that the outgoing and incoming scattering states, $\Psi_{\mathbf{a}}{}^{(+)}$ and $\Psi_{\mathbf{b}}^{(-)}$, can be calculated from the Lippmann–Schwinger equation (16.91) using the $E \to E \pm i\varepsilon$ prescription. The scattering functions are eigenfunctions of H,

$$H|\Psi_{\mathbf{a}}{}^{(+)}\,\rangle = E_a|\Psi_{\mathbf{a}}{}^{(+)}\,\rangle\,, \quad H|\Psi_{\mathbf{b}}{}^{(-)}\,\rangle = E_b|\Psi_{\mathbf{b}}{}^{(-)}\rangle\,,$$

but we have already noted that they are not all linearly independent. We shall use the Lippmann–Schwinger equation to derive their orthogonality and linear dependence relations.

From the formal solution (16.92) to the Lippmann–Schwinger equation, and the relation $[G(E^+)]^\dagger = G(E^-)$, we obtain

$$\langle\, \Psi_{\mathbf{a}}{}^{(+)}| = \langle\, \Phi_{\mathbf{a}}|[1 + G(E+i\varepsilon)\, V]^\dagger = \langle\, \Phi_{\mathbf{a}}|[1 + V\, G(E_a - i\varepsilon)]\,.$$

Hence we obtain

$$\begin{aligned} \langle\, \Psi_{\mathbf{a}}{}^{(+)}|\Psi_{\mathbf{b}}{}^{(+)}\,\rangle &= \langle\, \Phi_{\mathbf{a}}|\Psi_{\mathbf{b}}{}^{(+)}\,\rangle + \langle\, \Phi_{\mathbf{a}}|V(E_a - i\varepsilon - H)^{-1}|\Psi_{\mathbf{b}}{}^{(+)}\,\rangle \\ &= \langle\, \Phi_{\mathbf{a}}|\Psi_{\mathbf{b}}{}^{(+)}\,\rangle + (E_a - i\varepsilon - E_b)^{-1}\,\langle\, \Phi_{\mathbf{a}}|V|\Psi_{\mathbf{b}}{}^{(+)}\,\rangle\,. \end{aligned}$$

From (16.91) we have $|\Psi_{\mathbf{b}}{}^{(+)}\,\rangle = |\Phi_{\mathbf{b}}\,\rangle + (E_b + i\varepsilon - H_0)^{-1}\, V|\Psi_{\mathbf{b}}{}^{(+)}\,\rangle$, which we now substitute into the first term of the result above, obtaining

$$\begin{aligned}\langle \Psi_{\mathbf{a}}{}^{(+)}|\Psi_{\mathbf{b}}{}^{(+)}\rangle &= \langle \Phi_{\mathbf{a}}|\Phi_{\mathbf{b}}\rangle + \langle \Phi_{\mathbf{a}}|(E_b + i\varepsilon - H_0)^{-1}\,V|\Psi_{\mathbf{b}}{}^{(+)}\rangle \\ &\quad + (E_a - i\varepsilon - E_b)^{-1}\,\langle \Phi_{\mathbf{a}}|V|\Psi_{\mathbf{b}}{}^{(+)}\rangle \\ &= \langle \Phi_{\mathbf{a}}|\Phi_{\mathbf{b}}\rangle + \{(E_b + i\varepsilon - E_a)^{-1} \\ &\quad + (E_a - i\varepsilon - E_b)^{-1}\}\,\langle \Phi_{\mathbf{a}}|V|\Psi_{\mathbf{b}}{}^{(+)}\rangle \\ &= \langle \Phi_{\mathbf{a}}|\Phi_{\mathbf{b}}\rangle = (2\pi)^3\,\delta(\mathbf{k}_a - \mathbf{k}_b)\,\delta_{a,b}\,. \end{aligned} \tag{16.98}$$

[This normalization is a consequence of the choice $A = 1$ in (16.14), (16.42), and (16.46).] Thus we have shown that the outgoing scattering functions $\{\Psi_{\mathbf{a}}{}^{(+)}\}$ are mutually orthogonal. A similar calculation for the incoming scattering functions $\{\Psi_{\mathbf{b}}{}^{(-)}\}$ shows that

$$\langle \Psi_{\mathbf{a}}{}^{(-)}|\Psi_{\mathbf{b}}{}^{(-)}\rangle = (2\pi)^3\,\delta(\mathbf{k}_a - \mathbf{k}_b)\,\delta_{a,b}\,. \tag{16.99}$$

The linear dependence of the set $\{\Psi_{\mathbf{a}}{}^{(+)}\}$ on the set $\{\Psi_{\mathbf{b}}{}^{(-)}\}$ can be demonstrated by calculating the inner product of an incoming function with an outgoing function, $\langle \Psi_{\mathbf{b}}{}^{(-)}|\Psi_{\mathbf{a}}{}^{(+)}\rangle$. For the vector on the left, substitute $\langle \Psi_{\mathbf{b}}{}^{(-)}| = \langle \Phi_{\mathbf{b}}|[1 + G(E_b - i\varepsilon)\,V]^\dagger$, obtaining

$$\begin{aligned}\langle \Psi_{\mathbf{b}}{}^{(-)}|\Psi_{\mathbf{a}}{}^{(+)}\rangle &= \langle \Phi_{\mathbf{b}}|\Psi_{\mathbf{a}}{}^{(+)}\rangle + \langle \Phi_{\mathbf{b}}|V(E_b + i\varepsilon - H)^{-1}\,|\Psi_{\mathbf{a}}{}^{(+)}\rangle \\ &= \langle \Phi_{\mathbf{b}}|\Psi_{\mathbf{a}}{}^{(+)}\rangle + (E_b + i\varepsilon - E_a)^{-1}\,\langle \Phi_{\mathbf{b}}|V|\Psi_{\mathbf{a}}{}^{(+)}\rangle\,. \end{aligned}$$

We next substitute $|\Psi_{\mathbf{a}}{}^{(+)}\rangle = |\Phi_{\mathbf{a}}\rangle + (E_a + i\varepsilon - H_0)^{-1}\,V|\Psi_{\mathbf{a}}{}^{(+)}\rangle$ into the first term of this expression, obtaining

$$\begin{aligned}\langle \Psi_{\mathbf{b}}{}^{(-)}|\Psi_{\mathbf{a}}{}^{(+)}\rangle &= \langle \Phi_{\mathbf{b}}|\Phi_{\mathbf{a}}\rangle + \langle \Phi_{\mathbf{b}}|(E_a + i\varepsilon - H_0)^{-1}\,V|\Psi_{\mathbf{a}}{}^{(+)}\rangle \\ &\quad + (E_b + i\varepsilon - E_a)^{-1}\,\langle \Phi_{\mathbf{b}}|V|\Psi_{\mathbf{a}}{}^{(+)}\rangle \\ &= \langle \Phi_{\mathbf{b}}|\Phi_{\mathbf{a}}\rangle + \{(E_a + i\varepsilon - E_b)^{-1} \\ &\quad + (E_b + i\varepsilon - E_a)^{-1}\}\,\langle \Phi_{\mathbf{b}}|V|\Psi_{\mathbf{a}}{}^{(+)}\rangle \\ &= \langle \Phi_{\mathbf{b}}|\Phi_{\mathbf{a}}\rangle - i\,2\,\varepsilon\{(E_a - E_b)^2 + \varepsilon^2\}^{-1}\,\langle \Phi_{\mathbf{b}}|V|\Psi_{\mathbf{a}}{}^{(+)}\rangle\,. \end{aligned} \tag{16.100}$$

In the limit $\varepsilon \to 0$ this becomes

$$\begin{aligned}\langle \Psi_{\mathbf{b}}{}^{(-)}|\Psi_{\mathbf{a}}{}^{(+)}\rangle &= \langle \Phi_{\mathbf{b}}|\Phi_{\mathbf{a}}\rangle - i\,2\pi\,\delta(E_a - E_b)\,\langle \Phi_{\mathbf{b}}|V|\Psi_{\mathbf{a}}{}^{(+)}\rangle \\ &= (2\pi)^3\,\delta(\mathbf{k}_a - \mathbf{k}_b)\,\delta_{a,b} - i\,2\pi\,\delta(E_a - E_b)\,\langle \Phi_{\mathbf{b}}|T(E_a{}^+)|\Phi_{\mathbf{a}}\rangle\,. \end{aligned} \tag{16.101}$$

In the last line we have introduced the t matrix from (16.93).

The two sets of scattering functions $\{\Psi_{\mathbf{a}}{}^{(+)}\}$ and $\{\Psi_{\mathbf{b}}{}^{(-)}\}$ are linearly dependent on each other, and it is possible to express the members of one set as linear combinations of the other. Neither set is complete, since they span only the subspace of positive energy eigenfunctions of H, but both can be completed by including the bound states of H, $\{\Psi_n{}^{(B)}\}$, which span the negative energy subspace.

Let us define the S *matrix*:

$$S_{\mathbf{b},\mathbf{a}} = \langle \Psi_{\mathbf{b}}{}^{(-)} | \Psi_{\mathbf{a}}^{(+)} \rangle \,. \tag{16.102}$$

Then, in view of the orthogonality relation (16.99), the expression for the outgoing scattering functions in terms of the incoming functions is

$$|\Psi_{\mathbf{a}}{}^{(+)} \rangle = (2\pi)^{-3} \sum_b \int |\Psi_{\mathbf{b}}{}^{(-)} \rangle \, S_{\mathbf{b},\mathbf{a}} \, d^3k_b \,, \tag{16.103}$$

where the sum is over the discrete internal states. According to (16.101), $\Psi_{\mathbf{a}}{}^{(+)}$ and $\Psi_{\mathbf{b}}^{(-)}$ are orthogonal if $E_a \neq E_b$, so in fact only functions belonging to the same energy are mixed in (16.103), but it is inconvenient to indicate this explicitly in the notation.

Unitarity of the S matrix

Because the S matrix is the linear transformation between two orthogonal sets of functions, which span the same space, it follows that it must be unitary. This can be demonstrated more easily if we introduce an abbreviated notation:

$$\begin{aligned} \sum_{\mathbf{b}} &\leftrightarrow (2\pi)^{-3} \int d^3k_b \sum_b \,, \\ \mathbf{b} &\leftrightarrow (\mathbf{k}_b, b) \,, \\ \delta_{\mathbf{a},\mathbf{b}} &\leftrightarrow (2\pi)^3 \, \delta(\mathbf{k}_a - \mathbf{k}_b) \, \delta_{a,b} \,. \end{aligned} \tag{16.104}$$

Then (16.103) can be symbolically written as

$$|\Psi_{\mathbf{a}}{}^{(+)} \rangle = \sum_{\mathbf{b}} |\Psi_{\mathbf{b}}{}^{(-)} \rangle S_{\mathbf{b},\mathbf{a}} = \sum_{\mathbf{b}} |\Psi_{\mathbf{b}}{}^{(-)} \rangle \langle \Psi_{\mathbf{b}}{}^{(-)} | \Psi_{\mathbf{a}}{}^{(+)} \rangle$$

and the inverse relation can similarly be written as

$$|\Psi_{\mathbf{b}}{}^{(-)} \rangle = \sum_{\mathbf{a}} |\Psi_{\mathbf{a}}{}^{(+)} \rangle \langle \Psi_{\mathbf{a}}{}^{(+)} | \Psi_{\mathbf{b}}{}^{(-)} \rangle = \sum_{\mathbf{a}} |\Psi_{\mathbf{a}}{}^{(+)} \rangle \, (S^{-1})_{\mathbf{a},\mathbf{b}} \,.$$

Therefore we must have

$$(S^{-1})_{\mathbf{a},\mathbf{b}} = \langle \Psi_{\mathbf{a}}{}^{(+)}|\Psi_{\mathbf{b}}{}^{(-)}\rangle = (S_{\mathbf{b},\mathbf{a}})^* \,, \tag{16.105}$$

which is to say that the S matrix is unitary. Note that $S_{\mathbf{b},\mathbf{a}}$ is a *unitary matrix*, rather than the matrix representation of a unitary operator, because it has been defined only on the positive energy scattering functions, which are not a complete set if H has any bound states.

The S matrix is related to the t matrix, and hence to the scattering amplitudes, through (16.102) and (16.101):

$$S_{\mathbf{b},\mathbf{a}} = (2\pi)^3\,\delta(\mathbf{k}_a - \mathbf{k}_b)\,\delta_{a,b} - i\,2\pi\,\delta(E_a - E_b)\,\langle \Phi_{\mathbf{b}}|T(E_a{}^+)|\Phi_{\mathbf{a}}\rangle\,. \tag{16.106}$$

(Notice that the S matrix elements are defined only between states of equal total energy, whereas the t matrix has elements between any two states.) In the abbreviated notation (16.104), the S matrix is denoted as $S_{\mathbf{b},\mathbf{a}} = \boldsymbol{\delta}_{\mathbf{b},\mathbf{a}} - i\,2\pi\,\delta(E_a - E_b)\,T_{\mathbf{b},\mathbf{a}}$, with $T_{\mathbf{b},\mathbf{a}} = \langle \Phi_{\mathbf{b}}|T(E_a{}^+)|\Phi_{\mathbf{a}}\rangle$. The unitary matrix condition $\mathbf{S}^\dagger\,\mathbf{S} = 1$ becomes $\sum_c (S_{\mathbf{c},\mathbf{b}})^*\,S_{\mathbf{c},\mathbf{a}} = \boldsymbol{\delta}_{\mathbf{b},\mathbf{a}}$, from which we obtain

$$\begin{aligned}\boldsymbol{\delta}_{\mathbf{b},\mathbf{a}} &= \sum_{\mathbf{c}} [\boldsymbol{\delta}_{\mathbf{b},\mathbf{c}} + i\,2\pi\,\delta(E_b - E_c)\,(T_{\mathbf{c},\mathbf{b}})^*]\,[\boldsymbol{\delta}_{\mathbf{a},\mathbf{c}} - i\,2\pi\,\delta(E_a - E_c)\,T_{\mathbf{c},\mathbf{a}}]\\ &= \boldsymbol{\delta}_{\mathbf{b},\mathbf{a}} + i\,2\pi\,\delta(E_a - E_b)\,[(T_{\mathbf{a},\mathbf{b}})^* - T_{\mathbf{b},\mathbf{a}}]\\ &\quad + 4\pi^2 \sum_{\mathbf{c}} \delta(E_b - E_c)\,\delta(E_a - E_c)\,(T_{\mathbf{c},\mathbf{b}})^*\,T_{\mathbf{c},\mathbf{a}}\,.\end{aligned}$$

We may put $\delta(E_b - E_c)\,\delta(E_a - E_c) = \delta(E_a - E_b)\,\delta(E_c - E_a)$, since the effect of either part of δ functions is to require $E_a = E_b = E_c$. Therefore we obtain

$$\sum_{\mathbf{c}} \delta(E_c - E_a)\,(T_{\mathbf{c},\mathbf{b}})^*\,T_{\mathbf{c},\mathbf{a}} = (2\pi i)^{-1}\,[(T_{\mathbf{a},\mathbf{b}})^* - T_{\mathbf{b},\mathbf{a}}] \tag{16.107}$$

as the condition on the t matrix that is imposed by unitarity of the S matrix.

We shall evaluate (16.107) for $\mathbf{a} = \mathbf{b}$. On the left side of (16.107) we have

$$\begin{aligned}\sum_{\mathbf{c}} \delta(E_c - E_a)|T_{\mathbf{c},\mathbf{a}}|^2 &= (2\pi)^{-3} \sum_{c} \int d\Omega_{kc} \int k_c{}^2\,dk_c\,\delta(E_c - E_a)|T_{\mathbf{c},\mathbf{a}}|^2\\ &= (2\pi)^{-3} \sum_{c} \frac{\mu}{\hbar^2}\,k_c \int d\Omega_{kc}|T_{\mathbf{c},\mathbf{a}}|^2\,.\end{aligned} \tag{16.108}$$

In the last line, k_c takes the value that is required by the energy conservation condition: $\hbar^2 k_c{}^2/2\mu + e_c = \hbar^2 k_a{}^2/2\mu + e_a$. Now the differential cross section for scattering into channel c is $\sigma_{a\to c}(\Omega_{kc}) = (k_c/k_a)|f_{\mathbf{ac}}(\Omega_{kc})|^2 = (\mu/2\pi\hbar^2)^2(k_c/k_a)|T_{\mathbf{c,a}}|^2$, so (16.108) is equal to $(\hbar^2/2\pi\mu)k_a \sum_c \int d\Omega_{kc} \times \sigma_{a\to c}(\Omega_{kc}) = (\hbar^2/2\pi\mu)k_a\, \sigma_T(\mathbf{a})$, where $\sigma_T(\mathbf{a}) = \sum_c \int d\Omega_{kc}\, \sigma_{a\to c}(\Omega_{kc})$ is the *total cross section* for scattering from the initial state **a**, integrated over all scattering angles and summed over all channels. Putting $\mathbf{a} = \mathbf{b}$ on the right side of (16.107), we have $-\pi^{-1}\,\text{Im}\,[T_{\mathbf{b,a}}] \to (2\hbar^2/\mu)\,\text{Im}\,[f_{\mathbf{aa}}(\theta = 0)]$. Therefore

$$\sigma_T(\mathbf{a}) = \frac{4\pi}{k_a}\,\text{Im}\,[f_{\mathbf{aa}}(\theta = 0)]\,. \tag{16.109}$$

Stated in words, this relation says that the total cross section for scattering into all channels, both elastic and inelastic, is equal to $4\pi/k_a$ multiplied by the imaginary part of the elastic scattering amplitude in the forward direction. This relation, which we have derived from the unitary nature of the S matrix, is often called the *optical theorem*, because an analogous theorem for light scattering was known before quantum mechanics.

One consequence of the optical theorem is that a purely inelastic scatterer is impossible. For example, a perfectly absorbing target is impossible, and even a black hole must produce some amplitude for elastic scattering. It is easily verified that the phase shift expressions (16.29) and (16.31) for scattering by a central potential satisfy the optical theorem, even if the phase shifts are not calculated exactly. On the other hand, the Born approximation (16.60) does not satisfy the optical theorem.

Symmetries of the S matrix

It is clear from (16.106) that the S matrix carries the same information about scattering probabilities as does the t matrix or the scattering amplitude. However, the compact form (16.102) of the S matrix, $S_{\mathbf{b,a}} = \langle\, \Psi_{\mathbf{b}}{}^{(-)}|\Psi_{\mathbf{a}}{}^{(+)}\,\rangle$, makes it particularly convenient for studying the consequences of symmetry. The S matrix is a function of the Hamiltonian H, and therefore the S matrix is invariant under all the transformations that leave H invariant.

Consider the effect of *time reversal* (Sec. 13.3). The time reversal operation involves the taking of complex conjugates. Its effect is to transform $|\Psi_{\mathbf{a}}{}^{(+)}\,\rangle$ into $|\Psi_{T\mathbf{a}}{}^{(-)}\,\rangle$ and $|\Psi_{\mathbf{b}}{}^{(-)}\,\rangle$ into $|\Psi_{T\mathbf{b}}{}^{(+)}\,\rangle$. If $\mathbf{a} = (\mathbf{k}_a, a)$ then $T\mathbf{a} = (-\mathbf{k}_a, Ta)$, where a denotes the labels of the internal state and Ta denotes the labels of the time-reversed internal state. (Because the symbol T could represent the time reversal operator or the t matrix, we shall not use the t matrix in this discussion of symmetries.) Because the time reversal operator is anti-unitary, it follows that $\langle \Psi_{T\mathbf{b}}{}^{(+)} | \Psi_{T\mathbf{a}}{}^{(-)} \rangle = \langle \Psi_{\mathbf{b}}{}^{(-)} | \Psi_{\mathbf{a}}{}^{(+)} \rangle^*$ (see Problem 13.4). Therefore it follows that

$$S_{T\mathbf{a},T\mathbf{b}} = S_{\mathbf{b},\mathbf{a}} \,. \tag{16.110}$$

The meaning of this relation is illustrated in Fig. 16.4. "Time reversal" is more accurately described as *motion reversal.* The scattering event (b) in the figure is derived from the event (a) by interchanging the initial and the final states, and reversing all velocities and spins. Time reversal invariance implies that these two events have equal scattering amplitudes. Note that the equality is between the *amplitudes* rather than the cross sections. In view of the relation (16.43) between scattering amplitudes and cross sections, we have

$$k_a{}^2\, \sigma_{a\to b} = k_b{}^2\, \sigma_{Tb\to Ta} \,. \tag{16.111}$$

The effect of *space inversion* on the event (a) is to reverse the velocities and leave the spins unchanged, as shown in (c). The cross sections for the events (a) and (c) are equal.

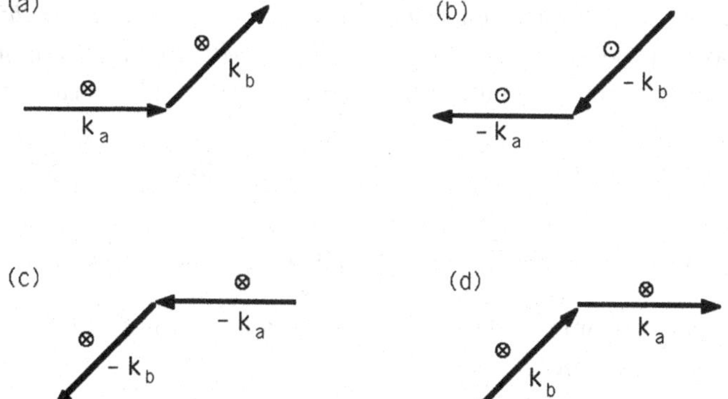

Fig. 16.4 (a) A collision event; (b) time-reversed collision; (c) space-inverted collision; (d) inverse collision. The transition probabilities of (a), (b), and (c) are equal.

The event (d) in the figure is the *inverse collision* of (a), obtained from (a) by interchanging the initial and the final states. We saw in Sec. 16.4 that the direct and inverse collision rates need not be equal, the difference being described as *skew scattering.* Thus, of the four collision events in Fig. 16.4, the cross sections for the first three are related by symmetry, but the fourth is not related. If, however, the particles were spinless, it is apparent from the figure that (d) could be derived from (b) by a rotation in the plane of the vectors.

Therefore a central potential that does not affect spin cannot produce skew scattering. For this common special case, the cross sections for the direct and inverse collisions will be equal. This is called *detailed balance*. But detailed balance does not hold generally.

16.6 *Scattering Resonances*

The scattering cross sections can exhibit a great variety of behaviors as a function of energy. One of the most striking is the appearance of a sharp peak superimposed on a smooth background. This occurs when one of the phase shifts passes rapidly through $\pi/2$. The nature and cause of this phenomenon are the subject of this section. We shall treat only scattering by a spherical potential, and shall neglect any internal degrees of freedom of the particles; however the phenomenon of resonant scattering also occurs in more general systems.

The phase shift δ_ℓ is determined by the logarithmic derivative of the partial wave function, $\gamma_\ell = (1/R_\ell)(dR_\ell/dr)$, as is shown by Eqs. (16.32)–(16.34), and so we must study the energy dependence of γ_ℓ. Consider the partial wave equation (16.23) for two energies, $E_1 = \hbar^2 {k_1}^2/2\mu$ and $E_2 = \hbar^2 {k_2}^2/2\mu$. The corresponding solutions are R_{ℓ,E_1} and R_{ℓ,E_2}. Multiplying the first equation by R_{ℓ,E_2} and the second equation by R_{ℓ,E_1}, and then subtracting, we obtain

$$({k_1}^2 - {k_2}^2)\, R_{\ell,E_1}\, R_{\ell,E_2} = R_{\ell,E_1} \frac{1}{r^2}\frac{d}{dr} r^2 \frac{d}{dr} R_{\ell,E_2} - R_{\ell,E_2} \frac{1}{r^2}\frac{d}{dr} r^2 \frac{d}{dr} R_{\ell,E_1}\,,$$

$$({k_1}^2 - {k_2}^2)\, R_{\ell,E_1}\, R_{\ell,E_2}\, r^2 = \frac{d}{dr}\left[r^2 \left[R_{\ell,E_1} \frac{d}{dr} R_{\ell,E_2} - R_{\ell,E_2} \frac{d}{dr} R_{\ell,E_1}\right]\right].$$

Integrating from $r = 0$ to $r = a$, the distance beyond which $V(r)$ is assumed to vanish, we obtain

$$({k_1}^2 - {k_2}^2) \int_0^a R_{\ell,E_1}\, R_{\ell,E_2}\, r^2\, dr$$
$$= a^2\, R_{\ell,E_1}(a)\, R_{\ell,E_2}(a) \left[\gamma_\ell(E_2, a) - \gamma_\ell(E_1, a)\right],$$

where $\gamma_\ell(E,a) = [R_{\ell,E}(a)]^{-1}\, [dR_{\ell,E}(r)/dr]|_{r=a}$. In the limit $E_1 - E_2 \to 0$ this becomes

$$\frac{\partial}{\partial E}\gamma_\ell(E,a) = \frac{-2\mu}{\hbar^2 a^2 |R_{\ell,E}(a)|^2} \int_0^a |R_{\ell,E}|^2\, r^2\, dr\,, \qquad (16.112)$$

from which it follows that $\partial\gamma_\ell/\partial E < 0$. The logarithmic derivative γ_ℓ is a monotonically decreasing function of E, except that it jumps discontinuously from $-\infty$ to $+\infty$ whenever $R_{\ell,E}(a)$ vanishes. Its qualitative behavior is similar to that of $\cot(ka)$.

For δ_ℓ to achieve the value $\pi/2$, which maximizes the contribution to the cross section from the partial wave ℓ, it is necessary for the denominator of (16.34) to vanish. Suppose this happens at the energy $E = E_r$. Then in a neighborhood of $E = E_r$ we may write, approximately, $\gamma_\ell \approx c - b(E - E_r)$, where $c = k\, n_\ell'(ka)/n_\ell(ka)$. We must have $b > 0$, since $\partial\gamma_\ell/\partial E < 0$, and it is clear that the approximation can be valid only if $n_\ell(ka) \neq 0$. In the neighborhood of $E = E_r$, Eq. (16.34) becomes

$$\begin{aligned} \tan(\delta_\ell) &\approx \frac{k\, j_\ell' - \gamma_\ell\, j_\ell}{n_\ell\, b(E - E_r)} \\ &\approx \frac{k\, j_\ell' - c\, j_\ell}{n_\ell\, b(E - E_r)} \\ &= \frac{k(j_\ell'\, n_\ell - n_\ell'\, j_\ell)}{(n_\ell)^2\, b(E - E_r)}\,, \end{aligned}$$

where for brevity we have omitted the argument ka of the Bessel functions and their derivatives. This expression can be simplified by means of the Wronskian relation, $j_\ell'(z)\, n_\ell(z) - n_\ell'(z)\, j_\ell(z) = -z^{-2}$, which follows directly from the differential equation satisfied by the Bessel functions. Thus we obtain

$$\tan(\delta_\ell) \approx \frac{\frac{1}{2}\Gamma}{E_r - E}\,, \tag{16.113}$$

where $\frac{1}{2}\Gamma = \{k\, a^2\, b[n_\ell(ka)]^2\}^{-1}$ and $\hbar^2 k^2/2\mu = E_r$. Without further approximation this yields

$$\sin(\delta_\ell)\, \exp(i\,\delta_\ell) \approx \frac{\Gamma}{2(E_r - E) - i\Gamma}\,, \tag{16.114}$$

which may be substituted into the expression (16.29) for the scattering amplitude. The contribution of this resonant partial wave ℓ to the total cross section is

$$\sigma_\ell = \frac{4\pi(2\ell + 1)}{k^2}\, \frac{\Gamma^2}{4(E_r - E)^2 + \Gamma^2}\,. \tag{16.115}$$

If Γ is small this term will produce a sharp narrow peak in the total cross section.

Decay of a resonant state

The physical nature of a resonant scattering state can be understood by examining its behavior in time. Instead of a stationary (monoenergetic) state, we now consider a time-dependent state involving a spectrum of energies that is much broader that Γ,

$$\Psi(\mathbf{r},t) = \int A(\mathbf{k})\, \Psi_{\mathbf{k}}{}^{(+)}(\mathbf{r})\, e^{-iEt/\hbar}\, d^3k\,, \tag{16.116}$$

where $\Psi_{\mathbf{k}}{}^{(+)}(\mathbf{r})$ is a stationary scattering state of the type we have previously been considering, and $E = \hbar^2 k^2/2\mu$. The function $A(\mathbf{k})$ should be nonzero only for values of $\mathbf{k}$ that are collinear with the incident beam. This state function can be divided into an incident wave and a scattered wave, in the manner of (16.12) and (16.14), and the scattered wave will be of the form

$$\Psi_s(\mathbf{r},t) \sim \int A(\mathbf{k})\, f_k(\theta,\phi)\, \frac{e^{ikr}}{r}\, e^{-iEt/\hbar}\, d^3\, k \tag{16.117}$$

in the limit of large r.

Suppose now that all phase shifts are small except the one that is resonant. Then the scattering amplitude will be dominated by that one value of ℓ, and using the resonance approximation (16.114) we obtain

$$\Psi_s(\mathbf{r},t) \sim (2\ell+1)P_\ell(\cos\theta)$$
$$\times \int A(\mathbf{k})\, \frac{e^{ikr}}{kr} \left\{\frac{\Gamma}{2(E_r - E) - i\Gamma}\right\} e^{-iEt/\hbar}\, d^3k\,.$$

Here θ is the angle of $\mathbf{r}$ relative to the incident beam. This integral can most conveniently be analyzed by going to polar coordinates and using $E = \hbar^2k^2/2\mu$ as a variable of integration, so we put

$$d^3k = k^2\, d\Omega_k\, dk = \frac{\mu}{\hbar^2}\, d\Omega_k\, k\, dE\,.$$

This yields

$$\Psi_s(\mathbf{r},t) \sim \frac{\mu}{\hbar^2}\,(2\ell+1)\, P_\ell(\cos\theta)\, \frac{F(r,t)}{r}\,, \tag{16.118}$$

where

$$F(r,t) = \int_0^\infty \alpha(E)\,\Gamma\, \frac{\exp\,[i(kr - Et/\hbar)]}{2(E_r - E) - i\Gamma}\, dE\,, \tag{16.119}$$

with

$$\alpha(E) = \int A(\mathbf{k})\, d\Omega_k\,.$$

The precise time dependence of $F(r,t)$ is determined by the details of the initial state through the function $\alpha(E)$, and can be quite complicated. We have assumed that $\alpha(E)$ is a smooth function of energy, nearly constant over an energy range Γ, and so it is reasonable to replace $\alpha(E)$ by $\alpha(E_r)$ in the integral. In the resonance approximation the integral is dominated by contributions in the energy range $E_r \pm \Gamma$. Therefore we replace k in the exponential by its Taylor series, $k \approx k_r + (E - E_r)/\hbar v_r$, where $E_r = \hbar^2 {k_r}^2/2\mu$ and $v_r = \hbar k_r/\mu$. Introducing a dimensionless variable of integration, $z = (E - E_r)/\Gamma$, and a retarded time $\tau = t - r/v_r$, we can rewrite Eq. (16.119) as

$$F(r,t) = -\alpha(E_r)\,\Gamma\, \exp(ik_r r - iE_r t/\hbar) \int_{-E_r/\Gamma}^{\infty} \frac{\exp(-i\tau\Gamma z/\hbar)}{2z + i}\, dz\,. \quad (16.120)$$

If $\Gamma \ll E_r$, the lower limit can be replaced by $-\infty$. The integral can then be evaluated for positive τ by closing the contour of integration with an infinite semicircle in the lower half of the complex z plane. From the residue of the pole at $z = -i/2$ we obtain the time dependence $\exp(-\tau\Gamma/2\hbar)$. For negative τ the contour must be closed in the upper half-plane, where there are no poles, and so the integral vanishes. Thus we have determined the time dependence of the scattered wave (16.118) at large distances to be

$$\begin{aligned} \Psi_s(\mathbf{r},t) &\propto e^{-\Gamma t/2\hbar} \qquad \text{for } t > \frac{r}{v_r}\,, \\ \Psi_s(\mathbf{r},t) &= 0 \qquad \text{for } t < \frac{r}{v_r}\,. \end{aligned}$$

It is zero before $t = r/v_r$ because that is the time needed for propagation from the scattering center to the point of detection. For times greater than this, the detection probability goes like $|\Psi_s|^2 \propto e^{-\Gamma t/\hbar}$. Thus we see that resonant scattering provides an example of approximately exponential decay, such as was discussed in Sec. 12.2.

Virtual bound states

The physical picture of a scattering resonance, which we derive from the above analysis, is of a particle being temporarily captured in the scattering potential in a *virtual bound state* whose mean lifetime is $\hbar/\Gamma$. It is possible to exhibit a closer connection between bound states and resonances. Suppose that the potential supports a bound state at the negative energy $E = -E_B$. As the strength of the potential is reduced, the binding energy E_B will decrease and eventually vanish. As the potential strength is further reduced, a resonance, or virtual bound state, appears at positive energy. Further reduction in the potential strength results in Γ increasing, so that the virtual bound state has so short a lifetime that it is no longer significant.

We shall illustrate this connection only for $E = 0$, which is the boundary between bound states and scattering states. It is apparent from (16.34) that in the limit $k \to 0$ we have $\tan(\delta_\ell) \to 0$ for almost all values of the logarithmic derivative γ_ℓ. The exception occurs if the denominator vanishes, in which case the phase shift has the zero energy limit $\pi/2$, and we have a zero energy resonance. In this case we must have $\gamma_\ell = k\, n_\ell'(ka)/n_\ell(k_a) \to -(\ell+1)/a$ in the limit $k \to 0$.

A bound state function for negative energy must match onto the exponentially decaying solution of the free particle wave equation. These functions are just the spherical Bessel functions evaluated for imaginary values of $k = i\kappa$, such that $\hbar^2 k^2/2\mu = -\hbar^2\kappa^2/2\mu = -E_B \le 0$. It is well known that the Hankel function $h_\ell(z) = j_\ell(z) + i n_\ell(z)$ is proportional to e^{iz} for large z. Therefore $h_\ell(i\kappa r)$ is proportional to $e^{-\kappa r}$ for large r. Its logarithmic derivative is $\gamma_\ell = i\kappa\, h_\ell'(i\kappa a)/h_\ell(i\kappa a)$, which has the limit $\gamma_\ell \to -(l+1)/a$ when $\kappa \to 0$. We have thus shown that the conditions for a zero energy resonance and a zero energy bound state are identical. Of course this zero energy bound state may not be square-integrable over all space, and so may not be a genuine bound state. Its significance is in its being the intermediate case between genuine bound states and resonance states, which we now see to be closely related.

16.7 *Diverse Topics*

We present here some examples that apply and illustrate various aspects of the theory developed in the preceding sections.

General behavior of phase shifts

The general behavior of phase shifts as function of energy and potential strength can be illustrated by the example of the *square well potential*,

$$\begin{aligned} V(r) &= -V_0 \quad (r < a)\,, \\ &= 0 \qquad (r > a)\,. \end{aligned} \tag{16.121}$$

The solution to the radial wave equation (16.23) for $r \le a$ is $R_\ell(r) = c\, j_\ell(\alpha r)$, where $\alpha^2 = (2\mu/\hbar^2)(E + V_0)$. Its logarithmic derivative at $r = a$ is $\gamma_\ell \equiv R_\ell'(a)/R_\ell(a) = \alpha\, j_\ell'(\alpha a)/j_\ell(\alpha a)$, from which the tanget of the phase shift δ_ℓ is calculated by means of (16.34). In Fig. 16.5 the phase shifts and total scattering cross sections for several square wells are plotted against k, which

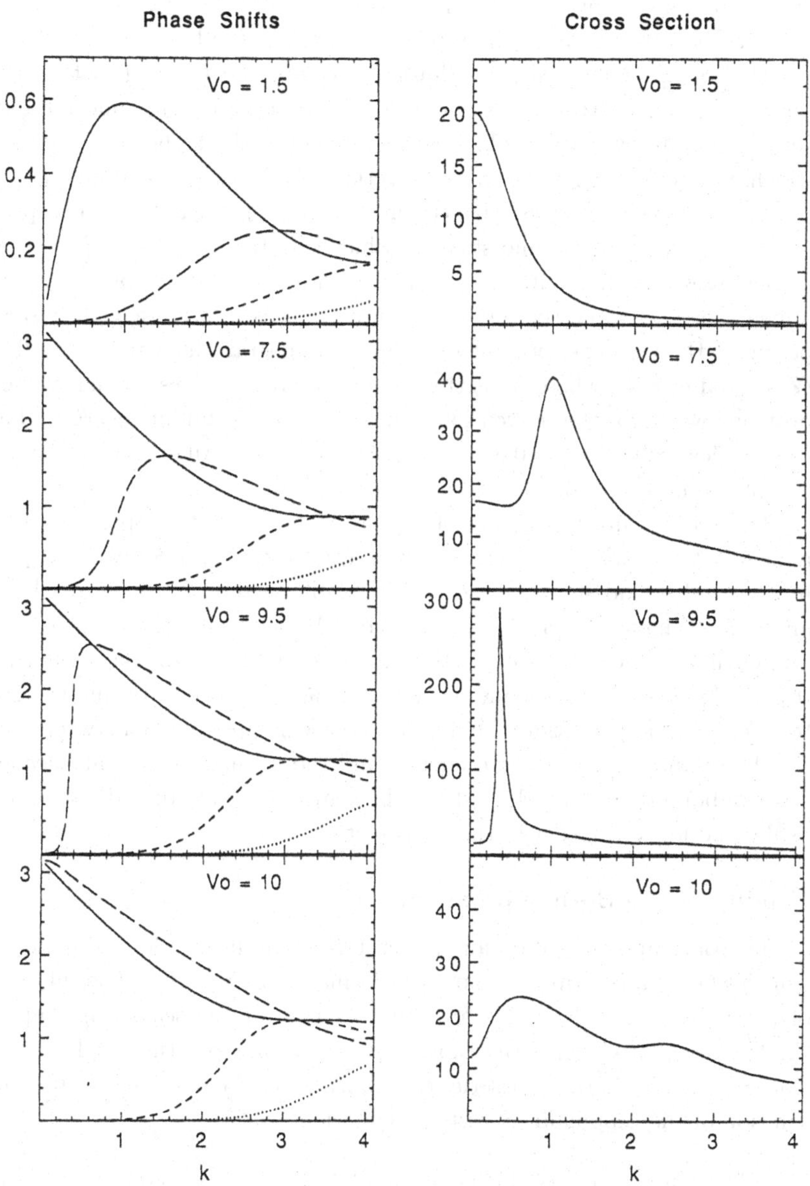

Fig. 16.5 Phase shifts and total cross sections for several square well potentials of radius $a = 1$. Units: $\hbar^2/2\mu = 1$. Key to phase shifts: $\ell = 0$, solid line; $\ell = 1$, long dashes; $\ell = 2$, short dashes; $\ell = 3$, dots.

is a more convenient variable than the energy $E = \hbar^2 k^2/2\mu$. The potential strength V_0 increases from the top of the figure to the bottom.

At the top of Fig. 16.5, we illustrate a potential ($V_0 = 1.5$) that has no bound states. All phase shifts rise from $\delta_\ell = 0$ at zero energy and fall back to zero at infinite energy. The larger the value of ℓ, the higher is the energy at which δ_ℓ has its strength. The cross section is smooth and structureless.

It was shown in the previous section that $\tan(\delta_\ell) = 0$ at $E = 0$, except when there is a zero energy bound state, in which case $\tan(\delta_\ell) = \infty$ at $E = 0$. As V_0 increases, we reach a critical value at which a bound state for a particular ℓ appears at $E = 0$, and the zero energy limit of δ_ℓ is $\pi/2$. For a slightly stronger potential the zero energy limit of δ_ℓ is π. In general, the zero energy limit of δ_ℓ is equal to πN_ℓ, where N_ℓ is the number of bound states for that value of ℓ, provided we adopt the convention that δ_ℓ vanishes at infinite energy. The first such critical value for $\ell = 0$ is $V_0\, a^2 = (\hbar^2/2\mu)(\pi/2)^2$. All examples in Fig. 16.5 except the first have one bound state with $\ell = 0$.

The second and third rows of Fig. 16.5 show the development of a resonance for $\ell = 1$, which is associated with the capture and decay of the particle in a virtual bound state. For $V_0 = 7.5$ the phase shift δ_1 barely reaches $\pi/2$, and the resonance is very broad in energy. When V_0 has increased to 9.5 the phase shift δ_1 rises very steeply through $\pi/2$, and the resonance is much narrower. Moreover, the resonance occurs at a lower energy, reflecting the greater tendency of the potential to bind. At a critical value of V_0, between 9.5 and 10, the resonance reaches zero energy, and the virtual bound state becomes a genuine bound state. For $V_0 = 10$ we have $\delta_1 = \pi$ at $k = 0$, and the resonance peak is no longer present in the cross section.

Validity of the Born approximation

The Born approximation for the scattering amplitude, derived in Sec. 16.4, can also be obtained from the operator formalism of Sec. 16.5. The substitution of (16.86) into (16.93) yields an infinite series for the scattering amplitude, the first term of which is the Born approximation (sometimes called the *first* Born approximation). That series can alternatively be obtained by an iterative solution of the Lippmann–Schwinger equation (16.91):

$$|\Psi_a{}^{(+)}\rangle = |\Phi_a\rangle + G_0(E^+)\,V|\Phi_a\rangle + G_0(E^+)\,V\,G_0(E^+)\,V|\Phi_a\rangle + \cdots\,, \tag{16.122}$$

which is then substituted into (16.93). [Notice the similarity of this series to Eq. (10.100) of Brillouin–Wigner perturbation theory. The main difference is that there we treated a discrete spectrum, whereas now we are dealing with the continuum.] A sufficient condition for the Born approximation to be accurate is that the higher order terms of (16.122) be small compared with the leading term.

The first order term of (16.122) for a central potential $V(r)$, in coordinate representation, is equal to

$$f_1(\mathbf{r}) = \int G_0(\mathbf{r}, \mathbf{r}'; E^+)\, V(r')\, e^{i\mathbf{k}\cdot\mathbf{r}'}\, d^3r' \,,$$

where $G_0(\mathbf{r}, \mathbf{r}'; E^+)$ is given by (16.95). The Born approximation will be accurate if $|f_1(\mathbf{r})| \ll 1$. We can most easily evaluate $f_1(\mathbf{r})$ at the point $r = 0$, which should be representative of the region where $f_1(\mathbf{r})$ is largest.

$$\begin{aligned} f_1(0) &= -\frac{\mu}{2\pi\hbar^2} \int \frac{e^{-ikr'}}{r'}\, V(r')\, e^{i\mathbf{k}\cdot\mathbf{r}'}\, d^3r' \\ &= -\frac{2\mu}{\hbar^2 k} \int_0^\infty e^{-ikr'}\, V(r')\, \sin(kr')\, dr' \,. \end{aligned} \tag{16.123}$$

If $V(r)$ does not change sign, the largest value of (16.123) will occur for $k = 0$. This yields the condition $(2\mu/\hbar^2) \int |V(r')| r'\, dr' \ll 1$, which would ensure the validity of the Born approximation for all values of k. Expressed as an order of magnitude, this condition can be written as $|V_0|\mu a^2/\hbar^2 \ll 1$, where V_0 measures the strength of the potential and a is its range. This condition is very restrictive, and it is seldom useful. A much more useful condition is obtained from the fact that (16.123) becomes small at large k, not only because of the factor k^{-1}, but also because of oscillations of the integrand. If $ka \gg 1$ there will be many oscillations within the range of $V(r)$, which will reduce the value of the integral. Although a precise estimate is difficult to obtain, it is clear that the Born approximation will become accurate in the high energy limit, and it is in that regime that it is most useful.

Multiple scattering

Suppose that the scattering potential is a sum of identical potentials centered on different atoms, $V(\mathbf{r}) = \sum_i v(\mathbf{r} - \mathbf{R}_i)$, with $\mathbf{R}_i$ being the position of the ith atom. The series (16.86) for the t matrix is then of the form

$$T = \sum_i v_i + \sum_i \sum_j v_i\, G_0\, v_j + \sum_i \sum_j \sum_m v_i\, G_0\, v_j\, G_0\, v_m + \cdots \,, \tag{16.124}$$

where $v_i = v(\mathbf{r} - \mathbf{R}_i)$. We would like to describe the total scattering process as a series of multiple scattering from the various atoms. But (16.124) cannot be so interpreted because of the terms with $i = j$, or $j = m$, etc. These do not represent "repeated scattering by the same atom", for no such process exists. They are actually an artifact of expanding the scattering amplitude from a single atom in powers of the atomic potential. Let us define the t matrix of atom i, which is the t matrix that would exist if only the potential v_i were present:

$$t_i = v_i + v_i\, G_0\, v_i + v_i\, G_0\, v_i\, G_0\, v_i + \cdots . \tag{16.125}$$

Then the complete t matrix of the system can be written as

$$T = \sum_i t_i + \sum_{i \neq j}\sum t_i\, G_0\, t_j + \sum_{i \neq j \neq m}\sum\sum t_i\, G_0\, t_j\, G_0\, t_m + \cdots . \tag{16.126}$$

The restriction on the summations is that adjacent atoms in a term must be distinct. (Note that $i = m$ is allowed in the third term.) The terms of this series can indeed be interpreted contributions to the total scattering amplitude from multiple scattering from the various atoms.

According to (16.93) the scattering amplitude is equal to $-\mu/2\pi\hbar^2$ multiplied by

$$\langle\, \Phi_{\mathbf{k}'}|T|\Phi_{\mathbf{k}}\,\rangle = \iint \exp(-i\mathbf{k}'\cdot \mathbf{r}')\, \langle \mathbf{r}'|T|\mathbf{r}\,\rangle\, \exp(i\mathbf{k}\cdot \mathbf{r}) d^3r' d^3r$$

where $\hbar\mathbf{k}$ is the initial momentum of the incident particle, and $\hbar\mathbf{k}'$ is its final momentum. The simplest approximation is to include only the first term of (16.126). Now the contribution of the ith atom is

$$\begin{aligned}\langle\, \Phi_{\mathbf{k}'}|t_i|\Phi_{\mathbf{k}}\,\rangle &= \iint \exp(-i\mathbf{k}'\cdot \mathbf{r}')\, \langle\, \mathbf{r}' - \mathbf{R}_i|t_0|\mathbf{r} - \mathbf{R}_i\,\rangle \exp(i\mathbf{k}\cdot \mathbf{r}) d^3r' d^3r \\ &= \exp[i(\mathbf{k} - \mathbf{k}')\cdot \mathbf{R}_i] \iint \exp(-i\mathbf{k}'\cdot \mathbf{r}')\langle\, \mathbf{r}'|t_0|\mathbf{r}\,\rangle\, \exp(i\mathbf{k}\cdot \mathbf{r}) d^3r' d^3r \\ &= \exp[i(\mathbf{k} - \mathbf{k}')\cdot \mathbf{R}_i]\, \langle\, \Phi_{\mathbf{k}'}|t_0|\Phi_{\mathbf{k}}\,\rangle ,\end{aligned}$$

where t_0 is the t matrix of an atom located at the origin of coordinates. Hence the first term of (16.126) yields

$$\langle\, \Phi_{\mathbf{k}'}|T|\Phi_{\mathbf{k}}\,\rangle \approx \langle\, |\Phi_{\mathbf{k}'}|t_0|\Phi_{\mathbf{k}}\,\rangle \sum_j \exp[i(\mathbf{k} - \mathbf{k}')\cdot \mathbf{R}_j] . \tag{16.127}$$

The scattering probability, which is the absolute square of the amplitude, depends upon the positions of the atoms through the factor

$$\left|\sum_j \exp[i(\mathbf{k}-\mathbf{k}')\cdot\mathbf{R}_j]\right|^2 = \sum_j\sum_m \exp[i(\mathbf{k}-\mathbf{k}')\cdot(\mathbf{R}_j-\mathbf{R}_m)]\,.$$

Thus it is possible to obtain information about the relative positions of the atoms by means of scattering measurements. This technique is very useful for determining the structures of solids and liquids.

The inverse scattering problem

In all of our theory and examples, we have proceeded from an assumed knowledge of the scattering interaction to a calculation of the scattering cross sections. In practice, one often wants to infer the interaction from observed scattering data, this being called the *inverse problem* of scattering theory. The mathematical theory of the inverse scattering problem is too lengthy and complex to present here, so we shall only summarize the main results for the scattering of spinless particles by a central potential.

The first problem is to deduce the phase shifts from the differential cross section. If the scattering amplitude were known, we could easily obtain the phase shifts by expanding it in Legendre polynomials, as in (16.29). But the amplitude is complex, and experiment yields only its magnitude but not its phase. From the unitarity condition (16.107), it is possible to deduce an integral equation relating to the phase. But it is usually more practical to fit the differential cross section data to a model involving a small number of phase shifts as adjustable parameters, using as many (or, rather, as few) parameters as are needed to reproduce the data within experimental accuracy. In this sense, the phase shifts can be regarded as measurable quantities.

It is possible to determine $V(r)$ uniquely from a knowledge of $\delta_\ell(E)$ for all E and only one value of ℓ, provided there are no bound states. If there are N_ℓ bound states of angular momentum ℓ then the solution is not unique, and there is an N_ℓ-parameter family of potentials all of which produce the same $\delta_\ell(E)$ for all E. It is also possible to determine $V(r)$ from a knowledge of $\delta_\ell(E)$ for all ℓ at one value of E. Although in principle any value of E can be used, it is clear that a small value of E is unsuitable because all phase shifts beyond $\ell = 0$ or 1 will be too small to measure. For details of these theorems, the reader is referred to Newton (1982, Ch. 20) or Wu and Ohmura (1962, Ch. 1, Sec. G). These theorems are of considerable mathematical interest. They are useful for telling us how much information is required to determine the scattering potential. However, their practical utility is limited because, in practice, one knows only a finite number of phase shifts over a limited range of energy, and this does not allow one to apply either of the theorems.

Further reading for Chapter 16

Goldberger and Watson (1964) has long been regarded as the authoritative reference on scattering theory, although it has now been superseded to some extent by Newton (1982). Both of them are research tomes. The beginning student may find the treatment by Rodberg and Thaler (1967) to be more accessible. Wu and Ohmura (1962) is intermediate between the textbook and research levels.

Problems

16.1 Derive Eq. (16.5) from momentum and energy conservation.

16.2 Calculate the $\ell = 0$ phase shift for the repulsive δ shell potential, $V(r) = c\,\delta(r-a)$. Determine the conditions under which it will be approximately equal to the phase shift of a hard sphere of the same radius a, and note the conditions under which it may significantly differ from the hard sphere phase shift even though c is very large.

16.3 Show that $\Psi^{(+)}$ and $\Psi^{(-)}$, defined by (16.69), have the correct asymptotic forms, (16.42) and (16.46), respectively.

16.4 Use the Born approximation to calculate the differential cross section for scattering by the Yukawa potential, $V(r) = V_0\, e^{-\alpha r}/\alpha r$.

16.5 In Example 1 of Sec. 16.4 (spin–spin interaction), assume that the two particles are an electron and a proton, and add to H_0 the magnetic dipole interaction $-\mathbf{B}\cdot(\boldsymbol{\mu}_e + \boldsymbol{\mu}_p)$. Calculate the scattering cross sections in the Born Approximation, taking into account the fact that kinetic energy will not be conserved.

16.6 For Example 1 of Sec. 16.4 (without a magnetic field), assume that the phase shifts for the central potential $V_0(r)$ are known, and use the DWBA to calculate the additional scattering due to the spin–spin interaction $V_s(r)\boldsymbol{\sigma}^{(1)}\cdot\boldsymbol{\sigma}^{(2)}$. Does skew scattering occur?

16.7 Show that Example 2 of Sec. 16.4 (spin–orbit interaction) can be solved "exactly" by introducing the total angular momentum eigenfunctions (7.104) as basis functions, and computing a new set of phase shifts that depend upon both the orbital angular momentum ℓ and the total angular momentum j. The solution will be as "exact" as the computation of the phase shifts. [Goldberger and Watson (1964), Sec. 7.2.]

16.8 Use phase shifts to evaluate the total cross section in the low energy limit ($E \to 0$) for the square well potential: $V(r) = -V_0$ for $r < a, V(r) = 0$ for $r > a$.

16.9 (a) Calculate the differential cross section of the square well potential using the Born approximation. (b) Evaluate the total cross section in the low energy limit. Explain any differences between this result and that of Problem 16.8.

16.10 Express the mean lifetime, $\hbar/\Gamma$, of a virtual bound state of angular momentum ℓ in terms of the energy derivative of the phase shift δ_ℓ.

Chapter 17

Identical Particles

In this chapter we discuss the properties of systems of identical particles, following principles that were first expounded by Messiah and Greenberg (1964). Three successively stronger expressions of particle identity can be distinguished: (1) permutation symmetry of the Hamiltonian, which leads to degeneracies and selection rules, as does any symmetry; (2) permutation symmetry of *all* observables, which leads to a superselection rule; and (3) the symmetrization postulate, which restricts the states for a species of particle to be of a single symmetry type (either symmetric or antisymmetric). The stronger principles in this sequence cannot be deduced from the weaker principles, and some misleading arguments in the literature will be corrected.

17.1 *Permutation Symmetry*

All electrons are identical in their properties; the same is true of all protons, all neutrons, etc. Two physical situations that differ only by the interchange of identical particles are indistinguishable. One of the consequences of this fact is that *any physical Hamiltonian must be invariant under permutation of identical particles.*

Consider first a system of two identical particles. Basis vectors for the state space may be constructed by taking products of single particle basis vectors, $|\alpha\rangle|\beta\rangle \equiv |\alpha\rangle \otimes |\beta\rangle$. In this notation the order of the factors formally distinguishes the two particles, the eigenvalue α corresponding to the first particle and the eigenvalue β corresponding to the second particle. Any arbitrary vector in the two-particle state space can be expressed as a linear combination of these basis vectors, $|\Psi\rangle = \sum_{\alpha,\beta} c_{\alpha,\beta}|\alpha\rangle|\beta\rangle$. A state function in the two-particle configuration space will be of the form

$$\Psi(\mathbf{x}_1, \mathbf{x}_2) = (\langle \mathbf{x}_1| \otimes \langle \mathbf{x}_2|)|\Psi\rangle = \sum_{\alpha,\beta} c_{\alpha,\beta} \, \langle \mathbf{x}_1|\alpha\rangle \, \langle \mathbf{x}_2|\beta\rangle \, .$$

We next define a *permutation operator* P_{12} by the relation

$$P_{12}|\alpha\rangle|\beta\rangle = |\beta\rangle|\alpha\rangle\,. \tag{17.1}$$

Clearly P_{12} is its own inverse, and P_{12} is both unitary and Hermitian. Its effect on a two-particle state function is

$$P_{12}\,\Psi(\mathbf{x}_1,\mathbf{x}_2) = \Psi(\mathbf{x}_2,\mathbf{x}_1)\,. \tag{17.2}$$

If the Hamiltonian H is invariant under interchange of the two particles, it must be the case that $P_{12}H = HP_{12}$. It follows (Theorem 5, Sec. 1.3) that the operators H and P_{12} possess a complete set of common eigenvectors. Because $(P_{12})^2 = I$, the only eigenvalues of P_{12} are $+1$ and -1, and its eigenfunctions are either *symmetric,* $\Psi(\mathbf{x}_2,\mathbf{x}_1) = \Psi(\mathbf{x}_1,\mathbf{x}_2)$, or *antisymmetric,* $\Psi(\mathbf{x}_2,\mathbf{x}_1) = -\Psi(\mathbf{x}_1,\mathbf{x}_2)$, under interchange of the two particles. Hence it follows that for a system of two identical particles, the eigenvectors of H can be chosen to have either symmetry or antisymmetry under permutation of the particles.

The situation is more complicated if there are more than two particles, and the general principles can be illustrated by considering a system of three identical particles. Basis vectors for the state space will now be of the form $|\alpha\rangle|\beta\rangle|\gamma\rangle$. There are six distinct permutations of three objects, so we can define six different permutation operators. These are the *identity operator* I, the *pair interchange operators* P_{12}, P_{23}, and P_{31}, and the *cyclic permutations* P_{123} and $(P_{123})^2$. The effects of these operators on a typical basis vector are

$$\begin{aligned} &P_{12}|\alpha\rangle|\beta\rangle|\gamma\rangle = |\beta\rangle|\alpha\rangle|\gamma\rangle\,, \quad P_{23}|\alpha\rangle|\beta\rangle\gamma\rangle = |\alpha\rangle|\gamma\rangle|\beta\rangle\,,\\ &P_{31}|\alpha\rangle|\beta\rangle|\gamma\rangle = |\gamma\rangle|\beta\rangle|\alpha\rangle\,, \quad P_{123}|\alpha\rangle|\beta\rangle|\gamma\rangle = |\gamma\rangle|\alpha\rangle|\beta\rangle\,,\\ &\quad (P_{123})^2 = |\beta\rangle|\gamma\rangle|\alpha\rangle\,. \end{aligned} \tag{17.3}$$

It is easily verified that the six permutation operators are not mutually commutative, for example $P_{12}P_{23} \neq P_{23}P_{12}$. Therefore a complete set of common eigenvectors for these operators does not exist, and so it is not possible for every eigenvector of H to be symmetric or antisymmetric under pair interchanges. However, we can divide the vector space into *invariant subspaces,* which have the property that a vector in any invariant subspace is transformed by the permutation operators into another vector in the same subspace. The basis vector $|\alpha\rangle|\beta\rangle|\gamma\rangle$ (with α, β, and γ all unequal) and its permutations (17.3) span a six-dimensional vector space. This may be reduced into four invariant subspaces, spanned by the following vectors, all of which are orthogonal:

Symmetric:

$$6^{-1/2}\{\,|\alpha\rangle|\beta\rangle|\gamma\rangle + |\beta\rangle|\alpha\rangle|\gamma\rangle + |\alpha\rangle|\gamma\rangle|\beta\rangle + |\gamma\rangle|\beta\rangle|\alpha\rangle + |\gamma\rangle|\alpha\rangle|\beta\rangle + |\beta\rangle|\gamma\rangle|\alpha\rangle\,\}$$

Antisymmetric:

$$6^{-1/2}\{\,|\alpha\rangle|\beta\rangle|\gamma\rangle - |\beta\rangle|\alpha\rangle|\gamma\rangle - |\alpha\rangle|\gamma\rangle|\beta\rangle - |\gamma\rangle|\beta\rangle|\alpha\rangle + |\gamma\rangle|\alpha\rangle|\beta\rangle + |\beta\rangle|\gamma\rangle|\alpha\rangle\,\}$$

Partially symmetric:

$$12^{-1/2}\,\{\,2|\alpha\rangle|\beta\rangle|\gamma\rangle + 2|\beta\rangle|\alpha\rangle|\gamma\rangle - |\alpha\rangle|\gamma\rangle|\beta\rangle - |\gamma\rangle|\beta\rangle|\alpha\rangle - |\gamma\rangle|\alpha\rangle|\beta\rangle - |\beta\rangle|\gamma\rangle|\alpha\rangle\,\}$$

$$2^{-1}\,\{\,0+0 - |\alpha\rangle|\gamma\rangle|\beta\rangle + |\gamma\rangle|\beta\rangle|\alpha\rangle + |\gamma\rangle|\alpha\rangle|\beta\rangle - |\beta\rangle|\gamma\rangle|\alpha\rangle\,\}$$

Partially symmetric:

$$2^{-1}\,\{\,0+0 - |\alpha\rangle|\gamma\rangle|\beta\rangle + |\gamma\rangle|\beta\rangle|\alpha\rangle - |\gamma\rangle|\alpha\rangle|\beta\rangle + |\beta\rangle|\gamma\rangle|\alpha\rangle\,\}$$

$$12^{-1/2}\,\{\,2|\alpha\rangle|\beta\rangle|\gamma\rangle - 2|\beta\rangle|\alpha\rangle|\gamma\rangle + |\alpha\rangle|\gamma\rangle|\beta\rangle + |\gamma\rangle|\beta\rangle|\alpha\rangle - |\gamma\rangle|\alpha\rangle|\beta\rangle - |\beta\rangle|\gamma\rangle|\alpha\rangle\,\}.$$

In these expressions, we have always written the basis vectors in the same order as they appear in (17.3), with zeros as place holders where necessary. The symmetric subspace is invariant under all permutations. The antisymmetric subspace changes sign under the pair interchanges P_{12}, P_{23}, and P_{31}, and is unchanged by the other permutations. In general, the action of permutation operators on the vectors in a partially symmetric subspace is to transform them into linear combinations of each other; however, under P_{12} the first member in each subspace is even, and the second member is odd.

Because the Hamiltonian commutes with the permutation operators, it is possible to form the eigenvectors of H so that each eigenvector is constructed from basis vectors that belong to only one of these invariant subspaces. Thus the stationary states may be classified according to their symmetry type under permutations; moreover this symmetry type will be conserved by any permutation-invariant interaction. This is so because $H\Psi$ has the same permutation symmetry as does Ψ, since H is permutation-invariant, and hence $\partial\Psi/\partial t = (i\hbar)^{-1}\,H\Psi$ must also have the same symmetry as Ψ. Therefore the symmetry type does not change. These conclusions can easily be generalized to any number of particles by means of group representation theory.

17.2 *Indistinguishability of Particles*

If a set of particles are indistinguishable, then their Hamiltonian must be unchanged by permutations of the particles. However, the converse is not true.

The Hamiltonian $H = ({P_e}^2 + {P_p}^2)/2M - e^2/r$ of a positronium atom, which consists of an electron and a positron, is invariant under interchange of the two particles. Therefore all of the conclusions of the previous section apply to positronium. But of course an electron and a positron are not identical particles, and they can be distinguished by applying an electric or a magnetic field.

Following Messiah and Greenberg (1964), we state the principle of indistinguishability of identical particles: *Dynamical states that differ only by a permutation of identical particles cannot be distinguished by any observation whatsoever*. This principle implies the permutation symmetry of the Hamiltonian, which was discussed in the previous section, but it also implies much more. Let A be an operator that represents an observable dynamical variable, and let $|\Psi\rangle$ represent a state of a system of identical particles. The state obtained by interchanging particles i and j is described by the vector $P_{ij}|\Psi\rangle$. Then according to the principle of indistinguishability we must have

$$\langle\Psi|A|\Psi\rangle = \langle\Psi|\,(P_{ij})^\dagger\, AP_{ij}|\Psi\rangle\,, \tag{17.4}$$

and this must hold for any vector $|\Psi\rangle$. Therefore we deduce (Problem 1.12) that $A = (P_{ij})^\dagger\, AP_{ij}$. Since $P_{ij} = (P_{ij})^\dagger = (P_{ij})^{-1}$, it follows that

$$P_{ij}\; A = AP_{ij}\,. \tag{17.5}$$

Since A represents an arbitrary observable, we have shown that *all physical observables must be permutation-invariant.*

An example of a dynamical variable that satisfies (17.5) is a component of the total spin, $S_x = \sum_j s_x^{(j)}$, where the sum is over all identical particles in the system. On the other hand, a spin component of one particular particle, ${s_x}^{(1)}$, is not permutation-invariant, and so is not observable according to our criterion. This does not mean that the spin of a single particle cannot be measured. If, for example, there were particles localized in the neighborhoods of $\mathbf{x}_1, \mathbf{x}_2, \ldots, \mathbf{x}_j, \ldots$, then "the spin of the particle located at $\mathbf{x}_1$" would be a permutation-invariant observable. It is only the attachment of labels to distinguish the individual particles themselves that is forbidden by the principle of indistinguishability.

The Hamiltonian H is itself an observable, and must be permutation-invariant. Thus all of the consequences of Sec. 17.1, such as the classification of states by symmetry type and the conservation of symmetry type, follow from the principle of indistinguishability. But in addition to those properties, which we may term "selection rules", there is a *superselection rule* which states that *interference between states of different permutation symmetry is not observable.* For symmetric and antisymmetric states, this can be shown by the same methods that were used in Sec. 7.6 to derive the $R(2\pi)$ superselection rule. Let $|s\rangle$ be a symmetric vector, $P_{ij}|s\rangle = |s\rangle$, and let $|a\rangle$ be an antisymmetric vector, $P_{ij}|a\rangle = -|a\rangle$. For any observable A we have, from (17.5),

$$\langle s|P_{ij}\, A|a\rangle = \langle s|AP_{ij}|a\rangle\,,$$
$$\langle s|A|a\rangle = -\langle s|A|a\rangle\,.$$

This is possible only if $\langle s|A|a\rangle = 0$. Now consider a superposition state,

$$|\Psi\rangle = |s\rangle + c|a\rangle\,,$$

where c is a complex constant with $|c| = 1$. The average of the arbitrary observable A in this state is

$$\langle\Psi|A|\Psi\rangle = \langle s|A|s\rangle + \langle a|A|a\rangle\,,$$

which is independent of the phase of c because $\langle s|A|a\rangle = 0$. Thus even if the state vector was a superposition of symmetric and antisymmetric components, no interference would be observed. Using the methods of group representation theory, Messiah and Greenberg (1964) generalize this result to apply to all permutation symmetry types, including the partially symmetric states that are neither symmetric nor antisymmetric under all permutations.

17.3 *The Symmetrization Postulate*

It was shown in the first section of this chapter that the invariance of the Hamiltonian under permutation of identical particles implies that state vectors can be classified according to their permutation symmetry type. This is a mathematical deduction from the principles of quantum mechanics. In the second section we introduced the *principle of indistinguishability*, which implies that no interference can occur between states of different permutation symmetry. This principle cannot be deduced from the other principles of quantum mechanics, although it could be argued that it merely defines what we mean by calling particles "identical". We now introduce an even stronger principle, which asserts that the states of a particular species of particles can only be of one permutation symmetry type. The *symmetrization postulate* states that:

(a) Particles whose spin is an *integer* multiple of $\hbar$ have only *symmetric* states. (These particles are called *bosons*.)
(b) Particles whose spin is a *half odd-integer* multiple of $\hbar$ have only *antisymmetric* states. (These particles are called *fermions*.)
(c) Partially symmetric states do not exist. (Nevertheless they give rise to the name *paraparticles*.)

The superselection rule deduced in Sec. 17.2 is trivialized by the symmetrization postulate, since obviously no interference between symmetry types is possible if only one symmetry type exists. The three parts of this postulate cannot be deduced from the other principles of quantum mechanics, so we shall examine their consequences and the empirical evidence that supports them.

[[Many books contain arguments that purport to derive one or more parts of the symmetrization postulate from other principles of quantum mechanics. A typical argument begins with the assertion that *permutation of identical particles must not lead to a different state.* (This is stronger than the *principle of indistinguishability*, which asserts only that such a permutation must not lead to any *observable* differences.) Hence it is asserted that an allowable state vector must satisfy $P_{ij}|\Psi\rangle = c|\Psi\rangle$. Since $(P_{ij})^2 = 1$, it follows that $c = \pm 1$, and so the argument concludes that only symmetric and antisymmetric states are permitted.

Implicit in this argument is the assumption that a state must be represented by a one-dimensional vector space, i.e. by a state vector with at most its overall phase being arbitrary. But this is equivalent to excluding by fiat the partially symmetric state vectors, which belong to multidimensional invariant subspaces. So it is practically equivalent to the assumption of (c), which was to be proven. If one drops the assumption that a state must be represented by a one-dimensional vector space, the conclusion no longer follows. Consider an n-dimensional permutation-invariant subspace. (Examples with $n = 2$ were given in Sec. 17.1.) From it we construct a state operator ρ,

$$\rho = n^{-1} \sum_{i=1}^{n} |u_i\rangle\langle u_i| \,,$$

with the sum being over all the basis vectors of the the invariant subspace. Clearly the state described by ρ is not changed by permutation of identical particles, since $P_{ij}\,\rho P_{ij} = \rho$.

Another common argument claims that the special properties of the states of identical particles, such as their restriction to be either symmetric or antisymmetric, are related to the *indeterminacy principle.* According to this argument, identical particles could be distinguished in classical mechanics by continuously following them along their trajectories. But in quantum mechanics the indeterminacy relation (8.33) does not allow position and momentum to both be sharp in any state. Therefore we cannot identify separate trajectories, and so the argument concludes that we cannot distinguish the particles. However, the pragmatic indistinguishability that is deduced from this argument implies nothing about the symmetry of the state vector. The derivation of the indeterminacy relations in Sec. 8.4 uses no property of the state vector or state operator except its existence. Even if we used an absurd state vector, having the wrong symmetry and violating the Schrödinger equation, we would still not violate the indeterminacy relations. Therefore the indeterminacy relations tell us nothing about the properties of the state vector.]]

We now examine the empirical consquences of the symmetrization postulate. Consider the three-particle antisymmetric function given in Sec. 17.1. $\Psi_{\alpha\beta\gamma} = \{|\alpha\rangle|\beta\rangle|\gamma\rangle - |\beta\rangle|\alpha\rangle|\gamma\rangle - |\alpha\rangle|\gamma\rangle|\beta\rangle - |\gamma\rangle|\beta\rangle|\alpha\rangle + |\gamma\rangle|\alpha\rangle|\beta\rangle + |\beta\rangle|\gamma\rangle|\alpha\rangle\}/\sqrt{6}$. If we put $\alpha = \beta$ we obtain $\Psi_{\beta\beta\gamma} = 0$. A similar result clearly holds for an antisymmetrized product state vector for any number of particles. This is the basis of the *Pauli exclusion principle*, which asserts that in a system of identical fermions no more than one particle can have exactly the same single particle quantum numbers. The exclusion principle forms the basis of the theory of atomic structure and atomic spectra, and so is very well established. Thus we have strong empirical evidence that electrons are fermions.

The *rotational spectra of diatomic molecules* provides evidence about the permutation symmetry of nucleons. Consider the molecule H_2, which consists of two protons and two electrons. The spin of a proton is $\frac{1}{2}$ (in units of $\hbar$), and the set of two-proton spin states comprises a singlet with total spin $S = 0$ and a triplet with total spin $S = 1$. The Clebsch–Gordan coefficients for constructing these states are given by (7.99). If we denote the single particle spin states by $|+\rangle$ and $|-\rangle$, then the singlet state vector is $(|+\rangle|-\rangle - |-\rangle|+\rangle)/\sqrt{2}$, and the members of the triplet are $|+\rangle|+\rangle, (|+\rangle|-\rangle + |-\rangle|+\rangle)/\sqrt{2}$, and $|-\rangle|-\rangle$. Transitions between the singlet and triplet states are very rare under ordinary conditions, and over a time scale of a day or so it is possible to regard the singlet and triplet states of H_2 as two different stable molecules. The two-proton state function will be of the form $\Psi(\mathbf{X}, \mathbf{r})|\text{spin}\rangle$, where $\mathbf{X} = (\mathbf{x}_1 + \mathbf{x}_2)/2$ is the CM mass coordinate, $\mathbf{r} = \mathbf{x}_1 - \mathbf{x}_2$ is the relative coordinate of the protons, and $|\text{spin}\rangle$ may be either a singlet or a triplet. Because of rotational symmetry, the function $\Psi(\mathbf{X}, \mathbf{r})$ may be an eigenfunction of the relative orbital angular momentum of the protons. According to the symmetrization postulate, the full state function must change sign when the protons are interchanged. Since the singlet spin state is odd under permutation, it must be accompanied by an orbital function $\Psi(\mathbf{X}, \mathbf{r})$ that is even in $\mathbf{r}$. Similarly, the triplet spin state is unchanged by permutation, so it must be accompanied by an orbital function that is odd in $\mathbf{r}$. Therefore the protons in the *singlet* state (called "para-H_2", not to be confused with the hypothetical paraparticles that violate the symmetrization postulate) can have only *even* values of orbital angular momentum L, while the protons in the *triplet* state (called "ortho-H_2") can have only *odd* values of L. The rotational kinetic energy is $\hbar^2 L(L+1)/2I$, where I is the moment of inertia of the molecule. Thus the energy levels, and more importantly the differences between energy levels, will not be the same in the singlet and in the triplet state H_2.

These predictions of the symmetrization postulate are confirmed by molecular spectroscopy. The rotational energy levels of singlet and triplet state H_2 molecules also have a dramatic effect on the specific heat of hydrogen gas [see Wannier (1966), Sec. 11.4]. Thus we have good evidence that protons are fermions. Similar phenomena exist for other homonuclear diatomic molecules. Some of these nuclei, such as ^{14}N, contain an odd number of neutrons, and hence provide evidence that neutrons are fermions.

For evidence about the permutation symmetry of other particles such as mesons and hyperons, for which one cannot produce stable many-particle states, we refer to Messiah and Greenberg (1964). It is necessary to search for a reaction that is forbidden by the symmetrization postulate and is *not* forbidden by any other symmetry or selection rule. Observation of such a reaction would contradict the symmetrization postulate; conversely the absence of such reactions supports it.

Scattering of identical particles can also provide evidence for the symmetrization postulate. The two scattering events shown in Fig. 17.1 cannot be distinguished if the particles are identical, and so the differential cross section for this process will include both of them.

The usual scattering wave function has the asymptotic form

$$\psi(\mathbf{r}) \sim e^{i\mathbf{k}\cdot\mathbf{r}} + f(\theta)\,\frac{e^{ikr}}{r}\,,$$

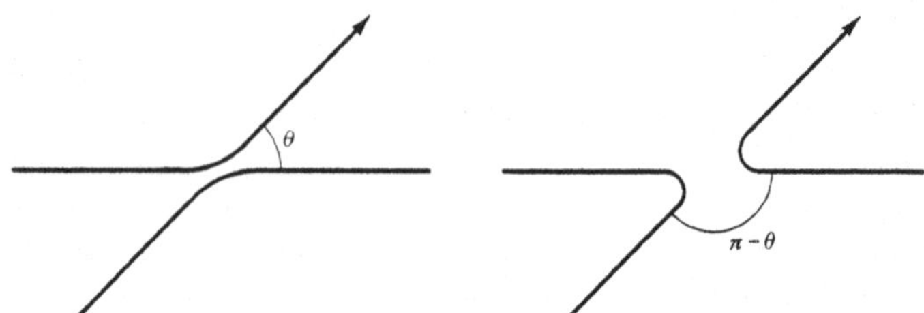

Fig. 17.1 Indistinguishable scattering events.

where **r** is the relative coordinate of the two particles. The differential cross section is then obtained as the square of the scattering amplitude: $\sigma(\theta) = |f(\theta)|^2$. But if the two particles are identical, it is necessary for the spatial wave function to be symmetric or antisymmetric under interchange of particles ($\mathbf{r} \to -\mathbf{r}$), according to the permutation symmetry of the spin state. After symmetrization/antisymmetrization of the spatial wave function, the scattering amplitude (defined as the coefficient of e^{ikr}/r) will be $f(\theta) \pm f(\pi - \theta)$. Thus the differential cross section will be

$$\begin{aligned}\sigma(\theta) &= |f(\theta) \pm f(\pi - \theta)|^2 \\ &= |f(\theta)|^2 + |f(\pi - \theta)|^2 \ \pm \ 2\,\mathrm{Re}\,[f^*(\theta)\, f(\pi - \theta)]\,. \end{aligned} \tag{17.6}$$

The plus sign applies for spin $\frac{1}{2}$ particles in the singlet state. The minus sign applies for spin $\frac{1}{2}$ particles in the triplet state, or for spinless particles. The first and second terms of (17.6) are just the differential cross sections that would be calculated for the two events in Fig. 17.1 if the particles were distinguishable. The third term, which is a consequence of the symmetrization postulate, manifests itself most clearly as an enhancement (+ sign) or diminution (− sign) of the cross section near $\theta = \pi/2$.

17.4 *Creation and Annihilation Operators*

The symmetrization postulate restricts the states of a species of particles to a single permutation symmetry type, either symmetric or antisymmetric under pair interchanges. This greatly simplifies the theory of many-particle states, and allows us to introduce an elegant formalism involving *creation* and *annihilation* operators. This formalism is not restricted to systems with a fixed number of particles. Instead, it treats the number of particles as a dynamical variable, and it treats states having any number of particles. This is a useful technique even in the theory of stable particles, for which the number of particles is constant; moreover it is easily extended to describe the physical creation and annihilation of particles that occur at high energies. However we shall not be treating such applications in this book.

The orthonormal basis vectors of the state space (known as *Fock space*) consist of: the vacuum or no-particle state, $|0\rangle$; a complete set of one-particle state vectors, $\{|\phi_\alpha\rangle : (\alpha = 1, 2, 3, \ldots)\}$, where the label α is actually an abbreviation for all the quantum numbers needed to specify a unique state; a complete set of two-particle state vectors; a complete set of three-particle state vectors; etc. However, these *complete* sets of many-particle states contain *only* vectors of the correct permutation symmetry, and hence they are *complete* within the appropriate permutation-invariant subspace. There is no need for them to span the full vector space considered in Sec. 17.1, which is physically irrelevant in light of the symmetrization postulate. There are technical differences between the formalisms for *fermions* and for *bosons*, which make it convenient to treat the two cases separately. Nevertheless, we shall see that a very strong analogy exists between the two formalisms, and many results will apply to both cases.

Fermions

We define the *creation operator* ${C_\alpha}^\dagger$ by the relations

$$ {C_\alpha}^\dagger|0\rangle = |\alpha\rangle \equiv |\phi_\alpha\rangle\,, \tag{17.7a}$$

$$ {C_\alpha}^\dagger|\beta\rangle = {C_\alpha}^\dagger\,{C_\beta}^\dagger|0\rangle = |\alpha\beta\rangle = -|\beta\alpha\rangle\,, \tag{17.7b}$$

$$ {C_\alpha}^\dagger|\beta\gamma\rangle = {C_\alpha}^\dagger\,{C_\beta}^\dagger\,{C_\gamma}^\dagger|0\rangle = |\alpha\beta\gamma\rangle\,, \tag{17.7c}$$

etc.

These vectors are defined to be antisymmetric under interchange of adjacent arguments, and thus $|\alpha\beta\gamma\rangle = -|\alpha\gamma\beta\rangle = |\gamma\alpha\beta\rangle$. The antisymmetric three-particle example given Sec. 17.1 is equal to $|\alpha\beta\gamma\rangle$. In coordinate representation, these vectors become

$$\langle \mathbf{x}|\alpha\rangle = \phi_\alpha(\mathbf{x})\,,$$

$$\langle \mathbf{x}_1, \mathbf{x}_2|\alpha\beta\rangle = \frac{\phi_\alpha(\mathbf{x}_1)\,\phi_\beta(\mathbf{x}_2) - \phi_\beta(\mathbf{x}_1)\,\phi_\alpha(\mathbf{x}_2)}{\sqrt{2}}\,.$$

We shall refer to the function $\phi_\alpha(\mathbf{x})$ as an *orbital*, and we shall say that in the vector $|\alpha\beta\gamma\rangle$ the α, β, and γ orbitals are *occupied*, while all other orbitals are unoccupied. The infinite sequence of equations (17.7) can be summarized by the formula

$$C_\alpha{}^\dagger|\cdots\rangle = |\alpha\cdots\rangle\,, \tag{17.8}$$

where the string denoted as "$\cdots$" (which may be of any length, or null) is the same on both sides of the equation.

If we operate with $C_\alpha{}^\dagger$ on a vector in which the α orbital is occupied, we formally obtain $C_\alpha{}^\dagger|\alpha\cdots\rangle = |\alpha\alpha\cdots\rangle$. Since the vector $|\alpha\alpha\cdots\rangle$ must change sign upon interchange of its first two arguments, we have $|\alpha\alpha\cdots\rangle = -|\alpha\alpha\cdots\rangle$, and therefore

$$C_\alpha{}^\dagger|\alpha\cdots\rangle = 0\,. \tag{17.9}$$

Thus the *Pauli exclusion principle* is automatically satisfied, it being impossible for an orbital to be more than singly occupied.

The operator $C_\alpha{}^\dagger$ is fully defined by (17.8) and (17.9), from which the properties of its adjoint, $C_\alpha = (C_\alpha{}^\dagger)^\dagger$, can be deduced. From (17.8) and (17.9) we have

$$\langle\alpha\cdots|C_\alpha{}^\dagger|\cdots(\sim\alpha)\rangle = 1\,,$$

$$\langle\psi|C_\alpha{}^\dagger|\cdots(\sim\alpha)\rangle = 0\,, \quad \text{if } \langle\psi|\alpha\cdots\rangle = 0\,.$$

In these two lines, the string denoted by "$\cdots$" is the same in both instances within the same line. The notation $|\cdots(\sim\alpha)\rangle$ signifies that the α orbital is unoccupied. From (17.9) we have

$$\langle\alpha\cdots|C_\alpha = 0\,.$$

The three relations above yield, respectively,

$$\langle(\sim\alpha)\cdots|C_\alpha|\alpha\cdots\rangle = 1\,, \tag{17.10}$$

$$\langle(\sim\alpha)\cdots|C_\alpha|\psi\rangle = 0\,, \quad \text{if } \langle\psi|\alpha\cdots\rangle = 0\,, \tag{17.11}$$

$$\langle\alpha\cdots|C_\alpha|\psi\rangle = 0\,. \tag{17.12}$$

Applying these relations with $|\psi\rangle = |0\rangle$, we deduce from (17.11) that the vector $C_\alpha|0\rangle$ is orthogonal to any basis vector in which the α orbital is unoccupied, and from (17.12) it follows that $C_\alpha|0\rangle$ is orthogonal to any basis vector in which the α orbital is occupied. Therefore we have

$$C_\alpha|0\rangle = 0\,. \tag{17.13}$$

Applying (17.12) with $|\psi\rangle = |\alpha\rangle$, we deduce that vector $C_\alpha|\alpha\rangle$ is orthogonal to all basis vectors in which the α orbital is occupied. From (17.11) we deduce that $C_\alpha|\alpha\rangle$ is orthogonal to all but one basis vector in which tha α orbital is unoccupied, the single exception being $\langle 0|C_\alpha|\alpha\rangle = 1$. Therefore we have

$$C_\alpha|\alpha\rangle = |0\rangle\,. \tag{17.14}$$

By means of a similar argument we deduce that

$$C_\alpha|\alpha\cdots\rangle = |\cdots(\sim\alpha)\rangle\,. \tag{17.15}$$

Finally we examine the case of $|\psi\rangle = |\cdots(\sim\alpha)\rangle$, for which (17.11) and (17.12) imply that

$$C_\alpha|\cdots(\sim\alpha)\rangle = 0\,. \tag{17.16}$$

Thus we see that the effect of the operator C_α is to empty the α orbital if it is occupied, and to destroy the vector if that orbital is unoccupied. Hence C_α may be called an *annihilation operator.*

To summarize, the *creation operator* ${C_\alpha}^\dagger$ adds a particle to the α orbital if it is empty, and the *annihilation operator* C_α removes a particle from the α orbital if it is occupied. Otherwise these operators yield zero.

If the creation operator ${C_\alpha}^\dagger$ acts twice in succession on an arbitrary vector, it would create a doubly occupied orbital, ${C_\alpha}^\dagger\,{C_\alpha}^\dagger|\psi\rangle = |\alpha\alpha\psi\rangle$, and so the result must vanish, as was pointed out in deriving (17.9). Since the initial vector $|\psi\rangle$ was arbitrary, we have the operator equality

$${C_\alpha}^\dagger\,{C_\alpha}^\dagger = 0\,. \tag{17.17}$$

The adjoint of this equation is

$$C_\alpha\,C_\alpha = 0\,. \tag{17.18}$$

Consider next the operator combination ${C_\alpha}^\dagger\,{C_\beta}^\dagger + {C_\beta}^\dagger\,{C_\alpha}^\dagger$ (which is called the *anticommutator* of ${C_\alpha}^\dagger$ and ${C_\beta}^\dagger$) acting on an arbitrary vector:

$$\begin{aligned}({C_\alpha}^\dagger\,{C_\beta}^\dagger + {C_\beta}^\dagger\,{C_\alpha}^\dagger)|\psi\rangle &= |\alpha\beta\psi\rangle + |\beta\alpha\psi\rangle \\ &= |\alpha\beta\psi\rangle - |\alpha\beta\psi\rangle = 0\,.\end{aligned}$$

Thus we obtain the operator relations

$${C_\alpha}^\dagger\,{C_\beta}^\dagger + {C_\beta}^\dagger\,{C_\alpha}^\dagger = 0\,, \tag{17.19}$$

$$C_\alpha C_\beta + C_\beta C_\alpha = 0\,. \tag{17.20}$$

Finally, we consider the anticommutator of a creation operator and an annihilation operator, $C_\alpha\, C_\beta{}^\dagger + C_\beta{}^\dagger\, C_\alpha$. For $\alpha \neq \beta$ it is apparent from our previous results that this operator will yield zero if either the α orbital is empty or the β orbital is occupied. Hence we need only consider its effect on a vector of the form $|\alpha\cdots(\sim\beta)\rangle$.

$$\begin{aligned}(C_\alpha C_\beta{}^\dagger + C_\beta{}^\dagger\, C_\alpha)|\alpha\cdots(\sim\beta)\rangle &= C_\alpha|\beta\alpha\cdots\rangle + C_\beta{}^\dagger|\cdots(\sim\alpha,\sim\beta)\rangle \\ &= -C_\alpha|\alpha\beta\cdots\rangle + C_\beta{}^\dagger|\cdots(\sim\alpha,\sim\beta)\rangle \\ &= -|\beta\cdots(\sim\alpha)\rangle + |\beta\cdots(\sim\alpha)\rangle = 0\,.\end{aligned}$$

For $\alpha = \beta$ we consider separately the cases of the α orbital being occupied or empty:

$$(C_\alpha C_\alpha{}^\dagger + C_\alpha{}^\dagger\, C_\alpha)|\alpha\cdots\rangle = 0 + C_\alpha{}^\dagger|\cdots(\sim\alpha)\rangle = |\alpha\cdots\rangle\,,$$

$$(C_\alpha C_\alpha{}^\dagger + C_\alpha{}^\dagger\, C_\alpha)|\cdots(\sim\alpha)\rangle = C_\alpha|\alpha\cdots\rangle + 0 = |\cdots(\sim\alpha)\rangle\,.$$

Thus it is apparent that $C_\alpha C_\alpha{}^\dagger + C_\alpha{}^\dagger\, C_\alpha$ is the identity operator. These last few results are summarized by the equation

$$C_\alpha C_\beta{}^\dagger + C_\beta{}^\dagger C_\alpha = \delta_{\alpha\beta} I\,. \tag{17.21}$$

All of the Fock basis vectors are eigenvectors of the operator $C_\alpha{}^\dagger\, C_\alpha$. It is easily verified that if the α orbital is empty the operator $C_\alpha{}^\dagger\, C_\alpha$ has eigenvalue 0, and if the α orbital is occupied the operator $C_\alpha{}^\dagger\, C_\alpha$ has eigenvalue 1. Thus $C_\alpha{}^\dagger\, C_\alpha$ functions as the *number operator* for the α orbital. The *total number operator* is therefore equal to

$$N = \sum_\alpha C_\alpha{}^\dagger\, C_\alpha\,. \tag{17.22}$$

A vector that is an arbitrary linear combination of these basis vectors need not be an eigenvector of N, in which case there will be a probability distribution for N, as there is for any dynamical variable.

Change of basis. The creation and annihilation operators have been defined with respect to a particular set of single particle basis functions, $C_\alpha{}^\dagger|0\rangle = |\alpha\rangle$ corresponding to the function $\phi_\alpha(\mathbf{x})$. Let us introduce another set of creation and annihilation operators, $b_j{}^\dagger$ and b_j, with $b_j{}^\dagger|0\rangle = |j\rangle$ corresponding to the function $f_j(\mathbf{x})$. The two sets of functions $\{\phi_\alpha(\mathbf{x})\}$ and $\{f_j(\mathbf{x})\}$ are both complete and orthonormal, and the members of one set can be expressed as linear combinations of the other:

$$f_j(\mathbf{x}) = \sum_\alpha \phi_\alpha(\mathbf{x})\langle\alpha|j\rangle\,,$$

or, equivalently,

$$b_j{}^\dagger|0\rangle = \sum_\alpha C_\alpha{}^\dagger|0\rangle\langle\alpha|j\rangle\,. \tag{17.23}$$

The new creation and annihilation operators must also satisfy the anticommutation relations (17.19), (17.20), and (17.21), since these characterize the essential properties of creation and annihilation operators. All of these requirements are satisfied by the linear transformation,

$$b_j{}^\dagger = \sum_\alpha C_\alpha{}^\dagger\,\langle\alpha|j\rangle\,,\ b_j = \sum_\alpha \langle j|\alpha\rangle C_\alpha\,. \tag{17.24}$$

As an example of such a change of basis, we consider the family of operators that create position eigenvectors,

$$\psi^\dagger(\mathbf{x})|0\rangle = |\mathbf{x}\rangle\,. \tag{17.25}$$

Applying (17.24) to this case, we obtain

$$\psi^\dagger(\mathbf{x}) = \sum_\alpha C_\alpha{}^\dagger\,\langle\alpha|\mathbf{x}\rangle\,,\ \psi(\mathbf{x}) = \sum_\alpha \langle\mathbf{x}|\alpha\rangle C_\alpha\,.$$

Now $\langle\mathbf{x}|\alpha\rangle = \phi_\alpha(\mathbf{x})$ is just the original basis function in coordinate repesentation, and hence we have

$$\psi^\dagger(\mathbf{x}) = \sum_\alpha [\phi_\alpha(\mathbf{x})]^* C_\alpha{}^\dagger\,,\ \psi(\mathbf{x}) = \sum_\alpha \phi_\alpha(\mathbf{x}) C_\alpha\,. \tag{17.26}$$

These new operators, which create and annihilate at a point in space, are often called the *field operators.* The product $\psi^\dagger(\mathbf{x})\psi(\mathbf{x})$ is the *number density* operator (not to be confused with the occasional use of the term "density matrix" as a synonym for "state operator"), and the total number operator (17.22) is equal to

$$N = \int \psi^\dagger(\mathbf{x})\psi(\mathbf{x})d^3x\,. \tag{17.27}$$

It should be noted that by introducing these "field" operators, we have not made a transition to quantum field theory, but have merely introduced another representation for many-particle quantum mechanics. However, this representation provides a close mathematical analogy between particle quantum mechanics and quantum field theory.

Bosons

The construction of the Fock space basis vectors proceeds very similarly in this case as it did for fermions, except that now the multiparticle states must be *symmetric* under interchange of particles. This implies that multiple occupancy of orbitals is now possible, since, for example, $|\alpha\rangle \otimes |\alpha\rangle \otimes |\alpha\rangle$ is acceptable as a symmetric three-particle state vector. So whereas we could label the antisymmetric states of fermions by merely specifying the occupied orbitals, we must now also specify the degree of occupancy. If the single particle basis vectors consist of the set of orbitals $\{|\phi_\alpha\rangle : (\alpha = 1, 2, 3, \ldots)\}$, we may denote a many-boson state vector as $|n_1, n_2, n_3, \ldots\rangle$ where the nonnegative integer n_α is the occupancy of the orbital ϕ_α. (This notation might also have been used for fermion states, in which case we would have restricted n_α to be 0 or 1. However, a different notation was more convenient in that case.)

We now define *creation operators* with the following properties:

$$\begin{aligned} a_\alpha{}^\dagger|0\rangle &= |\phi_\alpha\rangle = |0, 0, \ldots, n_\alpha = 1, 0, \ldots\rangle\,, \\ a_\alpha{}^\dagger|n_1, n_2, \ldots, n_\alpha, \ldots\rangle &\propto |n_1, n_2, \ldots, n_\alpha + 1, \ldots\rangle\,. \end{aligned} \tag{17.28}$$

Since these vectors are symmetric under permutation of particles, we must have $a_\alpha{}^\dagger a_\beta{}^\dagger = a_\beta{}^\dagger a_\alpha{}^\dagger$. It follows from arguments similar to those used for fermions that the adjoint operator, $a_\alpha = (a_\alpha{}^\dagger)^\dagger$, functions as an *annihilation operator*, with the properties

$$\begin{aligned} &a_\alpha|\phi_\alpha\rangle = |0\rangle\,, \\ &a_\alpha|n_1, n_2, \ldots, n_\alpha, \ldots\rangle \propto |n_1, n_2, \ldots, n_\alpha - 1, \cdots\rangle\,,\ (n_\alpha > 0)\,, \\ &a_\alpha|n_1, n_2, \ldots, n_\alpha = 0, \ldots\rangle = 0\,. \end{aligned} \tag{17.29}$$

The unspecified proportionality factor in these equations is fixed by requiring the product $a_\alpha{}^\dagger\, a_\alpha$ to serve as the *number* operator for the α orbital:

$$a_\alpha{}^\dagger\, a_\alpha|n_1, n_2, \ldots, n_\alpha, \ldots\rangle = n_\alpha|n_1, n_2, \ldots, n_\alpha, \ldots\rangle\,. \tag{17.30}$$

Thus from the relation

$$(\langle n_1, n_2, \ldots, n_\alpha, \ldots |a_\alpha{}^\dagger)(a_\alpha|n_1, n_2, \ldots, n_\alpha, \ldots\rangle) = n_\alpha$$

we obtain

$$a_\alpha|n_1, n_2, \ldots, n_\alpha, \ldots\rangle = (n_\alpha)^{1/2}|n_1, n_2, \ldots, n_\alpha - 1, \ldots\rangle\,. \tag{17.31}$$

(The arbitrary phase factor has been set equal to 1, since that is the simplest choice.) The single equation (17.31) embodies all three of the equations (17.29).

We can now determine the proportionality factor in (17.28), which we rewrite as

$$a_\alpha{}^\dagger|n_1, n_2, \ldots, n_\alpha, \ldots\rangle = c|n_1, n_2, \ldots, n_\alpha + 1, \ldots\rangle\,. \tag{17.32}$$

Operating with a_α and using (17.31), we obtain

$$a_\alpha a_\alpha{}^\dagger|n_1, n_2, \ldots, n_\alpha, \ldots\rangle = (n_\alpha + 1)^{1/2} c|n_1, n_2, \ldots, n_\alpha, \ldots\rangle\,.$$

Operating again with $a_\alpha{}^\dagger$ and using (17.32), we obtain

$$a_\alpha{}^\dagger a_\alpha a_\alpha{}^\dagger|n_1, n_2, \ldots, n_\alpha, \ldots\rangle = c^2(n_\alpha + 1)^{1/2}|n_1, n_2, \ldots, n_\alpha + 1, \ldots\rangle\,. \tag{17.33}$$

But the left side of this equation can alternatively be evaluated using (17.30), obtaining

$$(a_\alpha{}^\dagger a_\alpha)a_\alpha{}^\dagger|n_1, n_2, \ldots, n_\alpha, \ldots\rangle = (n_\alpha + 1)c|n_1, n_2, \ldots, n_\alpha + 1, \ldots\rangle\,. \tag{17.34}$$

Equating (17.33) and (17.34), we obtain $c = (n_\alpha + 1)^{1/2}$, and hence

$$a_\alpha{}^\dagger|n_1, n_2, \ldots, n_\alpha, \ldots\rangle = (n_\alpha + 1)^{1/2}|n_1, n_2, \ldots, n_\alpha + 1, \ldots\rangle\,. \tag{17.35}$$

From (17.31) and (17.35) we deduce the commutation relation

$$a_\alpha a_\beta{}^\dagger - a_\beta{}^\dagger a_\alpha = \delta_{\alpha\beta} I\,, \tag{17.36}$$

which complements the previously determined commutation relations

$$a_\alpha{}^\dagger a_\beta{}^\dagger - a_\beta{}^\dagger a_\alpha{}^\dagger = a_\alpha a_\beta - a_\beta a_\alpha = 0\,. \tag{17.37}$$

Comparing with (17.19), (17.20), and (17.21), we see that the essential difference between the creation and annihilation operators for *fermions* and those for *bosons* is that the former satisfy *anticommutation relations* while the latter satisfy *commutation relations.* We also note that boson creation and annihilation operators are mathematically isomorphic to the raising and lowering operators of a harmonic oscillator, which were introduced in Sec. 6.1.

Representation of operators

Much of the formalism of creation and annihilation operators is the same for bosons as for fermions. For example, the linear transformation (17.24) for a change of basis is applicable to both cases. The expression of arbitrary dynamical variables in terms of creation and annihilation operators is essentially the same for bosons and fermions. We shall demonstrate it explicitly for fermions because their anticommutation relations require more care about + or − signs than is necessary for bosons.

The simplest dynamical variables are those that are additive over the particles. Some examples of *additive one-body operator* are:

Momentum $\sum_{i=1}^{n} \mathbf{P}_i$

Kinetic energy $\sum_{i=1}^{n} -\frac{\hbar^2}{2M} \nabla_i{}^2$

External potential $\sum_{i=1}^{n} W(\mathbf{x}_i)$

General form $$R = \sum_{i=1}^{n} R_i \tag{17.38}$$

The conventional form of such an operator is a sum of operators, each of which acts only on one individual labeled particle. But this labeling has no significance for identical particles. The representation of an additive one-body operator in terms of creation and annihilation operators is

$$R = \sum_{\alpha} \sum_{\beta} \langle \phi_\alpha | R_1 | \phi_\beta \rangle C_\alpha{}^\dagger C_\beta \,. \tag{17.39}$$

It has advantage of not referring to fictitiously labeled particles, and its form does not depend on the number of particles. We shall prove the equivalence of (17.38) and (17.39) by demonstrating that they have the same matrix elements between any pair of n-particle state vectors.

We first show that the form (17.39) is invariant under change of basis by considering a similar operator defined in another basis,

$$R' = \sum_j \sum_k \langle f_j|R_1|f_k\rangle\, b_j{}^\dagger\, b_k\,.$$

Using the substitution (17.24), we obtain

$$\begin{aligned} R' &= \sum_j \sum_k \sum_\alpha \sum_\beta C_\alpha{}^\dagger\, \langle\phi_\alpha|f_j\rangle\langle f_j|R_1|f_k\rangle\langle f_k|\phi_\beta\rangle C_\beta \\ &= \sum_\alpha \sum_\beta \langle\phi_\alpha|R_1|\phi_\beta\rangle C_\alpha{}^\dagger\, C_\beta = R\,. \end{aligned}$$

Since the form (17.39) is independent of the basis, as has just been shown, we may choose any convenient basis in which to demonstrate the equivalence of (17.38) and (17.39). Therefore we choose the new basis functions $\{|f_k\rangle\}$ to diagonalize the single particle operator R_1: $R_1|f_k\rangle = r_k|f_k\rangle$, which yields $R = \sum_k r_k\, b_k{}^\dagger\, b_k$. In this basis, the diagonal matrix elements of the operator R are equal to $\sum_k r_k n_k$, where n_k is the occupancy of the orbital f_k, and the nondiagonal matrix elements are zero. This is clearly in agreement with the matrix elements of (17.38), provided the number of particles $\sum_k n_k = n$ is definite.

The next kind of dynamical variable that must be considered is an *additive pair operator*, of which the interaction potential is the most important example:

$$V = \sum\sum_{i<j} v(\mathbf{x}_1, \mathbf{x}_j) = \frac{1}{2} \sum\sum_{i \neq j} v(\mathbf{x}_i, \mathbf{x}_j)\,. \tag{17.40}$$

We assume no self-interaction, so there are no terms in this expression having $i = j$. The representation of this operator in terms of creation and annihilation operators is

$$V = \frac{1}{2} \sum_\alpha \sum_\beta \sum_\gamma \sum_\delta v_{\alpha\beta,\gamma\delta} C_\alpha{}^\dagger\, C_\beta{}^\dagger\, C_\delta C_\gamma\,. \tag{17.41}$$

(Note that the order of the last two operators, $C_\delta C_\gamma$, is reversed relative to the order of the indices in the matrix element. This is necessary for anticommuting fermion operators.) The matrix element in the above expression is calculated for unsymmetrized product vectors,

$$v_{\alpha\beta,\gamma\delta} = \iint \phi_\alpha{}^*(\mathbf{x}_1)\phi_\beta{}^*(\mathbf{x}_2) v(\mathbf{x}_1, \mathbf{x}_2)\phi_\gamma(\mathbf{x}_1)\phi_\delta(\mathbf{x}_2) d^3x_1 d^3x_2\,. \tag{17.42}$$

It has certain symmetries:

$$v_{\gamma\delta,\alpha\beta} = (v_{\alpha\beta,\gamma\delta})^*\,, \quad v_{\beta\alpha,\delta\gamma} = v_{\alpha\beta,\gamma\delta}\,.$$

The matrix element between antisymmetric two-particle state vectors, of the form $|\alpha\beta\rangle = (|\phi_\alpha\rangle|\phi_\beta\rangle - |\phi_\beta\rangle|\phi_\alpha\rangle)/\sqrt{2}$, is

$$\langle\alpha\beta|v(\mathbf{x}_1,\mathbf{x}_2)|\gamma\delta\rangle = v_{\alpha\beta,\gamma\delta} - v_{\alpha\beta,\delta\gamma}\,. \tag{17.43}$$

Using these antisymmetric matrix elements, we may write (17.41) as

$$V = \frac{1}{4}\sum_\alpha\sum_\beta\sum_\gamma\sum_\delta \langle\alpha\beta|v|\gamma\delta\rangle C_\alpha{}^\dagger\, C_\beta{}^\dagger\, C_\delta C_\gamma\,. \tag{17.44}$$

To prove the equivalence of the forms (17.40) and (17.44), we choose a representation that diagonalizes the pair interaction $\langle\alpha\beta|v|\gamma\delta\rangle$, so that we obtain $\langle\alpha\beta|v|\gamma\delta\rangle = \langle\alpha\beta|v|\gamma\delta\rangle\delta_{\alpha\beta,\gamma\delta}$. Here we have introduced a variant of the Kronecker delta, with the property that $\delta_{\alpha\beta,\gamma\delta} = 1$ if and only if $|\alpha\beta\rangle$ and $|\gamma\delta\rangle$ describe the *same state*, and $\delta_{\alpha\beta,\gamma\delta} = 0$ otherwise. Note that the vector $|\beta\alpha\rangle = -|\alpha\beta\rangle$ describes the same state as does $|\alpha\beta\rangle$, so that $\delta_{\alpha\beta,\alpha\beta} = \delta_{\alpha\beta,\beta\alpha} = 1$. In this diagonal representation (17.44) reduces to

$$\begin{aligned} V &= \frac{1}{4}\sum_\alpha\sum_\beta\{\langle\alpha\beta|v|\alpha\beta\rangle C_\alpha{}^\dagger\, C_\beta{}^\dagger\, C_\beta C_\alpha + \langle\alpha\beta|v|\beta\alpha\rangle C_\alpha{}^\dagger\, C_\beta{}^\dagger\, C_\alpha C_\beta\} \\ &= \frac{1}{2}\sum_\alpha\sum_\beta \langle\alpha\beta|v|\alpha\beta\rangle C_\alpha{}^\dagger\, C_\alpha C_\beta{}^\dagger\, C_\beta\,. \end{aligned}$$

In simplifying this expression, we have relied on the antisymmetry of the states to ensure that terms with $\alpha = \beta$ do not occur. The diagonal matrix element of this operator is equal to $\sum_\alpha\sum_\beta \frac{1}{2}n_\alpha n_\beta\langle\alpha\beta|v|\alpha\beta\rangle$. Since $\frac{1}{2}n_\alpha n_\beta$ is just the number of pairs of particles occupying the orbitals α and β, it is apparent that the above form for V is equivalent to the sum-over-pairs form (17.40).

Example

Very commonly the pair interaction depends only on the relative separation of the two particles, $v(\mathbf{x}_1,\mathbf{x}_2) = v(\mathbf{x}_2 - \mathbf{x}_1)$. This translational invariance of the interaction makes it convenient to choose momentum eigenfunctions as basis vectors, $\phi_\alpha(\mathbf{x}) = (2\pi)^{-3/2}\exp(i\mathbf{k}\cdot\mathbf{x})$. Introducing the central and relative coordinates, $\mathbf{R} = (\mathbf{x}_2 + \mathbf{x}_1)/2$ and $\mathbf{r} = \mathbf{x}_2 - \mathbf{x}_1$, the unsymmetrical matrix element (17.42) becomes

$$
\begin{aligned}
v_{\alpha\beta,\gamma\delta} &= (2\pi)^{-6} \iint \exp\left\{i\left[-\mathbf{k}_\alpha\cdot(\mathbf{R}-\tfrac{1}{2}\mathbf{r}) - \mathbf{k}_\beta\cdot(\mathbf{R}+\tfrac{1}{2}\mathbf{r})\right.\right.\\
&\qquad \left.\left.+\mathbf{k}_\gamma\cdot(\mathbf{R}-\tfrac{1}{2}\mathbf{r}) + \mathbf{k}_\delta\cdot(\mathbf{R}+\tfrac{1}{2}\mathbf{r})\right]\right\} v(\mathbf{r})d^3r d^3R\\
&= (2\pi)^{-3}\int \exp[-i\mathbf{R}\cdot(\mathbf{k}_\alpha+\mathbf{k}_\beta-\mathbf{k}_\gamma-\mathbf{k}_\delta)]d^3R\\
&\qquad \times (2\pi)^{-3}\int \exp[-i\tfrac{1}{2}\mathbf{r}\cdot(\mathbf{k}_\gamma-\mathbf{k}_\alpha+\mathbf{k}_\beta-\mathbf{k}_\delta)]d^3r\\
&= \delta(\mathbf{k}_\alpha+\mathbf{k}_\beta-\mathbf{k}_\gamma-\mathbf{k}_\delta)\tilde{v}(\mathbf{q})\,, \qquad (17.45)
\end{aligned}
$$

with $\mathbf{q} = \mathbf{k}_\gamma - \mathbf{k}_\alpha = \mathbf{k}_\beta - \mathbf{k}_\delta$ and $\tilde{v}(\mathbf{q}) = (2\pi)^{-3}\int \exp(-i\mathbf{q}\cdot\mathbf{r})v(r)d^3r$. The relations among the variables are illustrated in Fig. 17.2. The initial and final momentum (in units of $\hbar$) of the first particle are $\mathbf{k}_\gamma$ and $\mathbf{k}_\alpha$, the initial and final momentum of the second particle are $\mathbf{k}_\delta$ and $\mathbf{k}_\beta$, and the momentum transferred by the interaction between the particles is $\mathbf{q}$.

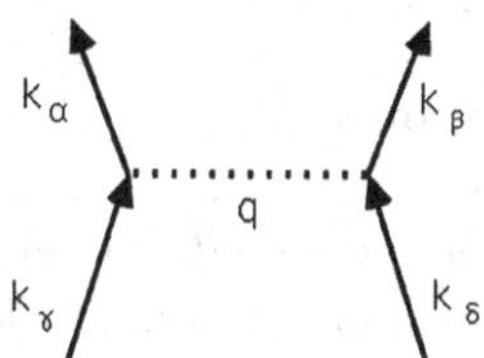

Fig. 17.2 Graphical representation of the interaction.

Wick's theorem

This theorem is very useful in performing calculations involving creation and annihilation operators. Before stating the theorem, we must define two preliminary notions. The *normal product* of a set of creation and annihilation operators is a product of those operators reordered so that all creation operators are to the left of all annihilation operators, multiplied by a factor (-1) for every pair interchange of fermion operators that is required to produce the reordering. Thus, for example, we have

$$N(a_\alpha a_\alpha{}^\dagger) = a_\alpha{}^\dagger\, a_\alpha \qquad \text{(bosons)}\,,$$

$$N(C_\alpha C_\alpha{}^\dagger) = -C_\alpha{}^\dagger\, C_\alpha \qquad \text{(fermions)}\,,$$

$$
\begin{aligned}
N(C_\alpha C_\beta C_\alpha{}^\dagger\, C_\beta{}^\dagger\, C_\gamma) &= C_\alpha{}^\dagger\, C_\beta{}^\dagger\, C_\alpha C_\beta C_\gamma\\
&= -C_\beta{}^\dagger\, C_\alpha{}^\dagger\, C_\alpha C_\beta C_\gamma\,.
\end{aligned}
$$

There may be several orderings all of which are *normal*, as in the last example, but all of them are equivalent. Operators within a normal product may be reordered as if all boson operators were commutative and all fermion operators were anticommutative. Indeed, any difference between the values of an ordinary product and the normal product of the same operators is due to nonvanishing commutators (for bosons) or anticommutators (for fermions).

The *contraction* of a pair of operators is defined to be their vacuum matrix element,

$$\langle AB\rangle_0 = \langle 0|AB|0\rangle\,. \tag{17.46}$$

It is easily verified that the only nonvanishing contractions involve a creation operator and an annihilation operator for the same orbital in antinormal order, i.e.

$$\langle C_\alpha {C_\alpha}^\dagger\rangle_0 = \langle 0|C_\alpha {C_\alpha}^\dagger|0\rangle \neq 0\,,$$

$$\langle a_\alpha {a_\alpha}^\dagger\rangle_0 = \langle 0|a_\alpha {a_\alpha}^\dagger|0\rangle \neq 0\,.$$

All other contractions are equal to zero.

We can now state *Wick's theorem:*

An ordinary product of any finite number of creation and annihilation operators is equal to the sum of normal products from which $0, 1, 2, \ldots$ *contractions have been removed in all possible ways.*

A few examples will make this abstract statement more comprehensible. For two operators Wick's theorem becomes

$$AB = N(AB) + \langle AB\rangle_0\,. \tag{17.47}$$

This is evidently true if A and B are both creation operators, or if A and B are both annihilation operators. In either case, we have $AB = N(AB)$ and the vacuum matrix element is zero. The same is true if A is a creation operator and B is an annihilation operator. If A is an annihilation operator and B is a creation operator, we may write

$$CC^\dagger = -C^\dagger C + [C, C^\dagger]_+ \quad \text{for fermions}\,,$$

$$aa^\dagger = a^\dagger a + [a, a^\dagger]_- \qquad \text{for bosons}\,.$$

(Here $[A,B]_+ \equiv AB + BA$ debotes the anticommutator, and $[A,B]_- \equiv AB - BA$ denotes the commutator.) But the terms $[C, C^\dagger]_+$ and $[a, a^\dagger]_-$ are multiples of the identity operator, and so they are equal to their vacuum matrix elements. Thus we obtain

$$CC^\dagger = -C^\dagger C + \langle 0|[C, C^\dagger]_+|0\rangle = -C^\dagger C + \langle 0|CC^\dagger|0\rangle = -C^\dagger C + \langle CC^\dagger\rangle_0 \,,$$
$$aa^\dagger = a^\dagger a + \langle 0|[a, a^\dagger]_-|0\rangle = a^\dagger a + \langle 0|aa^\dagger|0\rangle = a^\dagger a + \langle aa^\dagger\rangle_0 \,,$$

which confirms (17.47). Therefore we have proven Wick's theorem for two operators.

Consider next Wick's theorem applied to a product of four fermion operators:

$$\begin{aligned} ABCD = N(ABCD) & \\ & + N(AB)\langle CD\rangle_0 - N(AC)\langle BD\rangle_0 + N(BC)\langle AD\rangle_0 \\ & + N(AD)\langle BC\rangle_0 - N(BD)\langle AC\rangle_0 + N(CD)\langle AB\rangle_0 \\ & + \langle AB\rangle_0\langle CD\rangle_0 - \langle AC\rangle_0\langle BD\rangle_0 + \langle AD\rangle_0\langle BC\rangle_0 \,. \end{aligned} \tag{17.48}$$

The minus signs are due to the permutations of the operators that are needed to remove the contracted operators from the remaining factors. A similar expression without minus signs applies to four boson operators.

A general proof of Wick's theorem is given in App. C. It proceeds by induction, assuming the theorem true for n operators and proving that it must then be true for $n+1$ operators. The proof uses *only* the following properties of the operators:

(a) Any annihilation operator acting on the vector $|0\rangle$ yields zero;
(b) The commutator of two boson operators and the anticommutator of two fermion operators are multiples of the identity operator.

This will allow us to make useful generalizations of the theorem. Some applications of the creation and annihilation operator formalism to systems of fermions will be given in Ch. 18, and the boson operators will be used in Ch. 19.

Further reading for Chapter 17

The theoretical and experimental status of the symmetrization postulate is treated in considerable detail by Messiah and Greenberg, *Phys. Rev.* **B136**, 248–267 (1964). Of particular interest is the difficulty in experimentally disentangling the consequences of various symmetry principles. In the main text of this paper the authors cite the absence of decays of K_2^0 mesons into two pions as evidence that pions are bosons. Between the times of submission and publication of the paper, those two-pion decays were observed, and were generally interpreted as a violation of CP invariance ["charge conjugation" combined with "parity" (space inversion)] in weak interactions. In a note-added-in-proof, Messiah and Greenberg point out that the experiments could also be interpreted as evidence that pions are not bosons.

In their Resource Letter SS-1, *The Spin-Statistics Connection*, Curceanu, Gillaspy and Hilborn (2012) provide a review of the voluminous literature on this subject.

Problems

17.1 Prove that the formulas (17.22) and (17.27) for the number operator are equal.

17.2 The proof of equivalence of the forms (17.40) and (17.44) for an additive pair interaction that was given in the text used specific properties of fermion operators. Show that the equivalence also holds for bosons.

17.3 Show that the total number operator commutes with the Hamiltonian of a system of particles that interact via arbitrary pair interactions.

17.4 To what extent is a bound pair of fermions equivalent to a boson? (Some specific points to consider are: the permutation symmetry of many-particle states under interchange of fermion pairs; the commutation relations of the operators that create and annihilate bound pair states.)

17.5 Show that the anticommutation relations of the field operators (17.26) are $\psi(\mathbf{x})\psi(\mathbf{x}') + \psi(\mathbf{x}')\psi(\mathbf{x}) = 0$ and

$$\psi^\dagger(\mathbf{x})\psi(\mathbf{x}') + \psi(\mathbf{x}')\psi^\dagger(\mathbf{x}) = \delta(\mathbf{x} - \mathbf{x}') \,.$$

17.6 Show that for two particles of spin s the ratio of the number of symmetric states to the number of antisymmetric states is $(s + 1)/s$.

Chapter 18

Many-Fermion Systems

Most of the problems and applications of quantum mechanics that we considered prior to Ch. 17 effectively involved only single particle states. This was due in part to a deliberate selection of simple problems. In other cases the degrees of freedom could be decoupled, resulting in an effective one-particle problem. The separation of the two-particle scattering problem into CM and relative coordinates is a good example of this decoupling. In this chapter we shall consider some many-body problems that cannot be reduced to effective one-body problems. The new features that emerge are of both practical and fundamental significance.

18.1 *Exchange*

The term *exchange* is commonly used for those properties of many-fermion systems that are a direct consequence of the antisymmetry of the state function under interchange of particles. Consider the simplest example, a system of two free electrons, whose Hamiltonian in coordinate representation is $H = -(\hbar^2/2M)({\nabla_1}^2 + {\nabla_2}^2)$. Because H contains no coupling between the two electrons, the eigenvalue equation $H\Psi(\mathbf{x}_1, \mathbf{x}_2) = E\Psi(\mathbf{x}_1, \mathbf{x}_2)$ can be separated into a pair of single particle equations, with the eigenfunctions of H having the product form $\Psi(\mathbf{x}_1, \mathbf{x}_2) = \exp(i\mathbf{k} \cdot \mathbf{x}_1) \exp(i\mathbf{k}' \cdot \mathbf{x}_2)$. If the symmetrization postulate could be ignored, these product eigenfunctions would describe stationary states of the system, and the position probability density $|\Psi(\mathbf{x}_1, \mathbf{x}_2)|^2$ would exhibit no correlation between the positions of the two particles.

However, the symmetrization postulate requires the total state function to change sign when the two electrons are interchanged. So, instead of the product functions, we must take an appropriately symmetrized linear combination of them,

$$\begin{aligned}\Psi_\pm(\mathbf{x}_1, \mathbf{x}_2) &= 2^{-1/2} \left[e^{i\mathbf{k}\cdot\mathbf{x}_1} e^{i\mathbf{k}'\cdot\mathbf{x}_2} \pm e^{i\mathbf{k}\cdot\mathbf{x}_2} e^{i\mathbf{k}'\cdot\mathbf{x}_1} \right] \\ &= 2^{-1/2} e^{i(\mathbf{k}+\mathbf{k}')\cdot\mathbf{R}} \left[e^{i(\mathbf{k}-\mathbf{k}')\cdot\mathbf{r}/2} \pm e^{-i(\mathbf{k}-\mathbf{k}')\cdot\mathbf{r}/2} \right], \qquad (18.1)\end{aligned}$$

where $\mathbf{R} = (\mathbf{x}_1 + \mathbf{x}_2)/2$ is the center of mass, and $\mathbf{r} = \mathbf{x}_1 - \mathbf{x}_2$ is the separation between the electrons. The plus sign must be chosen with the antisymmetric *singlet* spin state, and the minus sign must be chosen with the symmetric *triplet* spin state. The position probability density is $|\Psi_+(\mathbf{x}_1, \mathbf{x}_2)|^2 = 2|\cos[(\mathbf{k}-\mathbf{k}')\cdot\mathbf{r}/2]|^2$ for the singlet state, and $|\Psi_-(\mathbf{x}_1, \mathbf{x}_2)|^2 = 2|\sin[(\mathbf{k}-\mathbf{k}')\cdot\mathbf{r}/2]|^2$ for the triplet state. Thus, as a consequence of antisymmetry, the positions of the two electrons are correlated, even though the Hamiltonian contains no interaction between them.

The Fermi sea

Let us now consider an arbitrary number of noninteracting electrons. Without increasing the complexity of the problem, we may assume the electrons to be bound in an external potential $W(\mathbf{x})$ which yields a set of single particle energy eigenfunctions $\{\phi_j(\mathbf{x})\}$ (called "orbitals"). These may be used as basis functions, with corresponding creation operators $C_\alpha{}^\dagger|0\rangle = |\phi_\alpha\rangle$. Here we use a composite label, $\alpha = (j, \sigma)$, where j labels the eigenfunction $\phi_j(\mathbf{x})$ and $\sigma = \pm 1$ labels the eigenvalue of the spin component σ_z. The ground state of a system of N noninteracting electrons is a "Fermi sea" in which the N lowest single particle energy levels are filled:

$$|F\rangle = \prod_{\alpha \leq N} C_\alpha{}^\dagger|0\rangle \,. \tag{18.2}$$

(The condition $\alpha \leq N$ is to be interpreted symbolically to mean that α ranges over the N lowest single particle energy levels. Assuming that the energy of an electron does not depend on the direction of its spin, the condition could also be written as $j \leq N/2$, provided N is even.)

The correlation functions can be calculated with the help of the number density operators. It is useful to introduce *spin density operators*,

$$n_\sigma(\mathbf{x}) = \boldsymbol{\psi}_\sigma{}^\dagger(\mathbf{x})\boldsymbol{\psi}_\sigma(\mathbf{x}) \,, \tag{18.3}$$

where we have defined a spin-dependent field operator, in generalization of (17.26), as

$$\boldsymbol{\psi}_\sigma(\mathbf{x}) = \sum_j \phi_j(\mathbf{x}) C_{j\sigma} \,. \tag{18.4}$$

For the "Fermi sea" ground state, the single particle distribution is $\langle F|n_\sigma(\mathbf{x})|F\rangle$, and the pair distribution is $\langle F|n_\sigma(\mathbf{x}_1)n_{\sigma'}(\mathbf{x}_2)|F\rangle$. These quantities can be evaluated by means of Wick's theorem.

The calculation is made easier if we perform a *particle/hole* transformation. We define a new set of creation and annihilation operators, $\{b_\alpha{}^\dagger, b_\alpha\}$:

$$b_\alpha{}^\dagger = C_\alpha\,, \quad b_\alpha = C_\alpha{}^\dagger\,, \quad \text{if } \alpha \text{ is occupied in } |F\rangle\,, \tag{18.5a}$$

$$b_\alpha{}^\dagger = C_\alpha{}^\dagger\,, \quad b_\alpha = C_\alpha\,, \quad \text{if } \alpha \text{ is empty in } |F\rangle\,. \tag{18.5b}$$

We say that the new creation operator $b_\alpha{}^\dagger$ creates a *hole* if α corresponds to an occupied energy level in the Fermi sea, and that it creates a *particle* if the α orbital is outside the Fermi sea. The new annihilation operator has the property

$$b_\alpha|F\rangle = 0 \tag{18.6}$$

for all α. Thus the Fermi sea ground state is analogous to a vacuum state for the particle/hole operators. This makes it possible to apply *Wick's theorem* (Sec. 17.4) with the following modifications:

(1) *Normal ordering* is taken with respect to the particle/hole operators, with $b_\alpha{}^\dagger$ taken to the left of b_β:
(2) *Contractions* are defined as matrix elements in the "Fermi sea" ground state,

$$\langle AB\rangle_F = \langle F|AB|F\rangle\,. \tag{18.7}$$

The only nonvanishing contraction is $\langle b_\alpha b_\alpha{}^\dagger\rangle_F$. Since the field operators and other dynamical variables are naturally expressed in terms of the original electron creation and annihilation operators, $C_\alpha{}^\dagger$ and C_α, it is useful to express the nonvanishing contractions in terms of them:

$$\langle C_\alpha{}^\dagger\, C_\alpha\rangle_F = 1\,, \quad \text{for } \alpha \le N \;\; (\alpha \text{ in the Fermi sea})\,, \tag{18.8a}$$

$$\langle C_\alpha\, C_\alpha{}^\dagger\rangle_F = 1\,, \quad \text{for } \alpha \le N \;\; (\alpha \text{ outside of the Fermi sea})\,. \tag{18.8b}$$

All other contractions are zero.

The *one-particle distribution function* is

$$\begin{aligned}\langle F|n_\sigma(\mathbf{x})|F\rangle &= \langle F|\psi_\sigma{}^\dagger(\mathbf{x})\,\psi_\sigma(\mathbf{x})|F\rangle \\ &= \sum_j\sum_k \phi^*{}_j(\mathbf{x})\,\phi_k(\mathbf{x})\,\langle\, F|C_{j\sigma}{}^\dagger\, C_{k\sigma}|F\,\rangle \\ &= \sum_{j\le N/2} |\phi_j(\mathbf{x})|^2\,.\end{aligned} \tag{18.9}$$

(For convenience, we assume that N is even.)

The *two-particle distribution function* is

$$\langle F|n_\sigma(\mathbf{x}_1)n_{\sigma'}(\mathbf{x}_2)|F\rangle = \langle F|\psi_\sigma{}^\dagger(\mathbf{x}_1)\psi_\sigma(\mathbf{x}_1)\psi_{\sigma'}{}^\dagger(\mathbf{x}_2)\psi_{\sigma'}(\mathbf{x}_2)|F\rangle\,. \tag{18.10}$$

Direct evaluation of this expression from the definitions (18.2) and (18.4) would be tedious, but it is made almost trivial by the particle/hole transformation and Wick's theorem. The product of the four field operators is expressed by Wick's theorem in terms of normal products and contractions [see Eq. (17.48)]. The Fermi ground state matrix element of any normal product is zero by virtue of (18.6), so only the fully contracted terms survive. Thus we obtain

$$\begin{aligned}\langle F|n_\sigma(\mathbf{x}_1)n_{\sigma'}(\mathbf{x}_2)|F\rangle &= \langle\psi_\sigma{}^\dagger(\mathbf{x}_1)\,\psi_\sigma(\mathbf{x}_1)\rangle_F\,\langle\psi_{\sigma'}{}^\dagger(\mathbf{x}_2)\,\psi_{\sigma'}(\mathbf{x}_2)\rangle_F\\ &\quad - \langle\psi_\sigma{}^\dagger(\mathbf{x}_1)\,\psi_{\sigma'}(\mathbf{x}_2)\rangle_F\,\langle\psi_{\sigma'}{}^\dagger(\mathbf{x}_2)\,\psi_\sigma(\mathbf{x}_1)\rangle_F\,. \end{aligned}\tag{18.11}$$

There are two cases to consider. For electrons of opposite spin orientation ($\sigma = -\sigma'$), the second term in (18.11) is zero. [None of the nonvanishing contractions (18.8) can occur in it because $\sigma \neq \sigma'$.] Therefore we have

$$\langle F|n_\uparrow(\mathbf{x}_1)n_\downarrow(\mathbf{x}_2)|F\rangle = \langle F|n_\uparrow(\mathbf{x}_1)|F\rangle\,\langle F|n_\downarrow(\mathbf{x}_2)|F\rangle\,. \tag{18.12}$$

There is no correlation between the positions of electrons with opposite spin orientation.

For electrons of parallel spin ($\sigma = \sigma'$), both terms of (18.11) contribute. The first term is the product of the one-particle distributions, as before. The first factor of the second term becomes

$$\begin{aligned}\langle\psi_\sigma{}^\dagger(\mathbf{x}_1)\psi_\sigma(\mathbf{x}_2)\rangle_F &= \sum_j\sum_k[\phi_j(\mathbf{x}_1)]^*\,\phi_k(\mathbf{x}_2)\,\langle F|C_{j\sigma}{}^\dagger\,C_{k\sigma}|F\rangle\\ &= \sum_{j\le N/2}\,[\phi_j(\mathbf{x}_1)]^*\,\phi_j(\mathbf{x}_2)\,.\end{aligned}$$

Hence the pair distribution for electrons with parallel spins is

$$\begin{aligned}\langle F|n_\uparrow(\mathbf{x}_1)n_\uparrow(\mathbf{x}_2)|F\rangle &= \sum_{j\le N/2}\;\sum_{k\le N/2}\,\{|\phi_j(\mathbf{x}_1)|^2\,|\phi_k(\mathbf{x}_2)|^2\\ &\quad - [\phi_j(\mathbf{x}_1)]^*\,\phi_j(\mathbf{x}_2)\,[\phi_k(\mathbf{x}_2)]^*\,\phi_k(\mathbf{x}_1)\}\,.\end{aligned}\tag{18.13}$$

The positions of the electrons are correlated as a consequence of the antisymmetry of the state, and it is apparent that the pair distribution vanishes in the limit $\mathbf{x}_1 \to \mathbf{x}_2$. Although the particles have been described as electrons, these results clearly apply to any particle of spin $\frac{1}{2}$. In particular, they would apply to protons and neutrons in a nucleus if the interparticle interaction were neglected.

The exchange interaction

For our next example, we consider two atoms, each having one electron. The nuclei of the atoms will be regarded as fixed force centers located at the positions $\mathbf{R}_A$ and $\mathbf{R}_B$, so this becomes a two-particle problem. The Hamiltonian of the two-electron system is

$$H = \sum_{j=1}^{2} \left\{ \frac{-\hbar^2}{2M} \nabla_j{}^2 + W(\mathbf{x}_j - \mathbf{R}_A) + W(\mathbf{x}_j - \mathbf{R}_B) \right\} + V(\mathbf{x}_1 - \mathbf{x}_2) , \quad (18.14)$$

where $W(\mathbf{x} - \mathbf{R}_A)$ is the interaction between an electron and the nucleus at $\mathbf{R}_A$, and $V(\mathbf{x}_1 - \mathbf{x}_2) = e^2/|\mathbf{x}_1 - \mathbf{x}_2|$ is the interaction between electrons.

We wish to take the isolated atoms as a starting point, and to treat the interaction between them as a perturbation. Therefore it would seem reasonable to separate the Hamiltonian into two terms, $H = H_0 + H'$, with

$$H_0 = -\frac{\hbar^2}{2M} \nabla_1{}^2 + W(\mathbf{x}_1 - \mathbf{R}_A) - \frac{\hbar^2}{2M} \nabla_2{}^2 + W(\mathbf{x}_2 - \mathbf{R}_B) , \quad (18.15)$$

$$H' = W(\mathbf{x}_2 - \mathbf{R}_A) + W(\mathbf{x}_1 - \mathbf{R}_B) + V(\mathbf{x}_1 - \mathbf{x}_2) . \quad (18.16)$$

The eigenfunctions of H_0 are products of atomic eigenfunctions,

$$H_0\, \phi_j(\mathbf{x}_1 - \mathbf{R}_A)\, \phi_k(\mathbf{x}_2 - \mathbf{R}_B) = (\varepsilon_j + \varepsilon_k)\, \phi_j(\mathbf{x}_1 - \mathbf{R}_A)\, \phi_k(\mathbf{x}_2 - \mathbf{R}_B) , \quad (18.17)$$

with

$$\left\{ -\frac{\hbar^2}{2M} \nabla^2 + W(\mathbf{x}) \right\} \phi_j(\mathbf{x}) = \varepsilon_j\, \phi_j(\mathbf{x}) , \quad (18.18)$$

so H_0 does indeed describe two isolated atoms. Unfortunately this particular separation of the Hamiltonian leads to technical difficulties because the terms H_0 and H' are not symmetric under permutation of the two electrons (although of course their sum H is symmetric). Therefore the symmetrized function $\phi_j(\mathbf{x}_1 - \mathbf{R}_A)\phi_k(\mathbf{x}_2 - \mathbf{R}_B) \pm \phi_j(\mathbf{x}_2 - \mathbf{R}_A)\phi_k(\mathbf{x}_1 - \mathbf{R}_B)$ is not an eigenfunction of H_0.

Instead of perturbation theory, we shall use the variational method with the trial function suggested by the above argument:

$$\psi_\pm(\mathbf{x}_1, \mathbf{x}_2) = \phi_A(1)\phi_B(2) \pm \phi_B(1)\phi_A(2) . \quad (18.19)$$

Here we have introduced an abbreviated notation, $\phi_A(1) = \phi(\mathbf{x}_1 - \mathbf{R}_A)$, $\phi_B(2) = \phi(\mathbf{x}_2 - \mathbf{R}_B)$, etc., where ϕ denotes the atomic orbital that is normally occupied in the isolated atom. The energies of the symmetric and antisymmetric states are then given by

$$E_\pm = \frac{\langle\psi_\pm|H|\psi_\pm\rangle}{\langle\psi_\pm|\psi_\pm\rangle}\,. \tag{18.20}$$

If the atomic orbitals are conventionally normalized, $\langle\phi_A|\phi_A\rangle = \langle\phi_B|\phi_B\rangle = 1$, we have

$$\langle\psi_\pm|\psi_\pm\rangle = 2(1 \pm |\langle\phi_A|\phi_B\rangle|^2)\,. \tag{18.21}$$

Then, using the permutation symmetry of H, we see that the energies reduce to

$$E_\pm = \frac{\langle\phi_A(1)\phi_B(2)|H|\phi_A(1)\phi_B(2)\rangle \pm \langle\phi_A(1)\phi_B(2)|H|\phi_B(1)\phi_A(2)\rangle}{1 \pm |\langle\phi_A|\phi_B\rangle|^2}\,. \tag{18.22}$$

A detailed evaluation of this expression is not difficult in principle, but the result involves many terms. [See Pauling and Wilson (1935), Sec. 43, for details, and for calculations with other trial functions.] The most interesting result of the calculation is the difference between the energies of the spatially symmetric and antisymmetric states. In the limit of small overlap between the atoms, for which we may neglect $|\langle\phi_A|\phi_B\rangle| \ll 1$, we obtain

$$E_+ - E_- \approx 2K\,, \tag{18.23}$$

$$\begin{aligned} K &= \langle\phi_A(1)\phi_B(2)|V|\phi_B(1)\phi_A(2)\rangle \\ &= \iint \phi^*(\mathbf{x}_1 - \mathbf{R}_A)\phi^*(\mathbf{x}_2 - \mathbf{R}_B)V(\mathbf{x}_1 - \mathbf{x}_2) \\ &\quad \times \phi(\mathbf{x}_1 - \mathbf{R}_B)\phi(\mathbf{x}_2 - \mathbf{R}_A)\,d^3x_1\,d^3x_2\,. \end{aligned} \tag{18.24}$$

This energy difference, whose existence is due entirely to the permutation symmetry of the state function, is called the *exchange interaction*. It should be emphasized that this is not a new kind of interaction, but is really only a manifestation of the Coulomb interaction between the two electrons. The average of the Coulomb interaction in one of the states $\psi_\pm$ is $\langle V\rangle = \langle\psi_\pm|V|\psi_\pm\rangle/\langle\psi_\pm|\psi_\pm\rangle$. In the approximation of small atomic overlap, this becomes $\langle V\rangle \approx C \pm K$, where

$$C = \iint |\phi(\mathbf{x}_1 - \mathbf{R}_A)|^2\phi(\mathbf{x}_2 - \mathbf{R}_B)|^2 V(\mathbf{x}_1 - \mathbf{x}_2)d^3x_1\,d^3x_2\,. \tag{18.25}$$

Now C would be the average Coulomb interaction energy if the positions of the two electrons were statistically independent (i.e. if the joint probability density for $\mathbf{x}_1$ and $\mathbf{x}_2$ was equal to the product of the single particle probability densities). But, in fact, the positions of the two electrons are correlated as a consequence of the symmetrization postulate, and hence their average Coulomb energy is not C, but rather $C \pm K$.

The functions ψ_+ and ψ_- [Eq. (18.19)] are only approximate eigenfunctions of the Hamiltonian (18.14). But because H is permutation-invariant, the exact eigenfunctions will also be of the symmetric or antisymmetric form. Hence we can define exact energies E_+ and E_-. A spatially symmetric function must correspond to the antisymmetric *singlet* spin state, and a spatially antisymmetric function must correspond to the symmetric *triplet* spin state, and hence we may describe E_+ as the *singlet state energy* and E_- as the *triplet state energy*. Now the scalar product of the Pauli spin operators for the two electrons has the following eigenvalues (Problem 7.9):

$$\boldsymbol{\sigma}^{(1)}\cdot\boldsymbol{\sigma}^{(2)}\,|\text{singlet}\rangle = -3\,|\text{singlet}\rangle\,,$$

$$\boldsymbol{\sigma}^{(1)}\cdot\boldsymbol{\sigma}^{(2)}\,|\text{triplet}\rangle = +1\,|\text{triplet}\rangle\,.$$

Therefore, within the four-dimensional subspace spanned by these state vectors, we can use the *effective Hamiltonian*

$$H_{\text{eff}} = a - b\boldsymbol{\sigma}^{(1)}\cdot\boldsymbol{\sigma}^{(2)}\,, \tag{18.26}$$

with $a = (E_+ + 3E_-)/4$, $b = (E_+ - E_-)/4$. (Note that $b \approx \frac{1}{2}K$ in the approximation of small overlap between the atoms.) This form of the exchange interaction is commonly used in the study of magnetism. If $b > 0$ the triplet state has the lowest energy and ferromagnetism (parallel spins) is favored. If $b < 0$ antiferromagnetism is favored. It is important to realize that this interaction, which is responsible for magnetism in matter, is not a magnetic interaction. The true Hamiltonian of the system is (18.14), which does not involve the spins of the electrons. The use of the spin-dependent effective Hamiltonian (18.26) is possible only because the symmetrization postulate correlates a symmetric spatial state function with an antisymmetric spin state function, and vice versa.

18.2 *The Hartree–Fock Method*

In the previous section, we described the states of an N-fermion system by vectors of the form

$$|\Psi\rangle = C^{\dagger}{}_1 C^{\dagger}{}_2 \cdots C^{\dagger}{}_N|0\rangle\,, \tag{18.27}$$

which are the simplest state vectors that are antisymmetric under interchange of particles. The basis vectors, $|\phi_\alpha\rangle = C_\alpha{}^\dagger|0\rangle$, were chosen to represent the single particle states in the presence of the external fields (or approximations to such states). Thus the N-particle state vector $|\Psi\rangle$ takes account of the symmetrization postulate and the external fields, but omits the effect of the interparticle interactions. The Hartree–Fock (HF) method retains the simple form (18.27) while using the variational principle to choose the best single particle basis vectors $\{|\phi_\alpha\rangle\}$. Therefore it takes account of the interactions as well as possible under the restriction imposed by the form (18.27).

A change in the basis vectors that preserves their orthonormal character can be effected by a unitary matrix, the vector $|\phi_\alpha\rangle$ being replaced by $\sum_\beta |\phi_\beta\rangle u_{\beta\alpha}$. The creation operator $C_\alpha{}^\dagger$ transforms the same way as does the basis vector $|\phi_\alpha\rangle$. An infinitesimal unitary matrix has the form $u_{\beta\alpha} = \delta_{\beta\alpha} + i\eta_{\beta\alpha}$, where the small quantities $\eta_{\beta\alpha}$ form a Hermitian matrix. Thus the infinitesimal variations of the vector $|\Psi\rangle$ consist of substitutions of the form

$$C_\alpha{}^\dagger \to C_\alpha{}^\dagger + i\sum_\beta C_\beta{}^\dagger \eta_{\beta\alpha} \tag{18.28}$$

for each of the creation operators. To the first order in $\eta_{\beta\alpha}$, one of the independent variations in the vector $|\Psi\rangle$ — call it $|\delta\Psi\rangle_{\beta\alpha}$ — will be obtained by replacing a particular operator $C_\alpha{}^\dagger$ in (18.27) by $iC_\beta{}^\dagger\eta_{\beta\alpha}$. This is achieved by writing

$$|\delta\Psi\rangle_{\beta\alpha} = i\eta_{\beta\alpha} C_\beta{}^\dagger C_\alpha |\Psi\rangle\,, \tag{18.29}$$

the effect of the multiplication by C_α being to remove $C_\alpha{}^\dagger$ from (18.27). It is clear that α must correspond to an occupied one-fermion state, and β must correspond to an empty one-fermion state for this variation to be nonzero.

The variational method (Sec. 10.6) requires that $\langle H\rangle$ should be stationary to the first order in $|\delta\Psi\rangle$. The variation $|\delta\Psi\rangle$ has been obtained from a unitary transformation, which preserves $\langle\Psi|\Psi\rangle = 1$, and hence the variational condition becomes $\langle\Psi|H|\delta\Psi\rangle = 0$. In view of (18.29), this yields the set of conditions

$$\langle\Psi|HC_\nu{}^\dagger C_\mu|\Psi\rangle = 0\,, \tag{18.30}$$

where $|\Psi\rangle$ is as given by (18.27), μ labels any *occupied* one-fermion state, ν labels any *empty* one-fermion state, and H is equal to

$$H = \sum_\alpha\sum_\beta T_{\alpha\beta} C_\alpha{}^\dagger C_\beta + \frac{1}{2}\sum_\alpha\sum_\beta\sum_\gamma\sum_\delta V_{\alpha\beta,\gamma\delta} C_\alpha{}^\dagger C_\beta{}^\dagger C_\delta C_\gamma\,, \tag{18.31}$$

where $T_{\alpha\beta}$ is a matrix element of the one-body additive operators ($T = P^2/2M + W$ is the sum of kinetic energy plus any external potential), and $V_{\alpha\beta,\gamma\delta}$ is the matrix element of the pair-additive interaction between unsymmetrized product vectors (17.41).

The expression (18.30) can be evaluated by Wick's theorem if we make use of a *particle/hole* transformation of the form (18.5), with the vector $|\Psi\rangle$ now taking the place of $|F\rangle$. As was discussed in other examples in Sec. 18.1, only fully contracted terms in the normal product expansion of (18.30) will survive, and the only nonvanishing contractions are

$$\langle C_\mu{}^\dagger C_\mu \rangle_\Psi = 1 \ \text{ for } \mu \ \text{ occupied}\,,$$

$$\langle C_\nu C_\nu{}^\dagger \rangle_\Psi = 1 \ \text{ for } \nu \ \text{ empty}\,.$$

Substituting (18.31) into (18.30) and forming all possible contractions, we see that the variational condition reduces to

$$T_{\mu\nu} + \sum_{\lambda \le N} (v_{\lambda\mu,\lambda\nu} - v_{\lambda\mu,\nu\lambda}) = 0\,, \quad \text{for } \mu \text{ occupied and } \nu \text{ empty}\,, \tag{18.32}$$

where the sum over λ covers the occupied one-fermion states. This condition acquires a simpler form if we define an *effective one-particle Hamiltonian*,

$$H^{\mathrm{HF}} = T + V^{\mathrm{HF}}\,, \tag{18.33}$$

where the HF effective potential is defined as

$$V^{\mathrm{HF}}_{\mu\nu} = \sum_{\lambda \le N} [v_{\lambda\mu,\lambda\nu} - v_{\lambda\mu,\nu\lambda}]\,. \tag{18.34}$$

The variational condition (18.32) then asserts that the effective Hamiltonian H^{HF} has no matrix elements connecting occupied one-fermion states with empty states. This can be achieved by diagonalizing H^{HF}, i.e. by choosing the basis vectors to be the eigenvectors of H^{HF},

$$H^{\mathrm{HF}}|\phi_\alpha\rangle = \varepsilon_\alpha|\phi_\alpha\rangle\,. \tag{18.35}$$

The occupied states must correspond to those eigenvectors that have the lowest eigenvalues. Since the operator H^{HF} depends on its own eigenvectors through the sum over interaction matrix elements in (18.34), it follows that (18.35) is really a nonlinear equation that must be solved self-consistently.

The energy of an N-fermion HF state can also be calculated using Wick's theorem and the particle/hole transformation:

$$E = \langle\Psi|H|\Psi\rangle = \sum_{\mu\leq N} T_{\mu\mu} + \frac{1}{2}\sum_{\mu\leq N}\sum_{\lambda\leq N}(v_{\mu\lambda,\mu\lambda} - v_{\mu\lambda,\lambda\mu})\,. \tag{18.36}$$

The single particle eigenvalue of (18.35) is equal to

$$\begin{aligned}\varepsilon_\alpha &= \langle\phi_\alpha|H^{\rm HF}|\phi_\alpha\rangle \\ &= T_{\alpha\alpha} + V^{\rm HF}_{\alpha\alpha} = T_{\alpha\alpha} + \sum_{\lambda\leq N}(v_{\alpha\lambda,\alpha\lambda} - v_{\alpha\lambda,\lambda\alpha})\,.\end{aligned} \tag{18.37}$$

The relation between the total energy E and the single particle energy eigenvalues is apparently

$$E = \sum_{\mu\leq N}\left(\varepsilon_\mu - \frac{1}{2}V^{\rm HF}_{\mu\mu}\right). \tag{18.38}$$

The reason why the total energy is not merely equal to the sum of the single particle energies is implicit in (18.37), where it is apparent that ε_1 includes the interaction of particle 1 with all other particles, and ε_2 includes the interaction of particle 2 with all other particles, and so the energy of interaction between particles 1 and 2 is counted twice in the sum $\varepsilon_1 + \varepsilon_2$. The final term of (18.38) corrects for this double counting.

It is useful to rewrite some of these results in coordinate spin representation. A label such as μ then becomes (k,σ), where k labels the orbital function and σ labels the z component of spin. The single particle eigenfunctions are denoted $\phi_{k\sigma}(\mathbf{x}) = \langle\mathbf{x},\sigma|\phi_{k\sigma}\rangle$. The total energy (18.36) becomes

$$\begin{aligned}E = &\sum_{k\sigma}\int[\phi_{k\sigma}(\mathbf{x})]^*\left[\frac{-\hbar^2}{2M}\nabla^2 + W(\mathbf{x})\right]\phi_{k\sigma}(\mathbf{x})d^3x \\ &+\frac{1}{2}\sum_{k\sigma}\sum_{j\sigma'}\iint|\phi_{k\sigma}(\mathbf{x}_1)|^2 v(\mathbf{x}_1-\mathbf{x}_2)|\phi_{j\sigma'}(\mathbf{x}_2)|^2\,d^3x_1\,d^3x_2 \\ &-\frac{1}{2}\sum_{k\sigma}\sum_{j\sigma'}\iint[\phi_{k\sigma}(\mathbf{x}_1)\phi_{j\sigma'}(\mathbf{x}_2)]^* v(\mathbf{x}_1-\mathbf{x}_2) \\ &\times\delta_{\sigma,\sigma'}\,\phi_{j\sigma'}(\mathbf{x}_1)\phi_{k\sigma}(\mathbf{x}_2)d^3x_1\,d^3x_2\,.\end{aligned} \tag{18.39}$$

All of the sums are over the occupied single particle states. The condition that the value of E should be stationary with respect to variations in the form of

$[\phi_{k\sigma}(\mathbf{x})]^*$ (recall from Sec. 10.6 that ϕ and ϕ^* may be varied independently), subject to the constraint that $\int [\phi_{k\sigma}(\mathbf{x})]^* \phi_{k\sigma}(\mathbf{x}) d^3x$ is held fixed, yields the integro-differential equation

$$\left[-\frac{\hbar^2}{2M}\nabla^2 + W(\mathbf{x})\right] \phi_{k\sigma}(\mathbf{x}) + \sum_{j\sigma'} \int v(\mathbf{x}-\mathbf{x}')|\phi_{j\sigma'}(\mathbf{x}')|^2 d^3x' \, \phi_{k\sigma}(\mathbf{x})$$
$$- \sum_j \phi_{j\sigma}(\mathbf{x}) \int [\phi_{j\sigma}(\mathbf{x}')]^* \, v(\mathbf{x}-\mathbf{x}')\phi_{k\sigma}(\mathbf{x}') d^3x' = \varepsilon_k \, \phi_{k\sigma}(\mathbf{x}) \,. \quad (18.40)$$

This equation could also have been deduced from (18.35) by directly transforming to coordinate spin representation. Notice that the "exchange" term (the last one on the left side) connects only states of parallel spin. This integro-differential equation must be solved self-consistently. We may begin by guessing plausible eigenfunctions, from which the integrals are evaluated. The eigenvalue equation is then solved, and the new eigenfunctions are used to re-evaluate the integrals. This procedure is carried out iteratively until it converges to a self-consistent solution. (Much of quantum chemistry is based upon sophisticated computer programs that solve such problems for atoms and molecules.)

Example: Coupled harmonic oscillators

We consider a model that was proposed and solved by Moshinsky (1968): two particles bound in a parabolic potential centered at the origin, and interacting with each other through a harmonic oscillator force. The Hamiltonian of the system is

$$H = \frac{1}{2}(\mathbf{p}_1{}^2 + \mathbf{r}_1{}^2) + \frac{1}{2}(\mathbf{p}_2{}^2 + \mathbf{r}_2{}^2) + \frac{1}{2}K(\mathbf{r}_1 - \mathbf{r}_2)^2 \,. \quad (18.41)$$

This model can be solved exactly and also by the HF method, and so the accuracy of the HF approximation can be assessed.

An exact solution can be obtained by transforming to the coordinates and momenta of the normal modes, which is achieved by the transformation

$$\mathbf{R} = \frac{\mathbf{r}_1 + \mathbf{r}_2}{\sqrt{2}} \,, \quad \mathbf{r} = \frac{\mathbf{r}_1 - \mathbf{r}_2}{\sqrt{2}} \,,$$
$$\mathbf{P} = \frac{\mathbf{p}_1 + \mathbf{p}_2}{\sqrt{2}} \,, \quad \mathbf{p} = \frac{\mathbf{p}_1 - \mathbf{p}_2}{\sqrt{2}} \,. \quad (18.42)$$

(Notice that this differs by numerical factors from the familiar transformation to CM and relative coordinates. We use this form to agree with Moshinsky's notation. Notice also that it is a canonical transformation, preserving the commutation relations between coordinates and momenta.) In terms of these new variables, the Hamiltonian becomes

$$H = \frac{1}{2}(\mathbf{P}^2 + \mathbf{R}^2) + \frac{1}{2}[\mathbf{p}^2 + (1 + 2K)\mathbf{r}^2]\,, \tag{18.43}$$

which describes two uncoupled harmonic oscillators. Comparing with the standard form (6.1) of the harmonic oscillator Hamiltonian, we see that the two terms of (18.43) describe oscillators with mass $M = 1$, and that their angular frequencies are $\omega' = 1$ and $\omega'' = (1 + 2K)^{1/2}$, respectively. The exact ground state energy consists of the sum of the zero-point energies of the six degrees of freedom:

$$E_0 = \frac{3}{2}(\omega' + \omega'') = \frac{3}{2}[1 + (1 + 2K)^{1/2}] \tag{18.44}$$

(in units of $\hbar = 1$). The corresponding eigenfunction of H is

$$\Psi_0 = \pi^{-3/2}(1 + 2K)^{3/8} \exp\left(-\frac{1}{2}R^2\right) \exp\left[-\frac{1}{2}(1 + 2K)^{1/2}r^2\right]. \tag{18.45}$$

Since this function is symmetric under interchange of the two particles, it must be multiplied by the antisymmetric singlet spin state.

The HF state function for the singlet state has the form

$$\Psi_0{}^{\mathrm{HF}} = \phi(\mathbf{r}_1)\,\phi(\mathbf{r}_2)\,, \tag{18.46}$$

where $\phi(\mathbf{r}_1)$ is obtained by applying (18.40) to this problem. Because the two particles have oppositely directed spins, the exchange term drops out, and (18.40) becomes

$$\frac{1}{2}(\mathbf{p}_1{}^2 + \mathbf{r}_1{}^2)\,\phi(\mathbf{r}_1) + \int \frac{1}{2}K(\mathbf{r}_1 - \mathbf{r}_2)^2|\phi(\mathbf{r}_2)|^2 d^3r_2\,\phi(\mathbf{r}_1) = \varepsilon_0\,\phi(\mathbf{r}_1)\,. \tag{18.47}$$

Within the integral is the factor $(\mathbf{r}_1 - \mathbf{r}_2)^2 = r_1{}^2 + r_2{}^2 + 2\mathbf{r}_1\cdot\mathbf{r}_2$. The term involving $\mathbf{r}_1\cdot\mathbf{r}_2$ vanishes upon integration, so (18.47) reduces to

$$\frac{1}{2}[\mathbf{p}_1{}^2 + (1 + K)r_1{}^2)]\phi(\mathbf{r}_1) + \frac{1}{2}K\int r_2{}^2|\phi(\mathbf{r}_2)|^2 d^3r_2\,\phi(\mathbf{r}_1) = \varepsilon_0\,\phi(\mathbf{r}_1)\,. \tag{18.48}$$

The solution of this equation is

$$\phi(r_1) = \pi^{-3/4}(1+K)^{3/8} \exp\left[-\frac{1}{2}(1+K)^{1/2}{r_1}^2\right], \tag{18.49}$$

$$\varepsilon_0 = \frac{3}{2}(1+K)^{1/2}\frac{3K+2}{2K+2}. \tag{18.50}$$

[This can be obtained from the usual HF iterative procedure, which converges after one step, or by substituting the "intelligent guess" $\phi(\mathbf{r}_1) \propto \exp(-\alpha {r_1}^2)$ and solving for the parameter α.] Thus the HF state function is

$$\begin{aligned}\Psi_0{}^{\rm HF} &= \pi^{-3/2}(1+K)^{3/4}\exp\left[-\frac{1}{2}(1+K)^{1/2}({r_1}^2+{r_2}^2)\right] \\ &= \pi^{-3/2}(1+K)^{3/4}\exp\left[-\frac{1}{2}(1+K)^{1/2}(R^2+r^2)\right].\end{aligned} \tag{18.51}$$

The HF approximation to the ground state energy is

$$E_0{}^{\rm HF} = \langle\Psi_0{}^{\rm HF}|H|\Psi_0{}^{\rm HF}\rangle = 3(1+K)^{1/2}. \tag{18.52}$$

This last calculation is made easier by rewriting H in the form

$$H = \frac{1}{2}\left[\mathbf{p}_1{}^2 + (1+K)\mathbf{r}_1{}^2\right]^2 + \frac{1}{2}[\mathbf{p}_2{}^2 + (1+K)\mathbf{r}_2{}^2]^2 - K\mathbf{r}_1\cdot\mathbf{r}_2,$$

since the last term does not contribute to (18.52) because of symmetry.

A comparison between the exact and HF solutions is now possible. Clearly the two become identical when the interaction vanishes ($K = 0$), and so the approximation will be most accurate for small K. To the lowest order in K, we have

$$E_0 = E_0{}^{\rm HF} = 3\left(1+\frac{1}{2}K\right) + O(K^2).$$

The error of the HF approximation increases with K, but even for $K = 1$ we obtain $E_0{}^{\rm HF}/E_0 = 1.035$, which is quite good. The overlap between the exact and approximate ground state functions has been calculated and plotted by Moshinsky (1968) (see the erratum). The quantity $|\langle\Psi_0{}^{\rm HF}|\Psi_0\rangle|^2$ is a decreasing function of K, having the value 0.94 for $K = 1$. The parameter K characterizes the ratio of the strength of the interaction between the particles to the strength of the binding potential. Hence it is analogous to the parameter Z^{-1}

in an atom, where Z is the atomic number. This analogy suggests that the HF approximation should be quite good for atoms, improving as Z increases.

18.3 *Dynamic Correlations*

The antisymmetry of the state function under permutation of identical fermions leads to a correlation between the positions of two such particles whose spins are parallel, even if there is no interaction between the particles. An interaction among the particles causes their positions to be correlated, even if there were no symmetrization postulate. The combination of these two effects is referred to by the jargon words "exchange" and "correlation" — "exchange" referring to the effect of antisymmetry and "correlation" referring to the effect of the interaction. But since both effects lead to a kind of correlation, the interaction effect is sometimes distinguished as "dynamic correlations". The two effects are not additive, and so a separation of them into "exchange" and "correlation" is only conventional. The usual separation is to describe as "exchange" those effects that are included in the HF approximation, and as "dynamic correlations" those effects that cannot be represented in a state function of the form (18.27). This separation is natural, in as much as the HF approximation is the simplest many-body theory that respects the symmetrization postulate, but we shall see that it is not always useful.

The HF approximation is quite accurate for atoms, making it possible to regard dynamic correlations as a higher order correction. However, it is a very poor approximation for electrons in a metal, not merely inaccurate, but even pathological in some respects. To see how the HF approximation can serve so well in one case yet fail so badly in another, we shall treat dynamic correlations as a perturbation on the HF state. Write the Hamiltonian as $H = H^{\text{HF}} + H_1$, with $H^{\text{HF}} = T + V^{\text{HF}}$ and $H_1 = V - V^{\text{HF}}$. The HF effective Hamiltonian (18.33) has the form of a one-body additive operator:

$$H^{\text{HF}} = \sum_{\mu}\sum_{\nu}(T_{\mu\nu} + V_{\mu\nu}^{\text{HF}}){C_\mu}^\dagger C_\nu\,. \tag{18.53}$$

Its eigenvectors, solutions of

$$H^{\text{HF}}|{\Psi_m}^{\text{HF}}\rangle = {E_m}^{\text{HF}}|{\Psi_m}^{\text{HF}}\rangle\,,$$

are of the form (18.27). The interaction V [second term of (18.31)] is not a one-body operator, and so the perturbation

$$H_1 = V - V^{\text{HF}} \tag{18.54}$$

leads to perturbed eigenfunctions that are linear combinations of the eigenvectors of H^{HF}. The first order correction to the HF ground state can be formally obtained from (10.68), its order of magnitude being determined by the ratio $\langle \Psi_m{}^{\mathrm{HF}}|H_1|\Psi_0{}^{\mathrm{HF}}\rangle/(E_m{}^{\mathrm{HF}} - E_0{}^{\mathrm{HF}})$. If the energy denominator $E_m{}^{\mathrm{HF}} - E_0{}^{\mathrm{HF}}$ is not too small, the perturbation correction to the HF state will be small. This is usually true for atoms. But for electrons in bulk matter the spacing between the energy levels is very small, and they practically form a continuum. Thus we have no assurance that the perturbation of the HF ground state by the residual interaction H_1 will be small. This argument does not tell us how large the error of the HF approximation should be for electrons in a metal, but at least it warns us that the approximation cannot be trusted in such a case.

Two-electron atoms

The simplest problem involving dynamic correlations is the helium atom, which has two electrons. Ions such as H^- or Li^+ also have two electrons. If the motion of the nucleus is neglected, the Hamiltonian of the two-electron system becomes

$$H = -\frac{1}{2}\nabla_1{}^2 - \frac{Z}{r_1} - \frac{1}{2}\nabla_2{}^2 - \frac{Z}{r_2} + \frac{1}{r_{12}}, \tag{18.55}$$

where r_1 and r_2 are the distances of the electrons from the fixed nucleus, and r_{12} is the distance between the electrons. We have chosen *atomic units* in which $\hbar = e = M_e = 1$. The atomic unit of energy is $M_e e^4/\hbar^2 \approx 27.2$ eV (electron volts). (Unfortunately the *atomic unit* and the *Rydberg unit*, $\mathrm{Ry} = M_e e^4/2\hbar^2 \approx 13.6$ eV, seem to be equally common in the literature of atomic physics, so one must beware of factors of 2. On my bookshelf there is a report by a well-known quantum chemist in which one-electron energy levels are expressed in Ry while the total atomic energies are in a.u.!)

The lowest energy eigenfunction of (18.55), $\Psi_0(\mathbf{r}_1, \mathbf{r}_2)$, is symmetric under permutation of the electronic coordinates, and so must correspond to the antisymmetric spin singlet. The variational method (Sec. 10.6) is the most powerful and convenient way to attack this problem. We shall compare the results of several different trial functions. According to the variational theorem, the approximate energy will be an upper bound to the true lowest eigenvalue, and so the lowest of the approximate values will be the best.

If the interaction between the electrons were neglected, the two-electron state function would be the product of hydrogen-like states for each electron, appropriately rescaled for the atomic number Z. Therefore our first trial function (unnormalized) is

$$\psi(\mathbf{r}_1, \mathbf{r}_2) = e^{-\alpha r_1} e^{-\alpha r_2} = e^{-\alpha(r_1+r_2)}. \tag{18.56}$$

The parameter α will be varied to obtain the best approximate energy, rather than fixing it at the value $\alpha = Z$, which would be obtained by scaling the hydrogenic function. The variational method was applied to the hydrogen atom in Sec. 10.6, and we may adapt that calculation to obtain (in atomic units)

$$\frac{\langle\psi|(-\frac{1}{2}\nabla_1{}^2)|\psi\rangle}{\langle\psi|\psi\rangle} = \frac{\alpha^2}{2},$$

$$\frac{\langle\psi|(-Z/r_1)|\psi\rangle}{\langle\psi|\psi\rangle} = -Z\alpha.$$

The electronic interaction term can be evaluated with the help of the well-known identity

$$\frac{1}{r_{12}} = \frac{1}{r_1}\sum_\ell \left(\frac{r_2}{r_1}\right)^\ell P_\ell(\cos\theta), \quad r_1 > r_2,$$

$$\frac{1}{r_{12}} = \frac{1}{r_2}\sum_\ell \left(\frac{r_1}{r_2}\right)^\ell P_\ell(\cos\theta), \quad r_1 < r_2,$$

where θ is the angle between $\mathbf{r}_1$ and $\mathbf{r}_2$. Because the function ψ does not depend on θ, it is clear that only the $\ell = 0$ term will contribute to the average interaction energy, which can easily be evaluated to be

$$\left\langle \frac{1}{r_{12}} \right\rangle = \frac{5\alpha}{8}.$$

Adding all terms, we have

$$\langle H\rangle = \frac{\langle\psi|H|\psi\rangle}{\langle\psi|\psi\rangle} = \alpha^2 - 2Z\alpha + \frac{5\alpha}{8}. \tag{18.57}$$

The minimum energy is obtained for

$$\alpha = Z - \frac{5}{16} \tag{18.58}$$

and its value in atomic units is

$$E = \langle H\rangle_{\min} = -\left(Z - \frac{5}{16}\right)^2. \tag{18.59}$$

If the interaction between the electrons had been neglected, the value of α would have been Z, corresponding to the hydrogen-like ground state for a nucleus of charge Ze. The smaller value (18.58) can be understood as a screening of the nucleus by the electrons. If one of the electrons is instantaneously closer to the nucleus, then the more distant electron will experience the attraction of the net charge $(Z-1)e$. In fact, both of the electrons are in motion, and on the average the net attraction corresponds to approximately $(Z-5/16)e$.

An obvious improvement over the previous approximation would be

$$\psi(\mathbf{r}_1,\mathbf{r}_2)=\phi(r_1)\,\phi(r_2)\,, \tag{18.60}$$

where the best possible function $\phi(r)$ is determined by the HF equation (18.40). Because the spins of the electrons are oppositely directed in the singlet state, the exchange term of the HF equation does not contribute. The factored form of (18.60) implies that correlations between the electrons are not taken into account in this approximation, so it will serve as a reference point from which to judge the importance of dynamic correlations. The HF equation can be solved numerically, yielding a total energy of $E=-2.86168$ (a.u.) for He $(Z=2)$. [This number was obtained from two different computer programs, one of which integrated (18.40) numerically, and the other expressed $\phi(r)$ as a linear combination of several basis functions and thereby converted (18.40) into a matrix equation.]

To improve on the HF approximation, we must introduce correlations into the trial function. The ground state function $\Psi_0(\mathbf{r}_1,\mathbf{r}_2)$ actually depends on only the three distances that form the sides of the triangle whose corners are the two electrons and the nucleus, r_1, r_2, and r_{12}. It is more convenient to use the variables $s=r_1+r_2$, $t=r_2-r_1$, and $u=r_{12}$. By 1930 E. A. Hylleraas had carried out a series of calculations using functions of the form

$$\psi(s,t,u)=e^{-\alpha s}\,p(s,t,u)\,, \tag{18.61}$$

where $p(s,t,u)$ is a power series in its variables. Only even powers of t are permitted because the function must be symmetric under permutation of electrons. Some of his results are summarized in the following table. In view of the fact that Hylleraas worked long before the invention of the digital computer, his work is very impressive. Modern computations have made only small improvements to his results. The results in the table are taken from the book by Pauling and Wilson (1935), and from the review paper by Hylleraas (1964). Hylleraas points out that a computational error was responsible for one of his

Variational calculations of the binding energy of two-electron systems.

Trial function (unnormalized)	Energy (Ry) H^-	He
$e^{-\alpha r_1}\,e^{-\alpha r_2} = e^{-\alpha s}$	−0.94531	−5.69531
$\phi(r_1)\,\phi(r_2)$		−5.72336
$e^{-\alpha s}\times$ (3 terms)	−1.0506	−5.8048
$e^{-\alpha s}\times$ (6 terms)		−5.80648
$e^{-\alpha s}\times$ (12 terms)	−1.05284	
Best modern value	−1.05550	−5.807449
Experiment	−1.055	−5.80744

results (line 10 in Table 29.1 of Pauling and Wilson) being lower than the experimental value.

The rows in the table above the dashed line do not include correlations, whereas those below the dashed line do. Even though the correlation effect on the total energy is small, it is clearly significant. More precise calculations than these are possible, and it then becomes necessary to take into account the motion of the nucleus and certain relativistic effects.

The stability of the H^- ion is determined by the difference between its total energy and the energy of a neutral hydrogen atom plus a free electron, which is −1 Ry. If the energy of the H^- ion were greater than −1 Ry, it would spontaneously eject an electron and go to the state of lowest energy. We see from the results in the table that the negative ion is only marginally stable, and that its stability is due to the correlation between electrons.

Electrons in a metal

We have just seen that the HF approximation provides a useful starting point for atoms, the corrections due to dynamic correlations being small, although not negligible. However, the HF approximation yields a qualitatively incorrect description of the behavior of the conduction electrons in a metal, as will now be demonstrated.

The simplest model of a metal is obtained by neglecting the periodic potential of the lattice, and regarding the conduction electrons as a fluid of charged particles confined within the interior of the metal. We choose the specimen to be a cube of side L. The specific boundary condition imposed on the state functions at the surface of the metal is not critical when the length L is large, and it is convenient to use periodic boundary conditions. The single particle state functions will then be plane waves (momentum eigenfunctions),

$$\phi_k(\mathbf{x}) = L^{-3/2}\, e^{i\mathbf{k}\cdot\mathbf{x}}\,, \tag{18.62}$$

with the three rectangular components of $\mathbf{k}$ being each an integer multiple of $2\pi/L$, in order to satisfy the periodic boundary condition.

Because of the translational invariance of this system, these momentum eigenfunctions also satisfy the HF equation (18.40), which now takes the form

$$-\frac{\hbar^2}{2M}\nabla^2\phi_{\mathbf{k}}(\mathbf{x}) + W(\mathbf{x})\phi_{\mathbf{k}}(\mathbf{x}) + 2\sum_{\mathbf{k}'}\int v(\mathbf{x}-\mathbf{x}')|\phi_{\mathbf{k}'}(\mathbf{x}')|^2 d^3x'\,\phi_{\mathbf{k}}(\mathbf{x})$$

$$-\sum_{\mathbf{k}'}\phi_{\mathbf{k}'}(\mathbf{x})\int \phi^*{}_{\mathbf{k}'}(\mathbf{x}')v(\mathbf{x}-\mathbf{x}')\phi_{\mathbf{k}}(\mathbf{x}')d^3x' = \varepsilon(k)\phi_{\mathbf{k}}(\mathbf{x})\,. \tag{18.63}$$

The sum over the occupied single particle states includes all values of the vector $\mathbf{k}$ such that $|\mathbf{k}| \le k_F$, where k_F is called the Fermi wave vector. The factor 2 multiplying the third term accounts for summing over both orientations of the electron spin. No such factor occurs in the fourth term because the spin orientations associated with $\phi_{\mathbf{k}}$ and $\phi_{\mathbf{k}'}$ must be the same in the exchange term. The third term of (18.63) is equivalent to the potential of a negative charge density $-2e\sum_{\mathbf{k}'}|\phi_{\mathbf{k}'}(\mathbf{x}')|^2$, which is the average charge density of the conduction electrons. This will be neutralized by the positive charge of the lattice, whose potential $W(\mathbf{x})$ makes up the second term. In our simplified model, we take $W(\mathbf{x})$ to be a constant, so the second and third terms of (18.63) cancel each other.

When (18.62) is substituted into (18.63), the fourth (exchange) term becomes

$$\begin{aligned}
&\frac{-1}{L^{9/2}}\sum_{\mathbf{k}'} e^{i\mathbf{k}'\cdot\mathbf{x}}\int e^{-i\mathbf{k}'\cdot\mathbf{x}'}\frac{e^2}{|\mathbf{x}'-\mathbf{x}|}e^{i\mathbf{k}\cdot\mathbf{x}'}d^3x' \\
&= -\frac{e^{i\mathbf{k}\cdot\mathbf{x}}}{L^{3/2}}\frac{1}{L^3}\sum_{\mathbf{k}'}\int e^{i(\mathbf{k}-\mathbf{k}')\cdot(\mathbf{x}'-\mathbf{x})}\frac{e^2}{|\mathbf{x}'-\mathbf{x}|}d^3x' \\
&= \varepsilon_x(k)\,\phi_{\mathbf{k}}(\mathbf{x})\,.
\end{aligned}$$

Here we have defined the *exchange energy* of the state $\phi_{\mathbf{k}}(\mathbf{x})$ as

$$\begin{aligned}\varepsilon_x(k) &= \frac{-1}{L^3}\sum_{\mathbf{k}'}\int e^{i(\mathbf{k}-\mathbf{k}')\cdot(\mathbf{x}'-\mathbf{x})}\frac{e^2}{|\mathbf{x}'-\mathbf{x}|}d^3x' \\ &= \frac{-1}{L^3}\sum_{\mathbf{k}'}\frac{4\pi e^2}{(\mathbf{k}-\mathbf{k}')^2}\,. \end{aligned} \tag{18.64}$$

Since L is very large, we may convert the sum into an integral, as in Eq. (5.9), and so obtain

$$\begin{aligned}\varepsilon_x(k) &= \frac{-1}{(2\pi)^3}\int_{k'\leq k_F}\frac{4\pi e^2}{|\mathbf{k}-\mathbf{k}'|^2}d^3k' \\ &= -\frac{2e^2}{\pi}k_F\, g\left(\frac{k}{k_F}\right)\end{aligned} \tag{18.65}$$

where

$$g(x) = \frac{1}{2} + \frac{1-x^2}{4x}\log\left|\frac{1+x}{1-x}\right|. \tag{18.66}$$

Therefore the single electron energy eigenvalues, as determined from the HF equation (18.63), are

$$\varepsilon(k) = \frac{\hbar^2k^2}{2M} + \varepsilon_x(k)\,. \tag{18.67}$$

This approximation for $\varepsilon(k)$ is unsatisfactory in many respects:

(a) The conduction bandwidth, $\varepsilon(k_F) - \varepsilon(0)$, is much too large. Indeed the *free electron* value, obtained from the first term of (18.67) alone, gives a much better result than does the HF approximation. [See Ashcroft and Mermin (1976), Fig. 17.1, for an illustration of this point.]

(b) One can easily verify that $d\varepsilon_x(k)/dk$ is infinite at $k = k_F$. Now the density of one-electron states is uniform in k space, and therefore the density of states per unit energy is proportional to $[d\varepsilon(k)/dk]^{-1}$. Thus the HF approximation predicts that the density of electronic states goes to zero at $k = k_F$. There is plenty of evidence from conductivity and specific heat data, as well as from spectroscopic measurements, that this is not true.

Evidently the HF approximation is not merely inaccurate, but is pathologically bad as a description of the dynamics of the conduction electrons in a metal. This is an extreme illustration of the fact that the variational method, of which the HF approximation is a particular case, is bound to yield as good an estimate as possible for the *total* energy, but it need not yield equally good results for other quantities.

In the physics of condensed matter, we have a situation in which the conventional division between exchange and correlation is inappropriate. The free electron model, which treats the electrons as independent noninteracting particles, is clearly a gross oversimplification, yet it yields better results than does the HF approximation. This can only mean that there is a cancellation taking place, and that the sum of exchange plus correlation effects is smaller than the exchange term of the HF approximation. The problem of dynamic correlations among electrons in a metal is too difficult to discuss in detail in this book. Overhauser (1971) has shown, by means of a simple model, that the logarithmic singularity in $\varepsilon_x(k)$ [Eq. (18.65)] is indeed canceled by the effects of dynamic correlations. The net effect of the Coulomb interaction between electrons on the single particle energy $\varepsilon(k)$ is not negligible, but it is much smaller than the misleading HF result.

18.4 *Fundamental Consequences for Theory*

The results of the previous section are important, not merely for their applications, but for what they imply about quantum theory. They provide evidence that the many-body Schrödinger equation is correct. This remark is not so trite as it may seem. In Sec. 4.2 we cautioned against too literal an interpretation of the notion of wave–particle duality, stressing that a system of N interacting particles does not correspond to N interacting waves in three-dimensional space, but rather to a single wave function in $3N$-dimensional configuration space. But that was merely a theoretical assertion, which was not justified by any empirical evidence at that time.

It is logically possible that the HF equation could have been the correct theory for N interacting particles. The HF equation satisfies all of the essential principles of quantum mechanics, including the symmetrization postulate. It reduces to the Schrödinger equation for $N = 1$, and most of the applications and experiments that we have considered may be described, at least approximately, as one-particle states. Thus, prior to the results of Sec. 18.3, someone might have maintained that the HF equation, rather than the Schrödinger equation, was true. But in the HF theory, N interacting particles are described by N interacting wave functions in three-dimensional space. Moreover the HF equation is nonlinear, and so would undermine much of the analysis of the measurement process in Ch. 9, which was based on the linearity of the Schrödinger equation of motion. Thus there is a tremendous conceptual difference between the HF equation and the Schrödinger equation. Therefore the experimental confirmations of the dynamic correlation effects that are predicted by the many-body Schrödinger equation are of fundamental significance for our understanding of quantum mechanics.

18.5 *BCS Pairing Theory*

The topic of this section is perhaps the most striking and important application of the quantum-mechanical many-body theory. It is the explanation of the remarkable phenomenon of *superconductivity* — the persistence of resistanceless electrical currents. The principal idea of this theory, introduced in 1957 by J. Bardeen, L. N. Cooper, and J. R. Schrieffer (BCS), is that an attractive interaction between electrons causes the system to condense into a state of correlated paris, whose energy is lower that of the weakly correlated normal ground state of a fermion system. The BCS state is stable because it is separated by an energy gap from its lowest excitations. We shall not discuss the detailed physics of superconductivity and the mechanisms that can lead to an effective attraction between electrons. (These involve a polarization of the material by the Coulomb force of an electron, and the subsequent attraction of another electron to that polarization.)

BCS ground state

The ground state of the BCS theory has the form

$$|\mathrm{BCS}\rangle = \prod_{\alpha>0} (u_\alpha + v_\alpha\, {C_\alpha}^\dagger {C_{-\alpha}}^\dagger)|0\rangle\,. \tag{18.68}$$

Here the label α signifies the quantum numbers of a one-electron state, $\alpha = (\mathbf{k}, \sigma)$, and $-\alpha$ signifies the time-reversed state, $-\alpha = (-\mathbf{k}, -\sigma)$. The restriction $\alpha > 0$ ensures that the pair $(\alpha, -\alpha)$ is included only once in the product. One may interpret the set of values in the range $\alpha > 0$ as including all values of $\mathbf{k}$ but only positive σ. The parameters u_α and v_α are real and must satisfy

$${u_\alpha}^2 + {v_\alpha}^2 = 1 \tag{18.69}$$

in order to ensure the normalization $\langle\mathrm{BCS}|\mathrm{BCS}\rangle = 1$. The values of u_α and v_α will be determined by the variational principle. It is apparent that the BCS state consists of correlated pairs of electrons, since the one-electron states α and $-\alpha$ are always associated together. However, it reduces to the HF state (18.27) in the limit: $u_\alpha = 0$, $v_\alpha = 1$ for $k_\alpha \le k_F$; $u_\alpha = 1$; $v_\alpha = 0$ for $k_\alpha > k_F$. We shall refer to this HF limit of the BCS state as the *trivial* case of (18.68). The criterion for the existence of the superconductivity will be that some nontrivial BCS state should have a lower energy than the HF state.

The number of particles in the BCS state (18.68) is not definite, since the state contains a superposition of components having any number of pairs of electrons. The use of a state containing a variable number of electrons is a matter of computational convenience, analogous to the use of the *grand canonical ensemble* in statistical thermodynamics. The minimization of the average energy $\langle H\rangle$ in the variational calculation should be subject to the constraint that the average number of electrons $\langle N\rangle$ be held constant. This constraint can most easily be handled by means of a Lagrange multiplier, which allows us to minimize instead the quantity $\langle H-\mu N\rangle$, with the Lagrange multiplier μ ultimately playing the role of a chemical potential.

Bogoliubov transformation

The calculation of the ground state energy and the excitation spectrum is simplified by introducing a canonical transformation (named after its inventor, N. N. Bogoliubov):

$$b_\alpha = u_\alpha C_\alpha - v_\alpha {C_{-\alpha}}^\dagger \,, \tag{18.70a}$$

$$b_\alpha{}^\dagger = u_\alpha C_\alpha{}^\dagger - v_\alpha C_{-\alpha} \,. \tag{18.70b}$$

For this transformation to be *canonical*, it must preserve the anticommutation relations, (17.19), (17.20), and (17.21), which may be written as $[C_\alpha, C_\beta]_+ = 0$, $[C_\alpha, C_\beta{}^\dagger]_+ = \delta_{\alpha\beta} I$ ($[A,B]_+ \equiv AB+BA$). From (18.70) and (18.69) we obtain

$$\begin{aligned}
[b_\alpha, b_\beta{}^\dagger]_+ &= u_\alpha u_\beta [C_\alpha, C_\beta{}^\dagger]_+ + v_\alpha v_\beta [C_{-\alpha}{}^\dagger, C_{-\beta}]_+ \\
&= (u_\alpha u_\beta + v_\alpha v_\beta)\delta_{\alpha\beta} I = \delta_{\alpha\beta} I \,, \\
[b_\alpha, b_\beta]_+ &= -u_\alpha v_\beta [C_\alpha, C_{-\beta}{}^\dagger]_+ - v_\alpha u_\beta [C_{-\alpha}{}^\dagger, C_\beta]_+ \\
&= -(u_\alpha v_\beta + v_\alpha u_\beta)\delta_{-\alpha\beta} I \,.
\end{aligned}$$

For this expression to have the canonical value (zero) for the case $\beta = -\alpha$, it is necessary that $u_\alpha v_{-\alpha} = -u_{-\alpha} v_\alpha$. Therefore we shall require that

$$u_\alpha = u_{-\alpha}\,, \quad v_\alpha = -v_{-\alpha}\,. \tag{18.71}$$

(The opposite choice of u_α odd and v_α even merely changes the phase of the vector $|\text{BSC}\rangle$, and so is not a real alternative.) It will be useful to rewrite the canonical transformation (18.70) in light of this result:

$$b_\alpha = u_\alpha C_\alpha - v_\alpha {C_{-\alpha}}^\dagger\,, \quad b_{-\alpha} = u_\alpha C_{-\alpha} + v_\alpha {C_\alpha}^\dagger\,, \tag{18.72a}$$

$${b_\alpha}^\dagger = u_\alpha {C_\alpha}^\dagger - v_\alpha C_{-\alpha}\,, \quad {b_{-\alpha}}^\dagger = u_\alpha {C_{-\alpha}}^\dagger + v_\alpha C_\alpha\,. \tag{18.72b}$$

These operators annihilate and create a kind of *quasiparticle* excitation. Its character is difficult to grasp intuitively, but it is apparently a linear combination of the "particle" and "hole" excitations that were discussed in Sec. 18.1. Indeed, in the trivial (HF) limit of the BCS state, these quasiparticles become either "particle" or "hole" excitations. The inverse of the transformation (18.72) will also be useful:

$$C_\alpha = u_\alpha b_\alpha + v_\alpha {b_{-\alpha}}^\dagger\,, \quad C_{-\alpha} = u_\alpha b_{-\alpha} - v_\alpha {b_\alpha}^\dagger\,, \tag{18.73a}$$

$${C_\alpha}^\dagger = u_\alpha {b_\alpha}^\dagger + v_\alpha b_{-\alpha}\,, \quad {C_{-\alpha}}^\dagger = u_\alpha {b_{-\alpha}}^\dagger - v_\alpha b_\alpha\,. \tag{18.73b}$$

It is easy to verify that the BCS state satisfies

$$b_\alpha |\text{BCS}\rangle = 0\,, \tag{18.74}$$

and hence it is the state of zero quasiparticles. This fact, and the fact that the quasiparticle operators satisfy the canonical anticommutation relations, allow us to adapt *Wick's theorem* (Sec. 17.4 and App. C) to the BCS state and the Bogoliubov transformation, much as we did for the Fermi sea state and the particle/hole transformation in Sec. 18.1. For present purposes we have:

(1) *Normal ordering* is taken with respect to the Bogoliubov quasiparticle operators, with ${b_\alpha}^\dagger$ taken to the left of b_α;
(2) The *contraction* of two operators X and Y is defined as the matrix element in the BCS state:

$$\langle XY\rangle_{\text{BCS}} = \langle \text{BCS}|XY|\text{BCS}\rangle\,. \tag{18.75}$$

The only nonvanishing contraction of the quasiparticle operators is $\langle b_\alpha {b_\alpha}^\dagger\rangle_{\text{BCS}} = 1$.

The contractions of the original electron creation and annihilation operators are obtained by using (18.73). Because the Bogoliubov transformation mixes creation and annihilation operators, it follows that two creation operators or two annihilation operators can have a nonvanishing contraction. Thus we have

$$\begin{aligned}\langle C_\alpha C_\beta \rangle_{\text{BCS}} &= u_\alpha v_\beta \, \langle b_\alpha b_{-\beta}{}^\dagger \rangle_{\text{BCS}} \\ &= u_\alpha v_\beta \delta_{-\alpha\beta} = -u_\alpha v_\alpha \delta_{-\alpha\beta}\,, \end{aligned} \tag{18.76}$$

$$\begin{aligned}\langle C_\alpha{}^\dagger C_\beta{}^\dagger \rangle_{\text{BCS}} &= v_\alpha u_\beta \, \langle b_{-\alpha} b_\beta{}^\dagger \rangle_{\text{BCS}} \\ &= v_\alpha u_\beta \delta_{-\alpha\beta} = u_\alpha v_\alpha \delta_{-\alpha\beta}\,, \end{aligned} \tag{18.77}$$

$$\langle C_\alpha C_\beta{}^\dagger \rangle_{\text{BCS}} = u_\alpha{}^2 \delta_{\alpha\beta}\,, \tag{18.78}$$

$$\langle C_\alpha{}^\dagger C_\beta \rangle_{\text{BCS}} = v_\alpha{}^2 \delta_{\alpha\beta}\,. \tag{18.79}$$

It follows from (18.79) that the average number of electrons in the state labeled $\alpha = (\mathbf{k}, \sigma)$ is

$$n_\alpha \equiv \langle C_\alpha{}^\dagger C_\alpha \rangle_{\text{BCS}} = v_\alpha{}^2\,. \tag{18.80}$$

The physical significance of the parameters u_α and v_α in (18.68) is now apparent: $v_\alpha{}^2$ is the average occupancy of the state $(\mathbf{k}, \sigma)$, and $u_\alpha{}^2 = 1 - v_\alpha{}^2$ is its average vacancy rate.

Energy minimization

The Hamiltonian of the system is

$$H = \sum_\alpha \sum_\beta T_{\alpha\beta} C_\alpha{}^\dagger C_\beta + \frac{1}{4} \sum_\alpha \sum_\beta \sum_\gamma \sum_\delta \langle \alpha\beta | V | \gamma\delta \rangle C_\alpha{}^\dagger C_\beta{}^\dagger C_\delta C_\gamma\,, \tag{18.81}$$

where $T_{\alpha\beta}$ is a matrix element of the kinetic energy plus any external potential, $T = P^2/2M + W$, and $\langle \alpha\beta | V | \gamma\delta \rangle = v_{\alpha\beta,\gamma\delta} - v_{\alpha\beta,\delta\gamma}$ is a matrix element of the interaction between antisymmetric states, as in (17.43). The average energy in the BCS ground state (18.68), $\langle \text{BCS}|H|\text{BCS}\rangle$, can easily be evaluated if we use Wick's theorem to rewrite H in terms of normal products and contractions, as in (17.48). Only the fully contracted terms will contribute to the ground state energy. Using Eq. (18.76) through Eq. (18.79) to evaluate the contractions, we obtain

$$
\begin{aligned}
\langle \text{BCS}|H|\text{BCS}\rangle &= \sum_\alpha T_{\alpha\alpha}\langle C_\alpha{}^\dagger C_\alpha\rangle_{\text{BCS}} \\
&\quad + \frac{1}{4}\sum_\alpha\sum_\beta \langle\alpha\beta|V|\alpha\beta\rangle\,\langle C_\alpha{}^\dagger C_\alpha\rangle_{\text{BCS}}\,\langle C_\beta{}^\dagger C_\beta\rangle_{\text{BCS}} \\
&\quad - \frac{1}{4}\sum_\alpha\sum_\beta \langle\alpha\beta|V|\beta\alpha\rangle\,\langle C_\alpha{}^\dagger C_\alpha\rangle_{\text{BCS}}\,\langle C_\beta{}^\dagger C_\beta\rangle_{\text{BCS}} \\
&\quad + \frac{1}{4}\sum_\alpha\sum_\gamma \langle\alpha,-\alpha|V|\gamma,-\gamma\rangle\,\langle C_\alpha{}^\dagger C_{-\alpha}{}^\dagger\rangle_{\text{BCS}}\,\langle C_{-\gamma}C_\gamma\rangle_{\text{BCS}} \\
&= \sum_\alpha T_{\alpha\alpha}{v_\alpha}^2 + \frac{1}{2}\sum_\alpha\sum_\beta\langle\alpha\beta|V|\alpha\beta\rangle {v_\alpha}^2\,{v_\beta}^2 \\
&\quad + \frac{1}{4}\sum_\alpha\sum_\gamma\langle\alpha,-\alpha|V|\gamma,-\gamma\rangle u_\alpha v_\alpha u_\gamma v_\gamma\,.
\end{aligned}
\tag{18.82}
$$

The first and second terms of the final expression above are very similar to the ground state energy in the HF approximation (18.36), and they become identical with it in the trivial limit ($u_\alpha = 0, v_\alpha = 1$ for occupied states; $u_\alpha = 1, v_\alpha = 0$ for empty states). The last term, which vanishes in the trivial limit, corresponds to the energy of pair correlations in the BCS state.

Since the number of electrons in the BCS state is not fixed, we subtract $\mu\langle\text{BCS}|N|\text{BCS}\rangle = \mu\sum_\alpha {v_\alpha}^2$ from the above energy, and vary u_α and v_α so as to minimize $\langle H\rangle - \mu\langle N\rangle$. From the relation (18.69), ${u_\alpha}^2 + {v_\alpha}^2 = 1$, it follows that $du_\alpha/dv_\alpha = -v_\alpha/u_\alpha$. Thus we obtain

$$
\begin{aligned}
\frac{\partial\langle H-\mu N\rangle}{\partial v_\alpha} &= 2v_\alpha(T_{\alpha\alpha}-\mu) + 2v_\alpha\sum_\beta\langle\alpha\beta|V|\alpha\beta\rangle{v_\beta}^2 \\
&\quad + \frac{1}{2}\sum_\gamma\frac{\langle\alpha,-\alpha|V|\gamma,-\gamma\rangle u_\gamma v_\gamma({u_\alpha}^2-{v_\alpha}^2)}{u_\alpha}\,.
\end{aligned}
\tag{18.83}
$$

The vanishing of this expression is the minimization condition.

The above condition will become easier to understand if we introduce two definitions. We first define

$$
\begin{aligned}
\Delta_\alpha &= -\frac{1}{2}\sum_\gamma\langle\alpha,-\alpha|V|\gamma,-\gamma\rangle u_\gamma v_\gamma \\
&= -\sum_{\gamma>0}\langle\alpha,-\alpha|V|\gamma,-\gamma\rangle u_\gamma v_\gamma\,.
\end{aligned}
\tag{18.84}
$$

This parameter (called the *gap parameter*, for reasons that will become apparent) characterizes the strength of the pair correlation energy, which is the most important feature of the BCS state. The minus sign is conventionally introduced because V will be negative for an attractive interaction. Second, we define

$$\varepsilon_\alpha = T_{\alpha\alpha} + \sum_\beta \langle\alpha\beta|V|\alpha\beta\rangle {v_\beta}^2 \,. \tag{18.85}$$

This is analogous to the HF single particle energy (18.37). It will be an eigenvalue of an effective one-electron Hamiltonian if we choose the single particle basis vectors $|\alpha\rangle$ so as to diagonalize the matrix

$$\begin{aligned}(H_1)_{\alpha\beta} &\equiv T_{\alpha\beta} + \sum_\gamma \langle\alpha\gamma|V|\beta\gamma\rangle {v_\gamma}^2 \\ &= \varepsilon_\alpha \delta_{\alpha\beta} \,.\end{aligned} \tag{18.86}$$

The operator H_1 is analogous to the effective Hamiltonian H^{HF} (18.33), and the basis vectors that diagonalize it are the best single particle vectors for this problem. (In a translationally invariant system, they will be momentum eigenvectors.) The eigenvalue ε_α is usually interpreted as the single particle energy in the *normal* (nonsuperconducting) state because it excludes the pairing energy term proportional to Δ_α, although this interpretation is only approximately correct because ${v_\gamma}^2$ implicitly depends upon Δ_α.

When these two definitions are substituted into (18.83), the minimization condition $\partial\langle H - \mu N\rangle/\partial v_\alpha = 0$ becomes

$$2u_\alpha v_\alpha(\varepsilon_\alpha - \mu) - \Delta_\alpha({u_\alpha}^2 - {v_\alpha}^2) = 0 \,. \tag{18.87}$$

From this equation and the normalization (18.69) we obtain

$${u_\alpha}^2 = \frac{1}{2}\left[1 + \frac{\varepsilon_\alpha - \mu}{[(\varepsilon_\alpha - \mu)^2 + {\Delta_\alpha}^2]^{1/2}}\right], \tag{18.88}$$

$${v_\alpha}^2 = \frac{1}{2}\left[1 - \frac{\varepsilon_\alpha - \mu}{[(\varepsilon_\alpha - \mu)^2 + {\Delta_\alpha}^2]^{1/2}}\right]. \tag{18.89}$$

[These results are easily verified by substitution into (18.87) and (18.69).] The average occupancy of the single particle state $|\alpha\rangle$, ${v_\alpha}^2$, is shown in Fig. 18.1. The chemical potential μ is chosen so that $\langle N\rangle = \sum_\alpha {v_\alpha}^2$ is the number of electrons in the system.

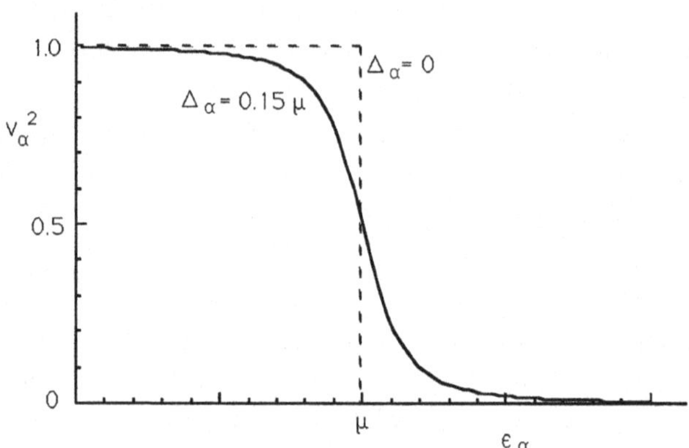

Fig. 18.1 Occupancy of single particle states versus energy [Eq. (18.89)].

Every quantity of interest may now be expressed in terms of the parameter Δ_α. From (18.88) and (18.89) it follows that

$$u_\alpha v_\alpha = \frac{\frac{1}{2}\Delta_\alpha}{[(\varepsilon_\alpha - \mu)^2 + {\Delta_\alpha}^2]^{1/2}} \,. \tag{18.90}$$

Using this result with (18.84) and (18.85), we can rewrite the BCS ground state energy (18.82) as

$$E_0 = \sum_\alpha \frac{1}{2}(T_{\alpha\alpha} + \varepsilon_\alpha){v_\alpha}^2 - \frac{1}{2}\sum_{\alpha>0} \frac{{\Delta_\alpha}^2}{[(\varepsilon_\alpha - u)^2 + {\Delta_\alpha}^2]^{1/2}} \,. \tag{18.91}$$

(Note that both v_α and Δ_α are odd with respect to α. We have defined them so as to be positive for $\alpha > 0$.) If $\Delta_\alpha = 0$, the ground state energy E_0 becomes equal to the HF energy (18.36). If there is a nontrivial solution with $\Delta_\alpha \neq 0$, then the BCS state will have a lower energy than the HF state.

The gap parameter Δ_α was defined by (18.84). With the substitution (18.90), that definition leads to the equation

$$\Delta_\alpha = -\frac{1}{2}\sum_{\gamma>0} \frac{\langle \alpha, -\alpha|V|\gamma, -\gamma\rangle \Delta_\gamma}{[(\varepsilon_\gamma - \mu)^2 + {\Delta_\gamma}^2]^{1/2}} \,, \tag{18.92}$$

which must be solved self-consistently for Δ_α. The trivial solution, $\Delta_\alpha = 0$, is always possible. Whether a nontrivial solution exists depends on the nature of the interaction. Clearly it is necessary that the interaction be attractive, in the sense that we must have $\langle\alpha,-\alpha|V|\gamma,-\gamma\rangle < 0$ for those matrix elements that dominate (18.92). In general, the solution of (18.92) can only be carried out numerically.

Elementary excitations

The elementary excitations from the BCS ground state consist of the quasiparticles that are generated by the operator ${b_\alpha}^\dagger$ [Eq. (18.70b)], the ground state $|\text{BCS}\rangle$ being the state of zero quasiparticles. The vector $|\text{BCS}\rangle$ describes an indefinite number of electrons, and it is (approximately) the ground state of the operator

$$G = H - \mu N\,. \tag{18.93}$$

This is the zero temperature Gibbs free energy, and its eigenvalues correspond to the work needed to transfer an electron into various states of the system from a reservoir with chemical potential μ. It can be shown (Problem 18.3) that the fluctuations in N are a negligible fraction of $\langle N\rangle$ in the limit as $\langle N\rangle$ becomes very large, and therefore the excitation spectrum of G will be very nearly the same as that of H, provided the single particle energies are measured relative to the chemical potential μ. Looking at the problem from another point of view, we note that the operators H and N commute, and therefore the diagonalization of G is equivalent to the diagonalization of H. Thus the excitation spectrum of H is equivalent to the excitation spectrum of G with $\langle N\rangle$ held constant. This equivalence is obscured by the Bogoliubov transformation, and it may be violated by any approximations that are introduced. Nevertheless, an approximate diagonalization of G is, in principle, as valid as an approximate diagonalization of H.

We shall use Wick's theorem to express G in terms of contractions and normal products of quasiparticle operators. Following (17.48) as a model, we obtain

$$G = G_0 + G_{11} + G_{20} + H_4\,, \tag{18.94}$$

where $G_0 = H_0 - \mu N_0$ contains no uncontracted operators, each term of $G_{11} = H_{11} - \mu N_{11}$ contains one creation and one annihilation operator, each term of $G_{20} = H_{20} - \mu N_{20}$ contains two creation or two annihilation operators, and H_4 contains four operators in normal order. The fully contracted terms have already been calculated. H_0 is equal to the BCS ground state energy, (18.82) or (18.91), and N_0 is equal to the average number of electrons in the state. The term H_4 represents an interaction between quasiparticles, and it will be neglected. Calculation of the other terms is straightforward but tedious. We obtain

$$G_{11} = \sum_{\alpha>0} \sum_{\beta>0} \{[(H_1)_{\alpha\beta} - \mu\delta_{\alpha\beta}](u_\alpha u_\beta - v_\alpha v_\beta) - g_{\alpha\beta}(u_\alpha v_\beta + v_\alpha u_\beta)\} \times ({b_\alpha}^\dagger b_\beta + {b_{-\beta}}^\dagger b_{-\alpha}), \tag{18.95}$$

$$G_{20} = \sum_{\alpha>0} \sum_{\beta>0} \{[(H_1)_{\alpha\beta} - \mu\delta_{\alpha\beta}](u_\alpha v_\beta + v_\alpha u_\beta) + g_{\alpha\beta}(u_\alpha u_\beta - v_\alpha v_\beta)\} \times ({b_\alpha}^\dagger {b_{-\beta}}^\dagger + b_{-\alpha} b_\beta), \tag{18.96}$$

where $(H_1)_{\alpha\beta}$ is as defined in (18.86), and we have introduced

$$g_{\alpha\beta} = \sum_{\gamma>0} \langle \alpha, -\beta | V | \gamma, -\gamma \rangle u_\gamma v_\gamma . \tag{18.97}$$

Earlier we argued that $(H_1)_{\alpha\beta}$ should be diagonal in momentum representation, so we will have $(H_1)_{\alpha\beta} = \varepsilon_\alpha \delta_{\alpha\beta}$. Writing explicitly the momentum and spin labels, we have $\alpha = (\mathbf{k}_\alpha\uparrow), -\beta = (-\mathbf{k}_\beta\downarrow)$, etc. Thus

$$\langle \alpha, -\beta | V | \gamma, -\gamma \rangle = \langle \mathbf{k}_\alpha\uparrow, -\mathbf{k}_\beta\downarrow | V | \mathbf{k}_\gamma\uparrow, -\mathbf{k}_\gamma\downarrow \rangle$$

will vanish as a consequence of momentum conservation unless $\mathbf{k}_\alpha - \mathbf{k}_\beta = 0$. Therefore $g_{\alpha\beta}$ will be diagonal, provided the interaction v is translationally invariant, and from (18.84) we obtain $g_{\alpha\beta} = -\Delta_\alpha \delta_{\alpha\beta}$. Thus Eq. (18.95) simplifies to

$$G_{11} = \sum_{\alpha>0} [(\varepsilon_\alpha - \mu)({u_\alpha}^2 - {v_\alpha}^2) + 2\Delta_\alpha u_\alpha v_\alpha]({b_\alpha}^\dagger b_\alpha + {b_{-\alpha}}^\dagger b_{-\alpha}) \tag{18.98}$$

and Eq. (18.96) becomes

$$G_{20} = \sum_{\alpha>0} [2(\varepsilon_\alpha - \mu) u_\alpha v_\alpha - \Delta_\alpha({u_\alpha}^2 - {v_\alpha}^2)]({b_\alpha}^\dagger {b_{-\alpha}}^\dagger + b_{-\alpha} b_\alpha) .$$

Now the minimization condition (18.87) implies that G_{20} vanishes identically; therefore, with the approximation of neglecting H_4, we find that $G \approx G_0 + G_{11}$ is diagonal and its spectrum can be obtained trivially. (This method of diagonalization would not have succeeded if we had worked with H rather than $G = H - \mu N$.)

The excitation spectrum is given directly by G_{11}, which we may write in the form

$$G_{11} = \sum_{\alpha>0} E_\alpha ({b_\alpha}^\dagger b_\alpha + {b_{-\alpha}}^\dagger b_{-\alpha}) . \tag{18.99}$$

Now ${b_\alpha}^\dagger b_\alpha$ is a quasiparticle number operator. A one-quasiparticle eigenvector of G_{11} is of the form

$$G_{11}({b_\beta}^\dagger|\text{BCS}\rangle) = E_\beta({b_\beta}^\dagger|\text{BCS}\rangle)\,.$$

In general, the excitation energy is just the sum of the energies E_α of the quasiparticles that are present. Comparing (18.99) with (18.98), we find that the excitation energy (illustrated in Fig. 18.2) is

$$\begin{aligned} E_\alpha &= (\varepsilon_\alpha - \mu)({u_\alpha}^2 - {v_\alpha}^2) + 2\Delta_\alpha u_\alpha v_\alpha \\ &= [(\varepsilon_\alpha - \mu)^2 + {\Delta_\alpha}^2]^{1/2}\,. \end{aligned} \tag{18.100}$$

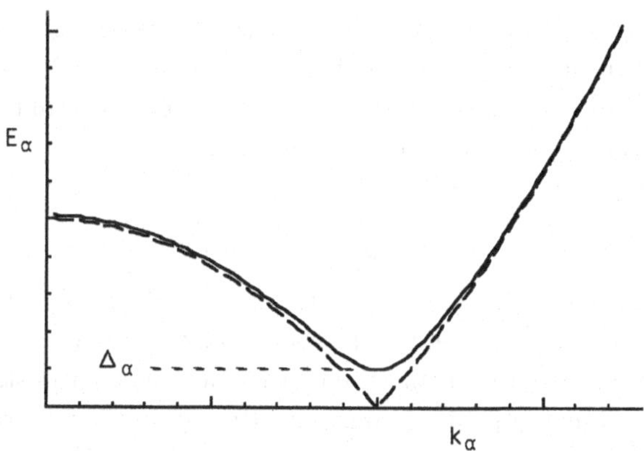

Fig. 18.2 Quasiparticle excitation energy (solid curve), according to Eq. (18.100). The dashed curve corresponds to $\Delta_\alpha = 0$.

This energy is easily interpreted in the limit $\Delta_\alpha \to 0$. There, for $\varepsilon_\alpha > \mu$, the quasiparticle is a "particle", and it requires energy $\varepsilon_\alpha - \mu$ to transfer it from the reservoir to the state α. For $\varepsilon_\alpha < \mu$, the quasiparticle is a "hole", and it requires energy $\mu - \varepsilon_\alpha$ to remove an electron from the system to the reservoir so as to create the "hole" in state α.

If Δ_α is not zero, the character of the quasiparticles is partly particle-like and partly hole-like. Because the minimum excitation energy is Δ_α, the BCS ground state is stable against small perturbations. The correlated pairs of electrons in the BCS ground state have zero total momentum, so this state has a "persistent current" of magnitude 0. However, one can readily construct a similar state that is displaced in momentum, so that all correlated pairs carry a nonzero momentum and a nonzero current. This state will have a higher energy than the zero-current ground state, but provided its additional energy per electron is small compared with Δ_α, this persistent current state will also be stable against the multitude of small perturbations that would produce electrical resistance in the nonsuperconducting state.

Simple model

Equation (18.92) for the gap parameter Δ_α can be solved for a simple but useful model. We assume that

$$\begin{aligned}\langle \alpha, -\alpha|V|\gamma, -\gamma\rangle &= -V_0 \quad \text{for } |\varepsilon_\alpha - \mu| < \omega\,,\\ &= 0 \qquad \text{for } |\varepsilon_\alpha - \mu| > \omega\,.\end{aligned} \tag{18.101}$$

Here V_0 is a positive constant characterizing the strength of the attractive interaction. Substituting this model interaction into (18.92), we find that $\Delta_\alpha = \Delta_0$ for $|\varepsilon_\alpha - \mu| < \omega$, $\Delta_\alpha = 0$ for $|\varepsilon_\alpha - \mu| > \omega$. The constant value of the gap parameter Δ_0 is determined by the condition

$$1 = \frac{1}{2} V_0 \int_{-\omega}^{\omega} \frac{n(x)}{(x^2 + {\Delta_0}^2)^{1/2}} dx\,. \tag{18.102}$$

Here $x = \varepsilon_\alpha - \mu$ is the energy relative to the chemical potential (or Fermi energy). In deriving this equation from (18.92), we have converted the sum into an integral by introducing the density of one-electron states, $n(x)$. If we further approximate $n(x)$ by a constant $n(0)$ within the integral, then (18.102) reduces to

$$1 = n(0)V_0 \sinh^{-1}\left(\frac{\omega}{\Delta_0}\right),$$

and hence the gap parameter is

$$\Delta_0 = \frac{\omega}{\sinh[1/n(0)V_0]}\,. \tag{18.103}$$

If the interaction strength V_0 is small, this becomes

$$\Delta_0 \approx 2\omega \exp\left(\frac{-1}{n(0)V_0}\right).$$

This formula is notable because it has an *essential singularity* at $V_0 = 0$, and therefore the results of the BCS theory could never have been obtained if the interaction V had been treated by perturbation theory. This is a valuable lesson to remember when studying the elegant systematic perturbation formalisms that have been developed for many-body theory and for quantum field theory. There may be phenomena that cannot be discovered by perturbation theory, even if it can be summed to arbitrarily high order.

Further reading for Chapter 18

Many-body theory is a very large subject. The books by March, Young, and Sampanthar (1967), and by Fetter and Walecka (1971) cover many aspects of it.

Problems

18.1 Evaluate the normalized pair correlation function for a Fermi sea of free electrons with parallel spins,

$$g_{\uparrow\uparrow}(\mathbf{x}_1, \mathbf{x}_2) = \langle F|n_\uparrow(\mathbf{x}_1)n_\uparrow(\mathbf{x}_2)|F\rangle / \langle F|n_\uparrow(\mathbf{x}_1)|F\rangle \, \langle F|n_\uparrow(\mathbf{x}_2)|F\rangle \,.$$

[This can be done by evaluating (18.13) with the orbitals $\phi_j(\mathbf{x})$ being plane waves.]

18.2 It was shown in Eq. (18.12) that in a system of N noninteracting electrons, there is no correlation between the positions of particles having opposite spin orientation. Yet Eq. (18.1) shows a correlation between two electrons in the singlet spin state. Resolve this apparent contradiction.

18.3 Evaluate mean square fluctuation in the number of particles, $\langle (N - \langle N\rangle)^2\rangle$, in the BCS state. Hence show that the relative fluctuation, $\langle (N - \langle N\rangle)^2\rangle / \langle N\rangle^2$, becomes negligible in the limit of a very large system.

18.4 Generalize the virial theorem (Problem 10.8) to a many-body system, and hence show that the average kinetic and potential energies of a system of particles that interact by Coulomb forces are related by $\langle V\rangle = -2\langle T\rangle$.

Chapter 19

Quantum Mechanics of the Electromagnetic Field

The development of electromagnetic theory from Coulomb's law to Maxwell's equations is accompanied by a change in the concept of the *field* from a merely passive agent that mediates the interactions between particles to a dynamical system in its own right. One may expect the dynamics of the EM field to be governed by quantum mechanics, as is the dynamics of particles. This expectation, however, cannot be justified *a priori*, and we shall need to seek experiments which can distinguish between classical and quantum electrodynamics.

We shall develop the theory in a form that is most convenient for application to quantum optics, because that subject is rich in experiments that illustrate the distinctive features of quantum electromagnetism, as well as being of considerable practical importance. Since the optical reflectors and cavities define a natural frame of reference for the description of the experiments, we need not pay such close attention to Lorentz invariance as is necessary in those formulations of quantum electrodynamics that are designed for application to particle physics. Although our theory will in fact be Lorentz-invariant in its content (except for specific approximations), its form will not be kept manifestly Lorentz-invariant.

In some respects the quantized EM field behaves as a system of bosons, although these particle-like excitations, called *photons*, play a less fundamental role than is suggested by some informal accounts of the subject. In particular, we shall see the quantum EM field is not describable as merely a gas of photons.

19.1 *Normal Modes of the Field*

Maxwell's equations for the electric and magnetic fields, **E** and **B**, in empty space are

$$\boldsymbol{\nabla} \times \mathbf{E} = -\frac{1}{c}\frac{\partial \mathbf{B}}{\partial t}, \tag{19.1}$$

$$\nabla \times \mathbf{B} = \frac{1}{c}\frac{\partial \mathbf{E}}{\partial t}, \tag{19.2}$$

$$\nabla \cdot \mathbf{E} = 0, \tag{19.3}$$

$$\nabla \cdot \mathbf{B} = 0. \tag{19.4}$$

Although the values of **E** and **B** at each point in space are the fundamental dynamical variables of the theory, it is convenient to decompose the fields into normal modes before attempting a quantum-mechanical description.

By taking the curl of (19.1) and the time derivative of (19.2), we obtain $\nabla \times (\nabla \times \mathbf{E}) = -c^{-2}\,\partial^2\mathbf{E}/(\partial t)^2$. Using the vector identity $\nabla \times (\nabla \times \mathbf{v}) = \nabla(\nabla \cdot \mathbf{v}) - \nabla^2\mathbf{v}$ and (19.3), we then obtain the *wave equation* for **E**,

$$\nabla^2\mathbf{E} - \frac{1}{c^2}\frac{\partial^2\mathbf{E}}{\partial t^2} = 0. \tag{19.5}$$

An identical wave equation holds for **B**.

The solution of the wave equation (19.5) can be facilitated by representing the electric field as a sum of *mode functions*, $\mathbf{u}_m(\mathbf{x})$, which are defined by the following eigenvalue equation and subsidiary conditions:

$$\nabla^2\mathbf{u}_m(\mathbf{x}) = -{k_m}^2\,\mathbf{u}_m(\mathbf{x}), \tag{19.6a}$$

$$\nabla \cdot \mathbf{u}_m(\mathbf{x}) = 0, \tag{19.6b}$$

$$\hat{\mathbf{n}} \times \mathbf{u}_m(\mathbf{x}) = 0 \text{ on any conducting surface}, \tag{19.6c}$$

where $\hat{\mathbf{n}}$ is the unit normal to the surface. The latter condition is imposed because the tangential component of **E** must vanish on a conducting surface. It can readily be shown that Eqs. (19.6) describe a Hermitian eigenvalue problem, and so the mode functions that correspond to unequal eigenvalues must be orthogonal. Hence we may choose the set of mode functions to satisfy the orthonormality condition

$$\int \mathbf{u}_{m'}(\mathbf{x}) \cdot \mathbf{u}_m(\mathbf{x})\,d^3x = \delta_{m',m}\,. \tag{19.7}$$

We now represent the electric field as a sum of mode functions,

$$\mathbf{E}(\mathbf{x},t) = \sum_m f_m(t)\,\mathbf{u}_m(\mathbf{x}), \tag{19.8}$$

and substitute this series into the wave equation. Since the mode functions are linearly independent, the coefficients of each mode must separately add up to zero in order to satisfy (19.5). Thus we obtain a separate equation for the amplitude of each mode,

$$\frac{d^2}{dt^2} f_m(t) + c^2 k_m^2 f_m(t) = 0 . \tag{19.9}$$

This is the equation of motion for a harmonic oscillator with angular frequency $\omega_m = c\, k_m$.

The magnetic field can be determined from the electric field by means of (19.1). Substitution of (19.8) into (19.1) yields

$$\mathbf{B}(\mathbf{x}, t) = \sum_m h_m(t)\, \boldsymbol{\nabla} \times \mathbf{u}_m(\mathbf{x}) , \tag{19.10}$$

with

$$\frac{dh_m(t)}{dt} = -c\, f_m(t) . \tag{19.11}$$

The curl of a mode function, $\boldsymbol{\nabla} \times \mathbf{u}_m$, has many useful properties. Since the divergence of a curl is zero, it follows that Maxwell's equation (19.4) is automatically satisfied by (19.10). From the boundary condition (19.6c), it follows that $\oint_C \mathbf{u}_m \cdot d\boldsymbol{\ell} = \int\int_S (\boldsymbol{\nabla} \times \mathbf{u}_m) \cdot d\mathbf{S} = 0$, where C is the closed curve bounding any portion S of a conducting surface. Hence it follows that

$$\hat{\mathbf{n}} \cdot (\boldsymbol{\nabla} \times \mathbf{u}_m) = 0 , \tag{19.12}$$

where $\hat{\mathbf{n}}$ is a unit vector normal to the conducting surface. Thus the dynamic boundary condition for the magnetic field, $\hat{\mathbf{n}} \cdot \mathbf{B} = 0$, is automatically satisfied. Moreover, the curls of the mode functions satisfy an orthogonality relation,

$$\int (\boldsymbol{\nabla} \times \mathbf{u}_{m'}) \cdot (\boldsymbol{\nabla} \times \mathbf{u}_m)\, d^3x = {k_m}^2\, \delta_{m',m} . \tag{19.13}$$

To prove this relation, we integrate the identity

$$(\boldsymbol{\nabla} \times \mathbf{u}_{m'}) \cdot (\boldsymbol{\nabla} \times \mathbf{u}_m) = \mathbf{u}_{m'} \cdot \boldsymbol{\nabla} \times (\boldsymbol{\nabla} \times \mathbf{u}_m) + \boldsymbol{\nabla} \cdot [\mathbf{u}_{m'} \times (\boldsymbol{\nabla} \times \mathbf{u}_m)]$$

over the volume of the system. Because of the eigenvalue equation (19.6a), the integral of the first term on the right hand side becomes

$${k_m}^2 \int \mathbf{u}_{m'} \cdot \mathbf{u}_m\, d^3x = {k_m}^2\, \delta_{m',m} .$$

The divergence theorem can be used to convert the volume integral of the second term into a surface integral, which vanishes by virtue of the boundary conditions (19.6c) and (19.12).

Substitution of (19.8) and (19.10) into (19.2) leads to the relation

$$\frac{df_m(t)}{dt} = c\,{k_m}^2\,h_m(t)\,. \tag{19.14}$$

When combined with (19.11), this yields

$$\frac{d^2}{dt^2}\,h_m(t) + c^2\,{k_m}^2\,h_m(t) = 0\,, \tag{19.15}$$

which has the same form as (19.9).

This analysis into normal modes has shown that the electromagnetic field is dynamically equivalent to an infinite number of independent harmonic oscillators. This result will be used in the next section to obtain a quantum-mechanical description of the field.

19.2 *Electric and Magnetic Field Operators*

To identify appropriate operators for the electric and magnetic fields, we first obtain the Hamiltonian by calculating the total energy of the EM field:

$$\begin{aligned} H_{\text{EM}} &= (8\pi)^{-1} \int \left(E^2 + B^2\right)\, d^3x \\ &= (8\pi)^{-1} \sum_{m',m} f_{m'}\, f_m \int \mathbf{u}_{m'} \cdot \mathbf{u}_m\, d^3x \\ &\quad + (8\pi)^{-1} \sum_{m',m} h_{m'}\, h_m \int (\boldsymbol{\nabla} \times \mathbf{u}_{m'}) \cdot (\boldsymbol{\nabla} \times \mathbf{u}_m)\, d^3x \\ &= (8\pi)^{-1} \sum_m \left({f_m}^2 + {k_m}^2\,{h_m}^2\right). \end{aligned} \tag{19.16}$$

The Hamiltonian for a set of harmonic oscillators, each having unit mass, is

$$H_{\text{osc}} = \sum_m \tfrac{1}{2}\left({P_m}^2 + {\omega_m}^2\,{Q_m}^2\right), \tag{19.17}$$

with Q_m being the coordinate and $P_m = dQ_m/dt$ being the momentum or velocity. If we use (19.14) to eliminate h_m in (19.16), then H_{EM} will have the form of (19.17) provided we make the identification

$$Q_m \leftrightarrow \frac{f_m}{2\,\omega_m\,\sqrt{\pi}}\,. \tag{19.18}$$

On the other hand, if we use (19.11) to eliminate f_m in (19.16), then H_{EM} will also have the form of (19.17), but with the alternative identification

$$Q_m \leftrightarrow \frac{h_m}{2\,c\sqrt{\pi}}\,. \tag{19.19}$$

The strategy is to use the known forms of the operators for a harmonic oscillator to deduce appropriate operators for the EM field. But the ambiguity associated with these two possible identifications must be resolved before we can be sure that we have the correct operators.

From the results of Sec. 6.1, the *Hamiltonian operator* for a system of independent oscillators is of the form

$$H = \sum_m \hbar\omega_m \left({a_m}^\dagger\, a_m + \frac{1}{2} \right), \tag{19.20}$$

where the raising and lowering operators, ${a_m}^\dagger$ and a_m, satisfy the commutation relation

$$\left[a_{m'}, {a_m}^\dagger\right] = \delta_{m',m}\,. \tag{19.21}$$

We shall assume that this form holds for the Hamiltonian of the modes of the EM field. Using (6.4), (6.5), (6.8), and (6.9), the position and momentum operators for the normal mode oscillators are

$$Q_m = \left(\frac{\hbar}{2\,\omega_m}\right)^{1/2} \left({a_m}^\dagger + a_m\right), \tag{19.22}$$

$$P_m = i \left(\frac{\hbar\,\omega_m}{2}\right)^{1/2} \left({a_m}^\dagger - a_m\right). \tag{19.23}$$

(The masses of the oscillators have been set equal to 1. This amounts only to a choice of units.) Now either of the identifications (19.18) and (19.19) will give us forms for the electric and magnetic field operators. But which choice is correct? We shall investigate both choices.

Case (i). If we choose (19.18), then we obtain

$$f_m = \left(2\,\omega_m \sqrt{\pi}\right)\, Q_m = (2\pi\,\hbar\,\omega_m)^{1/2} \left({a_m}^\dagger + a_m\right).$$

The *electric field operator*, obtained from (19.8), is then

$$\mathbf{E}(\mathbf{x}, t) = \sum_m (2\pi\,\hbar\,\omega_m)^{1/2} \left\{{a_m}^\dagger(t) + a_m(t)\right\}\, \mathbf{u}_m(\mathbf{x})\,. \tag{19.24}$$

Note that $\mathbf{E}(\mathbf{x},t)$ is now an *operator field*; that is to say, an electric field operator is defined at each space–time point $(\mathbf{x},t)$. The space coordinate $\mathbf{x}$ is not an operator. The time dependence of these operators is interpreted in the sense of the Heisenberg picture (Sec. 3.7).

Case (ii). If we choose (19.19), then we obtain $h_m = (2\,c\sqrt{\pi})\,Q_m$. Then from (19.11), it follows that

$$f_m(t) = \frac{-1}{c}\,\frac{dh_m(t)}{dt} = -\left(2\sqrt{\pi}\right)\,\frac{dQ_m}{dt}$$

$$= -\left(2\sqrt{\pi}\right)\,P_m = -i(2\pi\,\hbar\,\omega_m)^{1/2}\,\left({a_m}^\dagger - a_m\right).$$

With this identification, (19.8) yields

$$\mathbf{E}(\mathbf{x},t) = \sum_m -i(2\pi\,\hbar\,\omega_m)^{1/2}\,\left\{{a_m}^\dagger(t) - a_m(t)\right\}\,\mathbf{u}_m(\mathbf{x})\,. \tag{19.25}$$

The apparent conflict between (19.24) and (19.25) is now easily resolved. The Heisenberg equation of motion (3.73) for the operator a_m is

$$\frac{d}{dt}\,a_m(t) = \frac{i}{\hbar}\,[H, a_m(t)] = -i\,\omega_m\,a_m(t)\,.$$

Its solution is $a_m(t) = a_m \exp(-i\,\omega_m t)$. Similarly, we obtain ${a_m}^\dagger(t) = {a_m}^\dagger \exp(i\,\omega_m t)$. Because of this simple time dependence of the raising and lowering operators, it is apparent that (19.25) differs from (19.24) only in the phases of the modes. Since the initial phase of a mode is arbitrary, there is no physically significant difference between case (i) and case (ii). In case (i) the electric field is analogous to the position of the oscillator. In case (ii) it is analogous to the momentum. Since the Hamiltonian of a harmonic oscillator is symmetric with respect to position and momentum (provided they are expressed in suitable units), the two analogies lead to the same physical results.

Henceforth we shall use only case (i), and (19.24) will be the appropriate form for the electric field operator. The corresponding form for the *magnetic field operator* is found from (19.10) to be

$$\mathbf{B}(\mathbf{x},t) = \sum_m ic\left(\frac{2\pi\,\hbar}{\omega_m}\right)^{1/2}\,\left\{{a_m}^\dagger(t) - a_m(t)\right\}\,\boldsymbol{\nabla}\times\mathbf{u}_m(\mathbf{x})\,. \tag{19.26}$$

Comparing the way in which the raising and lowering operators enter (19.24) and (19.26) with the way they enter (19.22) and (19.23), we may say that, roughly speaking, the electric field is analogous to the position and the magnetic field is analogous to the momentum of an oscillator.

[[The operators for the electric and magnetic fields have been obtained here by *analogy* with those for a mechanical oscillator. The operators for the dynamical variables of particles were deduced in Ch. 3 from the symmetry transformations of space–time. It is worth pointing out that the principle of invariance under transformations such as $\mathbf{Q} \to \mathbf{Q} + \mathbf{a}$, which certainly holds for the dynamics of a particle, has no apparent justification for the abstract oscillator of a field mode. Therefore our identification of the operators for the fields must be regarded as a plausible hypothesis, rather than a deduction from established principles. Perhaps the weaker status of the field operators should be kept in mind when we encounter some difficulties in the next section.]]

Complex basis functions

Since the electric and magnetic fields are real, it is natural to chose the basis functions $\mathbf{u}_m(\mathbf{x})$ to be real, as we have done above. However, it is sometimes convenient to use complex basis functions. The complex exponential $e^{i\mathbf{k}\cdot\mathbf{x}}$ is the most common example of such a function, but we shall present the necessary mathematics in a more general form.

For the introduction of complex basis functions to be possible, it is necessary that for every frequency there should be two independent mode functions. For the frequency ω_k we denote the two degenerate mode functions as $\mathbf{c}_k(\mathbf{x})$ and $\mathbf{s}_k(\mathbf{x})$ $(k > 0)$. We denote the corresponding creation operators as ${a_c}^\dagger$ and a_s. We now define a complex mode function,

$$\mathbf{e}_k(\mathbf{x}) = [\mathbf{c}_k(\mathbf{x}) + i\,\mathbf{s}_k(\mathbf{x})]\sqrt{\tfrac{1}{2}}\,, \quad (k > 0)\,, \tag{19.27}$$

so that $\mathbf{c}_k = \sqrt{\frac{1}{2}}(\mathbf{e}_k + {\mathbf{e}_k}^*)$ and $\mathbf{s}_k = \sqrt{\frac{1}{2}}(\mathbf{e}_k - {\mathbf{e}_k}^*)/i$. It is convenient to define

$$\mathbf{e}_{-k}(\mathbf{x}) = {\mathbf{e}_k}^*(\mathbf{x}) = [\mathbf{c}_k(\mathbf{x}) - i\,\mathbf{s}_k(\mathbf{x})]\sqrt{\tfrac{1}{2}}\,, \quad (k > 0)\,. \tag{19.28}$$

The most common example of these functions is $\mathbf{c}_k(\mathbf{x}) = (2/\Omega)^{1/2}\mathbf{u}\cos(\mathbf{k}\cdot\mathbf{x})$, $\mathbf{s}_k(\mathbf{x}) = (2/\Omega)^{1/2}\,\mathbf{u}\,\sin(\mathbf{k}\cdot\mathbf{x})$, and $\mathbf{e}_k(\mathbf{x}) = \mathbf{u}\,e^{i\mathbf{k}\cdot\mathbf{x}}/\sqrt{\Omega}$, where $\mathbf{u}$ is a polarization vector, and Ω is the volume over which the functions are normalized. The symbolic restriction $k > 0$ is to be interpreted as restricting the vector $\mathbf{k}$ to a half-space, since the replacement of $\mathbf{k}$ by $-\mathbf{k}$ does not yield independent mode functions. It is convenient to think in terms of these functions, although the mathematics does not require us to be so specific.

Apart from the numerical factor $(2\pi\,\hbar\,\omega_k)^{1/2}$, the contribution of the two degenerate modes to the electric field operator (19.24) is

$$\begin{aligned}&\left(a_c{}^\dagger + a_c\right)\,\mathbf{c}_k(\mathbf{x}) + \left(a_s{}^\dagger + a_s\right)\,\mathbf{s}_k(\mathbf{x})\\ &\quad = \mathbf{e}_k(\mathbf{x})\left(\frac{a_c{}^\dagger - i\,a_s{}^\dagger}{\sqrt{2}} + \frac{a_c - i\,a_s}{\sqrt{2}}\right) + \mathbf{e}_k{}^*(\mathbf{x})\left(\frac{a_c{}^\dagger + i\,a_s{}^\dagger}{\sqrt{2}} + \frac{a_c + i\,a_s}{\sqrt{2}}\right).\end{aligned} \tag{19.29}$$

It is now convenient to define

$$a_k^\dagger = \frac{a_c{}^\dagger + i\,a_s{}^\dagger}{\sqrt{2}}\,, \quad a_{-k}{}^\dagger = \frac{a_c{}^\dagger - i\,a_s{}^\dagger}{\sqrt{2}}\,, \tag{19.30}$$

from which it follows that

$$a_k = \frac{a_c - i\,a_s}{\sqrt{2}}\,, \quad a_{-k} = \frac{a_c + i\,a_s}{\sqrt{2}}\,. \tag{19.31}$$

Then (19.29) becomes

$$\begin{aligned}&\left(a_c{}^\dagger + a_c\right)\,\mathbf{c}_k(\mathbf{x}) + \left(a_s{}^\dagger + a_s\right)\,\mathbf{s}_k(\mathbf{x})\\ &\quad = \left(a_k{}^\dagger + a_{-k}\right)\,\mathbf{e}_k{}^*(\mathbf{x}) + \left(a_{-k}{}^\dagger + a_k\right)\,\mathbf{e}_k(\mathbf{x})\,.\end{aligned} \tag{19.32}$$

Defining $\omega_{-k} = \omega_k$ and using (19.28), we may extend the range of k to negative values and rewrite (19.24) as

$$\mathbf{E}(\mathbf{x},t) = \sum_{\mathbf{k}} (2\pi\,\hbar\,\omega_k)^{1/2}\,\left\{a_k{}^\dagger(t)\,\mathbf{e}_k{}^*(\mathbf{x}) + a_k(t)\,\mathbf{e}_k(\mathbf{x})\right\}\,. \tag{19.33}$$

19.3 *Zero-Point Energy and the Casimir Force*

The Hamiltonian operator (19.20) for the EM field has the form of a harmonic oscillator for each mode of the field. As was shown in Sec. 6.1, the lowest energy of a harmonic oscillator is $\frac{1}{2}\hbar\omega$. Since there are infinitely many modes of arbitrarily high frequency in any finite volume, it follows that there should be an infinite zero-point energy in any volume of space. Needless to say, this conclusion is unsatisfactory.

In order to gain some appreciation for the magnitude of the zero-point energy, we shall calculate the zero-point energy in a rectangular cavity due to those field modes whose frequency is less than some cutoff ω_c. The *mode functions* $\mathbf{u}(\mathbf{x})$, solutions of (19.6) for a cavity of dimensions $L_1 \times L_2 \times L_3$, have the vector components

$$\begin{aligned} u_x &= A_1 \cos(k_1 x) \sin(k_2 y) \sin(k_3 z)\,, \\ u_y &= A_2 \sin(k_1 x) \cos(k_2 y) \sin(k_3 z)\,, \\ u_z &= A_3 \sin(k_1 x) \sin(k_2 y) \cos(k_3 z)\,. \end{aligned} \tag{19.34}$$

They may be labeled by a wave vector $\mathbf{k}$ whose components are

$$k_1 = \frac{n_l\,\pi}{L_1}\,, \quad k_2 = \frac{n_2\,\pi}{L_2}\,, \quad k_3 = \frac{n_3\,\pi}{L_3}\,, \tag{19.35}$$

with n_1, n_2, and n_3 being nonnegative integers in order to satisfy the boundary condition (19.6c). The frequency of the mode is $\omega_k = c\sqrt{{k_1}^2 + {k_2}^2 + {k_3}^2}$. At least two of the integers must be nonzero, otherwise the mode function would vanish identically. The amplitudes of the three components of (19.34) are related by the divergence condition (19.6b), which requires that

$$A_1 k_1 + A_2 k_2 + A_3 k_3 = 0\,. \tag{19.36}$$

This condition can be written as $\mathbf{A} \cdot \mathbf{k} = 0$, from which it is clear that there are two linearly independent *polarizations* (directions of $\mathbf{A}$) for each $\mathbf{k}$, and hence there are two independent modes for each set of positive integers (n_1, n_2, n_3) in (19.35). If one of the integers is zero, it is clear from (19.34) that two of the components of $\mathbf{u}(\mathbf{x})$ will vanish, so there is only one mode in this exceptional case.

If the dimensions of the cavity are large, the allowed values of $\mathbf{k}$ approximate a continuum, and the density of modes in the positive octant of k space is $2\Omega/\pi^3$. Here $\Omega = L_1\,L_2\,L_3$ is the volume of the cavity. The *zero-point energy density* for all modes of frequency less that ω_c is then given by

$$\frac{2}{\Omega}\sum_{\mathbf{k}} \frac{1}{2}\hbar\omega_k = \frac{2}{8\pi^3} \underset{k<k_c}{\rightarrow} \int \frac{1}{2}\hbar ck\; 4\pi k^2\, dk = \frac{\hbar c {k_c}^4}{8\pi^2}\,, \tag{19.37}$$

with $k_c = 2\pi/\lambda_c = \omega_c/c$. [The sum in (19.37), over the allowed vectors $\mathbf{k}$ in the positive octant, has been approximated by one eighth of an integral over the full sphere.]

The factor ${k_c}^4$ indicates that this energy density is dominated by the high-frequency, short-wavelength modes. Taking a minimum wavelength of

$\lambda_c = 0.4 \times 10^{-6}$ m, so as to include the visible light spectrum, yields a zero-point energy density of 23 J/m^3. This may be compared with energy density produced by a 100 W light bulb at a distance of 1 m, which is 2.7×10^{-8} J/m^3. Of course it is impossible to extract any of the zero-point energy, since it is the minimum possible energy of the field, and so our inability to perceive that large energy density is not incompatible with its existence. Indeed, since most experiments detect only energy differences, and not absolute energies, it is often suggested that the troublesome zero-point energy of the field should simply be omitted and the Hamiltonian (19.20) should be replaced by

$$H = \sum_m \hbar\omega_m \, {a_m}^\dagger \, a_m \,.$$

One phenomenon that depends on absolute energies is gravitation, since all forms of mass–energy act as gravitational sources. Of course an infinite energy density is not acceptable, but there is reason to believe that quantum-gravitational effects may lead to a minimum wavelength for all fields no greater than the Planck length, $(\hbar G/c^3)^{1/2} = 1.6 \times 10^{-33}$ cm. According to general relativity, the geometry of the universe will be curved into closure if the mass density of the universe exceeds $\rho_c \approx 5 \times 10^{-30}$ g/cm^3 [see Misner, Thorne, and Wheeler (1973), Ch. 27]. The known mass density is a factor of 10 or 20 smaller than this, but some people speculate that dark matter may bring the total up to ρ_c. Suppose that the zero-point energy were responsible for this "missing mass". A minimum wavelength, $\lambda_c = 2\pi/k_c$, might be estimated by setting $\rho_c c^2$ equal to (19.37), obtaining $\lambda_c \approx 0.02$ cm. Needless to say, this is much too large a value for a minimum wavelength!

Although these calculations suggest that the zero-point energy of the electromagnetic field cannot be real, an argument by Casimir (1948) leads to the opposite conclusion. We consider a large cavity of dimensions $L \times L \times L$ bounded by conducting walls (Fig. 19.1). A conducting plate is inserted at a distance R from one of the yz faces ($R \ll L$). The new boundary condition at $x = R$ alters the energy (or frequency) of each field mode. Following Casimir, we shall calculate the energy shift as a function of R.

Let W_X denote the electromagnetic energy within a cavity whose length in the x direction is X. The change in the energy due to the insertion of the plate at $x = R$ will be

$$\Delta W = W_R + W_{L-R} - W_L \,. \tag{19.38}$$

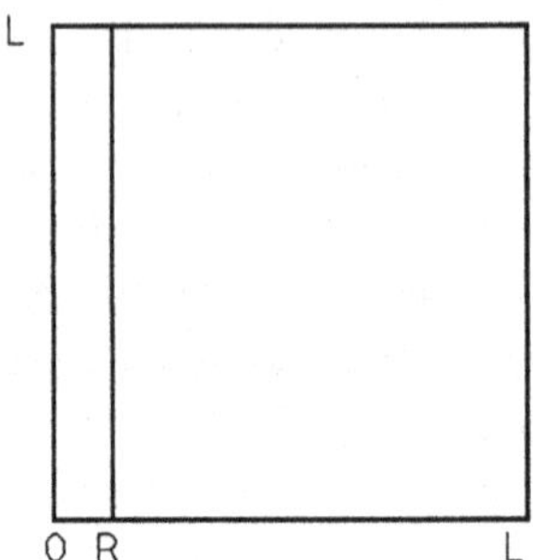

Fig. 19.1 A large cavity with a plate inserted near one end.

Each of these three terms is infinite, but the difference will turn out to be finite. Each mode has a zero-point energy of $\frac{1}{2}\hbar\omega = \frac{1}{2}\hbar ck$, and the total energy is the sum of zero-point energies for all modes. If the linear dimensions of the cavity are very large we can replace the sum over discrete modes by an integral. Thus we have, formally,

$$W_L = \frac{2\,L^3}{\pi^3} \iiint_0^\infty \tfrac{1}{2}\,\hbar\,ck\,dk_x\,dk_y\,dk_z\,,$$

$$W_{L-R} = \frac{2\,L^2(L-R)}{\pi^3} \iiint_0^\infty \tfrac{1}{2}\,\hbar\,ck\,dk_x\,dk_y\,dk_z\,,$$

with $k = \sqrt{{k_x}^2 + {k_y}^2 + {k_z}^2}$. The discreteness of k_x must be taken into account in calculating W_R, so we have

$$W_R = \sum_{n=0}^{\infty} 2\,\theta_n\,\frac{L^2}{\pi^2} \iint_0^\infty \tfrac{1}{2}\,\hbar\,ck\,dk_y\,dk_z\,,$$

with $k = \sqrt{(n\pi/R)^2 + {k_y}^2 + {k_z}^2}$, and $\theta_n = 1$ for $n > 0$, $\theta_n = \frac{1}{2}$ for $n = 0$. The factor θ_n must be included because there are two polarization states for $n > 0$ but only one for $n = 0$. (This single exceptional point does not affect the values of the integrals.)

Because the sum and integrals are divergent it is necessary to introduce a smooth *cutoff* function, $f(k/k_c)$, having the properties

$$f\left(\frac{k}{k_c}\right) \to 1\,, \quad k \ll k_c\,, \tag{19.39a}$$

$$f\left(\frac{k}{k_c}\right) \to 0\,, \quad k \gg k_c\,. \tag{19.39b}$$

This is not merely a mathematical artifice. For sufficiently high frequencies ($\omega \gg ck_c$), a metal plate will not behave as a conductor, and so will not affect the EM field at those frequencies. Thus the plate at $x = R$ will have no effect on the energies of very high frequency modes, and hence they will not contribute to ΔW. The detailed form of $f(k/k_c)$, and the value of k_c depend upon the nature of the material, but these details do not affect ∇W in the lowest order approximation.

Combining the above results, we have

$$\Delta W = \hbar c \frac{L^2}{\pi^2} \left\{ \sum_{n=0}^{\infty} \theta_n \, g\left(\frac{n\pi}{R}\right) - \frac{R}{\pi} \int_0^{\infty} g(k_x) \, dk_x \right\}, \tag{19.40}$$

where

$$g(k_x) = \iint_0^{\infty} k \, f\left(\frac{k}{k_c}\right) \, dk_y \, dk_z$$

and $k = \sqrt{k_x{}^2 + k_y{}^2 + k_z{}^2}$. The integral expression for $g(k_x)$ can be simplified by means of a few substitutions. First we introduce $\kappa = \sqrt{k_y{}^2 + k_z{}^2}$ and use polar coordinates in the yz plane, to obtain

$$g(k_x) = \frac{4\pi}{8} \int_0^{\infty} \sqrt{k_x{}^2 + \kappa^2} \, f\left(\frac{\sqrt{k_x{}^2 + \kappa^2}}{k_c}\right) \kappa \, d\kappa \, .$$

We next introduce dimensionless variables n and α in place of $k_x = n\pi/R$ and $\kappa = \alpha\pi/R$:

$$g(k_x) = \frac{\pi}{2} \frac{\pi^3}{R^3} \int_0^{\infty} \sqrt{n^2 + \alpha^2} \, f\left(\frac{\pi\sqrt{n^2 + \alpha^2}}{R \, k_c}\right) \alpha \, d\alpha \, .$$

Finally we substitute $w = n^2 + \alpha^2$, $dw = 2\alpha \, d\alpha$, and obtain

$$g(k_x) = \frac{\pi^4 \, F(n)}{4R^3} \, ,$$

with

$$F(n) = \int_{n^2}^{\infty} \sqrt{w} \, f\left(\frac{\pi\sqrt{w}}{R \, k_c}\right) \, dw \, . \tag{19.41}$$

Thus (19.40) becomes

$$\Delta W = \frac{\hbar c \, L^2 \, \pi^4}{4\pi^2 \, R^3} \left\{ \sum_{n=0}^{\infty} \theta_n \, F(n) - \int_0^{\infty} F(n) \, dn \right\}. \tag{19.42}$$

The discrete sum in the expression (19.42) is a common numerical approximation to the integral, known as the trapezoidal rule. The difference between the sum and the integral may be estimated by means of the *Euler–Maclaurin* formula,

$$\sum_{n=0}^{\infty} \theta_n F(n) - \int_0^{\infty} F(n)\, dn = \frac{-1}{6 \times 2!} F'(0) + \frac{1}{30 \times 4!} F'''(0) - \cdots . \tag{19.43}$$

From (19.41) we obtain

$$\begin{aligned} F'(n) &= -2n^2 f\left(\frac{\pi n}{Rk_c}\right), \\ F''(n) &= -4n f\left(\frac{\pi n}{Rk_c}\right) - 2n^2 f'\left(\frac{\pi n}{Rk_c}\right)\left(\frac{\pi}{Rk_c}\right), \\ F'''(n) &= -4 f\left(\frac{\pi n}{Rk_c}\right) - 8n f'\left(\frac{\pi n}{Rk_c}\right)\left(\frac{\pi}{Rk_c}\right) \\ &\quad - 2n^2 f''\left(\frac{\pi n}{Rk_c}\right)\left(\frac{\pi}{Rk_c}\right)^2, \end{aligned}$$

where the prime notation on a function indicates the derivative with respect to its own argument. Because of (19.39a), this yields $F'(0) = 0$ and $F'''(0) = -4$. Therefore the shift of the electromagnetic zero-point energy due to the conducting plate inserted at $x = R$ is

$$\Delta W = -\hbar c \frac{\pi^2}{720} \frac{L^2}{R^3} . \tag{19.44}$$

It is clear from the above calculations that the contributions of the higher derivative terms that are omitted from (19.43) will all be proportional to some power of $1/Rk_c$, and therefore the result (19.44) will be valid provided $Rk_c \gg 1$. Equivalently, we may say that the result holds for $R \gg \lambda_c$, where $\lambda_c = 2\pi/k_c$ is the wavelength of radiation at the cutoff frequency ω_c.

In the limit $L \to \infty$, the energy shift produces a force per unit area between the conducting plates at $x = 0$ and $x = R$:

$$F = -\frac{1}{L^2} \frac{\partial \Delta W}{\partial R} = -\frac{\hbar c}{240} \frac{\pi^2}{R^4} . \tag{19.45}$$

The minus sign indicates that the force is attractive.

The Casimir force (19.45) is difficult to measure. The surfaces must be flat and clean, and free from any electrostatic charge. However, the measurements carried out by Sparnaay (1958) confirm Casimir's theoretical prediction, the greatest uncertainty being in the determination of the distance between the surfaces (of order 10^{-6} m). These results indicate that the notion of a zero-point or vacuum energy is not spurious, and cannot be discarded. We are left with a conundrum that has never been adequately resolved: an infinite zero-point energy density is nonsensical, but finite parts of it yield experimentally confirmed consequences. (In the above calculation only field modes whose frequencies are less than ω_c contribute significantly to the result.)

[[The infinities of quantum field theory, of which we have here seen only the first, are somewhat of a hidden scandal. Most physicists seem content to ignore them because there are procedures (the so-called renormalization theory) which allow us to avoid the infinities in many practical cases. One prominent physicist who was not complacent about the infinities was Dirac. On the last page of his book he says of renormalization theory, "the rules ... do not fit in with the logical foundations of quantum mechanics. They should therefore not be considered as a satisfactory solution of the difficulties." In the final sentence of his book Dirac states, "The difficulties, being of a profound character, can be removed only by some drastic change in the foundations of the theory, probably a change as drastic as the passage from Bohr's orbit theory to the present quantum mechanics."]]

19.4 *States of the EM Field*

It was shown in Sec. 19.2 that the electromagnetic field can be analyzed into normal modes, each of which is dynamically equivalent to a harmonic oscillator. Thus the EM field equivalent to a denumerably infinite set of independent harmonic oscillators, and a state for the EM field can be specified in terms of the states of the modes.

Photon number eigenstates

The Hamiltonian of the mth mode of the EM field is $H_m = \hbar\omega_m({a_m}^\dagger\, a_m + \frac{1}{2})$. Its eigenvectors, denoted as $|n_m\rangle$ with $n_m = 0, 1, 2, \ldots$, are the eigenvectors of the number operator $N_m = {a_m}^\dagger\, a_m$. The field Hamiltonian (19.20) is the sum of the mode Hamiltonians, and its eigenvectors are of the form

$$|\{n\}\rangle \equiv |n_1, n_2, \ldots, n_m, \ldots\rangle = |n_1\rangle \otimes |n_2\rangle \otimes \cdots |n_m\rangle \otimes \cdots . \tag{19.46}$$

This is similar in form to the state vectors for a system of bosons, which were treated in Sec. 17.4. Therefore the mth mode of this state is described as containing n_m *photons*. These elementary excitations of the EM field behave in many respects like particles, carrying energy and momentum. However, the analogy is incomplete, and it is not possible to replace the EM field by a gas of photons.

[[A very similar situation occurs in the theory of harmonic vibrations of a crystalline solid. The coupled motions of the atoms may by analyzed into independent normal modes, each of which is treated as a quantized harmonic oscillator. The elementary excitations of these normal modes carry energy and momentum, and are called *phonons*. When introducing this term, J. Frenkel (1932) added the following remark (p. 267): "It is not in the least intended to convey the impression that such phonons have a real existence. On the contrary, the possibility of their introduction rather serves to discredit the belief in the real existence of photons."]]

In a state with definite photon numbers, the electric and magnetic fields are indefinite and fluctuating. The probability distributions for the electric and magnetic fields in such a state are analogous to the distributions for the position and momentum of an oscillator in an energy eigenstate. A simple calculation using the operator (19.24) shows that the average electric field is zero in a photon number eigenstate:

$$\langle \mathbf{E}(\mathbf{x},t) \rangle = \langle n_1, n_2, \ldots, n_m, \ldots | \mathbf{E}(\mathbf{x},t) | n_1, n_2, \ldots, n_m, \ldots \rangle = 0 \,. \tag{19.47}$$

The mean square of the electric field amplitude is

$$\begin{aligned}\langle |\mathbf{E}|^2 \rangle &= 2\pi\hbar \sum_m \sum_n (\omega_m\,\omega_n)^{1/2}\, \mathbf{u}_m \cdot \mathbf{u}_n \,\langle \left({a_m}^\dagger + a_m\right)\left({a_n}^\dagger + a_n\right) \rangle \\ &= 2\pi\hbar \sum_m \omega_m |\mathbf{u}_m(\mathbf{x})|^2 \,\langle n_m | \left({a_m}^\dagger + a_m\right)^2 | n_m \rangle \\ &= \sum_m 2\pi\hbar\,\omega_m |\mathbf{u}_m(\mathbf{x})|^2\,(2n_m + 1) \,. \end{aligned} \tag{19.48}$$

The sum over all modes is infinite. If $n_m = 0$ for all m, this is the vacuum state, with the infinite zero-point energy that was discussed in Sec. 19.3.

This divergence problem can often be circumvented (but *not* solved) by recognizing that a particular experiment will effectively couple to the EM field only over some finite bandwidth. In quantum optics we usually excite only a finite number of modes, or even a single mode. The mean square field associated with the mth mode, averaged over the normalization volume Ω, as well as over the ensemble described by the state vector, is

$$\Omega^{-1}\int\langle\,|\mathbf{E}_m(\mathbf{x})|^2\,\rangle\,d^3x = \Omega^{-1}\int\langle\,|\mathbf{B}_m(\mathbf{x})|^2\,\rangle\,d^3x$$
$$= \frac{4\pi}{\Omega}\,\hbar\omega_m\left(n_m+\tfrac{1}{2}\right).$$

The result can be understood in terms of an equal division of the mode energy, $\hbar\omega_m(n_m+\frac{1}{2})$, between the energy densities of the electric and magnetic fields.

Coherent states — Mathematical relations

Photon number eigenvectors form a complete orthonormal basis for the state space, but of course not all states of the EM field need to be eigenstates of the photon number operators. It is possible to define eigenvectors of the electric field (analogous to position eigenvectors of a mechanical oscillator), but the fluctuations of the magnetic field would be unbounded in such states. Of much greater use are the *coherent states*, in which the fluctuations of both the electric and magnetic fields are comparably small.

We shall first treat a single mode of the EM field. This is equivalent to a single harmonic oscillator, and it is convenient to use the language appropriate to an oscillator and then to apply the results to the EM field using the relations developed in Sec. 19.2. The field mode oscillator has unit mass, angular frequency ω, creation and annihilation operators $a^\dagger$ and a, and ground state $|0\,\rangle$, with the property

$$a|0\,\rangle = 0\,. \tag{19.49}$$

In the ground state, the position and momentum distributions have zero mean,

$$\langle\,0|Q|0\,\rangle = \langle\,0|P|0\,\rangle = 0\,, \tag{19.50}$$

and the variances satisfy the indeterminacy relation (8.33) as an equality,

$$\langle\,0|Q^2|0\,\rangle\,\langle\,0|P^2|0\,\rangle = \frac{\hbar^2}{4}\,. \tag{19.51}$$

The latter result follows from the fact that the ground state function, the case $n=0$ of (6.32), is an example of a minimum uncertainty state function (8.35).

We now introduce a unitary operator,

$$D(z) = \exp\left(z\, a^\dagger - z^* a\right), \tag{19.52}$$

where z is a complex parameter whose significance will soon be determined. Using the relations (19.22) and (19.23), we may express $D(z)$ in terms of the position and momentum operators of the oscillator:

$$\begin{aligned} D(z) &= \exp\left\{(\text{Re } z)\left(a^\dagger - a\right) + i(\text{Im } z)\left(a^\dagger + a\right)\right\} \\ &= \exp\left\{\frac{i}{\hbar}\left(p_0 Q - q_0 P\right)\right\} \\ &= D(q_0, p_0)\,, \end{aligned} \tag{19.53}$$

where

$$q_0 = \left(\frac{2\hbar}{\omega}\right)^{1/2} \text{Re}\, z\,, \tag{19.54}$$

$$p_0 = (2\hbar\omega)^{1/2}\, \text{Im}\, z\,. \tag{19.55}$$

We shall use the notations $D(z)$ and $D(q_0, p_0)$ interchangeably. The operator $D(q_0, p_0)$ is a *displacement operator in phase space*, with the properties

$$D(q_0, p_0)\, Q\, D^{-1}(q_0, p_0) = Q - q_0\, I\,, \tag{19.56}$$

$$D(q_0, p_0)\, P\, D^{-1}(q_0, p_0) = P - p_0\, I\,. \tag{19.57}$$

These relations are easily derived from the identity (Problem 3.3)

$$e^Y X\, e^{-Y} = X + [Y, X] + \frac{1}{2}[Y, [Y, X]] + \cdots. \tag{19.58}$$

Substituting $Y = (i/\hbar)(p_0 Q - q_0 P)$ and $X = Q$, we obtain (19.56). Substituting $X = P$, we obtain (19.57).

The *coherent state vectors* are obtained by displacing the ground state of the oscillator in phase space,

$$|\, z\,\rangle = D(z)|\, 0\,\rangle\,, \tag{19.59}$$

where $z = (\omega/2\hbar)^{1/2}\, q_0 + i(1/2\hbar\,\omega)^{1/2} p_0$ may be any complex number. Because the operator $D(z)$ is unitary, it is clear that $\langle\, z|z\,\rangle = \langle\, 0|0\,\rangle = 1$. The average position in a coherent state is

$$\begin{aligned}\langle z|Q|z\rangle &= \langle 0|D^{\dagger}(z)\,Q\,D(z)|0\rangle = \langle 0|D^{-1}(z)\,Q\,D(z)|0\rangle \\ &= \langle 0|D(-z)\,Q\,D^{-1}(-z)|0\rangle \\ &= \langle 0|(Q+q_0\,I)|0\rangle \\ &= q_0\,. \end{aligned} \tag{19.60}$$

Similarly, we can show that the average momentum is

$$\langle z|P|z\rangle = p_0\,. \tag{19.61}$$

In fact, the entire position and momentum probability distributions are those of the ground state, diaplaced from the origin of phase space to the point (q_0, p_0).

The coherent state vectors are eigenvectors of the annihilation operator. To show this we operate on (19.49) with $D(z)$:

$$\begin{aligned}0 &= D(z)\,a|0\rangle \\ &= D(z)\,a\,D^{-1}(z)\,D(z)|0\rangle \\ &= (a - zI)|z\rangle\,. \end{aligned}$$

[In the last step we have use the identity (19.58) with $Y = za^{\dagger} - z^{*}a$ and $X = a$.] Thus we have

$$a|z\rangle = z|z\rangle\,. \tag{19.62}$$

Since the operator a is not Hermitian, the eigenvalue z need not be real, and the eigenvector is right-handed only. The dual equation is

$$\langle z|a^{\dagger} = z^{*}\langle z|\,. \tag{19.63}$$

It follows that any normally ordered function of creation and annihilation operators can be trivially evaluated in a coherent state:

$$\langle z|f\left(a^{\dagger}\right)\,g(a)|z\rangle = f(z^{*})\,g(z)\,. \tag{19.64}$$

A coherent state $|z\rangle$ can be expanded in terms of the number eigenvectors $\{|n\rangle\}$ with the help of the Baker–Hausdorff identity (Problem 3.4), from which it follows that

$$D(z) \equiv e^{za^{\dagger} - z^{*}a} = \exp\left(-\tfrac{1}{2}|z|^2\right)\,e^{za^{\dagger}}\,e^{-z^{*}a}\,. \tag{19.65}$$

Thus we have

$$\begin{aligned}|z\rangle = D(z)|0\rangle &= \exp(-\tfrac{1}{2}|z|^2)\ e^{za^\dagger}\ e^{-z^*a}|0\rangle \\ &= \exp(-\tfrac{1}{2}|z|^2)\ e^{za^\dagger}|0\rangle\,, \quad (\text{using } a|0\rangle = 0)\,, \\ &= \exp(-\tfrac{1}{2}|z|^2) \sum_{n=0}^{\infty} \frac{z^n}{n!}\,(a^\dagger)^n\,|0\rangle\,.\end{aligned}$$

Using (6.17) we obtain our desired expansion,

$$|z\rangle = \exp(-\tfrac{1}{2}|z|^2) \sum_{n=0}^{\infty} \frac{z^n}{(n!)^{1/2}}\,|n\rangle\,. \tag{19.66}$$

The set of coherent state vectors $\{|z\rangle\}$ can be used as basis vectors, but its properties are very different from those of the familiar orthonormal sets. Although the coherent state vectors have the usual normalization, $\langle z|z\rangle = 1$, they are *not orthogonal*. Using (19.66) we deduce that

$$\begin{aligned}\langle z|z'\rangle &= \exp\left[-\frac{1}{2}(|z|^2 + |z'|^2)\right] \sum_n \sum_m \langle n|m\rangle \frac{(z^*)^n\,(z')^m}{(n!\,m!)^{1/2}} \\ &= \exp\left[-\frac{1}{2}(|z|^2 + |z'|^2) + z^*z'\right].\end{aligned} \tag{19.67a}$$

The absolute value of the inner product is

$$|\langle z|z'\rangle| = \exp\left(-\frac{1}{2}|z - z'|^2\right), \tag{19.67b}$$

from which we see that, although no two coherent states are orthogonal, the overlap between them is very small if they are reasonably far apart in the complex z plane.

The coherent state vectors obey a *completeness* relation,

$$\pi^{-1} \int |z\rangle\langle z|\,d^2z = I\,, \tag{19.68}$$

where I is the identity operator. The integration is over the area of the complex z plane. If $z = x + iy = r\,e^{i\theta}$, then $d^2z = dx\,dy = r\,d\theta\,dr$. To prove (19.68), we once again use the expansion (19.66):

$$\begin{aligned}\int |z\rangle\langle z|\, d^2z &= \int \exp\left(-|z|^2\right) \sum_n \sum_m |n\rangle\langle m| \frac{z^n\,(z^*)^m}{(n!\,m!)^{1/2}}\, d^2z \\ &= \int_0^\infty \int_0^{2\pi} \exp\left(-r^2\right) \sum_n \sum_m |n\rangle\langle m| \frac{r^{n+m}}{(n!\,m!)^{1/2}}\, e^{i(n-m)\theta}\, r\, d\theta\, dr \\ &= 2\pi \sum_n |n\rangle\langle n| \int_0^\infty \exp(-r^2)\, \frac{r^{2n+1}}{n!}\, dr \\ &= \pi \sum_n |n\rangle\langle n| = \pi\, I\,,\end{aligned}$$

which proves the completeness relation. Using this relation, we can express an arbitrary vector $|\psi\rangle$ as a linear combination of the set $\{|z\rangle\}$:

$$|\psi\rangle = I|\psi\rangle = \pi^{-1} \int |z\rangle\langle z|\psi\rangle\, d^2z\,.$$

However, this representation is not unique, because the set $\{|z\rangle\}$ is *overcomplete*; that is to say, it is *not linearly independent*. The vector $|z\rangle$ can be expressed as a linear combination of other vectors in the set $\{|z\rangle\}$:

$$|z\rangle = I|z\rangle = \pi^{-1} \int |z'\rangle\langle z'|z\rangle\, d^2z'\,.$$

One important consequence of this overcompleteness is that the usual formula for the *trace* of an operator does not apply in the coherent state basis:

$$\mathrm{Tr}(A) \neq \int \langle z|A|z\rangle\, d^2z\,. \tag{19.69}$$

Coherent states — Physical properties

A coherent state of the EM field is obtained by specifying a coherent state for each of the mode oscillators of the field. Thus the coherent state vector will have the form

$$|\{z\}\rangle \equiv |z_1, z_2, \ldots, z_m, \ldots\rangle = |z_1\rangle \otimes |z_2\rangle \otimes \cdots |z_m\rangle \otimes \cdots\,. \tag{19.70}$$

It is parameterized by a denumberably infinite sequence of complex numbers. The electric field operator (19.24) in the Heisenberg picture may be written as

$$\mathbf{E}(\mathbf{x},t) = \sum_m (2\pi\,\hbar\,\omega_m)^{1/2}\, \{a_m{}^\dagger\, e^{i\omega_m t} + a_m\, e^{-i\omega_m t}\}\, \mathbf{u}_m(\mathbf{x})\,, \tag{19.71}$$

where the time dependence has been removed from the creation and annihilation operators into explicit factors. It follows from (19.64) that the average electric field in the coherent state (19.70) is

$$\begin{aligned}\langle \mathbf{E}(\mathbf{x},t)\rangle &= \langle\{z\}|\mathbf{E}(\mathbf{x},t)|\{z\}\rangle \\ &= \sum_m (2\pi\,\hbar\,\omega_m)^{1/2}\,\{z_m{}^*\,e^{i\omega_m t} + z_m\,e^{-i\omega_m t}\}\,\mathbf{u}_m(\mathbf{x})\,. \end{aligned} \tag{19.72}$$

This is exactly the same form as a normal mode expansion of a classical solution of Maxwell's equations, with the parameter z_m representing the amplitude of a classical field mode. However, it must be emphasized that, in spite of this similarity, a coherent state of the quantized EM field is not equivalent to a classical field. Equation (19.72) gives only the *average* field in the state, and there are also the characteristic quantum fluctuations. A coherent state provides a good description of the EM field produced by a laser. (We shall not treat laser theory in this book.)

We have already seen that there are nonvanishing zero-point fluctuations of a field mode in its ground state (represented by the vector $|n=0\,\rangle = |z=0\,\rangle$), and that the sum of these fluctuations over all modes leads to an infinite mean square fluctuation. To avoid that complication, we shall calculate the fluctuations of a single field mode. We use the notation $\mathbf{E}_m(\mathbf{x},t)$ to indicate the mth term of the operator (19.71). The average of the square of the field is

$$\begin{aligned}\langle\,|\mathbf{E}_m(\mathbf{x},t)|^2\rangle &= \langle\, z_m|\mathbf{E}_m\cdot\mathbf{E}_m|z_m\,\rangle \\ &= \langle\, z_m|\Big\{\left(a_m{}^\dagger\right)^2\,e^{i2\omega_m t} + (a_m)^2\,e^{-i2\omega_m t} \\ &\quad + a_m{}^\dagger\,a_m + a_m\,a_m{}^\dagger\Big\}|z_m\,\rangle 2\pi\,\hbar\,\omega_m\,|\mathbf{u}_m(\mathbf{x})|^2 \\ &= \left\{(z_m{}^*)^2\,e^{i2\omega_m t} + (z_m)^2\,e^{-i2\omega_m t} + z_m{}^* z_m + (z_m{}^* z_m + 1)\right\} \\ &\quad \times 2\pi\,\hbar\,\omega_m\,|\mathbf{u}_m(\mathbf{x})|^2\,. \end{aligned} \tag{19.73}$$

(Here we have used the commutation relation $a_m\,a_m{}^\dagger = a_m{}^\dagger\,a_m{+}1$ to rearrange the last term into normal order.) The mean square fluctuation is

$$\begin{aligned}\langle\,\Delta(\mathbf{E}_m)^2\rangle &= \langle\,|\mathbf{E}_m(\mathbf{x},t)|^2\,\rangle - \langle\,\mathbf{E}(\mathbf{x},t)\,\rangle^2 \\ &= 2\pi\,\hbar\,\omega_m\,|\mathbf{u}_m(\mathbf{x})|^2\,. \end{aligned} \tag{19.74}$$

This is independent of z_m, and is equal to the mean square fluctuation in the ground state. This result was to be expected, since the coherent state was obtained by displacing the ground state in phase space.

The ratio of the rms fluctuation to the mean field, $\langle \Delta(\mathbf{E}_m)^2 \rangle^{1/2} / \langle |\mathbf{E}_m| \rangle$, is of order $|z_m|^{-1}$, and so the fractional fluctuation of the field becomes negligible in the large amplitude limit, $|z_m| \to \infty$. In this way the classical limit of the EM field is reached.

The *photon number distribution* for each mode in a coherent state can be determined directly from the expansion (19.66), from which we obtain

$$\text{Prob}(n|z) = |\langle n|z \rangle|^2 = \frac{|z|^{2n}\, e^{-|z|^2}}{n!} \,. \tag{19.75}$$

The probability of finding a total of n photons in the field is apparently governed by the *Poisson distribution.* The averages $\langle n \rangle$ and $\langle n^2 \rangle$ can be obtained from the properties of this well-known distribution (see Problem 1.17). However we shall calculate them directly from the quantum-mechanical operators. The average number of photons contained in a field mode in a coherent state is

$$\langle n \rangle = \langle z|a^\dagger\, a|z \rangle = z^* z = |z|^2 \,. \tag{19.76}$$

The average of n^2 is

$$\begin{aligned}
\langle n^2 \rangle &= \langle z|a^\dagger a\, a^\dagger a|z \rangle \\
&= z^* \, \langle z|a\, a^\dagger|z \rangle \, z \\
&= |z|^2 \, \langle z|\, (a^\dagger\, a + 1)\, |z \rangle \\
&= |z|^2 \, \left(|z|^2 + 1\right) .
\end{aligned} \tag{19.77}$$

Thus the mean square fluctuation in the photon number of this mode is

$$\begin{aligned}
\langle (n - \langle n \rangle)^2 \rangle &= \langle n^2 \rangle - \langle n \rangle^2 \\
&= |z|^2 \,.
\end{aligned} \tag{19.78}$$

The fluctuations in the fields in a coherent state are as small as possible. We see that this entails a fluctuation in the number of photons, and that the magnitude of the photon number fluctuation increases as the field amplitude z increases. However, the relative fluctuation of the photon number is

$$\frac{\langle (n - \langle n \rangle)^2 \rangle^{1/2}}{\langle n \rangle} = \frac{1}{|z|} = \frac{1}{\langle n \rangle^{1/2}} \,, \tag{19.79}$$

which becomes small in the large amplitude (classical) limit.

The photon number eigenstates (19.46) and the coherent states (19.70) are not the only possible states of the EM field. Other pure states are described by superpositions of these vectors, and nonpure states are represented by state operators that are formally mixtures of these basic states. We shall not develop any formal theory of such general states, but some examples will be treated in the following sections.

19.5 *Spontaneous Emission*

In Sec. 12.6 we treated atomic radiation in a model in which the EM field was regarded as an external classical field. (An "external" field is one that acts on the atom but is not affected by it.) That model describes the changes of the atomic state corresponding to absorption and to stimulated emission of radiation, but is unable to predict spontaneous emission. However, it was possible to relate the probability of spontaneous emission to that of stimulated emission by means of a statistical argument due to Einstein. Now we shall show that the phenomenon of spontaneous emission emerges simply from a more complete theory in which both the atom and the EM field are treated as quantum-mechanical systems.

The Hamiltonian of our system is of the form

$$H = H_{\text{at}} + H_{\text{em}} + H_{\text{int}} \,. \tag{19.80}$$

Here H_{at} is the Hamiltonian of the atom, and H_{em} is the Hamiltonian of the EM field (19.20). These two operators commute because they operate on separate degrees of freedom. If there were no interaction between the atom and the field, the stationary state vectors of the system would be of the form $|\Psi\rangle = |\text{atom}\rangle \otimes |\text{ field}\rangle$, which is an eigenvector of both H_{at} and H_{em}. However, the interaction term H_{int} does not commute with H_{at} or H_{em}, and so a product eigenvector of H_{at} and H_{em} is not an eigenvector of H and does not represent a stationary state. In general, the interaction must be expressed in terms of the vector potential, which is now represented by an operator like those for the electric and magnetic fields. The gauge problem, discussed in Sec. 12.6, also occurs in quantum field theory, but fortunately the solutions discussed earlier are applicable here too. We shall confine our treatment to the *electric dipole approximation*, which is valid whenever the wavelength of the radiation is very much larger that the diameter of the atom. Thus we shall take the interaction operator to be

$$H_{\text{int}} \approx H_D = -\mathbf{D} \cdot \mathbf{E} \,, \tag{19.81}$$

where **D** is the dipole moment operator of the atom, and **E** is the electric field operator (19.71) evaluated at the position of the atom.

We shall calculate the spontaneous transition rate in lowest order perturbation theory using Fermi's rule (12.62):

$$R_s = \frac{2\pi}{\hbar}\,|\langle\,\Psi_f|H_{\text{int}}|\Psi_i\,\rangle|^2\,n(\varepsilon_f)\,. \tag{19.82}$$

That formula was derived under the assumption that the perturbation acted only for a duration T, with the limit $T \to \infty$ then being taken. Presumably we may use it for an interaction that is always present and cannot be switched on and off. The factor $n(\varepsilon_f)$ is the density of final states per unit energy. Our system now includes the EM field, whose states are continuous, as well as the atom whose states are discrete, so $n(\varepsilon_f)$ will be the density of photon states. If the field modes are confined within a cavity of volume Ω, there will be one allowed value of $\mathbf{k}$ in a portion of k space whose volume is $(2\pi)^3/\Omega$. (This is most easily verified for a cubic cavity with periodic boundary conditions, but the result is actually independent of the shape and the particular boundary condition.) Thus the density of photon states per unit energy is

$$\begin{aligned} n(\varepsilon_f) &= 2\times\Omega(2\pi)^{-3}\times 4\pi k^2\times\frac{dk}{d\varepsilon} \\ &= \frac{\Omega\,\omega^2}{\pi^2\,\hbar c^3}\,. \end{aligned} \tag{19.83}$$

The initial factor of 2 is for the two polarization states; the next factors are the density of states per unit k space times that volume element. We have introduced the angular frequency $\omega = ck$, and the photon energy $\varepsilon = \hbar\omega$.

The initial state we take to be $|\Psi_i\,\rangle = |i\,\rangle\otimes|0\rangle$, where $|i\,\rangle$ is the initial atomic state, and $|0\,\rangle$ is zero-photon state of the EM field. The final state in the formula (19.82) will be of the form $|\Psi_f\,\rangle = |f\,\rangle\otimes|n_m = 1\,\rangle$, where $|f\,\rangle$ is the final state of the atom, and $|n_m = 1\,\rangle$ is a one-photon state such that the energy conservation condition, $\varepsilon_i - \varepsilon_f = \hbar\omega$, is satisfied. [The energy conservation condition is implicit in (19.82), and it appeared explicitly in the derivation of Fermi's rule (12.62).] Thus the matrix element squared is

$$|\langle\,\Psi_f|H_{\text{int}}|\Psi_i\,\rangle|^2 = 2\pi\,\hbar\omega|\langle\,f|\mathbf{D}\cdot\mathbf{u}_m|i\,\rangle|^2\,.$$

Substituting for the various factors in (19.82), we obtain

$$R_S = \frac{4\omega^3}{\hbar c^3}\,|\langle\,f|\mathbf{D}|i\,\rangle|^2\,\frac{\Omega}{3}\,|\mathbf{u}_m|^2\,. \tag{19.84}$$

The factor 1/3 comes from the angular average of $(\mathbf{D}\cdot\mathbf{u}_m)^2$. If the modes functions $\mathbf{u}_m(\mathbf{x})$ are plane waves, as is appropriate to an atom radiating into empty space, we will have $|\mathbf{u}_m|^2 = \Omega^{-1}$. Then the spontaneous emission rate becomes

$$R_s = \frac{4\omega^3}{3\hbar c^3}\,|\langle f|\mathbf{D}|i\rangle|^2 . \tag{19.85}$$

The reader can verify that this is equal to the value of Einstein's coefficient A_{fi} in (12.78) if we substitute the value of B_{fi} that is implicit in (12.75). Thus Einstein's intuitive calculation is justified by modern quantum field theory. Our calculation made use of the electric dipole approximation, and so applies only to transitions between atomic states for which there is a nonzero matrix element of the dipole moment operator. The generalization to higher order multipoles involves only technical complications, but does not affect the principle of the calculation.

Enhancement and inhibition of spontaneous radiation

Since the formula (19.85) involves only fundamental constants and properties of the atomic states, there is a tendency to regard the spontaneous emission probability of an atomic state (and the closely analogous radioactive decay probability of an unstable nucleus) as being an intrinsic property of the atom (or nucleus) that is fundamentally uncontrollable. But, in fact, the spontaneous emission probability depends not only on the nature of the atomic state, but also on the properties of the surrounding vacuum field fluctuations. This can be seen in the geometrical factor $(\Omega/3)|\mathbf{u}_m|^2$ in (19.84). If $|\mathbf{u}_m(\mathbf{x})|^2$ is not uniform in space, one can enhance or inhibit the spontaneous emission probability by locating the atom at a position where $|\mathbf{u}_m(\mathbf{x})|^2$ is greater or less than its average value of Ω^{-1}. If the vacuum field is anisotropic, it will not be appropriate to replace $(\mathbf{D}\cdot\mathbf{u}_m)^2$ by its angular average $(1/3)|\mathbf{D}|^2\,|\mathbf{u}_m|^2$, and the spontaneous emission probability may depend on the polarization direction of the radiation.

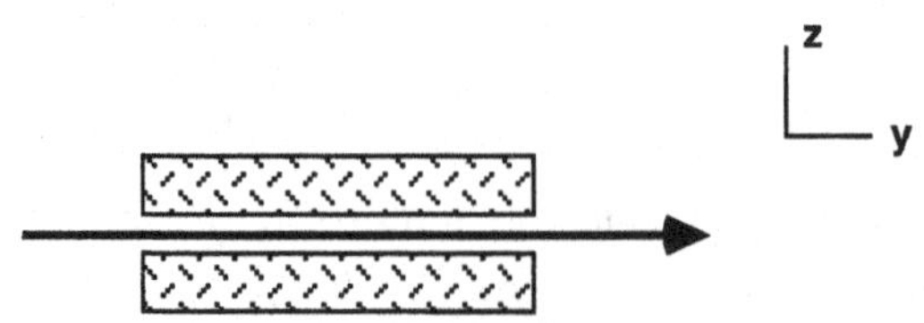

Fig. 19.2 Mirrors inhibit spontaneous emission from an atomic beam.

The most spectacular effect is obtained by placing the atom in a small optical cavity in which there are no field modes whose frequency satisfies the condition $\hbar\omega = \varepsilon_i - \varepsilon_f$. As shown in Fig. 19.2, a beam of atoms of atoms in an excited state is directed between two closely spaced mirrors. The reflecting surfaces are parallel to the xy plane. Let us suppose that the excited state has angular momentum quantum numbers $\ell = 1,\ m = 1$ (in the original experiments much larger values were used), and that the ground state has $\ell = 0$, $m = 0$. The atoms can be oriented by means of an electric or magnetic field in the z direction. From elementary considerations of symmetry (or from the Wigner–Eckart theorem if a more formal argument is preferred), it is apparent that the matrix element of the dipole moment operator between the excited and the ground state will vanish for D_z, and hence the electric field of the radiation must lie in the xy plane. Because the tangential component of the electric field vanishes on the reflecting surfaces (assumed to be perfect conductors), the z dependence of the mode functions between the mirrors must be proportional to $\sin(n\pi z/d)$, where d is the separation between the reflecting surfaces and z is the distance from the lower surface. The angular frequency of such a mode is $\omega = c({k_x}^2 + {k_y}^2 + {k_z}^2)^{1/2}$, with $k_z = n\pi/d$ (n is a positive integer). Therefore there are no modes of suitable polarization whose frequencies are less than $\omega_c = \pi c/d$. Hence, if energies of the atomic states satisfy $\varepsilon_i - \varepsilon_f > \hbar\omega_c$, then the effect of the mirrors will be to "turn off" the spontaneous emission. In the original experiments of Hulet, Hilfer, and Kleppner (1985), the spontaneous emission rate was decreased (and the lifetime of the excited state increased) by a factor of 20 by this technique.

19.6 *Photon Detectors*

In later sections we shall discuss experiments that involve photon counting and photon correlations. It is therefore necessary to understand the principle of the photoelectric detector. This analysis will also serve to motivate the introduction of the correlation functions of the electric field, which turn out to be closely related to photoelectric detection.

Reduced to its simplest essence, a photoelectric detector consists of an atom that can be ionized by absorption of a photon. (In practice the signal of a single electron must be amplified in order to be detectable, but we shall not consider the necessary technique in detail.) We shall calculate the probability of ionization in first order perturbation theory, and so we write the Hamiltonian of the system as

$$H = H_0 + V\,,$$

where

$$H_0 = H_{\text{at}} + H_{\text{em}}$$

is the sum of the Hamiltonians of the isolated atom and the EM field. For the interaction between them, we use the electric dipole approximation,

$$V = -\mathbf{D}\cdot\mathbf{E}\,,$$

which was discussed in the previous section.

It is convenient to transform to the *interaction picture*, which is obtained from the standard Schrödinger picture by the unitary transformation

$$\begin{aligned}|\Psi_I(t)\rangle &= e^{iH_0t/\hbar}|\Psi(t)\rangle\,,\\ V_I(t) &= e^{iH_0t/\hbar}\,V\,e^{-iH_0t/\hbar} = -\mathbf{D}_I(t)\cdot\mathbf{E}_I(t)\,.\end{aligned} \tag{19.86}$$

It is readily shown (Problem 3.10) that the equation satisfied by the state vector in the interaction picture is

$$V_I(t)|\Psi_I(t)\rangle = i\,\hbar\,\frac{\partial}{\partial t}\,|\Psi_I(t)\rangle\,, \tag{19.87}$$

and thus this transformation draws attention to the effects of the interaction. Since the two terms of H_0 commute, we have

$$\mathbf{E}_I(t) = e^{iH_0t/\hbar}\,\mathbf{E}\,e^{-iH_0t/\hbar} = e^{iH_{\text{em}}t/\hbar}\,\mathbf{E}\,e^{-iH_{\text{em}}t/\hbar}\,,$$

where H_{em} is given by (19.20). Therefore the interaction picture operator $\mathbf{E}_I(t)$ is the same operator (19.24), which we previously described as "Heisenberg picture" when our system consisted of only the EM field. Similarly the atomic dipole moment operator is

$$\mathbf{D}_I(t) = e^{iH_0t/\hbar}\,\mathbf{D}\,e^{-iH_0t/\hbar} = e^{iH_{\text{at}}t/\hbar}\,\mathbf{D}\,e^{-iH_{\text{at}}t/\hbar}\,.$$

Let $|a\rangle$ and $|b\rangle$ be eigenvectors of H_{at} with energy eigenvalues ε_a and ε_b, respectively. Then the matrix element of the dipole moment operator is

$$\langle b|\mathbf{D}_I(t)|a\rangle = e^{i\omega_{ba}t}\,\langle b|\mathbf{D}|a\rangle\,, \tag{19.88}$$

where $\hbar\omega_{ba} = \varepsilon_b - \varepsilon_a$ and $\langle b|\mathbf{D}|a\rangle$ is the matrix element in the Schrödinger picture.

From (19.87) it follows that the time dependence of the state vector is given by

$$|\Psi_I(t)\rangle = |\Psi_I(0)\rangle + (i\hbar)^{-1}\int_0^t V_I(t')|\Psi_I(t')\rangle\, dt'$$

$$\approx |\Psi_I(0)\rangle + (i\hbar)^{-1}\int_0^t V_I(t')\, dt'\,|\Psi_I(0)\rangle\,, \qquad (19.89)$$

where the first line is exact, and second line is correct to the first order in V. We assume that at $t = 0$ the detector atom is prepared in a state that is not correlated with the state of the EM field, so the initial state vector of the system is

$$|\Psi_I(0)\rangle = |a\rangle \otimes |\psi_i\rangle \equiv |a;\psi_i\rangle\,.$$

Since our objective is to measure something about the field, we must allow the initial state of the field $|\psi_i\rangle$ to be arbitrary. Let $|b;\psi_f\rangle \equiv |b\rangle \otimes |\psi_f\rangle$ denote some possible final state, orthogonal to $|a;\psi_i\rangle$. The transition amplitude to this final state, to the first order, is

$$\langle b;\psi_f|\Psi_I(t)\rangle = (i\hbar)^{-1}\int_0^t \langle b;\psi_f|V_I(t')|a;\psi_i\rangle\, dt'$$

$$= -(i\hbar)^{-1}\int_0^t e^{i\omega_{ba}t'}\langle b|\mathbf{D}|a\rangle\cdot\langle\psi_f|\mathbf{E}(t')|\psi_i\rangle\, dt'\,. \qquad (19.90)$$

Here and henceforth, we omit the subscript I from the electric field operator, since it is in fact the same operator that we denoted as $\mathbf{E}$ in previous sections.

The electric field operator (19.71) consists of two parts, which are called the *negative* and *position frequency* components,

$$\mathbf{E}(t) = \mathbf{E}^{(-)}(t) + \mathbf{E}^{(+)}(t)\,, \qquad (19.91)$$

where

$$\mathbf{E}^{(-)}(t) = \sum_m (2\pi\,\hbar\,\omega_m)^{1/2}\,\mathbf{u}_m(\mathbf{x})\,{a_m}^{\dagger}\,e^{i\omega_m t}\,, \qquad (19.92)$$

$$\mathbf{E}^{(+)}(t) = \sum_m (2\pi\,\hbar\,\omega_m)^{1/2}\,\mathbf{u}_m(\mathbf{x})\,a_m\,e^{-i\omega_m t}\,. \qquad (19.93)$$

The negative frequency part contains the creation operators, and the positive frequency part contains the annihilation operators. (Since the conventional frequency factor is $e^{-i\omega t}$ with ω positive, we must define the positive frequency part $\mathbf{E}^{(+)}$ to contain the exponential with the minus sign. One must also avoid confusing the superscript $^{(+)}$ with the Hermitian conjugate sign † in the creation operator. To get the notational conventions right, just remember that everything is the opposite of what would be suggested by visual association!)

It is apparent that the transition amplitude (19.90) involves the Fourier component of the field at the frequency ω_{ba}. If the atom had only two states, it would function as a detector of radiation only at this frequency. In an actual photoelectric detector, the initial state $|a\rangle$ is the ground state of the atom, and the final state $|b\rangle$ of the ionized atom is in a continuum, and therefore we have $\omega_{ba} > 0$. Thus the integral in (19.90) is dominated by components of the electric field whose frequencies are near ω_{ba}. The contributions of the negative frequency components of the field operator to the integrand are rapidly oscillating as a function of t', and their net contribution to the integral is very small. In any optical measurement the observation time will be much longer than the period of oscillation of the radiation, so that $\omega_{ba}t \gg 1$, and it will be a good approximation to neglect the contribution of the negative frequency components to (19.90). Thus only the annihilation operator (positive frequency) components of the electric field operator (19.91) will contribute significantly to the ionization probability. This is intuitively understandable from the fact that an atom in its ground state can only absorb radiation, annihilating a photon, but cannot emit radiation. Therefore it will be very good approximation to *replace the electric field operator* $\mathbf{E}(t)$ *by its positive frequency part* $\mathbf{E}^{(+)}(t)$. Thus instead of (19.90) we will now have

$$\langle b;\psi_f|\Psi_I(t)\rangle = -(i\hbar)^{-1}\int_0^t e^{i\omega_{ba}t'}\sum_{\nu=1}^{3}\langle b|D_\nu|a\rangle\langle\psi_f|E_\nu{}^{(+)}(t')|\psi_i\rangle\,dt'\,. \tag{19.94}$$

The square of the amplitude (19.94) gives the probability of a transition to a particular final state of the system (atom + field). But only the state of the atom, and not the state of the field, will be detected, since it is only through its effect on matter that we obtain information about the field. Therefore we sum the transition probability over all final states of the field. Let $M(b)$ be the probability that a photoelectron, excited to state b, will be registered by the counters. (This factor allows for the necessary amplification and electronics that we are not considering in detail.) Then the probability at time t that such an event has occurred and been registered by our apparatus is

$$P_b(t) = M(b)\sum_f |\langle b;\psi_f|\Psi_I(t)\rangle|^2\,. \tag{19.95}$$

This is the probability that an electronic transition $a \to b$ has occurred and been detected, regardless of the final state of the field. The dependence of this probability on the EM field will clearly be through the quantity

$$\sum_f \langle \psi_f | E_\mu{}^{(+)}(t'') | \psi_i \rangle^* \langle \psi_f | E_\nu{}^{(+)}(t') | \psi_i \rangle$$

$$= \sum_f \langle \psi_i | E_\mu{}^{(-)}(t'') | \psi_f \rangle \langle \psi_f | E_\nu{}^{(+)}(t') | \psi_i \rangle$$

$$= \langle \psi_i | E_\mu{}^{(-)}(t'') E_\nu{}^{(+)}(t') | \psi_i \rangle . \tag{19.96}$$

Here we have used the relation $[E_\mu{}^{(+)}]^\dagger = E_\mu{}^{(-)}$. A *broadband detector* does not discriminate the final state b of the electron, so the probability that it has been registered by the time t is

$$P(t) = \int P_b(t)\, n(\varepsilon_b)\, d\varepsilon_b ,$$

where $n(\varepsilon_b)$ is the density of states available to the photoelectron at the energy $\varepsilon_b = \varepsilon_a + \hbar\omega_{ba}$. Combining (19.94), (19.95), and (19.96), we can write the *detection probability* as

$$P(t) = \int_0^t \int_0^t \sum_\mu \sum_\nu s_{\nu\mu}(t'-t'') \langle E_\mu{}^{(-)}(t'')\, E_\nu{}^{(+)}(t') \rangle \, dt'\, dt'' . \tag{19.97}$$

It involves a function that characterizes the detector, and a function that depends on the state of the electric field.

The properties of the detector are summarized in the *sensitivity function*,

$$s_{\nu\mu}(t'-t'') = \int M(b)\, \hbar^{-2} \langle b|D_\nu|a\rangle \langle b|D_\mu|a\rangle^*\, e^{i\omega_{ba}(t'-t'')}\, n(\varepsilon_b)\, d\varepsilon_b , \tag{19.98}$$

which determines the selectivity of the detector to the frequency and polarization of the radiation.

The state of the field enters through the *correlation function* $\langle E_\mu{}^{(-)}(t'') E_\nu{}^{(+)}(t') \rangle$. If the initial state of the field is the pure state $|\psi_i\rangle$, as was assumed above, the correlation function is

$$\langle E_\mu{}^{(-)}(t'')\, E_\nu{}^{(+)}(t') \rangle = \langle \psi_i | E_\mu{}^{(-)}(t'')\, E_\nu{}^{(+)}(t') | \psi_i \rangle . \tag{19.99}$$

If the initial state of the field is not a pure state, but is instead described by a state operator of the form $\rho_{\text{em}} = \sum_i w_i |\psi_i\rangle\langle\psi_i|$, then the correlation function will be

$$\langle E_\mu{}^{(-)}(t'')\, E_\nu{}^{(+)}(t') \rangle = \text{Tr}\{\rho_{\text{em}}\, E_\mu{}^{(-)}(t'')\, E_\nu{}^{(+)}(t')\} . \tag{19.100}$$

The general form (19.97) of the detection probability is valid regardless of the nature of the state of the field. Since ${E_\nu}^{(+)}(t')$ contains only annihilation operators, the field correlation function will vanish identically in the vacuum state. More generally, it will contain no contribution from the zero-point fluctuations of any field mode that is in its ground state. (This is not to say that vacuum field fluctuations have no physical effects, but only that they cannot be detected by absorption of photons.)

It is apparent from (19.97) that the probability of absorbing a photon at some instant t does not depend merely upon the field at time t, but upon the fields over some range of times that depends on the nature of the absorbing device. If the integrand of (19.98) was independent of frequency (or energy) except for the explicit factor $e^{i\omega_{ba}(t'-t'')}$, then the sensitivity function $s_{\nu\mu}(t'-t'')$ would be proportional to $\delta(t'-t'')$, and the probability would not involve the correlation of fields at two different times. This ideal limit is strictly impossible because, as was stated earlier, only positive frequencies ($\omega_{ba} > 0$) correspond to absorption of photons, and all frequencies from $-\infty$ to $+\infty$ would be needed to make up a delta function of time. However this limitation of principle can often be practically overcome. Let us write the sensitivity fucntion (19.98) as

$$s_{\nu\mu}\,(t'-t'') = \int e^{i\omega(t'-t'')}\, s_{\nu\mu}(\omega)\, d\omega\,,$$

where $s_{\nu\mu}(\omega)$ is called the *frequency response function* of the detector. Then the detection probability (19.97) can be written as

$$P(t) = \sum_\mu \sum_\nu \int_0^t dt' \int d\omega\, s_{\nu\mu}(\omega) \int_0^t e^{i\omega(t'-t'')}\, \langle\, {E_\mu}^{(-)}(t'')\, E_\nu^{(+)}(t')\rangle\, dt''\,.$$

The integral furthest to the right in this expression will usually be appreciably greater than zero only over some finite frequency range, which we call the *bandwidth* of the radiation. Only the values of $s_{\nu\mu}(\omega)$ over this bandwidth will influence the value of $P(t)$. Therefore, if the frequency response of the detector is nearly constant over the bandwidth of the radiation, we may replace $s_{\nu\mu}(\omega)$ with that constant value, $s_{\nu\mu}$, and formally extend the range of integration over ω from $-\infty$ to $+\infty$. It is apparent that this is equivalent to replacing the sensitivity function $s_{\nu\mu}(t'-t'')$ by $s_{\nu\mu}\,\delta(t'-t'')$. With this approximation, we obtain a *detection rate* (probability per unit time) equal to

$$R(t) \equiv \frac{dP(t)}{dt} = \sum_\mu \sum_\nu s_{\nu\mu}\,\langle\, {E_\mu}^{(-)}(t)\, {E_\nu}^{(+)}(t)\rangle\,, \qquad (19.101)$$

which depends only on the electric field correlation function at one time. One should remember that this expression, which will be used henceforth, is valid only under the assumption that *the bandwidth of the detector is greater than the bandwidth of the radiation.* The sensitivity function $s_{\nu\mu}(t' - t'')$ must be nonvanishing for a time interval of order $|t'-t''| \sim \tau$, where τ is the reciprocal of the bandwidth of the detector. The derivation of (19.101) will be valid provided $\langle E_\mu{}^{(-)}(t+\delta t)\, E_\nu{}^{(+)}(t)\rangle$ is approximately constant for $|\delta t| < \tau$.

Detection of n photons

We can generalize the above analysis to treat the detection of n photons at n different space–time points. Without going into details, it should be clear that this involves the positive frequency components of the electric field at the positions of the n different atoms that will make up our idealized detector, and the transition amplitude will involve $\langle \psi_f|E^{(+)}(\mathbf{x}_n, t_n)\cdots E^{(+)}(\mathbf{x}_1, t_1)|\psi_i\rangle$. (For simplicity we omit the polarization subscripts.) When this amplitude is squared and summed over all final states of the field, it will yield

$$\sum_f \langle \psi_i|E^{(-)}(\mathbf{x}_1, t_1)\cdots E^{(-)}(\mathbf{x}_n, t_n)|\psi_f\rangle\langle \psi_f|E^{(+)}(\mathbf{x}_n, t_n)\cdots E^{(+)}(\mathbf{x}_1, t_1)|\psi_i\rangle$$

$$= \langle \psi_i|E^{(-)}(\mathbf{x}_1, t_1)\cdots E^{(-)}(\mathbf{x}_n, t_n)\, E^{(+)}(\mathbf{x}_n, t_n)\cdots E^{(+)}(\mathbf{x}_1, t_1)|\psi_i\rangle\,.$$

Thus the probability of detecting n photons in coincidence (or delayed coincidence) is directly related to a higher order correlation function of the electric field.

Semiclassical theory

We shall be seeking experiments that can distinguish quantum electrodynamics from the semiclassical theory, which treats quantum-mechanical matter interacting with a classical EM field. The response of a photoelectric detector to a classical EM field can be obtained by applying time-dependent perturbation theory to an atom perturbed by a classical electric field. The essential parts of this calculation were already carried out in Sec. 12.6, where we calculated the transition rate for absorption of energy by an atom. The separation of the electric field into positive and negative frequency components occurred naturally in Sec. 12.5 when we distinguished the conditions for *resonant absorption* and *resonant emission.*

The positive and negative frequency components of a real time-dependent field $E_\nu(t)$ can be formally defined through the Fourier integral

$$E_\nu(\omega) = (2\pi)^{-3} \int E_\nu(t)\, e^{i\omega t}\, dt\,.$$

The Fourier transform must satisfy the reality condition $E_\nu(-\omega) = [E_\nu(\omega)]^*$. The positive and negative frequency components are defined to be

$$E_\nu{}^{(+)}(t) = \int_0^\infty E_\nu(\omega)\, e^{-\omega t}\, d\omega\,, \tag{19.102}$$

$$E_\nu{}^{(-)}(t) = \int_{-\infty}^0 E_\nu(\omega)\, e^{-i\omega t}\, d\omega$$

$$= \int_0^\infty [E_\nu(\omega)]^*\, e^{i\omega t}\, d\omega\,. \tag{19.103}$$

If there were no noise, the electric field correlation function $\langle\, E_\mu{}^{(-)}(t)\, E_\nu{}^{(+)}(t)\rangle$ for a classical field would simply be equal to the product of the fields, $E_\mu{}^{(-)}(t)\, E_\nu{}^{(+)}(t)$. The ensemble average brackets $\langle\cdots\rangle$ now are interpreted as an average over the probability distribution of the noise. With this definition, the formulas (19.97) and (19.101) remain valid.

19.7 *Correlation Functions*

We saw in the previous section that the probability of absorbing one or more photons is determined by certain correlations of the EM field. We shall now define the field correlation functions generally, and show how they are useful in various optical experiments. It is convenient to adopt an abbreviated notation, writing $E_\mu(\mathbf{x}_n, t_n) = E(x_n)$. The single label x_n is now an alias for the space, time, and polarization variables $(\mathbf{x}_n, t_n, \mu)$. We define *the correlation function of degree* n to be

$$G^{(n)}(x_1, \ldots, x_n; x_{n+1}, \ldots, x_{2n})$$
$$= \langle\, E^{(-)}(x_1) \cdots E^{(-)}(x_n)\, E^{(+)}(x_{n+1}) \cdots E^{(+)}(x_{2n})\rangle\,. \tag{19.104}$$

(Some authors call this correlation function $G^{(2n)}$.) Notice that the operators are in normal order, with all creation operators to the left of all annihilation operators. The average is calculated from the state vector or the state operator in the usual way. In the classical theory, the electric field operators are replaced by classical fields, and the average is over the appropriate ensemble to account for noise fluctuations.

It was shown in the previous section that the probability of detecting a photon at each of the space–time points $x_1, \ldots, x_n$ is proportional to the diagonal correlation function $G^{(n)}(x_1,\ldots,x_n;x_n,\ldots,x_1)$, with the factor of proportionality depending on the sensitivity of the detector. If our only means of measuring the EM field is photon counting, then the diagonal correlation functions are the only measurable quantities. However, if we can sample and combine the fields from two or more space–time points, then interference between them in effect allows us to measure nondiagonal correlations. This is illustrated in Fig. 19.3, where two signals are extracted from points x_1 and x_2 in the optical cavity and combined at a detector. The signal at the detector will be $E_d = E(x_1) + E(x_2)$, and so the photon detection rate at the detector will be proportional to $\langle\, E_d{}^{(-)}\, E_d{}^{(+)}\rangle = G^{(1)}(x_1;x_1) + G^{(1)}(x_2;x_2) + G^{(1)}(x_1;x_2) + G^{(1)}(x_2;x_1)$. Thus the nondiagonal correlations can be determined. This interpretation assumes that signals can be extracted without significantly perturbing the original field distribution. Strictly speaking, the introduction of the mirrors to sample the field introduces new boundary conditions and sets up new mode functions satisfying these new boundary conditions, and strictly speaking, we really measure only the diagonal correlation function at the position of the detector, $G^{(1)}(x_d;x_d)$. However, the two points of view are practically equivalent in many cases.

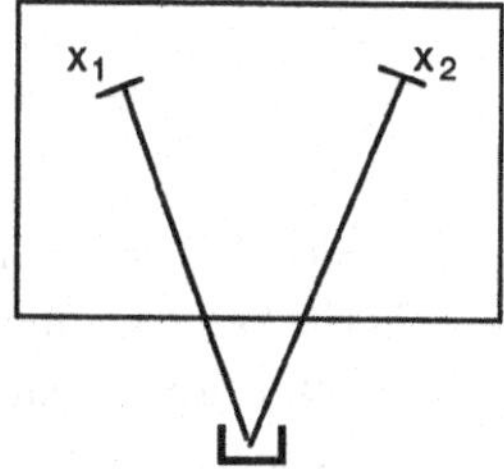

Fig. 19.3 Extraction of signals from points x_1 and x_2 so as to measure $G^{(1)}(x_1;x_2)$.

Several mathematical properties of these correlation functions are easily proven:

(a) $G^{(n)}(x_1,\ldots,x_n;x_{n+1},\ldots,x_{2n})$ is invariant under permutation of the variables $\{x_1,\ldots,x_n\}$ among themselves, and of $\{x_{n+1},\ldots,x_{2n}\}$ among themselves. This is so because all creation operators commute among themselves, as do all annihilation operators.

(b)
$$G^{(n)}(x_{2n},\ldots,x_{n+1};x_n,\ldots,x_1) = [G^{(n)}(x_1,\ldots,x_n;x_{n+1},\ldots,x_{2n})]^* \,. \tag{19.105}$$

Taking $n = 1$ as an example, this follows from the fact that $E^{(-)}(x_2) E^{(+)}(x_1) = [E^{(-)}(x_1)\, E^{(+)}(x_2)]^\dagger$.

(c) The diagonal correlation function is nonnegative:

$$G^{(n)}(x_1, \ldots, x_n; x_n, \ldots, x_1) \geq 0\,. \tag{19.106}$$

This follows from the fact that $\mathrm{Tr}(\rho A^\dagger A\} \geq 0$ for any operator ρ. In classical theory it follows from $E^{(-)}(x)\, E^{(+)}(x) = |E^{(+)}(x)|^2 \geq 0$.

(d)
$$G^{(n)}(x_1, \ldots, x_n; x_n, \ldots, x_1)\, G^{(n)}(x_{n+1}, \ldots, x_{2n}; x_{2n}, \ldots, x_{n+1}) \\ \geq |G^{(n)}(x_1, \ldots, x_n; x_{n+1}, \ldots, x_{2n})|^2\,. \tag{19.107}$$

The proof is similar to that for the Schwarz inequality.

First order correlations: Interference

The rate of detecting photons (of a certain polarization that is not explicitly indicated in the notation) at the space–time point $x = (\mathbf{x}, t)$ is proportional to

$$G^{(1)}(x; x) = \langle\, E^{(-)}(\mathbf{x}, t)\, E^{(+)}(\mathbf{x}, t) \rangle \\ = \sum_n \sum_m 2\pi\, \hbar (\omega_n\, \omega_m)^{1/2}\, u_n(\mathbf{x})\, u_m(\mathbf{x})\, e^{i(\omega_n - \omega_m)t}\, \langle\, a_n{}^\dagger\, a_m \,\rangle\,.$$

The spatial form of the interference pattern is given by the product of the mode functions, $u_n(\mathbf{x})\, u_m(\mathbf{x})$, and is the same as in classical electromagnetic theory. However, the amplitude of the interference pattern reflects the quantum state through the quantity $\langle\, a_n{}^\dagger\, a_m \,\rangle$. Clearly at least two modes must be excited in order for interference to occur.

A simple but very useful model consists of *two plane wave modes*. In this model, the only modes excited above the ground state are $e^{i\mathbf{k}_1 \cdot \mathbf{x}}$ and $e^{i\mathbf{k}_2 \cdot \mathbf{x}}$, with $|\mathbf{k}_1| = |\mathbf{k}_2| = \omega/c$. Since field modes in their ground state do not contribute to the photon detection probability; we need to consider only those terms of the field operators that correspond to excited modes. Thus we may substitute

$$E^{(+)}(\mathbf{x}, t) = C\, (a_1\, e^{i\mathbf{k}_1 \cdot \mathbf{x}} + a_2\, e^{i\mathbf{k}_2 \cdot \mathbf{x}})\, e^{-i\omega t}\,, \\ E^{(-)}(\mathbf{x}, t) = C\, (a_1{}^\dagger\, e^{-i\mathbf{k}_1 \cdot \mathbf{x}} + a_2{}^\dagger\, e^{-i\mathbf{k}_2 \cdot \mathbf{x}})\, e^{i\omega t}\,, \tag{19.108}$$

where several constants have been absorbed into the factor C. The photon detection rate for this model will be proportional to

$$
\begin{aligned}
G^{(1)}(\mathbf{x}, t; \mathbf{x}, t) &= C^2 \left\{ \langle a_1{}^\dagger a_1 \rangle + \langle a_2{}^\dagger a_2 \rangle \right. \\
&\quad \left. + \langle a_1{}^\dagger a_2 \rangle e^{-i(\mathbf{k}_1 - \mathbf{k}_2) \cdot \mathbf{x}} + \langle a_2{}^\dagger a_1 \rangle e^{i(\mathbf{k}_1 - \mathbf{k}_2) \cdot \mathbf{x}} \right\} \\
&= C^2 \left\{ \langle a_1{}^\dagger a_1 \rangle + \langle a_2{}^\dagger a_2 \rangle \right. \\
&\quad \left. + 2 |\langle a_1{}^\dagger a_2 \rangle| \cos[(\mathbf{k}_1 - \mathbf{k}_2) \cdot \mathbf{x} - \phi] \right\} , \qquad (19.109)
\end{aligned}
$$

where the phase ϕ comes from $\langle a_1{}^\dagger a_2 \rangle = |\langle a_1{}^\dagger a_2 \rangle| \, e^{i\phi}$.

Some experimental realizations of the two-plane-wave model are shown in Fig. 19.4. In the top picture, double slit diffraction, the field modes on the right are really cylindrical waves, but far from the slits they may be locally approximated by plane waves. In all three of the pictures, the model does not apply throughout all space, but it is a good local approximation in the region

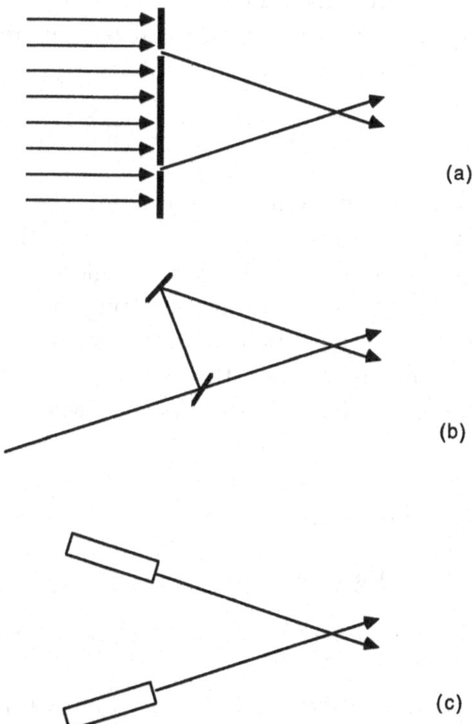

Fig. 19.4 Examples of the two-mode model: (a) double slit diffraction; (b) beam splitting and recombination; (c) interference between two laser beams.

of overlap between the two beams. According to Eq. (19.109), the photon detection rate may vary periodically in space in the direction of $\mathbf{k}_1 - \mathbf{k}_2$, which is the vertical direction in the picture.

Example (i): One-photon state

Consider a state vector for the EM field of the form

$$|\Psi_1\rangle = \alpha|1,0\rangle + \beta|0,1\rangle\,, \quad \left(|\alpha|^2 + |\beta|^2 = 1\right), \tag{19.110}$$

where the vector $|1,0\rangle = {a_1}^\dagger|0\rangle$ describes one photon in mode 1 and $|0,1\rangle = {a_2}^\dagger|0\rangle$ describes one photon in mode 2. The vacuum state is denoted as $|0\rangle$. The state vector $|\Psi_1\rangle$ is an eigenvector of the total photon number operator $N = \sum_m {a_m}^\dagger a_m$, even though the number of photons in each mode is indefinite. One can also write $|\Psi_1\rangle = b^\dagger|0\rangle$, where the creation operator is $b^\dagger = \alpha {a_1}^\dagger + \beta {a_2}^\dagger$. Evaluating (19.109) for the state $|\Psi_1\rangle$, we obtain a photon detection probability proportional to

$$G^{(1)}(\mathbf{x},t;\mathbf{x},t) = C^2\left\{|\alpha|^2 + |\beta|^2 + 2\,\mathrm{Re}[\beta^*\,\alpha\, e^{i(\mathbf{k}_1-\mathbf{k}_2)\cdot\mathbf{x}}]\right\}. \tag{19.111}$$

The interference pattern, identical in form to that of classical optics, exists even for a one-photon state. Of course this interference pattern cannot be observed by detecting a single photon. We must perform an ensemble of measurements, preparing the state and detecting the photon in each repetition, and the statistical distribution of the detected photons will take the form of an interference pattern. The method of preparing a one-photon state will be discussed in the next section.

Example (ii): Two-photon state

Consider now the state vector

$$|\Psi_2\rangle = \alpha|2,0\rangle + \beta|1,1\rangle + \gamma|0,2\rangle\,, \quad \left(|\alpha|^2 + |\beta|^2 + |\gamma|^2 = 1\right), \tag{19.112}$$

where the three component vectors describe two photons in mode 1, one photon in each mode, and two photons in mode 2, respectively. Evaluating (19.109) for this state vector, with the help of (6.16) and (6.20), yields a photon detection probability proportional to

$$G^{(1)}(\mathbf{x},t;\mathbf{x},t) = C^2 \left\{ 2(|\alpha|^2 + |\beta|^2 + |\gamma|^2) + 2\,\mathrm{Re}\left[\sqrt{2}\,(\beta^*\,\alpha + \gamma^*\,\beta)\; e^{i(\mathbf{k}_1-\mathbf{k}_2)\cdot\mathbf{x}}\right] \right\}. \quad (19.113)$$

The interference pattern is generally similar to that in Example (i), but two particular cases deserve special attention. For $\alpha = \gamma = 0$, $\beta = 1$, corresponding to exactly one photon in each mode, there is no interference. For $\beta = 0$, $\alpha \neq 0$, $\gamma \neq 0$ the interference pattern also disappears. However we shall see that higher order correlations exist in these states.

Example (iii): Independent laser beams

The EM field produced by a single mode laser is described by a coherent state vector $|z_m\rangle$. The state of the field produced by two intersecting laser beams, shown in Fig. 19.4(c), is therefore described by the state vector

$$|\Psi\rangle = |z_1\rangle \otimes |z_2\rangle\,. \quad (19.114)$$

This product form is appropriate because the lasers are independent, and each laser excites one mode of the field. Equation (19.109) can easily be evaluated using (19.64), obtaining

$$G^{(1)}(\mathbf{x},t;\mathbf{x},t) = C^2\{|z_1|^2 + |z_2|^2 + 2\,\mathrm{Re}[z_2{}^*\,z_1\,e^{i(\mathbf{k}_1-\mathbf{k}_2)\cdot\mathbf{x}}]\}\,. \quad (19.115)$$

This result is derived under the assumption that the frequencies of the lasers are equal and state. If we consider the possibility that their frequencies might differ, then the space-dependent exponential should be multiplied by the time-dependent factor $e^{-i(\omega_1-\omega_2)t}$. Although we may have $\omega_1 - \omega_2 = 0$ on the average, the two frequencies will be subject to independent random fluctuations, which will produce a random drift of the interference pattern. Thus the interference pattern can be observed only if a sufficient number of photons can be collected before the phase factor $e^{-i(\omega_1-\omega_2)t}$ drifts too much. The interference between independent lasers has been observed, but it requires careful experimental technique.

[[On page 9 of his textbook, Dirac states, “Each photon then interferes only with itself. Interference between two different photons never occurs.” There is no evidence in the context that he intended that remark to be very deep or profound, much less controversial. It seems intended only to indicate, rather loosely and metaphorically, the direction in which quantum theory would proceed. Nevertheless, some people have treated it as an oracular pronouncement, and have argued as to whether or not it is strictly true. In view of the casual manner in which Dirac made the assertion, that degree of scholastic scrutiny seems misplaced.

Let us examine Dirac's statement in the light of the three examples above. Interference in a one-photon state is clearly compatible with Dirac's statement. The absence of interference in the two-photon state $|1,1\rangle$ is consistent with his statement that interference between two different photons does not occur. But what about Example (iii), which seems to involve interference between photons from two different lasers? Some people were led by Dirac's statement to predict that such interference could not occur. (Presumably they were unaware that their prediction would have led them to contradict classical wave theory.) In this case Dirac's overworked metaphor ceases to be helpful. Strictly speaking, it is not photons that interfere, neither with themselves nor with each other, but rather the interference pattern is in the electromagnetic field. Recall that it is the electric and magnetic fields that are the fundamental dynamical variables of the theory, and it is these (and not positions and momenta of photons) that are represented by quantum-mechanical operators. The photon enters the theory as a secondary quantity, namely as an elementary excitation of the field. We have here an example in which the primary nature of the fields is emphasized, and it is not helpful to regard the field as merely a stream of particles.]]

Second order correlations

If two detectors are placed at $\mathbf{x}_1$ and $\mathbf{x}_2$, the probability of both detecting a photon is proportional to the second order correlation function,

$$G^{(2)}(x_1, x_2; x_2, x_1) = \langle\, E^{(-)}(x_1)\, E^{(-)}(x_2)\, E^{(+)}(x_2)\, E^{(+)}(x_1)\rangle\,. \qquad (19.116)$$

Adopting the two-plane-wave mode model, we substitute (19.108) for the field operators. The resulting expression for $G^{(2)}$ has 16 terms. If we specialize to the state vector $|1,1\,\rangle$, which corresponds to one photon in each mode, only four of them survive and we obtain

$$\langle 1,1|E^{(-)}(x_1)\,E^{(-)}(x_2)\,E^{(+)}(x_2)\,E^{(+)}(x_1)|1,1\rangle$$
$$= C^4\,\langle a_1{}^\dagger\, a_2{}^\dagger\, a_2\, a_1\rangle\; 2\{1+\cos[(\mathbf{k}_1-\mathbf{k}_2)\cdot(\mathbf{x}_1-\mathbf{x}_2)]\}$$
$$= C^4\, 2\{1+\cos[(\mathbf{k}_1-\mathbf{k}_2)\cdot(\mathbf{x}_1-\mathbf{x}_2)]\}\,. \tag{19.117}$$

We have already seen from (19.113) (for the case $\alpha = \gamma = 0$) that $G^{(1)}(x;x)$ is constant for this state. Therefore the photon detection probability for any one detector will be independent of position, and no ordinary diffraction pattern will be observed. Nevertheless we see from (19.117) that the two photons are correlated and an interference-like pattern appears in the second order correlation function. Indeed there are values of the separations $(\mathbf{x}_1 - \mathbf{x}_2)$ between detectors for which the joint probability goes to zero. These correlations in a two-photon state have been observed by Ghosh and Mandel (1987), who point out that the correlations are stronger by a factor of 2 than those predicted by classical optics.

Quantum beats

In Sec. 12.4 we discussed quantum beats as an interesting time-dependent phenomenon that exhibits a striking departure from exponential decay. Since we could give only a semiclassical treatment in Sec. 12.4, we now briefly reexamine the phenomenon with a quantum-mechanical treatment of the EM field. We are interested in first order correlations, but as a function of time rather than space.

Referring to Fig. 12.1, we have an atom with ground state $|a\rangle$ and two closely spaced excited states $|b\rangle$ and $|c\rangle$. The energies of these atomic states (neglecting the interaction with the EM field) are $\varepsilon_a, \varepsilon_b$, and ε_c. For simplicity, we shall consider only two modes of the field: mode 1 having frequency $\omega_1 = (\varepsilon_b - \varepsilon_a)/\hbar$, and mode 2 having frequency $\omega_2 = (\varepsilon_c - \varepsilon_a)/\hbar$.

Suppose that the initial state of the atom + field system is $|b\rangle \otimes |0,0\rangle$. Because of the interaction between the atom and the field, this is not a stationary state, and in the course of time it will evolve into the linear combination $\beta(t)|b\rangle \times |0,0\rangle + \alpha_1(t)|a\rangle \otimes |1,0\rangle$, with $|\beta|^2 + |\alpha_1|^2 = 1$. The coefficients $\beta(t)$ and $\alpha_1(t)$ are smoothly varying functions of t, which could be calculated from (19.89). (The interaction picture is used here.) The quantity $|\alpha_1(t)|^2$ represents the probability that a photon of frequency ω_1 has been emitted spontaneously. Alternatively let us suppose that the initial state of the system is $|c\rangle \otimes |0,0\rangle$. This would evolve into the linear combination $\gamma(t)|c\rangle \otimes |0,0\rangle + \alpha_2(t)|a\rangle \otimes |0,1\rangle$, with $|\gamma|^2 + |\alpha_2|^2 = 1$, and $|\alpha_2(t)|^2$ being the probability that a photon of frequency ω_2 has been emitted spontaneously.

In the actual problem of interest, the initial state of the system is

$$|\Psi_I(0)\rangle = (B|b\,\rangle + C|c\,\rangle) \otimes |0,0\rangle \tag{19.118}$$

with $|B|^2 + |C|^2 = 1$. Since this is a linear combination of the two cases considered above, the state of the system after a time t will be

$$\begin{aligned}|\Psi_I(t)\rangle = {} & B\,\beta(t)|b\,\rangle \otimes |0,0\rangle + B\,\alpha_1(t)|a\,\rangle \otimes |1,0\rangle \\ & + C\,\gamma(t)|c\,\rangle \otimes |0,0\rangle + C\,\alpha_2(t)|a\,\rangle \otimes |0,1\rangle\,. \end{aligned} \tag{19.119}$$

(In reality, there will be many more terms corresponding to atomic decays involving other field modes that the two that we have considered, but these four terms are sufficient to illustrate the phenomenon of interest.) The probability of detecting a photon at the time t is proportional to $G^{(1)}(\mathbf{x},t;\mathbf{x},t)$, which we shall now write as $G^{(1)}(t;t)$ since we keep the detector fixed in space. Only two modes need to be retained in the field operators, so we may write

$$\begin{aligned} E^{(+)}(t) &\propto a_1\, e^{-i\omega_1 t} + a_2\, e^{-i\omega_2 t}\,, \\ E^{(-)}(t) &\propto {a_1}^\dagger\, e^{i\omega_1 t} + {a_2}^\dagger\, e^{i\omega_2 t}\,. \end{aligned}$$

In the state (19.119), the first order correlation function is

$$\begin{aligned} G^{(1)}(t;t) &= \langle\Psi_I(t)|E^{(-)}(t)\, E^{(+)}(t)|\Psi_I(t)\rangle \\ &\propto |B\alpha_1(t)|^2 + |C\alpha_2(t)|^2 + 2\,\mathrm{Re}\{BC^*\,\alpha_1(t)[\alpha_2(t)]^*\, e^{i(\omega_1-\omega_2)t}\}\,. \end{aligned} \tag{19.120}$$

Thus the photon detection probability will be modulated at the beat frequency $\omega_1-\omega_2$, and hence the phenomenon is called *quantum beats*. Note that the phase of the quantum beat depends on the relative phase of the constants B and C which determine the initial state (19.118). Therefore, in order to observe the beats, it is essential that one be able to prepare the initial state of the atom with a well-defined relative phase between the two components. If the phase were to fluctuate, this would in effect average over the phase of BC^*, and the beats would not be seen.

19.8 *Coherence*

The difference between coherent and incoherent radiation is, roughly speaking, due to the absence or presence of noise fluctuations. But since no quantum state can be entirely free of fluctuations, this informal notion of coherence is not apparently applicable. A precise definition of coherence that is applicable in both classical and quantum theories was given in a pair of papers by R. J. Glauber (1963). According to this definition, *coherence* is not a single condition but an infinite sequence of conditions.

Let us recall the definition (19.104) of the nth order correlation function,

$$G^{(n)}(x_1, \ldots, x_n; x_{n+1}, \ldots, x_{2n})$$
$$= \langle E^{(-)}(x_1) \cdots E^{(-)}(x_n) E^{(+)}(x_{n+1} \cdots E^{(+)}(x_{2n}) \rangle \, .$$

Following Glauber, we say that the state of the field has *nth order coherence* if and only if there exists a function $f(x)$ such that

$$G^{(n)}(x_1, \ldots, x_n; x_{n+1}, \ldots, x_{2n}) = \prod_{j=1}^{n} f^*(x_j) \prod_{k=n+1}^{2n} f(x_k) \, , \qquad (19.121)$$

and similar relations hold for all correlations of lower degree. If this condition holds for *all* n, we say the state is *fully coherent*. A noiseless classical field is fully coherent, since then the average bracket $\langle \cdots \rangle$ is redundant and $f(x)$ is simply equal to the positive frequency part of the classical field. Since all quantum states contain fluctuations, it is not obvious that full coherence is possible in quantum mechanics. However, a quantum-mechanical "coherent" state $|z_1, \ldots, z_m, \ldots \rangle$ (19.70) is also fully coherent according to this definition. This is so because of the eigenvalue relation (19.62), $a_m |z_m\rangle = z_m |z_m\rangle$. Thus (19.121) holds with $f(x) = \mathcal{E}^{(+)}(x)$, where $\mathcal{E}^{(+)}(x)$ is obtained by replacing the annihilation operators $\{a_m\}$ in the electric field operator $E^{(+)}(x)$ with their eigenvalues, the complex numbers $\{z_m\}$.

A quantitative measure of the degree of coherence is provided by the *normalized coherence functions*,

$$g^{(n)}(x_1, \ldots, x_{2n}) = \frac{G^{(n)}(x_1, \ldots, x_n; x_{n+1}, \ldots, x_{2n})}{\prod_{k=1}^{2n} \left\{ G^{(1)}(x_k; x_k) \right\}^{1/2}} \, . \qquad (19.122)$$

A fully coherent state satisfies the condition $|g^{(n)}(x_1, \ldots, x_{2n}| = 1$ for all n and all values of the variables $\{x_k\}$. When this condition is satisfied only approximately, or only over some limited region, we may speak of the state as being approximately coherent, in some finite order, within some finite volume of space–time. Having thus given the term "coherent" a precise meaning, we have deprived its opposite, "incoherent", of its usefulness, since it makes much more sense to speak of various degrees of partial coherence than to lump them together under the label "incoherent".

First order coherence

If (19.121) is satisfied for $n = 1$, we have first order coherence,

$$G^{(1)}(x_1; x_2) = f^*(x_1)\, f(x_2)\,. \tag{19.123}$$

From this, it obviously follows that the inequality (19.107) for $n = 1$ holds as an equality,

$$G^{(1)}(x_1; x_1)\, G^{(1)}(x_2; x_2) = |G^{(1)}(x_1; x_2)|^2\,. \tag{19.124}$$

The converse is also true, i.e. (19.124) implies (19.123), although this is not obvious. [For the proof of this result and the next, see Klauder and Sudarshan (19.68), pp. 159–162]. First order coherence also implies that the higher order correlation functions must have the form

$$G^{(n)}(x_1, \ldots, x_n; x_{n+1}, \ldots, x_{2n}) = g_n \prod_{j=1}^{n} f^*(x_j) \prod_{k=n+1}^{2n} f(x_k)\,, \tag{19.125}$$

where g_n is independent of the variables $\{x_k\}$ and $g_n \geq 0$. By definition, we have $g_1 = 1$ in a first order coherent state, but for $n > 1$ the nonnegative values of g_n may be greater or less than 1. Substituting (19.125) into (19.122), we see that the normalized coherence function for a first order coherent state is independent of $\{x_k\} : g^{(n)}(x_1, \ldots, x_{2n}) = g_n$. Thus if the state is known to be coherent in the first order, the nature of its higher order coherence depends only upon the values of the constants g_2, g_3, etc.

If only one mode of the field is excited, then the numerator and denominator of (19.122) will contain identical space- and time-dependent factors, and hence the normalized coherence function reduces to a constant,

$$g^{(n)}(x_1, \ldots, x_{2n}) = g^{(n)} = \frac{\langle\, a^{\dagger\, n}\, a^n \,\rangle}{\langle\, a^\dagger\, a \,\rangle^n}\,, \tag{19.126}$$

where $a^\dagger$ and a are the creation and annihilation operators for the relevant field mode. The constant $g^{(n)}$ is defined for a field with only one mode excited, whereas g_n in (19.125) is defined for a first order coherent state. However, it is apparent from (19.126) that $g^{(1)} = 1$, and so a single mode field necessarily has first order coherence. A first order coherent state need not correspond to a single excited normal mode. However, if we generalize the notion of a *mode function* to an arbitrary linear combination of normal modes, of the form $f(\mathbf{x},t) = \sum_m c_m u_m(\mathbf{x})\, e^{-i\omega_m t}$, then it can be shown that any first order coherent state corresponds to the excitation of a single generalized mode. [Proof is given by Klauder and Sudarshan as part of their derivation of (19.125).]

We now consider some examples of partially coherent states.

(a) *Single mode m-photon state.* The normalized coherence function for a single mode field (19.126) can be evaluated by means of the identity (Problem 19.4)

$$a^{\dagger\, n} a^n = N(N-1)(N-2)\cdots(N-n+1)\,, \tag{19.127}$$

where $N = a^\dagger a$ is the number operator for the mode. Thus for the m-photon state, $N|m\,\rangle = m|m\,\rangle$, we obtain

$$\begin{aligned} g^{(n)} &= \frac{m!}{(m-n)!\,m^n}\,, \qquad (n \le m)\,, \\ &= 0\,, \quad (n > m)\,. \end{aligned} \tag{19.128}$$

Some values of this function are listed in the following table:

m photons	$g^{(1)}$	$g^{(2)}$	$g^{(3)}$	$g^{(4)}$
$m = 1$	1	0	0	0
2	1	$\frac{1}{2}$	0	0
3	1	$\frac{2}{3}$	$\frac{2}{9}$	0
4	1	$\frac{3}{4}$	$\frac{3}{8}$	$\frac{3}{32}$

A single mode state necessarily has $g^{(1)} = 1$, so all these photon number eigenstates for $m > 0$ have first order coherence. For fixed n, we have $g^{(n)} \to 1$ in the limit as the number of photons, m, becomes infinite. However, no photon number eigenstate can be fully coherent.

(b) *Filtered thermal radiation.* The state operator for blackbody thermal radiation is $\rho_\mathcal{T} = e^{-H_{\rm em}/k_BT}[\mathrm{Tr}(e^{-H_{\rm em}/k_BT})]^{-1}$. The frequency bandwidth of thermal radiation can be reduced by means of a narrow passband filter. The ultimate limit of filtering (not attainable in practice) would be a single field mode. The resultant state operator for the mode would be that of a harmonic oscillator in thermal equilibrium,

$$\rho = \frac{\sum_m e^{-\alpha m} |m\rangle\langle m|}{\sum_m e^{-\alpha m}}, \quad (\alpha = \hbar\omega/k_B T) . \tag{19.129}$$

The normalized coherence function (constant because this is a single mode field) is

$$\begin{aligned} g^{(n)} &= \frac{\mathrm{Tr}\left(\rho a^{\dagger\, n}\, a^n\right)}{\{\mathrm{Tr}\,(\rho a^\dagger\, a)\}^n} \\ &= \frac{\{\sum_m \langle m|a^{\dagger\, n}\, a^n|m\rangle\, e^{-\alpha m}\}\,\{\sum_m e^{-\alpha m}\}^{n-1}}{\{\sum_m m\, e^{-\alpha m}\}^n} . \end{aligned} \tag{19.130}$$

(The sums over m are from 0 to ∞.) This expression can be evaluated with the help of the identity (19.127). Putting $\lambda = e^{-\alpha}$, the first sum in the numerator of (19.130) can be written as

$$\begin{aligned} \sum_{m=n}^{\infty} \lambda^m \frac{m!}{(m-n)!} &= \lambda^n \frac{d^n}{d\lambda^n} \sum_{m=0}^{\infty} \lambda^m \\ &= \lambda^n \frac{d^n}{d\lambda^n} \frac{1}{1-\lambda} = \frac{n!\,\lambda^n}{(1-\lambda)^{n+1}} . \end{aligned}$$

The sum in the denominator of (19.130) is a special case of this result with $n = 1$, and the second sum in the numerator is another special case with $n = 0$. Therefore (19.130) simplifies to

$$g^{(n)} = n! . \tag{19.131}$$

We now have examples of states with $g^{(n)} = 1$ (coherent states), $g^{(n)} < 1$ (photon number eigenstates), and $g^{(n)} > 1$ (thermal states). There are qualitative differences among these three types of states, and it clearly would make no sense to lump together all states for which $g^{(n)} \neq 1$ under the label "incoherent". All of these states are coherent in the first order. Therefore the differences between them cannot be detected by means of ordinary interferometry, which measures the intensity (or photon-counting rate) distribution, since this depends only upon the first order correlations. We shall see later that these various radiation states can be distinguished in experiments that measure second-order correlations.

Coherence and monochromaticity

In practice, the improvement of the coherence of radiation sources is often closely linked with narrowing the frequency bandwidth. Hence there is a danger of confusion between the concepts of *coherent* radiation and *monochromatic* (monofrequency) radiation. A coherent state clearly need not be monochromatic. Coherence implies (in the first order) that

$$G^{(1)}(t_1, t_2) = f^*(t_1)\, f(t_2)\,, \tag{19.132}$$

with no restriction being imposed on the time dependence of $f(t)$. (Since we are concerned here with the time dependence, we shall omit space variables.) Suppose, however, that we have a *stationary state*, which is one that is invariant under displacements in time. Then the correlation function must satisfy $G^{(1)}(t_1 + \tau, t_2 + \tau) = G^{(1)}(t_1, t_2)$, and hence

$$G^{(1)}(t_1, t_2) = G^{(1)}(t_1 - t_2)\,. \tag{19.133}$$

A *coherent stationary state* must satisfy both of these conditions, and hence it satisfies

$$G^{(1)}(t_1 - t_2) = f^*(t_1)\, f(t_2)\,. \tag{19.134}$$

Let us write $f(t) = \exp[\alpha(t)]$. Then $f^*(t_1)\, f(t_2) = \exp[\alpha^*(t_1) + \alpha(t_2)]$. This will be a function of the difference $t_1 - t_2$ only if $\alpha(t)$ is a linear function of t with the coefficient of t being pure imaginary, i.e. of the form $\alpha(t) = a + i\omega$, with ω real. In this case $f(t)$ has only one frequency. Thus we have shown that a *coherent stationary state must be monochromatic.*

Coherent states versus Pure states

In addition to the distinction between *coherent* and *incoherent* (or, preferably, *partially coherent*) states of the EM field, there is the broad distinction between *pure* states and *nonpure* (*or mixed*) states. Both *coherent* states and *pure* states differ from their opposites by having smaller statistical fluctuations. (Recall the discussion in Sec. 8.4 of minimum uncertainty states, which were shown to necessarily be pure states.) In spite of this superficial similarity, the classes of pure states and of coherent states are not identical. Pure states need not be coherent, and coherent states need not be pure. Examples can be given in each of the four logical categories.

(i) *Coherent and pure.* The coherent state, denoted by the vector $|z\rangle$ in Sec. 19.4, is both pure and coherent. This is also true of the multimode state $|z_1, \ldots, z_m, \ldots\rangle$.

(ii) *Coherent but not pure.* Write the complex amplitude z of a coherent state in terms of its real magnitude and phase, $z = r\,e^{i\phi}$. Using a nonnegative weight function $w(\phi)$ we now construct a (nonpure) state operator,

$$\rho = \int w(\phi)|r\,e^{i\phi}\rangle\langle r\,e^{i\phi}|\,d\phi\,.$$

It is easily verified that the correlation functions for this nonpure state are identical with those in the pure coherent state $|z\rangle$. Therefore ρ describes a coherent nonpure state. (Note that the coherent state vectors $|r\,e^{i\phi}\rangle$ and $|r\,e^{i\phi'}\rangle$ are linearly independent if $\phi \neq \phi'$. The physically significant phase of the field mode amplitude z is not to be confused with the physically insignificant phase of the vector $|z\rangle$. $|r\,e^{i\phi}\rangle \neq e^{i\phi}|r\rangle$.)
A similar construction is possible for a multimode coherent state. The nonpure state described by the operator

$$\rho = \int w(\phi)\,|e^{i\phi}\,z_1,\ldots,e^{i\phi}\,z_m,\ldots\rangle\langle e^{i\phi}\,z_1,\ldots,e^{i\phi}\,z_m,\ldots|\,d\phi$$

has the same field correlation functions as does the pure coherent state $|z_1,\ldots,z_m,\ldots\rangle$. The essential point in this example is that only the overall phase of the mode amplitudes has statistical dispersion, but the *relative* phases of the modes are well defined. The correlations depend only upon relative phases.

(iii) *Pure but not coherent.* As was shown in Example (a) above, and especially by (19.128), the photon number eigenstates are not fully coherent. The single mode states are coherent only in the first order. The two-mode state $|1,1\rangle$, with one photon in each mode, is not coherent even in the first order. This can be verified by putting $\alpha = \gamma = 0$ in (19.113).

(iv) *Neither pure nor coherent.* There are numerous examples of such states, blackbody radiation being the most familiar example.

Classical theory

We have seen that quantum field theory presents us with a rich variety of states, whose qualitatively different properties can be distinguished experimentally. It is of interest to determine whether *classical* field theory could also account for the same phenomena. Any experiment that uses only one detector, and so measures only the first order correlation function, can in principle be described by classical field theory. We need only reinterpret the "photon detection probability" as a measure of the intensity of the classical radiation field. The discrete detection events (the clicks of the photoelectric counting device) are accounted for by the quantum nature of matter, with its discrete energy levels, and do not provide compelling evidence for quantization of the EM field. (Recall Secs. 12.1 and 12.6, which treated quantum-mechanical spin systems and atoms in a classical external EM field.) Only by considering second and higher order correlations can we distinguish between the predictions of classical and quantum field theories.

We shall first examine the normalized coherence function (19.122) for a *classical field* in the case where all the variables are equal: $x_1 = x_2 = \cdots = x_{2n} = x$. This special case can be expressed in terms of the intensity of the classical field, $X = E^{(-)}(x)\, E^{(+)}(x) = |E^{(+)}(x)|^2$, thus:

$$g^{(n)}(x) \equiv g^{(n)}(x,\ldots,x) = \frac{G^{(n)}(x,\ldots,x;x,\ldots,x)}{\left\{G^{(1)}(x;x)\right\}^n} = \frac{\langle X^n \rangle}{\langle X \rangle^n}\,. \tag{19.135}$$

[Note that $g^{(1)}(x) = 1$ by definition, but this does not imply first order coherence because it applies only to equal values of the arguments, $x_1 = \cdots = x_{2n}$.] The classical averages can be calculated from any arbitrary nonnegative probability density, $w(x) \geq 0$:

$$\langle X^n \rangle = \int_0^\infty X^n\, w(X)\, dX\,.$$

From these definitions, it follows that

$$\begin{aligned}
g^{(n+1)}(x) - g^{(n)}(x) &= \int_0^\infty \left[\frac{X}{\langle X \rangle} - 1\right] \frac{X^n}{\langle X \rangle^n}\, w(X)\, dX \\
&\geq \int_0^\infty \left[\frac{X}{\langle X \rangle} - 1\right] \frac{X^n}{\langle X \rangle^n}\, w(X)\, dX \\
&\quad - \int_0^\infty \left[\frac{X}{\langle X \rangle} - 1\right] \left[\frac{X^n}{\langle X \rangle^n} - 1\right] w(X)\, dX \\
&= \int_0^\infty \left[\frac{X}{\langle X \rangle} - 1\right] w(X)\, dX = 0\,.
\end{aligned}$$

Therefore classical field theory leads to a nondecreasing sequence,

$$1 = g^{(1)}(x) \leq g^{(2)}(x) \leq g^{(3)}(x) \cdots\,. \tag{19.136}$$

This contrasts with the photon number eigenstates, for which we have shown that $g^{(n)} < 1$ for $n > 2$. Thus any state containing only a finite number of photons has properties that are incompatible with classical field theory.

Photon bunching and antibunching

The probability of n photons being detected by n separate detectors is $G^{(n)}(x_1, \ldots, x_n; x_n, \ldots, x_1)$, apart from a factor that depends upon instrumental details. It follows from (19.121) that if a state is *coherent*, this probability will factor, $G^{(n)}(x_1, \ldots, x_n; x_n, \ldots, x_1) = G^{(1)}(x_1; x_1)\, G^{(1)}(x_2; x_2) \cdots G^{(1)}(x_n; x_n)$, indicating that the detections of the photons are statistically independent and uncorrelated. This conclusion applies to the response of detectors at different points in space, and also to one detector at n different times. The arrival times of photons in a coherent state of the field are uncorrelated events. This may seem counterintuitive, on the first encounter. After all, a coherent state with no fluctuations in the mode amplitudes would seem to be the most regular and least random state of the field, and one might expect the photons in such a regular field to arrive at uniform intervals. This paradox is resolved by recognizing that the relation between the amplitude of the field and the detection of a photon is a probabilistic relation. If the field amplitude fluctuates, one is likely to observe a burst of photons when the amplitude is high (bunching), and few or no photons when the amplitude is low. Therefore correlations among the photons is a characteristic of incoherence, and statistical independence is a characteristic of coherence. In fact, a periodic stream of photons would be an extreme case of antibunching.

The correlations of photons in time are most conveniently described by means of the following normalized correlation function [which is really a special case of the normalized coherence function (19.122):

$$g^{(2)}(t, t+\tau) = \frac{G^{(2)}(t, t+\tau; t+\tau, t)}{G^{(1)}(t; t)\, G^{(1)}(t+\tau; t+\tau)}. \tag{19.137}$$

This function is a measure of the probability of detecting another photon at a time τ later than the detection of the first photon, but divided by the single photon-counting rates so that its value is 1 for uncorrelated photons. For a stationary state, both the numerator and the denominator are independent of t, and we may write $g^{(2)}(t, t+\tau) = g^{(2)}(\tau)$.

Three possible behaviors of the correlation function $g^{(2)}(\tau)$ are shown in Fig. 19.5. If $g^{(2)}(\tau) > 1$ for small τ, then the photons tend to arrive close together: this is called *bunching*. If $g^{(2)}(\tau) < 1$ for small τ, then the photons tend to be separated; this is called *antibunching*. At sufficiently long time

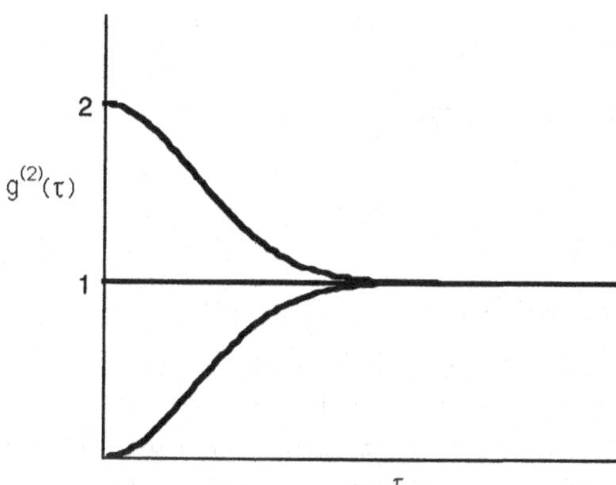

Fig. 19.5 Photon pair correlation function [defined in Eq. (19.137)], showing: bunching (upper curve), no correlation (middle curve), and antibunching (lower curve).

separations, the photons will not be correlated, and so $g^{(2)}(\tau) \to 1$ as $\tau \to \infty$. The shapes of the curves in the figure are only schematic, and it is possible for the correlation function to oscillate before reaching its asymptotic limit. For coherent radiation, the line is horizontal: $g^{(2)}(\tau) = 1$.

The detailed time dependence of $g^{(2)}(\tau)$ depends on the detailed frequency spectrum of the radiation. However the limit at $\tau = 0$, $g^{(2)}(0)$, is just the special case of the normalized coherence function $g^{(2)}(x_1, x_2, x_3, x_4)$, [Eq. (19.122)], in which all arguments are equal. This has already been evaluated for several states, being given by (19.126) if only one mode is excited. Putting $n = 2$ in (19.128), we find that for an m-photon state its value is $g^{(2)}(0) = 1 - 1/m$. For filtered thermal radiation, Eq. (19.131) yields $g^{(2)}(0) = 2$.

In the *classical* theory, the function (19.137) becomes a correlation function for the intensity:

$$g^{(2)}(t, t+\tau) = \frac{\langle X(t)\, X(t+\tau)\rangle}{\langle X(t)\rangle \langle X(t+\tau)\rangle}, \tag{19.138}$$

with $X(t) = |E^{(+)}(t)|^2$ being the intensity of the radiation. In the thermal state the magnitude of the electric field has a Gaussian distribution. This may be understood from the fact that $|E|^2$ is proportional to an energy density, and so the canonical Boltzmann distribution function would yield $\exp(-\beta|E|^2)$, with β being inversely proportional to the temperature. Since thermal equilibrium is a steady state, $g^{(2)}(t, t+\tau)$ in (19.138) does not depend upon t. For $\tau = 0$ it reduces to $g^{(2)}(0) = \langle X^2 \rangle / \langle X \rangle^2$. The averages in this expression are easily calculated from the integral

$$S_r = \int_0^\infty X^r \, e^{-\beta X} \, dX \,,$$

since it is apparent that $\langle X \rangle = S_1/S_0$ and $\langle X^2 \rangle = S_2/S_0$. The relevant values of the integral S_r are $S_0 = \beta^{-1}, S_1 = -\partial S_0/\partial\beta = \beta^{-2}, S_2 = -\partial S_1/\partial\beta = 2\beta^{-3}$. Therefore the classical theory yields $g^{(2)}(0) = 2$ for thermal radiation, in agreement with the quantum-theoretical value. Thus classical field theory is able to account for the phenomenon called *photon bunching*. But we have already shown in (19.136) that classical theory can never yield a value for $g^{(2)}(0)$ that is less than 1, and therefore classical theory cannot account for *antibunching*.

Experimental evidence for the bunching of photons from a thermal light source was first found by Hanbury-Brown and Twiss in 1956, and has been confirmed in several more recent experiments. The demonstration of antibunching proved to be more difficult, but Diedrich and Walther (1987) showed that the fluorescence radiation from a single atom exhibits a value for $g^{(2)}(0)$ that is near zero. The reason for antibunching is easy to understand. After an atom has just emitted a photon, it cannot emit another photon until it has been re-excited. Thus it is unlikely that the time interval between the emission of two photons will be less than the characteristic time for the atom to be excited from its ground state. It is important to be able to observe the radiation from a single atom, since if several atoms radiate at once it is possible for two different atoms to emit photons at an arbitrary time separation, thus obscuring the antibunching in the radiation from a single atom.

We have already noted that the arrival times of photons in a coherent state are statistically independent events. In a stationary coherent state, the photon detection probability per unit time would be a constant (call it λ), and the number of photons detected in a fixed time interval would be governed by the Poisson distribution (see Problem 1.17). The probability of finding n photons in the field at one instant of time is also governed by the Poisson distribution (19.75). Different states of the field have different photon distributions, which can be characterized, in part, by their *mean* $\langle n \rangle$ and *variance* $\sigma^2 = \langle (n - \langle n \rangle)^2 \rangle = \langle n^2 \rangle - \langle n \rangle^2$. For a single excited mode, the $\tau = 0$ limit of the photon pair correlation function is given by (19.126):

$$
\begin{aligned}
g^{(2)}(0) &= \frac{\langle a^{\dagger 2} a^2 \rangle}{\langle a^\dagger a \rangle^2} = \frac{\langle a^\dagger (a a^\dagger - 1) a \rangle}{\langle a^\dagger a \rangle^2} \\
&= \frac{\langle (a^\dagger a)^2 - a^\dagger a \rangle}{\langle a^\dagger a \rangle^2} = \frac{\langle n^2 \rangle - \langle n \rangle}{\langle n \rangle^2} \\
&= 1 + \frac{\sigma^2 - \langle n \rangle}{\langle n \rangle^2} . \qquad (19.139)
\end{aligned}
$$

For a coherent state (Poisson distribution), we have $\sigma^2 = \langle n \rangle$ [see (19.76) and (19.77)]. It is apparent that photon bunching $[g^{(2)}(0) > 1]$ is associated with a larger photon number variance, and that antibunching $[g^{(2)}(0) < 1]$ is associated with a reduced photon number variance, compared to the Poisson distribution. This so-called *sub-Poissonian* photon statistics has also been observed in the fluorescence radiation of a single atom. The extreme case (not yet realized in any experiment) of completely regular photon emission would have zero variance, and hence $g^{(2)}(0) = 0$.

The single-photon state

There is an obvious sense in which the one-photon state is the most distinctively quantum-mechanical, most anticlassical state of the EM field. This has long been recognized, and hence many experiments have been performed to verify the persistence of interference at light intensities so low that $\langle n \rangle < 1$, where $\langle n \rangle$ is the average number of photons present in the system at one time. Those experiments confirm that the form of the interference pattern is independent of the intensity, thus ruling out the notion that diffraction might be due to some cooperative interaction among photons. While we do not deny the value of those experiments, it should be emphasized that *the mere attenuation of light from a conventional source cannot yield an anticlassical state of the field.* If we attenuate laser light, which is described by the coherent state vector $|z\rangle$, we reduce the amplitude $|z|$ and hence reduce the average photon number, $\langle n \rangle = |z|^2$. But this does not change the coherence properties of the state, and a fully coherent state is compatible with classical field theory. If we attenuate thermal radiation, we will reduce the mean square electric field. But this will not alter the Gaussian form of the electric field distribution, which leads to the prediction $g^{(2)}(0) = 2$ and is compatible with the classical theory. Even though the condition $\langle n \rangle \ll 1$ may be achieved by such means, it will not produce a state of the field whose coherence properties are incompatible with classical field theory.

A clever method of producing a single photon state was devised by Grangier *et al.* (1986). Certain atomic excited states cannot decay directly to the ground state, but must decay via an intermediate state, emitting two photons of frequencies ω_1 and ω_2 within a very short time of each other. (This will occur if the excited state and the ground state both have angular momentum $J = 0$, in which case the dipole selection rule prohibits a direct transition, but allows a cascade through an intermediate state of $J = 1$.) Because the two photons have different frequencies, they can be separated into different directions by a diffraction grating. The detection of the first photon ($\hbar\omega_1$) is then a signal that the second photon ($\hbar\omega_2$) will be emitted a fraction of second later. The first photon can be used as a signal to turn on the detectors associated with an interferometer. By this method, it has been possible to confirm the existence of an interference pattern such as (19.111) in a single photon state.

19.9 *Optical Homodyne Tomography — Determining the Quantum State of the Field*

It is possible to determine the state of an ensemble of similarly prepared systems by measuring a sufficient number of dynamical variables (Sec. 8.2), although it may not be obvious what constitutes a sufficient number of measurements for any particular system. Smithey, Beck, Raymer, and Faridani (1993) have used Wigner's phase-space representation to devise a method for determining the state of a field mode.

The Wigner function $\rho_w(q,p)$ [Eq. (15.4)] is not a probability distribution, and is not directly observable; nevertheless, its marginal integrals, $\int \rho_w(q,p)\,dp$ and $\int \rho_w(q,p)\,dq$, are measurable, being the position and momentum distributions, respectively. Similar relations hold if we define new position and momentum variables by the linear canonical transformation

$$q_\phi = q\ \cos\phi + p\ \sin\phi\,, \tag{19.140a}$$

$$p_\phi = -q\ \sin\phi + p\ \cos\phi\,. \tag{19.140b}$$

The probability distribution for the new position q_ϕ is given by the integral of the Wigner function over the conjugate momentum p_ϕ:

$$P_\phi(q_\phi) = \int_{-\infty}^{\infty} \rho_w(q_\phi\ \cos\phi - p_\phi\ \sin\phi,\ q_\phi\ \sin\phi + p_\phi\ \cos\phi)\,dp_\phi\,. \tag{19.141}$$

This fact is not very useful unless q_ϕ is a measurable quantity, which is usually not the case for a mechanical particle. But it turns out that the analogous variable for a field mode is measurable.

A normal mode of a field is formally isomorphic to a harmonic oscillator, and the raising and lowering operators for the oscillator, $a^\dagger$ and a, are the photon creation and annihilation operators for the field mode. Equations (19.22) and (19.23) can be used to define canonical position momentum operators for the field mode, and hence the state of the field mode can be described by a Wigner function. For convenience, we choose units in which $\hbar$ and the angular frequency ω are equal to 1. In these units, the position and momentum operators for the mode are

$$q = \frac{a^\dagger + a}{\sqrt{2}}, \quad p = \frac{i\left(a^\dagger - a\right)}{\sqrt{2}}, \tag{19.142}$$

and the canonically transformed position operator (19.140a) is

$$q_\phi = \frac{a^\dagger\, e^{i\phi} + a\, e^{-i\phi}}{\sqrt{2}}. \tag{19.143}$$

Applying (19.24) to a single mode with $\omega = 1$, it is apparent that the electric field amplitude of the mode is proportional to q, and that q_ϕ is proportional to the amplitude of a rotated quadrature component of the electric field.

A rotated quadrature component of the electric field can be measured by *homodyne detection*, shown schematically in Fig. 19.6. The signal field is mixed with a local oscillator field by a 50–50 beam splitter, and the detectors D1 and D2 count the photons in the two output beams. The useful result R is the difference between the count rates of the two detectors.

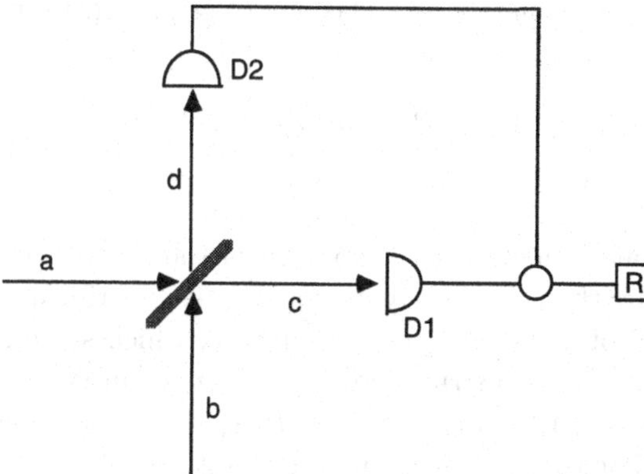

Fig. 19.6 Homodyne detection. The signal field a and the local oscillator field b are mixed by the beam splitter. The measurement result R is the difference between the photon count rates of detectors D1 and D2.

The output modes, c and d, are orthogonal linear combinations of the input modes, a and b. Therefore (with a suitable choice of the phases of the mode functions) the photon creation operators for the output modes will be related to those of the input modes thus:

$$c^\dagger = \frac{a^\dagger + b^\dagger}{\sqrt{2}}, \quad d^\dagger = \frac{-a^\dagger + b^\dagger}{\sqrt{2}}. \tag{19.144}$$

The photon number operators for detectors D1 and D2 are $N_1 = c^\dagger c$ and $N_2 = d^\dagger d$. In terms of the input modes, these are

$$N_1 = c^\dagger c = \tfrac{1}{2}\left(a^\dagger a + b^\dagger b + a^\dagger b + b^\dagger a\right), \tag{19.145}$$

$$N_2 = d^\dagger d = \tfrac{1}{2}\left(a^\dagger a + b^\dagger b - a^\dagger b - b^\dagger a\right). \tag{19.146}$$

The first and second terms in the parentheses on the right are the photon number operators for the signal and local oscillator, respectively, while the third and fourth terms are interference terms. The difference between the outputs of detectors D1 and D2 is represented by the operator

$$R = N_1 - N_2 = a^\dagger b + b^\dagger a. \tag{19.147}$$

The states of the signal and local oscillator are uncorrelated, and therefore we have $\langle a^\dagger b \rangle = \langle a^\dagger \rangle \langle b \rangle$. If we choose the state of the local oscillator to be a coherent state $|z\rangle$, with $z = r\, e^{i\phi}$, then we will have $\langle b \rangle = r\, e^{i\phi}$ and $\langle b^\dagger \rangle = r\, e^{-i\phi}$. If the amplitude r of the coherent state is large, the relative fluctuations of the field will be small, and the field in the coherent state will behave approximately as a classical electromagnetic field. Then (19.147) will become

$$\begin{aligned} R &= r(a^\dagger e^{i\phi} + a\, e^{-i\phi}) \\ &= \sqrt{2}\, r\, q_\phi. \end{aligned} \tag{19.148}$$

Thus the rotated quadrature component of the field mode, q_ϕ [Eq. (19.143)], is measured by the homodyne detector. In the experiment, the signal consists of a sequence of identically prepared pulses, which serve as an ensemble for which the probability distribution of q_ϕ, $P_\phi(q_\phi)$, can be determined.

From (19.141) it is apparent that $P_\phi(q_\phi)$ is a one-dimensional projection of the two-dimensional Wigner function $\rho_w(q,p)$ along the direction ϕ. The Wigner function can be reconstructed from its projections $P_\phi(q_\phi)$ along a large number of directions ϕ by a technique known as *tomography*. In medical science, tomography is used to construct a three-dimensional model of the interior of a patient's body from a large number of x–ray images, each of which is a two–dimensional projection of the tissue density. The mathematical transformation from $P_\phi(q_\phi)$ to $\rho_w(q,p)$ is the *inverse Radon transform* (Vogel and Risken, 1989), which is a one-to-one transformation. Thus the Wigner representation of the quantum state of the field mode is measurable. Any other representation of the state can be calculated from the Wigner function.

Further reading for Chapter 19

The literature on the Casimir force is reviewed by Lamoraux (1999). The theory can be generalized to nonzero temperatures, and to forces between atoms (Spruch, 1986). Knight and Allen (1983) give a concise introduction to the principles of quantum optics, as well as reprints of 17 original papers. The nontrivial nature of the quantum vacuum is investigated in detail by Milonni (1984, 1994). Silverman (1995) describes many ingenious experiments involving two-photon correlations and interference.

Problems

19.1 In the limit $\Omega \to \infty$, the normal modes of the EM field approach a continuum. Sum the expression (19.48) for the mean square zero-point field over the modes within a small but finite bandwidth $\Delta\omega$, and so obtain the value of $\langle E^2 \rangle$ per unit bandwidth.

19.2 If the initial state vector of a harmonic oscillator is the coherent state, $|\psi(0)\rangle = |z\rangle$, show that the state remains coherent, and the time evolution of $|\psi(t)\rangle$ corresponds to a classical orbit in (q_0, p_0) space.

19.3 Generalize the theory of spontaneous emission in Sec. 19.5 to treat stimulated emission and absorption. Do this by taking the initial state of the field to contain n photons, and calculating the transition probability to states of $n+1$ and $n-1$ photons. (Compare your result with that obtained in Sec. 12.6, where the field was not quantized.)

19.4 Prove the identity (19.127), $a^{\dagger\, n}\, a^n = N(N-1)(N-2)\cdots(N-n+1)$, where $N = a^\dagger\, a$.

19.5 Use the quantum theory to calculate the zero-time limit of the photon pair-correlation function, $g^{(2)}(0)$, for unfiltered blackbody radiation.

19.6 We have shown that the state $|\Psi\rangle = \alpha|2,0\rangle + \gamma|0,2\rangle$ of the two-plane-wave model exhibits no interference pattern in the (first order) photon detection probability $G^{(2)}(x_1;x_1)$. (The vector $|2,0\rangle$ represents two photons in mode $^{\#}1$, and $|0,2\rangle$ represents two photons in mode $^{\#}2$.)

Determine whether there is an interference pattern in the spatial correlation of photons by evaluating the second order correlation function $G^{(2)}(x_1, x_2; x_2, x_1)$.

19.7 The field produced by two fully coherent single mode lasers is described by the state vector $|z_1, z_2\rangle = |z_1\rangle \otimes |z_2\rangle$ or, equivalently, by the state operator $\rho = |z_1, z_2\rangle\langle z_1, z_2|$. Suppose that the two lasers are subject to independent phase fluctuations, described by the probability density $w(\phi_1)\, w(\phi_2)$, where $z_1 = |z_1|\, e^{i\phi_1}$ and $z_2 = |z_2|\, e^{i\phi_2}$. Investigate the effect of this noise on the interference pattern.

19.8 Investigate the effect of fluctuations in the amplitudes $|z_1|$ and $|z_2|$ on the interference pattern of the two lasers of the previous problem. What is the difference, if any, between the effects of phase fluctuations and amplitude fluctuations?

19.9 Evaluate the average number of photons $\langle n \rangle$ in a field mode of angular frequency ω in the thermal equilibrium state.

19.10 The operator $D(z) = \exp(z\, a^\dagger - z^* a)$, for a harmonic oscillator or a single field mode, is a *displacement* operator in phase space. This is apparent from (19.56) and (19.57). Determine the composition law for the product of two successive displacements in phase space, $D(z_1)\, D(z_2)$. Why is the result not simply equal to $D(z_1 + z_2)$?

19.11 This problem and the next two involve the *squeezing operator*, $S(\zeta) = \exp[\frac{1}{2}(\zeta\, aa - \zeta^*\, a^\dagger\, a^\dagger)]$, where ζ may be a complex number. Show that if ζ is replaced by a real number r, the effect of the unitary transformation $S(r)$ is to rescale the position operator Q of the oscillator by a constant factor, and to rescale the momentum operator P by the reciprocal of that factor. (Hint: use the result of Problem 3.3.)

19.12 Consider the action of the squeezing operator on a coherent state vector, $|z\rangle = D(z)|0\rangle$, for an oscillator or field mode. Show that for any real value of r, the vector $|r, z\rangle = S(r)|z\rangle$ describes a state of *minimum uncertainty* [in the sense that the indeterminacy relation (8.33) becomes an equality], but is not a *coherent state*. (It is known as a *squeezed state*.)

19.13 For the field mode (harmonic oscillator) Hamiltonian $H = \hbar\omega(a^\dagger\, a + \frac{1}{2})$, determine the time evolution of an initial state vector of the form $|\zeta, 0\rangle = S(\zeta)|0\rangle$, where ζ may be complex. For this nonstationary state (known as a *squeezed vacuum state*), calculate the variance of the electric field of the mode as a function of time. How does it compare the variance of the electric field in the vacuum?

Chapter 20

Bell's Theorem and Its Consequences

In this chapter we shall show that some simple results about correlations turn out to have very profound and puzzling consequences about the nature of the world, as it is described by quantum mechanics. Certain ideas that seem natural, and indeed almost inevitable, from the point of view of special relativity have consequences that are contradicted by quantum mechanics. The starting point of the investigation was an argument by Einstein, Podolsky, and Rosen in 1935, but it was not pursued until 30 years later, by J. S. Bell, who obtained much more significant results that had not been anticipated in any previous work. Bell's work has led to an interesting series of experiments, which have confirmed the numerical correctness of the predictions of QM, but have not made the implications of the results seem any less strange.

20.1 *The Argument of Einstein, Podolsky, and Rosen*

In 1935 Einstein, Podolsky, and Rosen (EPR) posed the question "Can quantum-mechanical description of reality be considered complete?" The meaning of the term *complete*, in this context, is specified by their requirement that in a complete theory "*every element of physical reality must have a counterpart in the physical theory*". As a sufficient condition for recognizing an *element of physical reality*, they proposed: "*If, without in any way disturbing a system, we can predict with certainty the value of a physical quantity, then there exists an element of physical reality corresponding to this physical quantity.*" Note that this is only a *sufficient condition* for recognizing the existence of an element of physical reality, and it should not be construed as a necessary condition or as a definition of an element of reality.

EPR then considered a system of two particles prepared in a state in which the relative position, $x_1 - x_2$, and the total momentum, $p_1 + p_2$, have definite values. (Since the operators for $x_1 - x_2$ and $p_1 + p_2$ commute, it follows that such an eigenstate exists.) After the state preparation has been completed, there is to be no interaction between the two particles. By measuring the position of particle #1 we can predict with certainty the position of particle #2. Since the second particle is spatially separated from the first and there is no interaction between them, the measurement on particle #1 does not disturb particle #2. Therefore, according to the criterion above, the position of particle #2 must be an element of reality. Alternatively, we could measure the momentum of particle #1 and predict with certainty the value of the momentum of particle #2. We therefore infer that the momentum of particle #2 is an element of reality. By hypothesis, there is no interaction between the two particles, so any measurement on particle #1 should have no physical effect on the condition of particle #2. It would be most unreasonable for the reality of the attributes of particle #2 to depend on operations that do not disturb it in any way. Therefore EPR concluded that, in this situation, the values of both the position x_2 and the momentum p_2 of particle #2 are elements of reality. Since the corresponding operators have no common eigenvectors that could describe sharp values for both of these elements of reality, it follows that the description of reality that is provided by the quantum-mechanical state vector is not *complete*, in the sense defined above.

[[The original version of the EPR argument made use of the notion of *reduction of the state vector* during measurement, not because the authors believed it to be true, but because their purpose was to criticize the then current interpretation of QM, of which that notion was a component. Several arguments against that notion have been given in this book (see Ch. 9 and Sec. 12.2). The version of the EPR argument given above does not employ *reduction of the state vector* in order to make it clear that the main thrust of their argument is still relevant even after the notion of state reduction has been discarded.]]

The EPR argument and its conclusion no longer seem so startling or controversial as they did in 1935. Indeed, the question of completeness is of only secondary interest. Bohm's theory of the quantum potential and associated quantal trajectories (discussed briefly in Sec. 14.3) is an example of a more complete description than the statistical state description of standard quantum theory. We are now accustomed to a hierarchy of theoretical models, each of which provides a more detailed description of reality than the one before: we may describe the atom as a point nucleus surrounded by electrons; the idealization of a point nucleus is then replaced by a system of protons and neutrons; the nucleons are then constructed out of quarks; the quarks themselves are not elementary particles, but only transitory excitations of an underlying field. There is no reason to believe that the latest stage of theory building represents a *complete* edifice.

The greater importance of the EPR argument is that it first confronted quantum mechanics with a principle of *locality*, which Einstein later expressed in the words "*The real factual situation of the system S_2 is independent of what is done with the system S_1, which is spatially separated from the former.*" This principle is motivated by special relativity, which prohibits instantaneous action at a distance. Such a principle was implicitly invoked in the EPR argument when it was asserted that a measurement on particle #1 cannot affect the condition of the spatially separated particle #2, since there is no interaction between the particles.

The *locality* principle may seem so abstract and metaphysical that one may be inclined to doubt that it can be experimentally tested, and it was not until several decades after the EPR argument that its empirical consequences were deduced.

20.2 *Spin Correlations*

The idealized experiment proposed by EPR is not a suitable model from which to design a real experiment. It is not practical to prepare their initial state (an eigenstate of relative position and total momentum), and even if it could be prepared, it would have only a transitory existence, since an eigenstate of relative position cannot be a stationary state. A more realistic experiment, illustrating the same principles, was proposed by Bohm. He considered a system of two atoms, each having spin $s = \frac{1}{2}$, prepared in a state of zero total spin. (Certain diatomic molecules have unstable excited states with the desired properties.) This singlet spin state vector for the two particles has the form

$$|\Psi_0\rangle = (\langle|+\rangle \otimes |-\rangle - |-\rangle \otimes |+\rangle) \sqrt{\tfrac{1}{2}}\,, \tag{20.1}$$

where the single particle vectors $|+\rangle$ and $|-\rangle$ denote "spin up" and "spin down" with respect to some coordinate system. Even though the orbital state is not stationary, the interactions do not involve spin and so the spin state will not change. The particles are allowed to separate, and when they are well beyond the range of interaction we can measure the z component of spin of particle #1. Because the total spin is zero, we can predict with certainty, and without in any way disturbing the second particle, that the z component of spin of particle #2 must have the opposite value. Thus the value of ${\sigma_z}^{(2)}$ is an element of reality, according to the EPR criterion. But the singlet state is invariant under rotation, and it has the same form (20.1) in terms of "spin up" and "spin down" vectors if the directions "up" and "down" are referred to the x axis, or y axis, or any other axis. Thus, following EPR, we may argue that the values of ${\sigma_x}^{(2)}, {\sigma_y}^{(2)}$, and any number of other spin components are also elements of reality, and hence that the quantum state description is not a complete description of physical reality.

Except for this restatement of the EPR argument in terms of a practicable experiment, no further progress was made until 1964, when it occurred to J. S. Bell to consider the correlations not only between the components of spin in the same spatial direction, such as ${\sigma_z}^{(1)}$ and ${\sigma_z}^{(2)}$, but also between components of spin in arbitrary directions. Let $\sigma_a \equiv \boldsymbol{\sigma}\cdot\hat{\mathbf{a}}$ denote the component of the Pauli spin operator in the direction of the unit vector $\hat{\mathbf{a}}$, and $\sigma_b \equiv \boldsymbol{\sigma}\cdot\hat{\mathbf{b}}$ denote the component in the direction of the unit vector $\hat{\mathbf{b}}$. If we measure the spin of particle #1 along the direction $\hat{\mathbf{a}}$ and the spin of particle #2 along the direction $\hat{\mathbf{b}}$, the results will be correlated, and for the singlet state the correlation is

$$\langle\Psi_0|\sigma_a \otimes \sigma_b|\Psi_0\rangle = -\cos(\theta_{ab})\,, \tag{20.2}$$

where θ_{ab} is the angle between the directions $\hat{\mathbf{a}}$ and $\hat{\mathbf{b}}$. This result can be calculated from the properties of the Pauli spin matrices by brute force. Alternatively, we can invoke the rotational invariance of the singlet state, and without loss of generality, choose $\hat{\mathbf{a}}$ to be in the z direction. Then the two terms of (20.1) each become eigenvectors of σ_a, and we obtain

$$\langle\Psi_0|\sigma_a \otimes \sigma_b|\Psi_0\rangle = \frac{1}{2}(\langle -|\sigma_b|-\rangle - \langle +|\sigma_b|+\rangle) = -\cos(\theta_{ab})\,.$$

This innocuous expression for spin correlations was shown by Bell to conflict with Einstein's *locality* principle.

We shall examine Bell's arguments in the next section, but first we show the existence of a conflict using a simple argument, similar to one introduced by N. Herbert (1975). We idealize the source of the singlet state as a generator of two correlated signals, and two spin-measuring devices are used as detectors of those signals. Detector A measures the component of the spin of particle #1 in the direction $\hat{\mathbf{a}}$, and detector B measures the component of the spin of particle #2 in the direction $\hat{\mathbf{b}}$. The message recorded by detector A is the value of $-{\sigma_a}^{(1)}$, and the message recorded by detector B is the value of ${\sigma_b}^{(2)}$. When the two detectors are aligned in the same direction ($\theta_{ab} = 0$), the two messages (strings of $+\,1$ and -1) will be identical because of the correlation (20.2). If detector B is rotated by an angle θ, the two messages will no longer agree, and the fractional rate of disagreement will be $d(\theta) = [1 - \cos(\theta)]/2$.

We now introduce a form of the *locality* postulate, assuming that any change in the message recorded by B is due only to the change in the orientation of B, and does not depend upon the orientation of the spatially separated detector A. Hence the rotation of B through the angle θ may be said to introduce an error rate $d(\theta)$ in the message that it records. (The term "error" is only figurative, since the whole setup is rotationally invariant, and there is no "correct" direction.) If detector A is also rotated through the angle θ (so that once again $\theta_{ab} = 0$), an error rate $d(\theta)$ is also introduced into message A, but the errors in the A and B messages exactly cancel, and the two messages agree. If detector B is rotated through a further angle θ, another set of errors will be introduced into message B, so that the disagreement rate will again be $d(\theta)$. Now consider the error rate in message B if the detector were originally rotated through the angle $2\theta, d(2\theta)$. This must be equivalent to the cumulative effect of the errors introduced by two steps of θ. However, some of those errors might cancel, so we have an inequality,

$$d(2\theta) \le 2d(\theta)\,. \tag{20.3}$$

This result has been deduced from the assumption that the error production is only local.

Although this result applies for all θ, a comparison with quantum mechanics is easiest at small angles. Suppose that $d(\theta) \propto \theta^\alpha$. Then the inequality yields $2^\alpha \le 2$, or $\alpha \le 1$. But for the singlet state, we have $d(\theta) = \frac{1}{2}[1 - \cos(\theta)]$, which for small θ becomes $d(\theta) \propto \theta^2$. Thus the predictions of quantum mechanics for this state are in conflict with the inequality that was derived from the *locality* assumption.

20.3 *Bell's Inequality*

The argument leading to Herbert's inequality (20.3) has the merit of being very brief. But since the conclusion — a conflict between quantum mechanics and locality — is so surprising and potentially far-reaching, it is important to seek more general arguments that can indicate more precisely the source of the conflict. The original arguments by J. S. Bell are more effective for this purpose. Here we shall follow, approximately, his second (1971) argument. Although inspired by the spin correlation model of Sec. 20.2, it is generalized so as to apply to systems other than spins.

We consider a two-component system and a pair of instruments that can measure a two-valued variable on each of the components. The components will for convenience be called particles, although no specific particle model is assumed. The possible results of a measurement are taken to be ± 1. Each instrument has a range of settings, corresponding to the possible orientations of a spin-measuring apparatus such as a Stern–Gerlach magnet (described in Sec. 9.1). These will be denoted as $\mathbf{a}$ for the first instrument and $\mathbf{b}$ for the second instrument. The result of a measurement may depend on the controllable parameters $\mathbf{a}$ and $\mathbf{b}$, and on any number of uncontrolled parameters denoted collectively as λ. The result $A(= \pm 1)$ of the measurement on the first particle may depend on the setting $\mathbf{a}$ of the first instrument and on the uncontrolled parameters λ. Therefore we assume that there is a function $A(\mathbf{a}, \lambda) = \pm 1$ which determines the result of the measurement on the first particle. Similarly we assume that there is a function $B(\mathbf{b}, \lambda) = \pm 1$ which determines the result of the measurement on the second particle. But, in accordance with Einstein's principle of *locality* (introduced in Sec. 20.1), we assume that the result of a measurement on the first particle does not depend on the setting $\mathbf{b}$ of the second instrument, and that the result of a measurement on the second particle does not depend on the setting $\mathbf{a}$ of the first instrument. Thus we exclude functions of the form $A(\mathbf{a}, \mathbf{b}, \lambda)$ and $B(\mathbf{a}, \mathbf{b}, \lambda)$. Nothing need be assumed about the uncontrollable parameters λ. They may be associated with the particles, with the instruments, with the environment, or jointly with all of these. It makes no difference to the argument.

It is possible, in principle, that the two measurements (including the setting up of instruments) could be carried out in spatially separated regions of space–time, so that no light signal could communicate the value of the setting of the first instrument to the region of the second instrument before the second measurement was completed. Thus our assumption that the result $B(\mathbf{b}, \lambda)$ of the measurement on the second particle is independent of the setting $\mathbf{a}$ of the first instrument seems well motivated by special relativity. The first surprise may be that any nontrivial physical conclusions can be drawn from the assumption that the result of one measurement should not depend on the setting of another distant instrument. It was Bell's great accomplishment to show not only that such a bland assumption has testable consequences, but also that it conflicts with the predictions of quantum mechanics.

We wish to study the correlation between the results of the measurements on the two particles. The uncontrollable parameters λ are subject to some probability distribution $\rho(\lambda)$, so for fixed settings of the instruments, the correlation function is of the form

$$C(\mathbf{a},\mathbf{b}) = \int A(\mathbf{a},\lambda)\, B(\mathbf{b},\lambda)\, \rho(\lambda)\, d\lambda\,, \tag{20.4}$$

where $A(\mathbf{a},\lambda) = \pm 1, B(\mathbf{b},\lambda) = \pm 1, \rho(\lambda) \geq 0$, and $\int \rho(\lambda)\, d\lambda = 1$. The problem of determining the properties of $C(\mathbf{a},\mathbf{b})$ is now reduced to a mathematical exercise. In fact, we shall not use the restrictions $A(\mathbf{a},\lambda) = \pm 1$ and $B(\mathbf{b},\lambda) = \pm 1$, but only the weaker restrictions

$$|A(\mathbf{a},\lambda)| \leq 1\,, \quad |B(\mathbf{b},\lambda)| \leq 1\,. \tag{20.5}$$

This will permit an important generalization of the result that we are about to derive.

We consider two alternative settings, $\mathbf{a}$ and $\mathbf{a}'$, for the first instrument, and two settings, $\mathbf{b}$ and $\mathbf{b}'$, for the second instrument. Then

$$\begin{aligned} C(\mathbf{a},\mathbf{b}) - C(\mathbf{a},\mathbf{b}') &= \int [A(\mathbf{a},\lambda)\, B(\mathbf{b},\lambda) - A(\mathbf{a},\lambda)\, B(\mathbf{b}',\lambda)]\, \rho(\lambda)\, d\lambda \\ &= \int [A(\mathbf{a},\lambda)\, B(\mathbf{b},\lambda)\, \{1 \pm\, A(\mathbf{a}',\lambda)\, B(\mathbf{b}',\lambda)\}]\, \rho(\lambda)\, d\lambda \\ &\quad - \int [A(\mathbf{a},\lambda)\, B(\mathbf{b}',\lambda)\, \{1 \pm\, A(\mathbf{a}',\lambda)\, B(\mathbf{b},\lambda)\}]\, \rho(\lambda)\, d\lambda\,. \end{aligned}$$

Using (20.5) we obtain

$$\begin{aligned} |C(\mathbf{a},\mathbf{b}) - C(\mathbf{a},\mathbf{b}')| &\leq \int [1 \pm\, A(\mathbf{a}',\lambda)\, B(\mathbf{b}',\lambda)]\, \rho(\lambda)\, d\lambda \\ &\quad + \int [1 \pm\, A(\mathbf{a}',\lambda)\, B(\mathbf{b},\lambda)]\, \rho(\lambda)\, d\lambda \\ &= 2 \pm\, [C(\mathbf{a}',\mathbf{b}') + C(\mathbf{a}',\mathbf{b})]\,, \end{aligned}$$

which can be written as

$$|C(\mathbf{a},\mathbf{b}) - C(\mathbf{a},\mathbf{b}')| + |C(\mathbf{a}',\mathbf{b}') + C(\mathbf{a}',\mathbf{b})| \leq 2\,. \tag{20.6}$$

This result is known as *Bell's inequality*.

The derivation of Bell's inequality made no use of quantum mechanics, but only some much simpler postulates, among which Einstein's locality principle was the most prominent. Hence it is not obvious whether quantum mechanics is consistent with the inequality. We shall compare (20.6) with the quantum-mechanical correlation in the singlet spin state (20.2). Since the state is spherically symmetric, the correlation function depends only upon the relative angle between the orientations of the spin-measuring instruments, $C(\mathbf{a},\mathbf{b}) = C(\theta_{ab})$. We choose the four directions $\mathbf{a}, \mathbf{b}, \mathbf{a}'$, and $\mathbf{b}'$ to be coplanar, with relative directions as shown in Fig. 20.1.

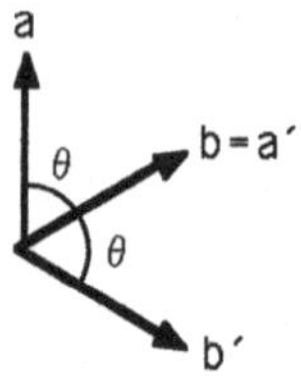

Fig. 20.1 Spin directions chosen to test Bell's inequality, Eq. (20.7).

Then Bell's inequality (20.6) reduces to

$$|C(\theta) - C(2\theta)| + |C(\theta) + C(0)| \leq 2\,. \tag{20.7}$$

If we were to substitute the quantum-mechanical value (20.2) for the spin correlation function in the singlet state, $C(\theta) = -\cos(\theta)$, we would obtain

$$2\,\cos(\theta) - \cos(2\theta) \leq 1\,.$$

But this inequality is violated for a wide range of θ. The maximum violation occurs for $\theta = \pi/3$, for which the expression on the left is equal to 3/2. Therefore quantum mechanics is in conflict with at least one of the assumptions that were used in the derivation of Bell's inequality. The proof that such a conflict with quantum mechanics exists is known as *Bell's theorem.*

Since Bell's inequality is violated by a large amount in the singlet state, it would seem that an experimental test would not be difficult. No measurements have yet been performed on a spontaneously dissociating molecule in the singlet state, but an analogous two-photon state has been studied (see Sec. 20.5). The most serious limitation of those experiments is the inefficiency of the detectors; many particles go undetected. To show what effect this has, we redefine the functions that describe the results of the measurements, $A(\mathbf{a}, \lambda)$ and $B(\mathbf{b}, \lambda)$, to have three possible values: $+1$, -1, and 0. The instruments can record only the values ± 1; the value $A = 0$ means that the first particle was not detected, and $B = 0$ means that the second particle was not detected. Bell's inequality (20.6) remains valid with this redefinition because the only properties of the functions $A(\mathbf{a}, \lambda)$ and $B(\mathbf{b}, \lambda)$ that were used in the derivation were $|A| \leq 1$ and $|B| \leq 1$.

The experimental value of the correlation is

$$C_{\text{exp}}(\mathbf{a}, \mathbf{b}) = \frac{N_{++} + N_{--} - N_{+-} - N_{-+}}{N}\,, \tag{20.8}$$

where N_{++} is the number of events for which both instruments recorded $+1, N_{+-}$ is the number of events for which the first instrument recorded $+ 1$ and the second instrument recorded -1, etc. The total number of pairs emitted by the source is $N = N_{++} + N_{--} + N_{+-} + N_{-+} + N_{+0} + N_{-0} + N_{0+} + N_{0-} + N_{00}$, where the subscript 0 means that the corresponding particle was not detected. The number of events N_{00} for which neither particle was detected is unknown, and so the true value $C_{\text{exp}}(\mathbf{a}, \mathbf{b})$ is unknown. If N_{00} is very large, the magnitude of $C_{\text{exp}}(\mathbf{a}, \mathbf{b})$ will be so small that it will automatically satisfy the inequality (20.6). But if we assume that *the detected particles are a statistically representative sample of the whole*, then we may compare the quantity

$$C'_{\text{exp}}(\mathbf{a}, \mathbf{b}) = \frac{N_{++} + N_{--} - N_{+-} - N_{-+}}{N_{++} + N_{--} + N_{+-} + N_{-+}} \tag{20.9}$$

with the theoretical predictions, and this quantity (or its analog in photon experiments) has usually been found to agree with quantum theory and to disagree with Bell's inequality.

It is possible to construct theoretical models for which the detected particles are not representative of the whole, and for which $C'_{\text{exp}}(\mathbf{a}, \mathbf{b})$ would agree with quantum theory but the true correlation function (20.8) would obey Bell's inequality. These models are somewhat artificial, but some theorists argue that the strange features of their models are more plausible than would be the consequences of rejecting Einstein's locality principle. It appears more likely that the question will be settled by designing better detectors than by further theoretical arguments.

20.4 *A Stronger Proof of Bell's Theorem*

As the theoretical significance of Bell's theorem became known, many other derivations of it were given. Often the same theorem can be proven from different sets of assumptions. A proof is considered stronger or weaker according as it invokes fewer or more assumptions. The proof of Bell's theorem to be given in this section is superior to that of the previous section in two respects. On the theoretical side, it eliminates the assumption of *determinism*. On the practical side, it leads to an inequality that refers only to detected results, and does not involve the number of undetected particles, thereby making experimental tests more feasible. The ideas were first published by Clauser, Horne, Shimony, and Holt (1969), and by Clauser and Horne (1974).

In the previous section, it was assumed that for fixed settings of the instruments, the results of the measurements $A(\mathbf{a}, \lambda)$ and $B(\mathbf{b}, \lambda)$ are *determined* by the uncontrollable parameters λ. (The simpler argument presented in Sec. 20.2 also implicitly contains such an assumption, but it is not so clearly formulated.) The question of whether the statistical distributions of quantum mechanics can be realized as averages over uncontrollable hidden variables was posed in the early days of quantum mechanics. That question was usually dismissed on the grounds that it is futile to speculate about things that are unobservable. Bell's work shows that attitude to be mistaken, since the hypothetical existence of hidden variables can have testable consequences, namely Bell's inequality. Bue the fact that the predictions of quantum mechanics are in the form of probabilities has led to a bias in favor of indeterminism. Hence one might reasonably suspect that the source of the conflict between Bell's inequality and quantum mechanics is in the assumption of a hidden determinism. That, however, is not the case.

The experimental arrangement envisaged by Clauser and Horne is shown schematically in Fig. 20.2. A source emits correlated pairs of particles, each

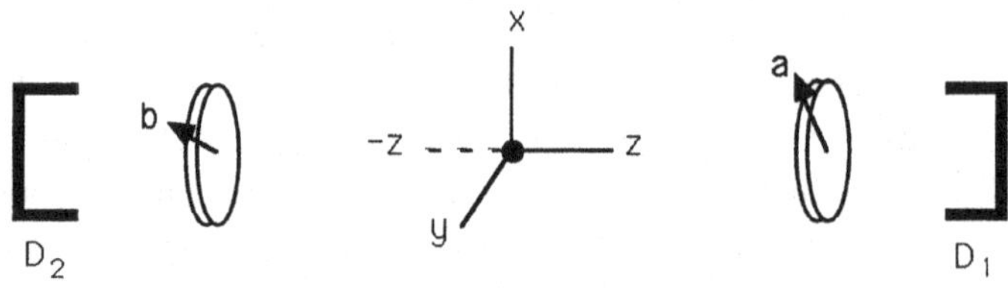

Fig. 20.2 Schematic apparatus for testing the Clauser–Horne inequality. A source at the origin emits a pair of correlated particles, each of which must pass through an analyzer (**a** or **b**) before reaching one of the detectors (D_1 or D_2).

of which must pass through an analyzer before reaching a detector. In the previously considered example of two spin $\frac{1}{2}$ particles, the analyzers would be Stern–Gerlach magnets. In the case of photons (to be treated in the next section), they are polarization filters. The detectors merely record counts (unlike those in Bell's analysis which were to record two possible results, $+1$ or -1).

We suppose that the probability of detector D_1 recording a count may depend on the setting of the analyzer **a** and on some uncontrollable parameters λ. This probability is denoted as $p_1(\mathbf{a}, \lambda)$. Likewise the probability of detector D_2 recording a count is $p_2(\mathbf{b}, \lambda)$. The values of the functions $p_1(\mathbf{a}, \lambda)$ and $p_2(\mathbf{b}, \lambda)$ are bounded between 0 and 1. Unlike the analysis in Sec. 20.3, we do not assume that the *outcome* (count or no count) at D_1 is determined by **a** and λ, but only that the *probability* of the outcome is determined by those parameters. The hypothesis of a hidden determinism could be recovered if we required the functions $p_1(\mathbf{a},\lambda)$ and $p_2(\mathbf{b},\lambda)$ to take on only the value 0 or 1, but such a restriction will be avoided here. The probability of a *coincidence*, i.e. of simultaneous detection by both D_1 or D_2, is denoted as $p_{12}(\mathbf{a},\mathbf{b},\lambda)$.

The uncontrolled parameters have some probability distribution $\rho(\lambda)$. Upon averaging over this distribution, we obtain the probability of detecting a count at D_1 to be

$$P_1(\mathbf{a}) = \int p_1(\mathbf{a},\lambda)\,\rho(\lambda)\,d\lambda\,. \tag{20.10}$$

Similarly the probability of detecting a count at D_2 is

$$P_2(\mathbf{b}) = \int p_2(\mathbf{b},\lambda)\,\rho(\lambda)\,d\lambda\,, \tag{20.11}$$

and the probability of detecting a coincidence is

$$P_{12}(\mathbf{a},\mathbf{b}) = \int p_{12}(\mathbf{a},\mathbf{b},\lambda)\,\rho(\lambda)\,d\lambda\,. \tag{20.12}$$

We now make use of Einstein's *locality* principle. It has, of course, implicitly been used through the assumption that $p_1(\mathbf{a},\lambda)$ does not depend on $\mathbf{b}$ and $p_2(\mathbf{b},\lambda)$ does not depend on $\mathbf{a}$. It is now invoked again, so that we may assume that

$$p_{12}(\mathbf{a},\mathbf{b},\lambda) = p_1(\mathbf{a},\lambda)\,p_2(\mathbf{b},\lambda)\,. \tag{20.13}$$

This factorization of the coincidence probability expresses the idea that there is no action at a distance between instrument #1 (consisting of analyzer $\mathbf{a}$ and detector D_1) and instrument #2 (analyzer $\mathbf{b}$ and detector D_2). Therefore the propensity of instrument #1 to detect a count should be independent of the setting of instrument #2, and vice versa. It should be emphasized that the assumption of (20.13) in no way denies the possibility that the result of one of the measurements could give information about the other. The parameters λ may contain information about both particles, and indeed the observable joint probability for the two measurements (20.12) may show correlations and need not factor.

To derive their principal result, Clauser and Horne first prove a lemma. If x, x', y, y', X, Y are real numbers such that $0 \le x, x' \le X$ and $0 \le y, y' \le Y$, then the following inequality holds:

$$-XY \le xy - xy' + x'y + x'y' - Yx' - Xy \le 0\,. \tag{20.14}$$

This awkward-looking inequality is established through examination of the various special cases. To prove the upper bound, we rewrite the quantity to be bounded as $(x-X)y+(y-Y)x'+(x'-x)y'$, which is clearly nonpositive for $x \geq x'$. In the case of $x < x'$ we rewrite it as

$$\begin{aligned} &x(y-y')+(x'-X)y+x'(y'-Y) \\ &\quad \leq x(y-y')+(x'-X)y+x(y'-Y)=(x'-X)y-x(y-Y)\leq 0\,. \end{aligned}$$

Thus the upper bound of (20.14) is proven. The lower bound will not be used in the experiment, so we omit the proof. This lemma is now applied by substituting $x = p_1(\mathbf{a}, \lambda), x' = p_1(\mathbf{a}', \lambda), y = p_2(\mathbf{b}, \lambda), y' = p_2(\mathbf{b}', \lambda), X = Y = 1$. Using (20.13), multiplying by $\rho(\lambda)$ and integrating over λ, we then obtain

$$-1 \leq P_{12}(\mathbf{a},\mathbf{b}) - P_{12}(\mathbf{a},\mathbf{b}') + P_{12}(\mathbf{a}',\mathbf{b}) + P_{12}(\mathbf{a}',\mathbf{b}') - P_1(\mathbf{a}') - P_2(\mathbf{b}) \leq 0\,, \tag{20.15}$$

which is known as the *Clauser–Horne inequality* (or CH inequality). It is very closely related to Bell's inequality, although the two are not precisely equivalent. The relation between them is discussed in detail by Clauser and Horne (1974).

The experimental data will consist of the number of counts by each detector, $N_1(\mathbf{a})$ and $N_2(\mathbf{b})$, and the number of coincidences $N_{12}(\mathbf{a},\mathbf{b})$, for various settings $\mathbf{a}$ and $\mathbf{b}$ of the analyzers. The appropriate comparison between theory and experiment is

$$\begin{aligned} P_1(\mathbf{a}) &\leftrightarrow \frac{N_1(\mathbf{a})}{N}\,, \\ P_2(\mathbf{b}) &\leftrightarrow \frac{N_2(\mathbf{b})}{N}\,, \\ P_{12}(\mathbf{a},\mathbf{b}) &\leftrightarrow \frac{N_{12}(\mathbf{a},\mathbf{b})}{N}\,, \end{aligned}$$

where N is the total number of pairs emitted by the source. Since the number of undetected particles is unknown, it would appear that a rigorous test of the CH inequality will encounter the same difficulty that was encountered with Bell's inequality in Sec. 20.3. But since the upper bound of (20.15) is zero, we can write an inequality involving only the ratio

$$\frac{P_{12}(\mathbf{a},\mathbf{b}) - P_{12}(\mathbf{a},\mathbf{b}') + P_{12}(\mathbf{a}',\mathbf{b}) + P_{12}(\mathbf{a}',\mathbf{b}')}{P_1(\mathbf{a}') + P_2(\mathbf{b})} \leq 1\,. \tag{20.16}$$

This may be compared with the experimental quantity

$$\frac{N_{12}(\mathbf{a},\mathbf{b}) - N_{12}(\mathbf{a},\mathbf{b}') + N_{12}(\mathbf{a}',\mathbf{b}) + N_{12}(\mathbf{a}',\mathbf{b}')}{N_1(\mathbf{a}') + N_2(\mathbf{b})}, \tag{20.17}$$

which is independent of the unknown number N. Thus an experimental test of the CH inequality is more easily achieved than is a test of the Bell inequality.

Following the example of the previous section, it is easy to obtain a conflict between the predictions of quantum mechanics for the singlet state and the CH inequality. (This is one of the problems at the end of the chapter.) Since no use was made of any form of determinism in the derivation of the CH inequality, the possible sources of the conflict have been correspondingly narrowed, with the *locality* postulate, as embodied in (20.13), appearing to be the most likely source.

20.5 *Polarization Correlations*

Most of the experiments that have been performed to test the Bell and CH inequalities are based on correlations between the polarizations of pairs of photons. Suitably correlated photons can be produced in several different processes. A commonly used source involves systems that decay from an initial state of angular momentum $J = 0$ to a final state of $J = 0$ by emitting a pair of photons. One example is positronium (an atom consisting of an electron and a positron), which annihilates predominantly into two photons. (Energy and momentum could not be conserved by annihilation into a single photon.) Another example is an atom in an excited state of zero angular momentum that decays by means of a $J = 0 \to J = 1 \to J = 0$ cascade, emitting two photons in rapid succession. In both of these examples, the initial angular momentum is zero, and so the final state of the two-photon system must have zero angular momentum; equivalently, it must be spherically symmetric.

The states of the electromagnetic field can conveniently be described by complex basis functions of the form $\hat{\mathbf{u}}e^{i\mathbf{k}\cdot\mathbf{x}}$, where the unit vector $\hat{\mathbf{u}}$ represents the polarization of the mode, and $\mathbf{k}$ is the propagation vector. A state with one photon in this mode will be denoted as $|\hat{\mathbf{u}},\mathbf{k}\rangle$. The condition (19.6b) requires that the polarization be transverse, $\hat{\mathbf{u}}\cdot\mathbf{k} = 0$, and so there are two linearly independent polarizations corresponding to each value of $\mathbf{k}$. The one-photon state vector $|\hat{\mathbf{u}},\mathbf{k}\rangle$ behaves as an ordinary three-vector in its dependence on the polarization vector $\hat{\mathbf{u}}$ (but *not* with respect to its dependence on $\mathbf{k}$). That is to say, for example, the one-photon state vector for a field mode polarized in the direction $\hat{\mathbf{u}} = \cos\theta\ \hat{\mathbf{u}}_1 + \sin\theta\ \hat{\mathbf{u}}_2$ is $|\hat{\mathbf{u}}, \mathbf{k}\rangle = \cos\theta|\hat{\mathbf{u}}_1, \mathbf{k}\rangle + \sin\theta|\hat{\mathbf{u}}_2, \mathbf{k}\rangle$. Thus we may think of the state vector $|\hat{\mathbf{u}}, \mathbf{k}\rangle$ as being proportional to $\hat{\mathbf{u}}$.

The most general state vector describing one photon in each of two modes is of the form

$$\sum c(\hat{\mathbf{u}}_1, \mathbf{k}_1; \hat{\mathbf{u}}_2, \mathbf{k}_2)\ |\hat{\mathbf{u}}_1, \mathbf{k}_1\rangle \otimes |\hat{\mathbf{u}}_2, \mathbf{k}_2\rangle\,, \tag{20.18}$$

where the sum is over all values of the propagation vectors and polarizations of the two modes. For the state vector to be rotationally invariant (i.e. to have zero total angular momentum), it is necessary for the coefficient $c(\hat{\mathbf{u}}_1, \mathbf{k}_1; \hat{\mathbf{u}}_2, \mathbf{k}_2)$ to be a scalar function of its arguments. It must also be linear in $\hat{\mathbf{u}}_1$ and $\hat{\mathbf{u}}_2$, these factors arising from the polarization dependence of $|\hat{\mathbf{u}}_1, \mathbf{k}_1\rangle \otimes |\hat{\mathbf{u}}_2, \mathbf{k}_2\rangle$. The number of possibilities is considerably reduced if we may work in the center-of-momentum frame, in which $\mathbf{k}_1 = \mathbf{k}, \mathbf{k}_2 = -\mathbf{k}$, and the magnitude $k = |\mathbf{k}|$ is fixed by conservation of energy. This is the case for the decay of positronium from a state of zero total momentum (spin singlet and orbital angular momentum $\ell = 0$). The only nontrivial scalars that can be constructed under these conditions are $\mathbf{k}\cdot(\hat{\mathbf{u}}_1 \times \hat{\mathbf{u}}_2)$ and $\hat{\mathbf{u}}_1\cdot\hat{\mathbf{u}}_2$, since the transversality condition requires that $\mathbf{k}\cdot\hat{\mathbf{u}}_1 = \mathbf{k}\cdot\hat{\mathbf{u}}_2 = 0$. Under space inversion, the first of these functions has odd parity, and the second has even parity. Both functions can be multiplied by an arbitrary function of k.

Positronium decay

The parity of a state of a particle–antiparticle system is opposite to that of a similar state of a two-particle system. This follows from Dirac's relativistic quantum theory, which treats both electron and positron states. [See Gasiorowicz (1966), p. 46.] Thus the ground state of positronium has negative parity, whereas the similar ground state of hydrogen has positive parity. Therefore the ground state of positronium decays into two photons in the odd parity state,

$$|\Psi_-\rangle = C \int \sum \mathbf{k}\cdot(\hat{\mathbf{u}}_1 \times \hat{\mathbf{u}}_2)\ |\hat{\mathbf{u}}_1, \mathbf{k}\rangle \otimes |\hat{\mathbf{u}}_2, -\mathbf{k}\rangle\ d\Omega_k\,, \tag{20.19}$$

where C is a normalization factor, $2\mathbf{k}$ is the relative momentum of the photons, the integral is over the directions of $\mathbf{k}$, and the sum is over the polarizations. The experimental setup, shown in Fig. 20.2, selects those values of $\mathbf{k}$ that are close to the z axis, so if we neglect the small spread in propagation directions, we may write the relevant part of the state vector as

$$|\Psi'_-\rangle = \left(|\hat{\mathbf{x}}_1, \mathbf{k}\rangle \otimes |\hat{\mathbf{y}}_2 - \mathbf{k}\rangle - |\hat{\mathbf{y}}_1, \mathbf{k}\rangle \otimes |\hat{\mathbf{x}}_2, -\mathbf{k}\rangle\right)\sqrt{\tfrac{1}{2}}\,, \tag{20.20}$$

where $\hat{\mathbf{x}}$ and $\hat{\mathbf{y}}$ are unit vectors in the x and y directions, respectively, and it is understood that $\mathbf{k}$ is in the z direction.

According to quantum theory, the probability that a photon in the state $|\hat{\mathbf{u}},\mathbf{k}\rangle$ will pass through an ideal polarization filter oriented in the direction $\hat{\mathbf{a}}$ is $|\cos(\theta_{au})|^2$, where θ_{au} is the angle between the directions $\hat{\mathbf{a}}$ and $\hat{\mathbf{u}}$. It is apparent that the polarizations of the two photons in the state $|\Psi'_-\rangle$ are correlated. If the first photon passes through a filter oriented in the x direction, then the second photon will pass through a filter oriented in the y direction but will not pass a filter oriented in the x direction. But the state $|\Psi'_-\rangle$ is invariant under rotations about the z axis, and so this correlation exists not only for the x and y directions, but for any pair of orthogonal directions in the xy plane. Let us next consider the first filter oriented in the x direction ($\hat{\mathbf{a}} = \hat{\mathbf{x}}$), and the second filter in some other direction $\hat{\mathbf{b}}$ in the xy plane. By inspection of (20.20), we see that the probability of the first photon passing through the $\hat{\mathbf{a}}$ filter is $\frac{1}{2}$, and the conditional probability of the second photon passing through the $\hat{\mathbf{b}}$ filter is $|\sin(\theta_{ab})|^2$ if the first photon passes. Thus the joint, or coincidence, probability is equal to $\frac{1}{2}|\sin(\theta_{ab})|^2$. But because of the rotational invariance of the state, this result cannot depend upon the absolute direction of $\hat{\mathbf{a}}$, but only upon the relative angle θ_{ab}. Thus the coincidence probability is equal to

$$[P_{12}(\mathbf{a},\mathbf{b})]_{qm} = \tfrac{1}{2}|\ \sin\ (\theta_{ab})|^2 \qquad (20.21)$$

for arbitrary directions of $\hat{\mathbf{a}}$ and $\hat{\mathbf{b}}$. We shall see that, for certain angles, this result violates the CH inequality (20.15).

$J = 0 \to 1 \to 0$ cascade

There are many atoms that have an excited state of angular momentum $J = 0$, which decays to an intermediate state of $J = 1$ by emitting a photon of angular frequency ω_1, and then reaches the $J = 0$ ground state by emitting a second photon of angular frequency ω_2. There is no net change of angular momentum or parity of the atom in this process. Let us adopt the frame of reference in which the initial linear momentum of the atom is zero. If the recoil of the atom could be neglected, it would follow that the final state of the two photons would have total linear and angular momentum equal to zero, and even parity. The only state with these properties is similar in form to (20.19) but with the even function $\hat{\mathbf{u}}_1{\cdot}\hat{\mathbf{u}}_2$ replacing the odd function $\mathbf{k}{\cdot}(\hat{\mathbf{u}}_1 \times \hat{\mathbf{u}}_2)$. However, the recoil of the atom may take up a significant amount of momentum, and so this simple argument is not valid.

Because of the recoil momentum of the atom, the directions of the photon momenta, $\hbar\mathbf{k}_1$ and $\hbar\mathbf{k}_2$, are not strongly correlated. In order to emphasize the transversality conditions, $\hat{\mathbf{u}}_1 \cdot \mathbf{k}_1 = 0$ and $\hat{\mathbf{u}}_2 \cdot \mathbf{k}_2 = 0$, we shall write the polarization vectors as

$$\hat{\mathbf{u}}_1 = \hat{\mathbf{u}}_{\perp 1} \equiv \hat{\mathbf{u}}_1 - \frac{\mathbf{k}_1(\mathbf{k}_1 \cdot \hat{\mathbf{u}}_1)}{{k_1}^2}\,,$$

$$\hat{\mathbf{u}}_2 = \hat{\mathbf{u}}_{\perp 2} \equiv \hat{\mathbf{u}}_2 - \frac{\mathbf{k}_2(\mathbf{k}_2 \cdot \hat{\mathbf{u}}_2)}{{k_2}^2}\,.$$

Now, in addition to the even parity scalar function $\hat{\mathbf{u}}_{\perp 1} \cdot \hat{\mathbf{u}}_{\perp 2}$ which we have already identified, there is another possibility of the form $(\hat{\mathbf{u}}_{\perp 1} \cdot \mathbf{k}_2)(\hat{\mathbf{u}}_{\perp 2} \cdot \mathbf{k}_1)$. An admixture of this second function in the state would diminish the correlation between the polarizations of the two photons. However, if we use the experimental arrangement shown in Fig. 20.2 to select $\mathbf{k}_1$ in the z direction and $\mathbf{k}_2$ in the $-z$ direction, then this second function vanishes, and the relevant part of the state vector becomes

$$\begin{aligned} |\Psi'_+\rangle &= \sum (\hat{\mathbf{u}}_{\perp 1} \cdot \hat{\mathbf{u}}_{\perp 2})\, |\hat{\mathbf{u}}_1, \mathbf{k}_1\rangle \otimes |\hat{\mathbf{u}}_2, \mathbf{k}_2\rangle \\ &= \big(|\hat{\mathbf{x}}_1, \mathbf{k}_1\rangle \otimes |\hat{\mathbf{x}}_2, \mathbf{k}_2\rangle + |\hat{\mathbf{y}}_1, \mathbf{k}_1\rangle \otimes |\hat{\mathbf{y}}_2, \mathbf{k}_2\rangle\big)\sqrt{\tfrac{1}{2}}\,. \end{aligned} \tag{20.22}$$

The sum in the first line is over the independent directions of the polarization vectors, and it is understood that $\mathbf{k}_1$ is in the z direction and $\mathbf{k}_2$ is in the $-z$ direction. It is apparent that the two photons in the state $|\Psi'_+\rangle$ have parallel linear polarizations. The joint probability that the first photon passes through the filter oriented in the direction $\hat{\mathbf{a}}$ and the second photon passes through the filter oriented in the direction $\hat{\mathbf{b}}$ (both in the xy plane) depends only on the relative angle θ_{ab}, and is equal to

$$[P_{12}(\mathbf{a}, \mathbf{b})]_{qm} = \tfrac{1}{2}|\cos(\theta_{ab})|^2\,. \tag{20.23}$$

Experimental tests

In the experiment depicted in Fig. 20.2, the directions of the polarization analyzers $\hat{\mathbf{a}}$ and $\hat{\mathbf{b}}$ lie in the xy plane. Since the states that we shall consider are invariant under rotations about the z axis, the CH inequality (20.16) may be simplified for these cases. The single-count probabilities P_1 and P_2 will now be independent of the orientation of the analyzers, and the coincidence detection probability will depend only on the relative angle, $P_{12}(\mathbf{a}, \mathbf{b}) = P_{12}(\theta_{ab})$. If we choose the alternative directions of the polarization analyzers as shown in Fig. 20.3, the CH inequality (20.16) becomes

$$S(\theta) \equiv \frac{3\,P_{12}(\theta) - P_{12}(3\theta)}{P_1 + P_2} \leq 1\,. \tag{20.24}$$

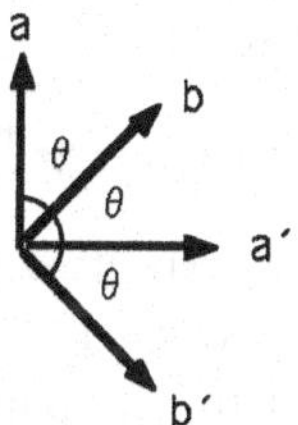

Fig. 20.3 Polarization directions chosen to test the Clauser–Horne inequality, Eq. (20.24).

If we substitute into (20.24) the ideal quantum-mechanical predictions for the state (20.22), $P_1 = P_2 = \frac{1}{2}$, $P_{12}(\theta) = \frac{1}{2}|\cos(\theta)|^2$, we obtain the result shown in Fig. 20.4. The CH inequality is violated whenever $S(\theta)$ exceeds 1. The maximum violation occurs at the angle $\theta = \pi/8$. Similar results hold for the state (20.20), with violations of the CH inequality occurring at a different range of angles.

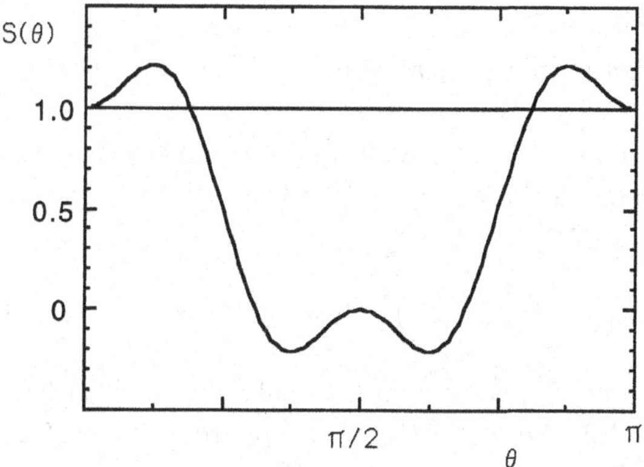

Fig. 20.4 The ideal value of $S(\theta)$ (20.24), calculated from quantum theory, violates the CH inequality whenever it exceeds 1.

The values of $S(\theta)$ shown in Fig. 20.4 are ideal in the sense that they ignore several essential features of an actual experiment. A more realistic analysis, taking account of the limitations of the apparatus, yields predictions of the forms

$$\begin{aligned}[P_1]_{\text{exp}} &= \tfrac{1}{2}\,\eta_1\, f_1\, \varepsilon_1{}^+ ,\\ [P_2]_{\text{exp}} &= \tfrac{1}{2}\,\eta_2\, f_2\, \varepsilon_2{}^+ ,\\ [P_{12}(\theta)]_{\text{exp}} &= \tfrac{1}{4}\,\eta_1\,\eta_2\, f_1\, g_{12}\,\{\varepsilon_1^+\varepsilon_2{}^+ + \varepsilon_1^-\varepsilon_2^- F\ \cos(2\theta)\}\,.\end{aligned} \tag{20.25}$$

The subscripts 1 and 2 refer to the two photons and the two halves of the apparatus in Fig. 20.2. The efficiencies of the detectors are η_1 and η_2. A polarization filter is described by the parameters $\varepsilon^+ = \varepsilon^M + \varepsilon^m$ and $\varepsilon^- = \varepsilon^M - \varepsilon^m$, where ε^M is the transmission coefficient for radiation polarized parallel to the axis of the filter, and ε^m is the transmission coefficient for radiation polarized perpendicular to the axis of the filter. f_1 is the probability that a photon of frequency ω_1 emitted by the source will enter the analyzer–detector instrument on the right, and f_2 is the similar probability for a photon of frequency ω_2 to enter the instrument on the left. These two parameters are determined by the acceptance angles of the instruments. The directional correlation between the momenta of the two photons determines the parameter g_{12}, which is the conditional probability that the second photon will enter the left instrument, given that the first photon enters the right instrument. Thus the probability that both members of a photon pair enter the two instruments is $f_1\, g_{12} = f_2\, g_{21}$. The parameter F is a measure of the degree of correlation between the polarizations of the photons in the initial state. If we put $\eta_1 = \eta_2 = \varepsilon_1{}^+ = \varepsilon_2{}^+ = \varepsilon_1{}^- = \varepsilon_2{}^- = F = f_1 = f_2 = g_{12} = 1$, we would recover the ideal quantum-mechanical result: $P_1 = P_2 = \frac{1}{2}, P_{12}(\theta) = [1+\cos(2\theta)]/4 = \frac{1}{2}|\cos(\theta)|^2$.

Let us assume, for simplicity, that the two instruments are identical, so that the subscripts 1 and 2 can be omitted from the parameters. Then substituting (20.25) into (20.24), we obtain

$$[S(\theta)]_{\text{expt}} = \frac{\eta g}{4\varepsilon^+}\,\{2(\varepsilon^+)^2 + (\varepsilon^-)^2 F[3\ \cos(2\theta) - \cos(6\theta)]\}\,. \tag{20.26}$$

Although the ideal quantum-mechanical value of $S(\theta)$ has a maximum value that exceeds 1.2 (see Fig. 20.4), it is apparent from (20.26) that the experimental value may fail to violate the CH inequality, $S(\theta) \leq 1$, merely because of the various instrumental parameters, and hence the experiment would not distinguish between quantum mechanics and the class of theories that obey the locality hypothesis of Bell's theorem. Each kind of experiment must be examined separately.

Positronium decay looks like a favorable case because the relative directions of the two photons are constrained by momentum conservation (hence $g \approx 1$), and the polarizations are strongly correlated as a consequence of angular momentum and parity conservation (hence $F \approx 1$). Unfortunately, efficient linear polarization filters do not exist for such high energy photons (hence $\varepsilon^- \ll 1$), and the polarization can only be inferred, indirectly and imprecisely, through Compton scattering of electrons. Hence values of $[S(\theta)]_{\text{expt}}$ greater than 1 cannot be obtained in this experiment.

The optical photons emitted in an *atomic cascade* are of much lower energies than those from positronium decay. Efficient polarization filters are readily available ($\varepsilon^- \equiv \varepsilon^M - \varepsilon^m > 0.95$). However, there are several other difficulties. Because of the variable recoil momentum of the atom, the directions of the two photons are not strongly correlated, and thus it is difficult to detect both members of a pair (hence g is small). If we increase the acceptance angles of the instruments so as to capture more pairs (increase g), then the directions of $\mathbf{k}_1$ and $\mathbf{k}_2$ need not be opposite, and the state vector will not be accurately given by (20.22). An admixture of the term proportional to $(\hat{\mathbf{u}}_{\perp 1}\cdot\mathbf{k}_2)(\hat{\mathbf{u}}_{\perp 2}\cdot\mathbf{k}_1)$ will be allowed, and the correlation between the polarizations of the photons will be diminished (F will be reduced). Lastly, the quantum efficiency of detectors of optical photons is not high enough. It has been shown (Garg and Mermin, 1987) that for an unambiguous demonstration of the violation of Bell's inequalities by quantum theory, we need detectors with efficiency $\eta > 0.83$. The highest reported efficiency is $\eta = 0.76$ (Kwiat *et al.*, 1993).

For these reasons, it is difficult to perform an experiment that discriminates strictly between quantum mechanics and the inequalities of Bell's theorem, and some supplementary assumptions have been invoked in order to draw conclusions from the feasible experiments. Those supplementary assumptions have been given in various forms, but in all cases their effect is to justify the conclusion that the photon pairs which are detected constitute a statistically representative sample of the whole ensemble of photons emitted by the source. [In Sec. 20.3 such an assumption allowed us to pass from (20.8) to (20.9).] If such an assumption is granted, then there are many experiments that confirm the predictions of quantum mechanics, such as (20.23), to a high degree of accuracy, and those predictions imply a violation of the Bell and CH inequalities. The need to invoke a supplementary assumption arises only because of technical limitations, not for any reason of principle, and there is every reason to expect that these limitations will be overcome in new experiments.

It has been proposed (Lo and Shimony, 1981) that one should study the spin correlations of an atomic pair that results from the dissociation of a diatomic molecule in a metastable singlet state. The efficiency of counting atoms approaches $\eta = 1$, but other technical difficulties make this a difficult experiment to realize.

Parametric down conversion of photons is an alternative source of correlated photon pairs. This is a nonlinear optical phenomenon in which a photon of frequency ω and wave vector $\mathbf{k}$ is converted into a pair of photons whose frequencies and wave vectors satisfy $\omega_1 + \omega_2 = \omega$ and $\mathbf{k}_1 + \mathbf{k}_2 = \mathbf{k}$. This method is superior to atomic cascade decay, in that the relative directions of the photons in the pair are better correlated, and the emission times of the photons are governed by the pumping of the nonlinear crystal, whereas the emission time is random for cascade decays. Shin and Alley (1988) have used this technique to demonstrate the violation of Bell's inequalities (subject, of course, to a supplementary assumption about the detection process).

It should be emphasized that, technical problems notwithstanding, there is no doubt that the ideal results of quantum mechanics are incompatible with the simple assumptions used to derive Bell's theorem.

20.6 *Bell's Theorem Without Probabilities*

In the preceding sections, Bell's theorem has been obtained as inequalities restricting correlations or probabilities, and the violation of these inequalities by quantum mechanics indicates that the set of assumptions (including Einstein's *locality* principle) used to derive the inequalities are not compatible with quantum mechanics. It seems peculiar that such a deep and fundamental conclusion should be accessible only through detailed quantitative calculations. In Fig. 20.4, for example, Bell's theorem follows from the fact that $S(\theta)$ exceeds 1 by a few percent over a rather small range of θ. In the simple, symmetric configurations for which the correlations can be deduced by qualitative reasoning, such as θ equal to π or $\pi/2$, the quantum-mechanical predictions obey the inequalities. The contradication can only be obtained through a fully quantitative calculation. Thus neither the theory nor the experiments provide any qualitative understanding of why quantum mechanics violates the intuitively reasonable principles that lead to Bell's theorem.

An important step forward was made by Greenberger, Horne, and Zeilinger, who were able to derive Bell's theorem without using probabilities or inequalities. This new method also connects Bell's theorem with another theorem, also proved by Bell, but usually attributed to Kochen and Specker. The derivation presented here closely follows that of Mermin (1993). A proof of the Kochen–Specker (KS) theorem will be given first, because it generalizes directly into the new proof of Bell's theorem.

The KS theorem

The KS theorem arises in the following context. It is well known that quantum theory does not predict the result of an individual measurement of a dynamical variable. An exception occurs if the state is an eigenstate of the dynamical variable being measured, in which case the corresponding eigenvalue is the uniquely predicted result. But, in general, only the probabilities of the various possibilities are predicted. Now, in the general case (not an eigenstate), *may we think of the dynamical variable as having a definite (but unknown) value before it is measured*? That is, given any quantum state of a system with dynamical variables $A, B, C, \ldots$, can we assign numerical values $v(A), v(B), v(C), \ldots$ to these observables? (The next stage of this program would be to construct a statistical ensemble of these hypothetical values that agrees with the quantum-mechanical probabilities, but we shall not be concerned with the probabilistic aspects here.)

The answer to the question in italics will be trivially affirmative if no conditions are imposed on the valuation function $v(A)$. But there are some conditions that clearly should be applied, the first being:

(i) The value $v(A)$ should be an eigenvalue of the operator A.

The second condition preserves some of the functional relations that hold among dynamical variables:

(ii) If a set of *mutually commuting* observables satisfies a functional relation of the form $f(A, B, C, \ldots) = 0$, then the same relation should be satisfied by the values, $f(v(A), v(B), v(C), \ldots) = 0$.

It is important that this condition be imposed only on commuting observables, otherwise trivial contradictions would arise. Consider, for example, the relation $\sigma_u = (\sigma_x + \sigma_y)\sqrt{\frac{1}{2}}$, where σ_u is the component of spin along an axis at an angle of 45° to the x and y axes. The eigenvalues of σ_x, σ_y, and σ_u are ± 1, and so cannot satisfy the inappropriate condition $v(\sigma_u) = [v(\sigma_x) + v(\sigma_y)]\sqrt{\frac{1}{2}}$. But if the operators are mutually commutative, then they possess a complete set of common eigenvectors, and their eigenvalues satisfy condition (ii). The content of assumption (ii) is then to extend an identity among eigenvalues, which holds for measured values in eigenstates, to the conjectured valuation functions in arbitrary states. I shall not present the arguments for and against the physical plausibility of this condition, since the KS theorem is being proved only as a precursor for the new proof of Bell's theorem.

The proof involves three independent spins of magnitude $\frac{1}{2}$, and makes use of the elementary properties of the Pauli spin operators, Eqs. (7.46) and (7.47). The ten observables that we need are shown in Fig. 20.5.

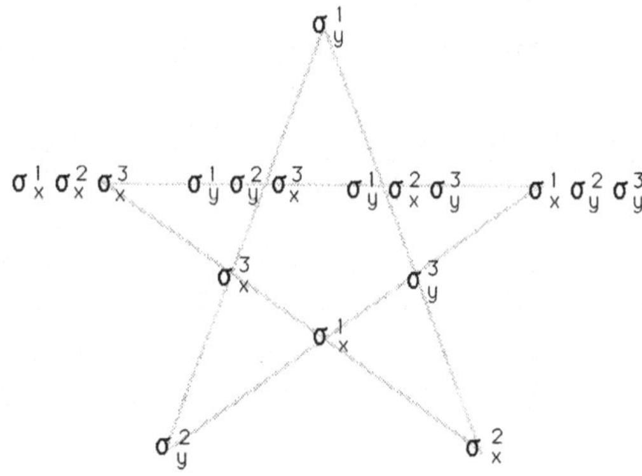

Fig. 20.5 These observables, lying on the sides of a pentagram, provide a proof of the KS theorem. The four on each line are mutually commutative. The product of the four on a line is 1, except for the horizontal line, where the product is -1.

The steps in the proof are as follows:

(a) The four operators on each of the five lines are mutually commutative.

This is obvious for all but the horizontal line, since the spin operators on different particles commute. For an interchange of any pair of the observables on the horizontal line, it follows from the anticommutation of different spin components on the same particle, and the fact that the interchange will involve an even number of such anticommutations.

(b) The product of the four operators on each line is 1, except for the horizontal line, where it is -1. These facts are easily verified.

(c) Since the values assigned to mutually commuting observables must obey the functional relations satisfied by the operators, it follows that the values assigned to the observables must obey (b). The product of the four values on a line must be 1, except for the horizontal line, where it must be -1.

(d) From (c) it follows that the product of the values over all five lines must be -1.

(e) But each observable will appear exactly twice in this product, since each lies on the intersection of two lines, and so the value of the product in (d) must be $+1$, not -1.

This is a contradiction. Therefore the assumed valuation satisfying (i) and (ii) must be impossible. This concludes the proof of the KS theorem.

The consequence of the KS theorem is often expressed by saying that the values of quantum-mechanical observables are *contextual*. This means that, in a given situation, the value of one of the observables will depend on what commuting set is being measured along with it. To have supposed, prior to the KS theorem, that the values of quantum-mechanical observables were noncontextual, may have been plausible. But such a belief would have been based only on a hope for simplicity, and not on any compelling argument. This situation changes when we move on to Bell's theorem.

Bell's theorem — new proof

The KS theorem may be summarized by saying that the assumption of noncontextual values for commutative quantum-mechanical observables leads to a contradiction. In the new proof of Bell's theorem, no such assumption is made. Instead, Einstein's locality principle is used to derive a limited degree of noncontextuality.

We now interpret the three vector operators, $\boldsymbol{\sigma}^i$, whose components are used in Fig. 20.5, as the spins of three distantly separated spin $\frac{1}{2}$ particles. Let us temporarily ignore the four observables on the horizontal line of Fig. 20.5. Each of the remaining six is a local observable on one of the particles. They belong to four sets (the nonhorizontal lines), each set consisting of one observable for each of the three particles. It can now be argued, invoking Einstein locality, that a local measurement on one of the particles cannot affect the value of an observable on either of the other two distant particles. Now each of these single particle observables belongs to two commutative sets (intersecting nonhorizontal lines) that differ in the choice of the observables belonging to the two distant particles. The locality principle asserts that the value of the first observable cannot depend on which spin component is or is not measured on the other distant particles. Hence we conclude that these six single particle observables should possess noncontextual values, this degree of noncontextuality being deduced from locality.

No such conclusion can be deduced for the four observables on the horizontal line, since they are not local observables on any of the particles. But since they are commutative, they possess a set of common eigenvectors. We therefore consider a special state: one of the eigenstates of these four operators. Each observable on the horizontal line now has a definite value, namely its eigenvalue. We next pick one of the nonhorizontal lines, and independently measure a local observable on each of the three particles, the product of the three measured values necessarily being equal to the eigenvalue of the nonlocal operator on that line. So far, we have done nothing but orthodox quantum theory. The next step is to invoke the locality argument given above, which says that the value obtained in a single local measurement at one particle cannot depend on which other distant particles are also measured. We thus obtain a unique, noncontextual value for each of ten observables in Fig. 20.5. These values will satisfy step (c) in the KS theorem: the product of the values on a nonhorizontal line will be 1, and the product of the eigenvalues on the horizontal line will be -1. We then proceed directly to steps (d) and (e), and derive the contradiction.

However, in this case, the noncontextuality of the measurement values was not assumed, but rather derived from locality. Therefore the contradiction is between quantum mechanics and the locality principle. Since the locality principle has strong physical motivation (unlike noncontextuality), the conclusion of Bell's theorem is much more impressive and surprising than is the conclusion of the KS theorem.

[[The first proof of Bell's theorem without probabilities or inequalities was presented at a conference by Greenberger, Horne, and Zeilinger, with a full exposition being published by Greenberger, Horne, Shimony, and Zeilinger (1990). It uses a four-particle system. The simpler proof given here, involving only three particles, is due to Mermin (1993), who also reviews the history of the problem. Hardy (1992, 1993) has devised a proof that uses only two spin $\frac{1}{2}$ particles. It is elaborated by Mermin (1994).

The KS theorem requires a state vector space of at least three dimensions, a system of spin 1 being the smallest for which the concept of *contextuality* can be formulated. (For spin $\frac{1}{2}$, there are no independent operators that commute with a spin operator, so the "context" of mutually commuting operators is trivial.) The theorem was proved by Bell (1966), who presented it as a version of a theorem by Gleason (who, however, had not developed its physical significance). Independently, Kochen and Specker (1967) provided the most celebrated proof, involving spin components in 117 distinct spatial directions. This *tour de force* reigned for over two decades before simpler proofs were found. Peres (1991) found an elegant proof using only 33 directions; J. Conway and S. Kochen (unpublished) have achieved 31. The use of a larger system allows a simpler proof. With a system of two spin $\frac{1}{2}$ particles (four-dimensional state vector space), Mermin (1993) gives an elegant proof using only nine observables.]]

20.7 *Implications of Bell's Theorem*

If the results of the preceding sections are taken at face value, they seem to imply that quantum mechanics is incompatible with Einstein's principle of locality: "*The real factual situation of the system S_2 is independent of what is done with the system S_1, which is spatially separated from the former.*" For the experimental setup shown in Fig. 20.2, this principle is implemented by the assumption that the result of a measurement performed with the instrument on the left is not affected by the setting of the instrument on the right, and vice versa. This seems to be a very reasonable assumption, since the instruments could be very far apart. The setting of the two instruments could, in principle, be carried out at a spacelike separation, so that within the measurement time it would be impossible for information about the setting of one instrument to be transmitted to the other instrument at any speed not exceeding the speed of light. So to deny the locality assumption would be a very extreme measure. Yet that assumption leads to Bell's inequality, which is violated by certain predictions of quantum theory. Such a radical conclusion cannot be accepted without the most rigorous scrutiny.

Is the contradiction due to some hypothesis other than locality?

Although we said above that Bell's theorem follows from the *locality* assumption, it is seldom the case that a theorem depends on only one hypothesis. So our conclusion should have been that *at least one* of the hypotheses used in the derivation of Bell's theorem is violated by quantum mechanics. Many assumptions, other than locality, that seem to be implicit in Bell's original argument have been identified, but in every case it has been possible to deduce a contradiction of quantum mechanics without that assumption. Some example are:

Determinism. In Sec. 20.3 it was assumed that the *result* of a measurement by one instrument was *determined* by the setting of that instrument and some uncontrolled parameters (denoted as λ). In Sec. 20.4 that assumption of a hidden determinism was relinquished, and we derived the theorem assuming only that the *probability* of the result, but not the result itself, was determined by the setting of the instrument and the uncontrolled parameters λ. Therefore determinism cannot be the cause of the contradiction.

Probability factorization. There has been some debate as to whether the factorization of the probability in (20.13) is justified by locality alone, or whether it requires some additional, stronger assumptions. If so, the contradiction might be blamed on those additional assumptions. The issue here is subtle, but fortunately it is now irrelevant, since the new proof in Sec. 20.6 does not make use of probability.

Counterfactual definiteness. This term (abbreviated to CFD) is used to describe the character of statements about "what would have occurred" in a measurement that we could have performed, but did not actually perform. CFD occurs in the EPR argument when they assert that if we had measured the position of particle #1 we could have learned the position x_2 of particle #2, and if we had measured the momentum of particle #1 we could have learned the momentum p_2 of particle #2. Although only one of these measurements can actually be carried out in a single case, the conclusion that both of the values x_2 and p_2 are well defined in nature is an instance of CFD. EPR do not assume CFD; rather, they deduce it from *locality* and some results taken from quantum mechanics.

In the proof of Bell's theorem in Sec. 20.6, CFD would imply that the measured values of the various local observables are not contextual. But, here also, no assumption of this form was needed, since the necessary degree of noncontextuality was deduced from locality.

H. P. Stapp (1985, 1988) has devoted considerable effort to reducing the number of assumptions needed to derive Bell's theorem. He dispenses entirely with the hidden parameters λ of Secs. 20.3 and 20.4, and works only with sequences of possible measurement results that are statistically compatible with quantum mechanics. From this he argues that quantum mechanics contradicts the *locality* postulate alone, without any assumption of hidden variables.

Are the experiments conclusive?

If we accept the theoretical arguments that quantum mechanics is incompatible with locality, the next question is whether the experiments are adequate for ruling out locality. We have already seen that, strictly speaking, they are not, because of inefficiencies of the detectors and other instrumental problems. However, the fact that those photon pairs that are detected are correlated in the manner predicted by quantum theory is certainly strong evidence for the correctness of those predictions. Although it is possible to devise local models that would obey Bell's inequality for ideal detectors, but which agree with quantum theory for the imperfect instruments presently available, such models seem rather contrived. This is especially true in view of the fact that the effect of the various systematic errors that experimentalists have studied is to *reduce* the coincidence detection rate. But quantum theory predicts a coincidence rate that is *greater* than is permitted by Bell's inequality.

Another concern is that most of the experiments were not carried out under one of the conditions specified by the locality postulate: that the settings of the two instruments be adjusted, and the two measurements carried out, in spacelike separated regions of space–time, so that it would be impossible for any light speed signal to "inform" one instrument about the setting of the other. It is under such conditions that the assumptions used to derive Bell's theorem are most compelling. To answer this objection Aspect, Dalibard, and Roger (1982) have carried out an experiment in which the instruments in Fig. 20.2 are rapidly switched between two polarizer orientations: $\mathbf{a}$ and $\mathbf{a}'$ on the right, and $\mathbf{b}$ and $\mathbf{b}'$ on the left. The switchings on the two sides are performed by two independent oscillators running at incommensurate frequenices, and presumably with independent phase drifts. The lifetime of the intermediate energy level in the cascade (5×10^{-9} sec) and the switching time between polarizers (10×10^{-9} sec) were both smaller than the time for a light signal to pass from one instrument to the other (40×10^{-9} sec). The polarization correlations were found to be the same as in experiments with static settings of the analyzers, and to agree with quantum theory.

Is quantum mechanics incompatible with relativity?

Einstein's *locality* postulate, which is the key to Bell's theorem, is strongly motivated by special relativity. Thus the conflict between quantum mechanics and locality suggests a deep incompatibility between quantum mechanics and relativity. We are now entering an area of uncertainty, and hence of controversy.

It is not valid to object that we have based our analysis on nonrelativistic quantum mechanics. In fact, only the properties of spin and polarization have been used, and these are essentially identical in both the relativistic and nonrelativistic theories.

The conclusion of Bell's theorem has, unfortunately, sometimes been summed up by saying that *local realistic* models/theories cannot agree with all of the predictions of quantum mechanics. The term "local realism" should be depricated because it is seriously misleading. It suggests that we have a particular choice of how to deal with the violation of Bell's inequalities: either give up *locality* and keep *realism*, or keep *locality* and give up *realism*. But a *local non-realistic* theory is an empty set, and probably a contradiction of terms. Without some degree of *realism* the distinction between *local* and *nonlocal* has no meaning. Users of the depricated phrase "local realism" seldom define what they mean by "realism". In response to this obscurity, Norsen (2007) has examined several versions of *realism* that are common in the philosophy of science, and has argued persuasively that giving up one or more of them would not be effective in saving *locality* in the face of Bell's theorem.

If quantum mechanics implies *nonlocality* – influences between distant regions that are not restricted by the speed of light – can we make use of them to send messages at superluminal speeds? No! Several people have shown that quantum correlations cannot be used to transmit messages at superluminal speeds. This has been analysed in detail by Shimony (1986), and independently by Ballentine and Jarrett (1987). Their analyses center on the factorization in Eq. (20.13) and its role in preventing superluminal signaling. However, critiques of the factorization in (20.13) do not affect the newer arguments in Sec. (20.6), which do not depend on probabilities or inequalities.

It is doubtful that the requirements of special relativity are exhausted by excluding superluminal signals. But it is not clear how there could be superluminal "influences" – so as to violate Bell's inequality and satisfy quantum mechanics – that in principle cannot be used as signals. [A hint is provided by Bohm's *quantum potential*, (Secs. 14.2 and 14.3), which depends on the entire arrangement of the apparatus, and so would in principle depend on the settings of both polarizers.] Whether or not there is a deep incompatibility between quantum mechanics and relativity remains unclear, but it is apparent that the concepts which seem natural in one theory may not seem natural in the other. It is truly remarkable that such deep questions should have arisen from the simple sinusoidal correlations between spins and between polarizations!

Further reading for Chapter 20

All of J. S. Bell's papers on this subject are reprinted in a single volume (Bell, 1987). The review by Clauser and Shimony (1978) is an excellent account of both theory and experiment up to that date. The first experimental test of the Bell inequality was by Freedman and Clauser (1972); more comprehensive experiments are reported by Aspect *et al.* (1981, 1982). Rarity and Tapster (1990) have tested a version of Bell's theorem that does not involve spin or polarization. The Resource Letter by Ballentine (1987) contains an annotated bibliography of papers on this subject, some of which are contained in the reprint book (Ballentine, 1989). Wheeler and Zurek (1983) also reprint some of the early papers on this subject.

Problems

20.1 Show that Herbert's inequality (20.3) is a special case of Bell's inequality (20.6).

20.2 For two $s = \frac{1}{2}$ particles in a singlet state, evaluate the left hand side of the CH inequality (20.16) for the configuration shown in Fig. 20.1. Evaluate it also for the configuration shown in Fig. 20.3. For which configuration does the greatest violation of the inequality occur? Can you devise a configuration for which the violation would be even greater?

20.3 Consider the two-photon state vector $|\mathbf{x}_1, \mathbf{k}\rangle \otimes |\mathbf{x}_2, -\mathbf{k}\rangle$, which is essentially the first term of (20.22). Although this state lacks rotational invariance about the direction of **k**, the correlation of the polarizations in this state bears some similarity to the correlation in the state (20.22). Evaluate the coincidence probability $[P_{12}(\mathbf{a}, \mathbf{b})]_{qm}$ for this state, and show that it obeys the CH inequality (20.16). Note that it must depend on the directions of both **a** and **b**, and not only on their relative angle.) For what orientations of the analyzers would the coincidence probability for this state be the same as that for the state (20.22)?

20.4 For a system of two particles with spin $s = \frac{1}{2}$, the vector $|\uparrow\downarrow\rangle_u \equiv |\uparrow\rangle_u \otimes |\downarrow\rangle_u$ describes a state in which the component of spin in the direction of the unit vector $\hat{\mathbf{u}}$ is positive for the first particle and negative for the second particle. Thus the total spin in the direction $\hat{\mathbf{u}}$ is zero. This state is not rotationally invariant, but we can construct from it a rotationally invariant state operator (not a pure state) by averaging over all directions, $\rho = (4\pi)^{-1} \int |\uparrow\downarrow\rangle_u \langle\uparrow\downarrow|_u \, d\Omega_u$. Calculate the spin correlation function $C(\theta_{ab}) = \mathrm{Tr}(\rho\, \sigma_a \otimes \sigma_b)$ for this state, and compare with the corresponding function for the singlet state, (20.2). Verify that for this state, unlike the singlet state, Bell's inequality is obeyed.

20.5 A classical body whose total angular momentum is zero breaks up into two fragments: fragment #1 carries angular momentum $\mathbf{J}$ and fragment #2 carries angular momentum $-\mathbf{J}$. The direction of the classical angular momentum vector $\mathbf{J}$ is not reproducible, and is described by a probability distribution $\rho(\mathbf{J})$. The two fragments separate. On the first we measure the sign of the component of its angular momentum in the direction $\mathbf{a}$, and on the second we measure the sign of its angular momentum component in the direction $\mathbf{b}$. The two results may be denoted as $\sigma_a = \text{sgn}(\mathbf{J}\cdot\mathbf{a})$ and $\sigma_b = \text{sgn}(-\mathbf{J}\cdot\mathbf{b})$, and their possible values are $+1$ and -1. Calculate the correlation function $\langle\sigma_a\sigma_b\rangle$ as a function of the angle between the directions of $\mathbf{a}$ and $\mathbf{b}$ if the probability distribution $\rho(\mathbf{J})$ is isotropic. Does it obey the Bell inequality?

Chapter 21

Quantum Information

21.1 *Quantum States as Carriers of Information*

Throughout this book we have stressed the role of the *quantum state* as a generator of *probabilities*. Since *information* is closely related to probability, it is a short step to consider the quantum state as a carrier of information. Quantum information theory is a relatively new, but rapidly developing, branch of quantum theory. It is quite impossible to give a full treatment of the subject in one chapter; indeed, a whole book would be needed for that task. The emphasis in this chapter will be on the basic principles, including the relationship of quantum information theory to the foundations of quantum mechanics. Some of the most important applications will be treated, but reference to the current research literature is necessary to keep up with the rapid progress in this area.

Bits and Qubits

The simplest example of an information carrier is a physical variable that can take on either of two states, usually labeled as 0 and 1. The information carried by this binary variable is called a *bit*. The word "bit" is commonly used to refer either to the physical variable that stores the information, or to the abstract information itself, with the appropriate interpretation usually being clear from the context.

An information storage device consisting of N independent 2-state registers can store N bits of information. Since it has $W = 2^N$ distinct states (equivalently, W possible stored messages), the maximum information content of the device is $\log_2(W)$ bits.

This maximum information content is realized if all of the possible states are equally probable. In a more general situation, the amount of information is determined by a theory due to Shannon (1948), expounded more fully by Shannon and Weaver (1949). Suppose the probability of occurrence of the ith state is p_i. The *entropy* of that situation is defined as

$$H = -\sum_{i=1}^{W} p_i \log_2(p_i)\,. \tag{21.1}$$

This quantity, called the *Shannon entropy*[h] or *information entropy*, is a measure of the uncertainty about which of the possible message states may occur. If a subsequent measurement reduces that uncertainty, thereby making some possible states more probable and others less probable, then the final entropy will be smaller than the initial entropy. The *information* given by that measurement is defined to be the reduction in the Shannon entropy,

$$I = H_{initial} - H_{final}\,. \tag{21.2}$$

If the measurement reveals a unique message state, then its probability becomes one, and p_i is zero for all other states. Then the final entropy is $H_{final} = 0$, and the information given by that measurement is

$$I = -\sum_{i=1}^{W} p_i \log_2(p_i)\,. \tag{21.3}$$

(That the same formula, (21.1) or (21.3), may represent either entropy or information, is a perennial source of confusion to novices!) The formula (21.3) should be interpreted as the *average* information, averaged over all possible messages. The arrival of an unexpected, improbable (small p_i) message carries a large amount of information ($\log_2(1/p_i)$ bits), but the probability of that happening is small. Conversely, if one particular message was a near certainty ($p_i \approx 1$) then the information received is nearly zero.

Just as we can formulate an abstract theory of *numbers* without regard to any actual physical things which the numbers may count, so we can formulate an abstract theory of *information* without regard to the particular physical devices that store the information. It is this generality and device-independence that gives information theory its power. Nevertheless, it is wise to remember the words of R. Landauer (1992), that "information is physical." This is not to deny that information theory is abstract, like mathematics. But it is a reminder that the material world is not made out of information; rather, the information is carried by the structures of material objects, and information is physically

[h]There is a story that Shannon asked von Neumann what to call this quantity, and vN replied, *Why not call it entropy? No one knows what entropy is anyway.* Whether historical or not, this choice inevitably gave rise to the question, *Is the information entropy related to the thermodynamic entropy?* The answer is a qualified *Yes*, but the arguments and counter arguments lie outside the scope of this chapter.

relevant only if there is a physical method of coupling to it. This is especially relevant when the material objects are essentially quantum mechanical.

A *bit* of information can be stored in any 2-level quantum system, whose basis states may be denoted as Hilbert space vectors $|0\rangle$ and $|1\rangle$. But in addition to these two basis vectors, the superposition principle allows a continuum of pure states, of the form

$$|\psi\rangle = c_0|0\rangle + c_1|1\rangle \tag{21.4}$$

where the complex coefficients satisfy $|c_0|^2+|c_1|^2 = 1$. All 2-dimensional Hilbert spaces are isomorphic, and so it is convenient to think of (21.4) as a state of a spin-$\frac{1}{2}$ system. The most general spin-$\frac{1}{2}$ state operator was given in Eq. (7.50) as

$$\varrho = \frac{1}{2}(\mathbf{1} + \mathbf{a} \cdot \boldsymbol{\sigma})\,. \tag{21.5}$$

If we identify the information-basis vectors $|0\rangle$ and $|1\rangle$ with the spin-up and spin-down eigenvectors of σ_z, then the σ-matrices will have the same appearance here as in angular momentum theory. The *pure* states of the form (21.5) are those for which the "polarization" vector $\mathbf{a}$ has unit length. This geometrical representation of the state space is called the *Bloch sphere.*

The pure states of the spin correspond to the 4π steradians of solid angle over which the direction of the vector $\mathbf{a}$ can range. If the smallest solid angle that can be resolved is $\delta\Omega$, then the number of distinguishable states of our 2-level quantum system is $W = 4\pi/\delta\Omega$, and so the latent information stored in one of its pure states, $I = \log_2(W)$, would appear to be potentially very large. Unfortunately, this large latent information is not accessible because (as we shall see in Section 21.2) it is not possible to determine an unknown quantum state by measurements on a single copy of that state. All that we can do is to measure the component of $\boldsymbol{\sigma}$ in any *one* chosen direction, which will yield exactly one bit (0 or 1) of information. Nevertheless, that additional latent information is stored in the quantum state, and the possibility of its manipulation is responsible for the unique applications of quantum information.

The above-mentioned formula, estimating the bit content of a quantum state from the resolution of solid angle, is not very useful because there is no way to directly couple to the "polarization" vector $\mathbf{a}$. It is much more useful to treat the 2-dimensional Hilbert space, spanned by the basis vectors $|0\rangle$ and $|1\rangle$, as an entity called a *qubit.* The name derives from a contraction of "quantum bit", regarded as an extension from the two discrete states, $|0\rangle$ and $|1\rangle$, which form a *bit*, to include all of their linear combinations. As was the case for the term *bit*, the term *qubit* can have two interpretations. Sometimes it refers to the physical object that supports a 2-dimensional Hilbert space of pure states, and other times it refers to the abstract information contained in the quantum state. The choice between these two interpretations is usual clear from the context.

Just as a *bit* may be realized in many different material devices, so there are also many physical realizations of a *qubit.* The spin analogy leads to an obvious choice: the spin of an atom in a trap, whose states can be manipulated by appropriately designed fields. Another important realization is provided by the polarization states of a photon. If the vertical and horizontal linear polarizations are taken to be the analogs of the eigenvectors of σ_z, then the linear polarizations at $+45$ and -45 degrees from the horizontal are the analogs of the eigenvectors of σ_x, and the left and right circular polarizations are the analogs of the eigenvectors of σ_y. Indeed, a qubit can, in principle, be realized by any pair of discrete energy levels between which transitions can be induced in a controlled, reproducible manner, provided that transitions to other states are negligible. However, it is essential that not only the relative amplitudes of the two terms in (21.4) be controlled, but also their relative phase. The phase can be particularly sensitive to small external perturbations, a loss of phase stability being called *decoherence.* In practice, the choice of a particular physical implementation of a qubit may be governed by the technical problem of minimizing decoherence.

Classical vs Quantum Information

The terms *classical information* and *quantum information* are commonly used to refer to the storage, processing, and transmission of information by *bits* and by *qubits*, respectively. This is a natural terminology. Nevertheless, it can give rise to misconceptions. It suggests a dichotomy between "classical" and "quantum" parts of the world. But this misrepresents the relationship between classical and quantum mechanics. It is not the case that the two theories are mutually exclusive (although some people argue to the contrary). Rather quantum mechanics is the more general and fundamental theory, from which classical mechanics emerges as a special limiting case. The nature of that limit (see Ch. 14) is quite subtle, and has not yet been fully worked out in detail, but there are no counter examples to the claim that classical mechanics emerges as a limit of quantum mechanics. And the relationship between classical and quantum information is not such a counter example.

Many books contain the misleading statement that the Shannon entropy (21.1) is "classical", and that for a quantum state ρ it must be replaced by the von Neumann entropy, $-\text{Tr}(\rho \log \rho)$. We shall see, however, that the Shannon entropy plays a ubiquitous role in quantum information. It is defined for any probability distribution, of any observable, and it measures the maximum amount of information that may be gained by measuring that observable. The vN entropy is, in fact, just the Shannon entropy calculated in the representation that diagonalizes the state operator ρ. The vN entropy plays an important role in many theorems, however most of them are beyond the scope of this chapter.

If, indeed, quantum mechanics is the fundamental theory, then *all* information-processing devices are quantum mechanical in their nature. If only the states $|0\rangle$ and $|1\rangle$ (or some other set of orthogonal states) are used to represent and store information, then we may say that we are dealing with *classical* information, the quantum mechanical nature of the device notwithstanding. If, on the other hand, superposition states (21.4) are used to store information, then we may say that we are dealing with *quantum* information. *Information theory* should become one subject, embracing the storage and processing of information by qubits as well as by the more limited method of bits.

21.2 *Some Quantum Information Theorems*

Although *quantum information theory* could be regarded as merely the application of quantum mechanics to devices that store, process, or transmit information, it has developed into a distinct sub-field of quantum theory. Its distinctive character derives from a body of information-related theorems, which build on each other, and which depend on only a few basic principles from the more general quantum mechanics. The relation of quantum information theory to quantum mechanics is somewhat analogous to the relation of thermodynamics to (classical or quantum) mechanics. Every result of thermodynamics could, in principle, be derived as a (highly probable) consequence of mechanics. But the derivation of a particular result as being, say, a consequence of the Second Law of Thermodynamics, is much more satisfying, since it avoids a lot of mechanical details which, apparently, are not necessary for the result to hold. Within its domain, quantum information theory presents similar advantages. Another similarity with thermodynamics is that several of the quantum-information theorems show that certain kinds of devices or processes are impossible.

Theorem 21.1 (No-cloning theorem). *It is impossible for any device to take an arbitrary (unknown) quantum state as input and produce that state and a copy of it as output.*

This theorem has already been proved in Section 8.1, where two different arguments are given. The first proof involved only the *linearity* of quantum equation of motion for state vectors. We wish our hypothetical device to take an input of the form $|\phi\rangle \otimes |\omega\rangle$ and produce an output of the form $|\phi\rangle \otimes |\phi\rangle$. Here $|\phi\rangle$ is the unknown state that we wish to copy, and $|\omega\rangle$ is an empty holder to receive the copy. (We use here a slightly different notation from that of Section 8.1, where the holder $|\omega\rangle$ was taken to be part of the apparatus, and was not explicitly indicated.) Now suppose that our device is successful in copying two different states, $|\phi_1\rangle$ and $|\phi_2\rangle$, so that

$$|\phi_1\rangle \otimes |\omega\rangle \to |\phi_1\rangle \otimes |\phi_1\rangle\,, \tag{21.6}$$

$$|\phi_2\rangle \otimes |\omega\rangle \to |\phi_2\rangle \otimes |\phi_2\rangle\,. \tag{21.7}$$

Suppose that we next try to copy the superposition state $(|\phi_1\rangle + |\phi_2\rangle)$. The two equations above, combined with *linearity*, imply

$$(|\phi_1\rangle + |\phi_2\rangle) \otimes |\omega\rangle \to |\phi_1\rangle \otimes |\phi_1\rangle + |\phi_2\rangle \otimes |\phi_2\rangle\,. \tag{21.8}$$

But this is not the desired result of copying the superposition state, which would be $(|\phi_1\rangle + |\phi_2\rangle) \otimes (|\phi_1\rangle + |\phi_2\rangle)$, therefore the supposed state-copying device has failed to copy faithfully.

The second proof invokes *unitarity*, which implies that the inner product of two state vectors must be constant if they both evolve under the same unitary time-evolution operator. Take the inner product of the left-hand sides of equations (21.6) and (21.7), and equate it to the inner product of the right-hand sides. This yields a contradiction unless $\langle\phi_1|\phi_2\rangle$ is either 0 or 1. A consequence of this argument is that if the input state vector $|\phi\rangle$ is not completely arbitrary, but rather is an arbitrary member of a particular orthogonal set of vectors, then copying of that partially unknown state is possible. This is the case for copying *classical* states, such as copying a computer file. So it is apparent that the *superposition principle* of quantum theory is responsible for the *no-cloning theorem.*

Theorem 21.2. *It is impossible to determine the unknown quantum state of a single individual system, by any measurement or sequence of measurements.*

The significance of this theorem needs some explanation. Section 8.2 is devoted to the topic of *state determination*, where it was shown how to determine an unknown state. But in that chapter, a *state* was identified with a reproducible procedure for its preparation, which therefore defined the *state* as referring to an *ensemble* of similarly prepared systems. So, for example, if you have a state-preparation apparatus for which the instruction manual is missing, Section 8.2 tells you how, in principle, you might go about determining what state it prepares. It would require many measurements on many copies of the state, so as to determine the statistical distributions of sufficiently many (often non-commuting) observables to reconstruct the state operator. But for this theorem, you have only one copy of the state – only one system that has been prepared in that unknown state.

The proof of the theorem is as follows. Suppose, contrary to the theorem, it was possible to determine a complete description of the previously unknown state of a single system. One could then use that information to prepare an arbitrarily large number of copies of that state. (Section 8.1 discusses how, in principle, to prepare a *known* state. Here "in principle" means without regard to the practical availability of the fields or interactions required.) But the possibility of copying an arbitrary unknown state is forbidden by the *no-cloning theorem*, therefore the determination of an unknown state by measurements on a single copy of it must be impossible.

The simplicity of this proof illustrates the strength of information theoretic methods. A direct proof of this theorem from the formalism of quantum mechanics would first require an abstract characterization of the most general possible measurements. A nontrivial argument would be needed to justify any particular abstract characterization. Second, it would require a mathematical proof that no such generalized measurement could achieve its goal. Even then, doubts might remain that some completely different measurement strategy, outside of the one considered in the abstract characterization, might yet succeed. Contrast that with the simplicity of the quantum-information theoretic proof just given!

The next theorem concerns the distinguishability of quantum states. The eigenvalues of an observable can be distinguished by measurement, and this is equivalent to distinguishing between the orthogonal eigenstates of the observable. Thus it is possible, in principle, to devise a filter that will separate out a collection of mutually orthogonal states. The Stern-Gerlach apparatus (Section 9.1) does this for spin-up and spin-down states. But what if we want to separate *non-orthogonal* states? The next theorem asserts that to be impossible.

Theorem 21.3. *It is impossible for any device to unambiguously distinguish, and reliably separate, non-orthogonal states.*

Suppose, hypothetically, that we have an instrument that can separate two states, $|\phi_1\rangle$ and $|\phi_2\rangle$, which may or may not be orthogonal. Then these two states must become correlated with two distinguishable final states of the instrument. (This is very similar to the characterization of a *measurement* in Chapter 9.) The effect of the instrument must be

$$|\phi_1\rangle \otimes |\chi_0\rangle \to |\phi_1\rangle \otimes |\chi_1\rangle \tag{21.9}$$

$$|\phi_2\rangle \otimes |\chi_0\rangle \to |\phi_2\rangle \otimes |\chi_2\rangle \tag{21.10}$$

where $|\chi_0\rangle$ is the initial state of the instrument, and $|\chi_1\rangle$ and $|\chi_2\rangle$ are its final states. For the device to achieve its intended purpose, it is essential that its two final states must be *distinguishable*, so we require that they be orthogonal, $\langle\chi_1|\chi_2\rangle = 0$. The passage from initial to final state is carried out by a unitary time-evolution operator, which preserves the values of inner products between state vectors, so we equate the inner product of the left-hand sides of the two equations above to the inner product of the right-hand sides. We obtain $\langle\phi_1|\phi_2\rangle = 0$, because the right-hand sides are orthogonal. Therefore our device is capable of distinguishing between *orthogonal* states, but distinguishing between *non-orthogonal* states is impossible.

Because the device is macroscopic, it is unrealistic to assume, as we did for simplicity, that the preparation of its initial state is so precise as to always yield exactly the same pure state $|\chi_0\rangle$. It would be better to describe the initial state of the instrument as a mixed state operator, ρ_0, and the corresponding final states as ρ_1 and ρ_2. But to ensure the distinguishability of the final states, we must again require them to be orthogonal, $\text{Tr}\rho_1\rho_2 = 0$ being the orthogonality condition for state operators. So, with very slight modifications, the argument reaches the same conclusion: that $|\phi_1\rangle$ and $|\phi_2\rangle$ can be distinguished only if they are orthogonal.

These theorems reveal fundamental differences between *qubit*-based quantum information and *bit*-based classical information. But because the conclusions of the theorems are negative, one might wonder whether quantum information is of any practical value. We shall soon see that it does, indeed, have value.

21.3 *Quantum Transmission of Information*

Any process of transmitting information from the sender to the receiver (with or without processing it in between) is commonly referred to as a *channel*. There is a growing mathematical theory of quantum information channels, much of which is beyond the scope of this chapter. But the basic concepts that define a quantum information channel are the familiar concepts of *state preparation* and *measurement*, with state preparation playing the role of the sender and measurement playing the role of the detector of information. The impossibility theorems of the preceding section might seem to suggest that quantum communication would be technically inferior to the well-developed methods of classical communication. But those impossibilities can be used to our advantage in protecting the security of information from intruders and spies. This gives rise to the technique of *quantum cryptography*, which will be the subject of the next section. But first we need to discuss some more elementary aspects of quantum information.

Relevant Kinds of Probability

Section 1.5 contains a brief discussion of various interpretations of probability. Here the proper role of *interpretation* is to say what the abstract notion of *probability* means in a particular application or context. It should not, but unfortunately often does, lead to the formation of mutually hostile schools of thought, each championing its own favorite interpretation. In the application of quantum information theory, there may be more than one kind (or *interpretation*) of probability that is relevant, and we need to be aware of the differences.

One of the most fundamental aspects of quantum mechanics is *statistical causality*, which is a form of causality weaker than determinism, to which the name *propensity* has been attached. The preparation of a state has a causal influence on the results of future measurements, but does not determine the individual outcome. The state determines only the probability, or more precisely, the *propensity* for obtaining a particular outcome. If the preparation is repeatable, you can repeat the preparation-measurement sequence as many times as you wish, thereby generating an *ensemble* of results. The *frequencies* of various outcomes in this ensemble can then be analyzed by statistical techniques. The *Law of Large Numbers*, Eq. (1.58), establishes the close connection between *propensity* and *frequency* (denoted p and f_n in (1.58)).

A typical *quantum probability* has the form

$$P(D|\rho) = \mathrm{Tr}(E_\omega \rho)\,. \tag{21.11}$$

Here ρ is the state operator (or *density matrix*), and E_ω is a projector onto a subset ω of the spectrum of the dynamical variable that is to be measured. On the left-hand side, $P(D|\rho)$ is the probability (propensity) to obtain the result

$D = (x \in \omega)$ (meaning that the measured value of x lies in ω), conditional on the state ρ having been prepared. For example, if we choose a pure state $\rho = |\psi\rangle\langle\psi|$, choose ω to be the interval $(x, x+\delta)$ of the position variable, and define E_ω to yield $E_\omega\psi(x) = \psi(x)$ for $x \in \omega$, $E_\omega\psi(x) = 0$ for x not in ω, we obtain the familiar expression for the probability density, $\lim_{\delta\to 0} P(D|\rho)/\delta = |\psi(x)|^2$. This typifies the usual role of probability in quantum mechanics. One begins with the preparation of the state, and then calculates the probabilities for the possible results of measurements. This specifically *quantum* probability is often referred to as *irreducible* because the quantum state provides all the information that is relevant to predicting the results of subsequent measurements.

Now let us consider a different kind of problem. Suppose we begin with some specific measurement result D, and we wish to infer the unknown state that produced it. This is clearly relevant to the transmission of quantum information, since the message is encoded in the prepared quantum state, and the receiver would like to determine what the state was in order to read the message. Theorem 21.2 asserts that if the state is entirely unknown, then it is not possible to determine it by any kind of measurement. But the situation is not hopeless because it is possible, nevertheless, to deduce *information* about the unknown state. Instead of the quantum probability $P(D|\rho)$, given by (21.11), we need the inverse probability, $P(\rho|D)$, which is the probability that the unknown state was ρ, given the data D that was obtained by the measurement. This may be calculated from *Bayes' theorem* (Eq. (1.51)), which, when applied to the current problem, yields

$$P(\rho|D\&C) = P(D|\rho\&C)\frac{P(\rho|C)}{P(D|C)}\,. \tag{21.12}$$

Here the symbol C denotes any other information that may be relevant.

The differences between the probabilities (21.11) and (21.12) should be noted. In (21.11) we have a particular quantum state ρ, which has a (non-deterministic) causal influence on the subsequent measurement results, and the repeatability of the state preparation leads to the generation of an ensemble of measurement results. But this *propensity/ensemble* interpretation is not applicable to (21.12). Here we have a particular measurement result, but there is only one state – it just happens to be unknown. Clearly the data D has no causal influence on the preparation of that state, so the propensity/ensemble interpretation is not applicable. By using (21.12), we are engaged in *inductive inference*, and (21.12) may be called an *inferential* probability. It usually does not allow us to determine the unknown state with certainty, but it allows us to infer that some candidates are more likely than others.

When we consider *cryptography* we will be concerned, not only to transmit information from the sender to the receiver, but also do deny that information to others. This will lead us to use yet another kind of probability, the so-called *subjective* or *personal* probability that describes a particular agent's *degree of belief.* If (21.12) is used to describe an agent's degree of belief about the identity of the unknown state ρ, the answer may differ from one agent to another because the prior information C that each possesses may be different. That the fundamental *quantum* probabilities (21.11) are *not* subjective may be seen from the fact that the quantum probability for a particular measurement result depends only on the preparation of the state ρ, which is an objective physical process, and not merely someone's belief.

Information Transmission – Examples

As a simple example of information transmission, consider a single qubit with basis vectors $|0\rangle$ and $|1\rangle$. From these two, we construct another pair of orthogonal vectors, $|+\rangle = (|0\rangle + |1\rangle)/\sqrt{2}$ and $|-\rangle = (|0\rangle - |1\rangle)/\sqrt{2}$. To make the example more concrete, we could take the physical carrier of the qubit to be the polarization states of a single photon, with $|0\rangle$ corresponding to vertical linear polarization and $|1\rangle$ corresponding to horizontal polarization. Then the pair $|+\rangle$ and $|-\rangle$ will correspond to orthogonal polarizations at angles $+45$ and -45 degrees from the horizontal.

Now we construct a machine that can prepare each of these states, $\{|0\rangle, |1\rangle, |+\rangle, |-\rangle\}$, and can be programmed to produce them in a sequence, which may be random, or may encode a message.

Example 1:

For the first example, the machine is programmed to produce a random sequence of the nonorthogonal states $|1\rangle$ and $|+\rangle$, with equal probability. We give this machine, as a "black box" state-preparation device, to a tester who knows only that it prepares some unknown state, and who is asked to determine what state it produces. Since the preparation is repeatable, he can generate an ensemble of photons, and by measuring the averages of various polarizations he can accurately estimate the density matrix. (From the spin representation of a general qubit state operator (21.5), it is apparent that measuring the averages $\langle\sigma_x\rangle$, $\langle\sigma_y\rangle$, and $\langle\sigma_z\rangle$ will be sufficient to determine the state.)

The programming of the machine will yield the particular mixed state

$$\rho = \frac{1}{2}\left(|1\rangle\langle 1| + |+\rangle\langle +|\right). \tag{21.13}$$

Expressed as a density matrix in the basis $\{|0\rangle, |1\rangle\}$, it is

$$\rho = \begin{pmatrix} 0.25 & 0.25 \\ 0.25 & 0.75 \end{pmatrix}. \tag{21.14}$$

Our tester will be able to determine, within statistical uncertainties, that the density matrix of the mixed state which this "black box" produces is (21.14). But any mixed state operator can be expressed as a mixture of pure states in infinitely many ways, and our tester will not be able to discover that this particular machine prepares the particular mixture described by the right-hand side of (21.13). This is so because all quantum probabilities are determined solely by the state operator, as in (21.11), and do not depend on the particular mixture that may have been used to form that state operator. So if our machine was sending a coded message, rather than a random sequence of $|1\rangle$ or $|+\rangle$, our tester could never discover it.

Example 2:

For the second example, the same machine, programmed to produce a random sequence of $|1\rangle$ and $|+\rangle$, is given to a second tester. But she is given the additional information that the outputs will be either $|1\rangle$ or $|+\rangle$, with equal probability, and she is given a new task. Can she identify in which events $|1\rangle$ was prepared, and in which $|+\rangle$ was prepared? If she can do this, then she would be able to discover any message that might be substituted for the nominally "random" sequence.

Theorem 21.3 asserts that it is impossible to distinguish between non-orthogonal states *unambiguously*. However, it is possible to perform measurements that provide some information about the identity of the individual states, $|1\rangle$ or $|+\rangle$. This can be done by a filtering measurement that unambiguously separates $|1\rangle$ from $|0\rangle$. (Alternatively, a similar filtering of $|+\rangle$ from $|-\rangle$ is possible, but since the experiment is fully symmetric with respect to the two choices of basis, only the first choice need be considered.)

The information that the measurement results will provide about the identity of the individual prepared states can be determined from Bayes' theorem (21.12), which we rewrite in a notation suited to this particular problem.

$$P(S_i|D\&C) = P(D|S_i\&C)\frac{P(S_i|C)}{P(D|C)}. \tag{21.15}$$

Here C denotes the new information (available to the second tester, but not to the first), that the machine prepares either of two possible states, with equal probability. We denote them as $S_1 = |1\rangle$ and $S_2 = |+\rangle$. The data D is the result of the measurement, either $D = 0$ or $D = 1$.

The factor $P(S_i|C)$ is the *prior probability* that the prepared state was S_i. Here the term *prior* means before the measurement that yields the data D. It can be obtained from the information C, which describes the operation of the preparation machine, to be

$$P(S_1|C) = P(S_2|C) = 0.5\,. \tag{21.16}$$

Similarly, the denominator $P(D|C)$ in (21.15) is the *prior probability* that the data D would be obtained, based only on the information C about the machine, but before performing the measurement. It can be obtained in two ways. Since the sum of the probabilities in (21.15) over the two possible states S_1 and S_2 must be one, it follows that

$$P(D|C) = \sum_{i=1}^{2} P(D|S_i \& C) P(S_i|C)\,. \tag{21.17}$$

Alternatively, we can use the density matrix (21.14), which was obtained from the information C, and evaluate $P(D|C)$ directly from it. Whence

$$P(D = 0|C) = 0.25\,, \quad P(D = 1|C) = 0.75\,. \tag{21.18}$$

The remaining factor on the right-hand side of (21.15) is the ordinary quantum probability for the possible results of a measurement performed on a given state S_i. It simplifies to $P(D|S_i \& C) = P(D|S_i)$ because the information C, that the unknown state is either $|1\rangle$ or $|+\rangle$, is redundant when the particular state is given. A simple calculation yields the values of $P(D|S_i)$ to be

$$\begin{aligned} P(0|S_1) &= 0\,, \quad P(0|S_2) = 0.5 \\ P(1|S_1) &= 1\,, \quad P(1|S_2) = 0.5\,. \end{aligned} \tag{21.19}$$

Finally, the probabilities $P(S_i|D\&C)$, that the prepared state was $|1\rangle$ or $|+\rangle$, conditioned on the measurement result $D = 0$ or $D = 1$, are

	$S_1 = \|1\rangle$	$S_2 = \|+\rangle$
$D = 0$	0	1
$D = 1$	$\frac{2}{3}$	$\frac{1}{3}$

(21.20)

Using these results, the tester should predict the prepared state to be $|+\rangle$ whenever she observes $D = 0$, and $|1\rangle$ whenever $D = 1$. She will be correct every time when $D = 0$, and 2/3 of the time when $D = 1$. From (21.18) we know that $D = 0$ will occur 25% of the time and $D = 1$ will occur 75% of the time, in a long run of the test, so her overall success rate will be 75%. Theorem 21.3 guarantees that no testing strategy can raise the success rate to 100%.

Note that in the case of $D = 0$, where the tester's success rate was 100%, what the measurement result actually told her is that the state was *not* $|1\rangle$. If she had not known that there were only two possible states (if, for example, she had suspected that there were three or four possibilities), then she could not have been certain that the state was $|+\rangle$ when $D = 0$. She could infer that it must be $|+\rangle$, only because there was no other possibility.

Finally, we can use (21.1) and (21.2) to calculate the amount of information that is given by the measurement. Although these equations use base-2 logarithms, it is easier to use natural logarithms to compute the entropies, and then divide by log(2) to convert the result into *bits*. Since the initially unknown state is one of two possibilities, with equal probability, the initial entropy is $H_{initial} = \log(2)$, which is 1 bit. When the measurement result is $D = 0$, the state must be $|+\rangle$, and the final entropy is $H_{D=0} = 0$. Hence the amount of information provided by this result is $I_{D=0} = H_{initial} - H_{D=0} = 1$ bit. When the result is $D = 1$, the final entropy is $H_{D=1} = -\left(\frac{2}{3}\log(\frac{2}{3}) + \frac{1}{3}\log(\frac{1}{3})\right)/\log(2) = 0.91830$ bits. Hence the information in this case is $I_{D=1} = 0.08170$ bits. Since, in the long run, the result $D = 0$ occurs 25% of the time and $D = 1$ occurs 75% of the time, the average amount of information per measurement will be $I_{av} = 0.25 \times 1 \text{ bit} + 0.75 \times 0.08170 \text{ bit} = 0.31128$ bits. Although this is considerably less than the 1 bit of information that would be needed to identify the unknown state unambiguously, it still provides a success rate that is much better than chance.

21.4 *Cryptography*

The essence of cryptography is that every character may be represented by some other character, so that the encrypted message is unreadable to anyone who does not have the decryption key. In order to show the advantage of *quantum* cryptography, we first examine two cases of ordinary cryptography, to see what difficulties they encounter.

Example 1: A Static Encryption Key

The first example is based on a real case that occurred in the Second World War, although this account of it is simplified. The German radio transmissions were encoded so that every character was represented by a different character.

The key for encryption and decryption was almost random, and was changed frequently. Nevertheless, the British were able to decode the German transmissions. One weakness lay in the fact that the encryption key was only *almost* random. There was one constraint – no character was ever allowed to represent itself. With a fully *random* key, a character would sometimes, purely by chance, be represented by itself (about once every 27 times, for a 27 character alphabet, counting "space" as a character). One might think that this constraint would make the encrypted message *harder* to read, but in fact the opposite is true.

Consider two random strings, each N characters long. If they are compared character by character, then by chance alone there might be a position where the corresponding characters would match. The probability that there would be *no matches* is $(26/27)^N$, which becomes small if N is large enough. Now the German messages often contained certain stock phrases near the beginning or end, "HEIL HITLER" being the most common. The probability that two random strings of 11 characters would have no matches is $(26/27)^{11} = 0.66$. The British decoders could begin by searching, near the beginning or end of the message, for an 11 character string having *no matches* with a corresponding character in the string "HEIL HITLER". Since the probability of a non-match occurring by chance is 0.66, it follows that they had 1 chance in 3 of getting a very easy decoding for seven characters (E,H,I,L,R,T,space). Other conventional decoding techniques enabled them to complete the decoding of the message.

Example 2: A Dynamic Encryption Key

The encryption scheme in the previous example has another weakness. Its key is static – the mapping from message characters into code characters is the same throughout the message. If one instance of the message character A was represented by the code character M, then all instances of A throughout the message would be represented by M, and all instances of M in the code would represent A. The message would be much harder to decrypt if the key were dynamic, that is, if it were to change in an apparently random manner as the message progressed.

This can be conveniently and effectively implemented at the binary level, where the message is a string of bits (0 or 1). First we generate a random key having the same length as the message, and then form the cryptogram by adding (modulo 2) the key bits to the corresponding message bits. The effect is to leave the message bit unchanged if the key bit is 0, and flip the message bit if the key bit is 1. Addition of the key to the cryptogram will recover the original message. In this scheme, two instances of the same character may be represented differently in the cryptogram. For example, in the standard 7-bit ASCII character set, the letter A is represented by the binary string 1000001. After encryption with a random key, two instances of A may be represented by two quite different binary strings. It can be proven that if the key is random (meaning, it has no detectable patterns), and if it is used only once, then the cryptogram is unbreakable, in the sense that no decrypting strategy is any more likely to reveal the message than is a lucky guess.

In spite of its theoretically unbreakable character, this scheme has an obvious practical limitation. The secure communication of the key from the sender to the recipient is just as difficult as would be the secure communication of the original plain text message! In practice, the potential senders and recipients must carry with them a prearranged key, which would have to be used repeatedly. The reuse of the same key imposes a pattern on the cryptograms, which can, in principle, be detected by statistical methods. Under repeated use the key ceases to be unbreakable.

Quantum Cryptography

The special features of quantum information can be used to securely communicate the random one-time-use encryption key that was needed in the preceding example. The method is due to Bennett and Brassard (1984).

Suppose that Alice and Bob[i] wish to communicate securely, using the polarization of photons as their medium of communication. The polarization states of a photon form a qubit, and any two orthogonal polarizations can represent a bit. Alice and Bob represent the bits of their key using either of the two orthogonal bases that were discussed in Section 21.3. When using the vertical/horizontal polarizations, the states $|0\rangle$ and $|1\rangle$ represent the values 0 and 1, respectively. When using the ± 45 degree polarizations, the states $|+\rangle$ and $|-\rangle$ represent the values 0 and 1. To create and send the secret key, Alice chooses randomly between these two bases, and sends a random sequence of 0's and 1's. Bob changes the orientation of his polarization filter randomly between the two bases, and measures the bit values of the photons. Both Alice and Bob must record their sequences of polarizer orientations and the bit values that they send or receive. After a sufficiently long sequence of bits has been transmitted, Alice and Bob reveal (through a public, unsecure channel) their

[i]It has become a convention of the quantum information community to call the communicating parties Alice and Bob. At the Perimeter Institute in Waterloo, Ontario, Canada, there are lecture rooms called the "Alice Room" and the "Bob Room". It's an in-joke.

sequences of polarizer orientations (but do not reveal their bit sequences). By chance alone, in about half of the cases their orientations will coincide and so Bob's measured bit will agree with Alice's transmitted bit. Those bits form their secret key. The transmission, coding and decoding of the message then follows as in the preceding example.

Suppose that an eavesdropper (named – you guessed it – Eve) has intercepted their transmission of the key. Eve intends to measure the polarization of each photon, and then send a photon of the same polarization to Bob so that he is none the wiser. But since Alice's choice of polarizer orientations is unknown to Eve, all she can do is guess and hope she is right. Only the cases in which it later turns out that Alice and Bob chose the same polarizer orientations will be used. By chance, Eve will also have made the same choice in about half of those cases, and for that half she will know the bits of the key. But consider, on the other hand, what happens when Alice and Bob both chose the vertical/horizontal orientation, while Eve chose the ± 45 degree orientation. Then Eve's measured bit value will be equally likely to be 0 or 1, uncorrelated to the actual value sent by Alice. Eve will then send a wrongly polarized state to Bob, and his measured bit value will be uncorrelated with the true value sent by Alice. As a result, when Bob receives the coded message and decodes it with his corrupted key, it will make no sense. He can then warn Alice that their key has been corrupted, and Eve's spying will be discovered.

This analysis has, for simplicity, assumed a perfectly noiseless transmission. But the communication protocol can be generalized to allow for noise, and the essential features are preserved. Subject only to statistical uncertainties, the presence of an eavesdropper can be detected, and the security of the message can be certified.

21.5 *Entanglement*

We have seen how distinctive and useful features of quantum information arise from the *superposition principle*, through the forming of linear combinations of the (classically discrete) bit states, $|0\rangle$ and $|1\rangle$. Even more unusual and powerful features arise from *entanglement*, which is an application of the superposition principle to systems of two or more components. The prototypical example of an entangled state is the *singlet* state of two spin-$\frac{1}{2}$ particles, or equivalently a pair of qubits,

$$\frac{1}{\sqrt{2}}\big(|\uparrow\rangle|\downarrow\rangle - |\downarrow\rangle|\uparrow\rangle\big) \equiv \frac{1}{\sqrt{2}}\big(|0\rangle|1\rangle - |1\rangle|0\rangle\big)\,. \tag{21.21}$$

(The spin-state and bit-state notations may be used interchangeably. We will use the former when intuition about spin is helpful.) This state gives no information about the orientation of a single spin (or the value of a single bit), but it tells us that the orientations of the two spins (and the values of the two bits) are opposite. The singlet state was the first example of a quantum state that violates the Bell inequalities (Section 20.3), which implies that the correlations in the quantum state (21.21) are stronger than would be possible in any classical theory.

Definition of Entanglement

In generalizing the notion of entanglement beyond the prototype (21.21), we need to do two things: provide a mathematical criterion for entanglement, and identify the physical phenomenon that this mathematics is supposed to describe. Both of these tasks encounter difficulties.

The definition that has been adopted is based on the mathematical notion of a *separable* state. A separable state is either the tensor product of state operators for each component, $\rho^a \otimes \rho^b$, (superscripts ^{a} and ^{b} label components), or a mixture of such product states. Hence a general *separable* state operator for a two-component system can be written in the form

$$\rho^{ab} = \sum_i c_i\, \rho_i^a \otimes \rho_i^b \tag{21.22}$$

where $0 < c_i \leq 1$, and $\sum_i c_i = 1$. An *entangled* state is then defined as one that cannot be expressed in this separable form, and a *non-entangled* state is identified as a separable state.

Identifying Entanglement

The definition of a non-entangled state requires only that it *can be* written in the form (21.22). But a mixed state can be written as a mixture of other states in infinitely many ways. A state may be represented in a form that does not look like (21.22), but to be sure that it is entangled we must *prove* that no representation of it in the form (21.22) is possible. It would be desirable to have an operational criterion that can identify entanglement.

To investigate this problem, let us take ρ^{ab} to be an arbitrary state (not necessarily of the form (21.22)) of a two-component system, and let $\rho^a = \mathrm{Tr}^b\, \rho^{ab}$ and $\rho^b = \mathrm{Tr}^a\, \rho^{ab}$ be the partial states of the two components. In Section 8.3 the full state and its partial states were classified by the categories: pure, non-pure (mixed), correlated, and uncorrelated (factored).

Pure States:

For a pure state, $\rho^{ab} = |\Psi^{ab}\rangle\langle\Psi^{ab}|$, we find in Table 8.2 that there are only two possibilities:

(i) Both ρ^a and ρ^b are pure states, and $\rho^{ab} = \rho^a \otimes \rho^b$. The state vector also has a factored form, $|\Psi^{ab}\rangle = |\psi^a\rangle \otimes |\phi^b\rangle$.
(ii) ρ^{ab} and $|\Psi^{ab}\rangle$ are not factorable, and neither ρ^a nor ρ^b is pure.

In case (i) the two-component state is *non-entangled*, while in case (ii) it is *entangled*. For pure states, the concepts of *correlation* and *entanglement* are equivalent. All pure correlated states are also entangled, and vice versa.

A practical test is as follows. It is assumed that ρ^{ab} is a non-negative Hermitian operator with unit trace, these being the conditions required for any state operator (Section 2.3). Then a necessary and sufficient condition for ρ^{ab} to be a pure state is that $\text{Tr}\big((\rho^{ab})^2\big) = 1$. If in addition, we find that $\text{Tr}\big((\rho^{a})^2\big) = 1$ then we have case (i), and the state is not entangled. If, on the other hand, we find that $\text{Tr}\big((\rho^{a})^2\big) < 1$ then we have case (ii), and the state is entangled.

It should be emphasized that *entanglement* is only defined relative to a particular decomposition of the system into components. Consider the ground state of the hydrogen atom. The Schrödinger equation can be separated in relative and center-of-mass variables, and in terms of those "components" the ground state is a factored, non-entangled state. But in terms of the electron and proton coordinates, the ground state is entangled. In quantum information theory, it is assumed that there is a natural separation of the system into components. It may be the spins of separate particles. It may be the polarizations of separate photons. It may be different qubits that serve as registers in a quantum computer.

Mixed States:

For mixed states, the concepts of *correlation* and *entanglement* are distinct. Since an uncorrelated state is factorable, all entangled states are correlated, by definition. But a non-entangled state can exhibit correlations between the components when its separable form (21.22) has at least two terms.

A useful criterion, introduced by Peres (1996), is based on the notion of a *partial transpose* of the density matrix of a two-component state. Choose an orthonormal set of product basis vectors, $\{|m\rangle^a \otimes |\mu\rangle^b\}$. The matrix elements of the state operator ρ will have the form $\rho_{m\mu,n\nu}$ where the latin indices refer to the first component and the greek indices refer to the second component.

The partial transpose of this density matrix with respect to its first component yields another matrix,

$$\sigma_{m\mu,n\nu} = \rho_{n\mu,m\nu} \,. \tag{21.23}$$

(Note that only the latin indices, m and n, have been transposed.) Peres then asked whether the new matrix $\sigma_{m\mu,n\nu}$ satisfies the conditions required of a density matrix. It is apparent that it has unit trace, and is Hermitian, like $\rho_{m\mu,n\nu}$. But it is not clear whether it is non-negative (meaning $\langle u|\sigma|u\rangle \geq 0$ for all vectors $|u\rangle$).

To see the effect of partial transposition on a *separable* density matrix we write out (21.22) as a matrix,

$$\rho^{ab}_{m\mu,n\nu} = \sum_i c_i \, (\rho^a_i)_{mn} \, (\rho^b_i)_{\mu\nu} \,. \tag{21.24}$$

The effect, in each term of the sum, is to replace the matrix $(\rho^a_i)_{mn}$ with its transpose (which, for a Hermitian matrix, is its complex conjugate), which has the same (non-negative) eigenvalues as the original matrix. Thus (21.24), being the sum of non-negative matrices, is itself a non-negative matrix. Therefore, we conclude that *for a density matrix to be separable, it is necessary that its partial transpose be non-negative.*

It was later proven (Horodecki 1996) that this condition is both *necessary and sufficient* if the Hilbert spaces of the two components of the system have dimensions $2 \otimes 2$ or $2 \otimes 3$. However the condition is not sufficient for higher dimensional spaces, in which there exist entangled states whose density matrices have non-negative partial transposes. So we still lack an operational method for identifying all entangled states, but the Peres criterion suffices for the important special case of two qubits.

[[The reader should be warned that some authors carelessly use the word "positive" (which means "greater than zero") when the proper term should be "non-negative" (greater than, or equal to zero). The term *positive partial transpose* (PPT), as used by Horodecki (2001) in an otherwise excellent article, should properly be *non-negative partial transpose.* Zero eigenvalues are allowed; negative eigenvalues are not. Their equations are correct; only their words are misleading. Peres (1996) used the correct terminology.

This is a good place to rant against the abominable term *positive semi-definite*, which is all too often used in similar contexts. Let us analyze it. "Positive" means "greater than zero". "Positive definite" means the same thing, but with an added rhetorical emphasis, "I really mean >0, not ≥0." "Semi-definite" has no real meaning, but the prefix "semi" nullifies the effect of "definite". The proper term is *non-negative.* A mathematical quantity can no more be "semi-definite" than a woman can be "semi-pregnant".]]

Physical Meaning of Entanglement

The mathematical definition of an *entangled state* is one whose density matrix is non-separable. But we would also like to know what *physical* attributes or phenomena constitute the qualitative nature of *entanglement.* Many classes of states can be defined by the values of some observables. For example, consider the class of states for which the average of the observable A is positive, $\mathrm{Tr}(A\rho) > 0$. It is apparent that if two or more states have this property, then so will *all mixtures* of those states. Furthermore, a mixture of *separable* states is also separable, as is apparent from the definition.

But *entanglement* does not behave like that. A mixture of two entangled states can be non-entangled. For example, consider a mixture of the two entangled pure spin states,

$$|\Psi_1\rangle = \big(|\uparrow\rangle\otimes|\uparrow\rangle \ + |\downarrow\rangle\otimes|\downarrow\rangle\big)/\sqrt{2} \tag{21.25}$$

$$|\Psi_2\rangle = \big(|\uparrow\rangle\otimes|\uparrow\rangle \ - |\downarrow\rangle\otimes|\downarrow\rangle\big)/\sqrt{2}\,. \tag{21.26}$$

The density matrix for $\rho_1 = |\Psi_1\rangle\langle\Psi_1|$, in the product basis, is

$$\rho_1 = \begin{pmatrix} \frac{1}{2} & 0 & 0 & \frac{1}{2} \\ 0 & 0 & 0 & 0 \\ 0 & 0 & 0 & 0 \\ \frac{1}{2} & 0 & 0 & \frac{1}{2} \end{pmatrix}. \tag{21.27}$$

The density matrix for $\rho_2 = |\Psi_2\rangle\langle\Psi_2|$ is similar, but with negative values in the off-diagonal elements. An equal mixture of these two entangled states has the diagonal density matrix

$$\rho = \mathrm{diag}\Big(\frac{1}{2},\ 0,\ 0,\ \frac{1}{2}\Big) \tag{21.28}$$

which is clearly a separable (non-entangled) state. That mixing a separable state with an entangled state would diminish the entanglement is not surprising. But in the above example we see two strongly entangled states, which upon mixing become non-entangled. Apparently, the physical nature of entanglement is more subtle than that of an ordinary quantum observable.

Entanglement is related to correlation, although the two concepts are not equivalent. It is not difficult to show that all separable states (21.22) obey the

Bell inequalities [*Hint for proof*: compare the structure of (21.22) with that of (20.4)], whereas the entangled spin-singlet state (21.21) violates the Bell inequalities. Thus it is tempting to identify the physical nature of entanglement with *correlations that are stronger than is possible in a classical theory.* However there are states, known as *Werner states*, which are entangled but nevertheless obey the Bell inequalities. The simplest example is a mixture of a singlet with a completely unpolarized state,

$$\rho_W = x\,|\Psi_S\rangle\langle\Psi_S| + \frac{1-x}{4}\,\mathbf{1}\,. \tag{21.29}$$

Here $|\Psi_S\rangle$ denotes the singlet state (21.21) and $\mathbf{1}$ is the identity operator. The partial transpose of the density matrix has a negative eigenvalue for $x > \frac{1}{3}$, hence by this criterion the state is entangled for $\frac{1}{3} < x \leq 1$. However the Bell inequality holds for $x < 1/\sqrt{2}$. So for x in the range from 0.33333 to 0.70711 the state (21.29) is entangled, yet satisfies the Bell inequality.

From the purely mathematical point of view, this merely tells us that the Bell inequality is less powerful than the partial-transpose as a criterion for distinguishing between separable and non-separable states. But the physical significance is less obvious. The example of Werner states seems to contradict the conjecture that entanglement is, in essence, the phenomenon of superclassical correlation. There might, however, be more subtle correlations that are not tested by the Bell inequalities.

The Werner state (21.29) for $x = 0.5$ is nonseparable, but does not violate the Bell inequalities. However, Sandu (1994) showed that it could, nevertheless, be used to accomplish the task of quantum teleportation of information (see the next Section), although with only half the success rate that can be achieved from a pure singlet. This supports the conjecture that more than one non-classical property is contained under the broad heading of entanglement.

Sandu (1995) later generalized the concept of the $x = 0.5$ Werner state from the original $2 \otimes 2$ system to higher dimensions, studying in detail the case of $5 \otimes 5$ (realizable as a pair of spin-2 particles). For this larger system, the possibilities for correlation measurements are much richer than for the $2 \otimes 2$ system, and he found that some of the measurements were capable of a classical explanation, while others were not. It has also been suggested that for systems of dimension greater than $2 \otimes 2$ or $2 \otimes 3$ we should distinguish between *distillable entanglement* and *bound entanglement* (see Horodecki 2001), with these playing quite different roles in quantum information. It seems likely that, by defining entanglement as *non-separability*, we are including more than one distinct physical phenomenon under one label. In their review of proposed measures of entanglement, Plenio and Virmani (2007) consider measures based on several different properties and operations.

21.6 *Teleportation of Quantum States*

The word *teleportation* suggests science fiction, but nothing like that is intended here. Consider how you might send a picture to a friend who lives far away. You could put it in an envelope and send it though the postal service, marked "Fragile, Do Not Bend". Or, after unfortunate experience with that method of transportation, you could scan the picture and send the digital file electronically, and your friend would use the information in that file to recreate a copy of the picture. That is the nature of teleportation. Instead of sending the material object, you send information from which the object can be recreated.

Similarly, if we want to transport a quantum state, we could simply ship the quantum object. But it would be more convenient if we could put the object into a quantum scanner, which would encode a complete description of the state in a digital file, which could be sent to the receiver, who could use the information to create a copy of the original state. But Theorem 21.2 tells us that such a quantum scanner is impossible, and so the teleportation of an arbitrary quantum state by an ordinary information channel is impossible. Bennett *et al.* (1993) showed, however, that by supplementing the classical information channel with a pair of particles in an entangled quantum state, the teleportation of an arbitrary quantum state becomes possible.

First we need to define an entangled basis for a system of two qubits (often called the *Bell states*):

$$|\Psi_+^{ab}\rangle = \left(|0^a\rangle|1^b\rangle + |1^a\rangle|0^b\rangle\right)/\sqrt{2} \tag{21.30}$$

$$|\Psi_-^{ab}\rangle = \left(|0^a\rangle|1^b\rangle - |1^a\rangle|0^b\rangle\right)/\sqrt{2} \tag{21.31}$$

$$|\Phi_+^{ab}\rangle = \left(|0^a\rangle|0^b\rangle + |1^a\rangle|1^b\rangle\right)/\sqrt{2} \tag{21.32}$$

$$|\Phi_-^{ab}\rangle = \left(|0^a\rangle|0^b\rangle - |1^a\rangle|1^b\rangle\right)/\sqrt{2}\,. \tag{21.33}$$

These states are orthogonal and normalized. They share with the *singlet* (the second member of the set) the property of maximally violating a Bell inequality. They also have the property that any one of the set can be transformed into any other by means of a unitary transformation on just one of the component particles. (In the realization of the qubits by spins, the unitary transformation is a rotation.)

Now suppose that Alice has a particle (labeled c), whose unknown state

$$|\psi^c\rangle = u|0^c\rangle + v|1^c\rangle, \quad \left(|u|^2 + |v|^2 = 1\right), \tag{21.34}$$

she wishes to teleport to Bob. Before beginning the process, Alice and Bob must obtain an entangled pair, prepared in the singlet state $|\Psi_-^{ab}\rangle$, which they will share. Alice takes particle a and Bob takes particle b.

The initial state of the three particles is

$$\begin{aligned}|\Psi^{abc}\rangle &= |\psi^c\rangle|\Psi_-^{ab}\rangle \\ &= \frac{1}{\sqrt{2}}\left(u|0^c\rangle|0^a\rangle|1^b\rangle - u|0^c\rangle|1^a\rangle|0^b\rangle + v|1^c\rangle|0^a\rangle|1^b\rangle - v|1^c\rangle|1^a\rangle|0^b\rangle\right).\end{aligned} \tag{21.35}$$

The products of the states of particles c and a can now be expressed in terms of the Bell states of c and a, yielding

$$\begin{aligned}|\Psi^{abc}\rangle = \frac{1}{2}\big[&|\Psi_+^{ca}\rangle(-u|0^b\rangle + v|1^b\rangle) \\ &+ |\Psi_-^{ca}\rangle(-u|0^b\rangle - v|1^b\rangle) \\ &+ |\Phi_+^{ca}\rangle(-v|0^b\rangle + u|1^b\rangle) \\ &+ |\Phi_-^{ca}\rangle(+v|0^b\rangle + u|1^b\rangle)\big].\end{aligned} \tag{21.36}$$

It should be noted that although the state of particle c is not entangled with the state of the singlet pair ab (see (21.35)), the state of c and a, considered as a unit, is entangled with the state of b (21.36). This illustrates once again that the state of a system may be considered entangled, or not, depending on how the system is divided into components.

The next step is for Alice to make a joint measurement on the two particles c and a, which distinguishes among the four Bell states of the ca pair in (21.36). In other words, it must be a measurement of some dynamical variable whose eigenstates are those Bell states. It is apparent from (21.36) that the Bell states of Alice's particles ca are correlated with the states of Bob's particle b. So if the result of Alice's measurement were to correspond to Ψ_+^{ca}, then the conditional state[j] of particle b would be $(-u|0^b\rangle + v|1^b\rangle)$, and so on. Moreover, these conditional states are each related to the initial (unknown) state (21.34) of particle c by a simple unitary transformation.

Denote the state vector $|\psi^c\rangle$ by the column $\begin{pmatrix} u \\ v \end{pmatrix}^c$. In the equations below we list the results of Alice's measurement with the corresponding conditional

[j] The term "conditional state" is used in analogy with the term "conditional probability." The state of a composite system yields a joint probability distribution for any set of measurements on each of the component parts of the system. If we focus on a particular result of the measurement on one part, the (appropriately renormalized) distribution for the other part is its conditional probability distribution. The conditional state is defined similarly, and the conditional probability can be calculated from it.

state of Bob's particle, preceded by the unitary transformation that he must perform to transform it into the desired state $\begin{pmatrix} u \\ v \end{pmatrix}^b$.

$$\Psi_+^{ca} \leftrightarrow \begin{bmatrix} -1 & 0 \\ 0 & 1 \end{bmatrix} \begin{pmatrix} -u \\ v \end{pmatrix}^b, \quad \Psi_-^{ca} \leftrightarrow \begin{bmatrix} -1 & 0 \\ 0 & -1 \end{bmatrix} \begin{pmatrix} -u \\ -v \end{pmatrix}^b, \tag{21.37}$$

$$\Phi_+^{ca} \leftrightarrow \begin{bmatrix} 0 & 1 \\ -1 & 0 \end{bmatrix} \begin{pmatrix} -v \\ u \end{pmatrix}^b, \quad \Phi_-^{ca} \leftrightarrow \begin{bmatrix} 0 & 1 \\ 1 & 0 \end{bmatrix} \begin{pmatrix} v \\ u \end{pmatrix}^b. \tag{21.38}$$

(It is essential that the unitary matrices do not depend on the unknown parameters u and v.) Alice now tells Bob the result of her measurement, and he can then use the appropriate unitary transformation to prepare his particle b into the original state (21.34) of particle c. This does not violate the *No-Cloning* theorem because the state of particle c has been changed by Alice's measurement, and so only one copy of the original state $|\psi^c\rangle$ exists. It has, in effect, been teleported from Alice's particle c to Bob's particle b.

Experimental Realization of Teleportation

To actually perform the teleportation of a quantum state, using the preceding protocall, there are three kinds of tasks that must be accomplished. First, the production of entangled states; second, the design of Alice's measurement to distinguish the four entangled Bell states; and third, the performance of Bob's final unitary transformations.

The first is quite well understood. In Section 20.5 we described three sources of photon-pairs with entangled polarization states: positronium decay, 2-photon decays of an excited atom, and parametric down conversion. The latter is now the most popular technique.

The third is also quite simple. If we ignore some overall phase factors that do not affect the final states, and use spins as the qubits, then in the first, third, and fourth cases of (21.37)–(21.38) we must rotate by π about the z, y, and x axes, respectively, while the second case needs no transformation. If the qubits are polarization states of photons, the necessary optical methods for transforming polarizations are also well-known.

Distinguishing the Bell States:

The second task, distinguishing the Bell states, is the most difficult. Indeed, it has been shown by Lütkenhaus *et al.* (1999) that it cannot be accomplished using only the linear operations of quantum optics, such as beam splitters and phase shifters. Nevertheless, such a measurement is possible.

Using the spin interpretation of qubits, the four Bell states are the common eigenvectors of the three commuting operators $\sigma_x \otimes \sigma_x$, $\sigma_y \otimes \sigma_y$ and $\sigma_z \otimes \sigma_z$. (That these operators are commutative follows from the anticommutative nature of the σ operators.) The eigenvalues of these operators, for each of the Bell states, are:

	$\sigma_x \otimes \sigma_x$	$\sigma_y \otimes \sigma_y$	$\sigma_z \otimes \sigma_z$
Ψ_+	$+1$	$+1$	-1
Ψ_-	-1	-1	-1
Φ_+	$+1$	-1	$+1$
Φ_-	-1	$+1$	$+1$

(21.39)

It is sufficient to measure any two of the three operators, since they satisfy the identity: $(\sigma_x \otimes \sigma_x)(\sigma_y \otimes \sigma_y)(\sigma_z \otimes \sigma_z) = -1$.

When we are concerned about the final state after a measurement, it is usually necessary that the measurement be of the *filtering* type (also called a *projective* measurement). These names refer to a measurement that separates out the various eigenvectors of the observable, but does not change any of the eigenvectors. (The Stern-Gerlach measurement is an example.) However, the requirements for teleportation (Section 21.6) are not quite that demanding. Alice needs to determine the Bell states of her particles c and a only for the purpose of determining the conditional state of Bob's particle b. Since her measurement will not interact with Bob's particle, it is immaterial whether her measurement preserves or destroys the Bell states. All that matters is that she determine the identity of her Bell state as it was *before* her measurement.

Suppose Alice chooses to measure $\sigma_z \otimes \sigma_z$ followed by $\sigma_x \otimes \sigma_x$. The first measurement determines whether the z-components of the two spins are parallel or anti-parallel (i.e. whether the Bell state is of the Ψ type or the Φ type). It is essential that this be done *without* measuring the actual values of σ_z for either of the particles. This is so because $\sigma_z \otimes I$ and $I \otimes \sigma_z$ do not commute with $\sigma_x \otimes \sigma_x$, and so measurement of either of them would change the state in a way that would spoil the outcome of the second measurement. But no such caution is needed for the second measurement. It is then necessary only to determine whether the x-components of the two spins are parallel or anti-parallel. This may be done by a careful measurement of $\sigma_x \otimes \sigma_x$ (without determining the values of σ_x for either particle), or by simply measuring σ_x on each particle. The latter method will, of course, destroy the Bell state, but that no longer matters. Only its identity before Alice's measurements is needed. So there is no doubt that the state teleportation procedure can be done in principle, and indeed, different experimenters have realized it by techniques that differ in detail.

Bouwmeester *et al.* (1997) used photon polarization states as qubits, but they were not able to distinguish all four of the Bell states. However, they could unambiguously identify the state Ψ_-^{ca}, which occurred in 25% of the cases, and in those cases they verified that the teleportation was successful. Boschi *et al.* (1998) also used photons, but in a different manner. Instead of using three photons for the qubits a, b, and c, they used the polarizations of two photons for two of the qubits, and the path through the interferometer taken by one of the photons (i.e. its **k**-vector) as the third qubit. Their entangled pair ab consisted of one photon whose path and polarization states were entangled. They were able to resolve all four of the Bell states, and so could successfully complete the teleportation in all cases. But since the entangled pair consisted of two different degrees of freedom of the same photon, it was not possible to spatially separate the members of the pair, and teleportation over a distance is not possible by this method. More recently, Nölleke *et al.* (2013) used a pair of atomic states of two single Rb atoms, each trapped in an optical cavity, as their entangled pair. Using quantum optical techniques, they were able to resolve all four of the Bell states, and to achieve teleportation over a distance of 21 meters.

Purifying Entanglement – Distillation of Singlets

The experimental realization of teleportation of quantum states leads to the question of its possible practical application. One difficulty is immediately evident. Other things being equal, it would be much simpler to merely transport the quantum object. But quantum states are fragile, and must be protected from even weak perturbations. There is no assurance that a quantum state, so transported via a noisy channel, would arrive intact. But the teleportation procedure seems to fare no better, since it requires the separation and transport of the members of an entangled pair, and any perturbations along the way would degrade the entanglement. However, Bennett *et al.* (1996) devised a method of *purifying* a source of degraded entangled pairs into near-perfect singlets. (We will follow them in using the language of spins, but will retain the information-basis notation, with the understanding that $|0\rangle \equiv |\uparrow\rangle$ and $|1\rangle \equiv |\downarrow\rangle$.)

Suppose that Alice and Bob have a plentiful source of singlet pairs, but as a result of noisy perturbations in transit, the particles arrive in a degraded mixed state. This fact would be revealed if they were to measure the spins of some pairs along the same axis, and were to discover that in some cases the spins were parallel (instead of anti-parallel, as expected for a singlet). The state of the degraded pairs can be written as a mixture of the original singlet and a noise term. If the effect of the noise is statistically isotropic, then the resultant state will be rotationally invariant, and can be written in the form of a Werner state (21.29). (Bennett *et al.* (1996) suggest that as a first step, Alice and Bob should subject the unknown noisy state to random rotations, thereby averaging out any anisotropies, and effectively converting the unknown state into a Werner state.)

The objective of the procedure is to increase the *singlet fraction* of the noisy mixed state ρ, defined as

$$F_S = \langle \Psi_-^{ab} | \rho | \Psi_-^{ab} \rangle \,. \tag{21.40}$$

For the Werner state, ρ_W (21.29), the singlet fraction is $F_S = (3x+1)/4$. (That it is not equal to x can be seen from the fact that the identity operator in (21.29) is equal to a sum of projection operators onto the singlet and the three triplet states of the pair of spins.) Bennett *et al.* describe three purification procedures, and compare their effectiveness. The procedures consist of Alice and Bob taking *two* pairs of particles, performing certain transformations on them, and measuring the spins on both members of one pair. If the test is favorable, the other pair is saved. If not, it is discarded. It is demonstrated that if the singlet fraction of the initial state is $F_S > 0.5$, then the subensemble of saved pairs will have a singlet fraction $F_S' > F_S$, and moreover, that iteration of the procedure can yield a singlet fraction that approaches one. The cost is the discarding of most of the original pairs. But all of the operations can be carried out locally by Alice and Bob, each on their own halves of a pair, with the results of their measurements being communicated classically to each other to decide mutually whether the test was a success. Thus, even if Alice and Bob are far separated, they can *distill* good singlets from the degraded pairs whose halves they receive, and so can reliably teleport quantum states, provided only that the singlet fraction of their noisy ensemble of pairs is not less than $F_S > 0.5$.

[[After the possibility of quantum state teleportation – transporting the information rather than the object – had been demonstrated, one of its inventors, Asher Peres, was asked by a reporter whether teleportation could transport the soul as well as the body. He relied, laconically, "No, not the body, just the soul." (Reported in his obituary, Physics Today, August 2005, pp. 65–66.)]]

21.7 *Quantum Information from Independent Pairs*

The predictions of quantum mechanics are statistical, and the standard scheme for testing them is given by the procedure: *state preparation – measurement – repeat until enough statistical data has been collected.* Since measurement usually changes the state, only one measurement can be done on a single copy of the state. It is therefore necessary to have a repeatable procedure for preparing the state, and hence to be able to create an ensemble of samples that are all prepared in that same state.

However, there is no reason why an individual sample must be "used up" by measurement before the next sample is prepared. So instead of having an ensemble of individual samples of the state ρ, we could equally well prepare an ensemble of pairs in the state $\rho \otimes \rho$, or an ensemble of triplets in the state $\rho \otimes \rho \otimes \rho$, etc. But would there be any advantage in doing that? Since there is no interaction or correlation between the members of a pair, or a triplet, one might expect the answer to be *No.*

But suppose that state preparation is expensive, and we can afford only a small number of samples of the state ρ. It would then be important to use those samples in the most efficient way, so as to obtain the maximum amount of information. It turns out that testing them as pairs may be more effective than testing them as individuals.

We have already seen (Example 2, Section 21.3) how information about an unknown state can be obtained from a measurement on a single particle. Peres and Wootters (1991) carried out a similar study for a pair of non-interacting particles, prepared independently in the same state. Two measurements can now be made, one on each particle. Surprisingly, it turns out that more information can be obtained from a joint measurement on the pair than by independent measurements on each particle separately.

We consider two spin-$\frac{1}{2}$ particles, both prepared in the same state. However, there are three possible directions for the polarization: vertical along the z-axis, or in the x-z plane tilted at ± 120 degrees from the z-axis. The three possible state vectors for the pair will be denoted as

$$|a\rangle = |\uparrow\rangle \otimes |\uparrow\rangle \tag{21.41}$$

$$|b\rangle = |\swarrow\rangle \otimes |\swarrow\rangle \tag{21.42}$$

$$|c\rangle = |\searrow\rangle \otimes |\searrow\rangle\,. \tag{21.43}$$

The set of possible states has 3-fold rotational symmetry in the x-z plane. The initial entropy is $H_0 = \log_2(3) = 1.58496$ bits, which is the minimum amount of information required to unambiguously identify the state.

Before considering the quantum measurements, it is instructive to consider a *classical analog*, consisting of a pair of classical spins that are always parallel, but which may be oriented in any one of the three directions above. The classical states will be denoted as $(\uparrow\uparrow)$, $(\swarrow\swarrow)$, and $(\searrow\searrow)$. Since classical measurements have none of the restrictions that apply to quantum measurements, we must impose some rules to make the comparison fair. The rules are: (i) only one measurement per particle; (ii) one component of spin in any chosen direction may be measured; (iii) only the sign (not the magnitude) of that component will be determined. With these rules, a classical measurement will yield a maximum of 1 bit of information, as is the case for a quantum measurement on a spin-$\frac{1}{2}$ particle. We begin by measuring the z-component of the first spin. If it is positive, the state must be $(\uparrow\uparrow)$, and we are finished. If it is negative the state $(\uparrow\uparrow)$ is excluded. We then measure the x-component of the second spin, which will distinguish between $(\swarrow\swarrow)$ and $(\searrow\searrow)$. So in the classical case we always succeed in identifying the prepared state.

Now let us return to the quantum system. We begin by measuring the spin of the first particle along the z-axis. If the result is negative, then the state $|a\rangle$ is excluded, as in the classical analog. A positive result is evidence for state $|a\rangle$, since that result is more likely to have been obtained from $|a\rangle$ than from $|b\rangle$ or $|c\rangle$, but all three states remain possibilities. We then have a choice. We can measure the z-component of the second spin. A positive result will strengthen the evidence for $|a\rangle$, or a negative result will exclude it. But neither result will discriminate between $|b\rangle$ and $|c\rangle$. Alternatively, we can measure the x-component of the second spin, which will yield information discriminating between $|b\rangle$ and $|c\rangle$, but will tell us nothing more about $|a\rangle$.

Peres and Wootters found that the greatest information gain was obtained as follows. Measure the z-component of the first spin. If the result is positive, measure the z-component of the second spin. If the result is negative, measure the x-component of the second spin. These two alternatives yield different amounts of total information from the two measurements. Averaging that information over all possible outcomes of the measurements (as was done in Example 2, Section 21.3) yields an average of 1.05228 bits.

However, a better result can be obtained from a single joint measurement on the particles, treated as a pair. Notice that the state $|a\rangle$ is a member of the *triplet*, which has total spin 1. The same is true for the states $|b\rangle$ and $|c\rangle$, since they are obtained from $|a\rangle$ by rotation. So these three linearly independent vectors span the 3-dimensional subspace of the spin-triplet. Inconveniently, they are not orthogonal, but three orthogonal vectors can be constructed.

Define

$$|A\rangle = \frac{1}{9}[54 + 8(18)^{1/2}]^{1/2}|a\rangle - \frac{1}{9}[18 - 4(18)^{1/2}]^{1/2}(|b\rangle + |c\rangle)\,. \qquad (21.44)$$

Similarly, define $|B\rangle$ and $|C\rangle$ by cyclic permutation. The three vectors $|A\rangle$, $|B\rangle$, and $|C\rangle$ are orthogonal and normalized, and so form a basis for the triplet-state subspace. The fourth basis vector in the Hilbert space of the 2-particle system is the singlet, which is orthogonal to the triplet subspace.

We now wish to measure, on the composite system, a dynamical variable whose eigenstates are $|A\rangle$, $|B\rangle$, and $|C\rangle$. One such variable is

$$|B\rangle\langle B| - |C\rangle\langle C| = \left(\frac{2}{3}\right)^{1/2} J_x - \left(\frac{1}{3}\right)^{1/2} (J_x J_z + J_z J_x)\,. \qquad (21.45)$$

Here $J_x = \frac{1}{2}(\sigma_x \otimes I + I \otimes \sigma_x)$ and $J_z = \frac{1}{2}(\sigma_z \otimes I + I \otimes \sigma_z)$ are components of the total spin of the 2-particle system. These operators will be represented by 4×4 matrices, but in our chosen basis they will consist of a 3×3 block and a 1×1 block, for the triplet and singlet states, respectively. But since all of our possible prepared states are in the triplet subspace, the singlet block is irrelevant, and we may work, effectively, with the 3×3 matrices from the triplet block. Indeed, it is as if we were dealing with an effective spin-1 system, upon which we might hope, in principle,[k] to perform a generalized Stern-Gerlach measurement. The information gain from this measurement, averaged over all three possible results, is 1.36907 bits. This is considerably greater than was obtained from separate measurements on each of the spin-$\frac{1}{2}$ particles.

That a single joint measurement on the pair yields more information than do separate measurements on each particle would not be surprising if the pair were prepared in a correlated state. But for this to happen when the two particles are prepared independently is very surprising. Peres and Wootters suggest that this situation can be regarded as the converse of Bell's theorem. In Bell's situation the particles are prepared in an entangled (non-factorable) state, and the measurements are carried out on each of the particles separately. Here the state is factorized, but the measurement is of a non-factorable observable (21.45).

21.8 *Measurable "in Principle"*

The basic postulates of quantum mechanics assert that observables are represented by Hermitian operators and pure states are represented by Hilbert-

[k]But see the next section about "in principle" measurability.

space vectors. But the converse propositions – that *all* Hermitian operators represent measurable quantities, and that *all* vectors describe realizable states – do not follow immediately from the postulates. Nor would it be justifiable to introduce those converse propositions as additional postulates. They might be consequences of the fundamental principles, or they might be false. Attitudes toward these converse propositions vary from optimism to skepticism. The introduction of the paper by Swift and Wright (1980) quotes from prominent exponents of both positions.

Within the quantum information community, optimism seems dominant, it being commonly assumed that any mathematically well-defined operation based on unitary transformations and projections can, in principle, be realized as a laboratory operation. This optimistic attitude is surprising, in view of the fact that some of the most striking results of quantum information theory are proofs that certain classically familiar operations, such as measuring and copying states, are impossible in the quantum domain. So the realizability "in principle" of some basic operations needs more careful examination.

Local Observables

The *local observables* here are the dynamical variables of a single particle (or compact object), such as its position, momentum, and the components of its spin. Since quantum information deals mainly with internal degrees of freedom like spin and polarization, we shall consider only finite dimensional Hilbert spaces.

The prototype of a quantum measurement is the Stern-Gerlach experiment, described schematically in Section 9.1, and more realistically by Swift and Wright (1980). The magnetic moment of the particle (proportional to its spin) couples to an external magnetic field. The gradient of the magnetic energy provides a force that separates the particle beam into separate beams corresponding to each eigenvalue of the spin component along the direction of the field. For a particle of spin $s = \frac{1}{2}$, the Hermitian matrices are 2×2 in size, and all such matrices are linear combinations of the unit matrix and the matrices for the three components of spin. Therefore, in this case, all Hermitian matrices do, indeed, describe measurable quantities.

For higher values of spin, the set of Hermitian matrices is much larger. For $s = 1$ the Hermitian matrices form a 9-parameter set, while the linear combinations of the unit matrix and the three spin components form only a 4-parameter set. Swift and Wright showed, however, that the matrices can be expressed as a sum of tensors that have the symmetries of dipole, quadrupole, octupole, etc. These in turn can couple to electric or magnetic fields of the same multipolar symmetry. Spatial gradients in the strengths of those fields will provide the necessary forces for a generalized Stern-Gerlach experiment to separate particles according to the eigenvalues of the multipole operator that is to be measured. Thus all the Hermitian matrices on the $(2s+1)$-dimensional Hilbert space of a spin-s particle correspond to dynamical variables that can, in principle, be measured.

Nonlocal Observables of Two Objects

By a *nonlocal observable* we mean a collective variable belonging jointly to a pair of objects that are not connected in any way. The example by Peres and Wootters (Section 21.7) showed that a joint measurement on a pair of independently prepared, non-interacting spin-$\frac{1}{2}$ particles could, in principle, yield more information than would measurements on the particles separately. The nonlocal observable in this case, Eq. (21.45), is a function of the total spin of the two particles. However, it is not obvious that such a nonlocal measurement can actually be performed.

Let us consider the measurement of observables related to the total spin of two independent particles. A spin-$\frac{1}{2}$ particle carries a magnetic dipole moment, so a component of total spin, $J_x = \frac{1}{2}(\sigma_x \otimes I + I \otimes \sigma_x)$, can be measured by an ordinary Stern-Gerlach measurement. If the two independent particles are sent through the magnetic field, they will each be deflected up or down (relative to the field gradient). There are three possible results: two particles go up, one up and one down, or two go down. These correspond to the three eigenvalues of J_x: $+1$, 0, or -1. If the two independent particles could be caught and bound together by a central force that acts only on their positions, and does not affect their spins, then in the "one up and one down" case the net deflecting force on the composite would be zero, and we would obtain the familiar 3-spot deflection pattern for a spin-1 particle. But binding them is not necessary to measure J_x because it is the sum of the (commuting) spins of the separate particles.

For a collective variable like $(J_x J_z + J_z J_x)$, in the second term of (21.45), the situation is quite different. It has quadrupolar symmetry, and if it possessed a physical quadrupole moment, then it might be measured by a generalized Stern-Gerlach measurement, as described by Swift and Wright (1980). Now individual spin-$\frac{1}{2}$ particles cannot possess an intrinsic quadrupole moment, as a consequence of the Wigner-Eckart theorem (Section 7.8). But just as a pair of (unequal) charges may produce a dipole, so a pair of dipoles may produce a quadrupole. However any such collective quadrupole moment of the two dipoles would depend on their relative positions and motions. If we could capture and bind the pair with a central force that does not affect the spin-states, we might produce a quadrupole, but it would depend on the *orbital state* of the two particles. Measuring it would give us information about the orbital state, but it would not measure $(J_x J_z + J_z J_x)$. It appears, therefore, that a nonlocal collective variable of this type cannot be measured by any generalized Stern-Gerlach measurement, contrary to the hopes of Peres and Wootters.

An alternative interpretation of the operator $(J_x J_z + J_z J_x)$ is possible. If we multiply it out, and use the anticommutation relation $\sigma_x \sigma_z = -\sigma_z \sigma_x$, then apart from constants we are left with the operator $\sigma_x \otimes \sigma_z + \sigma_z \otimes \sigma_x$, which can be interpreted as an anisotropic spin–spin interaction. If we imagine the possibility of suddenly turning on such an interaction, then we might create an energy in the form of the right-hand side of (21.45), and so regard it as being measurable "in principle". But that seems to be stretching the "principle" rather far.

21.9 *Quantum Computing*

The most ambitious application of quantum information is the design and building of a quantum computer. This is a subject in which theory is far ahead of experimental realization. It is a very active field of research, and any attempt to describe the state-of-the-art in a textbook would be futile, since relevant research papers are published every week. We shall, therefore, emphasize the basic principles, and give only brief accounts of a few important applications.

Quantum Operations (Quantum Gates)

The basic concepts involved in (classical) digital computing are the storage of information in discrete bits, and the processing of this information by the logical operations of AND, OR, and NOT. A quantum computer extends the information storage to qubits. The state vectors of a single qubit may be arbitrary linear superpositions of the bit states $|0\rangle$ and $|1\rangle$. They form a 2-dimensional complex Hilbert space $\mathcal{H}_2$. A quantum register consisting of n qubits has a 2^n-dimensional Hilbert space, $\mathcal{H}_{2^n} = \mathcal{H}_2 \otimes \cdots \otimes \mathcal{H}_2$ (n factors). The most general transformation of an n-qubit state is a unitary transformation on the 2^n-dimensional Hilbert space. By means of such unitary operations, one can transform a product state of the n qubits into an arbitrary entangled state, and vice versa. A theorem due to Barenco *et al.* (1995) demonstrates that any such general unitary transformation can be generated by a product of simpler operations of two kinds: *local unitary operations*, which act only on the states of a single qubit, and *controlled NOT gates* (CNOT gates), which act on pairs of qubits.

A CNOT operation is an interaction between two qubits, called the *control* and the *target*. Its action on the bit states is

$$\begin{aligned} |0\rangle \otimes |0\rangle \to |0\rangle \otimes |0\rangle \ , \quad |0\rangle \otimes |1\rangle \to |0\rangle \otimes |1\rangle \\ |1\rangle \otimes |0\rangle \to |1\rangle \otimes |1\rangle \ , \quad |0\rangle \otimes |1\rangle \to |1\rangle \otimes |0\rangle \,. \end{aligned} \tag{21.46}$$

The first bit is the control and the second bit is the target. If the control bit is $|0\rangle$ the target bit is unchanged. If the control bit is $|1\rangle$ the target bit is reversed.

The four product states used above form an orthonormal basis for the 2-qubit Hilbert space $\mathcal{H}_4$, and it is apparent that the CNOT transformation maps orthogonal states onto orthogonal states. Therefore it is induced by a unitary operator, which may be written as

$$U_{\text{CNOT}} = |0\rangle\langle 0| \otimes I + |1\rangle\langle 1| \otimes \sigma_x \,. \tag{21.47}$$

Here σ_x is (isomorphic to) a Pauli spin operator (Eq. (7.45)), whose effect is to interchange the bit states. In terms of the bit-state product basis, the CNOT operator becomes the matrix

$$U_{\text{CNOT}} = \begin{pmatrix} 1 & 0 & 0 & 0 \\ 0 & 1 & 0 & 0 \\ 0 & 0 & 0 & 1 \\ 0 & 0 & 1 & 0 \end{pmatrix} . \tag{21.48}$$

(From the upper left corner, the row and column labels are: 00, 01, 10, 11.)

The CNOT transformation (21.46) is a purely classical bit operation. The power of CNOT as a quantum operator is demonstrated by using a control qubit that is a superposition of bit states. Consider, for example, the effect of using the state $|+\rangle = (|0\rangle + |1\rangle)/\sqrt{2}$ as the control qubit:

$$U_{\text{CNOT}}|+\rangle \otimes |0\rangle = (|0\rangle \otimes |0\rangle + |1\rangle \otimes |1\rangle)/\sqrt{2}\,. \tag{21.49}$$

This is one of the entangled Bell states (21.30). The ability to prepare entangled states is very important in quantum computing, so the unitary CNOT gate is very useful.

Many other useful qubit operations can be defined. An example is the *controlled unitary* (CU) gate, defined by

$$\begin{aligned} |0\rangle \otimes |0\rangle \to |0\rangle \otimes |0\rangle \ , \quad |0\rangle \otimes |1\rangle \to |0\rangle \otimes |1\rangle \\ |1\rangle \otimes |0\rangle \to |1\rangle \otimes U|0\rangle \ , \quad |1\rangle \otimes |1\rangle \to |1\rangle \otimes U|1\rangle \,. \end{aligned} \tag{21.50}$$

If the control qubit is $|1\rangle$, the local (i.e. single qubit) unitary operation U is performed on the target qubit.

Another useful example is the *swap* gate, which swaps the states of the two qubits,

$$U_{\text{swap}}\,|\phi\rangle \otimes |\psi\rangle = |\psi\rangle \otimes |\phi\rangle\,. \tag{21.51}$$

In the usual basis, it is represented by the matrix

$$U_{\text{swap}} = \begin{pmatrix} 1 & 0 & 0 & 0 \\ 0 & 0 & 1 & 0 \\ 0 & 1 & 0 & 0 \\ 0 & 0 & 0 & 1 \end{pmatrix}. \tag{21.52}$$

Superpositions of the bit states are often needed. These can be obtained by the *Hadamard transformation*,

$$\begin{aligned} |0\rangle &\to (|0\rangle + |1\rangle)/\sqrt{(2)} \\ |1\rangle &\to (|0\rangle - |1\rangle)/\sqrt{(2)} \end{aligned} \tag{21.53}$$

which is generated by the single-qubit unitary operator

$$U_{\text{H}} = \frac{1}{\sqrt{2}}\begin{pmatrix} 1 & 1 \\ 1 & -1 \end{pmatrix}. \tag{21.54}$$

When one is designing a complex set of transformations to implement a quantum computation algorithm, the algebraic representation of the set of transformations becomes burdensome. A graphical representation, known as a *quantum circuit*, has been devised, which is much more convenient for practical programming of a quantum computer. We refer to more specialized works on quantum computing for this technique. Quantum circuits lie outside of the scope of a book on quantum mechanics, just as classical digital circuits lie outside of the scope of a book on semiconductors.

Some Impossible Quantum Operations

While quantum information processing can do things that could not be done classically, there are also many seemingly plausible operations that are impossible. Some of these were demonstrated in the theorems of Section 21.2. Several more have been proven by Pati (2002). Here are some examples of impossible operations.

The universal NOT gate: It is easy to define a NOT operation for the bit states in a chosen basis. It has the effect of replacing $|0\rangle$ by $|1\rangle$, and vice versa. The hypothetical *universal NOT* operation would take an arbitrary qubit state

$|\psi\rangle$ and replace it by its orthogonal complement $|\psi^{\perp}\rangle$. This would be a basis-independent version of NOT.

The impossibility of this operation is most easily seen by regarding the qubits as quantum spin-$\frac{1}{2}$ systems, whose states are described by the Bloch sphere representation (21.5). The hypothetical u-NOT operation must reverse the spin-component along every possible direction. Now all unitary transformations of a spin-$\frac{1}{2}$ system correspond to rotations of the Bloch sphere. A rotation by π about any axis would reverse all spin components in the plane perpendicular to the chosen axis, but it would not affect the component along the axis. Therefore, a unitary operation cannot achieve the reversal of all spin-components, and so it cannot implement the u-NOT operation. The reversal of all spin-components could be achieved by the *time-reversal* operator (Eq. (13.29)), but that is an anti-unitary operator. Since all time-evolution operators must be unitary, the time-reversal operation cannot be achieved in any dynamic process. Therefore the u-NOT gate is impossible.

No qubit complementing: The no-cloning theorem shows that the process $|\psi\rangle \otimes |\omega\rangle \rightarrow |\psi\rangle \otimes |\psi\rangle$, which would make a copy of an unknown quantum state, is impossible. The no-qubit-complementing theorem shows that a similar process, which would produce an arbitrary qubit and its orthogonal complement, $|\psi\rangle \otimes |\omega\rangle \rightarrow |\psi\rangle \otimes |\psi^{\perp}\rangle$, is impossible. This result is not surprising, in light of the impossibility of the u-NOT gate, but it is a different theorem, involving the generation of another qubit rather than the transformation of the input qubit.

Although the theorem can be proven directly from linearity and unitarity, a simple physical argument may be more insightful. Suppose that it were possible to create the orthogonal complement $|\psi^{\perp}\rangle$ of an arbitrary input qubit $|\psi\rangle$. Then we could set aside the original qubit $|\psi\rangle$, and repeat the complementing operation on the qubit $|\psi^{\perp}\rangle$. But its orthogonal complement would be a copy of the original qubit $|\psi\rangle$ (possibly multiplied by an unimportant phase factor). We would then have both the original qubit and a copy of it, in violation of the no-cloning theorem. Therefore the general qubit-complementing operation is impossible. (Note, however, that whereas the no-cloning theorem applies to quantum states of any dimensionality, this proof of the no-complementing theorem applies only to a 2-dimensional Hilbert space, since only in that case does the orthogonal complement of the orthogonal complement necessarily yield the original state.)

No universal superposition gate: The *superposition principle* is one of the most distinctive features of quantum mechanics, and superposition states are essential features of quantum information processing. The hypothetical universal superposition gate (u-SUP gate) would take an arbitrary qubit state and transform it into an equal-amplitude superposition of the input state and its orthogonal complement,[1]

$$|\psi\rangle \rightarrow (|\psi\rangle + |\psi^{\perp}\rangle)/\sqrt{2}\,. \tag{21.55}$$

(Since $|\psi^{\perp}\rangle$ is defined only by the condition $\langle\psi|\psi^{\perp}\rangle = 0$, its phase relative to $|\psi\rangle$ may be chosen arbitrarily. Hence only the + sign need be considered in this equation.) If the u-SUP gate can be implemented by a unitary transformation, then it must preserve the value of the inner product. Now (21.55) yields, for the inner product of two state vectors,

$$\langle\psi_1|\psi_2\rangle \rightarrow \frac{1}{2}\big(\langle\psi_1|\psi_2\rangle + \langle\psi_1^{\perp}|\psi_2^{\perp}\rangle + \langle\psi_1|\psi_2^{\perp}\rangle + \langle\psi_1^{\perp}|\psi_2\rangle\big)\,. \tag{21.56}$$

Let us express the qubits in terms of a standard basis,

$$|\psi_i\rangle = \alpha_i|0\rangle + \beta_i|1\rangle \tag{21.57}$$
$$|\psi_i^{\perp}\rangle = \alpha_i^*|1\rangle - \beta_i^*|0\rangle\ , \quad (i = 1, 2)\,. \tag{21.58}$$

Here the coefficients α_i and β_i are arbitrary, except for the normalization $|\alpha_i|^2 + |\beta_i|^2 = 1$. It is easy to verify that $\langle\psi_i|\psi_i^{\perp}\rangle = 0$, $\langle\psi_1^{\perp}|\psi_2\rangle = -\langle\psi_1|\psi_2^{\perp}\rangle^*$, and $\langle\psi_1^{\perp}|\psi_2^{\perp}\rangle = \langle\psi_1|\psi_2\rangle^*$. Hence (21.56) reduces to

$$\langle\psi_1|\psi_2\rangle \rightarrow \frac{1}{2}\big(\langle\psi_1|\psi_2\rangle + \langle\psi_1|\psi_2\rangle^* + \langle\psi_1|\psi_2^{\perp}\rangle - \langle\psi_1|\psi_2^{\perp}\rangle^*\big)\,. \tag{21.59}$$

If the coefficients α_i and β_i are real, then all of the inner products above will be real, and the value of $\langle\psi_1|\psi_2\rangle$ will be preserved. This is true for a significantly large number of states, and for those states the u-SUP operation can be implemented. But the inner products on the right-hand side of (21.59) cannot be made real for arbitrary choices of $|\psi_1\rangle$ and $|\psi_2\rangle$, so a truly *universal* u-SUP gate is impossible.

This conclusion seems paradoxical. Linearity is the foundation of the *superposition principle.* Yet we have used linearity to prove the impossibility of a linear superposition of an unknown qubit with its complement. It is as if the superposition principle were defeating itself! The resolution of the paradox is that for any specific qubit state, a superposition with its complement can, in principle, be produced. But there is no process by which such a superposition can be produced for an arbitrary, unknown state.

[1]This can be thought of as a basis-independent version of the Hadamard gate. Our definition differs in detail from that of Pati (2002), but the conclusion is the same.

Several of the impossibility theorems can be seen to be related, in the light of this paradox. If we could measure the unknown qubit without changing it, then this information could be used to superpose it with its complement. But, of course, such a measurement is impossible. It might appear that a photon in an unknown polarization state could be superposed with its complement by a simple procedure: Pass it through a beam splitter that does not affect the polarization; apply the u-NOT operation to one of the sub-beams to form the complement of the polarization state; and then recombine the sub-beams to form the superposition. But, of course, the u-NOT operation is impossible. By exposing such connections, quantum information theory gives us a deeper understanding of quantum mechanics.

Advantages and Applications of a Quantum Computer

To justify an investment in quantum computing, it should offer some advantages over ordinary digital computing. It is possible to solve the Schrödinger equation numerically on an ordinary computer. Since a quantum computer is a physical embodiment of the Schrödinger equation for its qubits, it follows that a quantum computer can be simulated by a classical computer. Therefore, anything that a quantum computer might do can also be done, in principle, by a classical computer. The advantage of a quantum computer lies in its ability to perform certain tasks using less resources of memory and computation time than would be required by a classical computer. We shall next examine some kinds of applications for which a quantum computer should have superior performance.

Quantum Simulations

A large class of quantum many-body systems can be represented by a lattice of quantum spins (qubits). The most obvious example is a magnetic system, in which the qubits correspond directly to the localized magnetic atoms. Suppose there are N such spins in the system of interest. Then its state vector belongs to a Hilbert space of $n_d = 2^N$ dimensions, and its state operator (density matrix) has dimension $n_d \times n_d$. The amount of memory needed to store the state in an ordinary digital computer will grow exponentially with the size N of the system. The equation of motion for the state vector will consist of n_d coupled ordinary differential equations, and the computing time required to solve them will grow as some polynomial in n_d, and thus exponentially with N. Hence, for more than a few spins in the system, the computational requirements are impractically large.

If, on the other hand, we can simulate this N-spin system on a quantum computer, then the storage requirements for the state vector are only N qubits, and we shall see that the time needed to compute its motion grows only as a polynomial in N. Thus a quantum computer should be superior to a classical computer in simulating the dynamics of a many-body quantum system.

The Hamiltonian, H, of the system is a function of the Pauli spin operators on each of the lattice sites, with the interactions being represented by products of spin operators on neighboring sites. The time-evolution of the state is given by the unitary operator

$$U(t) = e^{-iHt/\hbar} . \tag{21.60}$$

If we can arrange for the interactions among the qubits of a quantum computer to be isomorphic to the interactions between the spins in the physical specimen, then we will have an *analog* computer for the magnetic system. The advantage of measuring the analog rather than the actual magnetic system will, presumably, be a more convenient time scale and the possibility of measuring individual qubits, whereas measuring individual spins in a magnet is usually not possible.

For most quantum systems, the creation of an isomorphic analog is not practical, and the quantum computer must implement *digital* techniques. The possibility of such quantum computations was demonstrated by Lloyd (1996). The Hamiltonian of the interacting system of N spins has the form

$$H = \sum_{i=1}^{\ell} H_i \tag{21.61}$$

where each H_i is a *local* operator, consisting of either a Pauli spin operator on a single site, or a product of two Pauli operators on adjacent interacting sites. As in a classical digital computation, the time evolution of the system is approximated by evolving it forward locally over small time steps. The global time-development operator (21.60) may be approximated by a sequence of local unitary transformations,

$$e^{-iHt/\hbar} \approx \left(e^{-iH_1 t/\hbar n} ... e^{-iH_\ell t/\hbar n}\right)^n \tag{21.62}$$

with an error of order $O(\ell t^2/n)$, which can be made small by choosing n large enough. For typical local interactions, ℓ will be proportional to N, and, at worst, to N^2 if all pairs interact. Thus the computation can be performed in a number of steps the grows, at most, as a polynomial in N, in contrast with the exponential dependence on N for a classical computer. For a system of more than about 20 particles, a quantum computer should demonstrate a clear superiority.

Both analog and digital quantum computers have been demonstrated by Blatt and Roos (2012), using a pair of energy levels in each of a set of trapped atomic ions as qubits. So far, only systems of a few (< 10) qubits have been realized, but there is reasonable optimism that quantum computers, scalable to large numbers of qubits, can be achieved.

Numerical Quantum Computing

That the simulation of one quantum system (an interacting many-body system) by another (a quantum computer) is feasible, and that it may be superior to simulation by an ordinary digital computer, may not be too surprising. Less obvious is the usefulness of a quantum computer in performing numerical calculations that are unrelated to quantum mechanics.

The first problem is that of *error control.* Storing binary numbers in *bit* states is inherently stable against small errors. If the bit value '0' is represented by 0 volts and the value '1' is represented by 1 volt, then errors of magnitude smaller than 0.5 volts can be tolerated. The inaccurate signal is simply interpreted as '0' or '1' according to which of the proper voltages is nearest to it. A larger error that can flip a single bit can also be detected and corrected, using redundant coding and parity checks.

In contrast to discrete *bit* states, the states of a *qubit* form a continuum, and so are susceptible to a variety of small errors. The detection of errors requires measurement, which usually changes the state, and the *no-cloning* theorem prevents us from making a spare copy of the state, on which (possibly destructive) measurements can be safely performed. Nevertheless, error-correcting protocols for qubit states have been devised.

The first step is to introduce redundancy by representing the logical '0' and '1' states by multi-qubit code words, rather than by single qubits. As a simple example, let us represent the logical '0' by the 3-qubit state $|000\rangle = |0\rangle\otimes|0\rangle\otimes|0\rangle$, and the logical '1' by $|111\rangle = |1\rangle\otimes|1\rangle\otimes|1\rangle$. A general superposition of these logical '0' and '1' states will have the form

$$|\psi\rangle = \alpha|000\rangle + \beta|111\rangle\,. \tag{21.63}$$

Now suppose that a random error were to flip the bit state of one physical qubit, yielding one of the three possible states:

$$|\psi'\rangle = \alpha|100\rangle + \beta|011\rangle\ , \tag{21.64}$$

$$|\psi''\rangle = \alpha|010\rangle + \beta|101\rangle\ , \tag{21.65}$$

$$|\psi'''\rangle = \alpha|001\rangle + \beta|110\rangle\ , \tag{21.66}$$

depending on which of the three qubits was flipped. If this anomaly could be detected, then the erroneous qubit could be identified and corrected by a "majority-rule". But measurements on each of the qubits separately will not work, since those measurements will yield either 0 or 1 randomly (with probability either $|\alpha|^2$ or $|\beta|^2$), and will change the state in some unpredictable way. However, we can detect the error by measuring the two collective observables

$$\text{ZZI} = \sigma_z \otimes \sigma_z \otimes I \tag{21.67}$$

$$\text{IZZ} = I \otimes \sigma_z \otimes \sigma_z\ . \tag{21.68}$$

Here the Pauli operator σ_z has eigenvalues $+1$ or -1 for the basis vectors $|0\rangle$ and $|1\rangle$, respectively. It is easily verified that the four vectors $|\psi\rangle \cdots |\psi'''\rangle$ are eigenstates of these two operators, so it is possible to perform the measurements without changing the states. The eigenvalues of the operators ZZI and IZZ are either $+1$ or -1 for each of the states $|\psi\rangle \cdots |\psi'''\rangle$, and the pattern of the pair of eigenvalues uniquely identifies the particular state. With that information, we can then apply the appropriate bit-flip operation to correct the single-qubit error.

This simple 3-qubit coding scheme cannot detect errors in the relative phase and amplitudes of the two terms in the superposition state (21.63). However, a considerably more complicated scheme, in which the logical states '0' and '1' are represented by entangled states of seven qubits, can detect amplitude and phase errors, although the probability of detecting such errors varies with the magnitude of the error. Details of the technique are described by Barnett (2009), (Ch. 6.4 and App. M). Without such error-correction techniques, it is doubtful that numerical quantum computing can be practical.

Quantum Function Evaluation

An n-bit binary number can be stored in n qubits. For example, the binary number $a = 0101$ ($= 5$ in decimal notation) can be represented by the state $|a\rangle = |0\rangle \otimes |1\rangle \otimes |0\rangle \otimes |1\rangle$. An orthonormal basis for the 2^n dimensional Hilbert space of an n-qubit register is given by the vectors $\{|a\rangle\}, (a = 0, 2^n - 1)$. Now suppose that we wish to evaluate an integer-valued function $f(a)$ that takes values in the range $0 \leq f(a) \leq 2^n - 1$. If there is a unitary operator U such that

$$U|a\rangle = |f(a)\rangle \tag{21.69}$$

then the function $f(a)$ can be evaluated easily by a quantum computer. But since a unitary transformation is one-to-one, this will be possible only if the function $f(a)$ generates a permutation of the integers $0 < a < 2^n - 1$. The equation must fail if $f(a_1) = f(a_2)$ for some $a_1 \neq a_2$.

A more effective method is to introduce a second n-qubit register, prepared in the state $|b\rangle$, and perform the unitary transformation

$$U_f|a\rangle \otimes |b\rangle = |a\rangle \otimes |b \oplus f(a)\rangle\,. \tag{21.70}$$

Here "$\oplus$" denotes bitwise addition of the two strings modulo 2. It can be verified directly that this equation always transforms orthogonal vectors into orthogonal vectors, for all values of a and b, regardless of the form of $f(a)$. Hence, unlike (21.69), this transformation is unitary for any function $f(a)$. If we choose $b = 0$, then the second factor on the right side yields the value of $f(a)$.

However, the full power of the quantum computer is evident only when it operates on *superposition* states. Let the two registers be initialized to the n-bit state $|0\rangle \otimes |0\rangle \otimes \cdots \otimes |0\rangle$. Next we apply the Hadamard operator (21.54) to each qubit of the first register, so that it becomes

$$U_{\mathrm{H}}|0\rangle \otimes U_{\mathrm{H}}|0\rangle \otimes \cdots \otimes U_{\mathrm{H}}|0\rangle = 2^{-n/2}(|0\rangle + |1\rangle) \otimes (|0\rangle + |1\rangle) \otimes \cdots \otimes (|0\rangle + |1\rangle)\,. \tag{21.71}$$

When the multiplications on the right side are expanded, it becomes the sum of the bit-state representations of all of the 2^n binary numbers from 0 to $2^n - 1$. Thus the above state is equal to

$$|\phi\rangle = 2^{-n/2} \sum_{a'=0}^{2^n-1} |a'\rangle\,. \tag{21.72}$$

Applying the transformation U_f of (21.70) then yields

$$U_f|\phi\rangle \otimes |0\rangle = 2^{-n/2} \sum_{a'=0}^{2^n-1} U_f|a'\rangle \otimes |0\rangle = 2^{-n/2} \sum_{a'=0}^{2^n-1} |a'\rangle \otimes |f(a')\rangle\,. \tag{21.73}$$

Thus a single invocation of the unitary transformation U_f has resulted in the computation of $f(a)$ for *all* 2^n values of a. The number of quantum gate operations required to perform this computational feat is the same as that needed to calculate only one value in (21.70). This illustrates the potential of a quantum computer, utilizing the superposition principle, to achieve much greater speeds in the limit of large n than is possible for classical computer.

Unfortunately, it is not possible to simply read the many values of $f(a)$ from the final state of (21.73). One must perform a simultaneous measurement of the bit values of the two registers. The result will be a pair of numbers $(a', f(a'))$ for some random value of a'. Nevertheless, the result of (21.73) can be used as an intermediate step in a larger quantum computation, and the exponential (in n) speedup can actually be realized.

Quantum Fourier Transform

One of the most common numerical tasks that is encountered in applied mathematics is the evaluation of a *discrete Fourier transform* (DFT),

$$f_j = \frac{1}{\sqrt{N}} \sum_{k=0}^{N-1} e^{i2\pi jk/N}\, g_k\,. \tag{21.74}$$

One may think of the input data points $\{g_k : k = 0, N-1\}$ as being in the time domain, and the output data points $\{f_j : j = 0, N-1\}$ as being in the frequency domain, but that specific interpretation is not essential. Sometimes the DFT arises as an approximation to the continuous Fourier integral, and other times it occurs in its own right in discrete mathematics. But it is so common that efficient algorithms for its computation are important. For large values of N, the computation time to evaluate (21.74) for all values of j will be dominated by the N^2 multiplications inside the sum, and so the time taken by a straightforward evaluation of the DFT will scale as $O(N^2)$.

The *Fast Fourier Transform* (FFT) can be much faster, especially if the number of points is a power of 2, $N = 2^n$. In that case, the original DFT can be written as the sum of two smaller DFTs, each of length $N/2$. This trick can be applied recursively, until we reach the trivial DFT on only one point. The number of iterations of the trick that are needed is $n = \log_2 N$, and consequently, the time for the FFT scales as $O(N \log_2 N)$, which is much smaller than $O(N^2)$ for large N.

The possibility of a *Quantum Fourier Transform* (QFT) arises from the observation that the DFT (21.74) is, in fact, a unitary transformation. If $\{g_k\}$ and $\{f_j\}$ are regarded as column vectors, then (21.74) takes the form $f = Mg$, where M is a unitary matrix whose elements are

$$M_{jk} = \frac{1}{\sqrt{N}} e^{i2\pi jk/N}\,. \tag{21.75}$$

The action in Hilbert space of a unitary operator such as M can be viewed in two ways. Let $|\psi\rangle = \sum_k x_k|k\rangle$ be an arbitrary vector, written in terms of the basis vectors $\{|k\rangle\}$. (In practice, this basis will be the 2^n bit-state basis vectors for an n-qubit register in a quantum computer.) By operating with M we obtain

$$M|\psi\rangle = \sum_k x_k M|k\rangle = \sum_k x_k|\tilde{k}\rangle\,, \tag{21.76}$$

where the vectors $|\tilde{k}\rangle = M|k\rangle$ form a new orthonormal basis. Alternatively, we can write

$$M|\psi\rangle = \sum_j \sum_k |j\rangle\langle j|M|k\rangle x_k = \sum_j y_j|j\rangle\;, \tag{21.77}$$

where $\langle j|M|k\rangle = M_{jk}$, and $y_j = \sum_k M_{jk}\,x_k$.

In the first form, we apply the Fourier transform to the basis vectors, and keep the original expansion coefficients x_k. In the second form, we use the original basis vectors, and apply a DFT to the expansion coefficients. Since a quantum computer applies a physical unitary transformation to the qubit states, a QFT realizes the first form, and its output is a Fourier transform of the bit-state basis vectors of an n-qubit register.

The efficiency of the QFT depends on how the unitary transformation M can be realized in terms of the basic operations of local (single qubit) unitaries and CNOT gates. This problem has an elegant (but nontrivial) solution, and the number of operations required for a QFT scales as $O(n^2)$. For large n, this is much faster than the FFT, which requires a number of operations that scales as $O(N\log_2 N) = O(2^n\,n)$.

As was the case for the previous example of quantum function evaluation, the QFT does not allow us to read out the actual Fourier transform of a function, and a measurement of the bit values of the appropriate register would only yield the amplitude of a single, randomly chosen Fourier component. But fortunately the Fourier transform is often not the end of a computation, but merely an important intermediate state in a longer computation. So the QFT can be a valuable component of another quantum computational algorithm.

Shor's Factoring Algorithm

A long-standing problem in number theory is to devise efficient algorithms to compute the factors of a large integer N. This problem is also relevant to cryptography. In Section 21.4, Alice and Bob shared an encryption/decryption key that they must keep secret from the rest of the world. The encryption and decryption processes were the same: add the key bitwise (modulo 2) to the message. *Public-key* encryption uses a quite different strategy, in that the encryption and decryption processes are different. Bob will publish an encryption key, and by using it in a standard encryption algorithm, anyone in the world can send him a secret message. The decryption of the message uses a different key, which is known only to Bob.

No public-key encryption scheme can be unconditionally secure. When both the encryption algorithm and the encryption key are publically known, it is mathematically possible to work out the inverse (decrypting) transformation. Practical security derives from the fact that the determination of the inverse transformation is a hard computational problem. Time-sensitive information will be secure if its expiry time is shorter than the computation time needed to work out the decrypting transformation. The widely used RSA encryption scheme uses an integer, $N = pq$, that is the product of two large primes, p and q. Bob's chosen value of N forms part of his public encryption key. A knowledge of the prime factors p and q enables Bob to efficiently decrypt the messages that he receives, and if a spy were able to efficiently factor large integers, he would be able to crack Bob's code.

The most elementary way to find a factor of an integer N is to try all possible divisors, from 2 up to the largest integer not greater than $\sqrt{N}$. This will either find a factor of N, or prove that N is prime. The number of divisions needed, and hence the computation time, will scale as $O(N^{1/2})$. If N requires no more than n bits for its binary representation, then the computation time will scale as $O(2^{n/2})$, which is exponential in the bit-length of N. Other classical algorithms exist, the best of which scales as $O(2^x)$, with $x = (\log n)^{2/3}\, n^{1/3}$. This is an improvement over $x = n/2$, but the time still grows more rapidly than any polynomial in n.

It would take us too far into number theory to describe Shor's algorithm in detail.[m] The most computationally expensive part of the algorithm is the determination of the period of a certain computable periodic function, of the form $f(r) = y^r \mod N$, where $y < N$, and N is the large (n-bit) number to be factored. This is a natural job for the Fourier transform. Recall that the computation time for an FFT scales as $O(2^n\, n)$, whereas that of the QFT scales as $O(n^2)$. Thus the QFT has the potential of reducing the computation time for factorization from exponential to polynomial in the bit-length of the number to be factored. This, in turn, could break the security of RSA encryption.

[m] Barnett (2009) gives a readable account of Shor's algorithm in 10 pages, including an essential Appendix.

At the time of writing, no quantum computer has been built which is scalable to large numbers of qubits, although a prototype model has succeeded in implementing Shor's algorithm to factor $15 = 3 \times 5$.

Adiabatic Quantum Computation

All of the examples of quantum computing that have been described above are based on quantum gates, which implement a sequence of unitary transformations on the states of the qubits. Since these are continuous transformations, they will accumulate quantitative errors. Although error-correcting techniques are possible in principle, they are very expensive to implement. In an example given earlier, the logical "0" and "1" bit states are each represented by 7-qubit code words. To store the results of the necessary error-detecting measurements, another 3 qubits are required, making a total of 10 qubits for every bit of stored information.

An *adiabatic* quantum computer (AQC) uses a very different principle. Its qubits are described by a Hamiltonian with controllable parameters for the various interactions. These are initially set to yield some Hamiltonian whose ground state is well known and easy to prepare. Then the parameters are varied adiabatically, to produce another Hamiltonian whose ground state contains the solution of our problem. The adiabatic principle (Section 12.7) ensures that the system of qubits will remain in the ground state of the slowly varying Hamiltonian, thus providing stability to the process. Although the range of problems that can be solved by an AQC may be smaller than that which can, theoretically, be solved by a quantum-gate based computer, the practical development of the adiabatic computer is considerably more advanced than that of the quantum-gate based computer. An AQC with 512 qubits is in operation,[n] whereas quantum-gate based computers have so far achieved no more than a dozen qubits.

The architecture and testing of the AQC processor are described by Harris *et al.* (2010). The qubits are superconducting current loops, with the bit states $|0\rangle$ and $|1\rangle$ being the states of clockwise and counter-clockwise currents (or equivalently, positive and negative magnetic flux through the loop). The energy of the qubit as a function of magnetic flux has the form of a double-well potential, similar to Fig. 13.2. It is convenient to describe its states by the spin analogy. Accordingly, the eigenvectors of σ_z represent the bit states, $|\uparrow\rangle = |0\rangle$ and $|\downarrow\rangle = |1\rangle$. Physically, these are clockwise and counter-clockwise current eigenstates, which correspond to the states localized in the left and right wells

[n]Designed and manufactured by *D-Wave Systems, Inc.*, Burnaby, B.C., Canada.

of the potential. Their symmetric and anti-symmetric superpositions are the eigenvectors of σ_x. The energy difference between the symmetric and anti-symmetric states, and hence also the tunneling rate between the clockwise and counter-clockwise current states, can be controlled by varying the height of the barrier between the left and right wells.

The effective Hamiltonian of the processor has the form

$$H(t) = -\sum_i h_i \sigma_z^{(i)} + \sum_{i<j} K_{ij} \sigma_z^{(i)} \sigma_z^{(j)} - \Gamma(t) \sum_i \sigma_x^{(i)} , \tag{21.78}$$

which is a variant of the *Ising model*, a familiar model in the statistical mechanics of magnetism. The local fields $\{h_i\}$ and couplings $\{K_{ij}\}$ can be programmed individually, to define a specific problem. The uniform transverse field $\Gamma(t)$ is used to control the adiabatic process.

A typical optimization problem is equivalent to finding the lowest energy state of (21.78) with the control term $\Gamma(t) = 0$. The initial state is prepared by starting with a very large value of $\Gamma(0)$. In the spin analogy, this aligns all spins in the x-direction. Thus the initial state of the qubits will be $(|0\rangle + |1\rangle) \otimes (|0\rangle + |1\rangle) \otimes \cdots$ (omitting normalization). When multiplied out, this becomes a uniform superposition of all bit states (as was previously studied in Eq. (21.71)). The field $\Gamma(t)$ is then varied (sufficiently slowly for the adiabatic theorem to hold) until it is very small, and the desired final state is obtained. The pattern of bits that is defined by this ground state, and which is the solution to the optimization problem, can be read out by measuring the magnetic flux in each qubit.

A complex optimization problem often presents a complex landscape of the energy function, containing many valleys of local minima that are separated by high ridges. A conventional digital minimization program can get stuck in a local minimum, and be unable to escape and seek a lower minimum state. An AQC has an advantage, in that it can tunnel through a ridge in the energy landscape, and so it can be more effective in finding the true minimum. For an Ising model of N spins, the number of basis states is 2^N, and so for a model with arbitrary couplings, the computation time for a digital computer to find the ground state will grow exponentially with N. The time required for adiabatic quantum optimization will typically grow only as N or N^2, and so will achieve an exponential speedup for large N.

Practical application of this technique depends on being able to map the chosen problem onto an Ising model whose ground state yields the optimum solution. While it is not obvious how to do this in general, it can and has been done for a large range of important problems. Smelyanskiy *et al.* (2012) describe how it can be done for many of the artificial intelligence problems that arise in unmanned space exploration. Other applications include protein folding (which is very important in biophysics and genetics), and automatic image processing and recognition.

21.10 *Quantum Information and Quantum Foundations*

In addition to producing many interesting new results, and creating the possibility of powerful new technologies, quantum information (QI) theory provides new points of view from which to examine the foundations of quantum mechanics (FQM). In this section, I will abandon the professorial "we",[o] and write in the first person singular, to emphasize that the point of view expressed is mine.

Changes in ideas about FQM can be seen by examining some earlier writings. For example, in his scientific biography of Einstein, Abraham Pais wrote (1982), "This conclusion [of the Einstein-Podolsky-Rosen paper] has not affected subsequent developments in physics, and it is doubtful that it ever will." Although Pais was usually very perceptive, he seriously missed the mark here. Today the EPR paper is recognized as foundational in the study of *entanglement*, which is a central topic in QI, and the negative responses to EPR are seldom quoted. Entangled states of spatially separated systems (often called "EPR states", or "Schrödinger-cat states") were once regarded as theoretical embarassments ("monstrous states", according to some writers), and efforts were directed at finding a physical principle that would exclude them. Today entanglement is valued as a resource, which can be used in quantum communication, teleportation, and computing.

The question of whether there might be a deterministic, quasi-classical set of *hidden variables* (HV) underlying the statistical predictions of QM, was often dismissed as irrelevant (like the question of how many angels can dance on the head of a pin, according to Pauli). However, one of the proofs of Bell's theorem shows that the postulated existence of local HVs has testable consequences (the Bell inequalities). As is now well known, the experimental data indicates against local HVs, but the fact that their conjectured existence is experimentally testable demonstrates the scientific content of the conjecture. More recently, it was realized that the simulation of a quantum computer by a digital computer is, in effect, a (presumably nonlocal) deterministic HV model that reproduces the results of QM. Thus to determine the conditions for a

[o] "From these equations, **we** see that this conclusion follows," means "I see it, and if you have been paying attention, you should see it too."

classical HV model to succeed is equivalent to determining something about the classical computing power that is needed to equal that of a quantum computer. Hence the old notion of an HV substrate below QM has acquired interest as a byproduct of quantum information studies.

Interpretations of Quantum States

The debate about the appropriate interpretation of the *quantum state* concept has received new impetus from quantum information theory. Three dichotomies are often used to classify the various interpretations:

individual versus *ensemble*;
ontic versus *epistemic*;
objective versus *subjective.*

These three pairs of categories are related, but are not equivalent. Throughout this book I have expounded an interpretation that lies in the categories of *ensemble* and *objective.*

Individual versus Ensemble:

The core of the *ensemble* interpretation is that QM does not predict *events*, but only the *probabilities* of the various possible events. In particular, it does not predict the outcomes of individual measurements. But if the same state can be repeatably prepared, thus generating an ensemble of similarly prepared systems, then the probabilities predicted by QM can be compared with the statistics of the results of measurements on the ensemble of prepared systems. In Ch. 9 (esp. Section 9.3) the ensemble interpretation (B) is compared with a specific individual interpretation (A), with the conclusion strongly favoring (B). An *ensemble* interpretation asserts that a state vector $|\psi\rangle$ generates the probability distributions for any and all measurements that may be performed on an ensemble of similarly prepared systems. It usually goes on to deny that the notion of a state vector of an individual system has any relevance. This interpretation is supported, within QI theory, by Theorem 21.2, which says that it is impossible to determine an unknown quantum state by measurements on a single representative system that has been prepared in that state.

[[At this point, it is amusing to point out a deep irony in the history of QM. In the early days of quantum theory, Bohr and Heisenberg argued that the simultaneous measurement of position and momentum was impossible. They, and their followers, then took the further step of denying scientific legitimacy to concepts that, like simultaneous values of position and momentum, were not measurable. In the famous Einstein-Podolsky-Rosen paper (1935), EPR argued that the QM description of physical reality is not complete. Bohr, in his reply, argued for the opposite conclusion. Since "physical reality" certainly includes individual systems, Bohr was, in effect, arguing that a quantum state vector is a complete description of an individual system. But Theorem 21.2 proves that the state vector of an individual system is not measurable, so by Bohr's and Heisenberg's own principle, it should not be a legitimate physical concept. Thus quantum information theory has undercut the basis of Bohr's argument against EPR.]]

The unmeasurability of the state of an individual system is a significant argument against an individual interpretation of quantum states, but by itself, it is not fully decisive. As the example of local HVs and Bell's theorem shows, the existence of an entity that is not directly measurable can, nevertheless, have testable consequences. The particular individual interpretation (A) in Ch. 9 is, indeed, excluded by the arguments in that chapter, but other individual interpretations might be possible.

Quantum information offers temptations to adopt an individual-state interpretation. Consider, for example, the teleportation of Alice's particle state to Bob's particle. Can that be possible unless Alice's particle possesses a definite state of its own, which can be transmitted to Bob's particle? That individual state may be unknown, and is certainly invisible, but the temptation to think of it as real remains strong. R.F. Werner (2001) argues that, nevertheless, we should resist the temptation (see his Sections 2.3.4 and 2.4.1). He points out that the adoption of an individual-state interpretation can generate misleading intuitions that would violate some impossibility theorems. He stresses the *statistical interpretation* of quantum states (here equivalent to our *ensemble* interpretation), according to which a *state* is operationally identified with a method of preparing systems, and its testable content consists of a probability distribution for each observable. Thus, to say that Alice's particle state is teleported to Bob's particle, means that if the preparation of the initial state and the teleportation process are repeated many times, so as to generate an ensemble, then Alice's original state and Bob's received state will be statistically indistinguishable. There is no need to compare Alice's and Bob's invisible state vectors, nor is there even a need to postulate their existence as real physical entities. In spite of Werner's noble efforts, many authors (including some authors of chapters in the same volume) write as if individual systems each have their own invisible, individual state vectors.

Ontic versus Epistemic:

The classification of quantum-state interpretations as *ontic* or *epistemic* has been exploited more recently. The words *ontic* and *epistemic* have been imported into physics from some rather pedantic branches of philosophy, hence their meanings must be adapted to fit their new context. *Ontic* refers to that which is real and exists. Its meaning is similar to EPR's *elements of physical reality.* An ontic entity may or may not be observable, so the term is not synonymous with HV. In philosophy, *epistemic* refers to knowledge, but in QM its meaning should be broadened to refer to *information.* (*Knowledge* refers, more narrowly, to *subjective* information – a distinction that is often overlooked.)[p] Although *ensemble* interpretations are often regarded as *epistemic*, and *individual* interpretations as *ontic*, those connections are not absolutely necessary. An *ensemble* interpretation may be of the type *ψ-supplemented* (to be defined below). Likewise, there could be an *individual* interpretation in which ψ merely gives information about the system, and so is *epistemic.*

A more precise study of ontic and epistemic states is made possible by the introduction of an *ontological model* (Harrigan and Spekkens, 2010). It consists of two elements. The first is a postulated set of *ontic states* Λ, with a particular ontic state being denoted as λ. The structure of Λ encodes the kind of reality envisaged by the model, and the particular state λ is the cause (either deterministic or statistical) of events, such as the outcome of an individual measurement. Let x denote a particular outcome of the measurement M. Then if the ontic state is measurement-deterministic, it will determine the *outcome* of the measurement, $x = x(\lambda, M)$. If the causality is only statistical, then the ontic state will only determine the probability (more precisely, the *propensity*) of the outcome, $P(x|\lambda, M)$.

The second element, called the *epistemic state*, is a probability distribution on the space of ontic states. It describes the result of a state preparation. The preparation may not be capable of determining a particular ontic state, but presumably the repetition of a well-defined preparation Q will produce a well-defined statistical distribution of ontic states, described by the probability $P(\lambda|Q)$. The observed distribution of measurement outcomes is obtained by folding together these two probabilities. Denote by Q_ρ a preparation that yields the quantum state ρ. We require the model to reproduce the quantum

[p]The broadening of the definition of *epistemic* to refer to information, rather than merely to knowledge, is necessary if we follow the terminology of Harrigan and Spekkens, who define *ψ-epistemic* to include any ontological model that is outside of the category *ψ-ontic*. It is, however, a departure from the usage of *epistemic* in philosophy.

probabilities,

$$\int P(x|\lambda, M)\, P(\lambda|Q_\rho)\, d\lambda = \mathrm{Tr}(\rho E_x)\,, \tag{21.79}$$

where E_x is a projector corresponding to the measurement outcome x.

The next question is the status of the state vector ψ – is it ontic or epistemic? Harrigan and Spekkens identify three kinds of ontological models.

At one extrteme, we have *ψ-complete*, for which the ontic state is ψ itself. In this model, ψ gives a complete description of the individual system. The discredited interpretation (A) of Ch. 9 is an example of *ψ-complete.* Whether there are other, more acceptable, *ψ-complete* models is not known.

At the other extreme, we have *ψ-epistemic*, for which the role of ψ is only to provide the probability distribution for the ontic state λ. If the probability distributions associated with two different epistemic states overlap, then fixing the ontic state λ (which, by assumption, is a complete description of physical reality) would not be sufficient to determine the epistemic state. Thus a *ψ-epistemic* model provides a natural explanation of why it is impossible to unambiguously distinguish two non-orthogonal quantum states.

Intermediate between these two, is *ψ-supplemented*, for which ψ has both ontic and epistemic roles. Although ψ itself is ontic, it is an incomplete description of physical reality, and must be supplemented by some other ontic variables ω in order to provide a complete description of the ontic state $\lambda = (\psi, \omega)$. But ψ may also determine the probability distribution of ω, and so it has also an epistemic role.

Harrigan and Spekkens provide examples for each of these three kinds of models. They also choose to *define* the category *ψ-ontic* to include all models in which the quantum state ψ is uniquely determined by the ontic state. If ψ is one of the ontic variables, it is automatically *ψ-ontic.* Thus, in their terminology, both *ψ-complete* and *ψ-supplemented* are classified as *ψ-ontic.*

One can impose other conditions on ontological models, such as *locality* or *non-contextuality*, and thereby see Bell's theorem and the Kochen-Specker theorem from new points of view. Perhaps most importantly, with the use of ontological models, the question of *ontic* versus *epistemic* interpretations of the quantum state can be studied at a mathematical level, rather than only by philosophical debate.

Objective versus Subjective:

The pair of categories, *objective* and *subjective*, are often conflated with the pair, *ontic* and *epistemic*, but there are important differences between them.

The term *subjective* refers to some person's knowledge or belief, and *objective* refers to physically real properties that do not depend on any perceiving subject. A *ψ-ontic* model is necessarily *objective*, however a *ψ-epistemic* model need not be *subjective.*

This can be see in the example above, where the preparation yields a well-defined distribution of ontic states, $P(\lambda|Q_\rho)$. That distribution is an objective characteristic of the preparation procedure. Now if the operator understands the workings of the device, he will know the distribution that it produces. So $P(\lambda|Q_\rho)$, regarded now as a subjective probability, will also describe his knowledge. But if the operator does not understand the workings of the device, or if it is malfunctioning, then his knowledge will not include $P(\lambda|Q_\rho)$. This example illustrates the difference (and occasional similarity) between *objective information* and *subjective knowledge.* It also shows that the subjective aspect is peripheral, and often irrelevant.

However, in some applications of quantum information, such as quantum cryptography, subjective information is useful. The distinction between the roles of objective and subjective information can best be illustrated at the level of *probability*, to which information is closely related.

As was discussed in Section 1.5, there are several interpretations of probability, all of which are governed by the same mathematical formalism. Perhaps the most common is the *frequency* interpretation, according to which the equation $P(A|C) = p$ is interpreted to mean that in a long run of measurements, each preceded by the same preparation C, the outcome A will occur in (approximately) a fraction p of the cases.

The *propensity* interpretation is closely related, and can be regarded as a foundation for the frequency interpretation. Propensity is a form of causality that is weaker than determinism. In this interpretation, the value of $P(A|C)$ measures the tendency for the condition C to produce the event A. The connection of *propensity* to *frequency* is deduced in the *Law of Large Numbers* (Eq. (1.58)), which makes precise the idea that the observed frequency of the event A should be close to the propensity p. Both of these related interpretations of probability are *objective*, referring only to the physical condition C, and not to any agent's knowledge.

Another kind of probability is that used in *inductive inference*, which may be regarded as a generalization of deductive logic. In this use of probability, the function $P(A|C)$ is interpreted as the degree to which the *proposition* A is supported by the *evidence* C. If $P(A|C)$ were to take on only the values 0 or 1, then this *inferential* system would reduce to deductive logic. This system of inductive inference is also objective, since it uses only *evidence*, not opinion or belief, as the condition C.

A *subjective* version of inferential probability is obtained by interpreting $P(A|C)$ as some person's *degree of belief* in the proposition A, in the light of the information C. If all persons were fully rational, then presumably every person's subjective probability would agree with the objective inferential probability, with differences arising only because different persons may possess different information. Both the objective and the subjective versions are examined in detail in the book by Jaynes (2003).

Quantum States do not describe Knowledge or Belief

The following example involves both *objective* and *subjective* probabilities in essential ways. It demonstrates that the two concepts are distinct, and play different roles. The situation is a game in which the Operator prepares a photon in a certain state of linear polarization, but keeps his preparation secret. He tells Alice that the polarization is in one of two states, and he tells Bob that it is in one of four states. We will attempt to represent Alice's and Bob's knowledge of the photon's state in the mathematical language of quantum mechanics.

Let us denote the vertical and horizontal polarizations by the information-basis vectors, $|0\rangle$ and $|1\rangle$, and denote the ± 45 degree polarizations by the normalized sum and difference vectors, $|+\rangle = (|0\rangle + |1\rangle)/\sqrt{2}$ and $|-\rangle = (|0\rangle - |1\rangle)/\sqrt{2}$. The Operator tells Alice that the polarization state of the photon is either vertical or horizontal ($|0\rangle$ or $|1\rangle$). He tells Bob that it is one of the four states defined above, whose density matrices are

$$\rho_0 = \begin{pmatrix} 1 & 0 \\ 0 & 0 \end{pmatrix}, \quad \rho_1 = \begin{pmatrix} 0 & 0 \\ 0 & 1 \end{pmatrix}, \tag{21.80}$$

$$\rho_+ = \begin{pmatrix} 1/2 & 1/2 \\ 1/2 & 1/2 \end{pmatrix}, \quad \rho_- = \begin{pmatrix} 1/2 & -1/2 \\ -1/2 & 1/2 \end{pmatrix}. \tag{21.81}$$

The actual prepared state of the photon is ρ_0, but Alice and Bob do not know this, so they must use their incomplete knowledge to make predictions. Alice knows that the state is either ρ_0 or ρ_1, so she may combine them with equal probability, obtaining a "subjective density matrix"[q] (SDM),

$$\rho_s = \begin{pmatrix} 1/2 & 0 \\ 0 & 1/2 \end{pmatrix}. \tag{21.82}$$

Bob knows only that the prepared state is one of the four states above, so he may combine them with equal probability, to obtain the same SDM, ρ_s. Notice

[q] A subjective density matrix is an attempt to represent someone's knowledge by the formalism of QM. It should not be confused with a "real" density matrix, which describes a physical state preparation.

that, even though Alice and Bob have different information, they both obtain the same SDM. Nevertheless, their future predictions will not necessarily be the same.

Suppose the Operator now performs a measurement that distinguishes between vertical and horizontal polarizations. The result, of course, confirms that the polarization is vertical. We, the readers, knew this in advance, but to Alice and Bob it is new information. Alice will now correctly infer that the initial state must have been ρ_0, since the only other possibility, ρ_1, has been ruled out by the measurement. But Bob, who previously knew only that the initial state was one of four, can only reduce the number of possibilities to three. However, he can update his information using Bayes theorem (21.12), which is copied here for convenience.

$$P(\rho|D\&C) = P(D|\rho\&C)\frac{P(\rho|C)}{P(D|C)} \,. \tag{21.83}$$

Here C is Bob's prior information (that the initial state was one of the original four), D is the data from the measurement (that the initial state was not ρ_1), and ρ is one of the states that are still allowed by his new information. Of course $P(D|\rho\&C) = P(D|\rho)$.

Because the set of possible states allowed by C has discrete rotational symmetry, the prior probability $P(\rho|C)$ is the same for all four of the ρ's, hence $P(\rho|C) = 0.25$. Similarly, the prior probability for the two possible results of the polarization measurement, $D = 0$ and $D = 1$, is $P(D|C) = 0.5$. The probability of observing vertical polarization in the measurement, $P(D{=}0|\rho)$, for each of the initially possible states, is

$$P(0|\,\rho_0) = 1\,, \quad P(0|\,\rho_1) = 0\,, \quad P(0|\,\rho_+) = 0.5\,, \quad P(0|\,\rho_-) = 0.5\,. \tag{21.84}$$

Substituting all of these values into (21.83) yields Bob's posterior probability for each of the four possible initial states to be

$$P(\rho_0) = 0.5\,, \; P(\rho_1) = 0\,, \quad P(\rho_+) = 0.25\,, \quad P(\rho_-) = 0.25\,. \tag{21.85}$$

(Here the second argument of $P(\rho|\cdot)$ has been omitted for brevity.) Finally, Bob can construct his new SDM by mixing the four states, each weighted by its probability from (21.85),

$$\rho_B = \begin{pmatrix} 0.75 & 0 \\ 0 & 0.25 \end{pmatrix} . \tag{21.86}$$

The term "subjective density matrix" (SDM) has been enclosed in quotes because it does not function as an ordinary density matrix. The normal role of a density matrix (or state operator) ρ is to encode the information provided by the physical operation of state preparation. The statistics of future measurements can then be calculated from the usual formula for quantum probabilities, $\mathrm{Tr}(\rho E_x)$, where E_x is a projector characterizing the outcome of the measurement. This quantum state and the associated quantum probabilities are *objective*, since they make no reference to any person's knowledge. The SDM, on the other hand, is an attempt to represent someone's knowledge by a quantum-like density matrix.

Some writers have attempted an alternative interpretation of quantum states as encoding the *subjective knowledge* of an observer. But the SDM does not fulfill that role, as is shown by this simple game. Firstly, although Alice and Bob begin the game with different information about the prepared state, they assign the same SDM (21.82). Second, although they have the same SDM, they make different inferences after learning the result of the polarization measurement. Alice correctly identifies the prepared state, but Bob can only slightly sharpen his information. Evidently, the SDM does not describe the different states of knowledge possessed by Alice and by Bob. Third, Bob's final state of knowledge is given by the probabilities in (21.85). But there is no way that they can be recovered from his SDM (21.86), since a particular mixed-state density matrix can be written as a sum of other states in infinitely many ways. One might guess that the *diagonal* representation of ρ_B might have some special significance, compared to other representations. But that form (21.86) yields the mixture $\rho_B = 0.75\rho_0 + 0.25\rho_1$. This certainly *does not* correspond to Bob's knowledge, for he knows the horizontal polarization state ρ_1 has zero probability.

I conclude that the SDM does not correctly describe the players' knowledge, at any stage of this game. Consequently, *the interpretation of quantum states as representing knowledge is untenable,* even though there may be some cases in which it gives plausible results. A person may have knowledge *about* a quantum state, or about its preparation, but the state itself does not represent knowledge.

Summary

The growth of *quantum information* as a new branch of quantum theory has, as a side effect, stimulated a renewed interest in the foundations and interpretation of quantum mechanics. But it has not yet resolved all questions. We have seen how results from QI can be used to argue both *for* and *against* an *individual* interpretation of quantum states. However, it has introduced new ideas into an old subject, and has revealed the complexity of the issues.

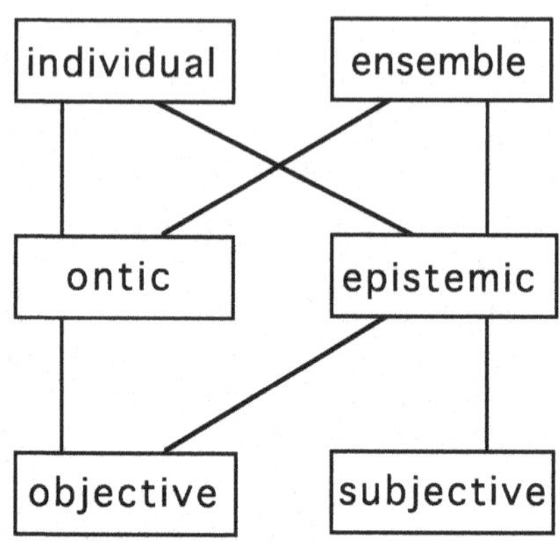

Fig. 21.1 Relations among various classes of models and interpretations. The lines between boxes indicate overlap of membership.

The relations among the various classes of models and interpretations are shown in Fig. 21.1. A line connecting two boxes indicates an overlap of membership between them. Notice, in particular, that all subjective interpretations are epistemic, but not all epistemic interpretations are subjective. Similarly, all ψ-ontic models are objective, but not all objective interpretations need be ψ-ontic. It is *not correct* to assert (as is all too often done) that if ψ is not an element of reality then it must represent knowledge. The range of alternatives is much greater.

A specific interpretation involves a choice of one category from each of the three rows in Fig. 21.1. In the top row, the most common *individual* interpretation (interpretation (A) in Ch. 9) has been rejected because it fails to give a satisfactory description of the measurement process. In the bottom row, the *subjective* interpretation of quantum states has been rejected because a density matrix fails to describe a person's knowledge of a quantum system. The second row, *ontic* versus *epistemic*, is the subject of active research. Some papers claim to exclude *ψ-epistemic* models, but they all introduce some additional assumptions that can be questioned. The issue cannot yet be regarded as settled.

Further reading for Chapter 21

Textbooks on Quantum Information:

Bennett, DiVincenzo, and Wootters (2009) – undergraduate level.

Barnett (2009) – postgraduate level.

General reference:

Alber *et al.* (2001) – Contains five chapters by different authors. Ch. 2 is a good introduction to the theory. Ch. 3 covers experimental aspects. Ch. 5 is a thorough account of entanglement theory.

Quantum Computing:

Barnett (2009) – an introduction to many aspects of quantum information, including *quantum circuits* and their uses.

Pittenger (2000) – quantum computing algorithms.

Nakahara and Ohmi (2008) – treats mathematical, programming, and hardware aspects of quantum computing.

Steeb and Hardy (2006) – a large collection of problems and solutions in quantum computing and quantum information.

ψ-ontic versus *psi-epistemic:*

Harrigan and Spekkens (2010) – the paper in which these concepts were introduced. Note that in later work, Spekkens does not rely on the concept of *separability*.

Pusey, Barrett and Rudolph (2012) – proves a theorem which seems to imply that *ψ-epistemic* models are impossible.

Lewis, Jennings, Barrett and Rudolph (2012) – shows that the previous paper used an additional (but plausible) assumption, without which the claimed theorem would be false.

Problems

21.1 Show that the entropy function (21.1) achieves its maximum value for a flat probability distribution, for which every p_i has the same value.

21.2 In Example 2 of Section 21.3, the tester's goal was to distinguish, as well as possible, between the non-orthogonal states $|1\rangle$ and $|+\rangle$. She chose a measurement that distinguished between $|1\rangle$ and $|0\rangle$, equivalent to measuring the observable $A = |1\rangle\langle 1| - |0\rangle\langle 0|$. Suppose that, instead, she had chosen to measure $B = |1\rangle\langle 1| - |+\rangle\langle +|$. What are the eigenvalues and eigenvectors of this operator? Carry out a similar calculation to Example 2, and obtain the final probabilities, analogous to the table (21.20). Does this new measurement strategy yield more or less information about the unknown state than did Example 2?

21.3 A Hermitian operator is invariant under the combined operations of transposition and complex conjugation. Therefore the transposition of a density matrix is equivalent to complex conjugation, and so is an *anti-linear* operator, like *time-reversal.*

(a) Consider the spin singlet state, $(|\uparrow\rangle|\downarrow\rangle - |\downarrow\rangle|\uparrow\rangle)/\sqrt{2}$. Construct its density matrix in the standard basis (labeled by $\uparrow\uparrow$, $\uparrow\downarrow$, $\downarrow\uparrow$, $\downarrow\downarrow$). Perform a partial time-reversal on the first component only. Verify that the result satisfies all the conditions required of a density matrix (especially non-negativity).

(b) Now perform a partial-transposition on the singlet density matrix. Is it different from the result of partial time-reversal in (a)? Is it non-negative?

21.4 Show that in the quantum teleportation protocol of Section 21.6, the entangled state $|\Phi_+^{ab}\rangle = (|0^a\rangle|0^b\rangle + |1^a\rangle|1^b\rangle)/\sqrt{2}$ may be used instead of the singlet Ψ_-^{ab}, with only minor changes to the protocol.

21.5 Add a third player, Carol, to the game that shows that quantum states do not represent knowledge. The Operator tells Carol that the prepared state is one of three: ρ_0, ρ_+, or ρ_-.

(a) What is her initial SDM?

(b) After the measurement shows that the prepared polarization cannot have been horizontal, what is her final SDM? Show that it does not represent her knowledge that the probability of the state ρ_1 is zero.

Appendix A

Schur's Lemma

Schur's lemma states that *a self-adjoint set of operators is irreducible if and only if any operator that commutes with all members of the set is a multiple of the identity operator.*

A *self-adjoint set* S is defined by the condition that if the operator T is a member of S then so is $T^\dagger$. To say that a set of operators is *irreducible* on a vector space V means that no subspace of V is invariant under the action of all operators in the set. If there is a subspace V_1 (other than V itself) such that if $|\phi\rangle \in V_1$ then also $T|\phi\rangle \in V_1$ for every operator T in S, we say that the set S is *reducible* (and also that S reduces V). Otherwise S is irreducible.

Let S be a self-adjoint set, and let S' be the set of operators that commute with all members of S. Thus if $R \in S'$ then $[T, R] = 0$ for all $T \in S$. Since S is a self-adjoint set, it follows that $R^\dagger$ is also a member of S'. Without loss of generality, we may consider only those operators in S' that are Hermitian, since an arbitrary member of S' is expressible as a linear combination of Hermitian operators in S' : $R = R_1 + iR_2$, with $R_1 = \frac{1}{2}(R + R^\dagger)$ and $R_2 = (R - R^\dagger)/2i$. Hence we take R to be Hermitian.

Let $\{|\phi_n\rangle\}$ be a complete orthogonal set of eigenvectors of R, with $R|\phi_n\rangle = r_n|\phi_n\rangle$. Define V_m to be the subspace spanned by those eigenvectors of R whose eigenvalue is r_m, and denote by $V_m^\perp$ the complementary subspace of vectors orthogonal to V_m. Since $[T, R] = 0$, we obtain $0 = \langle\phi_m|(TR - RT)|\phi_n\rangle = (r_n - r_m)\langle\phi_m|T|\phi_n\rangle$ for all T in the set S, where $|\phi_m\rangle \in V_m$ and $|\phi_n\rangle \in V_m^\perp$. Now if $r_n \neq r_m$ we would have $T|\phi_n\rangle$ orthogonal to $|\phi_m\rangle$. This would hold for all T in the set S and all of the eigenvectors $|\phi_n\rangle$ that span $V_m^\perp$, and hence the subspace $V_m^\perp$ would be invariant. But that is impossible if the set S is irreducible, so in this case we must have $r_n = r_m$ for all m and n. If R has only one distinct eigenvalue, then R is a multiple of the identity operator. Thus any operator R that commutes with all members of the irreducible set S can only be a multiple of the identity.

Conversely, if the set S were to reduce the space V into nontrivial invariant subspaces V_1 and $V^{\perp}$, we could choose two unequal numbers, r_1 and r_2, and define an operator R such that $R|\phi_m\rangle = r_1|\phi_m\rangle$ for any $|\phi_m\rangle \in V_1$ and $R|\phi_n\rangle = r_2|\phi_n\rangle$ for any $|\phi_n\rangle \in V^{\perp}$. Now $T|\phi_m\rangle \in V_1$ and $T|\phi_n\rangle \in V^{\perp}$ for every operator T in S, and hence it follows from the definition of R that $[T, R] = 0$. Thus we have an operator (not a multiple of the identity) that commutes with all members of the set S if S is not irreducible. So the lemma is proven.

Appendix B

Irreducibility of Q and P

For a single particle without internal degrees of freedom, we may use the coordinate representation (Ch. 4), in which the effect of the position operator Q is to multiply by the coordinate x, and the momentum operator P is the differential operator $-i\hbar\partial/\partial x$. (For simplicity we consider only one spatial dimension.) To prove that the set of operators $\{x, \partial/\partial x\}$ is *irreducible* in the sense of Schur's lemma, we shall show that any operator which commutes with both x and $\partial/\partial x$ must be a multiple of the identity.

Let M be an operator that commutes with both x and $\partial/\partial x$. Then

$$Mx\psi(x) = xM\psi(x) \quad \text{for all} \quad \psi(x)\,. \tag{B.1}$$

If M commutes with x, it must also commute with any function of x, $f(x)$, and hence $Mf(x)\psi(x) = f(x)M\psi(x)$. Choosing the particular function $\psi(x) = 1$, we obtain

$$Mf(x) = f(x)m(x)\,, \tag{B.2}$$

where by definition $m(x) = M1$ is the function that is produced by the operator M when acting on the particular function $\psi(x) = 1$. By hypothesis, we also have

$$\frac{\partial}{\partial x}M\psi(x) = M\frac{\partial\psi}{\partial x}\,. \tag{B.3}$$

Now (B.2) holds for any function $f(x)$, so it holds in particular for the functions $\psi(x)$ and $\partial\psi/\partial x$. Thus (B.3) yields

$$\frac{\partial}{\partial x}m(x)\psi(x) = m(x)\frac{\partial\psi}{\partial x}\,,$$

which is consistent only if $\partial m/\partial x = 0$. Therefore $m(x)$ is a constant, and the effect of the operator M is only to multiply by this constant. In other words, any operator M that commutes with both x and $\partial/\partial x$ must be a multiple of the identity, and so by Schur's lemma the set $\{x, \partial/\partial x\}$ is irreducible.

Appendix C

Proof of Wick's Theorem

This theorem, discussed in Section 17.4, will first be proved for generalized fermion operators. Let each of the operators $A_1, A_2, A_3, \ldots, A_n$ be either a creation operator $(C_\alpha{}^\dagger)$ or an annihilation operator (C_α). We assume that:

(a) There is a vector $|0\rangle$ such that $C_\alpha|0\rangle = 0$ for all α.
(b) The anticommutator of any two of the operators, $[A_j, A_k]_+ \equiv A_j A_k + A_j A_k$, is a multiple of the identity.

Define the *normal-ordered* product, $N(A_1, A_2, \ldots)$, of a set of operators as the product of those operators reordered so that all creation operators are to the left of all annihilation operators, multiplied by a factor (-1) for each pair interchange needed to produce the normal order. Define the *contraction* of two operators as $\langle A_1 A_2 \rangle = \langle 0|A_1 A_2|0\rangle$. Then the theorem states that any product of these operators is equal to its normal product expansion, which has the form

$$\begin{aligned} A_1 A_2 A_3 \cdots A_n &= N(A_1, A_2, A_3, \ldots, A_n) \\ &\quad + \underset{j<k}{\rightarrow} \sum\sum (-1)^{j+k-1} \langle A_j A_k \rangle \\ &\quad \times N(A_1, \ldots, A_{j-1}, A_{j+1}, \ldots, A_{k-1}, A_{k+1}, \ldots, A_n) \\ &\quad + \sum (-1)^P (2 \text{ contracted pairs}) \times (N\text{-product of } n-4 \text{ operators}) \\ &\quad + \sum (-1)^P (3 \text{ contracted pairs}) \times (N\text{-product of } n-6 \text{ operators}) \\ &\quad + \cdots \end{aligned} \tag{C.1}$$

Here P denotes the number of pair interchanges (modulo 2) needed to remove the contracted terms from the product. We shall prove the theorem by induction.

We first establish that

$$N(A_1, A_2, \ldots, A_n)A_{n+1} = N(A_1, A_2, \ldots, A_n, A_{n+1}) + \sum_{j=1}^{n}(-1)^{j+n}\langle A_j A_{n+1}\rangle N(A_1, \ldots, A_{j-1}, A_{j+1}, \ldots, A_n)\,. \quad \text{(C.2)}$$

If A_{n+1} is an annihilation operator the product on the left is already in normal order, and all of the contractions $\langle A_j A_{n+1}\rangle$ are zero by virtue of assumption (a), and so (C.2) is true in this case. If A_{n+1} is a creation operator, then we have

$$N(A_1, A_2, \ldots, A_n)A_{n+1} = (-1)^n A_{n+1}N(A_1, A_2, \ldots, A_n) + \sum_{j=1}^{n}(-1)^{j+n}[A_j, A_{n+1}]_+ N(A_1, \ldots, A_{j-1}, A_{j+1}, \ldots, A_n)\,.$$

But $N(A_1, A_2, \ldots, A_n, A_{n+1}) = (-1)^n A_{n+1}N(A_1, A_2, \ldots, A_n)$, and from (a) and (b) we have $\langle A_j A_{n+1}\rangle = [A_j, A_{n+1}]_+$, so (C.2) is true in this case too.

The effect of multiplying (C.1) on the right by A_{n+1} is to produce in every term of (C.1) a product of the form $N(A_1, A_2, \ldots)A_{n+1}$. According to the result (C.2), this is equal to

$$N(A_1, A_2, \ldots)A_{n+1} = N(A_1, A_2, \ldots, A_{n+1}) + \sum \text{(all terms containing a contraction involving } A_{n+1})\,.$$

Thus (C.1) remains true upon right multiplication by A_{n+1}, provided it was true for n operators. It was demonstrated in Section 17.4 that Wick's theorem is true for $n = 2$, so by induction the theorem is true for all $n \geq 2$.

The proof of Wick's theorem for boson operators follows the same pattern, except that anticommutators must be replaced by commutators, and the factors of (-1) for permutations do not appear.

[[Wick's theorem is often stated for a *time-ordered* product of operators, rather than for a general product as we have done. This is because of the context in which the theorem was first applied. Actually the notions of time dependence and time ordering are extraneous to the mathematics, and there is no gain in simplicity in return for the loss of generality. To apply the theorem to time-ordered products (not considered in this book), we simply substitute that particular product in place of the general product $A_1A_2A_3 \cdots A_n$ in (C.1).]]

Appendix D

Solutions to Selected Problems

1.10 Because of the inclusion relation $\Omega \subset \mathcal{H} \subset \Omega^{\times}$, we need only identify the smallest of the three spaces that contains the vector. (a) $\Omega^{\times}$. (b) $\mathcal{H}$. (c) $\Omega^{\times}$. (d) None of the three. To prove this, consider the function $\phi(x) = e^{-c|x|}$, which clearly belongs to Ω provided $c > 0$. For $f(x)$ to belong to $\Omega^{\times}$, it is necessary that $(\phi, f) = \int_{-\infty}^{\infty} \phi^*(x) f(x) dx$ be finite for all ϕ in Ω. But $\int_{-\infty}^{\infty} e^{-c|x|} e^{-ax} dx$ diverges if $a > c$, so e^{-ax} is not in $\Omega^{\times}$. (e) $\mathcal{H}$. For any $\varepsilon > 0$, there is a constant C such that $\log(1 + |x|) < C|x|^{\varepsilon}$ as $|x| \to \infty$. Hence it is easy to show that the function $\log(1 + |x|)/(1 + |x|)$ is square-integrable. (f) Ω. (g) Ω.

1.11 According to Theorem 1, Sec. 1.3, an operator H is Hermitian if $\langle\psi|H|\psi\rangle - \langle\psi|H|\psi\rangle^* = 0$ for all complex functions ψ. Therefore we shall calculate the quantity

$$R = \int \{\psi^* \nabla^2 \psi - \psi \nabla^2 \psi^*\} d\tau$$

within some volume, and determine the conditions under which it will vanish. Using the identity $\boldsymbol{\nabla} \cdot (\psi^* \boldsymbol{\nabla} \psi) = (\boldsymbol{\nabla}\psi)^* \cdot \boldsymbol{\nabla}\psi + \psi^* \nabla^2 \psi$, we obtain

$$\begin{aligned} R &= \int \boldsymbol{\nabla} \cdot (\psi^* \boldsymbol{\nabla} \psi - \psi^* \boldsymbol{\nabla} \psi) d\tau \\ &= \iint (\psi^* \boldsymbol{\nabla} \psi - \psi^* \boldsymbol{\nabla} \psi) \cdot d\mathbf{S}\,, \end{aligned}$$

where the last integral is over the closed surface bounding the volume of the first integration. Therefore we will have $R = 0$ for all ψ, and the operator ∇^2 will be Hermitian, if ψ and all functions in the linear space to which it belongs satisfy some boundary condition that ensures the vanishing of the surface integral. Some satisfactory boundary conditions for a finite volume are: (a) $\psi = 0$ on the bounding surface; (b) $\hat{\mathbf{n}} \cdot \nabla\psi = 0$, where $\hat{\mathbf{n}}$ is normal to the surface; (c) periodic boundary conditions on a surface whose opposite sides are parallel. If the volume is infinite, then we must have: (d) the integrand of the surface integral must go to zero more rapidly than r^{-2} in the limit as $r \to \infty$, so that the surface integral vanishes in the limit $r \to \infty$.

1.13 (a) The probability of life in the vicinity of some arbitrarily selected star is equal to $pqr = 10^{-6}$, assuming that the three conditions are independent.

(b) The probability P that life exists in the vicinity of at least one star is given by $P = 1 - P_0$, where P_0 is the probability that no stars have life about them. The probability of no life about some arbitrarily selected star is $1 - pqr$, so we have $P_0 = (1 - pqr)^N$. Thus we have

$$\log(P_0) = N \log(1 - pqr) \approx N(-pqr) = -10^{-5},$$

$$P_0 = e^{-pqrN} = \exp(-10^{-5}) \ll 1.$$

Hence $P = 1 - e^{-pqrN}$, which is very close to 1. Even a very rare event is almost certain to occur in a large enough sample.

1.17 Denote the probability that exactly n particles are emitted in the time interval t as $P_n(t)$. Since the average emission rate is λ particles per second and each emission event is independent of all others, the probability of an event must be the same in any short time interval of duration h. It seems intuitively clear that this probability should be equal to λh (neglecting corrections that vanish more rapidly than h), and that the probability of more than one event occurring within the interval h should be of higher order in h, and hence negligible in the limit $h \to 0$. (These simplifying assumptions will be confirmed in the final solution.)

Suppose that there are n events ($n > 0$) during the interval $t + h$. This could happen in two ways: there could have been n events in the interval t and none in the following interval h, or there could have been $n - 1$ events in the interval t and one in the interval h. Since the occurrence of events in the intervals t and h are independent, we have

$$P_n(t + h) = P_n(t)(1 - \lambda h) + P_{n-1}(t)\lambda h, \quad (n > 0).$$

Dividing by h and taking the limit $h \to 0$, we obtain

$$P_n{}'(t) = -\lambda P_n(t) + \lambda P_{n-1}(t), \quad (n > 0),$$

where $P_n{}'(t)$ is the derivative of $P_n(t)$ with respect to t. For $n = 0$ we have simply

$$P_0{}'(t) = -\lambda P_0(t).$$

These differential equations can easily be solved successively, subject to the boundary conditions $P_0(0) = 1$, $P_n(0) = 0$ for $n > 0$. The solutions are $P_0(t) = e^{-\lambda t}$, $P_1(t) = \lambda t e^{-\lambda t}$, and in general

$$P_n(t) = \frac{(\lambda t)^n e^{-\lambda t}}{n!}.$$

This is known as the *Poisson distribution.* It is apparent that it is correctly normalized: $\Sigma_n P_n(t) = 1$. The average number of events in the time interval t may now be calculated to be

$$\begin{aligned}\langle n\rangle &= \sum_n n P_n(t) = e^{-\lambda t} \sum_n n \frac{(\lambda t)^n}{n!} \\ &= e^{-\lambda t} \lambda t \frac{d}{d(\lambda t)} \sum_n \frac{(\lambda t)^n}{n!} = e^{-\lambda t} \lambda t \frac{d}{d(\lambda t)} e^{\lambda t} \\ &= \lambda t,\end{aligned}$$

which confirms our initial hypothesis. The Poisson distribution plays an important role in the study of photon statistics in Ch. 19.

2.5 (a) Not acceptable, has a negative eigenvalue; (b) pure state, with state vector $(3/5, 4/5)$; (c) pure state, with state vector $(|u\rangle + \sqrt{2}|v\rangle)/\sqrt{3}$; (d) not acceptable, has a negative eigenvalue; (e) acceptable, not a pure state.

3.3 Let $f(x) = e^{xA} B e^{-xA}$. We wish to expand this operator in powers of the parameter x:

$$f(x) = f(0) + x \left.\frac{df}{dx}\right|_{x=0} + \frac{x^2}{2} \left.\frac{d^2 f}{dx^2}\right|_{x=0} + \cdots$$

From the definition of $f(x)$, we obtain

$$\frac{df(x)}{dx} = Af(x) - f(x)A = [A, f(x)].$$

Differentiating again with respect to x, we obtain

$$\frac{d^2 f(x)}{dx^2} = \left[A, \frac{df(x)}{dx}\right] = [A, [A, f(x)]]\,.$$

Clearly this kind of result generalizes to any order of derivative. Since $f(0) = B$, the power series is

$$e^{xA} B e^{-xA} = B + [A, B]x + \frac{[A, [A, B]]x^2}{2} + \frac{[A, [A, [A, B]]]x^3}{6} + \cdots$$

3.4 Consider the operator $f(x) = e^{xA} e^{xB}$, where x is a parameter. Then

$$\begin{aligned} df(x)/dx &= A e^{xA} e^{xB} + e^{xA} B e^{xB} \\ &= (A + e^{xA} B e^{-xA}) f(x)\,. \end{aligned}$$

Using the result of Problem 3.3, and assuming that $[A, [A, B]] = 0$, we obtain

$$\frac{df(x)}{dx} = \{A + B + [A, B]x\} f(x)\,.$$

We now assume also that $[B, [A, B]] = 0$, so that $(A + B)$ and $[A, B]$ commute. The solution of the above differential equation then becomes

$$f(x) \equiv e^{xA} e^{xB} = \exp\left\{(A + B)x + \frac{[A, B]x^2}{2}\right\}.$$

Putting $x = 1$, we deduce that if $[A, [A, B]] = 0$ and $[B, [A, B]] = 0$ then

$$\begin{aligned} e^{(A+B)} &= e^A e^B e^{-[A,B]/2} \\ &= e^B e^A e^{[A,B]/2}\,, \end{aligned}$$

the last time being obtained by interchanging A and B.

3.7 The desired transformation, $U(\mathbf{v}, t) = \exp(i\mathbf{v} \cdot \mathbf{G}_t)$, is a combination of an instantaneous Galilei transformation, which affects the velocity operator but not the position operator, and a space displacement through the distance $\mathbf{v}t$. In view of the result of Problem 3.4, it seems appropriate to try $\mathbf{G}_t = M\mathbf{Q} - t\mathbf{P}$, which is the sum of the generators of the two transformations.

(We take $\hbar = 1$ for convenience.) We can now use the result of Problem 3.3 to calculate the effect of this transformation.

$$\exp(i\mathbf{v}\cdot\mathbf{G}_t)Q_\alpha\exp(-i\mathbf{v}\cdot\mathbf{G}_t) = Q_\alpha + [i\mathbf{v}\cdot\mathbf{G}_t, Q_\alpha] + \cdots$$

The commutator has the value $[i\mathbf{v}\cdot\mathbf{G}_t, Q_\alpha] = -[i\mathbf{v}\cdot\mathbf{P}t, Q_\alpha] = -v_\alpha tI$. Since this is a multiple of the identity operator, all the higher order commutator terms are zero, and hence

$$\exp(i\mathbf{v}\cdot\mathbf{G}_t)Q_\alpha \exp(-i\mathbf{v}\cdot\mathbf{G}_t) = Q_\alpha - v_\alpha tI\,.$$

Similarly we have

$$\exp(i\mathbf{v}\cdot\mathbf{G}_t)P_\alpha \exp(-i\mathbf{v}\cdot\mathbf{G}_t) = P_\alpha + [i\mathbf{v}\cdot\mathbf{G}_t, P_\alpha] + \cdots$$

The commutator has the value $[i\mathbf{v}\cdot\mathbf{G}_t, P_\alpha] = [iM\mathbf{v}\cdot\mathbf{Q}, P_\alpha] = -Mv_\alpha I$. Again all higher order commutator terms are zero, and hence

$$\exp(i\mathbf{v}\cdot\mathbf{G}_t)P_\alpha \exp(-i\mathbf{v}\cdot\mathbf{G}_t) = P_\alpha - Mv_\alpha I\,.$$

Dividing this equation by M yields the correct transformation for velocity operator $V_\alpha = P_\alpha/M$, so our choice of $\mathbf{G}_t = M\mathbf{Q} - t\mathbf{P}$ for the generator has proved to be correct.

We note in passing that $\mathbf{G}_t$ is *not* equal to the Heisenberg time-dependent operator obtained from the Schrödinger operator $\mathbf{G} = M\mathbf{Q}$. According to (3.72), that operator is $e^{iHt}M\mathbf{Q}e^{-iHt}$, with $H = P^2/2M$. Using Problem 3.3, we determine this Heisenberg operator to be $M\mathbf{Q} + t\mathbf{P}$.

4.5 Let us choose units such that $\hbar = 2M = 1$, and write the spherically symmetric state function as $\psi(r) = u(r)/r$. Then the stationary state Schrödinger equation, in spherical coordinates, becomes

$$u''(r) + [E - W(r)]u(r) = 0\,,$$

where $u''(r)$ is the second derivative of $u(r)$. For the potential $W(r) = C/r^n$ with $n > 2$, this differential equation has an *irregular singular point* at $r = 0$, and it is easily verified that $u(r)$ cannot behave as any power of r (positive, negative, or fractional) in the neighborhood of $r = 0$. Instead we shall try $u(r) = \exp(-a/r^\gamma)f(r)$, where $f(r)$ is a smooth function, and $\gamma > 0$. In the limit $r \to 0$ the dominant contribution to $u''(r)$ is $u''(r) \approx (a^2\gamma^2/r^{2\gamma+2})\exp(-a/r^\gamma)f(r)$. Collecting the coefficients of the dominant singular terms in the differential equation, we obtain

$$\frac{a^2\gamma^2}{r^{2\gamma+2}} - \frac{C}{r^n} = 0\,.$$

Therefore we must have $2\gamma + 2 = n$, and $a^2\gamma^2 = C$. Thus we have $\gamma = \frac{1}{2}n - 1$. (Since $\gamma > 0$ is a condition for validity of our solution, it is apparent that the case $n > 2$ is qualitatively different from that of $n \leq 2$.)

For $C > 0$ (repulsive potential) we obtain $a = C^{1/2}/\gamma > 0$. Thus $u(r)$ vanishes rapidly as $r \to 0$, yielding a physically acceptable solution.

For $C < 0$ (attractive potential) the parameter $a = i|a|$ is pure imaginary. Thus the state function is of the form $\psi(r) = A(r)e^{-S(r)}$, with amplitude $A(r) = f(r)/r$ and phase $S(r) = -i|a|/r^\gamma$. From (4.22b), $\mathbf{J} = A^2\boldsymbol{\nabla}S$, the radial probability flux goes as $J_r \propto 1/r^{\gamma+3}$. The integrated flux over a small sphere of radius r will be proportional to $1/r^{\gamma+1}$. For any $\gamma > 0$ this would correspond to a sink (or source) at $r = 0$, and so it is physically unacceptable. In physical terms, for $n > 2$ the attractive potential is so strong that any state would collapse into the center.

4.6 In one spatial dimension, the condition div $\mathbf{f} = 0$ reduces to $\partial f/\partial x = 0$. Hence the continuity equation for the flux $J(x,t)$ is also satisfied if we add to $J(x,t)$ an arbitrary function $f(t)$, provided only that $f(t)$ is independent of position. For any physically realizable state we must have $\Psi(x,t) \to 0$ and $J(x,t) \to 0$ as $|x| \to \infty$. Since $f(t)$ does not depend on x, this condition can be satisfied only if $f(t) \equiv 0$. Thus the usual identification of the probability flux is unique in one dimension, but this argument does not generalize to three dimensions.

4.10 The Lagrangian of a free particle is $\mathcal{L} = \frac{1}{2}mv^2$, where $v = dx/dt$. Since the velocity is constant along the classical path, we have $v = (x_2-x_1)/(t_2-t_1)$, and the action is $S = \int \mathcal{L}dt = \frac{1}{2}m(x_2-x_1)^2/(t_2-t_1)$. The de Broglie wavelength is $\lambda = mv/2\pi\hbar$, and hence the Feynman phase factor, $e^{iS/\hbar}$, is equal to $e^{i\pi(x_2-x_1)/\lambda}$.

Surprise? The quantity $(x_2 - x_1)/\lambda$ is equal to the number of wavelengths in the path. But one would expect each wavelength to contribute 2π to the phase, rather that only π as in the above answer.

The paradox is resolved by noting that we are dealing with a path in space–time, for which the action can be written as $\int(pdx - Edt)$. The first (momentum) term contributes the expected 2π per wavelength, and the second (energy) term cancels half of the first term.

5.4 For convenience we shall adopt units such that $\hbar = 2M = 1$. We shall also use box normalization in a cube of side L, so that our (discrete) momentum eigenvectors are orthonormal: $\langle \mathbf{k}'|\mathbf{k}\rangle = \delta_{\mathbf{k}',\mathbf{k}}$. In momentum representation the eigenvalue equation $(H - E)|\Psi\rangle = 0$ becomes

$$(k^2 - E)\langle \mathbf{k}|\Psi\rangle + \sum_{\mathbf{k}'} \langle \mathbf{k}|W|\mathbf{k}'\rangle \, \langle \mathbf{k}'|\Psi\rangle = 0\,.$$

The lattice periodicity of the potential W is expressed by the equation

$$U(\mathbf{R}_n) W U^\dagger(\mathbf{R}_n) = W\,,$$

where $U(\mathbf{R}_n) = \exp(-i\mathbf{P}\cdot\mathbf{R}_n)$ and $\mathbf{R}_n$ is given by (5.16). The periodicity of the potential implies that its matrix in momentum representation must satisfy

$$\begin{aligned}\langle \mathbf{k}|W|\mathbf{k}'\rangle &= \langle \mathbf{k}|U(\mathbf{R}_n)\, W U^\dagger(\mathbf{R}_n)|\mathbf{k}\rangle \\ &= \exp\left[i(\mathbf{k}' - \mathbf{k})\cdot\mathbf{R}_n\right] \langle \mathbf{k}|W|\mathbf{k}'\rangle\,.\end{aligned}$$

This is possible only if $\exp[i(\mathbf{k}' - \mathbf{k})\cdot\mathbf{R}_n] = 1$ for all lattice vectors $\mathbf{R}_n$. Therefore the matrix element $\langle \mathbf{k}|W|\mathbf{k}'\rangle$ must vanish unless $\mathbf{k}' - \mathbf{k} = \mathbf{G}$, where $\mathbf{G}$ is some reciprocal lattice vector. Thus the energy eigenvalue equation becomes

$$(k^2 - E)\langle \mathbf{k}|\Psi\rangle + \sum_{\mathbf{G}} \langle \mathbf{k}|W|\mathbf{k}+\mathbf{G}\rangle\langle \mathbf{k}+\mathbf{G}|\Psi\rangle = 0\,,$$

the sum being over all reciprocal lattice vectors. It is apparent that those components $\langle \mathbf{k}'|\Psi\rangle$ for which $\mathbf{k}' \neq \mathbf{k} + \mathbf{G}$ for any reciprocal lattice vector are decoupled from the equation, and so are irrelevant. Therefore we may conveniently set them equal to zero, in which case the eigenvector will be of the form

$$|\Psi\rangle = \sum_{\mathbf{G}} |\mathbf{k}+\mathbf{G}\rangle\, \langle \mathbf{k}+\mathbf{G}|\Psi\rangle\,.$$

In coordinate representation, this becomes

$$\Psi(\mathbf{x}) = \sum_{\mathbf{G}} \langle \mathbf{x}|\mathbf{k}+\mathbf{G}\rangle \, \langle \mathbf{k}+\mathbf{G}|\Psi\rangle$$

$$= \sum_{\mathbf{G}} \langle \mathbf{k}+\mathbf{G}|\Psi\rangle \, \frac{\exp\left[i(\mathbf{k}+\mathbf{G})\cdot\mathbf{x}\right]}{L^3} \,,$$

which is equivalent to the representation (5.23) for a Bloch type function. Thus even if we did not know Bloch's theorem, it would emerge naturally in the momentum representation solution of a periodic equation.

5.6 The stationary state Schrödinger equation in one dimension can be written as $(E - P^2/2M)|\Psi\rangle = W|\Psi\rangle$, where P is the momentum operator. In momentum representation this becomes

$$\left(E - \frac{p^2}{2M}\right) \Psi(p) = \int_{-\infty}^{\infty} \langle p|W|p'\rangle \, \Psi(p') \, dp' \,,$$

where p is an eigenvalue of P, and $\Psi(p) = \langle p|\Psi\rangle$. For a delta function potential, $W(x) = c\,\delta(x)$, we have $\langle p|W|p'\rangle = c/2\pi\hbar$, so the above integral equation reduces to the simple form

$$\left(E - \frac{p^2}{2M}\right) \Psi(p) = \frac{\alpha}{2M} \,,$$

where we have introduced

$$\alpha = \frac{Mc}{\pi\hbar} \int_{-\infty}^{\infty} \Psi(p) \, dp \,.$$

Anticipating that E will be negative for the ground state in an attractive potential ($c < 0$), we write $E = -\beta^2/2M$. The equation for $\Psi(p)$ then becomes $(p^2 + \beta^2)\Psi(p) = -\alpha$, which has the solution

$$\Psi(p) = -\frac{\alpha}{p^2 + \beta^2} \,.$$

Substituting this result into the definition of α above, we obtain $\alpha = -(Mc/\hbar\beta)\alpha$, which fixes the value of β and so determines the energy to be

$$E = -\frac{Mc^2}{2\hbar^2} \,.$$

The parameter α remains arbitrary, since it is only a normalization constant. The coordinate space state function $\Psi(x)$ can be obtained from $\Psi(p)$ by means of a Fourier transformation. Appropriately normalized, it is

$$\Psi(x) = \left(\frac{M|c|}{\hbar^2}\right)^{1/2} \exp\left(-\frac{M|cx|}{\hbar^2}\right).$$

[This problem and solution are due to M. Lieber, *Am. J. Phys.* **43**, 486–491 (1975).]

Question: Why was the solution for the state function *unique*, when it is usual to have a whole family of energy eigenfunctions?

7.2 The angular momentum and linear momentum operators have the forms $\mathbf{J} = \mathbf{x} \times (-i\hbar\boldsymbol{\nabla}) + \mathbf{S}$ and $\mathbf{P} = -i\hbar\boldsymbol{\nabla} + \mathbf{M}$, respectively. The internal contributions, $\mathbf{S}$ and $\mathbf{M}$, commute with the external terms, and so they must satisfy the same commutation relations as the total angular and linear moment operators:

$$[M_\alpha, M_\beta] = 0\,, \quad [S_\alpha, S_\beta] = i\hbar\varepsilon_{\alpha\beta\gamma}\, S_\gamma\,, \quad [S_\alpha, M_\beta] = i\hbar\varepsilon_{\alpha\beta\gamma}\, M_\gamma\,.$$

Because the components of $\mathbf{M}$ are commutative, we can choose a basis that diagonalizes the three matrices M_x, M_y, M_z. In that basis, we have $(M_\alpha)_{jk} = (M_\alpha)_j\ \delta_{jk}$, $(\alpha = x, y, z)$. From the third commutation relation, we obtain $(S_x)_{jk}(M_y)_k - (M_y)_j(S_x)_{jk} = i\hbar(M_z)_j\ \delta_{jk}$. Putting $j = k$ yields $(M_z)_j = 0$. Clearly a similar result holds for M_x and M_y, and therefore we have shown that $M_\alpha \equiv 0$. Thus there is no such thing as "internal linear momentum".

Since the external, or "orbital", contributions to linear and angular momentum satisfy the same commutation relations as do $\mathbf{M}$ and $\mathbf{S}$, one may ask why this argument cannot be extended to prove that $\mathbf{P} \equiv 0$. The crucial difference is that the internal momentum operator $\mathbf{M}$ must be represented by a *discrete* matrix, since it was assumed to operate in the same space as $\mathbf{S}$. However the momentum eigenvectors have infinite norms, and form a continuous basis. In that basis, the matrix element of the momentum operator is of the form $\langle \mathbf{k}'|P_\alpha|\mathbf{k}\rangle = \hbar k_\alpha\ \delta(\mathbf{k}' - \mathbf{k})$, and the matrix element of the position operator is even more highly singular. The diagonal matrix element of the commutator $[J_x, P_y]$ involves the difference of two infinite terms. However, one cannot validly conclude that $\infty - \infty = 0$, so one escapes the incorrect conclusion that $P_\alpha \equiv 0$.

7.16 The three-spin basis vectors, $|m_1\rangle \otimes |m_2\rangle \otimes |m_3\rangle$, where the m's take on the values $\pm\frac{1}{2}$, are eigenvectors of $S_z{}^{(1)}, S_z{}^{(2)}$, and $S_z{}^{(3)}$. We shall abbreviate them to $|\pm\rangle|\pm\rangle|\pm\rangle$, with the position of the factor indicating which particle it refers to. The total angular momentum operator is $\mathbf{J} = \mathbf{S}^{(1)} + \mathbf{S}^{(2)} + \mathbf{S}^{(3)}$. We wish to form the eigenvectors of $\mathbf{J}\cdot\mathbf{J}$ and J_z, whose eigenvalues (in units of $\hbar = 1$) are $J(J+1)$ and M, respectively.

Let us begin with the state in which all three spins are in the positive z direction, $|+\rangle|+\rangle|+\rangle$. This is an eigenvector of $\mathbf{J}\cdot\mathbf{J}$ and J_z, with both J and M taking the maximum value, 3/2, as can easily be verified, and so we shall write

$$|\tfrac{3}{2},\tfrac{3}{2}\rangle = |+\rangle|+\rangle|+\rangle\,.$$

Applying the lowering operator $J_- = S_-{}^{(1)} + S_-{}^{(2)} + S_-{}^{(3)}$ and using (7.15), we obtain

$$(3^{-1/2})\,|\tfrac{3}{2},\tfrac{1}{2}\rangle = |+\rangle|+\rangle|-\rangle \;+\; |+\rangle|-\rangle|+\rangle \;+\; |-\rangle|+\rangle|+\rangle\,.$$

Repeating this procedure twice more we obtain a set of four eigenvectors with $j = 3/2$.

To determine the remaining eigenvectors of $\mathbf{J}\cdot\mathbf{J}$ and J_z, we must combine the spins two at time using the Clebsch–Gordan coefficients, as in (7.90). Combining first spins of particles 2 and 3, we obtain

$$|J_{23}, M_{23}\rangle = \sum_{m_2,m_3} \left(\tfrac{1}{2},\tfrac{1}{2},m_2,m_3|J_{23},M_{23}\right)\,|m_2\rangle|m_3\rangle\,.$$

The relevant CG coefficients are given in (7.99), and so we have a triplet with $J_{23} = 1$ and a singlet with $J_{23} = 0$. These can now be combined with the spin of the first particle, yielding the total angular momentum eigenvectors for three spins,

$$|J_{23}; J, M\rangle = \sum_{m_1,M_{23}} \left(\tfrac{1}{2},J_{23},m_1,M_{23}|J,M\right)\,|m_1\rangle|J_{23},M_{23}\rangle\,.$$

The necessary CG coefficients can be obtained from (7.103). The $J = 3/2$ eigenvectors that were obtained above by a more direct method correspond to $J_{23} = 1$. The three-spin eigenvectors are summarized in the following table.

J	M	J_{23}	
3/2	3/2	1	$\vert +\rangle\vert +\rangle\vert +\rangle$
	1/2		$3^{-1/2}\,(\vert +\rangle\vert +\rangle\vert -\rangle + \vert +\rangle\vert -\rangle\vert +\rangle + \vert -\rangle\vert +\rangle\vert +\rangle)$
	−1/2		$3^{-1/2}\,(\vert +\rangle\vert -\rangle\vert -\rangle + \vert -\rangle\vert +\rangle\vert -\rangle + \vert -\rangle\vert -\rangle\vert +\rangle)$
	−3/2		$\vert -\rangle\vert -\rangle\vert -\rangle$
1/2	1/2	1	$6^{-1/2}\,(2\vert -\rangle\vert +\rangle\vert +\rangle - \vert +\rangle\vert +\rangle\vert -\rangle - \vert +\rangle\vert -\rangle\vert +\rangle)$
	−1/2		$6^{-1/2}\,(\vert -\rangle\vert +\rangle\vert -\rangle + \vert -\rangle\vert -\rangle\vert +\rangle - 2\vert +\rangle\vert -\rangle\vert -\rangle)$
1/2	1/2	0	$2^{-1/2}\,(\vert +\rangle\vert +\rangle\vert -\rangle - \vert +\rangle\vert -\rangle\vert +\rangle)$
	−1/2		$2^{-1/2}\,(\vert -\rangle\vert +\rangle\vert -\rangle - \vert -\rangle\vert -\rangle\vert +\rangle)$

This example illustrates the fact that when three or more angular momenta are combined, the total angular momentum quantum numbers J and M are insufficient to uniquely label the vectors, whereas they are sufficient when only two angular momenta are combined.

8.2 For a system of spin $s = 1$, the state operator ρ can be represented as a 3×3 matrix, which depends on eight independent real parameters after the conditions $\mathrm{Tr}\,\rho = 1$ and $\rho^\dagger = \rho$ have been imposed. The three data conditions, $\langle S_x \rangle = 0$, $\langle S_y \rangle = 0$, $\langle S_z \rangle = a$ (in units $\hbar = 1$), reduce the number of real free parameters to five, and the most general matrix satisfying all of these conditions can be written in the following form [using the standard representation (7.52) in which S_z is diagonal]:

$$\rho = \begin{bmatrix} \rho_{11} & \rho_{12} & \rho_{13} \\ {\rho_{12}}^* & 1 + a - 2\rho_{11} & -\rho_{12} \\ {\rho_{13}}^* & -{\rho_{12}}^* & \rho_{11} - a \end{bmatrix}.$$

The five parameters are ρ_{11} and the real and imaginary parts of ρ_{12} and ρ_{13}. This is not yet the solution to the problem, because the nonnegativeness condition has not yet been imposed on ρ, and its effect is different in the various special cases.

$\mathbf{a = 1}$: Clearly we must have $\rho_{11} = 1$, since no diagonal element of ρ may be negative. The nonnegativeness condition also implies that $\mathrm{Tr}\,\rho^2 \le 1$ (see Problem 2.4), which in turn implies that $\rho_{12} = \rho_{13} = 0$. This corresponds to the eigenvector of S_z with eigenvalue $+1$, and so is a pure state. The solution is unique, and the appearance of five free parameters was illusory.

$\mathbf{a < 1}$: In this case the solution does depend on five parameters, although their allowed ranges are restricted by several inequalities that are derivable from the nonnegativeness condition. The general solution is given in terms of more physically significant parameters in Eq. (8.9).

Pure states: A separate enumeration of these can be done, using their representation by state vectors. The most general three-dimensional state vector depends on four real parameters, since the norm is fixed and the absolute phase is irrelevant. We may write that vector as

$$|\psi\rangle = \begin{bmatrix} u\, e^{i\theta} \\ v\, e^{i\phi} \\ w \end{bmatrix}$$

with $u^2 + v^2 + w^2 = 1$. The data condition $\langle S_z \rangle = a$ implies that $u^2 - w^2 = a$. The conditions $\langle S_x \rangle = \langle S_y \rangle = 0$ imply that $uv\, e^{i(\theta-\phi)} = -vw\, e^{i\phi}$. This equation has two solutions: (i) $v = 0$; or (ii) $u = w$ and $\theta - \phi = \phi \pm \pi$. We must now examine the various special cases.

$\mathbf{a \neq 0}$: If $u^2 - w^2 = a \neq 0$ then $u \neq w$, and so we must have $v = 0$. This makes the phase ϕ irrelevant. The normalization condition becomes $u^2 + w^2 = 1$, and we obtain $u^2 = \frac{1}{2}(1+a), w^2 = \frac{1}{2}(1-a)$. Only the parameter θ remains free, and so we have a one-parameter family of pure states,

$$|\psi\rangle = \begin{bmatrix} e^{i\theta}\sqrt{\frac{1}{2}(1+a)} \\ 0 \\ \sqrt{\frac{1}{2}(1-a)} \end{bmatrix}.$$

$\mathbf{a = 1}$: In this limit, only the first element of the vector is nonzero, and so the phase θ has no significance. There is only one possible state in this case, as was already proven above.

$\mathbf{a = 0}$: Since $u^2 = w^2$ we can use solution (ii) with $v \neq 0$, $\phi = (\theta \pm \pi)/2$. Hence in this case we have a two-parameter family of pure states,

$$|\psi\rangle = \begin{bmatrix} e^{i\theta} u \\ \pm i\, e^{i\theta/2}\sqrt{1 - 2u^2} \\ u \end{bmatrix}.$$

8.9 If $|\psi\rangle$ is a common eigenvector of the operators A and B, and if $[A, B] = iC$, then it follows trivially that $|\psi\rangle$ is an eigenvector of C with eigenvalue zero, $C|\psi\rangle = 0$.

(a) Hence if $|\psi\rangle$ is a common eigenvector of the angular momentum operators L_x and L_y it must be the case that $L_z|\psi\rangle = 0$. By an extension of this argument, it follows that $|\psi\rangle$ can be a common eigenvector of L_x, L_y, and L_z only if $L_x|\psi\rangle = L_y|\psi\rangle = L_z|\psi\rangle = 0$. Hence it must also satisfy the relation $L^2|\psi\rangle = 0$. Therefore the only single particle eigenfunctions of L_x, L_y, and L_z must be of the form (in coordinate representation) $\psi(r) = f(r)\, Y_0{}^0(\theta,\phi) = f(r)/\sqrt{4\pi}$, with the radial function $f(r)$ being arbitrary.

(b) From the commutation relations $[L_x, P_y] = i\hbar P_z$, etc., it follows that any common eigenfunctions of $\mathbf{L}$ and $\mathbf{P}$ must satisfy $\mathbf{P}|\psi\rangle = 0$. The eigenfunctions of linear momentum have the form (in coordinate representation) $ce^{i\mathbf{k}\cdot\mathbf{x}}$, and therefore the unique solution to our problem is the unnormalizable function $\psi = c$ (a constant). Looking at the problem from a geometrical point of view, for ψ to be an eigenfunction of linear (angular) momentum with eigenvalue zero means that ψ must be invariant under displacements (rotations). Only a constant function is both translationally and rotationally invariant.

8.10 As in the derivation of (8.30), we minimize $\mathrm{Tr}(\rho TT^\dagger)$. But now we have $T = A_0 + i\omega B_0$ and $T^\dagger = A_0 - i\omega^* B_0$, and we must minimize with respect to both the real and imaginary parts of ω. The final result is a stronger inequality,

$$\Delta_A{}^2\,\Delta_B{}^2 \geq \left\{\mathrm{Tr}\left(\frac{\rho C}{2}\right)\right\}^2 + \left[\mathrm{Tr}\left(\frac{\rho\{A_0B_0 + B_0A_0\}}{2}\right)\right]^2,$$

where $A_0 = A - \langle A\rangle$, $B_0 = B - \langle B\rangle$, and $iC = AB - BA$. The second term on the right hand side of the inequality describes the correlation between the dynamical variables A and B, which must be zero in order to obtain a minimum uncertainty state.

9.5 Let the state vector for the detector be $|u_1\rangle$ or $|u_2\rangle$, according to whether the particle goes through the top hole or the bottom hole. Then the state vector for the system consisting of the particle and the detector will be $|\Psi\rangle = |\psi_1(x)\rangle \otimes |u_1\rangle + \psi_2(x)\rangle \otimes |u_2\rangle$. The partial state operator for the particle (p) is $\rho^{(p)} = \mathrm{Tr}^{(d)}(|\Psi\rangle\langle\Psi|)$, where the trace is over the state vector space of the detector (d). The position probability density is then given by the diagonal element of the density matrix,

$$\langle x|\rho^{(p)}|x\rangle = |\psi_1(x)|^2 + |\psi_2(x)|^2 + 2\,\mathrm{Re}\{[\psi_1(x)]^*\,\psi_2(x)\,\langle u_1|u_2\rangle\}\,.$$

If the detector is to discriminate unambiguously between the two holes, then its two possible final states must be orthogonal, $\langle u_1|u_2\rangle = 0$, and there will be no interference pattern. At the other extreme, if the detector is totally insensitive, $|u_1\rangle = |u_2\rangle$, the interference pattern will be unaffected. More interesting is the intermediate case of partial discrimination, $|\langle u_1|u_2\rangle| < 1$. In this case the probability of correctly inferring which hole the particle passed through is $1-\varepsilon^2$, where $\varepsilon = |\langle u_1|u_2\rangle|$ is proportional to the strength of the interference pattern. If we demand an unambiguous determination of which hole each particle passes through, then no interference pattern will be formed. But if we are satisfied with 90% confidence in determining which hole a particle passes through, then the interference pattern will persist with its strength reduced only by a factor of about 1/3.

9.6 If the input to the interferometer is a polarized spin-up state, the amplitude reaching the screen (via the lower path) will be $\psi_1(x)|\uparrow\rangle$, and the probability density will be $|\psi_1(x)|^2$. If the input to the interferometer is a polarized spin-down state, the amplitude reaching the screen (via the upper path and the spin flipper) will be $\psi_2(x)|\uparrow\rangle$, and the probability density will be $|\psi_2(x)|^2$. The functions $\psi_1(x)$ and $\psi_2(x)$ characterize the geometric distributions of the amplitudes in the two paths. Suppose that the Stern–Gerlach magnet is oriented to measure the z component of the spin of the particle on the left. The spins of the two particles must be opposite in the singlet state. Hence we deduce that the probability density on the screen, conditional on the result $\sigma_z = -1$ on the left, is $|\psi_1(x)|^2$; and the probability density on the screen, conditional on the result $\sigma_z = +1$ on the left, is $|\psi_2(x)|^2$. Since the two results $\sigma_z = +1$ and $\sigma_z = -1$ are equally probable, the probability density on the screen for the whole ensemble will be $[|\psi_1(x)|^2 + |\psi_2(x)|^2]/2$, which exhibits no interference.

Now let us rotate the Stern–Gerlach magnet so as to measure the x component of spin of the particle on the left. Recall that the relevant eigenvectors are $|\sigma_x = +1\rangle = (|\uparrow\rangle + |\downarrow\rangle)/\sqrt{2}$ and $|\sigma_x = -1\rangle = (|\uparrow\rangle - |\downarrow\rangle)/\sqrt{2}$. For a singlet state the result $\sigma_x = -1$ on the left implies the value $\sigma_x = +1$ for the particle emerging to the right, so it follows that the amplitude at the screen (after spin flip of the upper beam) will be $[\psi_1(x) + \psi_2(x)]|\uparrow\rangle/\sqrt{2}$, and the probability density, conditional on the result $\sigma_x = -1$ on the left, will be $\frac{1}{2}|\psi_1(x) + \psi_2(x)|^2$.

Similarly the probability density, conditional on the result $\sigma_x = +1$ on the left, will be $\frac{1}{2}|\psi_1(x) - \psi_2(x)|^2$. Since the two results are equally probable, the probability density on the screen for the whole ensemble will be $\left[\frac{1}{2}|\psi_1(x) + \psi_2(x)|^2 + \frac{1}{2}|\psi_1(x) - \psi_2(x)|^2\right]/2 = \left[|\psi_1(x)|^2 + |\psi_2(x)|^2\right]/2$. This is the same result as was obtained for the z orientation of the Stern–Gerlach magnet. Thus there is no paradox, since the pattern on the screen does not depend on a measurement that may or may not be performed on the other particle on the left.

But instead of merely looking at the total density on the screen, we could detect the pairs of particles in concidence. Then we would find those particles that were detected in coincidence with the result $\sigma_x = -1$ on the left to be distributed over the screen in the form of the interference pattern $\frac{1}{2}|\psi_1(x) + \psi_2(x)|^2$, and those particles that were detected in coincidence with the result $\sigma_x = +1$ on the left to be distributed over the screen in the form of the interference pattern $\frac{1}{2}|\psi_1(x) - \psi_2(x)|^2$. The modulations of these two patterns cancel out when they are added together.

The coincidence measurement would detect no interference pattern if σ_z was measured on the left.

10.8 In a bound state, which is an eigenstate of the Hamiltonian $H = P^2/2M + W$, the average of any dynamical variable will be independent of time. In particular, we must have $d\langle \mathbf{x} \cdot \mathbf{P} \rangle/dt = (i/\hbar)\langle [H, \mathbf{x} \cdot \mathbf{P}] \rangle = 0$. Using the result of Problem 4.1, we then obtain $2\langle T \rangle - \langle \mathbf{x} \cdot \nabla W \rangle = 0$. If W is proportional to r^n, we have $\langle \mathbf{x} \cdot \nabla W \rangle = n\langle W \rangle$, and hence $2\langle T \rangle = n\langle W \rangle$. This relation can be verified in the explicit solutions for the harmonic oscillator ($n = 2$) and the hydrogen atom ($n = -1$).

11.6 We use cylindrical coordinates (ρ, ϕ, z), where $\rho = \sqrt{x^2 + y^2}$ is the perpendicular distance from the z axis, and ϕ is the angle of rotation about the z axis. The cylindrical components of the vector potential are $A_\rho = 0, A_\phi = \frac{1}{2}B\rho, A_z = 0$. The energy eigenvalue equation then becomes

$$\frac{-\hbar^2}{2M}\left[\frac{\partial^2\Psi}{\partial\rho^2} + \frac{1}{\rho}\frac{\partial\Psi}{\partial\rho} + \frac{1}{\rho^2}\frac{\partial^2\Psi}{\partial\phi^2} + \frac{\partial^2\Psi}{\partial z^2}\right] + i\frac{\hbar qB}{2Mc}\frac{\partial\Psi}{\partial\rho} + \frac{q^2B^2}{8Mc^2}\rho^2\Psi = E\Psi\,.$$

We substitute $\Psi(\rho, \phi, z) = R(\rho)e^{im\phi}\, e^{ikz}$, and thereby reduce the partial differential equation to an ordinary differential equation,

$$\frac{d^2R}{d\rho^2} + \frac{1}{\rho}\frac{dR}{d\rho} + \left[\beta + \frac{q}{|q|}\frac{m}{{a_m}^2} - \frac{m^2}{\rho^2} - \frac{\rho^2}{4{a_m}^4}\right]R = 0\,,$$

where $\beta = (2ME/\hbar^2) - k^2$, and $a_m = (\hbar c/|q|B)^{1/2}$ is the magnetic length. The behavior of the solution in the limit $\rho \to 0$ can be determined by substituting $R(\rho) \approx \rho^\alpha$, which yields $\alpha = \pm m$. Since R^2 must be integrable, only the value $\alpha = |m|$ is acceptable. In the limit $\rho \to \infty$, it can easily be verified that the asymptotic behavior of the solution is dominated by a factor $\exp[\pm(\rho/2a_m)^2]$. Only the solution with the negative sign is acceptable, so our solution must be of the form

$$R(\rho) = \rho^{|m|}\,\exp\left[-\left(\frac{\rho}{2a_m}\right)^2\right]\,f(\rho)\,,$$

where $f(\rho)$ is a regular analytic function, which can be expressed in a power series:

$$f(\rho) = \sum_{n=0}^{\infty} c_n\,\rho^n, \text{ with } c_0 \neq 0\,.$$

From the differential equation, we obtain the recurrence relation

$$[n^2 + (2|m| + 4)n + 4|m| + 4]\,c_{n+2} + \left\{\beta + {a_m}^{-2}\left[\left(\frac{q}{|q|}\right)m - |m| - n - 1\right]\right\}c_n = 0\,.$$

(That only even values n are relevant could have been anticipated from the fact that the differential equation is invariant under the substitution $\rho \to -\rho$.) If the series does not terminate, the asymptotic ratio as $n \to \infty$ is $c_{n+2}/c_n \sim n{a_m}^2$, like the function $\exp(\rho^2/{a_m}^2)$. This unacceptable divergence at $\rho \to \infty$ is avoided only if the series terminates. This happens if the coefficient of c_n in the recurrence relation vanishes for some (even) integer $n = 2r$:

$$\beta{a_m}^2 + \left(\frac{q}{|q|}\right)m - |m| - 2r - 1 = 0\,.$$

The energy eigenvalue can now be obtained from β:

$$E = \frac{\hbar^2k^2}{2M} + \frac{1}{2}\hbar|\omega_c|\left(2r + 1 + |m| - \left(\frac{q}{|q|}\right)m\right),$$

where $\omega_c = qB/Mc$ is the cyclotron frequency, and the ranges of r

and m are $r = 0, 1, 2, 3, \ldots$; $m = 0, \pm 1, \pm 2, \pm 3, \ldots$. Notice that for fixed E, the allowed range of the angular momentum eigenvalue m is unbounded above (for $q > 0$), but is bounded below, in agreement with Eq. (11.38).

11.7 This problem is treated by Peshkin (1981), to whom we refer for details.

12.11 The perturbation is of the form

$$H_1(t) = e\mathbf{x} \cdot \mathbf{E} = \frac{r\,\cos(\theta) e A \tau}{t^2 + \tau^2}\,,$$

where the charge of the electron is $-e$. The first order transition amplitude is given by Eq. (12.51), modified to account for the fact that the perturbation acts from $-\infty$ to ∞, rather than from 0 to T. Thus the probability of excitation is

$$P_{\text{if}} = \hbar^{-2} \left| \int_{-\infty}^{\infty} \langle f|H_1(t)|i\rangle\, e^{i\omega t}\, dt \right|^2 .$$

The frequency ω is determined by the energy difference between the initial and final states, $\omega = (\varepsilon_f - \varepsilon_i)/\hbar$. The initial state is $|i\rangle = |100\rangle$. The first excited state is degenerate, but the matrix element is nonvanishing only for the final state $|f\rangle = |210\rangle$. The value of the matrix element for this case, obtained by integrating the relevant hydrogenic functions, is

$$\langle 210|r\,\cos(\theta)|100\rangle = \left(\frac{2^{7.5}}{3^5}\right) a_0\,.$$

Hence the probability is

$$P_{\text{if}} = \frac{2^{15}\, e^2\, A^2\, \tau^2\, a_0}{3^{10}\, \hbar^2} \left| \int_{-\infty}^{\infty} \frac{e^{i\omega t}}{t^2 + \tau^2}\, dt \right|^2 .$$

To evaluate the integral, we may close the contour by an infinite semicircle in the upper half of the complex t plane. Its value is $(\pi/\tau)\, e^{-\omega\tau}$. Therefore the excitation probability is

$$P_{\text{if}} = \frac{2^{15}\, e^2\, A^2\, \tau^2\, a_0}{3^{10}\, \hbar^2}\, e^{-2\omega\tau}\,.$$

Notice that P_{if} becomes exceedingly small if $\omega\tau \gg 1$, i.e. if the characteristic time of the perturbation τ is large. This is an example of the *adiabatic principle*, according to which a perturbation that varies smoothly and slowly in time leaves the state of the system unchanged.

13.4 Let A be an antilinear operator which satisfies $\| |\phi\rangle \| = \|A|\phi\rangle\|$ for all $|\phi\rangle$. Denote $|u'\rangle = A|u\rangle$ and $|v'\rangle = A|v\rangle$. We shall consider the vectors $|\phi\rangle = |u\rangle + |v\rangle$ and $|\phi'\rangle = A|\phi\rangle = |u'\rangle + |v'\rangle$.

$$\begin{aligned}\| |\phi\rangle \|^2 &= \langle\phi|\phi\rangle = \langle u|u\rangle + \langle v|v\rangle + \langle u|v\rangle + \langle v|u\rangle \\ &= \langle u|u\rangle + \langle v|v\rangle + 2\,\mathrm{Re}\,\langle u|v\rangle\,.\end{aligned}$$

By hypothesis this must be equal to

$$\| |\phi'\rangle \|^2 = \langle\phi'|\phi'\rangle = \langle u'|u'\rangle + \langle v'|v'\rangle + 2\,\mathrm{Re}\,\langle u'|v'\rangle\,,$$

and hence it follows that $\mathrm{Re}\langle u|v\rangle = \mathrm{Re}\,\langle u'|v'\rangle$.

We next consider the vectors $|\phi\rangle = |u\rangle + i|v\rangle$ and $|\phi'\rangle = A|\phi\rangle = |u'\rangle - i|v'\rangle$. This time the condition $\| |\phi\rangle \|^2 = \| |\phi'\rangle \|^2$ leads us to conclude that $\mathrm{Im}\langle u|v\rangle = -\mathrm{Im}\langle u'|v'\rangle$, and so we have $\langle u'|v'\rangle = \langle u|v\rangle^*$.

17.4 A fermion pair state is created by the operator $A^\dagger = C_\beta{}^\dagger\, C_\alpha{}^\dagger$, and is annihilated by the operator $A = C_\alpha C_\beta$. A simple calculation shows their commutation relation to be $AA^\dagger - A^\dagger A = 1 - C_\alpha{}^\dagger\, C_\alpha - C_\beta{}^\dagger\, C_\beta$. This is not equivalent to the correct boson commutation relation, $AA^\dagger - A^\dagger A = 1$, except when operating on a state vector for which both α and β orbitals are empty. Moreover $A^\dagger A^\dagger = 0$, indicating that no more than one such bound pair can occupy the same state, contrary to the behavior of real bosons. R. Penney [*J. Math. Phys.* **6**, 1031–1034 (1965)] has proved the more general result that one cannot construct boson creation and annihilation operators from a finite number of fermion operators.

18.3 The number operator is $N = \sum_\alpha C_\alpha{}^\dagger\, C_\alpha$, and it follows from (18.80) that in the BCS state one has $\langle N\rangle = \sum_\alpha {v_\alpha}^2$. The average of N^2 can readily be evaluated by means of Wick's theorem and Eqs. (18.76) through (18.79).

$$\begin{aligned}\langle N^2\rangle &= \sum_\alpha\sum_\beta \langle \mathrm{BCS}\,|C_\alpha{}^\dagger C_\alpha C_\beta{}^\dagger C_\beta|\,\mathrm{BCS}\rangle \\ &= \sum_\alpha\sum_\beta \{\langle C_\alpha{}^\dagger C_\alpha\rangle\,\langle C_\beta{}^\dagger C_\beta\rangle - \langle C_\alpha{}^\dagger C_\beta{}^\dagger\rangle\,\langle C_\alpha C_\beta\rangle \\ &\quad + \langle C_\alpha{}^\dagger C_\beta\rangle\langle C_\alpha C_\beta{}^\dagger\rangle\} \\ &= \sum_\alpha\sum_\beta \{{v_\alpha}^2 {v_\beta}^2 + (u_\alpha v_\alpha)^2\delta_{-\alpha\beta} + {u_\alpha}^2 {v_\alpha}^2\,\delta_{\alpha\beta}\} \\ &= \left\{\sum_\alpha {v_\alpha}^2\right\}^2 + 2\sum_\alpha {u_\alpha}^2 {v_\alpha}^2\,.\end{aligned}$$

Thus $\langle (N - \langle N \rangle)^2 \rangle = \langle N^2 \rangle - \langle N \rangle^2 = 2 \sum_\alpha {u_\alpha}^2 {v_\alpha}^2$. If we now use (18.88) and (18.89) and convert this sum into an integral, we will obtain an extensive quantity (a quantity of order $\langle N \rangle$). Therefore the relative mean square fluctuation, $\langle (N - \langle N \rangle)^2 \rangle / \langle N \rangle^2$, will go to zero as $\langle N \rangle$ becomes infinite.

19.7 & 19.8 The state operators that describe the noisy lasers in these two problems are special cases of the state operator

$$\rho = \iint w(r_1, \phi_1)\, w(r_1, \phi_1) |z_1, z_2\rangle\, \langle z_1, z_2|\, dr_1\, d\phi_1\, dr_2\, d\phi_2 \,.$$

Here $z_1 = r_1\, e^{i\phi_1}$, $z_2 = r_2\, e^{i\phi_2}$, and $w(r, \phi)$ is a nonnegative probability density that describes the noise fluctuations. Equation (19.115) gives the interference pattern for the pure state operator $|z_1, z_2\rangle\langle z_1, z_2|$. For this problem, that result must be averaged over the noise fluctuations. Thus we obtain

$$\begin{aligned} G^{(1)}(\mathbf{x}, \mathbf{x}) = C^2 & \iint w(r_1, \phi_1)\, w(r_2, \phi_2) \\ & \times \left\{ {r_1}^2 + {r_2}^2 + 2 r_1 r_2 \cos[(\mathbf{k}_1 - \mathbf{k}_2) \cdot \mathbf{x} + \phi_1 - \phi_2] \right\} \\ & \times dr_1\, d\phi_1\, dr_2\, d\phi_2 \,. \end{aligned}$$

It is apparent that small fluctuations in the phases ϕ_1 and ϕ_2 will tend to wash out the interference pattern, and if the fluctuations are so great that the phase difference $\phi_1 - \phi_2$ is uniformly distributed over a range of 2π, the interference pattern will be totally destroyed. On the other hand, fluctuations in the amplitudes r_1 and r_2 cannot destroy the interference pattern, but can only alter the *contrast* or *visibility* of the pattern. The usual measure of contrast is the ratio $(I_{\text{max}} - I_{\text{min}})/(I_{\text{max}} + I_{\text{min}})$, where I_{max} and I_{min} are the maximum and minimum intensities. In the absence of any fluctuation in phase or amplitude, this ratio is equal to $2 r_1 r_2 / ({r_1}^2 + {r_2}^2)$. Amplitude fluctuations cause this to be replaced by $2\langle r_1 r_2 \rangle / (\langle {r_1}^2 \rangle + \langle {r_2}^2 \rangle)$.

20.5 For the first fragment we measure σ_a, the sign of the projection of its angular momentum in the direction **a**. For the second fragment we measure σ_b, the sign of the projection of its angular momentum in the direction **b**. If the angular momentum of the first fragment is **J** and that of the second fragment is $-\mathbf{J}$, then σ_a will be positive if **J** points in the hemisphere centered on **a**, and σ_b will be positive if **J** points in the hemisphere opposite to **b**.

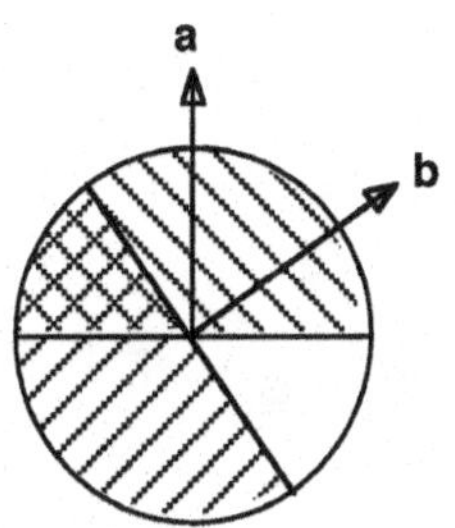

Let the angle between **a** and **b** be θ. As can be seen from the figure, σ_a and σ_b will both be positive if **J** lies in the intersection of those hemispheres. If the direction of **J** is uniformly distributed, then the probability that σ_a and σ_b will both be positive is $\theta/2\pi$. (Note that the angle is defined so that $0 \le \theta \le \pi$.) Similarly the probability that σ_a and σ_b will both be negative is also $\theta/2\pi$. Thus the probability that $\sigma_a \sigma_b = +1$ is θ/π, and the probability that $\sigma_a \sigma_b = -1$ is $1-\theta/\pi$. Hence the correlation function is $C(\theta) = \langle \sigma_a \sigma_b \rangle = (2\theta/\pi) - 1,\ (0 \le \theta \le \pi)$. The appropriate form of Bell's inequality is Eq. (20.7). It is satisfied as an equality, indicating that the correlation between the angular momenta of the two fragments is as strong as is possible for a system that obeys *locality*.

Bibliography

Abraham, R. and Marsden, J. E. (1978), *Foundations of Mechanics*, 2nd edn. (Benjamin, Reading, Massachusetts).

Aharonov, Y. and Bohm, D. (1959), "Significance of Electromagnetic Potentials in the Quantum Theory", *Phys. Rev.* **115**, 458–491.

Aharonov, Y., Massar, S., and Popescu, S. (2002), "Measuring Energy, Estimating Hamiltonians, and the Time-Energy Uncertainty Relation", *Phys. Rev. A* **66**, 052107.

Alber, T. *et al.* (2001), *Quantum Information: An Introduction to Basic Theoretical Concepts and Experiments* (Springer, Berlin, Heidelberg, New York).

Alstrom, P., Hjorth, P., and Mattuck, R. (1982), "Paradox in the Classical Treatment of the Stern-Gerlach Experiment", *Am. J. Phys.* **50**, 697–698.

Arens, R. and Babbitt, D. (1965), "Algebraic Difficulties of Preserving Dynamical Relations when Forming Quantum-Mechanical Operators", *J. Math. Phys.* **6**, 1071–1075.

Ashcroft, N. W. and Mermin, N. D. (1976), *Solid State Physics* (Holt, Rinehart, and Winston, New York).

Aspect, A., Grangier, P., and Roger, G. (1981), "Experimental Tests of Realistic Local Theories via Bell's Theorem", *Phys. Rev. Lett.* **47**, 460–463.

Aspect, A., Grangier, P., and Roger, G. (1982), "Experimental Realization of Einstein–Podolsky–Rosen–Bohm *Gedankenexperiment*: A New Violation of Bell's Inequality", *Phys. Rev. Lett.* **49**, 91–94.

Aspect, A., Dalibard, J., and Roger, G. (1982), "Experimental Test of Bell's Inequalities Using Time-Varying Analyzers", *Phys. Rev. Lett.* **49**, 1804–1807.

Badurek, G., Rauch, H., and Summhammer, J. (1983), "Time-Dependent Superposition of Spinors", *Phys. Rev. Lett.* **51**, 1015–1018.

Ballentine, L. E. (1970), "The Statistical Interpretation of Quantum Mechanics", *Rev. Mod. Phys.* **42**, 358–381.

Ballentine, L. E. (1972), "Einstein's Interpretation of Quantum Mechanics", *Am. J. Phys.* **40**, 1763–1771.

Ballentine, L. E. (1986), "Probability in Quantum Mechanics", *Am. J. Phys.* **54**, 883–889.

Ballentine, L. E. and Jarrett, J. P. (1987), "Bell's Theorem: Does Quantum Mechanics Contradict Relativity?", *Am. J. Phys.* **55**, 696–701.

Ballentine, L. E. (1987), "Resource letter IQM-2: Foundations of Quantum Mechanics Since the Bell Inequalities", *Am. J. Phys.* **55**, 785–791.

Ballentine, L. E. (1988), "What Do We Learn About Quantum Mechanics from the Theory of Measurement?", *Int. J. Theor. Phys.* **27**, 211–218.

Ballentine, L. E. (1990), "Limitations of the Projection Postulate", *Found. Phys.* **20**, 1329–1343.

Ballentine, L. E. and Huberman, M. (1977), "Theory of the Hall Effect in Simple Liquid Metals", *J. Phys. C: Solid State Phys.* **10**, 4991–5002.

Ballentine, L. E., Yang, Y., and Zibin, J. P. (1994), "Inadequacy of Ehrenfest's Theorem to Characterize the Classical Regime", *Phys. Rev. A* **50**, 2854–2859.

Ballentine, L. E. (2004), "Quantum-to-Classical Limit in a Hamiltonian System", *Phys. Rev. A* **70**, 032111.

Ballentine, L. E. (2007), "Objective and Subjective Probabilities in Quantum Mechanics", in *Quantum Theory: Reconsiderations of Foundations – 4*, AIP Conf. Proc. **962**, p. 28.

Ballentine, L. E. (2008), "Classicality without Decoherence: A Reply to Schlosshauer", *Found. Phys.* **38**, 916–922.

Band, W. and Park, J. L. (1970), "The Empirical Determination of Quantum States", *Found. Phys.* **1**, 133–144.

— (1971), "A General Method of Empirical State Determination in Quantum Physics", *Found. Phys.* **1**, 339–357.

— (1973), "Relations Among the Elements of the Density Matrix: 1. Definiteness Inequalities", *J. Math. Phys.* **14**, 551–553.

— (1979), "Quantum State Determination: Quorum for a Particle in One Dimension", *Am. J. Phys.* **47**, 188–191.

Barenco, A., Bennett, C. H., Cleve, R., DiVincenzo, D. P., Margolus, N., Shor, P., Sleator, T., Smolin, J. A., and Weinfurter, H. (1995), *Phys. Rev. A* **52**, 3457.

Bargmann, V. (1964), "Note on Wigner's Theorem on Symmetry Operators", *J. Math. Phys.* **5**, 862–868.

Barnett, S. M. (2009), *Quantum Information* (Oxford University Press, Oxford).

Bell, J. S. (1966), "On the Problem of Hidden Variables in Quantum Mechanics", *Rev. Mod. Phys.* **38**, 447–452.

Bell, J. S. (1987), *Speakable and Unspeakable in Quantum Mechanics* (Cambridge University Press, Cambridge).

Bell, J. S. (1990), "Against 'Measurement' ", in *Sixty-Two Years of Uncertainty*, ed. A. J. Miller (Plenum, New York), pp. 17–30.

Bennett, C. H. and Brassard, G. (1984), in *Proceedings of the IEEE International Conference on Computers, Systems and Signal Processing* (IEEE, New York).

Bennett, C. H., Brassard, G., Crépeau, C., Jozsa, R., Peres, A., and Wooters, W. K. (1993), "Teleporting an Unknown Quantum State via Dual Classical and Einstein–Podolsky–Rosen Channels", *Phys. Rev. Lett.* **70**, 1895.

Bennett, C. H., Brassard, G., Popescu, S., Schumacker, B., Smolin, J. A., and Wooters, W. K. (1996), "Purification of Noisy Entanglement and Faithful Teleportation via Noisy Channels", *Phys. Rev. Lett.* **76**, 722.

Bennett, C. H., DiVincenzo, D. P., and Wootters, W. K. (2009), *Quantum Information Theory* (Springer, London).

Berry, M. V. (1984), "Quantal Phase Factor Accompanying Adiabatic Changes", *Proc. R. Soc. Lond.* **A392**, 45–57.

Bethe, H. A. and Salpeter, E. E. (1957), *Quantum Mechanics of One-and Two-Electron Atoms* (Springer-Verlag, Berlin).

Biedenharn, L. C. and VanDam, H. (1965), *Quantum Theory of Angular Momentum* (Academic, New York).

Binnig, G., Rohrer, H., Gerber, C., and Weibel, E. (1982), "Vacuum Tunneling", *Physica* **109** & **110B**, 2075–2077.

Blatt, R. and Roos, C. F. (2012), "Quantum Simulations with Trapped Ions", *Nature Physics* **8**, 277.

Böhm, A. (1978), *The Rigged Hilbert Space and Quantum Mechanics*, Lecture Notes in Physics, Vol. 78 (Springer-Verlag, Berlin).

Bohm, D. (1952), "A Suggested Interpretation of the Quantum Theory in Terms of 'Hidden' Variables", *Phys. Rev.* **85**, 166–193.

Born, M. (1971), *The Born–Einstein Letters* (MacMillan, London).

Born, M. and Wolf, E. (1980), *Principles of Optics*, 6th edn. (Pergamon, Oxford).

Boschi, D., Branca, S., De Martini, F., Hardy, L., and Popescu, S. (1998), "Experimental Realization of Teleporting an Unknown Pure Quantum State via Dual Classical and Einstein–Podolsky–Rosen Channels", *Phys. Rev. Lett.* **80**, 1121.

Bouten, M. (1969), "On the Rotation Operators in Quantum Mechanics", *Physica* **42**, 572–580.

Bouwmeester, D., Pan, J.-W., Mattle, K., Eibl, M., Weinfurter, H., and Zeilinger, A. (1997), "Experimental Quantum Teleportation", *Nature* **390**, 575.

Brezger, B., Hackermüler, L., Uttenthaler, S., Petchinka, J., Arndt, M., and Zeilinger, A. (2002), "Matter-Wave Interferometer for Large Molecules", *Phys. Rev. Lett.* **88**, 100404.

Carnal, O. and Mlynek, J. (1991), "Young's Double-Slit Experiment with Atoms: a Simple Atom Interferometer", *Phys. Rev. Lett.* **66**, 2689–2692.

Casimir, H. B. G. (1948), "On the Attraction Between Two Perfectly Conducting Plates", *Proc. Kon. Ned. Akademie van Wetenschappen (Amsterdam)* **51**, 793–795.

Chambers, R. G. (1960), "Shift of an Interference Pattern by Enclosed Magnetic Flux", *Phys. Rev. Lett.* **5**, 3–5.

Cimmino, A., Opat, G. I., Klein, A. G., Kaiser, H., Werner, S. A., Arif, M., and Clothier, R. (1989), "Observation of the Topological Aharonov–Casher Phase Shift by Neutron Interferometry", *Phys. Rev. Lett.* **63**, 380–383.

Clauser, J. F. (1974), "Experimental Distinction Between the Quantum and Classical Field-Theoretic Predictions for the Photoelectric Effect", *Phys. Rev. D* **9**, 853–860.

Clauser, J. F., Horne, M. A., Shimony, A., and Holt, R. A. (1969), "Proposed Experiment to Test Local Hidden-Variable Theories", *Phys. Rev. Lett.* **23**, 880–884.

Clauser, J. F., and Horne, M. A. (1974), "Experimental Consequences of Objective Local Theories", *Phys. Rev. D* **10**, 526–535.

Clauser, J. F., and Shimony, A. (1978), "Bell's Theorem: Experimental Tests and Implications", *Rep. Prog. Phys.* **41**, 1881–1927.

Cohen, L. (1986), "Positive and Negative Joint Quantum Distributions", pp. 97–117, in *Frontiers of Nonequilibrium Statistical Physics*, eds. G. T. Moore and M. O. Scully (Plenum, New York).

Colella, R., Overhauser, A. W., and Werner, S. A. (1975), "Observation of Gravitationally Induced Quantum Interference", *Phys. Rev. Lett.* **34**, 1472–1474.

Cosma, G. (1983), "Comment on 'Direct Measurement of the Longitudinal Coherence Length of a Thermal Neutron Beam' ", *Phys. Rev. Lett.* **51**, 1105.

Cox, R. T. (1961), *The Algebra of Probable Inference* (Johns Hopkins Press, Baltimore).

Curceanu, C., Gillaspy, J. D., and Hilborn, R. C. (2012), "Resource Letter SS-1: The Spin-Statistics Connection", *Am. J. Phys.* **80**, 561–577.

Davisson, C. and Germer, L. (1927), "Diffraction of Electrons by a Crystal of Nickel", *Phys. Rev.* **30**, 705.

Diedrich, F. and Walther, H. (1987), "Nonclassical Radiation of a Single Stored Ion", *Phys. Rev. Lett.* **58**, 203–206.

Dirac, P. A. M. (1958), *The Principles of Quantum Mechanics*, 4th edn. (Clarendon, Oxford).

Draper, J. E. (1979), "Use of $|\Psi|^2$ and Flux to Simplify Analysis of Transmission Past Rectangular Barriers or Wells", *Am. J. Phys.* **47**, 525–530.

Draper, J. E. (1980), "Quantal Tunneling Through a Rectangular Barrier Using $|\Psi|^2$ and Flux", *Am. J. Phys.* **48**, 749–751.

Duane, W. (1923), "The Transfer in Quanta of Radiation Momentum to Matter", *Proc. Nat. Acad. Sci. (USA)* **9**, 158–164.

Edmonds, A. R. (1957), *Angular Momentum in Quantum Mechanics* (Princeton University Press, Princeton, New Jersey).

Einstein, A., Podolsky, B., and Rosen, N. (1935), "Can Quantum-Mechanical Description of Reality Be Considered Complete?", *Phys. Rev.* **47**, 777–780.

Einstein, A. (1949), in *Albert Einstein: Philosopher–Scientist*, ed. P. A. Schilpp (Harper & Row, New York).

Eisberg, R. and Resnick, R. (1985), *Quantum Physics* (Wiley, New York).

Fetter, A. L. and Walecka, J. D. (1971), *Quantum Theory of Many-Particle Systems* (McGraw-Hill, New York).

Feynman, R. P. (1948), "Space–Time Approach to Non-relativistic Quantum Mechanics", *Rev. Mod. Phys.* **20**, 367–387.

Feynman, R. P. and Hibbs, A. R. (1965), "*Quantum Mechanics and Path Integrals* (McGraw-Hill, New York).

Fine, T. L. (1973), *Theories of Probability, an Examination of Foundations* (Academic, New York).

Fischbach, E., Greene, G. L., and Hughes, R. J. (1991), "New Test of Quantum Mechanics: Is Planck's Constant Unique?", *Phys. Rev. Lett.* **66**, 256–259.

Fisher, G. P. (1971), "The Electric Dipole Moment of a Moving Magnetic Dipole", *Am. J. Phys.* **39**, 1528–1533.

Fonda, L., Ghirardi, G. C., and Rimini, A. (1978), "Decay Theory of Unstable Quantum Systems", *Rep. Prog. Phys.* **41**, 588–631.

Franck, J. and Hertz, G. (1914), *Verhandl. deut. physik. Ges.* **16**, 512.

Freedman, S. J. and Clauser, J. F. (1972), "Experimental Test of Local Hidden-Variable Theories", *Phys. Rev. Lett.* **28**, 938–941.

Frenkel, J. (1932), *Wave Mechanics, Elementary Theory* (Clarendon, Oxford).

Friedrich, H. and Wintgen, D. (1989), "The Hydrogen Atom in a Uniform Magnetic Field — An Example of Chaos", *Phys. Rep.* **183**, 37–79.

Gähler, R. and Zeilinger, A. (1991), "Wave-Optical Experiments with Very Cold Neutrons", *Am. J. Phys.* **59**, 316–324.

Garg, A. and Mermin, N. D. (1987), "Detector Inefficiencies in the Einstein–Podolsky–Rosen Experiment", *Phys. Rev. D* **35**, 3831–3835.

Gasiorowicz, S. (1966), *Elementary Particle Physics* (Wiley, New York).

Gel'fand, I. M. and Shilov, G. E. (1964), *Generalized Functions*, Vol. 1 (Academic, New York).

Gel'fand, I. M. and Vilenkin, N. Ya. (1964), *Generalized Functions*, Vol. 4 (Academic, New York).

Gerry, C. G. and Kiefer, J. (1988), "Numerical Path Integeration without Monté Carlo", *Am. J. Phys.* **56**, 1002–1005.

Ghosh, R. and Mandel, L. (1987), "Observation of Nonclassical Effects in the Inteference of Two Photons", *Phys. Rev. Lett.* **59**, 1903–1905.

Glauber, R. J. (1963a), "The Quantum Theory of Optical Coherence", *Phys. Rev.* **130**, 2529–2539.

Glauber, R. J. (1963b), "Coherent and Incoherent States of the Radiation Field", *Phys. Rev.* **131**, 2766–2788.

Goldberger, M. L. and Watson, K. M. (1964), *Collision Theory* (Wiley, New York).

Goldstein, H. (1980), *Classical Mechanics* (Addison-Wesley, Reading, Massachusetts).

Gould, P. L., Ruff, G. A., and Princhard, D. E. (1986), "Diffraction of Atoms by Light: The Near-Resonance Kapitza–Dirac Effect", *Phys. Rev. Lett.* **56**, 827–830.

Grangier, P., Roger, G., and Aspect, A. (1986), "A new Light on Single Photon Interferences", in *New Techniques and Ideas in Quantum Measurement Theory*, ed. D. M. Greenberger, *Ann. NY Acad. Sci.* **480**, 98–107.

Greenberger, D. M. and Overhauser, A. W. (1979), "Coherence Effects in Neutron Diffraction and Gravity Experiments", *Rev. Mod. Phys.* **51**, 43–78.

Greenberger, D. M. (1983), "The Neutron Interferometer as a Device for Illustrating the Strange Behaviour of Quantum Systems", *Rev. Mod. Phys.* **55**, 875–905.

Greenberger, D. M., Horne, M. A., Shimony, A., and Zeilinger, A. (1990), "Bell's Theorem Without Inequalities", *Am. J. Phys.* **58**, 1131–1143.

Greenberger, D. M., Horn, M. A., and Zeilinger, A. (1993), "Multi-particle Interferometry and the Superposition Principle", *Physics Today* **46**(8), 22–29.

Hagen, C. R. (1990), "Equivalence of Spin $\frac{1}{2}$ Aharonov–Bohm and Aharonov–Casher Effects", *Phys. Rev. Lett.* **64**, 2347–2349.

Haines, L. K. and Roberts, D. H. (1969), "One-Dimensional Hydrogen Atom", *Am. J. Phys.* **37**, 1145–1154.

Hardy, L. (1992), "Quantum Mechanics, Local Realistic Theories, and Lorentz-Invariant Realistic Theories", *Phys. Rev. Lett.* **68**, 2981–2984.

Hardy, L. (1993), "Nonlocality for Two Particles Without Inequalities for Almost All Entangled States", *Phys. Rev. Lett.* **71**, 1665–1668.

Harrigan, N. and Spekkens, R. W. (2010), "Einstein, Incompleteness, and the Epistemic View of Quantum States", *Found. Phys.* **40**, 125–157.

Harris, R. *et al.* (2010), "Experimental Investigation of an Eight-Qubit Unit Cell in a Superconducting Optimization Processor", *Phys. Rev. B* **82**, 024511.

Herbert, N. (1975), "Cryptographic Approach to Hidden Variables", *Am. J. Phys.* **43**, 315–316.

Hillery, M., O'Connell, R. F., Scully, M. O., and Wigner, E. P. (1984), "Distribution Functions in Physics: Fundamentals", *Phys. Rep.* **106**, 121–167.

Horodecki, M., Horodecki, P., and Horodecki, R. (1996), "Separability of Mixed States: Necessary and Sufficient Conditions", *Phys. Lett. A* **223**, 1–8.

Horodecki, M., Horodecki, P., and Horodecki, R. (2001), "Mixed-State Entanglement and Communication", in Alber (2001), pp. 151–195.

Hudson, R. L. (1974), "When Is the Wigner Quasi-probability Density Non-negative?", *Rep. Math. Phys.* **6**, 249–252.

Hudson, J. J., Kara, D. M., Smallman, I. J., Sauer, B. E., Tarbut, M. R., and Hinds, E. A. (2011), "Improved Measurement of the Shape of the Electron", *Nature (London)* **473**, 493–496.

Hughston, L. P., Joza, R., and Wootters, W. K. (1993), "A Complete Classification of Quantum Ensembles having a given Density Matrix", *Phys. Lett. A* **183**, 14–18.

Hulet, R. G., Hilfer, E. S., and Kleppner, D. (1985), "Inhibited Spontaneous Emission by a Rydberg Atom", *Phys. Rev. Lett.* **55**, 2137–2140.

Husimi, K. (1940), "Some Formal Properties of the Density Matrix", *Proc. Phys.-Math. Soc. Japan* **22**, 264–283.

Hylleraas, E. A. (1964), "The Schrödinger Two-Electron Atomic Problem", *Adv. Quantum Chem.* **1**, 1–33.

Iannuzzi, M., Merlo, V., Paciaroni, A., Petrillo, C., and Sacchetti, F. (2000), "A Coincidence Experiment of two Coherent Beams of Thermal Neutrons", *Found. Phys. Lett.* **13**, 1–9.

Jammer, M. (1996), *The Conceptual Development of Quantum Mechanics* (McGraw-Hill, New York); 2nd edn. (American Institute of Physics, New York, 1989).

Jammer, M. (1974), *The Philosophy of Quantum Mechanics* (John Wiley & Sons, New York).

Jauch, J. M. (1972), "On Bras and Kets", in *Aspects of Quantum Theory*, eds. A. Salam and E. P. Wigner (Cambridge University Press, Cambridge), pp. 137–167.

Jauch, W. (1993), "Heisenberg's Uncertainty Relation and Thermal Vibrations in Crystals", *Am. J. Phys.* **61**, 929–932.

Jaynes, E. T. (2003), *Probability Theory: The Logic of Science* (Cambridge University Press, Cambridge).

Jordan, T. F. (1969), *Linear Operators for Quantum Mechanics* (John Wiley & Sons, New York. Reprinted by T. F. Jordan, 2249 Dunedin Ave., Duluth, Minnesota).

Jordan, T. F. (1975), "Why $-i\nabla$ Is the Momentum", *Am. J. Phys.* **43**, 1089–1093.

Kac, M. (1959), *Statistical Independence in Probability, Analysis and Number Theory* (Mathematical Association of America and John Wiley & Sons).

Kaiser, H., Werner, S. A., and George, E. A. (1983), "Direct Measurement of the Longitudinal Coherence Length of a Thermal Neutron Beam", *Phys. Rev. Lett.* **50**, 560–563.

Kato (1949), *J. Phys. Soc. Japan* **4**, 334.

Keller, J. B. (1972), "Quantum Mechanical Cross Sections for Small Wavelengths", *Am. J. Phys.* **40**, 1035–1036.

Khalfin, L. A. (1958), "Contribution to the Decay Theory of a Quasistationary State", *Sov. Phys. JETP* **6**, 1053–1063.

Klauder, J. R. and Sudarshan, E. C. G. (1968), *Fundamentals of Quantum Optics* (Benjamin, New York).

Knight, P. L. and Alen, L. (1983), *Concepts of Quantum Optics* (Pergamon, Oxford).

Koba, D. H. and Smirl, A. L. (1978), "Gauge Invariant Formulation of the Interaction of Electromagnetic Radiation and Matter", *Am. J. Phys.* **46**, 624–633.

Koch, P. M., Galvez, E. J., van Leuwen, K. A. H., Moorman, L., Sauer, B. E., and Richards, D. (1989), "Experiments in Quantum Chaos: Microwave Ionization of Hydrogen Atoms", *Physica Scripta* **T26**, 51–58.

Kochen, S. and Specker, E. P. (1967), "The Problem of Hidden Variables in Quantum Mechanics", *J. Math. Mech.* **17**, 59–87.

Kwiat, P. G., Steinberg, Chiao, R. Y., Eberhard, P. H., and Petroff, M. D. (1993), "High-Efficiency Single-Photon Detectors", *Phys. Rev. A* **48**, R867–870.

Lam, C. S. and Varshni, Y. P. (1971), "Energies of s Eigenstates in a Static Screened Coulomb Potential", *Phys. Rev. A* **4**, 1875–1881.

Lamb, W. E. (1969), "An Operational Interpretation of Nonrelativistic Quantum Mechanics", *Physics Today* **22**(4), 23–28.

Landau, L. D. and Lifshitz, E. M. (1958), *Quantum Mechanics, Non-relativistic Theory* (Addison-Wesley, Reading, Massachusetts).

Landauer, R. (1992), *Information is Physical*, in Physics and Computation, 1992, Workshop on. IEEE (1993).

Landauer, R. (1996), "The Physical Nature of Information", *Phys. Lett. A* **217**, 188–193.

Lee, H.-W. (1995), "Theory and Applications of the Quantum Phase-Space Distribution Functions", *Phys. Rep.* **259**, 147–211.

Leggett, A. J. (1987), "Reflections on the Quantum Measurement Paradox", in *Quantum Implications*, ed. B. J. Hiley (Routledge & Kegan Paul, New York).

Levy-Leblond, J.-M. (1976), "Quantum Fact and Fiction: Clarifying Lande's Pseudo-paradox", *Am. J. Phys.* **44**, 1130–1132.

Lewis, P. G., Jennings, D., Barrett, J., and Rudolph, T. (2012), "Distinct Quantum States Can Be Compatible with a Single State of Reality", *Phys. Rev. Lett.* **109**, 150404.

Lloyd, S. (1996), "Universal Quantum Simulators", *Science* **273**, 1073.

Lo, T. K. and Shimony, A. (1981), "Proposed Molecular Test of Local Hidden-Variables Theories", *Phys. Rev. A* **23**, 3003–3012.

Lohman, B. and Weigold, E. (1981), "Direct Measurement of the Electron Momentum Probability Distribution in Atomic Hydrogen", *Phys. Lett.* **86A**, 139–141.

Lütkenhaus, N., Calsamiglia, J. and Suominen, K.-A. (1999), "Bell Measurements for Teleportation", *Phys. Rev. A* **59**, 3295.

March, N. H., Young, W. H., and Sampanthar, S. (1967), *The Many-Body Problem in Quantum Mechanics* (Cambridge University Press, Cambridge).

Margenau, H. and Cohen, L. (1967), "Probabilities in Quantum Mechanics", in *Quantum Theory and Reality*, ed. M. Bunge (Springer-Verlag, New York), pp. 71–89.

Mermin, N. D. (1993), "Hidden Variables and the Two Theorems of John Bell", *Rev. Mod. Phys.* **65**, 803–815.

Mermin, N. D. (1994), "What's Wrong with This Temptation?", *Physics Today* **47**(6), 9–10.

Merzbacher, E. (1970), *Quantum Mechanics*, 2nd edn. (John Wiley & Sons, New York).

Messiah, A. (1966), *Quantum Mechanics* (John Wiley & Sons, New York).

Messiah, A. and Greenberg, O. W. (1964), "Symmetrization Postulate and Its Experimental Foundation", *Phys. Rev.* **136**, B248–267.

Milonni, P. W. (1984), "Why Spontaneous Emission?", *Am. J. Phys.* **52**, 340–343.

Milonni, P. W. (1994), *The Quantum Vacuum* (Academic, Boston).

Milonni, P. W., Cook, R. J., and Ackeralt, J. R. (1989), "Natural Line Shape", *Phys. Rev. A* **40**, 3764–3768.

Misner, C. W., Thorne, K. S., and Wheeler, J. A. (1973), *Gravitation* (Freeman, San Francisco).

Mitchell, D. P. and Powers, P. N. (1936), "Bragg Reflection of Slow Neutrons", *Phys. Rev.* **50**, 486–487.

Morse, P. M. and Feshbach, H. (1953), *Methods of Theoretical Physics* (McGraw-Hill, New York).

Moshinsky, M. (1968), "How Good Is the Hartree–Fock Approximation?", *Am. J. Phys.* **36**, 52–53; erratum, p. 763.

Nakahara, M. and Ohmi, T. (2008), *Quantum Computing: From Linear Algebra to Physical Realizations* (CRC Press, Taylor & Francis Group, Boca Raton, Florida).

Newton, R. G. (1982), *Scattering Theory of Waves and Particles*, 2nd edn. (Springer, New York).

Newton, R. G. and Young, B. L. (1968), "Measurability of the Spin Density Matrix", *Ann. Phys.* **49**, 393–402.

Nölleke, C., Neuzner, A., Reiserer, A., Hahn, C., Rempe, G., and Ritter, R. (2013), "Efficient Teleportation between Remote Single-Atom Quantum Memories", *Phys. Rev. Lett.* **110**, 140403.

Norsen, T. (2007), "Against 'Realism'", *Found. Phys.* **37**, 311–339.

O'Connell, R. F. and Wigner, E. P. (1981), "Some Properties of a Non-negative Quantum-Mechanical Distribution Function", *Phys. Lett.* **85A**, 121–126.

O'Connell, R. F., Wang, L., and Williams, H. A. (1984), "Time Dependence of a General Class of Quantum Distribution Functions", *Phys. Rev. A* **30**, 2187–2192.

Overhauser, A. W. (1971), "Simplified Theory of Electron Correlations in Metals", *Phys. Rev. B* **3**, 1888–1897.

Pais, Abraham (1982), *Subtle Is the Lord: The Science and the Life of Albert Einstein* (Oxford University Press, Oxford).

Pati, A. K. (2002), "General Impossible Operations in Quantum Information", *Phys. Rev. A* **66**, 062319.

Pauling, L. and Wilson, E. B. (1935), *Introduction to Quantum Mechanics* (McGraw-Hill, New York).

Penney, R. (1965), "Bosons and Fermions", *J. Math. Phys.* **6**, 1031–1034.

Peres, A. (1991), "Two Simple Proofs of the Kochen–Specker Theorem", *J. Phys. A: Math. Gen.* **24**, L175–178.

Peres, A. and Wootters, W. K. (1991), "Optimal Detection of Quantum Information", Phys. Rev. Lett. **66**, 1119.

Peres, A. (1993), *Quantum Theory: Concepts and Methods* (Kluwer, Dordrecht).

Peres, A. (1996), "Separability Criterion for Density Matrices", *Phys. Rev. Lett.* **77**, 1413.

Perkins, D. H. (1982), *Introduction to High Energy Physics*, 2nd edn. (Addison-Wesley, Reading, Massachusetts).

Peshkin, M. (1981), "The Aharonov–Bohm Effect: Why It Cannot Be Eliminated from Quantum Mechanics", *Phys. Rep.* **80**, 375–386.

Pfeifer, P. (1995), "Generalized Time–Energy Uncertainty Relations and Bounds on Lifetimes of Resonances", *Rev. Mod. Phys.* **67**, 759–779.

Philippidis, C., Dewdney, C., and Hiley, B. J. (1979), "Quantum Interference and the Quantum Potential", *Nuovo Cimento* **52**, 15–28.

Pittenger, A. O. (2000), *An Introduction to Quantum Computing Algorithms* (Birkhäuser, Boston).

Platt, D. E. (1992), "A Modern Analysis of the Stern–Gerlach Experiment", *Am. J. Phys.* **60**, 306–308.

Plenio, M. B. and Virmani, S. (2007), "An Introduction to Entanglement Measurements", *Quantum Information and Computation* **7**, 1.

Popescu, S. (1964), "Bell's Inequalities versus Teleportation: What is Nonlocality?", *Phys. Rev. Lett.* **72**, 797.

Popescu, S. (1965), "Bell's Inequalities and Density Matrices: Revealing Hidden Nonlocality", *Phys. Rev. Lett.* **74**, 2619.

Popper, K. R. (1957), "The Propensity Interpretation of the Calculus of Probability, and the Quantum Theory", in *Observation and Interpretation*, ed. S. Korner (Butterworths, London), pp. 65–70.

Pritchard, D. E. (1992), "Atom Interferometers", in *Atomic Physics 13: Thirteenth International Conference on Atomic Physics*, Munich, Germany, 1992, eds. H. Walther, T. W. Hanch, and B. Neizert (American Institute of Physics, New York), pp. 185–199.

Pusey, M. F., Barrett, J., and Rudolph, T. (2012), "The Quantum State Cannot Be Interpreted Statistically", *Nature Phys.* **8**, 476.

Rarity, J. G. and Tapster, P. R. (1990), "Experimental Violation of Bell's Inequality Based on Phase and Momentum", *Phys. Rev. Lett.* **64**, 2495–2498.

Reck, M., Zeilinger, A., Bernstein, H. J., and Bertani, P. (1994), "Experimental Realization of Any Discrete Unitary Operator", *Phys. Rev. Lett.* **73**, 58–61.

Renyi, A. (1970), *Foundations of Probability* (Holden-Day, San Francisco).

Riesz, F. and Sz.-Nagy, B. (1955), *Functional Analysis* (Fredrick Ungar, New York).

Riisager, K. (1994), "Nuclear Halo States", *Rev. Mod. Phys.* **66**, 1105–1116.

Rodberg, L. S. and Thaler, R. M. (1967), *Introduction to the Quantum Theory of Scattering* (Academic, New York).

Rose, M. E. (1957), *Elementary Theory of Angular Momentum* (John Wiley & Sons, London).

Rotenberg, M., Bivins, R., Metropolis, N., and Wooten, J. K. (1959), *The 3-j and 6-j Symbols* (Technology Press, M.I.T., Cambridge, Massachusetts).

Rowe, E. G. P. (1987), "The Classical Limit of Quantum Mechanical Hydrogenic Radial Distributions", *Eur. J. Phys.* **8**, 81–87.

Sakuri, J. J. (1982), *Modern Quantum Mechanics* (Benjamin/Cummings, Menlo Park, California).

Sangster, K., Hinds, E. A., Barnett, S. M., and Riis, E. (1993), "Measurement of the Aharonov–Casher Phase in an Atomic System", *Phys. Rev. Lett.* **71**, 3641–3644.

Schiff, L. I. (1968), *Quantum Mechanics*, 3rd edn. (McGraw-Hill, New York).

Schlosshauer, M. (2008), "Classicality, the Ensemble Interpretation, and Decoherence", *Found. Phys.* **38**, 796–803.

Schulman, L. S. (1981), *Techniques and Applications of Path Integrals* (Wiley, New York).

Shannon, C. E. (1948), "A Mathematical Theory of Communication", *Bell Syst. Tech. J.* **27**, 623–656.

Shannon, C. E. and Weaver, W. (1949), *The Mathematical Theory of Communication* (University of Illinois, Urbana).

Shapere, A. and Wilczek, F. (1989), *Geometric Phases in Physics* (World Scientific, Singapore).

Shih, Y. H. and Alley, C. O. (1988), "New Type of Einstein–Podolsky–Rosen–Bohm Experiment Using Pairs of Light Quanta Produced by Optical Parametric Down Conversion", *Phys. Rev. Lett.* **26**, 2921–2924.

Shimony, A. (1986), "Events and Processes in the Quantum World", (esp. Secs. 12.3 and 12.4), in *Quantum Concepts in Space and Time*, eds. R. Penrose and C. J. Isham (Clarendon Press, Oxford).

Silverman, M. P. (1995), *More Than One Mystery* (Springer-Verlag, New York).

Smelyanskiy, V. N., Rieffel, E. G., Knysh, S. I., Williams, C. P., Johnson, M. W., Thom, M. C., Macready, W. G., and Pudenz, V. N. (2012), "A Near-Term Quantum Computing Approach for Hard Computational Problems in Space Exploration", arXiv:1204.2821v2.

Smithey, D. T., Beck, M., Raymer, N. G., and Faridani, A. (1993), "Measurement of the Wigner Distribution and the Density Matrix of a Light Mode Using Optical Homodyne Tomography: Application to Squeezed States and the Vacuum", *Phys. Rev. Lett.* **70**, 1244–1247.

Sparnaay, M. J. (1958), "Measurements of Attractive Forces Between Flat Plates", *Physica* **24**, 751–764.

Spruch, L. (1986), "Retarded, or Casimir, Long-Range Potentials", *Physics Today* **39**(11), 37–45.

Srinivas, N. D. (1982), "When Is a Hidden Variable Theory Compatible with Quantum Mechanics?", *Pramana* **19**, 159–173.

Stapp, H. P. (1985), "Bell's Theorem and the Foundations of Quantum Physics", *Am. J. Phys.* **53**, 306–317.

Stapp, H. P. (1988), "Quantum Nonlocality", *Found. Phys.* **18**, 427–447.

Staudenmann, J.-L., Werner, S. A., Colella, R., and Overhauser, A. W. (1980), "Gravity and Inertia in Quantum Mechanics", *Phys. Rev. A* **21**, 1419–1438.

Steeb, W.-H. and Hardy, Y. (2006), *Problems and Solutions in Quantum Computing and Quantum Information* (World Scientific, Singapore).

Stenholm, S. (1992), "Simultaneous Measurement of Conjugate Variables", *Ann. Phys. (NY)* **218**, 233–254.

Stern, A., Aharonov, A., and Imry, J. (1990), "Dephasing of Interference by a Back Reacting Environment", in *Quantum Coherence*, ed. J. S. Anandan (World Scientific, Singapore), pp. 201–219.

— (1990), "Phase Uncertainty and Loss of Interference", *Phys. Rev. A* **41**, 3436–3448.

Stillinger, F. H. and Herrick, D. R. (1975), "Bound States in the Continuum", *Phys. Rev. A* **11**, 446–454.

Stoler, D. and Newman, S. (1972), "Minimum Uncertainty and Density Matrices", *Phys. Lett.* **38A**, 433–434.

Summhammer, J., Badurek, G., Rauch, H., and Kischko, U. (1982), "Explicit Experimental Verification of Quantum Spin-State Superposition", *Phys. Lett.* **90A**, 110–112.

Swift, A. R. and Wright, R. (1980), "Generalized Stern-Gerlach Experiments and the Observability of Arbitrary Spin Operators", *J. Math. Phys.* **21**, 77–82.

Tinkham, M. (1964), *Group Theory and Quantum Mechanics* (McGraw-Hill, New York).

Tonomura, A., Umezaki, H., Matsuda, T., Osakabe, N., Endo, J., and Sugita, Y. (1983), "Electron Holography, Aharonov–Bohm Effect and Flux Quantization", in *Foundations of Quantum Mechanics in the Light of New Technology*, ed. S. Kamefuchi (Physical Society of Japan, Tokyo, 1984).

Tonomura, A., Endo, T., Matsuda, T., and Kawasaki, T. (1989), "Demonstration of Single-Electron Buildup of an Interference Pattern", *Am. J. Phys.* **57**, 117–120.

Trigg, G. L. (1971), *Crucial Experiments in Modern Physics* (Van Nostrand Reinhold, New York).

Uffink, J. (1993), "The Rate of Evolution of a Quantum State", *Am. J. Phys.* **61**, 935–936.

Vogel, K. and Risken, H. (1989), "Determination of Quasiprobability Distributions in Terms of Probability Distributions for the Rotated Quadrature Phase", *Phys. Rev.* **A40**, 2847–2849.

Walker, J. S. and Gathright, J. (1994), "Exploring One-Dimensional Quantum Mechanics with Transfer Matrices", *Am. J. Phys.* **62**, 408–422.

Wannier, G. H. (1966), *Statistical Physics* (Wiley, New York).

Weigert, S. (1992), "Pauli Problem for a Spin of Arbitrary Length: A Simple Method to Determine Its Wave Function", *Phys. Rev. A* **45**, 7688–7696.

Weigold, E. (1982), "Momentum Distributions: Opening Remarks", in AIP Conference Proceedings, No. 86, *Momentum Wave Functions – 1982*, pp. 1–4.

Werner, R. F. (2001), "Quantum Information Theory – an Invitation", in Alber (2001), pp. 14–57.

Wheeler, J. A. and Zurek, W. H. eds. (1983), *Quantum Theory and Measurement* (Princeton University Press, Princeton, New Jersey).

Wiebe, N. and Ballentine, L. E. (2005), "Quantum Mechanics of Hyperion", *Phys. Rev. A* **72**, 022109.

Wigner, E. P. (1932), "On the Quantum Correction for Thermodynamic Equilibrium", *Phys. Rev.* **40**, 749–759.

Wigner, E. P. (1971), "Quantum Mechanical Distribution Functions Revisited", in *Perspectives in Quantum Theory*, eds. W. Yourgrau and A. van der Merwe (Dover, New York), pp. 25–36.

Winter, R. G. (1961), "Evolution of a Quasistationary State", *Phys. Rev.* **123**, 1503–1507.

Wolsky, A. M. (1974), "Kinetic Energy, Size, and the Uncertainty Principle", *Am. J. Phys.* **42**, 760–763.

Wu, T.-Y. and Ohmura, T. (1962), *Quantum Theory of Scattering* (Prentice-Hall, Englewood Cliffs, New Jersey).

Zeilinger, A., Gähler, R., Shull, C. G., and Treimer, W. (1988), "Single- and Double-Slit Diffraction of Neutrons", *Rev. Mod. Phys.* **60**, 1067–1073.

Index

Printed in the United States
By Bookmasters